Statistical Pattern Recognition

Statistical Pattern Recognition

Third Edition

Andrew R. Webb • Keith D. Copsey
Mathematics and Data Analysis Consultancy, Malvern, UK

WILEY
A John Wiley & Sons, Ltd., Publication

This edition first published 2011

Registered office
John Wiley & Sons Ltd, The Atrium, Southern Gate, Chichester, West Sussex, PO19 8SQ, United Kingdom

For details of our global editorial offices, for customer services and for information about how to apply for permission to reuse the copyright material in this book please see our website at www.wiley.com.

Library of Congress Cataloging-in-Publication Data

Webb, A. R. (Andrew R.)
Statistical pattern recognition / Andrew R. Webb, Keith D. Copsey. – 3rd ed.
p. cm.
Includes bibliographical references and index.
ISBN 978-0-470-68227-2 (hardback) – ISBN 978-0-470-68228-9 (paper)
1. Pattern perception–Statistical methods. I. Copsey, Keith D. II. Title.
Q327.W43 2011
006.4–dc23 2011024957

A catalogue record for this book is available from the British Library.

HB ISBN: 978-0-470-68227-2
PB ISBN: 978-0-470-68228-9
ePDF ISBN: 978-1-119-95296-1
oBook ISBN: 978-1-119-95295-4
ePub ISBN: 978-1-119-96140-6
Mobi ISBN: 978-1-119-96141-3

Typeset in 10/12pt Times by Aptara Inc., New Delhi, India

To Rosemary,
Samuel, Miriam, Jacob and Ethan

Contents

Preface **xix**

Notation **xxiii**

1 Introduction to Statistical Pattern Recognition 1
1.1 Statistical Pattern Recognition 1
1.1.1 Introduction 1
1.1.2 The Basic Model 2
1.2 Stages in a Pattern Recognition Problem 4
1.3 Issues 6
1.4 Approaches to Statistical Pattern Recognition 7
1.5 Elementary Decision Theory 8
1.5.1 Bayes' Decision Rule for Minimum Error 8
1.5.2 Bayes' Decision Rule for Minimum Error – Reject Option 12
1.5.3 Bayes' Decision Rule for Minimum Risk 13
1.5.4 Bayes' Decision Rule for Minimum Risk – Reject Option 15
1.5.5 Neyman–Pearson Decision Rule 15
1.5.6 Minimax Criterion 18
1.5.7 Discussion 19
1.6 Discriminant Functions 20
1.6.1 Introduction 20
1.6.2 Linear Discriminant Functions 21
1.6.3 Piecewise Linear Discriminant Functions 23
1.6.4 Generalised Linear Discriminant Function 24
1.6.5 Summary 26
1.7 Multiple Regression 27
1.8 Outline of Book 29
1.9 Notes and References 29
Exercises 31

2 Density Estimation – Parametric 33
2.1 Introduction 33

2.2 Estimating the Parameters of the Distributions 34
2.2.1 Estimative Approach 34
2.2.2 Predictive Approach 35
2.3 The Gaussian Classifier 35
2.3.1 Specification 35
2.3.2 Derivation of the Gaussian Classifier Plug-In Estimates 37
2.3.3 Example Application Study 39
2.4 Dealing with Singularities in the Gaussian Classifier 40
2.4.1 Introduction 40
2.4.2 Naïve Bayes 40
2.4.3 Projection onto a Subspace 41
2.4.4 Linear Discriminant Function 41
2.4.5 Regularised Discriminant Analysis 42
2.4.6 Example Application Study 44
2.4.7 Further Developments 45
2.4.8 Summary 46
2.5 Finite Mixture Models 46
2.5.1 Introduction 46
2.5.2 Mixture Models for Discrimination 48
2.5.3 Parameter Estimation for Normal Mixture Models 49
2.5.4 Normal Mixture Model Covariance Matrix Constraints 51
2.5.5 How Many Components? 52
2.5.6 Maximum Likelihood Estimation via EM 55
2.5.7 Example Application Study 60
2.5.8 Further Developments 62
2.5.9 Summary 63
2.6 Application Studies 63
2.7 Summary and Discussion 66
2.8 Recommendations 66
2.9 Notes and References 67
Exercises 67

3 Density Estimation – Bayesian 70
3.1 Introduction 70
3.1.1 Basics 72
3.1.2 Recursive Calculation 72
3.1.3 Proportionality 73
3.2 Analytic Solutions 73
3.2.1 Conjugate Priors 73
3.2.2 Estimating the Mean of a Normal Distribution with Known Variance 75
3.2.3 Estimating the Mean and the Covariance Matrix of a Multivariate Normal Distribution 79
3.2.4 Unknown Prior Class Probabilities 85
3.2.5 Summary 87
3.3 Bayesian Sampling Schemes 87
3.3.1 Introduction 87

3.3.2 Summarisation 87
3.3.3 Sampling Version of the Bayesian Classifier 89
3.3.4 Rejection Sampling 89
3.3.5 Ratio of Uniforms 90
3.3.6 Importance Sampling 92
3.4 Markov Chain Monte Carlo Methods 95
3.4.1 Introduction 95
3.4.2 The Gibbs Sampler 95
3.4.3 Metropolis–Hastings Algorithm 103
3.4.4 Data Augmentation 107
3.4.5 Reversible Jump Markov Chain Monte Carlo 108
3.4.6 Slice Sampling 109
3.4.7 MCMC Example – Estimation of Noisy Sinusoids 111
3.4.8 Summary 115
3.4.9 Notes and References 116
3.5 Bayesian Approaches to Discrimination 116
3.5.1 Labelled Training Data 116
3.5.2 Unlabelled Training Data 117
3.6 Sequential Monte Carlo Samplers 119
3.6.1 Introduction 119
3.6.2 Basic Methodology 121
3.6.3 Summary 125
3.7 Variational Bayes 126
3.7.1 Introduction 126
3.7.2 Description 126
3.7.3 Factorised Variational Approximation 129
3.7.4 Simple Example 131
3.7.5 Use of the Procedure for Model Selection 135
3.7.6 Further Developments and Applications 136
3.7.7 Summary 137
3.8 Approximate Bayesian Computation 137
3.8.1 Introduction 137
3.8.2 ABC Rejection Sampling 138
3.8.3 ABC MCMC Sampling 140
3.8.4 ABC Population Monte Carlo Sampling 141
3.8.5 Model Selection 142
3.8.6 Summary 143
3.9 Example Application Study 144
3.10 Application Studies 145
3.11 Summary and Discussion 146
3.12 Recommendations 147
3.13 Notes and References 147
Exercises 148

4 Density Estimation – Nonparametric 150
4.1 Introduction 150
4.1.1 Basic Properties of Density Estimators 150

4.2 *k*-Nearest-Neighbour Method 152
4.2.1 *k*-Nearest-Neighbour Classifier 152
4.2.2 Derivation 154
4.2.3 Choice of Distance Metric 157
4.2.4 Properties of the Nearest-Neighbour Rule 159
4.2.5 Linear Approximating and Eliminating Search Algorithm 159
4.2.6 Branch and Bound Search Algorithms: kd-Trees 163
4.2.7 Branch and Bound Search Algorithms: Ball-Trees 170
4.2.8 Editing Techniques 174
4.2.9 Example Application Study 177
4.2.10 Further Developments 178
4.2.11 Summary 179
4.3 Histogram Method 180
4.3.1 Data Adaptive Histograms 181
4.3.2 Independence Assumption (Naïve Bayes) 181
4.3.3 Lancaster Models 182
4.3.4 Maximum Weight Dependence Trees 183
4.3.5 Bayesian Networks 186
4.3.6 Example Application Study – Naïve Bayes Text Classification 190
4.3.7 Summary 193
4.4 Kernel Methods 194
4.4.1 Biasedness 197
4.4.2 Multivariate Extension 198
4.4.3 Choice of Smoothing Parameter 199
4.4.4 Choice of Kernel 201
4.4.5 Example Application Study 202
4.4.6 Further Developments 203
4.4.7 Summary 203
4.5 Expansion by Basis Functions 204
4.6 Copulas 207
4.6.1 Introduction 207
4.6.2 Mathematical Basis 207
4.6.3 Copula Functions 208
4.6.4 Estimating Copula Probability Density Functions 209
4.6.5 Simple Example 211
4.6.6 Summary 212
4.7 Application Studies 213
4.7.1 Comparative Studies 216
4.8 Summary and Discussion 216
4.9 Recommendations 217
4.10 Notes and References 217
Exercises 218

5 Linear Discriminant Analysis 221
5.1 Introduction 221
5.2 Two-Class Algorithms 222
5.2.1 General Ideas 222

5.2.2 Perceptron Criterion 223
5.2.3 Fisher's Criterion 227
5.2.4 Least Mean-Squared-Error Procedures 228
5.2.5 Further Developments 235
5.2.6 Summary 235
5.3 Multiclass Algorithms 236
5.3.1 General Ideas 236
5.3.2 Error-Correction Procedure 237
5.3.3 Fisher's Criterion – Linear Discriminant Analysis 238
5.3.4 Least Mean-Squared-Error Procedures 241
5.3.5 Regularisation 246
5.3.6 Example Application Study 246
5.3.7 Further Developments 247
5.3.8 Summary 248
5.4 Support Vector Machines 249
5.4.1 Introduction 249
5.4.2 Linearly Separable Two-Class Data 249
5.4.3 Linearly Nonseparable Two-Class Data 253
5.4.4 Multiclass SVMs 256
5.4.5 SVMs for Regression 257
5.4.6 Implementation 259
5.4.7 Example Application Study 262
5.4.8 Summary 263
5.5 Logistic Discrimination 263
5.5.1 Two-Class Case 263
5.5.2 Maximum Likelihood Estimation 264
5.5.3 Multiclass Logistic Discrimination 266
5.5.4 Example Application Study 267
5.5.5 Further Developments 267
5.5.6 Summary 268
5.6 Application Studies 268
5.7 Summary and Discussion 268
5.8 Recommendations 269
5.9 Notes and References 270
Exercises 270

6 Nonlinear Discriminant Analysis – Kernel and Projection Methods 274
6.1 Introduction 274
6.2 Radial Basis Functions 276
6.2.1 Introduction 276
6.2.2 Specifying the Model 278
6.2.3 Specifying the Functional Form 278
6.2.4 The Positions of the Centres 279
6.2.5 Smoothing Parameters 281
6.2.6 Calculation of the Weights 282
6.2.7 Model Order Selection 284
6.2.8 Simple RBF 285

6.2.9 Motivation 286
6.2.10 RBF Properties 288
6.2.11 Example Application Study 288
6.2.12 Further Developments 289
6.2.13 Summary 290
6.3 Nonlinear Support Vector Machines 291
6.3.1 Introduction 291
6.3.2 Binary Classification 291
6.3.3 Types of Kernel 292
6.3.4 Model Selection 293
6.3.5 Multiclass SVMs 294
6.3.6 Probability Estimates 294
6.3.7 Nonlinear Regression 296
6.3.8 Example Application Study 296
6.3.9 Further Developments 297
6.3.10 Summary 298
6.4 The Multilayer Perceptron 298
6.4.1 Introduction 298
6.4.2 Specifying the MLP Structure 299
6.4.3 Determining the MLP Weights 300
6.4.4 Modelling Capacity of the MLP 307
6.4.5 Logistic Classification 307
6.4.6 Example Application Study 310
6.4.7 Bayesian MLP Networks 311
6.4.8 Projection Pursuit 313
6.4.9 Summary 313
6.5 Application Studies 314
6.6 Summary and Discussion 316
6.7 Recommendations 317
6.8 Notes and References 318
Exercises 318

7 Rule and Decision Tree Induction 322
7.1 Introduction 322
7.2 Decision Trees 323
7.2.1 Introduction 323
7.2.2 Decision Tree Construction 326
7.2.3 Selection of the Splitting Rule 327
7.2.4 Terminating the Splitting Procedure 330
7.2.5 Assigning Class Labels to Terminal Nodes 332
7.2.6 Decision Tree Pruning – Worked Example 332
7.2.7 Decision Tree Construction Methods 337
7.2.8 Other Issues 339
7.2.9 Example Application Study 340
7.2.10 Further Developments 341
7.2.11 Summary 342

7.3 Rule Induction 342
7.3.1 Introduction 342
7.3.2 Generating Rules from a Decision Tree 345
7.3.3 Rule Induction Using a Sequential Covering Algorithm 345
7.3.4 Example Application Study 350
7.3.5 Further Developments 351
7.3.6 Summary 351
7.4 Multivariate Adaptive Regression Splines 351
7.4.1 Introduction 351
7.4.2 Recursive Partitioning Model 351
7.4.3 Example Application Study 355
7.4.4 Further Developments 355
7.4.5 Summary 356
7.5 Application Studies 356
7.6 Summary and Discussion 358
7.7 Recommendations 358
7.8 Notes and References 359
Exercises 359

8 Ensemble Methods 361
8.1 Introduction 361
8.2 Characterising a Classifier Combination Scheme 362
8.2.1 Feature Space 363
8.2.2 Level 366
8.2.3 Degree of Training 368
8.2.4 Form of Component Classifiers 368
8.2.5 Structure 369
8.2.6 Optimisation 369
8.3 Data Fusion 370
8.3.1 Architectures 370
8.3.2 Bayesian Approaches 371
8.3.3 Neyman–Pearson Formulation 373
8.3.4 Trainable Rules 374
8.3.5 Fixed Rules 375
8.4 Classifier Combination Methods 376
8.4.1 Product Rule 376
8.4.2 Sum Rule 377
8.4.3 Min, Max and Median Combiners 378
8.4.4 Majority Vote 379
8.4.5 Borda Count 379
8.4.6 Combiners Trained on Class Predictions 380
8.4.7 Stacked Generalisation 382
8.4.8 Mixture of Experts 382
8.4.9 Bagging 385
8.4.10 Boosting 387
8.4.11 Random Forests 389
8.4.12 Model Averaging 390

8.4.13 Summary of Methods 396
8.4.14 Example Application Study 398
8.4.15 Further Developments 399
8.5 Application Studies 399
8.6 Summary and Discussion 400
8.7 Recommendations 401
8.8 Notes and References 401
Exercises 402

9 Performance Assessment 404
9.1 Introduction 404
9.2 Performance Assessment 405
9.2.1 Performance Measures 405
9.2.2 Discriminability 406
9.2.3 Reliability 413
9.2.4 ROC Curves for Performance Assessment 415
9.2.5 Population and Sensor Drift 419
9.2.6 Example Application Study 421
9.2.7 Further Developments 422
9.2.8 Summary 423
9.3 Comparing Classifier Performance 424
9.3.1 Which Technique is Best? 424
9.3.2 Statistical Tests 425
9.3.3 Comparing Rules When Misclassification Costs are Uncertain 426
9.3.4 Example Application Study 428
9.3.5 Further Developments 429
9.3.6 Summary 429
9.4 Application Studies 429
9.5 Summary and Discussion 430
9.6 Recommendations 430
9.7 Notes and References 430
Exercises 431

10 Feature Selection and Extraction 433
10.1 Introduction 433
10.2 Feature Selection 435
10.2.1 Introduction 435
10.2.2 Characterisation of Feature Selection Approaches 439
10.2.3 Evaluation Measures 440
10.2.4 Search Algorithms for Feature Subset Selection 449
10.2.5 Complete Search – Branch and Bound 450
10.2.6 Sequential Search 454
10.2.7 Random Search 458
10.2.8 Markov Blanket 459
10.2.9 Stability of Feature Selection 460
10.2.10 Example Application Study 462
10.2.11 Further Developments 462
10.2.12 Summary 463

10.3 Linear Feature Extraction 463
10.3.1 Principal Components Analysis 464
10.3.2 Karhunen–Loève Transformation 475
10.3.3 Example Application Study 481
10.3.4 Further Developments 482
10.3.5 Summary 483
10.4 Multidimensional Scaling 484
10.4.1 Classical Scaling 484
10.4.2 Metric MDS 486
10.4.3 Ordinal Scaling 487
10.4.4 Algorithms 490
10.4.5 MDS for Feature Extraction 491
10.4.6 Example Application Study 492
10.4.7 Further Developments 493
10.4.8 Summary 493
10.5 Application Studies 493
10.6 Summary and Discussion 495
10.7 Recommendations 495
10.8 Notes and References 496
Exercises 497

11 Clustering 501
11.1 Introduction 501
11.2 Hierarchical Methods 502
11.2.1 Single-Link Method 503
11.2.2 Complete-Link Method 506
11.2.3 Sum-of-Squares Method 507
11.2.4 General Agglomerative Algorithm 508
11.2.5 Properties of a Hierarchical Classification 508
11.2.6 Example Application Study 509
11.2.7 Summary 509
11.3 Quick Partitions 510
11.4 Mixture Models 511
11.4.1 Model Description 511
11.4.2 Example Application Study 512
11.5 Sum-of-Squares Methods 513
11.5.1 Clustering Criteria 514
11.5.2 Clustering Algorithms 515
11.5.3 Vector Quantisation 520
11.5.4 Example Application Study 530
11.5.5 Further Developments 530
11.5.6 Summary 531
11.6 Spectral Clustering 531
11.6.1 Elementary Graph Theory 531
11.6.2 Similarity Matrices 534
11.6.3 Application to Clustering 534
11.6.4 Spectral Clustering Algorithm 535
11.6.5 Forms of Graph Laplacian 535

11.6.6 Example Application Study 536
11.6.7 Further Developments 538
11.6.8 Summary 538
11.7 Cluster Validity 538
11.7.1 Introduction 538
11.7.2 Statistical Tests 539
11.7.3 Absence of Class Structure 540
11.7.4 Validity of Individual Clusters 541
11.7.5 Hierarchical Clustering 542
11.7.6 Validation of Individual Clusterings 542
11.7.7 Partitions 543
11.7.8 Relative Criteria 543
11.7.9 Choosing the Number of Clusters 545
11.8 Application Studies 546
11.9 Summary and Discussion 549
11.10 Recommendations 551
11.11 Notes and References 552
Exercises 553

12 Complex Networks 555
12.1 Introduction 555
12.1.1 Characteristics 557
12.1.2 Properties 557
12.1.3 Questions to Address 559
12.1.4 Descriptive Features 560
12.1.5 Outline 560
12.2 Mathematics of Networks 561
12.2.1 Graph Matrices 561
12.2.2 Connectivity 562
12.2.3 Distance Measures 562
12.2.4 Weighted Networks 563
12.2.5 Centrality Measures 563
12.2.6 Random Graphs 564
12.3 Community Detection 565
12.3.1 Clustering Methods 565
12.3.2 Girvan–Newman Algorithm 568
12.3.3 Modularity Approaches 570
12.3.4 Local Modularity 571
12.3.5 Clique Percolation 573
12.3.6 Example Application Study 574
12.3.7 Further Developments 575
12.3.8 Summary 575
12.4 Link Prediction 575
12.4.1 Approaches to Link Prediction 576
12.4.2 Example Application Study 578
12.4.3 Further Developments 578
12.5 Application Studies 579

12.6 Summary and Discussion 579
12.7 Recommendations 580
12.8 Notes and References 580
Exercises 580

13 Additional Topics 581
13.1 Model Selection 581
13.1.1 Separate Training and Test Sets 582
13.1.2 Cross-Validation 582
13.1.3 The Bayesian Viewpoint 583
13.1.4 Akaike's Information Criterion 583
13.1.5 Minimum Description Length 584
13.2 Missing Data 585
13.3 Outlier Detection and Robust Procedures 586
13.4 Mixed Continuous and Discrete Variables 587
13.5 Structural Risk Minimisation and the Vapnik–Chervonenkis Dimension 588
13.5.1 Bounds on the Expected Risk 588
13.5.2 The VC Dimension 589

References **591**

Index **637**

Preface

This book provides an introduction to statistical pattern recognition theory and techniques. Most of the material presented in this book is concerned with discrimination and classification and has been drawn from a wide range of literature including that of engineering, statistics, computer science and the social sciences. The aim of the book is to provide descriptions of many of the most useful of today's pattern processing techniques including many of the recent advances in nonparametric approaches to discrimination and Bayesian computational methods developed in the statistics literature and elsewhere. Discussions provided on the motivations and theory behind these techniques will enable the practitioner to gain maximum benefit from their implementations within many of the popular software packages. The techniques are illustrated with examples of real-world applications studies. Pointers are also provided to the diverse literature base where further details on applications, comparative studies and theoretical developments may be obtained.

The book grew out of our research on the development of statistical pattern recognition methodology and its application to practical sensor data analysis problems. The book is aimed at advanced undergraduate and graduate courses. Some of the material has been presented as part of a graduate course on pattern recognition and at pattern recognition summer schools. It is also designed for practitioners in the field of pattern recognition as well as researchers in the area. A prerequisite is a knowledge of basic probability theory and linear algebra, together with basic knowledge of mathematical methods (for example, Lagrange multipliers are used to solve problems with equality and inequality constraints in some derivations). Some basic material (which was provided as appendices in the second edition) is available on the book's website.

Scope

The book presents most of the popular methods of statistical pattern recognition. However, many of the important developments in pattern recognition are not confined to the statistics literature and have occurred where the area overlaps with research in machine learning. Therefore, where we have felt that straying beyond the traditional boundaries of statistical pattern recognition would be beneficial, we have done so. An example is the inclusion of some rule induction methods as a complementary approach to rule discovery by decision tree induction.

Most of the methodology is generic – it is not specific to a particular type of data or application. Thus, we exclude preprocessing methods and filtering methods commonly used in signal and image processing.

Approach

The approach in each chapter has been to introduce some of the basic concepts and algorithms and to conclude each section on a technique or a class of techniques with a practical application of the approach from the literature. The main aim has been to introduce the basic concept of an approach. Sometimes this has required some detailed mathematical description and clearly we have had to draw a line on how much depth we discuss a particular topic. Most of the topics have whole books devoted to them and so we have had to be selective in our choice of material. Therefore, the chapters conclude with a section on the key references. The exercises at the ends of the chapters vary from 'open book' questions to more lengthy computer projects.

New to the third edition

Many sections have been rewritten and new material added. The new features of this edition include the following:

- A new chapter on Bayesian approaches to density estimation (Chapter 3) including expanded material on Bayesian sampling schemes and Markov chain Monte Carlo methods, and new sections on Sequential Monte Carlo samplers and Variational Bayes approaches.
- New sections on nonparametric methods of density estimation.
- Rule induction.
- New chapter on ensemble methods of classification.
- Revision of feature selection material with new section on stability.
- Spectral clustering.
- New chapter on complex networks, with relevance to the high-growth field of social and computer network analysis.

Book outline

Chapter 1 provides an introduction to statistical pattern recognition, defining some terminology, introducing supervised and unsupervised classification. Two related approaches to supervised classification are presented: one based on the use of probability density functions and a second based on the construction of discriminant functions. The chapter concludes with an outline of the pattern recognition cycle, putting the remaining chapters of the book into context. Chapters 2, 3 and 4 pursue the density function approach to discrimination. Chapter 2 addresses parametric approaches to density estimation, which are developed further in Chapter 3 on Bayesian methods. Chapter 4 develops classifiers based on nonparametric schemes, including the popular *k* nearest neighbour method, with associated efficient search algorithms.

Chapters 5–7 develop discriminant function approaches to supervised classification. Chapter 5 focuses on linear discriminant functions; much of the methodology of this chapter (including optimisation, regularisation, support vector machines) is used in some of the nonlinear methods described in Chapter 6 which explores kernel-based methods, in particular, the radial basis function network and the support vector machine, and projection-based methods (the multilayer perceptron). These are commonly referred to as neural network methods. Chapter 7 considers approaches to discrimination that enable the classification function to be cast in the form of an interpretable rule, important for some applications.

Chapter 8 considers ensemble methods – combining classifiers for improved robustness. Chapter 9 considers methods of measuring the performance of a classifier.

The techniques of Chapters 10 and 11 may be described as methods of exploratory data analysis or preprocessing (and as such would usually be carried out prior to the supervised classification techniques of Chapters 5–7, although they could, on occasion, be post-processors of supervised techniques). Chapter 10 addresses feature selection and feature extraction – the procedures for obtaining a reduced set of variables characterising the original data. Such procedures are often an integral part of classifier design and it is somewhat artificial to partition the pattern recognition problem into separate processes of feature extraction and classification. However, feature extraction may provide insights into the data structure and the type of classifier to employ; thus, it is of interest in its own right. Chapter 11 considers unsupervised classification or *clustering* – the process of grouping individuals in a population to discover the presence of structure; its engineering application is to vector quantisation for image and speech coding. Chapter 12 on complex networks introduces methods for analysing data that may be represented using the mathematical concept of a graph. This has great relevance to social and computer networks.

Finally, Chapter 13 addresses some important diverse topics including model selection.

Book website

The website www.wiley.com/go/statistical_pattern_recognition contains supplementary material on topics including measures of dissimilarity, estimation, linear algebra, data analysis and basic probability.

Acknowledgements

In preparing the third edition of this book we have been helped by many people. We are especially grateful to Dr Gavin Cawley, University of East Anglia, for help and advice. We are grateful to friends and colleagues (past and present, from RSRE, DERA and QinetiQ) who have provided encouragement and made comments on various parts of the manuscript. In particular, we would like to thank Anna Skeoch for providing figures for Chapter 12; and Richard Davies and colleagues at John Wiley for help in the final production of the manuscript. Andrew Webb is especially thankful to Rosemary for her love, support and patience.

Andrew R. Webb
Keith D. Copsey

Notation

Some of the more commonly used notation is given below. We have used some notational conveniences. For example, we have tended to use the same symbol for a variable as well as a measurement on that variable. The meaning should be obvious from context. Also, we denote the density function of x as $p(x)$ and y as $p(y)$, even though the functions differ. A vector is denoted by a lower case quantity in bold face, and a matrix by upper case. Since pattern recognition is very much a multidisciplinary subject, it is impossible to be both consistent across all chapters and consistent with the commonly used notation in the different literatures. We have adopted the policy of maintaining consistency as far as possible within a given chapter.

p, d	number of variables
C	number of classes
n	number of measurements
n_j	number of measurements in the jth class
ω_j	label for class j
$X_1, \ldots, X_p$	p random variables
$x_1, \ldots, x_p$	measurements on variables, $X_1, \ldots, X_p$
$\mathbf{x} = (x_1, \ldots, x_p)^T$	measurement vector
$\boldsymbol{X} = [\boldsymbol{x}_1, \ldots, \boldsymbol{x}_n]^T$	$n \times p$ data matrix
$\boldsymbol{X} = \begin{bmatrix} x_{11} & \ldots & x_{1p} \\ \vdots & \ddots & \vdots \\ x_{n1} & \ldots & x_{np} \end{bmatrix}$	
$P(\boldsymbol{x}) = \text{prob}(X_1 \leq x_1, \ldots, X_p \leq x_p)$	
$p(\boldsymbol{x}) = \partial P / \partial \boldsymbol{x}$	probability density function
$p(\boldsymbol{x} \mid \omega_j)$	probability density function of class j
$p(\omega_j)$	prior probability of class j
$\boldsymbol{\mu} = \int \boldsymbol{x} p(\boldsymbol{x}) d\boldsymbol{x}$	population mean
$\boldsymbol{\mu}_j = \int \boldsymbol{x} p(\boldsymbol{x} \mid \omega_j) d\boldsymbol{x}$	mean of class j, $j = 1, \ldots, C$
$\boldsymbol{m} = (1/n) \sum_{i=1}^{n} \boldsymbol{x}_i$	sample mean
$\boldsymbol{m}_j = (1/n_j) \sum_{i=1}^{n} z_{ji} \boldsymbol{x}_i$	sample mean of class j, $j = 1, \ldots, C$; $z_{ji} = 1$ if $\boldsymbol{x}_i \in \omega_j$, 0 otherwise; n_j-number of patterns in ω_j, $n_j = \sum_{i=1}^{n} z_{ji}$

$\hat{\mathbf{\Sigma}} = \frac{1}{n}\sum_{i=1}^{n}(\boldsymbol{x}_i - \boldsymbol{m})(\boldsymbol{x}_i - \boldsymbol{m})^T$	sample covariance matrix (maximum likelihood estimate)
$n/(n-1)\hat{\mathbf{\Sigma}}$	sample covariance matrix (unbiased estimate)
$\hat{\mathbf{\Sigma}}_j = \frac{1}{n_j}\sum_{i=1}^{n} z_{ji}(\boldsymbol{x}_i - \boldsymbol{m}_j)(\boldsymbol{x}_i - \boldsymbol{m}_j)^T$	sample covariance matrix of class j (maximum likelihood estimate)
$\boldsymbol{S}_j = \frac{n_j}{n_j-1}\hat{\mathbf{\Sigma}}_j$	sample covariance matrix of class j (unbiased estimate)
$\boldsymbol{S}_W = \sum_{j=1}^{C}\frac{n_j}{n}\hat{\mathbf{\Sigma}}_j$	pooled within class sample covariance matrix
$\boldsymbol{S} = \frac{n}{n-C}\boldsymbol{S}_W$	pooled within class sample covariance matrix (unbiased estimate)
$\boldsymbol{S}_B = \sum\frac{n_j}{n}(\boldsymbol{m}_j - \boldsymbol{m})(\boldsymbol{m}_j - \boldsymbol{m})^T$	sample between class matrix
$\boldsymbol{S}_B + \boldsymbol{S}_W = \hat{\mathbf{\Sigma}}$	
$\Vert\boldsymbol{A}\Vert^2 = \sum_{ij}A_{ij}^2$	
$N(\boldsymbol{m}, \mathbf{\Sigma})$	normal (or Gaussian) distribution, mean $\boldsymbol{m}$, covariance matrix $\mathbf{\Sigma}$
$N(\boldsymbol{x}; \boldsymbol{m}, \mathbf{\Sigma})$	probability density function for the normal distribution, mean $\boldsymbol{m}$, covariance matrix $\mathbf{\Sigma}$, evaluated at $\boldsymbol{x}$
$\mathrm{E}[Y\vert X]$	expectation of Y given X
$I(\theta)$	indicator function, $I(\theta) = 1$ if $\theta = \texttt{true}$ else 0

1

Introduction to statistical pattern recognition

Statistical pattern recognition is a term used to cover all stages of an investigation from problem formulation and data collection through to discrimination and classification, assessment of results and interpretation. Some of the basic concepts in classification are introduced and the key issues described. Two complementary approaches to discrimination are presented, namely a decision theory approach based on calculation of probability density functions and the use of Bayes theorem, and a discriminant function approach.

1.1 Statistical pattern recognition

1.1.1 Introduction

We live in a world where massive amounts of data are collected and recorded on nearly every aspect of human endeavour: for example, banking, purchasing (credit-card usage, point-of-sale data analysis), Internet transactions, performance monitoring (of schools, hospitals, equipment), and communications. The data come in a wide variety of diverse forms – numeric, textual (structured or unstructured), audio and video signals. Understanding and making sense of this vast and diverse collection of data (identifying patterns, trends, anomalies, providing summaries) requires some automated procedure to assist the analyst with this ‘data deluge’. A practical example of pattern recognition that is familiar to many people is classifying email messages (as spam/not spam) based upon message header, content and sender.

Approaches for analysing such data include those for signal processing, filtering, data summarisation, dimension reduction, variable selection, regression and classification and have been developed in several literatures (physics, mathematics, statistics, engineering, artificial intelligence, computer science and the social sciences, among others). The main focus of this book is on pattern recognition procedures, providing a description of basic techniques

Statistical Pattern Recognition, Third Edition. Andrew R. Webb and Keith D. Copsey.

together with case studies of practical applications of the techniques on real-world problems. A strong emphasis is placed on the statistical theory of discrimination, but clustering also receives some attention. Thus, the main subject matter of this book can be summed up in a single word: 'classification', both supervised (using class information to design a classifier – i.e. discrimination) and unsupervised (allocating to groups without class information – i.e. clustering). However, in recent years many complex datasets have been gathered (for example, 'transactions' between individuals – email traffic, purchases). Understanding these datasets requires additional tools in the pattern recognition toolbox. Therefore, we also examine developments such as methods for analysing data that may be represented as a graph.

Pattern recognition as a field of study developed significantly in the 1960s. It was very much an interdisciplinary subject. Some people entered the field with a real problem to solve. The large number of applications ranging from the classical ones such as automatic character recognition and medical diagnosis to the more recent ones in *data mining* (such as credit scoring, consumer sales analysis and credit card transaction analysis) have attracted considerable research effort with many methods developed and advances made. Other researchers were motivated by the development of machines with 'brain-like' performance, that in some way could operate giving human performance.

Within these areas significant progress has been made, particularly where the domain overlaps with probability and statistics, and in recent years there have been many exciting new developments, both in methodology and applications. These build on the solid foundations of earlier research and take advantage of increased computational resources readily available nowadays. These developments include, for example, kernel-based methods (including support vector machines) and Bayesian computational methods.

The topics in this book could easily have been described under the term *machine learning* that describes the study of machines that can adapt to their environment and learn from example. The machine learning emphasis is perhaps more on computationally intensive methods and less on a statistical approach, but there is strong overlap between the research areas of statistical pattern recognition and machine learning.

1.1.2 The basic model

Since many of the techniques we shall describe have been developed over a range of diverse disciplines, there is naturally a variety of sometimes contradictory terminology. We shall use the term 'pattern' to denote the p-dimensional data vector $\boldsymbol{x} = (x_1, \ldots, x_p)^T$ of measurements (^{T} denotes vector transpose), whose components x_i are measurements of the features of an object. Thus the features are the variables specified by the investigator and thought to be important for classification. In discrimination, we assume that there exist C groups or *classes*, denoted $\omega_1, \ldots, \omega_C$ and associated with each pattern $\boldsymbol{x}$ is a categorical variable z that denotes the class or group membership; that is, if $z = i$, then the pattern belongs to $\omega_i, i \in \{1, \ldots, C\}$.

Examples of patterns are measurements of an acoustic waveform in a speech recognition problem; measurements on a patient made in order to identify a disease (diagnosis); measurements on patients (perhaps subjective assessments) in order to predict the likely outcome (prognosis); measurements on weather variables (for forecasting or prediction); sets of financial measurements recorded over time; and a digitised image for character recognition. Therefore, we see that the term 'pattern', in its technical meaning, does not necessarily refer to structure within images.

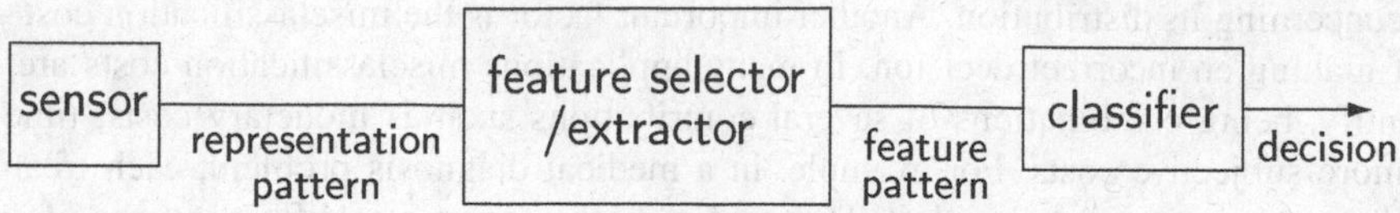

Figure 1.1 Pattern classifier.

The main topic in this book may be described by a number of terms including *pattern classifier design* or *discrimination* or *allocation rule design*. Designing the rule requires specification of the parameters of a pattern classifier, represented schematically in Figure 1.1, so that it yields the optimal (in some sense) response for a given input pattern. This response is usually an estimate of the class to which the pattern belongs. We assume that we have a set of patterns of known class $\{(\boldsymbol{x}_i, z_i), i = 1, \ldots, n\}$ (the *training* or *design* set) that we use to design the classifier (to set up its internal parameters). Once this has been done, we may estimate class membership for a pattern $\boldsymbol{x}$ for which the class label is unknown. Learning the model from a training set is the process of *induction*; applying the trained model to patterns of unknown class is the process of *deduction*.

Thus, the uses of a pattern classifier are to provide:

- A descriptive model that explains the difference between patterns of different classes in terms of features and their measurements.
- A predictive model that predicts the class of an unlabelled pattern.

However, we might ask why do we need a predictive model? Cannot the procedure that was used to assign labels to the training set measurements also be used for the test set in classifier operation? There may be several reasons for developing an automated process:

- to remove humans from the recognition process – to make the process more reliable;
- in banking, to identify good risk applicants before making a loan;
- to make a medical diagnosis without a post mortem (or to assess the state of a piece of equipment without dismantling it) – sometimes a pattern may only be labelled through intensive examination of a subject, whether person or piece of equipment;
- to reduce cost and improve speed – gathering and labelling data can be a costly and time consuming process;
- to operate in hostile environments – the operating conditions may be dangerous or harmful to humans and the training data have been gathered under controlled conditions;
- to operate remotely – to classify crops and land use remotely without labour-intensive, time consuming, surveys.

There are many classifiers that can be constructed from a given dataset. Examples include decision trees, neural networks, support vector machines and linear discriminant functions. For a classifier of a given type, we employ a learning algorithm to search through the parameter space to find the model that best describes the relationship between the measurements and class labels for the training set. The form derived for the pattern classifier depends on a number of different factors. It depends on the distribution of the training data, and the assumptions made concerning its distribution. Another important factor is the misclassification cost – the cost of making an incorrect decision. In many applications misclassification costs are hard to quantify, being combinations of several contributions such as monetary costs, time and other more subjective costs. For example, in a medical diagnosis problem, each treatment has different costs associated with it. These relate to the expense of different types of drugs, the suffering the patient is subjected to by each course of action and the risk of further complications.

Figure 1.1 grossly oversimplifies the pattern classification procedure. Data may undergo several separate transformation stages before a final outcome is reached. These transformations (sometimes termed preprocessing, feature selection or feature extraction) operate on the data in a way that, usually, reduces its dimension (reduces the number of features), removing redundant or irrelevant information, and transforms it to a form more appropriate for subsequent classification. The term *intrinsic dimensionality* refers to the minimum number of variables required to capture the structure within the data. In speech recognition, a preprocessing stage may be to transform the waveform to a frequency representation. This may be processed further to find formants (peaks in the spectrum). This is a *feature extraction* process (taking a possibly nonlinear combination of the original variables to form new variables). *Feature selection* is the process of selecting a subset of a given set of variables (see Chapter 10). In some problems, there is no automatic feature selection stage, with the feature selection being performed by the investigator who 'knows' (through experience, knowledge of previous studies and the problem domain) those variables that are important for classification. In many cases, however, it will be necessary to perform one or more transformations of the measured data.

In some pattern classifiers, each of the above stages may be present and identifiable as separate operations, while in others they may not be. Also, in some classifiers, the preliminary stages will tend to be problem specific, as in the speech example. In this book, we consider feature selection and extraction transformations that are not application specific. That is not to say the methods of feature transformation described will be suitable for any given application, however, but application-specific preprocessing must be left to the investigator who understands the application domain and method of data collection.

1.2 Stages in a pattern recognition problem

A pattern recognition investigation may consist of several stages enumerated below. Not all stages may be present; some may be merged together so that the distinction between two operations may not be clear, even if both are carried out; there may be some application-specific data processing that may not be regarded as one of the stages listed below. However, the points below are fairly typical.

1. Formulation of the problem: gaining a clear understanding of the aims of the investigation and planning the remaining stages.
2. Data collection: making measurements on appropriate variables and recording details of the data collection procedure (ground truth).

3. Initial examination of the data: checking the data, calculating summary statistics and producing plots in order to get a feel for the structure.
4. Feature selection or feature extraction: selecting variables from the measured set that are appropriate for the task. These new variables may be obtained by a linear or nonlinear transformation of the original set (feature extraction). To some extent, the partitioning of the data processing into separate feature extraction and classification processes is artificial, since a classifier often includes the optimisation of a feature extraction stage as part of its design.
5. Unsupervised pattern classification or clustering. This may be viewed as exploratory data analysis and it may provide a successful conclusion to a study. On the other hand, it may be a means of preprocessing the data for a supervised classification procedure.
6. Apply discrimination or regression procedures as appropriate. The classifier is designed using a training set of exemplar patterns.
7. Assessment of results. This may involve applying the trained classifier to an independent *test set* of labelled patterns. Classification performance is often summarised in the form of a confusion matrix:

		True class		
		ω_1	ω_2	ω_3
Predicted class	ω_1	e_{11}	e_{12}	e_{13}
	ω_2	e_{21}	e_{22}	e_{23}
	ω_3	e_{31}	e_{32}	e_{33}

where e_{ij} is the number of patterns of class ω_j that are predicted to be class ω_i. The accuracy, a, is calculated from the confusion matrix as

$$a = \frac{\sum_i e_{ii}}{\sum_{ij} e_{ij}}$$

and the error rate is $1 - a$.

8. Interpretation.

The above is necessarily an iterative process: the analysis of the results may generate new hypotheses that require further data collection. The cycle may be terminated at different stages: the questions posed may be answered by an initial examination of the data or it may be discovered that the data cannot answer the initial question and the problem must be reformulated.

The emphasis of this book is on techniques for performing the steps 4, 5, 6 and 7.

1.3 Issues

The main topic that we address in this book concerns classifier design: given a training set of patterns of known class, we seek to use those examples to design a classifier that is optimal for the expected operating conditions (the test conditions).

There are a number of very important points to make about this design process.

Finite design set

We are given a *finite* design set. If the classifier is too complex (there are too many free parameters) it may model noise in the design set. This is an example of *overfitting*. If the classifier is not complex enough, then it may fail to capture structure in the data. An illustration of this is the fitting of a set of data points by a polynomial curve (Figure 1.2). If the degree of the polynomial is too high then, although the curve may pass through or close to the data points thus achieving a low fitting error, the fitting curve is very variable and models every fluctuation in the data (due to noise). If the degree of the polynomial is too low, the fitting error is large and the underlying variability of the curve is not modelled (the model *underfits* the data). Thus, achieving optimal performance on the design set (in terms of minimising some error criterion perhaps) is not required: it may be possible, in a classification problem, to achieve 100% classification accuracy on the design set but the *generalisation performance* – the expected performance on data representative of the true operating conditions (equivalently the performance on an infinite test set of which the design set is a sample) – is poorer than could be achieved by careful design. Choosing the 'right' model is an exercise in *model selection*. In practice we usually do not know what is structure and what is noise in the data. Also, training a classifier (the procedure of determining its parameters) should not be considered as a separate issue from model selection, but it often is.

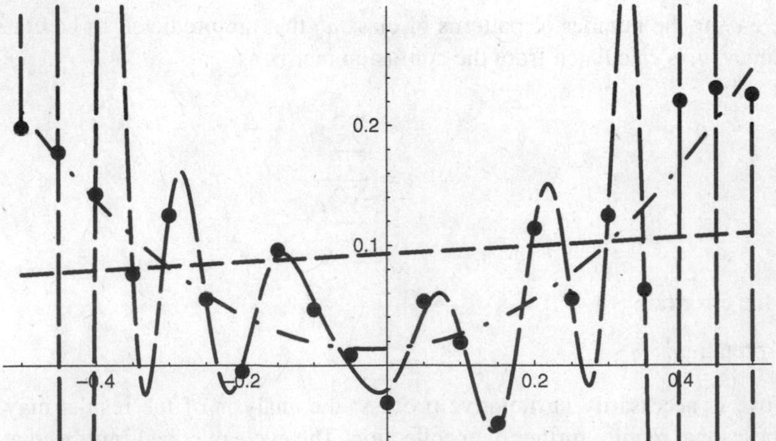

Figure 1.2 Fitting a curve to a noisy set of samples: the data samples are from a quadratic function with added noise; the fitting curves are a linear fit, a quadratic fit and a high-degree polynomial.

Optimality
A second point about the design of optimal classifiers concerns the word 'optimal'. There are several ways of measuring classifier performance, the most common being error rate, although this has severe limitations (see Chapter 9). Other measures, based on the closeness of the estimates of the probabilities of class membership to the true probabilities, may be more appropriate in many cases. However, many classifier design methods usually optimise alternative criteria since the desired ones are difficult to optimise directly. For example, a classifier may be trained by optimising a square-error measure and assessed using error rate.

Representative data
Finally, we assume that the training data are representative of the test conditions. If this is not so, perhaps because the test conditions may be subject to noise not present in the training data, or there are changes in the population from which the data are drawn (population drift), then these differences must be taken into account in the classifier design.

1.4 Approaches to statistical pattern recognition

There are two main divisions of classification: *supervised classification* (or discrimination) and *unsupervised classification* (sometimes in the statistics literature simply referred to as classification or clustering).

The problem we are addressing in this book is primarily one of *supervised pattern classification*. Given a set of measurements obtained through observation and represented as a pattern vector $\boldsymbol{x}$, we wish to assign the pattern to one of C possible classes, ω_i, $i = 1, \ldots, C$. A *decision rule* partitions the measurement space into C regions, Ω_i, $i = 1, \ldots, C$. If an observation vector is in Ω_i then it is assumed to belong to class ω_i. Each class region Ω_i may be multiply connected – that is, it may be made up of several disjoint regions. The boundaries between the regions Ω_i are the *decision boundaries* or *decision surfaces*. Generally, it is in regions close to these boundaries where the highest proportion of misclassifications occurs. In such situations, we may reject the pattern or withhold a decision until further information is available so that a classification may be made later. This option is known as the *reject option* and therefore we have $C + 1$ outcomes of a decision rule (the reject option being denoted by ω_0) in a C class problem: $\boldsymbol{x}$ belongs to ω_1 or ω_2 or ... or ω_C or withhold a decision.

In unsupervised classification, the data are not labelled and we seek to find groups in the data and the features that distinguish one group from another. Clustering techniques, described further in Chapter 11, can also be used as part of a supervised classification scheme by defining prototypes. A clustering scheme may be applied to the data for each class separately and representative samples for each group within the class (the group means for example) used as the prototypes for that class.

In the following section we introduce two approaches to discrimination that will be explored further in later chapters. The first assumes a knowledge of the underlying class-conditional probability density functions (the probability density function of the feature vectors for a given class). Of course, in many applications these will usually be unknown and must be estimated from a set of correctly classified samples termed the *design* or *training* set. Chapters 2, 3 and 4 describe techniques for estimating the probability density functions explicitly.

The second approach introduced in the next section develops decision rules that use the data to estimate the decision boundaries directly, without explicit calculation of the probability density functions. This approach is developed in Chapters 5 and 6 where specific techniques are described.

1.5 Elementary decision theory

Here we introduce an approach to discrimination based on knowledge of the probability density functions of each class. Familiarity with basic probability theory is assumed.

1.5.1 Bayes' decision rule for minimum error

Consider C classes, $\omega_1, \ldots, \omega_C$, with *a priori* probabilities (the probabilities of each class occurring) $p(\omega_1), \ldots, p(\omega_C)$, assumed known. If we wish to minimise the probability of making an error and we have no information regarding an object other than the class probability distribution then we would assign an object to class ω_j if

$$p(\omega_j) > p(\omega_k) \quad k = 1, \ldots, C; k \neq j$$

This classifies all objects as belonging to one class: the class with the largest prior probability. For classes with equal prior probabilities, patterns are assigned arbitrarily between those classes.

However, we do have an *observation vector* or *measurement vector* $\boldsymbol{x}$ and we wish to assign an object to one of the C classes based on the measurements $\boldsymbol{x}$. A decision rule based on probabilities is to assign $\boldsymbol{x}$ (here we refer to an object in terms of its measurement vector) to class ω_j if the probability of class ω_j given the observation $\boldsymbol{x}$, that is $p(\omega_j|\boldsymbol{x})$, is greatest over all classes $\omega_1, \ldots, \omega_C$. That is, assign $\boldsymbol{x}$ to class ω_j if

$$p(\omega_j|\boldsymbol{x}) > p(\omega_k|\boldsymbol{x}) \quad k = 1, \ldots, C; k \neq j \tag{1.1}$$

This decision rule partitions the measurement space into C regions $\Omega_1, \ldots, \Omega_C$ such that if $\boldsymbol{x} \in \Omega_j$ then $\boldsymbol{x}$ belongs to class ω_j. The regions Ω_j may be disconnected.

The *a posteriori* probabilities $p(\omega_j|\boldsymbol{x})$ may be expressed in terms of the *a priori* probabilities and the class conditional density functions $p(\boldsymbol{x}|\omega_i)$ using Bayes' theorem as

$$p(\omega_i|\boldsymbol{x}) = \frac{p(\boldsymbol{x}|\omega_i)p(\omega_i)}{p(\boldsymbol{x})}$$

and so the decision rule (1.1) may be written: assign $\boldsymbol{x}$ to ω_j if

$$p(\boldsymbol{x}|\omega_j)p(\omega_j) > p(\boldsymbol{x}|\omega_k)p(\omega_k) \quad k = 1, \ldots, C; k \neq j \tag{1.2}$$

This is known as Bayes' rule for *minimum error*.

For two classes, the decision rule (1.2) may be written

$$l_r(\boldsymbol{x}) = \frac{p(\boldsymbol{x}|\omega_1)}{p(\boldsymbol{x}|\omega_2)} > \frac{p(\omega_2)}{p(\omega_1)} \text{ implies } \boldsymbol{x} \in \text{class } \omega_1$$

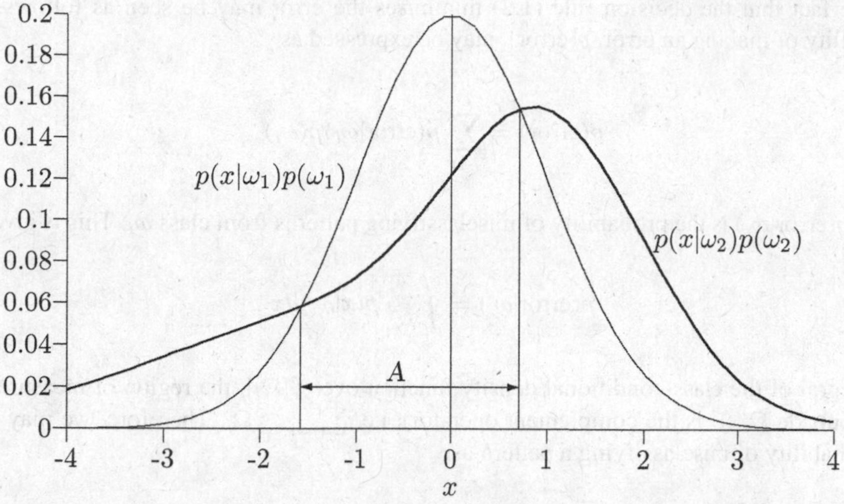

Figure 1.3 $p(x|\omega_i)p(\omega_i)$, for classes ω_1 and ω_2: for x in region A, x is assigned to class ω_1.

The function $l_r(\boldsymbol{x})$ is the *likelihood ratio*. Figures 1.3 and 1.4 give a simple illustration for a two-class discrimination problem. Class ω_1 is normally distributed with zero mean and unit variance, $p(\boldsymbol{x}|\omega_1) = N(\boldsymbol{x}; 0, 1)$. Class ω_2 is a *normal mixture* (a weighted sum of normal densities) $p(\boldsymbol{x}|\omega_2) = 0.6N(\boldsymbol{x}; 1, 1) + 0.4N(\boldsymbol{x}; -1, 2)$. Figure 1.3 plots $p(\boldsymbol{x}|\omega_i)p(\omega_i), i = 1, 2$, where the priors are taken to be $p(\omega_1) = 0.5, p(\omega_2) = 0.5$. Figure 1.4 plots the likelihood ratio $l_r(\boldsymbol{x})$ and the threshold $p(\omega_2)/p(\omega_1)$. We see from this figure that the decision rule (1.2) leads to a disconnected region for class ω_2.

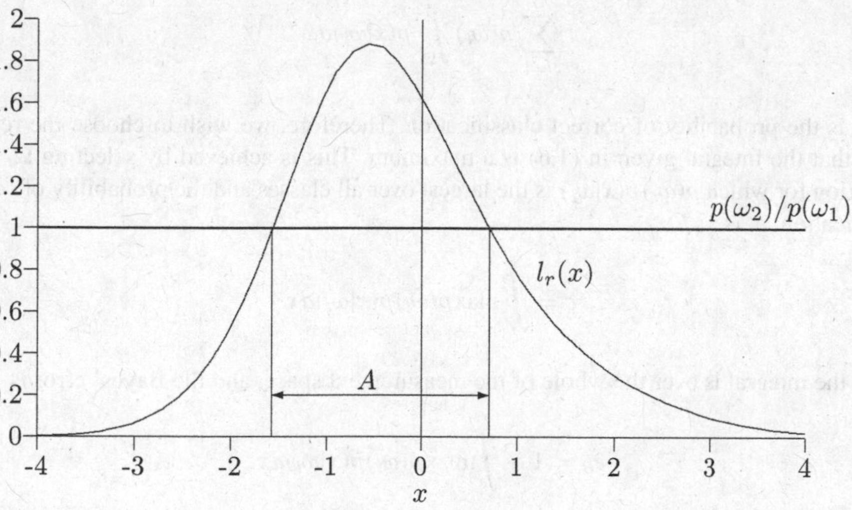

Figure 1.4 Likelihood function: for x in region A, x is assigned to class ω_1.

The fact that the decision rule (1.2) minimises the error may be seen as follows. The probability of making an error, $p(\text{error})$, may be expressed as

$$p(\text{error}) = \sum_{i=1}^{C} p(\text{error}|\omega_i)p(\omega_i) \tag{1.3}$$

where $p(\text{error}|\omega_i)$ is the probability of misclassifying patterns from class ω_i. This is given by

$$p(\text{error}|\omega_i) = \int_{\mathcal{C}[\Omega_i]} p(\boldsymbol{x}|\omega_i)d\boldsymbol{x} \tag{1.4}$$

the integral of the class-conditional density function over $\mathcal{C}[\Omega_i]$, the region of measurement space outside Ω_i ($\mathcal{C}$ is the complement operator), i.e. $\sum_{j=1, j\neq i}^{C} \Omega_j$. Therefore, we may write the probability of misclassifying a pattern as

$$\begin{aligned} p(\text{error}) &= \sum_{i=1}^{C} \int_{\mathcal{C}[\Omega_i]} p(\boldsymbol{x}|\omega_i)p(\omega_i)d\boldsymbol{x} \\ &= \sum_{i=1}^{C} p(\omega_i)\left(1 - \int_{\Omega_i} p(\boldsymbol{x}|\omega_i)d\boldsymbol{x}\right) \\ &= 1 - \sum_{i=1}^{C} p(\omega_i) \int_{\Omega_i} p(\boldsymbol{x}|\omega_i)d\boldsymbol{x} \end{aligned} \tag{1.5}$$

from which we see that minimising the probability of making an error is equivalent to maximising

$$\sum_{i=1}^{C} p(\omega_i) \int_{\Omega_i} p(\boldsymbol{x}|\omega_i)d\boldsymbol{x} \tag{1.6}$$

which is the probability of correct classification. Therefore, we wish to choose the regions Ω_i so that the integral given in (1.6) is a maximum. This is achieved by selecting Ω_i to be the region for which $p(\omega_i)p(\boldsymbol{x}|\omega_i)$ is the largest over all classes and the probability of correct classification, c, is

$$c = \int \max_i p(\omega_i)p(\boldsymbol{x}|\omega_i)d\boldsymbol{x} \tag{1.7}$$

where the integral is over the whole of the measurement space, and the Bayes' error is

$$e_B = 1 - \int \max_i p(\omega_i)p(\boldsymbol{x}|\omega_i)d\boldsymbol{x} \tag{1.8}$$

This is illustrated in Figures 1.5 and 1.6.

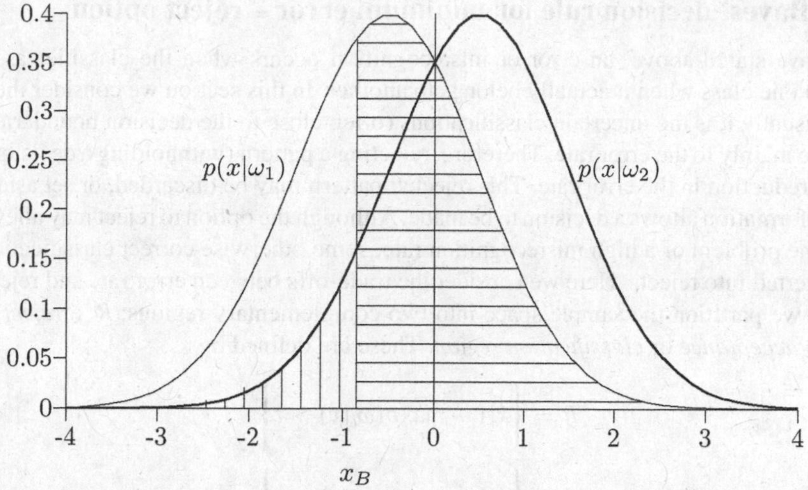

Figure 1.5 Class-conditional densities for two normal distributions.

Figure 1.5 plots the two distributions $p(\mathbf{x}|\omega_i)$, $i = 1, 2$ (both normal with unit variance and means ± 0.5), and Figure 1.6 plots the functions $p(\mathbf{x}|\omega_i)p(\omega_i)$ where $p(\omega_1) = 0.3$, $p(\omega_2) = 0.7$. The Bayes' decision boundary defined by the point where $p(\mathbf{x}|\omega_1)p(\omega_1) = p(\mathbf{x}|\omega_2)p(\omega_2)$ (Figure 1.6) is marked with a vertical line at x_B. The areas of the hatched regions in Figure 1.5 represent the probability of error: by Equation (1.4), the area of the horizontal hatching is the probability of classifying a pattern from class 1 as a pattern from class 2 and the area of the vertical hatching the probability of classifying a pattern from class 2 as class 1. The sum of these two areas, weighted by the priors [Equation (1.5)], is the probability of making an error.

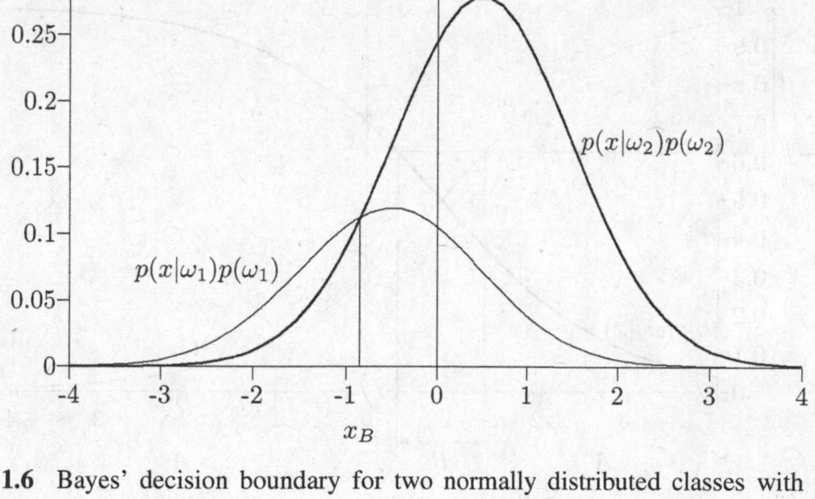

Figure 1.6 Bayes' decision boundary for two normally distributed classes with unequal priors.

1.5.2 Bayes' decision rule for minimum error – reject option

As we have stated above, an error or misrecognition occurs when the classifier assigns a pattern to one class when it actually belongs to another. In this section we consider the reject option. Usually it is the uncertain classifications (often close to the decision boundaries) that contribute mainly to the error rate. Therefore, rejecting a pattern (withholding a decision) may lead to a reduction in the error rate. This rejected pattern may be discarded, or set aside until further information allows a decision to be made. Although the option to reject may alleviate or remove the problem of a high misrecognition rate, some otherwise correct classifications are also converted into rejects. Here we consider the trade-offs between error rate and reject rate.

First, we partition the sample space into two complementary regions: R, a *reject region* and A, an *acceptance* or *classification region*. These are defined by

$$R = \left\{ \boldsymbol{x} | 1 - \max_i p(\omega_i|\boldsymbol{x}) > t \right\}$$

$$A = \left\{ \boldsymbol{x} | 1 - \max_i p(\omega_i|\boldsymbol{x}) \leq t \right\}$$

where t is a threshold. This is illustrated in Figure 1.7 using the same distributions as those in Figures 1.5 and 1.6.

The smaller the value of the threshold t then the larger is the reject region R. However, if t is chosen such that

$$1 - t \leq \frac{1}{C}$$

or equivalently,

$$t \geq \frac{C-1}{C}$$

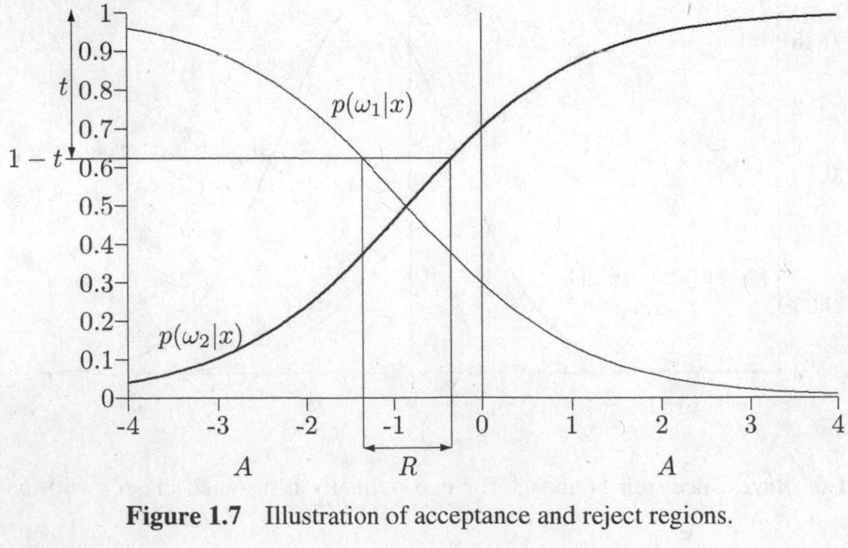

Figure 1.7 Illustration of acceptance and reject regions.

where C is the number of classes, then the reject region is empty. This is because the minimum value which $\max_i p(\omega_i|\boldsymbol{x})$ can attain is $1/C$ [since $1 = \sum_{i=1}^{C} p(\omega_i|\boldsymbol{x}) \leq C \max_i p(\omega_i|\boldsymbol{x})$], when all classes are equally likely. Therefore, for the reject option to be activated, we must have $t < (C-1)/C$.

Thus, if a pattern $\boldsymbol{x}$ lies in the region A, we classify it according to the Bayes' rule for minimum error [Equation (1.2)]. However, if $\boldsymbol{x}$ lies in the region R, we reject $\boldsymbol{x}$ (withhold a decision).

The probability of correct classification, $c(t)$, is a function of the threshold, t, and is given by Equation (1.7), where now the integral is over the acceptance region, A, only

$$c(t) = \int_A \max_i \left[p(\omega_i)p(\boldsymbol{x}|\omega_i)\right] d\boldsymbol{x}$$

and the unconditional probability of rejecting a measurement, r, also a function of the threshold t, is the probability that it lies in R:

$$r(t) = \int_R p(\boldsymbol{x})d\boldsymbol{x} \tag{1.9}$$

Therefore, the error rate, e (the probability of accepting a point for classification and incorrectly classifying it), is

$$\begin{aligned} e(t) &= \int_A \left(1 - \max_i p(\omega_i|\boldsymbol{x})\right) p(\boldsymbol{x})d\boldsymbol{x} \\ &= 1 - c(t) - r(t) \end{aligned}$$

Thus, the error rate and reject rate are inversely related. Chow (1970) derives a simple functional relationship between $e(t)$ and $r(t)$ which we quote here without proof. Knowing $r(t)$ over the complete range of t allows $e(t)$ to be calculated using the relationship

$$e(t) = -\int_0^t s\,dr(s) \tag{1.10}$$

The above result allows the error rate to be evaluated from the reject function for the Bayes' optimum classifier. The reject function can be calculated using *unlabelled* data and a practical application of the above result is to problems where labelling of gathered data is costly.

1.5.3 Bayes' decision rule for minimum risk

In the previous section, the decision rule selected the class for which the *a posteriori* probability, $p(\omega_j|\boldsymbol{x})$, was the greatest. This minimised the probability of making an error. We now consider a somewhat different rule that minimises an expected *loss* or risk. This is a very important concept since in many applications the costs associated with misclassification depend upon the true class of the pattern and the class to which it is assigned. For example, in a medical diagnosis problem in which a patient has back pain, it is far worse to classify a patient with severe spinal abnormality as healthy (or having mild back ache) than the other way round.

We make this concept more formal by introducing a loss that is a measure of the cost of making the decision that a pattern belongs to class ω_i when the true class is ω_j. We define a loss matrix $\mathbf{\Lambda}$ with components

$$\lambda_{ji} = \text{cost of assigning a pattern } \boldsymbol{x} \text{ to } \omega_i \text{ when } \boldsymbol{x} \in \omega_j$$

In practice, it may be very difficult to assign costs. In some situations, λ may be measured in monetary units that are quantifiable. However, in many situations, costs are a combination of several different factors measured in different units – money, time, quality of life. As a consequence, they are often a subjective opinion of an expert. The *conditional risk* of assigning a pattern $\boldsymbol{x}$ to class ω_i is defined as

$$l^i(\boldsymbol{x}) = \sum_{j=1}^{C} \lambda_{ji} p(\omega_j|\boldsymbol{x})$$

The average risk over region Ω_i is

$$\begin{aligned} r^i &= \int_{\Omega_i} l^i(\boldsymbol{x})\, p(\boldsymbol{x}) d\boldsymbol{x} \\ &= \int_{\Omega_i} \sum_{j=1}^{C} \lambda_{ij} p(\omega_i|\boldsymbol{x}) p(\boldsymbol{x}) d\boldsymbol{x} \end{aligned}$$

and the overall expected cost or *risk* is

$$r = \sum_{i=1}^{C} r^i = \sum_{i=1}^{C} \int_{\Omega_i} \sum_{j=1}^{C} \lambda_{ji} p(\omega_j|\boldsymbol{x}) p(\boldsymbol{x}) d\boldsymbol{x} \tag{1.11}$$

The above expression for the risk will be minimised if the regions Ω_i are chosen such that if

$$\sum_{j=1}^{C} \lambda_{ji} p(\omega_j|\boldsymbol{x}) p(\boldsymbol{x}) \leq \sum_{j=1}^{C} \lambda_{jk} p(\omega_j|\boldsymbol{x}) p(\boldsymbol{x}) \quad k = 1, \ldots, C \tag{1.12}$$

then $\boldsymbol{x} \in \Omega_i$. This is the *Bayes' decision rule for minimum risk*, with Bayes' risk, r^*, given by

$$r^* = \int_{\boldsymbol{x}} \min_{i=1,\ldots,C} \sum_{j=1}^{C} \lambda_{ji} p(\omega_j|\boldsymbol{x}) p(\boldsymbol{x}) d\boldsymbol{x}$$

One special case of the loss matrix $\mathbf{\Lambda}$ is the *equal cost* loss matrix for which

$$\lambda_{ij} = \begin{cases} 1 & i \neq j \\ 0 & i = j \end{cases}$$

Substituting into (1.12) gives the decision rule: assign x to class ω_i if

$$\sum_{j=1}^{C} p(\omega_j|x)p(x) - p(\omega_i|x)p(x) \leq \sum_{j=1}^{C} p(\omega_j|x)p(x) - p(\omega_k|x)p(x) \quad k = 1, \ldots, C$$

that is,

$$p(x|\omega_i)p(\omega_i) \geq p(x|\omega_k)p(\omega_k) \quad k = 1, \ldots, C$$

implies that $x \in$ class ω_i; this is the Bayes' rule for minimum error.

1.5.4 Bayes' decision rule for minimum risk – reject option

As with the Bayes' rule for minimum error, we may also introduce a reject option, by which the reject region, R, is defined by

$$R = \left\{ x \left| \min_i l^i(x) > t \right. \right\}$$

where t is a threshold. The decision is to accept a pattern x and assign it to class ω_i if

$$l^i(x) = \min_j l^j(x) \leq t$$

and to reject x if

$$l^i(x) = \min_j l^j(x) > t$$

This decision is equivalent to defining a reject region Ω_0 with a constant conditional risk

$$l^0(x) = t$$

so that the Bayes' decision rule is: assign x to class ω_i if

$$l^i(x) \leq l^j(x) \quad j = 0, 1, \ldots, C$$

with Bayes' risk

$$r^* = \int_R tp(x)dx + \int_A \min_{i=1,\ldots,C} \sum_{j=1}^{C} \lambda_{ji} p(\omega_j|x)p(x)dx \tag{1.13}$$

1.5.5 Neyman–Pearson decision rule

An alternative to the Bayes' decision rules for a two-class problem is the Neyman–Pearson test. In a two-class problem there are two possible types of error that may be made in the decision process. We may classify a pattern of class ω_1 as belonging to class ω_2 or a pattern from class ω_2 as belonging to class ω_1. Let the probability of these two errors be ϵ_1 and ϵ_2, respectively, so that

$$\epsilon_1 = \int_{\Omega_2} p(\boldsymbol{x}|\omega_1)\, d\boldsymbol{x} = \text{error probability of Type I}$$

and

$$\epsilon_2 = \int_{\Omega_1} p(\boldsymbol{x}|\omega_2)\, d\boldsymbol{x} = \text{error probability of Type II}$$

The Neyman–Pearson decision rule is to minimise the error ϵ_1 subject to ϵ_2 being equal to a constant, ϵ_0, say.

If class ω_1 is termed the positive class and class ω_2 the negative class, then ϵ_1 is referred to as the *false negative rate*: the proportion of positive samples incorrectly assigned to the negative class; ϵ_2 is the *false positive rate*: the proportion of negative samples classed as positive.

An example of the use of the Neyman–Pearson decision rule is in radar detection where the problem is to detect a signal in the presence of noise. There are two types of error that may occur; one is to mistake noise for a signal present. This is called a *false alarm*. The second type of error occurs when a signal is actually present but the decision is made that only noise is present. This is a *missed detection*. If ω_1 denotes the signal class and ω_2 denotes the noise then ϵ_2 is the probability of false alarm and ϵ_1 is the probability of missed detection. In many radar applications, a threshold is set to give a fixed probability of false alarm and therefore the Neyman–Pearson decision rule is the one usually used.

We seek the minimum of

$$r = \int_{\Omega_2} p(\boldsymbol{x}|\omega_1)\, d\boldsymbol{x} + \mu \left\{ \int_{\Omega_1} p(\boldsymbol{x}|\omega_2)\, d\boldsymbol{x} - \epsilon_0 \right\}$$

where μ is a Lagrange multiplier[1] and ϵ_0 is the specified false alarm rate. The equation may be written

$$r = (1 - \mu\epsilon_0) + \int_{\Omega_1} \{\mu p(\boldsymbol{x}|\omega_2)\, d\boldsymbol{x} - p(\boldsymbol{x}|\omega_1)\, d\boldsymbol{x}\}$$

This will be minimised if we choose Ω_1 such that the integrand is negative, i.e.

$$\text{if } \mu p(\boldsymbol{x}|\omega_2) - p(\boldsymbol{x}|\omega_1) < 0, \text{ then } \boldsymbol{x} \in \Omega_1$$

or, in terms of the likelihood ratio,

$$\text{if } \frac{p(\boldsymbol{x}|\omega_1)}{p(\boldsymbol{x}|\omega_2)} > \mu, \text{ then } \boldsymbol{x} \in \Omega_1 \tag{1.14}$$

---

[1] The method of Lagrange's undetermined multipliers can be found in most textbooks on mathematical methods, for example Wylie and Barrett (1995).

Thus the decision rule depends only on the within-class distributions and ignores the *a priori* probabilities.

The threshold μ is chosen so that

$$\int_{\Omega_1} p(\boldsymbol{x}|\omega_2)\,d\boldsymbol{x} = \epsilon_0,$$

the specified false alarm rate. However, in general μ cannot be determined analytically and requires numerical calculation.

Often, the performance of the decision rule is summarised in a receiver operating characteristic (ROC) curve, which plots the true positive against the false positive (that is, the probability of detection $[1 - \epsilon_1 = \int_{\Omega_1} p(\boldsymbol{x}|\omega_1)\,d\boldsymbol{x}]$ against the probability of false alarm $[\epsilon_2 = \int_{\Omega_1} p(\boldsymbol{x}|\omega_2)\,d\boldsymbol{x})]$ as the threshold μ is varied. This is illustrated in Figure 1.8 for the univariate case of two normally distributed classes of unit variance and means separated by a distance, d. All the ROC curves pass through the (0, 0) and (1, 1) points and as the separation increases the curve moves into the top left corner. Ideally, we would like 100% detection for a 0% false alarm rate and curves that are closer to this the better.

For the two-class case, the minimum risk decision [see Equation (1.12)] defines the decision rules on the basis of the likelihood ratio ($\lambda_{ii} = 0$):

$$\text{if } \frac{p(\boldsymbol{x}|\omega_1)}{p(\boldsymbol{x}|\omega_2)} > \frac{\lambda_{21}p(\omega_2)}{\lambda_{12}p(\omega_1)}, \text{ then } \boldsymbol{x} \in \Omega_1 \tag{1.15}$$

The threshold defined by the right-hand side will correspond to a particular point on the ROC curve that depends on the misclassification costs and the prior probabilities.

In practice, precise values for the misclassification costs will be unavailable and we shall need to assess the performance over a range of expected costs. The use of the ROC curve as a tool for comparing and assessing classifier performance is discussed in Chapter 9.

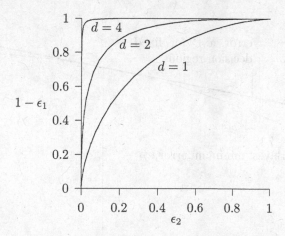

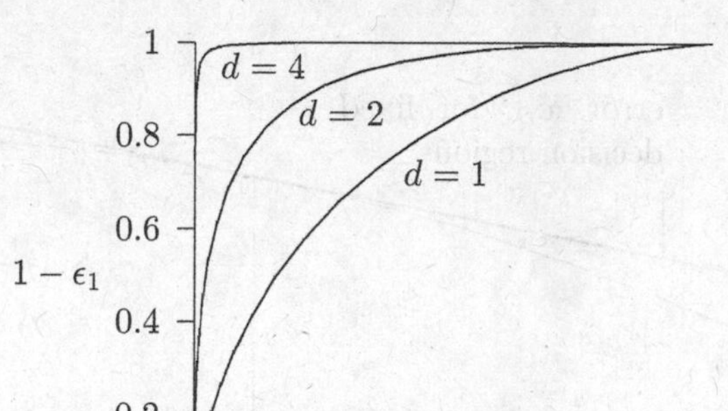

Figure 1.8 Receiver operating characteristic for two univariate normal distributions of unit variance and separation, d; $1 - \epsilon_1 = \int_{\Omega_1} p(\boldsymbol{x}|\omega_1)\,d\boldsymbol{x}$ is the true positive (the probability of detection) and $\epsilon_2 = \int_{\Omega_1} p(\boldsymbol{x}|\omega_2)\,d\boldsymbol{x}$ is the false positive (the probability of false alarm).

1.5.6 Minimax criterion

The Bayes' decision rules rely on a knowledge of both the within-class distributions and the prior class probabilities. However, situations may arise where the relative frequencies of new objects to be classified are unknown. In this situation a *minimax* procedure may be employed. The name *minimax* is used to refer to procedures for which either the maximum expected loss *or* the maximum of the error probability is a minimum. We shall limit our discussion below to the two-class problem and the minimum error probability procedure.

Consider the Bayes' rule for minimum error. The decision regions Ω_1 and Ω_2 are defined by

$$p(\boldsymbol{x}|\omega_1)p(\omega_1) > p(\boldsymbol{x}|\omega_2)p(\omega_2) \text{ implies } \boldsymbol{x} \in \Omega_1 \tag{1.16}$$

and the Bayes' minimum error, e_B, is

$$e_B = p(\omega_2)\int_{\Omega_1} p(\boldsymbol{x}|\omega_2)\,d\boldsymbol{x} + p(\omega_1)\int_{\Omega_2} p(\boldsymbol{x}|\omega_1)\,d\boldsymbol{x} \tag{1.17}$$

where $p(\omega_2) = 1 - p(\omega_1)$.

For *fixed* decision regions Ω_1 and Ω_2, e_B is a linear function of $p(\omega_1)$ (we denote this function $\tilde{e}_B$) attaining its maximum on the region [0, 1] either at $p(\omega_1) = 0$ or $p(\omega_1) = 1$. However, since the regions Ω_1 and Ω_2 are also dependent on $p(\omega_1)$ through the Bayes' decision criterion (1.16), the dependency of e_B on $p(\omega_1)$ is more complex, and not necessarily monotonic.

If Ω_1 and Ω_2 are fixed [determined according to (1.16) for some specified $p(\omega_i)$], the error given by (1.17) will only be the Bayes' minimum error for a particular value of $p(\omega_1)$, say p_1^* (Figure 1.9).

For other values of $p(\omega_1)$, the error given by (1.17) must be greater than the minimum error. Therefore, the optimum curve touches the line at a tangent at p_1^* and is concave down at that point.

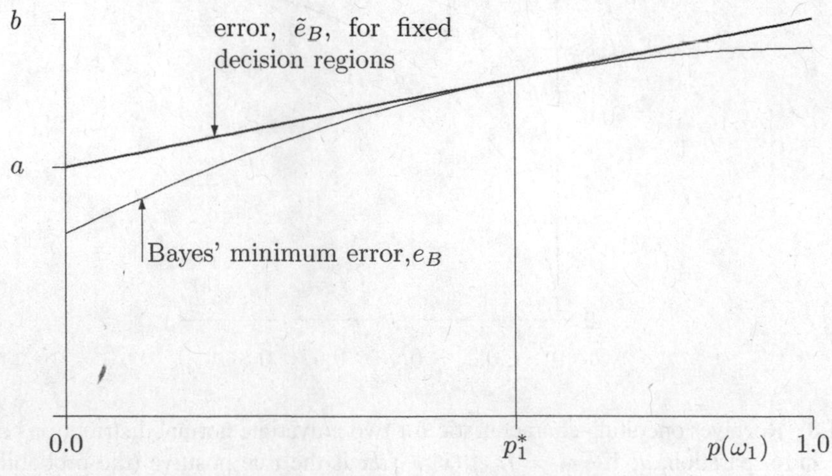

Figure 1.9 Minimax illustration.

The minimax procedure aims to choose the partition Ω_1, Ω_2, or equivalently the value of $p(\omega_1)$ so that the maximum error (on a test set in which the values of $p(\omega_i)$ are unknown) is minimised. For example, in Figure 1.9, if the partition were chosen to correspond to the value p_1^* of $p(\omega_1)$, then the maximum error which could occur would be a value of b if $p(\omega_1)$ were actually equal to unity. The minimax procedure aims to minimise this maximum value, i.e. minimise

$$\max\{\tilde{e}_B(0), \tilde{e}_B(1)\}$$

or minimise

$$\max\left\{\int_{\Omega_2} p(\boldsymbol{x}|\omega_1)\,d\boldsymbol{x}, \int_{\Omega_1} p(\boldsymbol{x}|\omega_2)\,d\boldsymbol{x}\right\}$$

This is a minimum when

$$\int_{\Omega_2} p(\boldsymbol{x}|\omega_1)\,d\boldsymbol{x} = \int_{\Omega_1} p(\boldsymbol{x}|\omega_2)\,d\boldsymbol{x} \tag{1.18}$$

which is when $a = b$ in Figure 1.9 and the line $\tilde{e}_B(p(\omega_1))$ is horizontal and touches the Bayes' minimum error curve at its peak value.

Therefore, we choose the regions Ω_1 and Ω_2 so that the probabilities of the two types of error are the same. The minimax solution may be criticised as being over-pessimistic since it is a Bayes' solution with respect to the least favourable prior distribution. The strategy may also be applied to minimising the maximum risk. In this case, the risk is

$$\int_{\Omega_1} [\lambda_{11} p(\omega_1|\boldsymbol{x}) + \lambda_{21} p(\omega_2|\boldsymbol{x})]\, p(\boldsymbol{x})d\boldsymbol{x} + \int_{\Omega_2} [\lambda_{12} p(\omega_1|\boldsymbol{x}) + \lambda_{22} p(\omega_2|\boldsymbol{x})]\, p(\boldsymbol{x})d\boldsymbol{x}$$

$$= p(\omega_1)\left[\lambda_{11} + (\lambda_{12} - \lambda_{11})\int_{\Omega_2} p(\boldsymbol{x}|\omega_1)d\boldsymbol{x}\right]$$

$$+ p(\omega_2)\left[\lambda_{22} + (\lambda_{21} - \lambda_{22})\int_{\Omega_1} p(\boldsymbol{x}|\omega_2)d\boldsymbol{x}\right]$$

and the boundary must therefore satisfy

$$\lambda_{11} - \lambda_{22} + (\lambda_{12} - \lambda_{11})\int_{\Omega_2} p(\boldsymbol{x}|\omega_1)\,d\boldsymbol{x} - (\lambda_{21} - \lambda_{22})\int_{\Omega_1} p(\boldsymbol{x}|\omega_2)\,d\boldsymbol{x} = 0$$

For $\lambda_{11} = \lambda_{22}$ and $\lambda_{21} = \lambda_{12}$, this reduces to the condition (1.18).

1.5.7 Discussion

In this section we have introduced a decision theoretic approach to classifying patterns. This divides up the measurement space into decision regions and we have looked at various strategies for obtaining the decision boundaries. The optimum rule in the sense of minimising the error is the Bayes' decision rule for minimum error. Introducing the costs of making incorrect decisions leads to the Bayes' rule for minimum risk. The theory developed assumes that the *a priori* distributions and the class-conditional distributions are known. In a real-world task, this is unlikely to be so. Therefore approximations must be made based on the data available. We consider techniques for estimating distributions in Chapters 2, 3 and 4. Two alternatives to the Bayesian decision rule have also been described, namely the Neyman–Pearson decision rule (commonly used in signal processing applications) and the minimax rule. Both require knowledge of the class-conditional probability density functions. The ROC curve characterises the performance of a rule over a range of thresholds of the likelihood ratio.

We have seen that the error rate plays an important part in decision making and classifier performance assessment. Consequently, estimation of error rates is a problem of great interest in statistical pattern recognition. For given fixed decision regions, we may calculate the probability of error using Equation (1.5). If these decision regions are chosen according to the Bayes' decision rule [Equation (1.2)], then the error is the *Bayes' error rate* or *optimal error rate*. However, regardless of how the decision regions are chosen, the error rate may be regarded as a measure of a given decision rule's performance.

Calculation of the Bayes' error rate (1.8) requires complete knowledge of the class conditional density functions. In a particular situation, these may not be known and a classifier may be designed on the basis of a training set of samples. Given this training set, we may choose to form *estimates* of the distributions (using some of the techniques discussed in Chapters 2 and 3) and thus, with these estimates, use the Bayes decision rule and estimate the error according to (1.8).

However, even with accurate estimates of the distributions, evaluation of the error requires an integral over a multidimensional space and may prove a formidable task. An alternative approach is to obtain bounds on the optimal error rate or distribution-free estimates. Further discussion of methods of error rate estimation is given in Chapter 9.

1.6 Discriminant functions

1.6.1 Introduction

In the previous section, classification was achieved by applying the Bayesian decision rule. This requires knowledge of the class-conditional density functions, $p(\boldsymbol{x}|\omega_i)$ (such as normal distributions whose parameters are estimated from the data – see Chapter 2), or nonparametric density estimation methods (such as kernel density estimation – see Chapter 4). Here, instead of making assumptions about $p(\boldsymbol{x}|\omega_i)$, we make assumptions about the forms of the *discriminant functions*.

A discriminant function is a function of the pattern $\boldsymbol{x}$ that leads to a classification rule. For example, in a two-class problem, a discriminant function $h(\boldsymbol{x})$ is a function for which

$$\begin{aligned} h(\boldsymbol{x}) > k &\Rightarrow x \in \omega_1 \\ h(\boldsymbol{x}) < k &\Rightarrow x \in \omega_2 \end{aligned} \tag{1.19}$$

for constant k. In the case of equality $[h(\boldsymbol{x}) = k]$, the pattern $\boldsymbol{x}$ may be assigned arbitrarily to one of the two classes. An optimal discriminant function for the two-class case is

$$h(\boldsymbol{x}) = \frac{p(\boldsymbol{x}|\omega_1)}{p(\boldsymbol{x}|\omega_2)}$$

with $k = p(\omega_2)/p(\omega_1)$. Discriminant functions are not unique. If f is a monotonic function then

$$g(\boldsymbol{x}) = f(h(\boldsymbol{x})) > k' \Rightarrow \boldsymbol{x} \in \omega_1$$
$$g(\boldsymbol{x}) = f(h(\boldsymbol{x})) < k' \Rightarrow \boldsymbol{x} \in \omega_2$$

where $k' = f(k)$ leads to the same decision as (1.19).

In the C group case we define C discriminant functions $g_i(\boldsymbol{x})$ such that

$$g_i(\boldsymbol{x}) > g_j(\boldsymbol{x}) \Rightarrow \boldsymbol{x} \in \omega_i \qquad j = 1, \ldots, C, j \neq i$$

That is, a pattern is assigned to the class with the largest discriminant. Of course, for two classes, a single discriminant function

$$h(\boldsymbol{x}) = g_1(\boldsymbol{x}) - g_2(\boldsymbol{x})$$

with $k = 0$ reduces to the two-class case given by (1.19).

Again, we may define an optimal discriminant function as

$$g_i(\boldsymbol{x}) = p(\boldsymbol{x}|\omega_i)p(\omega_i)$$

leading to the Bayes' decision rule, but as we showed for the two-class case, there are other discriminant functions that lead to the same decision.

The essential difference between the approach of the previous section and the discriminant function approach described here is that the form of the discriminant function is specified and is not imposed by the underlying distribution. The choice of discriminant function may depend on prior knowledge about the patterns to be classified or may be a particular functional form whose parameters are adjusted by a training procedure. Many different forms of discriminant function have been considered in the literature, varying in complexity from the linear discriminant function (in which g is a linear combination of the x_i) to multiparameter nonlinear functions such as the multilayer perceptron.

Discrimination may also be viewed as a problem in *regression* (see Section 1.7) in which the dependent variable, y, is a class indicator and the regressors are the pattern vectors. Many discriminant function models lead to estimates of $\mathrm{E}[y|\boldsymbol{x}]$, which is the aim of regression analysis (though in regression y is not necessarily a class indicator). Thus, many of the techniques we shall discuss for optimising discriminant functions apply equally well to regression problems. Indeed, as we find with feature extraction in Chapter 10 and also clustering in Chapter 11 similar techniques have been developed under different names in the pattern recognition and statistics literature.

1.6.2 Linear discriminant functions

First of all, let us consider the family of discriminant functions that are linear combinations of the components of $\boldsymbol{x} = (x_1, \ldots, x_p)^T$,

$$g(\boldsymbol{x}) = \boldsymbol{w}^T \boldsymbol{x} + w_0 = \sum_{i=1}^{p} w_i x_i + w_0 \tag{1.20}$$

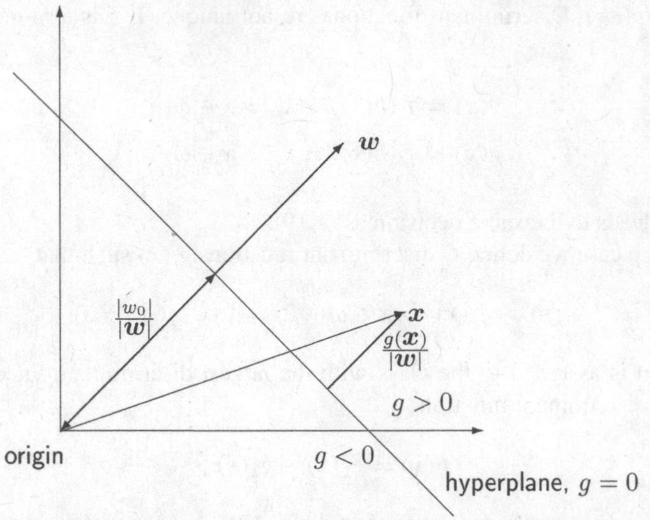

Figure 1.10 Geometry of linear discriminant function given by Equation (1.20).

This is a *linear discriminant function*, a complete specification of which is achieved by prescribing the *weight vector* $\mathbf{w}$ and *threshold weight* w_0. Equation (1.20) is the equation of a hyperplane with unit normal in the direction of $\mathbf{w}$ and a perpendicular distance $|w_0|/|\mathbf{w}|$ from the origin. The value of the discriminant function for a pattern $\boldsymbol{x}$ is a measure of the perpendicular distance from the hyperplane (Figure 1.10).

A linear discriminant function can arise through assumptions of normal distributions for the class densities, with equal covariance matrices (see Chapter 2). Alternatively, without making distributional assumptions, we may impose the form of the discriminant function to be linear and determine its parameters (see Chapter 5).

A pattern classifier employing linear discriminant functions is termed a *linear machine* (Nilsson, 1965), an important special case of which is the *minimum-distance classifier*. Suppose we are given a set of prototype points $\boldsymbol{p}_1, \ldots, \boldsymbol{p}_C$, one for each of the C classes $\omega_1, \ldots, \omega_C$. The minimum-distance classifier assigns a pattern $\boldsymbol{x}$ to the class ω_i associated with the nearest point $\boldsymbol{p}_i$. For each point, the squared Euclidean distance is

$$|\boldsymbol{x} - \boldsymbol{p}_i|^2 = \boldsymbol{x}^T\boldsymbol{x} - 2\boldsymbol{x}^T\boldsymbol{p}_i + \boldsymbol{p}_i^T\boldsymbol{p}_i$$

and minimum-distance classification is achieved by comparing the expressions $\boldsymbol{x}^T\boldsymbol{p}_i - \frac{1}{2}\boldsymbol{p}_i^T\boldsymbol{p}_i$ and selecting the largest value. Thus, the linear discriminant function is

$$g_i(\boldsymbol{x}) = \boldsymbol{w}_i^T\boldsymbol{x} + w_{i0}$$

where

$$\mathbf{w}_i = \boldsymbol{p}_i$$
$$w_{i0} = -\frac{1}{2}|\boldsymbol{p}_i|^2$$

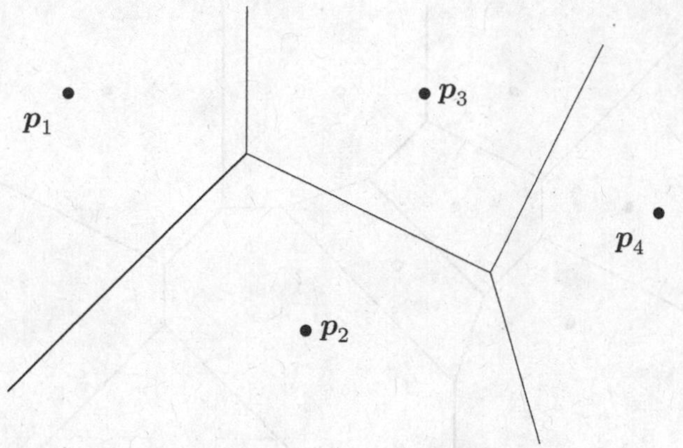

Figure 1.11 Decision regions for a minimum-distance classifier.

Therefore, the minimum-distance classifier is a linear machine. If the prototype points, $\boldsymbol{p}_i$, are the class means, then we have the nearest class mean classifier. Decision regions for a minimum-distance classifier are illustrated in Figure 1.11. Each boundary is the perpendicular bisector of the lines joining the prototype points of regions that are contiguous. Also, note from Figure 1.11 that the decision regions are convex (that is, two arbitrary points lying in the region can be joined by a straight line that lies entirely within the region). In fact, decision regions of a linear machine are always convex. Thus, the two class problems, illustrated in Figure 1.12, although separable, cannot be separated by a linear machine. Two generalisations that overcome this difficulty are piecewise linear discriminant functions and generalised linear discriminant functions.

1.6.3 Piecewise linear discriminant functions

This is a generalisation of the minimum-distance classifier to the situation in which there is more than one prototype per class. Suppose there are n_i prototypes in class ω_i, $\boldsymbol{p}_i^1, \ldots, \boldsymbol{p}_i^{n_i}$, $i = 1, \ldots, C$. We define the discriminant function for class ω_i to be

$$g_i(\boldsymbol{x}) = \max_{j=1,\ldots,n_i} g_i^j(\boldsymbol{x})$$

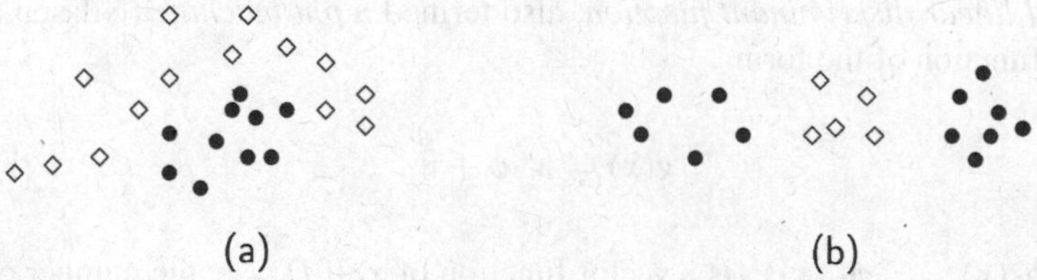

Figure 1.12 Two examples of groups not separable by a linear discriminant.

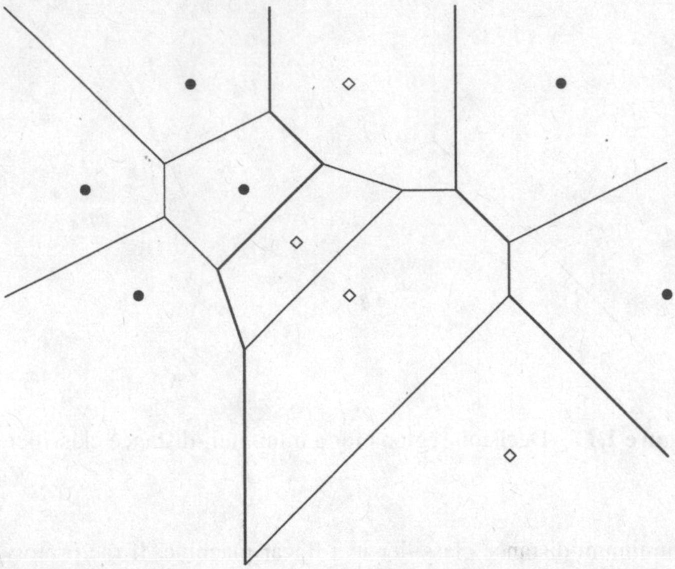

Figure 1.13 Dirichlet tessellation (comprising nearest-neighbour regions for a set of prototypes) and the decision boundary (thick lines) for two classes.

where g_i^j is a subsidiary discriminant function, which is linear and is given by

$$g_i^j(\boldsymbol{x}) = \boldsymbol{x}^T \boldsymbol{p}_i^j - \frac{1}{2}\boldsymbol{p}_i^{jT} \boldsymbol{p}_i^j \qquad j = 1, \ldots, n_i; i = 1, \ldots, C$$

A pattern $\boldsymbol{x}$ is assigned to the class for which $g_i(\boldsymbol{x})$ is largest; that is, to the class of the nearest prototype vector. This partitions the space into $\sum_{i=1}^{C} n_i$ regions known as the Dirichlet tessellation of the space. When each pattern in the training set is taken as a prototype vector, then we have the nearest-neighbour decision rule of Chapter 4. This discriminant function generates a piecewise linear decision boundary (Figure 1.13).

Rather than using the complete design set as prototypes, we may use a subset. Methods of reducing the number of prototype vectors (edit and condense) are described in Chapter 4, along with the nearest-neighbour algorithm. Clustering schemes may also be employed.

1.6.4 Generalised linear discriminant function

A *generalised linear discriminant function*, also termed a *phi machine* (Nilsson, 1965), is a discriminant function of the form

$$g(\boldsymbol{x}) = \boldsymbol{w}^T \boldsymbol{\phi} + w_0$$

where $\boldsymbol{\phi} = (\phi_1(\boldsymbol{x}), \ldots, \phi_D(\boldsymbol{x}))^T$ is a vector function of $\boldsymbol{x}$. If $D = p$, the number of variables, and $\phi_i(\boldsymbol{x}) = x_i$, then we have a linear discriminant function.

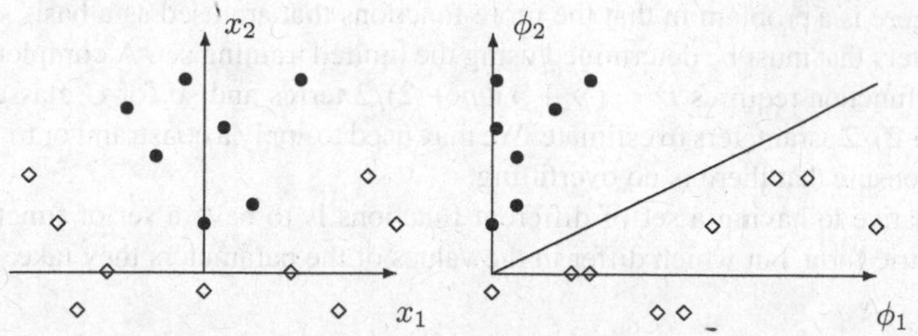

Figure 1.14 Nonlinear transformation of variables may permit linear discrimination.

The discriminant function is linear in the functions ϕ_i, not in the original measurements x_i. As an example, consider the two-class problem of Figure 1.14.

A linear discriminant function will not separate the two classes (denoted by the • and ◇ symbols in the left-hand illustration), even though they are separable. However, if we make the transformation

$$\phi_1(\boldsymbol{x}) = x_1^2$$
$$\phi_2(\boldsymbol{x}) = x_2$$

then the classes can be separated in the ϕ-space by a straight line as shown in the right-hand illustration. Similarly, disjoint classes can be transformed into a ϕ-space in which a lincar discriminant function could scparatc thc classcs (provided that they are separable in the original space).

The problem, therefore, is simple. Make a good choice for the functions $\phi_i(\boldsymbol{x})$, then use a linear discriminant function to separate the classes. But, how do we choose ϕ_i? Specific examples are shown in Table 1.1.

Table 1.1 Discriminant functions, ϕ.

Discriminant function	Mathematical form, $\phi_i(\boldsymbol{x})$
Linear	$\phi_i(\boldsymbol{x}) = x_i, i = 1, \ldots, p$
Quadratic	$\phi_i(\boldsymbol{x}) = x_{k_1}^{l_1} x_{k_2}^{l_2}, i = 1, \ldots, (p+1)(p+2)/2 - 1$ $l_1, l_2 = 0 \text{ or } 1; k_1, k_2 = 1, \ldots, p$ l_1, l_2 not both zero
νth order polynomial	$\phi_i(\boldsymbol{x}) = x_{k_1}^{l_1} \ldots x_{k_\nu}^{l_\nu}, i = 1, \ldots, \binom{p+\nu}{\nu} - 1$ $l_1, \ldots, l_\nu = 0 \text{ or } 1; k_1, \ldots, k_\nu = 1, \ldots, p$ l_i not all zero
Radial basis function	$\phi_i(\boldsymbol{x}) = \phi(\lvert \boldsymbol{x} - \boldsymbol{v}_i \rvert)$ for centre $\boldsymbol{v}_i$ and function ϕ
Multilayer perceptron	$\phi_i(\boldsymbol{x}) = f(\boldsymbol{x}^T \boldsymbol{v}_i + v_{i0})$ for direction $\boldsymbol{v}_i$ and offset v_{i0}. f is the logistic function, $f(z) = 1/(1 + \exp(-z))$

Clearly there is a problem in that the more functions that are used as a basis set, then the more parameters that must be determined using the limited training set. A complete quadratic discriminant function requires $D = (p + 1)(p + 2)/2$ terms and so for C classes there are $C(p+1)(p+2)/2$ parameters to estimate. We may need to apply a constraint or to 'regularise' the model to ensure that there is no overfitting.

An alternative to having a set of different functions is to have a set of functions of the same parametric form, but which differ in the values of the parameters they take,

$$\phi_i(\boldsymbol{x}) = \phi(\boldsymbol{x}; \boldsymbol{v}_i)$$

where $\boldsymbol{v}_i$ is a set of parameters. Different models arise depending on the way the variable $\boldsymbol{x}$ and the parameters $\boldsymbol{v}$ are combined. If

$$\phi(\boldsymbol{x}; \boldsymbol{v}) = \phi(|\boldsymbol{x} - \boldsymbol{v}|)$$

that is, ϕ is a function only of the magnitude of the difference between the pattern $\boldsymbol{x}$ and the weight vector $\boldsymbol{v}$, then the resulting discriminant function is known as a *radial basis function*. On the other hand, if ϕ is a function of the scalar product of the two vectors

$$\phi(\boldsymbol{x}; \boldsymbol{v}) = \phi(\boldsymbol{x}^T\boldsymbol{v} + v_0)$$

then the discriminant function is known as a *multilayer perceptron*. It is also a model known as projection pursuit. Both the radial basis function and the multilayer perceptron models can be used in regression.

In these latter examples, the discriminant function is no longer linear in the parameters. Specific forms for ϕ for radial basis functions and for the multilayer perceptron models will be given in Chapter 6.

1.6.5 Summary

In a multiclass problem, a pattern $\boldsymbol{x}$ is assigned to the class for which the discriminant function is the largest. A linear discriminant function divides the feature space by a hyperplane whose orientation is determined by the weight vector $\boldsymbol{w}$ and distance from the origin by the weight threshold w_0. The decision regions produced by linear discriminant functions are convex.

A piecewise linear discriminant function permits nonconvex and disjoint decision regions. Special cases are the nearest-neighbour and nearest class mean classifier.

A generalized linear discriminant function, with fixed functions ϕ_i, is linear in its parameters. It permits nonconvex and multiply connected decision regions (for suitable choices of ϕ_i). Radial basis functions and multilayer perceptrons can be regarded as generalised linear discriminant functions with flexible functions ϕ_i whose parameters must be determined or specified using the training set.

The Bayes' decision rule is optimal (in the sense of minimising classification error) and with sufficient flexibility in our discriminant functions we ought to be able to achieve optimal performance in principle. However, we are limited by a finite number of training samples and also, once we start to consider parametric forms for the ϕ_i, we lose the simplicity and ease of computation of the linear functions.

1.7 Multiple regression

Many of the techniques and procedures described within this book are also relevant to problems in *regression*, the process of investigating the relationship between a dependent (or response) variable Y and predictor (also referred to as regressor, measurement and independent) variables $X_1, \ldots, X_p$; a regression function expresses the expected value of Y in terms of $X_1, \ldots, X_p$ and model parameters. Regression is an important part of statistical pattern recognition and, although the emphasis of the book is on discrimination, practical illustrations are sometimes given on problems of a regression nature.

The discrimination problem itself is one in which we are attempting to predict the values of one variable (the class variable) given measurements made on a set of predictor variables (the pattern vector, $\boldsymbol{x}$). In this case, the response variable is categorical.

Regression analysis is concerned with predicting the mean value of the response variable given measurements on the predictor variables and assumes a model of the form,

$$E[y|\boldsymbol{x}] \triangleq \int y p(y|\boldsymbol{x}) dy = f(\boldsymbol{x}; \boldsymbol{\theta})$$

where f is a (possibly nonlinear) function of the measurements $\boldsymbol{x}$ and $\boldsymbol{\theta}$, a set of parameters of f. For example,

$$f(\boldsymbol{x}; \boldsymbol{\theta}) = \theta_0 + \boldsymbol{\theta}^T \boldsymbol{x},$$

where $\boldsymbol{\theta} = (\theta_1, \ldots, \theta_p)^T$, is a model that is linear in the parameters and the variables. The model

$$f(\boldsymbol{x}; \boldsymbol{\theta}) = \theta_0 + \boldsymbol{\theta}^T \boldsymbol{\phi}(\boldsymbol{x}),$$

where $\boldsymbol{\theta} = (\theta_1, \ldots, \theta_D)^T$ and $\boldsymbol{\phi} = (\phi_1(\boldsymbol{x}), \ldots, \phi_D(\boldsymbol{x}))^T$ is a vector of nonlinear functions of $\boldsymbol{x}$, is linear in the parameters but nonlinear in the variables. *Linear regression* refers to a regression model that is linear in the parameters, but not necessarily the variables.

Figure 1.15 shows an illustrative regression summary. For each value of x, there is a population of y values that varies with x. The solid line connecting the conditional means, $E[y|x]$, is the *regression line*. The dotted lines either side represent the spread of the conditional distribution (± 1 standard deviation from the mean).

It is assumed that the difference (commonly referred to as an error or residual), ϵ_i, between the measurement on the response variable and its predicted value conditional on the measurements on the predictors,

$$\epsilon_i = y_i - E[y|\boldsymbol{x}_i]$$

is an unobservable random variable. A normal model for the errors is often assumed,

$$p(\epsilon) = \frac{1}{\sqrt{(2\pi)}\sigma} \exp\left(-\frac{1}{2}\frac{\epsilon^2}{\sigma^2}\right)$$

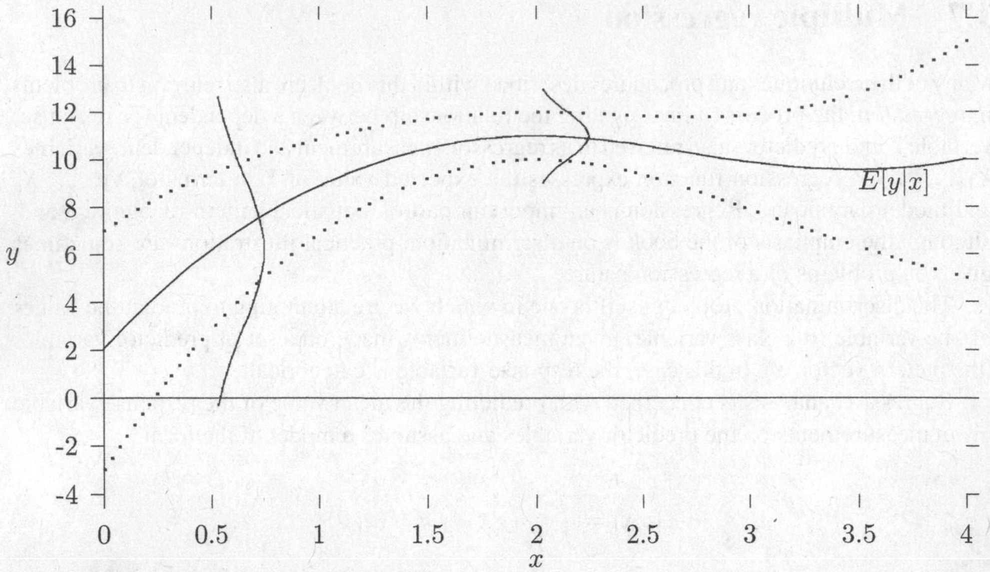

Figure 1.15 Population regression line (solid line) with representation of spread of conditional distribution (dotted lines) for normally distributed error terms, with variance depending on *x*.

That is,

$$p(y_i|\boldsymbol{x}_i, \boldsymbol{\theta}) = \frac{1}{\sqrt{(2\pi)}\sigma} \exp\left(-\frac{1}{2\sigma^2}(y_i - f(\boldsymbol{x}_i; \boldsymbol{\theta}))^2\right)$$

Given a set of data $\{(y_i, \boldsymbol{x}_i), i = 1, \ldots, n\}$, the maximum likelihood estimate of the model parameters (the value of the parameters for which the data are 'most likely'), $\boldsymbol{\theta}$, is that for which

$$p(\{(y_i, \boldsymbol{x}_i)\}|\boldsymbol{\theta})$$

is a maximum. Assuming independent samples, this amounts to determining the value of $\boldsymbol{\theta}$ for which the commonly used least squares error,

$$\sum_{i=1}^{n}(y_i - f(\boldsymbol{x}_i; \boldsymbol{\theta}))^2 \tag{1.21}$$

is a minimum (see the exercises at the end of the chapter).

For the linear model, procedures for estimating the parameters are described in Chapter 5.

1.8 Outline of book

The aim of this book is to provide a comprehensive account of statistical pattern recognition techniques with emphasis on methods and algorithms for discrimination and classification. In recent years there have been many developments in multivariate analysis techniques, particularly in nonparametric methods for discrimination and classification including kernel methods, but also the use of pattern recognition techniques applied to complex datasets such as those represented by a network. These are described in this book as extensions to the basic methodology developed over the years.

This chapter has presented some basic approaches to statistical pattern recognition. Supplementary material on probability theory and data analysis can be found on the book's website.

Chapters 2, 3 and 4 describe basic approaches to supervised classification via Bayes' rule and estimation of the class-conditional densities. Chapter 2 considers normal-based models. Chapter 3 develops these models to allow for uncertainty in model parameters. Chapter 4 addresses nonparametric approaches to density estimation.

Chapters 5–7 take a discriminant function approach to supervised classification. Chapter 5 describes algorithms for linear discriminant functions. Chapter 6 considers kernel-based approaches for constructing nonlinear discriminant functions, namely radial basis functions and support vector machine methods and alternative, projection-based methods, the multilayer perceptron neural network. Chapter 7 describes approaches that result in interpretable rules, often required for some applications to provide insight into the classification process.

Chapter 8 introduces the concept of classifier combination: can improvement be achieved with an ensemble of classifiers? Some classifiers may perform well in one part of the data space, other classifiers in another part. How should they be combined?

Chapter 9 addresses the important topic of performance assessment: how good is your designed classifier and how well does it compare with competing techniques?

Chapters 10 and 11 consider techniques that may form part of an exploratory data analysis. Chapter 10 describes methods of feature selection and extraction, both linear and nonlinear. Chapter 11 addresses *unsupervised* classification or clustering. Chapter 12 considers datasets that may be represented as complex networks. Many of the techniques employed for the analysis of such datasets are part of the pattern recognition literature presented in this book.

Finally, Chapter 13 covers additional topics on pattern recognition including model selection.

1.9 Notes and references

There was a growth of interest in techniques for automatic pattern recognition in the 1960s. Many books appeared in the early 1970s, some of which are still very relevant today and have been revised and reissued. More recently, there have been books detailing developments in pattern recognition, particularly neural network methods and kernel methods.

A very good introduction is provided by the book of Hand (1981a). Perhaps a little out of date now, it provides nevertheless a very readable account of techniques for discrimination and classification written from a statistical point of view and is to be recommended. Two of the main textbooks on statistical pattern recognition are those by Fukunaga (1990) and

Devijver and Kittler (1982). Written with an engineering emphasis, Fukunaga's book provides a comprehensive account of the most important aspects of pattern recognition, with many examples, computer projects and problems. Devijver and Kittler's book covers the nearest-neighbour decision rule and feature selection and extraction in some detail, though not at the expense of other important areas of statistical pattern recognition. It contains detailed mathematical accounts of techniques and algorithms, treating some areas in depth.

Another important textbook is that by Duda *et al.* (2001). This presents a thorough account of the main topics in pattern recognition. Other books that are an important source of reference material are those by Young and Calvert (1974), Tou and Gonzales (1974) and Chen (1973). Also, good accounts are given by Andrews (1972), a more mathematical treatment, and Therrien (1989), an undergraduate text.

Books that describe the 'neural network' aspects of developments in pattern recognition and their relationship to the more traditional methods include those of Haykin (1994), who provides a comprehensive treatment of neural networks, and Bishop (1995) who provides an excellent introduction to neural network methods from a statistical pattern recognition perspective. Ripley's (1996) account provides a thorough description of pattern recognition from within a statistical framework. It includes neural network methods, approaches developed in the field of machine learning, advances in statistical techniques as well as development of more traditional pattern recognition methods and gives valuable insights into many techniques gained from practical experience. Hastie *et al.* (2001) provide a thorough description of modern techniques in pattern recognition. Other books that deserve a mention are those by Schalkoff (1992) and Pao (1989).

Bishop (2007) provides an excellent introduction to pattern recognition, particularly recent developments and details of Bayesian computational methods.

The treatment of pattern recognition by Theodoridis and Koutroumbas (2009) is a comprehensive account, with similar goals to this book but with greater emphasis on unsupervised methods. Each chapter is supported by MATLAB code [see also the books by Nabney (2001), Theodoridis *et al.* (2010) and van der Heiden *et al.* (2004)].

Hand (1997) gives a short introduction to pattern recognition techniques and the central ideas in discrimination but places greater emphasis on the comparison and assessment of classifiers.

A more specialised treatment of discriminant analysis and pattern recognition is the book by McLachlan (1992a). This is a very good book. It is not an introductory textbook, but provides a thorough account of developments in discriminant analysis. Written from a statistical perspective, the book is a valuable source of reference of theoretical and practical work on statistical pattern recognition and is to be recommended for researchers in the field.

Comparative treatments of pattern recognition techniques (statistical, neural and machine learning methods) are provided in the volume edited by Michie *et al.* (1994) who report on the outcome of the *Statlog* project. Technical descriptions of the methods are given, together with the results of applying those techniques to a wide range of problems. This volume provides the most extensive comparative study available. More than 20 different classification procedures were considered for about 20 datasets.

Books on data mining often give good treatments of pattern recognition, including both supervised and unsupervised classification (Tan *et al.*, 2005; Witten and Frank, 2005; Han and Kamber, 2006).

There are many other books on pattern recognition. Some of those treating more specific parts (such as clustering) are cited in the appropriate chapters of this book. In addition, most textbooks on multivariate analysis devote some attention to discrimination and classification. These provide a valuable source of reference and are cited elsewhere in the book. There are also pattern recognition books for specialist applications, for example, medical imaging (Meyer-Baese, 2003) and forensics (Keppel *et al.*, 2006).

Exercises

In some of the exercises, it will be necessary to generate samples from a multivariate normal density with mean $\boldsymbol{\mu}$ and covariance matrix $\boldsymbol{\Sigma}$, denoted $N(\boldsymbol{\mu}, \boldsymbol{\Sigma})$. Many computer packages offer routines for this. However, it is a simple matter to generate samples from a normal distribution with unit variance and zero mean (Press *et al.*, 1992). Given a vector $\boldsymbol{Y}_i$ of such samples, then the vector $\boldsymbol{U}\boldsymbol{\Lambda}^{1/2}\boldsymbol{Y}_i + \boldsymbol{\mu}$ has the required distribution, where $\boldsymbol{U}$ is the matrix of eigenvectors of the covariance matrix and $\boldsymbol{\Lambda}^{1/2}$ is a diagonal matrix whose diagonal elements are the square roots of the corresponding covariance matrix eigenvalues.

1. Consider two multivariate normally distributed classes,

$$p(\boldsymbol{x}|\omega_i) = \frac{1}{(2\pi)^{p/2}|\boldsymbol{\Sigma}_i|^{1/2}} \exp\left\{-\frac{1}{2}(\boldsymbol{x} - \boldsymbol{\mu}_i)^T \boldsymbol{\Sigma}_i^{-1}(\boldsymbol{x} - \boldsymbol{\mu}_i)\right\}$$

   with means $\boldsymbol{\mu}_1$ and $\boldsymbol{\mu}_2$ and equal covariance matrices, $\boldsymbol{\Sigma}_1 = \boldsymbol{\Sigma}_2 = \boldsymbol{\Sigma}$. Show that the logarithm of the likelihood ratio is linear in the feature vector $\boldsymbol{x}$. What is the equation of the decision boundary?

2. Determine the equation of the decision boundary for the more general case of $\boldsymbol{\Sigma}_1 = \alpha\boldsymbol{\Sigma}_2$, for scalar α (normally distributed classes as in Exercise 1). In particular, for two univariate distributions, $N(0, 1)$ and $N(1, 1/4)$, show that one of the decision regions is bounded and determine its extent.

3. For the distributions in Exercise 1, determine the equation of the minimum risk decision boundary for the loss matrix,

$$\boldsymbol{\Lambda} = \begin{pmatrix} 0 & 2 \\ 1 & 0 \end{pmatrix}$$

4. Consider two multivariate normally distributed classes [ω_2 with mean $(-1, 0)^T$ and ω_1 with mean $(1, 0)^T$, and identity covariance matrix]. For a given threshold μ [see Equation (1.14)] on the likelihood ratio, determine the regions Ω_1 and Ω_2 in a Neyman–Pearson rule.

5. Consider three bivariate normal distributions, $\omega_1, \omega_2, \omega_3$ with identity covariance matrices and means $(-2, 0)^T$, $(0, 0)^T$ and $(0, 2)^T$. Show that the decision boundaries are piecewise linear. Now define a class, A, being the *mixture* of ω_1 and ω_3,

$$p_A(\boldsymbol{x}) = 0.5p(\boldsymbol{x}|\omega_1) + 0.5p(\boldsymbol{x}|\omega_3)$$

   and class B as bivariate normal with identity covariance matrix and mean $(a, b)^T$, for some a, b. What is the equation of the Bayes' decision boundary? Under what conditions is it piecewise linear?

6. Consider two uniform distributions with equal priors

$$p(x|\omega_1) = \begin{cases} 1 & \text{when } 0 \le x \le 1 \\ 0 & \text{otherwise} \end{cases}$$

$$p(x|\omega_2) = \begin{cases} \frac{1}{2} & \text{when } \frac{1}{2} \le x \le \frac{5}{2} \\ 0 & \text{otherwise} \end{cases}$$

Show that the reject function is given by

$$r(t) = \begin{cases} \frac{3}{8} & \text{when } 0 \le t \le \frac{1}{3} \\ 0 & \text{when } \frac{1}{3} \le t \le 1 \end{cases}$$

Hence calculate the error rate by integrating (1.10).

7. Reject option. Consider two classes, each normally distributed with means $x = 1$ and $x = -1$ and unit variances; $p(\omega_1) = p(\omega_2) = 0.5$. Generate a test set and use it (without using class labels) to estimate the reject rate as a function of the threshold t. Hence, estimate the error rate for no rejection. Compare with the estimate based on a labelled version of the test set. Comment on the use of this procedure when the true distributions are unknown and the densities have to be estimated.

8. The area of a sphere of radius r in p dimensions, S_p, is

$$S_p = \frac{2\pi^{\frac{p}{2}} r^{p-1}}{\Gamma(p/2)}$$

where Γ is the gamma function [$\Gamma(1/2) = \pi^{1/2}$, $\Gamma(1) = 1$, $\Gamma(x+1) = x\Gamma(x)$]. Show that the probability of a sample, $\boldsymbol{x}$, drawn from a zero-mean normal distribution with covariance matrix $\sigma^2 \boldsymbol{I}$ ($\boldsymbol{I}$ is the identity matrix) and having $|\boldsymbol{x}| \le R$ is

$$\int_0^R S_p(r) \frac{1}{(2\pi\sigma^2)^{p/2}} \exp\left(-\frac{r^2}{2\sigma^2}\right) dr$$

Evaluate this numerically for $R = 2\sigma$ and for $p = 1, \ldots, 10$. What do the results tell you about the distribution of normal samples in high-dimensional spaces?

9. In a two-class problem, let the cost of misclassifying a class ω_1 pattern be C_1 and the cost of misclassifying a class ω_2 pattern be C_2. Show that the point on the ROC curve that minimises the risk has gradient

$$\frac{C_2 p(\omega_2)}{C_1 p(\omega_1)}$$

10. Show that under the assumption of normally distributed residuals, the maximum likelihood solution for the parameters of a linear model is equivalent to minimising the sum-square error (1.21).

2

Density estimation – parametric

A discrimination rule may be constructed through explicit estimation of the class-conditional density functions and the use of Bayes' rule. One approach is to assume a simple parametric model for the density functions and to estimate the parameters of the model using an available training set. The Gaussian classifier and its variants are introduced. The more powerful approach of mixture models is then presented.

2.1 Introduction

In Chapter 1 we considered the basic theory of pattern classification. All the information regarding the density functions $p(\boldsymbol{x}|\omega_i)$ was assumed known. In practice, this knowledge is often unavailable or only partially known. Therefore, the next question that we must address is the estimation of the density functions themselves. If we can assume some parametric form for the distribution, perhaps obtained from theoretical considerations, or an assessment of the problem domain, then the problem reduces to one of estimating a finite number of parameters. Often the parametric form is chosen for convenience. In this chapter, special consideration is given to the normal (also referred to as Gaussian) distribution which leads to algorithms for the Gaussian classifier. The second major focus is on mixture models, which provide more general modelling capabilities.

We described in Chapter 1 how the minimum error decision is based on the probability of class membership $p(\omega_j|\boldsymbol{x})$, which using Bayes' theorem may be written

$$p(\omega_j|\boldsymbol{x}) = p(\omega_j)\frac{p(\boldsymbol{x}|\omega_j)}{p(\boldsymbol{x})} \tag{2.1}$$

The probability density function $p(\boldsymbol{x})$ is the same for all classes [in fact $p(\boldsymbol{x}) = \sum_{j=1}^{C} p(\omega_j)p(\boldsymbol{x}|\omega_j)$, where C is the number of classes]. Thus, assuming that the prior probabilities $p(\omega_j)$ are known, then in order to make a decision we need to estimate the class-conditional densities $p(\boldsymbol{x}|\omega_j)$, for all classes.

Statistical Pattern Recognition, Third Edition. Andrew R. Webb and Keith D. Copsey.

Estimation of the density $p(\boldsymbol{x}|\omega_j)$ is based on a sample of observations $\mathcal{D}_j = \{\boldsymbol{x}_1^j, \ldots, \boldsymbol{x}_{n_j}^j\}$ ($\boldsymbol{x}_i^j \in \mathbb{R}^d$) from class ω_j. In this chapter and the next we consider *parametric* approaches to density estimation. In the parametric approach, we assume that the class-conditional density for class ω_j is of a known form but has an unknown parameter, or set of parameters, $\boldsymbol{\theta}_j$, and we write this as $p(\boldsymbol{x}|\boldsymbol{\theta}_j)$. The alternative nonparametric approach to density estimation that we consider in Chapter 4 does not assume a simple functional form for the density.

2.2 Estimating the parameters of the distributions

We now introduce two approaches to estimating the parameters of a parametric class-conditional density, namely the *estimative approach* and the *predictive* or *Bayesian approach*. The focus in this chapter is on the estimative approach, with the Bayesian approach covered in Chapter 3.

2.2.1 Estimative approach

In the *estimative approach* to parametric density estimation we use an estimate of the parameter $\boldsymbol{\theta}_j$ in the parametric density. Thus we take

$$p(\boldsymbol{x}|\omega_j) = p(\boldsymbol{x}|\hat{\boldsymbol{\theta}}_j) \tag{2.2}$$

where $\hat{\boldsymbol{\theta}}_j = \hat{\boldsymbol{\theta}}_j(\mathcal{D}_j)$ is an estimate of the parameter $\boldsymbol{\theta}_j$ based on the data sample $\mathcal{D}_j$. A different data sample $\mathcal{D}_j$ would give rise to a different estimate $\hat{\boldsymbol{\theta}}_j$, but the estimative approach does not take into account this sampling variability.

The techniques and classifiers considered in this chapter use *maximum likelihood estimation* procedures to obtain the parameter estimates. In maximum likelihood estimation we seek to find the parameters $\hat{\boldsymbol{\theta}}_j$ that maximise the likelihood function defined using the data sample $\mathcal{D}_j$, i.e. we seek $\hat{\boldsymbol{\theta}}_j$ such that

$$L(\hat{\boldsymbol{\theta}}_j; \mathcal{D}_j) = \max_{\boldsymbol{\theta}_i} L(\boldsymbol{\theta}_i; \mathcal{D}_i)$$

where

$$L(\boldsymbol{\theta}_j; \mathcal{D}_j) = p(\mathcal{D}_j|\boldsymbol{\theta}_j) = p\left(\boldsymbol{x}_1^j, \ldots, \boldsymbol{x}_{n_j}^j|\boldsymbol{\theta}_j\right) \tag{2.3}$$

is the likelihood function (i.e. the probability density of the data measurements given the specific value $\boldsymbol{\theta}_j$ for the distribution parameters).

If the data measurement vectors making up the data sample are independent, then the likelihood function (2.3) can be written as a product of the known class-conditional densities for ω_j

$$L(\boldsymbol{\theta}_j; \mathcal{D}_j) = p(\mathcal{D}_j|\boldsymbol{\theta}_j) = \prod_{i=1}^{n_j} p\left(\boldsymbol{x}_i^j|\boldsymbol{\theta}_j\right) \tag{2.4}$$

The validity of the assumption of independence may depend on the manner in which the data sample was collected. If the data sample consists of sensor measurements (e.g. camera imagery) then noise correlations may occur between successive data vectors if the sampling rate is high, and therefore the data would not be independent. However, it is common to proceed with an independent likelihood assumption even if there are expected to be correlations between data vectors. This is primarily due to the difficulty of estimating such correlations.

Under the independence assumption the maximum likelihood estimation problem becomes one of optimising a known function [Equation (2.4)]. Typically, logarithms will be taken so that we seek to maximise the log-likelihood function

$$\log(L(\boldsymbol{\theta}_j; \mathcal{D}_j)) = \sum_{i=1}^{n_j} \log\left(p\left(\boldsymbol{x}_i^j|\boldsymbol{\theta}_j\right)\right)$$

which due to the strictly increasing nature of the logarithm is equivalent to maximising the likelihood function.

For some class-conditional densities (e.g. normal, as we shall see in the next section) an analytic solution for the optimal parameters is available. If an analytic solution is not available we can use numerical techniques such as gradient ascent or the Nelder–Mead method to maximise the likelihood function (details of such algorithms are available in Press *et al.*, 1992). An iterative optimisation scheme is used for optimising the parameters of mixture model class-conditional densities (Section 2.5).

2.2.2 Predictive approach

An alternative approach to parametric density estimation is the *predictive* or *Bayesian approach*, which is covered in Chapter 3. We write

$$p(\boldsymbol{x}|\omega_j) = \int p(\boldsymbol{x}|\boldsymbol{\theta}_j)p(\boldsymbol{\theta}_j|\mathcal{D}_j)d\boldsymbol{\theta}_j \tag{2.5}$$

where $p(\boldsymbol{\theta}_j|\mathcal{D}_j)$ is the Bayesian posterior density function for $\boldsymbol{\theta}_j$ based on a prior $p(\boldsymbol{\theta}_j)$ and the data $\mathcal{D}_j$. Thus, we admit that we do not know the true value of $\boldsymbol{\theta}_j$, and instead of taking a single estimate, we take a weighted sum of the densities $p(\boldsymbol{x}|\boldsymbol{\theta}_j)$, weighted by the distribution $p(\boldsymbol{\theta}_j|\mathcal{D}_j)$ (Aitchison *et al.*, 1977). This predictive approach is usually more complicated than the estimative approach (both in classifier design, and in application) and may be regarded as making allowance for the sampling variability of the estimate of $\boldsymbol{\theta}_j$.

2.3 The Gaussian classifier

2.3.1 Specification

Perhaps the most widely used classifier is that in which the class conditional densities are modelled using the normal (Gaussian) distribution.

Normal (Gaussian) distribution

The probability density function for a normal (Gaussian) distribution with mean μ and variance σ^2 is

$$p(x|\mu, \sigma^2) = N(x; \mu, \sigma^2) = \frac{1}{\sqrt{2\pi\sigma^2}} \exp\left(-\frac{(x-\mu)^2}{2\sigma^2}\right) \quad (2.6)$$

The probability density function for a multivariate normal (Gaussian) distribution with mean $\boldsymbol{\mu}$ and covariance matrix $\boldsymbol{\Sigma}$ (a symmetric positive semi-definite matrix) is

$$p(\boldsymbol{x}|\boldsymbol{\mu}, \boldsymbol{\Sigma}) = N(\boldsymbol{x}; \boldsymbol{\mu}, \boldsymbol{\Sigma}) = \frac{1}{(2\pi)^{d/2}|\boldsymbol{\Sigma}|^{1/2}} \exp\left\{-\frac{1}{2}(\boldsymbol{x}-\boldsymbol{\mu})^T \boldsymbol{\Sigma}^{-1}(\boldsymbol{x}-\boldsymbol{\mu})\right\} \quad (2.7)$$

where d is the dimensionality of the data.

We model a data vector from class ω_j as being drawn from a normal distribution with mean vector $\boldsymbol{\mu}_j$ and covariance matrix $\boldsymbol{\Sigma}_j$. The class conditional density is then given by

$$\begin{aligned} p(\boldsymbol{x}|\omega_j) &= N(\boldsymbol{x}; \boldsymbol{\mu}_j, \boldsymbol{\Sigma}_j) \\ &= \frac{1}{(2\pi)^{d/2}|\boldsymbol{\Sigma}_j|^{1/2}} \exp\left\{-\frac{1}{2}(\boldsymbol{x}-\boldsymbol{\mu}_j)^T \boldsymbol{\Sigma}_j^{-1}(\boldsymbol{x}-\boldsymbol{\mu}_j)\right\} \end{aligned}$$

Classification is achieved by assigning a pattern $\boldsymbol{x}$ to the class for which the posterior class probability, $p(\omega_j|\boldsymbol{x})$, is the greatest, or equivalently $\log(p(\omega_j|\boldsymbol{x}))$ is the greatest. Using Equations (2.1) and (2.2), and the normal modelling of the class conditional densities above, we have

$$\begin{aligned} \log(p(\omega_j|\boldsymbol{x})) &= \log(p(\boldsymbol{x}|\omega_j)) + \log(p(\omega_j)) - \log(p(\boldsymbol{x})) \\ &= -\frac{1}{2}(\boldsymbol{x}-\boldsymbol{\mu}_j)^T \boldsymbol{\Sigma}_j^{-1}(\boldsymbol{x}-\boldsymbol{\mu}_j) - \frac{1}{2}\log(|\boldsymbol{\Sigma}_j|) \\ &\quad -\frac{d}{2}\log(2\pi) + \log(p(\omega_j)) - \log(p(\boldsymbol{x})) \end{aligned}$$

Since $p(\boldsymbol{x})$ is the same for all classes, the discriminant rule is assign $\boldsymbol{x}$ to ω_i if $g_i > g_j$, for all $j \neq i$, where

$$g_j(\boldsymbol{x}) = \log(p(\omega_j)) - \frac{1}{2}\log(|\boldsymbol{\Sigma}_j|) - \frac{1}{2}(\boldsymbol{x}-\boldsymbol{\mu}_j)^T \boldsymbol{\Sigma}_j^{-1}(\boldsymbol{x}-\boldsymbol{\mu}_j) \quad (2.8)$$

Classifying a pattern $\boldsymbol{x}$ on the basis of the values of $g_j(\boldsymbol{x})$, $j = 1, \ldots, C$, gives the *normal-based quadratic discriminant function* (McLachlan, 1992a).

In the *estimative approach*, the quantities $\boldsymbol{\mu}_j$ and $\boldsymbol{\Sigma}_j$ in the above are replaced by estimates based on a training set. The estimates are obtained using maximum likelihood estimation given a set of (assumed independent) data samples from each class. Suppose that we have a set of samples $\{\boldsymbol{x}_1, \ldots, \boldsymbol{x}_n\}, \boldsymbol{x}_i \in \mathbb{R}^d$ from class ω_j. Then the maximum likelihood estimate of the mean for class ω_j is

$$\hat{\boldsymbol{\mu}}_j = \frac{1}{n} \sum_{i=1}^{n} \boldsymbol{x}_i, \tag{2.9}$$

the sample mean vector, and the maximum likelihood estimate for the covariance matrix is

$$\hat{\boldsymbol{\Sigma}}_j = \frac{1}{n} \sum_{i=1}^{n} (\boldsymbol{x}_i - \hat{\boldsymbol{\mu}}_j)(\boldsymbol{x}_i - \hat{\boldsymbol{\mu}}_j)^T, \tag{2.10}$$

the (biased) sample covariance matrix (we show how these estimates can be derived in the next subsection). Since the maximum likelihood estimate of the covariance matrix is biased $[E(\hat{\boldsymbol{\Sigma}}_j) = \frac{n-1}{n} \boldsymbol{\Sigma}_j$, see the exercises at the end of the chapter] it is common practice to replace it with an unbiased estimate

$$\hat{\boldsymbol{\Sigma}}_j = \frac{1}{n-1} \sum_{i=1}^{n} (\boldsymbol{x}_i - \hat{\boldsymbol{\mu}}_j)(\boldsymbol{x}_i - \hat{\boldsymbol{\mu}}_j)^T$$

Substituting the estimates of the means and the covariance matrices (termed the 'plug-in estimates') of each class into (2.8) gives the *Gaussian classifier* or quadratic discrimination rule: assign $\boldsymbol{x}$ to ω_i if $g_i > g_j$, for all $j \neq i$, where

$$g_j(\boldsymbol{x}) = \log(p(\omega_j)) - \frac{1}{2}\log(|\hat{\boldsymbol{\Sigma}}_j|) - \frac{1}{2}(\boldsymbol{x} - \hat{\boldsymbol{\mu}}_j)^T \hat{\boldsymbol{\Sigma}}_j^{-1} (\boldsymbol{x} - \hat{\boldsymbol{\mu}}_j) \tag{2.11}$$

If the training data have been gathered by collecting measurements in the operational (deployment) environment, then a plug-in estimate for the prior probability, $p(\omega_j)$, is $n_j / \sum_i n_i$, where n_j is the number of patterns in class ω_j. Other common choices are to use uniform prior probabilities $p(\omega_j) = 1/C$, or expert specified prior probabilities.

We may apply the Gaussian classifier (quadratic discrimination rule) to classify data vectors (e.g. members of a separate test set, if available). However, problems will occur in the Gaussian classifier if any of the matrices $\hat{\boldsymbol{\Sigma}}_j$ is singular. This is because the singularity will prevent the inversion of the covariance matrix in the discriminant rule (2.11), and will also prevent calculation of the logarithm of the determinant of the covariance matrix in the discriminant rule, since the determinant of a singular matrix is zero. Approaches for dealing with this are the subject of the next section.

2.3.2 Derivation of the Gaussian classifier plug-in estimates

To derive the intuitively reasonable plug-in estimates [Equations (2.9) and (2.10)] for the Gaussian classifier, consider the likelihood function for the set of samples, $\{\boldsymbol{x}_1, \ldots, \boldsymbol{x}_n\}$, $\boldsymbol{x}_i \in \mathbb{R}^d$, drawn independently from a normal distribution characterised by $\boldsymbol{\theta} = (\boldsymbol{\mu}, \boldsymbol{\Sigma})$. We have

$$L(\boldsymbol{\theta}; \boldsymbol{x}_1, \ldots, \boldsymbol{x}_n) = \prod_{i=1}^{n} \frac{1}{(2\pi)^{d/2} |\boldsymbol{\Sigma}|^{1/2}} \exp\left\{ -\frac{1}{2}(\boldsymbol{x}_i - \boldsymbol{\mu})^T \boldsymbol{\Sigma}^{-1} (\boldsymbol{x}_i - \boldsymbol{\mu}) \right\}$$

Maximisation of this function is easier if we replace Σ by $\Psi = \Sigma^{-1}$, the inverse of the covariance matrix (known as the *precision* matrix). We then have

$$\log(L(\boldsymbol{\mu}, \boldsymbol{\Psi}|\boldsymbol{x}_1, \ldots, \boldsymbol{x}_n)) = -\frac{nd}{2}\log(2\pi) + \frac{n}{2}\log(|\boldsymbol{\Psi}|) - \frac{1}{2}\sum_{i=1}^{n}(\boldsymbol{x}_i - \boldsymbol{\mu})^T\boldsymbol{\Psi}(\boldsymbol{x}_i - \boldsymbol{\mu})$$

where we have made use of the following matrix determinant property

$$|\boldsymbol{A}^{-1}| = |\boldsymbol{A}|^{-1}$$

Differentiating $\log(L)$ with respect to $\boldsymbol{\mu}$ and $\boldsymbol{\Psi}$ gives[1]

$$\frac{\partial \log(L)}{\partial \boldsymbol{\mu}} = \frac{1}{2}\sum_{i=1}^{n}\boldsymbol{\Psi}(\boldsymbol{x}_i - \boldsymbol{\mu}) + \frac{1}{2}\sum_{i=1}^{n}\boldsymbol{\Psi}^T(\boldsymbol{x}_i - \boldsymbol{\mu}) = \sum_{i=1}^{n}\boldsymbol{\Psi}(\boldsymbol{x}_i - \boldsymbol{\mu}) \qquad (2.12)$$

and

$$\frac{\partial \log(L)}{\partial \boldsymbol{\Psi}} = \frac{n}{2}\boldsymbol{\Psi}^{-1} - \frac{1}{2}\sum_{j=1}^{n}(\boldsymbol{x}_i - \boldsymbol{\mu})(\boldsymbol{x}_i - \boldsymbol{\mu})^T \qquad (2.13)$$

To derive Equation (2.13), we have used the result that

$$\frac{\partial |\boldsymbol{A}|}{\partial \boldsymbol{A}} = [\mathrm{adj}(\boldsymbol{A})]^T = |\boldsymbol{A}|(\boldsymbol{A}^{-1})^T$$

so that

$$\frac{\partial \log(|\boldsymbol{\Psi}|)}{\partial \boldsymbol{\Psi}} = |\boldsymbol{\Psi}|(\boldsymbol{\Psi}^{-1})^T\frac{1}{|\boldsymbol{\Psi}|} = (\boldsymbol{\Psi}^{-1})^T = \boldsymbol{\Psi}^{-1}$$

and also[2]

$$(\boldsymbol{x}_i - \boldsymbol{\mu})^T\boldsymbol{\Psi}(\boldsymbol{x}_i - \boldsymbol{\mu}) = \sum_{r=1}^{d}\sum_{s=1}^{d}(x_{ir} - \mu_r)\Psi_{r,s}(x_{is} - \mu_s)$$

which makes clear that

$$\frac{\partial (\boldsymbol{x}_i - \boldsymbol{\mu})^T\boldsymbol{\Psi}(\boldsymbol{x}_i - \boldsymbol{\mu})}{\partial \Psi_{r,s}} = (x_{ir} - \mu_r)(x_{is} - \mu_s)$$

---

[1] Differentiation with respect to a d-dimensional vector means differentiating with respect to each component of the vector. This gives a set of d equations which may be expressed as a vector equation. Similarly for differentiating with respect to a matrix.

[2] Using notation where x_{ir} is the rth component of the ith data measurement $\boldsymbol{x}_i$, μ_r is the rth component of the mean vector $\boldsymbol{\mu}$ and $\Psi_{r,s}$ is the component in the rth row and sth column of the matrix $\boldsymbol{\Psi}$.

Equating (2.12) to zero (and multiplying both sides by Ψ^{-1}) gives the maximum likelihood estimate of the mean [i.e. Equation (2.9)]. Equating (2.13) to zero gives

$$\Psi^{-1} = \frac{1}{n}\sum_{i=1}^{n}(\boldsymbol{x}_i - \boldsymbol{\mu})(\boldsymbol{x}_i - \boldsymbol{\mu})^T$$

Hence, replacing $\boldsymbol{\mu}$ by its maximum likelihood estimate, the maximum likelihood estimate of the covariance matrix is as specified in Equation (2.10). Note that since the maximisation gives rise to a valid covariance matrix (assuming a suitable amount of data) it does not matter that we did not perform a constrained maximisation (with the constraint enforcing the fact that the solution must be a positive semi-definite symmetric matrix).

2.3.3 Example application study

The problem

The purpose of this study is to investigate the feasibility of predicting the degree of recovery of patients entering hospital with severe head injury using data collected shortly after injury (Titterington *et al.*, 1981).

Summary

Titterington *et al.* (1981) report the results of several classifiers, each designed using different training sets. In this example, results of the application of a quadratic discriminant rule (2.11) are presented.

The data

The dataset comprises measurements on patients entering hospital with a head injury with a minimum degree of brain damage. Measurements are made on six categorical variables: `Age`, grouped into decades 0–9, 10–19, ..., 60–69, 70+; `EMV` score, relating to eye, motor and verbal responses to stimulation grouped into seven categories; `MRP`, a summary of the motor responses in all four limbs, graded 1 (nil) to 7 (normal); `Change`, the change in neurological function over the first 24 h, graded 1 (deteriorating) to 3 (improving); `Eye Indicant`, a summary of eye movement scores, graded 1 (bad) to 3 (good); `Pupils`, the reaction of pupils to light, graded 1 (nonreacting) or 2 (reacting). There are 500 patients in the training and test sets, distributed over three classes related to the predicted outcome (dead or vegetative; severe disability; and moderate disability or good recovery). The number of patterns in each of the three classes for the training and the test sets are: training – 259, 52, 189; test – 250, 48, 202. Thus there is an uneven class distribution. Also, there are many missing values. These have been substituted by class means on training and population means on test. Further details of the data are given by Titterington *et al.* (1981).

The model

The data in each class are modelled using a normal distribution leading to the discriminant rule (2.11).

Training procedure

Training consists of calculating the quantities $\{\hat{\boldsymbol{\mu}}_j, \hat{\boldsymbol{\Sigma}}_j, j = 1, \ldots, C\}$, the maximum likelihood estimates of the mean and covariance matrix for each class from the data. The prior class

Table 2.1 Left, confusion matrix for training data; right, results for the test data.

		True class					True class		
		1	2	3			1	2	3
Predicted class	1	209	22	15	Predicted class	1	188	19	29
	2	0	1	1		2	3	1	2
	3	50	29	173		3	59	28	171

probabilities are taken to be $p(\omega_j) = n_j/n, j = 1, \ldots, C$. A numerical procedure must be used to calculate the inverse covariance matrix, $\hat{\mathbf{\Sigma}}_j^{-1}$, and the determinant, $|\hat{\mathbf{\Sigma}}_j|$ for each class. Once calculated, these quantities are substituted into Equation (2.11) to give C functions, $g_j(\boldsymbol{x})$.

For each pattern, $\boldsymbol{x}$, in the training and test set, $g_j(\boldsymbol{x})$ is calculated and $\boldsymbol{x}$ assigned to the class for which the corresponding discriminant function, $g_j(\boldsymbol{x})$, is the largest.

Results

Results on training and test sets for a Gaussian classifier (quadratic rule) are given in Table 2.1 as misclassification matrices or *confusion matrices*. Note that class 2 is nearly always classified incorrectly as class 1 or class 3.

2.4 Dealing with singularities in the Gaussian classifier

2.4.1 Introduction

In problems where the data are from multivariate normal distributions with different covariance matrices, there may be insufficient data to obtain good estimates of the class covariance matrices. As was noted in the description of the Gaussian classifier (quadratic discrimination rule), problems will occur if any of the estimates of the covariance matrices $\hat{\mathbf{\Sigma}}_j$ are singular, since this will prevent calculation of the discriminant rule [Equation (2.11)]. There are several common approaches for dealing with this, which we discuss over the next few subsections. The notation and data measurements are the same as in the previous section.

2.4.2 Naïve Bayes

One of the simplest approaches to avoiding singular covariance matrices is to use diagonal covariance matrices; that is, set all off-diagonal terms of $\hat{\mathbf{\Sigma}}_j$ to zero. Since we are considering multivariate normal distributions, this is equivalent to making the assumption that the features making up each data vector are statistically independent (given the class to which the data vector belongs). This gives

$$p(\boldsymbol{x}|\omega_j) = \prod_{l=1}^{d} N\left(x_l; \hat{\mu}_{jl}, \hat{\sigma}_{j,l}^2\right)$$

where $\hat{\mu}_{jl}$ is the lth component of $\hat{\boldsymbol{\mu}}_j$ (i.e. the lth component of the mean of the data samples from class ω_j), $\hat{\sigma}^2_{j,l}$ is the lth diagonal element of $\hat{\boldsymbol{\Sigma}}_j$ (i.e. the variance of the lth component of the data samples from class ω_j)

$$\hat{\sigma}^2_{j,l} = \frac{1}{n-1}\sum_{i=1}^{n}(x_{ij} - \hat{\mu}_{jl})^2$$

and $N(x, \mu, \sigma^2)$ is the probability density function of the univariate normal distribution with mean μ and variance σ^2. The discriminant rule is assign $\boldsymbol{x}$ to ω_i if $g_i > g_j$, for all $j \neq i$, where

$$g_j(\boldsymbol{x}) = \log(p(\omega_j)) - \sum_{l=1}^{d}\log(\hat{\sigma}_{j,l}) - \frac{1}{2}\sum_{l=1}^{d}\frac{(x_l - \hat{\mu}_{jl})^2}{\hat{\sigma}^2_{j,l}}$$

The assumption of statistical independence between the features making up each data vector is made also in the nonparametric naïve Bayes classifier that is considered in Chapter 4. However, whilst the diagonal covariance matrix Gaussian classifier models each class-conditional component as an independent univariate normal distribution, a general naïve Bayes classifier can consider any univariate distribution for a given component.

2.4.3 Projection onto a subspace

Another approach is to project the data to a subspace in which $\hat{\boldsymbol{\Sigma}}_j$ is nonsingular, perhaps using a principal components analysis (PCA) (see Chapter 10), and then to use the Gaussian classifier in the reduced dimension space. Such an approach is assessed by Schott (1993) and linear transformations for reducing dimensionality are discussed in Chapter 10.

2.4.4 Linear discriminant function

Even in problems where the data are from multivariate normal distributions with different covariance matrices, sampling variability may mean that it is better to assume equal covariance matrices (i.e. assume that the class covariance matrices $\boldsymbol{\Sigma}_1, \ldots, \boldsymbol{\Sigma}_C$ are all the same). In this case the discriminant function (2.8) simplifies[3] and the discriminant rule becomes: assign $\boldsymbol{x}$ to ω_i if $g_i > g_j$, for all $j \neq i$, where g_j is (the plug-in estimate of) the linear discriminant

$$g_j(\boldsymbol{x}) = \log(p(\omega_j)) - \frac{1}{2}\hat{\boldsymbol{\mu}}_j^T \boldsymbol{S}_W^{-1}\hat{\boldsymbol{\mu}}_j + \boldsymbol{x}^T\boldsymbol{S}_W^{-1}\hat{\boldsymbol{\mu}}_j \tag{2.14}$$

where $\hat{\boldsymbol{\mu}}_j$ is the sample mean vector for class ω_j and $\boldsymbol{S}_W$ is the estimate for the common class covariance matrix. This is the *normal-based linear discriminant function*. The maximum

[3] The simplification arises from ignoring terms that are constant for all classes.

likelihood estimate for the common class covariance matrix is the pooled within-class sample covariance matrix

$$S_W = \sum_{j=1}^{C} \frac{n_j}{n} \hat{\boldsymbol{\Sigma}}_j \tag{2.15}$$

where n_j is the number of training data vectors for class ω_j and n is the total number of training data vectors (i.e. across all classes). The unbiased estimate is given by

$$\frac{n}{n-C} S_W$$

2.4.4.1 Special cases

For the special case of two classes, the rule (2.14) may be written:

- assign $\boldsymbol{x}$ to class ω_1 if

$$\boldsymbol{w}^T \boldsymbol{x} + w_0 > 0$$

- else assign $\boldsymbol{x}$ to class ω_2

where in the above

$$\boldsymbol{w} = S_W^{-1} (\hat{\boldsymbol{\mu}}_1 - \hat{\boldsymbol{\mu}}_2) \tag{2.16}$$

$$w_0 = -\log\left(\frac{p(\omega_2)}{p(\omega_1)}\right) - \frac{1}{2}(\hat{\boldsymbol{\mu}}_1 + \hat{\boldsymbol{\mu}}_2)^T \boldsymbol{w} \tag{2.17}$$

A special case of (2.14) for any number of classes occurs when the matrix S_W is taken to be the identity matrix and the class priors $p(\omega_i)$ are equal. This is the *nearest class mean classifier*: assign $\boldsymbol{x}$ to class ω_i if:

$$-2\boldsymbol{x}^T \hat{\boldsymbol{\mu}}_i + \hat{\boldsymbol{\mu}}_i^T \hat{\boldsymbol{\mu}}_i < -2\boldsymbol{x}^T \hat{\boldsymbol{\mu}}_j + \hat{\boldsymbol{\mu}}_j^T \hat{\boldsymbol{\mu}}_j \text{ for all } j \neq i$$

2.4.4.2 Discussion

The linear discriminant rule (2.14) is quite robust to departures from the equal covariance matrix assumptions (Wahl and Kronmal, 1977; O'Neill, 1992), and may give better performance than the optimum quadratic discriminant rule for normally distributed classes when the true covariance matrices are unknown and the sample sizes are small. However, it is better to use the quadratic rule if the sample size is sufficient.

2.4.5 Regularised discriminant analysis

Regularised discriminant analysis (RDA) was proposed by Friedman (1989) for small-sample, high-dimensional datasets as a means of overcoming the degradation in performance of the quadratic discriminant rule in such conditions. Two parameters are involved: λ, a complexity parameter providing an intermediate between a linear and a quadratic discriminant rule, and γ, a shrinkage parameter for covariance matrix updates.

The covariance matrix estimate $\hat{\boldsymbol{\Sigma}}_j$ for class ω_j [Equation (2.10)] in the quadratic discriminant rule (2.11) is replaced by a linear combination, $\boldsymbol{\Sigma}_j^\lambda$, of the sample covariance matrix $\hat{\boldsymbol{\Sigma}}_j$ and the pooled covariance matrix $\boldsymbol{S}_W$ [Equation (2.15)], as follows

$$\boldsymbol{\Sigma}_j^\lambda = \frac{(1-\lambda)n_j\hat{\boldsymbol{\Sigma}}_j + \lambda n \boldsymbol{S}_W}{(1-\lambda)n_j + \lambda n} \tag{2.18}$$

where $0 \le \lambda \le 1$, n_j is the number of training data vectors for class ω_j, and n is the total number of training data vectors (i.e. across all classes).

At the extremes of $\lambda = 0$ and $\lambda = 1$ we have covariance matrix estimates that lead to the quadratic discriminant rule and the linear discriminant rule, respectively

$$\boldsymbol{\Sigma}_j^\lambda = \begin{cases} \hat{\boldsymbol{\Sigma}}_j & \lambda = 0 \\ \boldsymbol{S}_W & \lambda = 1 \end{cases}$$

The second parameter γ is used to *regularise* the sample class covariance matrix further beyond that provided by (2.18),

$$\boldsymbol{\Sigma}_j^{\lambda,\gamma} = (1-\gamma)\boldsymbol{\Sigma}_j^\lambda + \gamma c_j(\lambda)\boldsymbol{I}_d \tag{2.19}$$

where $\boldsymbol{I}_d$ is the $d \times d$ identity matrix and

$$c_j(\lambda) = \mathrm{Tr}\{\boldsymbol{\Sigma}_j^\lambda\}/d,$$

the average eigenvalue of $\boldsymbol{\Sigma}_j^\lambda$.

The matrix $\boldsymbol{\Sigma}_j^{\lambda,\gamma}$ is then used as the plug-in estimate of the covariance matrix in the normal-based discriminant rule: assign $\boldsymbol{x}$ to ω_i if $g_i > g_j$, for all $j \neq i$, where

$$g_j(\boldsymbol{x}) = \log(p(\omega_j)) - \frac{1}{2}\log\left(|\boldsymbol{\Sigma}_j^{\lambda,\gamma}|\right) - \frac{1}{2}(\boldsymbol{x}-\hat{\boldsymbol{\mu}}_j)^T\left(\boldsymbol{\Sigma}_j^{\lambda,\gamma}\right)^{-1}(\boldsymbol{x}-\hat{\boldsymbol{\mu}}_j)$$

Friedman's approach is to choose the values of λ and γ to minimise a *cross-validated* estimate of future misclassification cost (that is, a robust estimate of the error rate using a procedure called cross-validation – see Chapter 9) – this approach is referred to as RDA. The strategy is to evaluate the misclassification risk on a grid of points ($0 \le \lambda, \gamma \le 1$) and to choose the optimal values of λ and γ to be those grid point values with smallest estimated risk. In order to reduce the computational cost of calculating each covariance matrix in the cross-validation procedure, a strategy that updates a covariance matrix when one observation is removed from a dataset may be employed.

2.4.5.1 Robust estimates

Robust estimates of the covariance matrices (see Chapter 13) may also be incorporated in the analysis. Instead of using $\mathbf{\Sigma}_j^\lambda$ in (2.19), Friedman (1989) proposes $\tilde{\mathbf{\Sigma}}_j^\lambda$, given by

$$\tilde{\mathbf{\Sigma}}_j^\lambda = \frac{(1-\lambda)\tilde{\mathbf{S}}_j + \lambda\sum_{k=1}^{C}\tilde{\mathbf{S}}_k}{(1-\lambda)W_j + \lambda\sum_{k=1}^{C}W_k}$$

where

$$\tilde{\mathbf{S}}_j = \sum_{i=1}^{n} z_{ji}w_i(\mathbf{x}_i - \tilde{\boldsymbol{\mu}}_j)(\mathbf{x}_i - \tilde{\boldsymbol{\mu}}_j)^T$$

$$\tilde{\boldsymbol{\mu}}_j = \sum_{i=1}^{n} z_{ji}w_i\mathbf{x}_i/W_j$$

$$W_j = \sum_{i=1}^{n} z_{ji}w_i$$

where the w_i ($0 \leq w_i \leq 1$) are weights associated with each observation and $z_{ji} = 1$ if $\mathbf{x}_i \in$ class ω_j and 0 otherwise. For $w_i = 1$ for all i, $\tilde{\mathbf{\Sigma}}_j^\lambda = \mathbf{\Sigma}_j^\lambda$.

2.4.5.2 Discussion

Friedman (1989) assesses the effectiveness of RDA on simulated and real datasets. He finds that model selection based on the cross-validation procedure performs well, and that fairly accurate classification can be achieved with RDA for a small ratio of the number of observations (n) to the number of variables (d). He finds that RDA has the potential for dramatically increasing the power of discriminant analysis when sample sizes are small and the number of variables is large.

2.4.6 Example application study

The problem
Face recognition, with relevance to face-based video indexing, biometric identity authentication and surveillance (Lu *et al.*, 2003).

Summary
The study uses RDA for image-based face recognition, after the images have been projected onto a lower dimension subspace. Special focus is applied to the *small sample size problem*, since the number of available training samples is small compared with the dimensionality of the feature space.

The data
The data were taken from the Face Recognition Technology (FERET) database (Phillips *et al.*, 2000). The subset used consisted of 606 grey-scale images of 49 subjects, each having more

than 10 samples. The imagery covered a range of illuminations and facial expressions. Data preprocessing was applied, including translation, rotation and scaling.

The model

The initial dimensionality reduction projection is based on a variant (L.F. Chen *et al.*, 2000) of linear discriminant analysis (see Chapter 5) designed for situations when the number of samples is smaller than the dimensionality of the samples. RDA is used on the resulting projected vectors.

Training procedure

Six experimental set-ups were considered, varying in the number of training examples for each subject (from 2 to 7 examples). In each experiment the test data consisted of all images apart from those used as training data. Rather than selecting the RDA parameters (λ, γ) using a validation stage, test results were provided for a range of RDA parameters (a grid of points). Results were compared with those obtained using the *Eigenfaces method* of Turk and Pentland (1991), a popular face recognition procedure in which PCA (see Chapter 10) is used to reduce the dimensionality of the images, and the nearest class mean classifier is applied to the resulting reduced dimension vectors.

Results

The performance from the optimal choice of RDA parameters was better than that from the Eigenfaces method. However, selection of these best RDA parameters without the aid of test data was not considered.[4]

2.4.7 Further developments

There have been several investigations of the robustness of the linear and quadratic discriminant rules to certain types of non-normality (Lachenbruch *et al.*, 1973; Chingánda and Subrahmaniam, 1979; Ashikaga and Chang, 1981; Balakrishnan and Subrahmaniam, 1985). They can be greatly affected by non-normality and, if possible, variables should be transformed to approximate normality before applying the rules.

There are several intermediate covariance matrix structures (Flury, 1987) that may be considered without making the restrictive assumption of equality of covariance matrices used in the linear discriminant function. These include diagonal, but different, covariance matrices; common principal components; and proportional covariance matrices models.

Aeberhard *et al.* (1994) report an extensive simulation study on eight statistical classification methods applied to problems when the number of observations is less than the number of variables. They found that out of the techniques considered, RDA was the most powerful, being outperformed by linear discriminant analysis only when the class covariance matrices were identical and for a large training set size. Reducing the dimensionality by feature extraction methods generally led to poorer results. However, Schott (1993) finds that dimension reduction prior to quadratic discriminant analysis can substantially reduce

[4] Later experiments by Lu *et al.* (2005) indicate that good test set performance is maintained when a separate validation set is used to optimise the RDA parameters.

misclassification rates for small sample sizes. It also decreases the sample sizes necessary for quadratic discriminant analysis to be preferred over linear discriminant analysis. Alternative approaches to the problem of discriminant analysis with singular covariance matrices are described by Krzanowski *et al.* (1995).

Further simulations have been carried out by Rayens and Greene (1991) who compare RDA with an approach based on an empirical Bayes framework for addressing the problem of unstable covariance matrices (see also Greene and Rayens, 1989). Aeberhard *et al.* (1993) propose a modified model selection procedure for RDA and Celeux and Mkhardi (1992) present a method of RDA for discrete data. Expressions for the shrinkage parameter are proposed by Loh (1995) and Mkhadri (1995).

An alternative regularised Gaussian discriminant analysis approach is proposed by Bensmail and Celeux (1996). Termed EDDA (eigenvalue decomposition discriminant analysis), it is based on the reparametrisation of the covariance matrix of a class in terms of its eigenvalue decomposition. Fourteen different models are assessed and results compare favourably with RDA. Raudys (2000) considers a similar development.

Hastie *et al.* (1995) cast the discrimination problem as one of regression using *optimal scaling* and use a penalised regression procedure (regularising the within-class covariance matrix). In situations where there are many highly correlated variables, their procedure offers promising results.

Extensions of linear and quadratic discriminant analysis to datasets where the patterns are curves or functions are developed by James and Hastie (2001).

2.4.8 Summary

Linear and quadratic discriminants (or equivalently Gaussian classifiers) are widely used methods of supervised classification and are supported by many statistical packages. Problems occur when the covariance matrices are close to singular and when class boundaries are nonlinear. The latter case occurs when the actual class distributions are highly non-Gaussian. The former case can be overcome by regularisation. This can be achieved by imposing structure on the covariance matrices, pooling/combining covariance matrices or by adding a penalty term to the within-class scatter. Friedman (1989) proposes a scheme (RDA) that includes a combination of covariance matrices and the addition of a penalty term.

2.5 Finite mixture models

2.5.1 Introduction

Finite mixture models have received wide application, being used to model non-Gaussian distributions, including those with multiple modes. They are particularly suitable for modelling distributions where the measurements arise from separate groups, but individual membership is unknown. A finite mixture model is a distribution with probability density function of the form

$$p(\boldsymbol{x}) = \sum_{j=1}^{g} \pi_j p(\boldsymbol{x}; \boldsymbol{\theta}_j) \tag{2.20}$$

where

- g is the number of mixture components (often referred to as the *model order*);
- $\pi_j \geq 0$ are the mixture component probabilities (also referred to as mixing proportions), which satisfy $\sum_{j=1}^{g} \pi_j = 1$;
- $p(\boldsymbol{x}; \boldsymbol{\theta}_j)$, $j = 1, \ldots, g$, are the *component density* functions (each of which depends on a parameter vector $\boldsymbol{\theta}_j$).

The component densities may be of different parametric forms and can be specified using knowledge of the data generation process, if available. In the normal (or Gaussian) mixture model, each component is a multivariate normal distribution, i.e. $p(\boldsymbol{x}; \boldsymbol{\theta}_j)$ is the probability density for the multivariate normal distribution with mean vector $\boldsymbol{\mu}_j$ and covariance matrix $\boldsymbol{\Sigma}_j$, so that $\boldsymbol{\theta}_j = \{\boldsymbol{\mu}_j, \boldsymbol{\Sigma}_j\}$. The normal mixture model therefore has probability density function

$$p(\boldsymbol{x}) = \sum_{j=1}^{g} \pi_j N(\boldsymbol{x}; \boldsymbol{\mu}_j, \boldsymbol{\Sigma}_j)$$

Even if the data are not known to be built up from a series of normal components, a normal mixture model with a sufficient number of components may be a suitable approximation to the data distribution. Figure 2.1 displays the probability density functions for some example

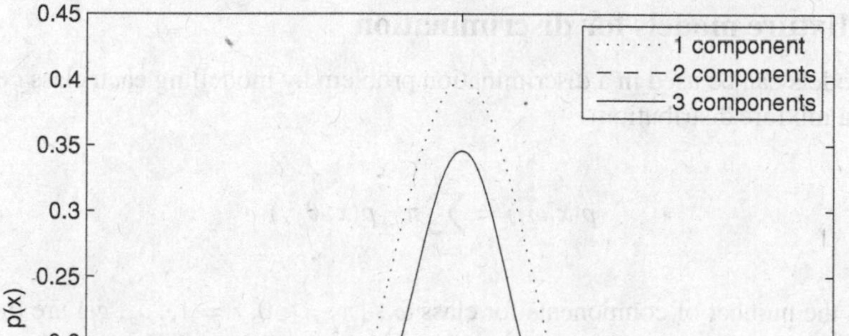

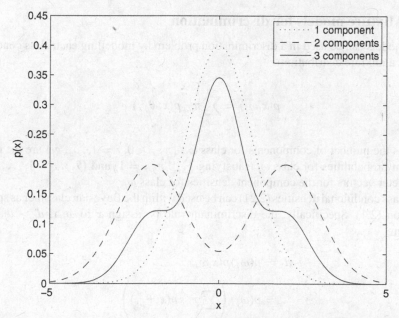

Figure 2.1 Examples of univariate normal mixture model probability density functions. The dotted line is a single component mixture $[(\mu, \sigma) = (0, 1)]$, the dashed line is a two-component mixture $[\pi = (0.5, 0.5), (\mu_1, \sigma_1) = (-2, 1), (\mu_2, \sigma_2) = (2, 1)]$ and the solid line is a three-component mixture $[\pi = (0.2, 0.6, 0.2), (\mu_1, \sigma_1) = (-2, 0.7), (\mu_2, \sigma_2) = (0, 0.7), (\mu_3, \sigma_3) = (2, 0.7)]$.

univariate normal mixture distributions. We can see that using normal distributed components, we can obtain both multimodal distributions (the two-component example) and highly non-Gaussian distributions (the three-component example).

As methods of density estimation, mixture models provide more flexible models than the simple normal-based models considered earlier in the chapter, and when used as class conditional densities they provide improved discrimination in some circumstances.

The modelling capability of mixture models may become clearer when one considers how to draw samples from a mixture model distribution. Samples can be generated using the following procedure:

- Sample the mixture component j according to the component probability vector $(\pi_1, \ldots, \pi_g)$ [i.e. so that $p(j = i) = \pi_i, i = 1, \ldots, g$]. To do this we draw a random variable $u \in U(0, 1)$ from the continuous uniform distribution on the interval [0, 1], and select the mixture component j satisfying $\sum_{k=1}^{j-1} \pi_k < u \leq \sum_{k=1}^{j} \pi_k$.
- Then sample the measurement from the selected component density $p(\boldsymbol{x}; \boldsymbol{\theta}_j)$. For a normal mixture model, this would involve drawing a sample from the distribution $N(\boldsymbol{\mu}_j, \boldsymbol{\Sigma}_j)$.

By repeating the above procedure we obtain independent samples from the mixture model. In words, we are drawing random samples from a population consisting of g separate groups, with the proportions of samples from each group given by the component probabilities $\pi_1, \ldots, \pi_g$.

2.5.2 Mixture models for discrimination

Mixture models can be used in a discrimination problem by modelling each class conditional density as a mixture distribution

$$p(\boldsymbol{x}|\omega_j) = \sum_{r=1}^{g_j} \pi_{j,r} p(\boldsymbol{x}; \boldsymbol{\theta}_{j,r}) \tag{2.21}$$

where g_j is the number of components for class ω_j, $\{\pi_{j,r} \geq 0, r = 1, \ldots, g_j\}$ are the mixture component probabilities for class ω_j (satisfying $\sum_{r=1}^{g_j} \pi_{j,r} = 1$) and $\{\boldsymbol{\theta}_{j,r}, r = 1, \ldots, g_j\}$ are the parameter vectors for the component densities for class ω_j.

The class conditional densities (2.21) can be used within the Bayesian classifier as specified in Equation (2.1). Specifically, the discriminant rule is: assign $\boldsymbol{x}$ to ω_i if $d_i > d_j$, for all $j \neq i$, where

$$\begin{aligned} d_j &= p(\omega_j) p(x|\omega_j) \\ &= p(\omega_j) \left(\sum_{r=1}^{g_j} \pi_{j,r} p(\boldsymbol{x}; \boldsymbol{\theta}_{j,r}) \right) \end{aligned} \tag{2.22}$$

The training phase of the mixture model classifier consists of estimating the parameters of each mixture model. This is done separately for each class in turn. In the *estimative approach* the estimated parameters are used directly as replacements for the parameters $\pi_{j,r}$ and $\boldsymbol{\theta}_{j,r}$ in (2.22).

2.5.3 Parameter estimation for normal mixture models

For a normal mixture model with pre-specified model order g, the parameters to be estimated are the mixture component probabilities $\{\pi_1, \ldots, \pi_g\}$ and the mixture component parameters $\boldsymbol{\theta}_j = \{\boldsymbol{\mu}_j, \boldsymbol{\Sigma}_j\}$, $j = 1, \ldots, g$. The most prevalent technique for optimising the parameters given a set of n independent observations $\{\boldsymbol{x}\} = \{\boldsymbol{x}_1, \ldots, \boldsymbol{x}_n\}$ (the training data) is an iterative procedure known as the Expectation Maximisation (EM) algorithm. The EM algorithm seeks to maximise the likelihood function

$$L(\pi_1, \ldots, \pi_g, \boldsymbol{\mu}_1, \ldots, \boldsymbol{\mu}_g, \boldsymbol{\Sigma}_1, \ldots, \boldsymbol{\Sigma}_g) = \prod_{i=1}^{n} \left\{ \sum_{j=1}^{g} \pi_j N(\boldsymbol{x}_i; \boldsymbol{\mu}_j, \boldsymbol{\Sigma}_j) \right\} \tag{2.23}$$

A full description and motivation of the EM algorithm is provided in Section 2.5.6. In this section we state the procedure for estimating the parameters of a normal mixture model.

2.5.3.1 Initialisation

The EM algorithm is an iterative procedure that requires an initial set of values for the parameters. Typically, a clustering algorithm such as k-means clustering is used to provide reasonable initial values. Clustering algorithms are considered in more detail in Chapter 11. For the purposes of clarity a k-means initialisation procedure for normal mixture models is provided in Algorithm 2.1. As described, the k-means algorithm can terminate naturally, when no further changes are made to the cluster assignments, or can be set to stop when a maximum number of iterations have taken place (reducing computational expense). Note that the provided algorithm often leads to a local optimum clustering solution, and is therefore not recommended as a clustering algorithm if the clusters are to be used with no subsequent processing (see Chapter 11).

2.5.3.2 Iterative EM procedure

Algorithm 2.2 states the EM procedure for normal mixture models. The procedure iteratively applies two steps to refine estimates of model parameters: the E-step and the M-step. Full motivations of these steps are provided in Section 2.5.6 as part of a generic description of the EM algorithm. In more detail, the EM algorithm for normal mixtures alternates between the E-step of estimating the w_{ij} [Equation (2.24)] and the M-step of calculating $\hat{\pi}_j$, $\hat{\boldsymbol{\mu}}_j$ and $\hat{\boldsymbol{\Sigma}}_j$ ($j = 1, \ldots, g$) given the values of w_{ij} [Equations (2.25), (2.26) and (2.27)]. The process iterates until convergence of the likelihood, i.e. until the increase in the likelihood from one iteration to the next falls below a pre-specified small threshold. The monitored likelihoods are calculated by using the parameter estimates within (the logarithm of) Equation (2.23). For computational considerations the procedure may be terminated when a maximum allowed number of iterations has passed.

As with the initialisation procedure described in Algorithm 2.1, it is usual to impose constraints on each estimated covariance matrix [Equation (2.27)], by adding a suitable multiple of the identity matrix if necessary (e.g. if the smallest eigenvalue falls below a threshold). This ensures that all covariance matrices can be inverted, and reduces the risk of one of the components collapsing down onto one of the training data vectors (i.e. a component mean equal to one of the data vectors, and the corresponding covariance matrix tending to the

Algorithm 2.1 k-means initialisation procedure for normal mixture models.

1. Initialisation: Pick g measurements from $\{\boldsymbol{x}_1, \ldots, \boldsymbol{x}_n\}$ randomly without replacement, and use these as initial values for the component mean vectors $\{\boldsymbol{\mu}_j, j = 1, \ldots, g\}$.
2. Iterative step:
   - For $i = 1, \ldots, n$, determine $\psi_i \in \{1, \ldots, g\}$ such that:

     $$|\boldsymbol{x}_i - \boldsymbol{\mu}_{\psi_i}|^2 = \min_{j=1,\ldots,g} |\boldsymbol{x}_i - \boldsymbol{\mu}_j|^2$$

     (i.e. find the closest centre to each training data vector).
   - If any value ψ_i, $i = 1, \ldots, n$, has changed from the previous iteration, recalculate the component means as follows:

     $$\boldsymbol{\mu}_j = \frac{1}{\sum_{i=1}^{n} I(\psi_i = j)} \sum_{i=1}^{n} I(\psi_i = j)\boldsymbol{x}_i$$

     with $I(r = j)$ the indicator function, taking value one if $r = j$, and 0 otherwise. If none of the ψ_i have changed from the previous iteration then move forward to step 4.
3. If the maximum number of iterations has been exceeded then move to step 4, otherwise repeat step 2.
4. Initialise the mixture component covariance matrices as follows:

   $$\boldsymbol{\Sigma}_j = \frac{1}{\sum_{i=1}^{n} I(\psi_i = j)} \sum_{i=1}^{n} I(\psi_i = j)(\boldsymbol{x}_i - \boldsymbol{\mu}_j)(\boldsymbol{x}_i - \boldsymbol{\mu}_j)^T$$

   Subsequent use of the EM algorithm is likely to be more efficient if the initial variances within the covariance matrices are not too small. It is therefore common practice to impose minimum values for each variance component (i.e. each diagonal element of each covariance matrix), or a minimum eigenvalue for each covariance matrix. This can be done by adding a multiple of the identity matrix to the covariance estimates. Indeed, some practitioners set all the initial covariance matrices to the scaled identity matrix.
5. Initialise the mixture component probabilities as follows:

   $$\pi_j = \frac{1}{n} \sum_{i=1}^{n} I(\psi_i = j)$$

zero matrix). Pre-specified constraints on the covariance matrices are considered in the next subsection.

The EM procedure is very easy to implement but the convergence rate can be poor depending on the data distribution and the initial estimates for the parameters. To reduce the risk of being trapped in local maxima of the likelihood function, it is common to repeat the training process a few times, each time using different initial estimates for the parameters. In such cases the final model is that giving rise to the largest likelihood.

Algorithm 2.2 Iterative EM procedure for normal mixture models.

Suppose that at the end of iteration $m \geq 0$ we have the following parameter estimates:

$$\{\pi_j^{(m)}, j = 1, \ldots, g\}, \quad \{\boldsymbol{\mu}_j^{(m)}, j = 1, \ldots, g\}, \text{ and } \{\boldsymbol{\Sigma}_j^{(m)}, j = 1, \ldots, g\}$$

(iteration 0 corresponds to the initialisation stage). The $(m+1)$th iteration of the EM algorithm is then as follows:

1. E-step: Calculate

$$w_{ij} = \frac{\pi_j^{(m)} N\left(\boldsymbol{x}_i; \boldsymbol{\mu}_j^{(m)}, \boldsymbol{\Sigma}_j^{(m)}\right)}{\sum_k \pi_k^{(m)} N\left(\boldsymbol{x}_i; \boldsymbol{\mu}_k^{(m)}, \boldsymbol{\Sigma}_k^{(m)}\right)} \tag{2.24}$$

for $i = 1, \ldots, n$ and $j = 1, \ldots, g$.

2. M-step: For $j = 1, \ldots, g$, update the estimates for the component parameters π_j, $\boldsymbol{\mu}_j$ and $\boldsymbol{\Sigma}_j$ in turn as follows:

$$\hat{\pi}_j^{(m+1)} = \frac{1}{n} \sum_{i-1}^{n} w_{ij} \tag{2.25}$$

$$\hat{\boldsymbol{\mu}}_j^{(m+1)} = \frac{\sum_{i=1}^n w_{ij} \boldsymbol{x}_i}{\sum_{i=1}^n w_{ij}} = \frac{1}{n \hat{\pi}_j^{(m+1)}} \sum_{i=1}^{n} w_{ij} \boldsymbol{x}_i \tag{2.26}$$

$$\hat{\boldsymbol{\Sigma}}_j^{(m+1)} = \frac{1}{n \hat{\pi}_j^{(m+1)}} \sum_{i=1}^{n} w_{ij} \left(\boldsymbol{x}_i - \hat{\boldsymbol{\mu}}_j^{(m+1)}\right) \left(\boldsymbol{x}_i - \hat{\boldsymbol{\mu}}_j^{(m+1)}\right)^T \tag{2.27}$$

2.5.4 Normal mixture model covariance matrix constraints

As mentioned above, problems can occur when one or more of the component covariance matrices is singular or near-singular. Such singularities prevent calculation of the w_{ij} within the E-step and would also prevent calculation of the discriminant rule [Equation (2.22)], since these both require the covariance matrices to be inverted. This problem is more likely to occur when the number of components is large relative to the number of training data vectors, when the dimensionality of the data is large, or when the mixture components are not well separated (the latter being hard to tell until the procedure is started). A more principled solution to this problem than the addition of a multiple of the identity matrix to near singular covariance matrices, is to impose constraints on the covariance matrix structure. Common constraints are:

- diagonal covariance matrices;
- spherical covariance matrices;
- a common covariance matrix across mixture components.

In these cases, the EM update equation for the covariance matrix [Equation (2.27)] is modified, but the rest of the EM procedure remains unchanged.

Diagonal covariance matrices

These ignore the cross-correlation terms in the covariance matrices, so that only the variances are being considered. The covariance matrices take the form $\mathbf{\Sigma}_j = \text{Diag}(\sigma_{j,1}^2, \ldots, \sigma_{j,d}^2)$, where d is the dimensionality of the data, and $\text{Diag}(a_1, \ldots, a_d)$ indicates a $d \times d$ diagonal matrix with lth diagonal element a_l. The EM covariance matrix update equation for the diagonal terms is

$$\hat{\sigma}_{j,l}^{2\,(m+1)} = \frac{1}{n\hat{\pi}_j^{(m+1)}} \sum_{i=1}^{n} w_{ij} \left(x_{i,l} - \hat{\mu}_{j,l}^{(m+1)} \right)^2$$

where $j = 1, \ldots, g$ indexes the mixture component, and $l = 1, \ldots, d$ the dimension.

Spherical covariance matrices

The spherical covariance matrix constraint models the covariances as multiples of the identity matrix, i.e. using a single variance σ_j^2 per mixture component, so that $\mathbf{\Sigma}_j = \sigma_j^2 I_{d,d}$, with $I_{d,d}$ the $d \times d$ identity matrix. The EM single variance update equation for the jth mixture component is

$$\hat{\sigma}_j^{2\,(m+1)} = \frac{1}{nd\hat{\pi}_j^{(m+1)}} \sum_{i=1}^{n} w_{ij} \left(\boldsymbol{x}_i - \hat{\boldsymbol{\mu}}_j^{(m+1)} \right)^T \left(\boldsymbol{x}_i - \hat{\boldsymbol{\mu}}_j^{(m+1)} \right)$$

Common covariance matrix

The third case of using a common covariance matrix across all mixture components gives rise to the following EM covariance matrix update equation

$$\hat{\mathbf{\Sigma}}^{(m+1)} = \frac{1}{n} \sum_{j=1}^{g} \sum_{i=1}^{n} w_{ij} \left(\boldsymbol{x}_i - \hat{\boldsymbol{\mu}}_j^{(m+1)} \right) \left(\boldsymbol{x}_i - \hat{\boldsymbol{\mu}}_j^{(m+1)} \right)^T$$

Further covariance matrix constraints

Other constraints on the covariance matrix structure may be considered. Celeux and Govaert (1995) propose a parametrisation of the covariance matrix that covers several different conditions, ranging from equal spherical clusters (covariance matrices equal and proportional to the identity matrix) to different covariance matrices for each cluster.

Hastie and Tibshirani (1996) consider the use of mixture models for classification. They allow each component within a class to have its own mean vector, but set a common covariance matrix across all mixture components and across all classes. This is a further way of restricting the number of parameters to be estimated.

2.5.5 How many components?

Choosing the number of components in a mixture model is not a trivial problem and depends on many factors including the actual distribution of the data being modelled, the shape of clusters, and the amount of available training data.

In a discrimination context, we can train a number of different mixture model classifiers, each using a different number of mixture components, and apply each to a separate test set. The model order can then be selected as that giving rise to the best discrimination performance on this test set. Note that the test set is being used as part of the training procedure (more properly termed a *validation set*) and error rates quoted using these data will be optimistically biased.

An alternative approach is to apply model selection criteria separately to each class-conditional mixture model (i.e. consider the fit of each class-conditional distribution, rather than the overall discrimination performance). A number of different model selection criteria have been proposed. Comparative studies of model selection criteria in the literature include that by Bozdogan (1993), who compared several information-theoretic criteria on simulated data consisting of overlapping and nonoverlapping clusters of different shape and compactness.

A popular criterion is the Akaike Information Criterion (AIC) (Akaike, 1974) given by

$$AIC = -2\log(L(\Psi)) + 2k_d$$

where $L(\Psi)$ is the model likelihood evaluated on the data [Equation (2.23) for normal mixture models, calculated for the estimated parameters], and k_d is the number of parameters in the model. When selecting between multiple models, the model with the smallest AIC is selected. The AIC penalises over-fitting using large models (which result in larger log-likelihood values) through the k_d term. However, it is often found to overestimate the number of components (as noted in McLachlan and Peel, 2000).

Another common criterion is the Bayesian Information Criterion (BIC), proposed by Schwarz (1978) and given by

$$BIC = -2\log(L(\Psi)) + k_d\log(n)$$

$L(\Psi)$ and k_d are as for the AIC, and n is the number of independent data vectors used to calculate the model likelihood. The model with the smallest BIC is selected. For $n \geq 8$ the model order penalty term in the BIC is larger than that in the AIC, and therefore the BIC is less likely to overestimate the number of components. The BIC is considered to have better theoretical motivations than the AIC for mixture models (McLachlan and Peel, 2000).

Celeux and Soromenho (1996) propose an entropy criterion, evaluated as a by-product of the EM algorithm, and compare its performance with several other criteria. Their model selection criterion involves minimising

$$C(g) = \frac{E(g)}{\log(L(g)) - \log(L(1))}, \quad g > 1$$

where $g > 1$ is the number of mixture components, $L(k)$ ($k \geq 1$) is the likelihood for a model with k components, and $E(g)$ is an entropy term

$$E(g) = \sum_{j=1}^{g}\sum_{i=1}^{n} w_{ij}\log(w_{ij})$$

where

$$w_{ij} = \frac{\pi_j p(\boldsymbol{x}_i|\boldsymbol{\theta}_j)}{\sum_{k=1}^{g} \pi_k p(\boldsymbol{x}_i|\boldsymbol{\theta}_k)}$$

calculated in the E-step of the EM algorithm [Equation (2.24)]. The criterion attempts to combine quality of fit (evidenced by large likelihoods), with selection of a mixture model whose components are well separated (i.e. clustering) giving rise to small values for the entropy term. It is therefore not suited to mixture models containing components that have similar mean vectors, but varying covariance matrices. A rule of thumb (McLachlan and Peel, 2000) is to define $C(1) = 1$.

Wolfe (1971) proposes a modified likelihood ratio test for testing between two mixture models with different numbers of components, in which the null hypothesis $g = g_0$ is tested against the alternative hypothesis that $g = g_1$. The quantity

$$-\frac{2}{n}\left(n - 1 - d - \frac{g_1}{2}\right)\log(\lambda)$$

where λ is the likelihood ratio [$\lambda = L(\Psi_0)/L(\Psi_1)$, with $L(\Psi_0)$ the likelihood under the null hypothesis, and $L(\Psi_1)$ the likelihood under the alternative hypothesis], is tested as a chi-square random variable with the degrees of freedom, χ, being twice the difference in the number of parameters in the two hypotheses (Everitt *et al.*, 2001), excluding mixing proportions. For components of a normal mixture with arbitrary covariance matrices,

$$\chi = 2(g_1 - g_0)\frac{d(d+3)}{2}$$

and with common covariance matrices (the case that was studied by Wolfe, 1971), $\chi = 2(g_1 - g_0)d$. The modification over a standard likelihood ratio test is needed, because the regularity conditions needed for the standard likelihood ratio test [i.e. for $-2\log(\lambda)$ to asymptotically have a chi-square distribution with degrees of freedom equal to the difference in the number of parameters] do not hold when comparing mixture models with a different number of components [see McLachlan and Peel (2000) for details and an example].

Wolfe's test has been investigated by Everitt (1981) and Anderson (1985). For the common covariance structure, Everitt finds that, for testing $g = 1$ against $g = 2$ in a two-component mixture, the test is appropriate if the number of observations is at least 10 times the number of variables. McLachlan and Basford (1988) recommend that Wolfe's modified likelihood ratio test be used as a guide to structure rather than rigidly interpreted.

Bayesian approaches to learning mixture model parameters can incorporate the model order as an unknown variable to be learnt. The underlying modelling techniques are discussed in Chapter 3. A seminal paper by Richardson and Green (1997)[5] developed a reversible jump Markov chain Monte Carlo (RJMCMC) algorithm for joint model selection and parameter estimation in univariate normal mixture models. The work has been extended to low-dimensionality multivariate distributions (Marrs, 1998), but currently is still considered inappropriate for application to high-dimensional data. Other fully Bayesian approaches, such

[5] The associated correction is in Richardson and Green (1998).

as the Markov birth–death process approach in Stephens (2000) (see also Stephens, 1997), have been proposed, but again are more suitable to low-dimensional data modelling (e.g. univariate and bivariate vectors).

2.5.6 Maximum likelihood estimation via EM

2.5.6.1 Introduction

We now return to the general problem of estimating the parameters of a mixture model. There are three sets of parameters to estimate: the value of g (i.e. the number of components), the values of π_j, and the parameters $\boldsymbol{\theta}_j$ defining the component densities. Selection of the number of mixture components has been discussed in Section 2.5.5, so in this section we focus on estimation of the the mixture model parameters for a fixed model order g.

Given a set of n independent observations $\{\boldsymbol{x}\} = \{\boldsymbol{x}_1, \ldots, \boldsymbol{x}_n\}$, the likelihood function for the mixture distribution $p(\boldsymbol{x}) = \sum_{j=1}^{g} \pi_j p(\boldsymbol{x}; \boldsymbol{\theta}_j)$ is

$$L(\boldsymbol{\Psi}) = \prod_{i=1}^{n} \left\{ \sum_{j=1}^{g} \pi_j p(\boldsymbol{x}_i|\boldsymbol{\theta}_j) \right\} \tag{2.28}$$

where $\boldsymbol{\Psi}$ denotes the set of parameters $\{\pi_1, \ldots, \pi_g; \boldsymbol{\theta}_1, \ldots, \boldsymbol{\theta}_g\}$ and we now denote the dependence of the component densities on their parameters as $p(\boldsymbol{x}|\boldsymbol{\theta}_j)$. In general, it is not possible to maximise the likelihood function through solving $\partial L/\partial \boldsymbol{\Psi} = 0$ explicitly for the parameters of the model. Instead, iterative schemes must be employed. The most popular approach for maximising the likelihood $L(\boldsymbol{\Psi})$ is to use a general class of iterative procedures known as EM algorithms, introduced in the context of missing data estimation by Dempster *et al.* (1977), though it had appeared in many forms previously.

2.5.6.2 Formulation as incomplete data problem

We suppose that we have a set of 'incomplete' data vectors $\{\boldsymbol{x}\} = \{\boldsymbol{x}_1, \ldots, \boldsymbol{x}_n\}$, and that we wish to maximise the likelihood function $L(\boldsymbol{\Psi}) = p(\{\boldsymbol{x}\}|\boldsymbol{\Psi})$. Let $\{\boldsymbol{y}\}$ denote a typical 'complete' version of $\{\boldsymbol{x}\}$, that is, each vector $\boldsymbol{x}_i$ is augmented by the 'missing' values z_i so that $\boldsymbol{y}_i^T = (\boldsymbol{x}_i^T, z_i^T)$. In the finite mixture model case, the augmentation is naturally the component labels for the data vectors. In the use of the EM procedure for more general missing data problems there may be many possible vectors $\boldsymbol{y}_i$ in which we can embed $\boldsymbol{x}_i$.

Let the likelihood of $\{\boldsymbol{y}\}$ be $p(\{\boldsymbol{y}\}|\boldsymbol{\Psi})$ whose form we know explicitly. The likelihood $p(\{\boldsymbol{x}\}|\boldsymbol{\Psi})$ is obtained from $p(\{\boldsymbol{y}\}|\boldsymbol{\Psi})$ by integrating over all possible $\{z\}$ alongside which the set $\{\boldsymbol{x}\}$ is embedded

$$\begin{aligned} L(\boldsymbol{\Psi}) = p(\{\boldsymbol{x}\}|\boldsymbol{\Psi}) &= \int p(\{\boldsymbol{x}\}, \{z\}|\boldsymbol{\Psi}) dz \\ &= \int \prod_{i=1}^{n} p(\boldsymbol{x}_i, z_i|\boldsymbol{\Psi}) dz_1 \ldots dz_n \end{aligned}$$

For the finite mixture model case, the complete data vector $\boldsymbol{y}$ is the observation augmented by a component label; that is, $\boldsymbol{y}^T = (\boldsymbol{x}^T, z^T)$, where z is an indicator vector of length g with a 1 in the kth position if $\boldsymbol{x}$ is in group (i.e. component) k and zeros elsewhere. The likelihood of $\boldsymbol{y}$ is

$$p(\boldsymbol{y}|\boldsymbol{\Psi}) = p(\boldsymbol{x}, z|\boldsymbol{\Psi}) = p(\boldsymbol{x}|z, \boldsymbol{\Psi})p(z|\boldsymbol{\Psi}) = p(\boldsymbol{x}|\boldsymbol{\theta}_k)\pi_k$$

which may be written as

$$p(\boldsymbol{y}|\boldsymbol{\Psi}) = \prod_{j=1}^{g} \left[p(\boldsymbol{x}|\boldsymbol{\theta}_j)\pi_j\right]^{z_j} \tag{2.29}$$

since z_j is zero except for $j = k$, for which it is one. The likelihood of $\boldsymbol{x}$ is

$$p(\boldsymbol{x}|\boldsymbol{\Psi}) = \sum_{\text{all possible z values}} p(\boldsymbol{y}|\boldsymbol{\Psi}) = \sum_{j=1}^{g} \pi_j p(\boldsymbol{x}|\boldsymbol{\theta}_j)$$

which is a mixture distribution. Thus, we may interpret mixture data as incomplete data where the missing values are the component labels.

2.5.6.3 EM procedure

The EM procedure generates a sequence of estimates of $\boldsymbol{\Psi}$, $\{\boldsymbol{\Psi}^{(m)}\}$, from an initial estimate $\boldsymbol{\Psi}^{(0)}$ and consists of two steps:

1. E-step: Evaluate:

$$Q(\boldsymbol{\Psi}, \boldsymbol{\Psi}^{(m)}) \triangleq \mathrm{E}\left[\log(p(\{\boldsymbol{y}\}|\boldsymbol{\Psi}))|\{\boldsymbol{x}\}, \boldsymbol{\Psi}^{(m)}\right]$$

   i.e. the expectation of the complete data log-likelihood, conditional on the observed data, $\{\boldsymbol{x}\}$, and the current value of the parameters, $\boldsymbol{\Psi}^{(m)}$. We have

$$\begin{aligned} Q(\boldsymbol{\Psi}, \boldsymbol{\Psi}^{(m)}) &= \mathrm{E}\left[\log\left(\prod_i p(\boldsymbol{x}_i, z_i|\boldsymbol{\Psi})\right)\Bigg|\{\boldsymbol{x}\}, \boldsymbol{\Psi}^{(m)}\right] \\ &= \mathrm{E}\left[\sum_i \log(p(\boldsymbol{x}_i, z_i|\boldsymbol{\Psi}))\Big|\{\boldsymbol{x}\}, \boldsymbol{\Psi}^{(m)}\right] \\ &= \int \sum_i \log(p(\boldsymbol{x}_i, z_i|\boldsymbol{\Psi}))p(\{z\}|\{\boldsymbol{x}\}, \boldsymbol{\Psi}^{(m)})dz_1 \ldots dz_n \end{aligned}$$

2. M-step: Find $\boldsymbol{\Psi} = \boldsymbol{\Psi}^{(m+1)}$ that maximises $Q(\boldsymbol{\Psi}, \boldsymbol{\Psi}^{(m)})$. Often the solution for the M-step may be obtained in closed form. Otherwise, numerical optimisation procedures can be used (Press *et al.*, 1992).

When iterated from an initial set of parameter estimates $\mathbf{\Psi}^{(0)}$, the above procedure is such that the likelihoods of interest satisfy

$$L\{\mathbf{\Psi}^{(m+1)}\} \geq L\{\mathbf{\Psi}^{(m)}\} \tag{2.30}$$

so that they are monotonically increasing (i.e. adopting the procedure leads to estimates with better likelihood values).

2.5.6.4 Why the EM procedure works

To see that the likelihoods arising from the iterative EM procedure are monotonically increasing we first define

$$\begin{aligned} H\left(\mathbf{\Psi}, \mathbf{\Psi}^{(m)}\right) &= \mathrm{E}\left[\log\left(p(\{z\}\,|\{\boldsymbol{x}\}, \mathbf{\Psi})\right)|\{\boldsymbol{x}\}, \mathbf{\Psi}^{(m)}\right] \\ &= \int \sum_i \log\left(p(z_i|\{\boldsymbol{x}\}, \mathbf{\Psi})\right) p\left(z_i|\{\boldsymbol{x}\}, \mathbf{\Psi}^{(m)}\right) dz_i \end{aligned}$$

so that $Q(\mathbf{\Psi}, \mathbf{\Psi}^{(m)})$ differs from the log-likelihood at $\mathbf{\Psi}$ by $H(\mathbf{\Psi}, \mathbf{\Psi}^{(m)})$, i.e.

$$\log(L(\mathbf{\Psi})) = Q(\mathbf{\Psi}, \mathbf{\Psi}^{(m)}) - H(\mathbf{\Psi}, \mathbf{\Psi}^{(m)}) \tag{2.31}$$

The validity of (2.31) can be seen by first noting that $p(\{\boldsymbol{x}\}|\mathbf{\Psi}))$ does not depend on $\{z\}$, and is therefore unchanged by taking its expectation with respect to $\{z\}$

$$\log(p(\{\boldsymbol{x}\}|\mathbf{\Psi})) = \mathrm{E}[\log(p(\{\boldsymbol{x}\}|\mathbf{\Psi}))|\{\boldsymbol{x}\}, \mathbf{\Psi}^{(m)}]$$

Then making use of the formula for conditional probabilities

$$p(\{y\}|\mathbf{\Psi}) = p(\{\boldsymbol{x}\}, \{z\}|\mathbf{\Psi}) = p(\{z\}|\{\boldsymbol{x}\}, \mathbf{\Psi})p(\{\boldsymbol{x}\}|\mathbf{\Psi})$$

we have

$$\begin{aligned} \log(p(\{\boldsymbol{x}\}|\mathbf{\Psi})) &= \mathrm{E}\left[\log\left(\frac{p(\{y\}|\mathbf{\Psi})}{p(\{z\}|\{\boldsymbol{x}\}, \mathbf{\Psi})}\right)\Big|\{\boldsymbol{x}\}, \mathbf{\Psi}^{(m)}\right] \\ &= \mathrm{E}[\log(p(\{y\}|\mathbf{\Psi}))|\{\boldsymbol{x}\}, \mathbf{\Psi}^{(m)}] - \mathrm{E}[\log\left(p(\{z\}|\{\boldsymbol{x}\}, \mathbf{\Psi})\right)|\{\boldsymbol{x}\}, \mathbf{\Psi}^{(m)}] \\ &= Q(\mathbf{\Psi}, \mathbf{\Psi}^{(m)}) - H(\mathbf{\Psi}, \mathbf{\Psi}^{(m)}) \end{aligned}$$

which is the relationship in Equation (2.31).

Examining the decomposition of the log-likelihood in Equation (2.31) we can see that a sufficient condition for the relationship (2.30) to hold is

$$Q(\mathbf{\Psi}^{(m+1)}, \mathbf{\Psi}^{(m)}) \geq Q(\mathbf{\Psi}^{(m)}, \mathbf{\Psi}^{(m)}) \tag{2.32}$$

$$H(\mathbf{\Psi}^{(m+1)}, \mathbf{\Psi}^{(m)}) \leq H(\mathbf{\Psi}^{(m)}, \mathbf{\Psi}^{(m)}) \tag{2.33}$$

Now the M-step of the EM procedure ensures that (2.32) holds. Hence it remains to show the relationship (2.33). In fact, the more general relationship $H(\boldsymbol{\Psi}, \boldsymbol{\Psi}^{(m)}) \leq H(\boldsymbol{\Psi}^{(m)}, \boldsymbol{\Psi}^{(m)})$ holds. Writing

$$\begin{aligned}\Upsilon(\boldsymbol{\Psi}, \boldsymbol{\Psi}^{(m)}) &= H(\boldsymbol{\Psi}, \boldsymbol{\Psi}^{(m)}) - H(\boldsymbol{\Psi}^{(m)}, \boldsymbol{\Psi}^{(m)}) \\ &= \mathrm{E}\left[\log\left(\frac{p(\{z\}|\{\boldsymbol{x}\}, \boldsymbol{\Psi})}{p(\{z\}|\{\boldsymbol{x}\}, \boldsymbol{\Psi}^{(m)})}\right)\Big|\{\boldsymbol{x}\}, \boldsymbol{\Psi}^{(m)}\right]\end{aligned}$$

we can make use of Jensen's inequality applied to the logarithmic function

$$\mathrm{E}[\log(T)] \leq \log(\mathrm{E}[T])$$

to obtain

$$\begin{aligned}\Upsilon(\boldsymbol{\Psi}, \boldsymbol{\Psi}^{(m)}) &\leq \log\left(\mathrm{E}\left[\left(\frac{p(\{z\}|\{\boldsymbol{x}\}, \boldsymbol{\Psi})}{p(\{z\}|\{\boldsymbol{x}\}, \boldsymbol{\Psi}^{(m)})}\right)\Big|\{\boldsymbol{x}\}, \boldsymbol{\Psi}^{(m)}\right]\right) \\ &= \log\left(\int_z \left(\frac{p(\{z\}|\{\boldsymbol{x}\}, \boldsymbol{\Psi})}{p(\{z\}|\{\boldsymbol{x}\}, \boldsymbol{\Psi}^{(m)})}\right) p(\{z\}|\{\boldsymbol{x}\}, \boldsymbol{\Psi}^{(m)})dz\right) \\ &= \log\left(\int_z p(\{z\}|\{\boldsymbol{x}\}, \boldsymbol{\Psi})dz\right) = \log(1) = 0\end{aligned}$$

proving the desired inequality [Equation (2.33)], and hence the monotonically increasing nature of the likelihoods obtained from the parameter estimates provided by the EM procedure.

2.5.6.5 Application of the EM algorithm to mixture models

Let us now consider the steps of the EM algorithm applied to mixture models (previously stated without justification for normal mixture models in Section 2.5.3). Extending (2.29) to multiple independent vectors we have

$$p(\{\boldsymbol{y}\}|\boldsymbol{\Psi}) = p(\boldsymbol{y}_1, \ldots \boldsymbol{y}_n|\boldsymbol{\Psi}) = \prod_{i=1}^{n}\prod_{j=1}^{g}[p(\boldsymbol{x}_i|\boldsymbol{\theta}_j)\pi_j]^{z_{ji}}$$

where z_{ji} are the indicator variables, $z_{ji} = 1$ if pattern $\boldsymbol{x}_i$ is in group j, zero otherwise. Hence

$$\begin{aligned}\log(p(\boldsymbol{y}_1, \ldots \boldsymbol{y}_n|\boldsymbol{\Psi})) &= \sum_{i=1}^{n}\sum_{j=1}^{g} z_{ji}\log(p(\boldsymbol{x}_i|\boldsymbol{\theta}_j)\pi_j) \\ &= \sum_{i=1}^{n} \boldsymbol{z}_i^T \boldsymbol{l} + \sum_{i=1}^{n} \boldsymbol{z}_i^T \boldsymbol{u}_i(\boldsymbol{\theta})\end{aligned}$$

where:

- the vector $\boldsymbol{l}$ has jth component $\log(\pi_j)$;
- $\boldsymbol{u}_i(\boldsymbol{\theta})$ has jth component $\log(p(\boldsymbol{x}_i|\boldsymbol{\theta}_j))$;
- z_i has components $z_{ji}, j = 1, \dots, g$,

The likelihood of $(\boldsymbol{x}_1, \dots, \boldsymbol{x}_n)$ is $L(\boldsymbol{\Psi})$ given by Equation (2.28). The steps in the basic EM iteration are:

1. E-step: Form Q given by

$$\begin{aligned} Q(\boldsymbol{\Psi}, \boldsymbol{\Psi}^{(m)}) &= \mathrm{E}\left[\log(p(\boldsymbol{y}_1, \dots, \boldsymbol{y}_n|\boldsymbol{\Psi}))|\{\boldsymbol{x}\}, \boldsymbol{\Psi}^{(m)}\right] \\ &= \sum_{i=1}^{n} \boldsymbol{w}_i^T \boldsymbol{l} + \sum_{i=1}^{n} \boldsymbol{w}_i^T \boldsymbol{u}_i(\boldsymbol{\theta}) \end{aligned}$$

   where

$$\boldsymbol{w}_i = \mathrm{E}\left[z_i|\boldsymbol{x}_i, \boldsymbol{\Psi}^{(m)}\right]$$

   with jth component the probability that $\boldsymbol{x}_i$ belongs to component j given the current parameter estimates, $\boldsymbol{\Psi}^{(m)}$:

$$w_{ij} = \frac{\pi_j^{(m)} p\left(\boldsymbol{x}_i|\boldsymbol{\theta}_j^{(m)}\right)}{\sum_k \pi_k^{(m)} p\left(\boldsymbol{x}_i|\boldsymbol{\theta}_k^{(m)}\right)} \tag{2.34}$$

2. M-step: This consists of maximising Q with respect to $\boldsymbol{\Psi}$. Consider the parameters π_j, $\boldsymbol{\theta}_j$ in turn. Maximising Q with respect to π_j (subject to the constraint that $\sum_{i=1}^{g} \pi_i = 1$) leads to the equation

$$\sum_{i=1}^{n} w_{ij} \frac{1}{\pi_j} - \lambda = 0$$

   obtained by differentiating $Q - \lambda(\sum_{j=1}^{g} \pi_j - 1)$ with respect to π_j, where λ is a Lagrange multiplier. The constraint $\sum_{j=1}^{g} \pi_j = 1$ gives $\lambda = \sum_{j=1}^{g} \sum_{i=1}^{n} w_{ij} = n$ and we have the estimate of π_j as

$$\hat{\pi}_j = \frac{1}{n} \sum_{i=1}^{n} w_{ij} \tag{2.35}$$

   For normal mixture models, $\boldsymbol{\theta}_j = (\boldsymbol{\mu}_j, \boldsymbol{\Sigma}_j)$ and we have

$$Q(\boldsymbol{\Psi}, \boldsymbol{\Psi}^{(m)}) = \sum_{i=1}^{n} \boldsymbol{w}_i^T \boldsymbol{l} + \sum_{j=1}^{g} \sum_{i=1}^{n} w_{ij} \log(N(\boldsymbol{x}_i; \boldsymbol{\mu}_j, \boldsymbol{\Sigma}_j)) \tag{2.36}$$

Differentiating Q with respect to $\boldsymbol{\mu}_j$ and equating to zero gives

$$\sum_{i=1}^{n} w_{ij}(\boldsymbol{x}_i - \boldsymbol{\mu}_j) = 0$$

which gives the new estimate for $\boldsymbol{\mu}_j$ as

$$\hat{\boldsymbol{\mu}}_j = \frac{\sum_{i=1}^{n} w_{ij}\boldsymbol{x}_i}{\sum_{i=1}^{n} w_{ij}} = \frac{1}{n\hat{\pi}_j}\sum_{i=1}^{n} w_{ij}\boldsymbol{x}_i \tag{2.37}$$

Differentiating Q with respect to $\boldsymbol{\Sigma}_j^{-1}$ and equating to zero gives

$$\hat{\boldsymbol{\Sigma}}_j = \frac{\sum_{i=1}^{n} w_{ij}\left(\boldsymbol{x}_i - \hat{\boldsymbol{\mu}}_j\right)\left(\boldsymbol{x}_i - \hat{\boldsymbol{\mu}}_j\right)^T}{\sum_{i=1}^{n} w_{ij}}$$

$$= \frac{1}{n\hat{\pi}_j}\sum_{i=1}^{n} w_{ij}\left(\boldsymbol{x}_i - \hat{\boldsymbol{\mu}}_j\right)\left(\boldsymbol{x}_i - \hat{\boldsymbol{\mu}}_j\right)^T \tag{2.38}$$

Equations (2.35), (2.37) and (2.38) are as the earlier stated normal mixture model update formulae of Section 2.5.3 [Equations (2.25), (2.26) and (2.27)].

2.5.7 Example application study

The problem

The practical application concerns the automatic recognition of ships using high-resolution radar measurements of the radar cross-section of targets (Webb, 2000).

Summary

This is a straightforward mixture model approach to discrimination, with maximum likelihood estimates of the parameters obtained via the EM algorithm. The mixture component distributions are taken to be gamma distributions.

The data

The data consist of radar range profiles (RRPs) of ships of seven class types. An RRP describes the magnitude of the radar reflections of the ship as a function of distance from the radar. The profiles are sampled at 3 m spacing and each RRP comprises 130 measurements. RRPs are recorded from all aspects of a ship as the ship turns through 360°. There are 19 data files and each data file comprises between 1700 and 8800 training patterns. The data files are divided into train and test sets. Several classes have more than one rotation available for training and testing.

The model

The density of each class is modelled using a mixture model. Thus, we have

$$p(\boldsymbol{x}) = \sum_{j=1}^{g} \pi_j p(\boldsymbol{x}|\boldsymbol{\theta}_j)$$

where $\boldsymbol{\theta}_j$ represents the set of parameters of mixture component j. An independence model[6] is assumed for each mixture component, $p(\boldsymbol{x}|\boldsymbol{\theta}_j)$, therefore

$$p(\boldsymbol{x}|\boldsymbol{\theta}_j) = \prod_{s=1}^{130} p(x_s|\theta_{js})$$

and the univariate factor, $p(x_s|\theta_{js})$, is modelled as a gamma distribution[7] with parameters $\theta_{js} = (m_{js}, \mu_{js})$

$$p(x_s|\theta_{js}) = \frac{m_{js}}{(m_{js}-1)!\mu_{js}} \left(\frac{m_{js}x_s}{\mu_{js}}\right)^{m_{js}-1} \exp\left(-\frac{m_{js}x_s}{\mu_{js}}\right)$$

where m_{js} is the order parameter for variable x_s of mixture component j and μ_{js} is the mean. Thus for each mixture component, there are two parameters associated with each dimension. We denote by $\boldsymbol{\theta}_j$ the set $\{\theta_{js}, s = 1, \ldots, 130\}$, the parameters of component j. The gamma distribution is chosen from physical considerations – it has special cases of Rayleigh scattering and a nonfluctuating target. Also, empirical measurements have been found to be gamma-distributed.

Training procedure

Given a set of n assumed independent observations (in a given class), the likelihood function is

$$L(\boldsymbol{\Psi}) = \prod_{i=1}^{n} \sum_{j=1}^{g} \pi_j p(\boldsymbol{x}_i|\boldsymbol{\theta}_j) \tag{2.39}$$

where $\boldsymbol{\Psi}$ represents all the parameters of the model, $\boldsymbol{\Psi} = \{\boldsymbol{\theta}_j, \pi_j, j = 1, \ldots, g\}$.

An EM approach to maximum likelihood is taken. If $\{\boldsymbol{\theta}_k^{(m)}, \pi_k^{(m)}\}$ denotes the estimate of the parameters of the kth component at the mth stage of the iteration, then the E-step estimates w_{ij}, the probability that $\boldsymbol{x}_i$ belongs to component j given the current estimates of the parameters as in Equation (2.34)

$$w_{ij} = \frac{\pi_j^{(m)} p(\boldsymbol{x}_i|\boldsymbol{\theta}_j^{(m)})}{\sum_k \pi_k^{(m)} p(\boldsymbol{x}_i|\boldsymbol{\theta}_k^{(m)})}$$

The M-step leads to the estimate of the mixture weights, π_j as in Equation (2.35)

$$\hat{\pi}_j = \frac{1}{n}\sum_{i=1}^{n} w_{ij}$$

---

[6] Note that this is independence of the elements of the data vector, conditional on the mixture component, and is unrelated to the independence of data vectors.

[7] There are several parametrisations of a gamma distribution. We present the one used in the study.

The equation for the mean can be obtained as

$$\hat{\boldsymbol{\mu}}_j = \frac{\sum_{i=1}^{n} w_{ij}\boldsymbol{x}_i}{\sum_{i=1}^{n} w_{ij}} = \frac{1}{n\hat{\pi}_j}\sum_{i=1}^{n} w_{ij}\boldsymbol{x}_i$$

but the equation for the gamma-order parameters, m_{js}, cannot be solved in closed form. We have

$$\nu(m_{js}) = -\frac{\sum_{i=1}^{n} w_{ij}\log(x_{is}/\mu_{js})}{\sum_{i=1}^{n} w_{ij}}$$

where $\nu(m) = \log(m) - \psi(m)$ and $\psi(m)$ is the digamma function, i.e. the derivative of the logarithm of the gamma function [$\psi(m) = \frac{d}{dm}\log(\Gamma(m))$, $\Gamma(m) = (m-1)!$ for $m \geq 1$ an integer].

Thus, for the gamma mixture problem, an EM approach may be taken, but a numerical root-finding routine must be used within the EM loop for the gamma distribution order parameters.

The number of mixture components per class was varied between 5 and 110 and the model for each ship determined by minimising a penalised likelihood criterion (the AIC, which is the likelihood penalised by a complexity term – see Section 2.5.5). This resulted in between 50 and 100 mixture components per ship. The density function for each class was constructed using (2.39). The class priors were taken to be equal. The model was then applied to a separate test set and the error rate estimated.

2.5.8 Further developments

Other procedures for maximising the mixture model likelihood function include Newton–Raphson iterative schemes (Hasselblad, 1966) and simulated annealing (Ingrassia, 1992).

Jamshidian and Jennrich (1993) propose an approach for accelerating the EM algorithm based on a generalised conjugate gradients numerical optimisation scheme (see also Jamshidian and Jennrich, 1997). Other competing numerical schemes for normal mixtures are described by Everitt and Hand (1981) and Titterington *et al.* (1985). Lindsay and Basak (1993) describe a method of moments approach to multivariate normal mixtures that may be used to initialise an EM algorithm.

Further extensions to the EM algorithm are given by Meng and Rubin (1992, 1993). The SEM (supplemented EM) algorithm is a procedure for computing the asymptotic variance–covariance matrix. The ECM (expectation/conditional maximisation) algorithm is a procedure for implementing the M-step when a closed-form solution is not available (i.e. for component distributions that are not Gaussian), replacing each M-step by a sequence of conditional maximisation steps. An alternative gradient algorithm for approximating the M-step is presented by Lange (1995) and the algorithm is further generalised to the ECME (ECM Either) algorithm by Liu and Rubin (1994).

Developments to data containing groups of observations with longer than normal tails are described by Peel and McLachlan (2000), who develop a mixture of t distributions model, with parameters determined using the ECM algorithm.

One of the main issues with likelihood optimisation in mixture models is that there is a multitude of 'useless' global maxima (Titterington *et al.*, 1985). In particular, if the mean of one of the mixture components is specified to be one of the training data points, then the likelihood tends to infinity as the variance of the component centred on that data point tends to zero. Similarly, if sample points are close together, then there will be high local maxima of the likelihood function and it appears that the maximum likelihood procedure fails for this class of mixture models. However, provided that we do not allow the variances to tend to zero, perhaps by imposing an equality constraint on the covariance matrices, then the maximum likelihood method is still viable (Everitt and Hand, 1981). Equality of covariance matrices may be a rather restrictive assumption in many applications. Convergence to parameter values associated with singularities is more likely to occur with small sample sizes and when components are not well separated.

Another problem with a mixture model approach is that there may be many local minima of the likelihood function and several initial configurations may have to be tried before a satisfactory estimate is produced. It is worthwhile trying several initialisations, since agreement between the resulting estimates lends more weight to the chosen solution. Celeux and Govaert (1992) describe approaches for developing the basic EM algorithm to overcome the problem of local optima.

Bayesian estimation procedures are available, most commonly utilising Markov chain Monte Carlo (MCMC) algorithms (Section 3.4), see e.g. Gilks *et al.* (1996) and Roeder and Wasserman (1997). Fully Bayesian approaches, estimating both the number of components and the component parameters have also been proposed, as highlighted in Section 2.5.5.

2.5.9 Summary

Modelling using normal mixtures is a simple and powerful way of developing the normal model to nonlinear discriminant functions. Even the restrictive assumption of a common covariance matrix for mixture components enables nonlinear decision boundaries to be found. The EM algorithm provides an appealing scheme for parameter estimation, and there have been various extensions accelerating the technique. The EM algorithm is not limited to normal mixtures, and can be utilised for learning the parameters of mixture components with a wide range of densities. A mixture model may also be used to partition a given dataset by modelling the dataset using a mixture, and assigning data samples to the group for which the probability of membership is the greatest. The use of mixture models in this (clustering) context is discussed further in Chapter 11.

2.6 Application studies

The application of the normal-based linear and quadratic discriminant rules covers a wide range of problems. These include the areas of:

- Medical research. Aitchison *et al.* (1977) compare predictive and estimative approaches to discrimination. Harkins *et al.* (1994) use a quadratic rule for the classification of red cell disorders. Hand (1992) reviews statistical methodology in medical research, including discriminant analysis (see also Jain and Jain, 1994). Stevenson (1993) discusses the role of discriminant analysis in psychiatric research. Borini and Guimarães (2003) use a quadratic rule for noninvasive classification of liver disease based upon patient responses and blood measurements. Bouveyron *et al.* (2009) consider the use of linear and quadratic discriminant functions, and also mixture models, for detecting cervical cancer.

- Machine vision. Magee *et al.* (1993) use a Gaussian classifier to discriminate bottles based on five features derived from images of the bottle tops.
- Target recognition. Kreithen *et al.* (1993) develop a target and clutter discrimination algorithm based on multivariate normal assumptions for the class distributions. Liu *et al.* (2008) use RDA to distinguish airports from surroundings in forward-looking infrared (FLIR) images.
- Spectroscopic data. Krzanowski *et al.* (1995) consider ways of estimating linear discriminant functions when covariance matrices are singular and analyse data consisting of infrared reflectance measurements.
- Radar. Haykin *et al.* (1991) evaluate a Gaussian classifier on a clutter classification problem. Lee *et al.* (1994) develop a classifier for polarimetric synthetic aperture radar (SAR) imagery based on the Wishart distribution. Solberg *et al.* (1999) use normal distributions within a procedure for detecting oil spills in SAR imagery.
- As part of a study by Aeberhard *et al.* (1994), RDA was one of eight discrimination techniques (including linear and quadratic discriminant analysis) applied to nine real datasets. An example is the wine dataset – the results of a chemical analysis of wines grown in the same region of Italy, but derived from different varieties. On all the real datasets, RDA performed best overall.
- Biometric recognition. Thomaz *et al.* (2004) consider linear, quadratic and regularised discriminant analysis for face recognition, face-expression recognition and fingerprint classification. They also propose a scheme for combining covariance matrices, based upon *maximising entropy*. Lee *et al.* (2010) utilise RDA within an approach for recognising facial expressions (e.g. happiness and disgust). RDA is used as the classifier within a *boosting* procedure (boosting, described in Chapter 8, is a sequential procedure that can improve classification performance by combining multiple instances of a classifier). The RDA parameters are optimised using *particle swarm optimisation*, a population-based search heuristic motivated by observance of birds flocking to a promising position (Kennedy and Eberhart, 1995).
- Conservation. Bavoux *et al.* (2006) use a Gaussian classifier to determine the sex of raptors based upon six measurements (body mass, bill length, wing chord length, tarsus length and width, and tail length).
- Microarray data analysis. Guo *et al.* (2007) develop a regularised version of a normal-distribution classifier, and apply it to microarray data analysis.

Comparative studies of normal-based models with other discriminant methods can be found in the papers by Curram and Mingers (1994), Bedworth *et al.* (1989) (on a speech recognition problem) and Aeberhard *et al.* (1994).

Applications of mixture models include:

- Speech recognition. Rabiner *et al.* (1985), and Juang and Rabiner (1985) describe a hidden Markov model approach to isolated digit recognition in which the probability density function associated with each state of the Markov process is a normal mixture model.

- Speech verification. Reynolds *et al.* (2000) use normal mixture models for speaker verification within conversational telephone speech.
- Face detection and tracking. In a study of face recognition (McKenna *et al.*, 1998), data characterising each subject's face (20 and 40 dimensional feature vectors) are modelled as a normal mixture, with component parameters estimated using the EM procedure. Classification is performed by using these density estimates in Bayes' rule.
- Intrusion detection in computer networks. Bahrololum and Khaleghi (2008) use normal mixture models to detect network attacks based on features derived from TCP dump data.
- Computer security. Hosseinzadeh and Krishnan (2008) consider how computer users can be recognised via keystroke patterns (e.g. the latency between key presses). Keystroke features extracted for a user are modelled by a normal mixture model.
- Handwritten character recognition. Revow *et al.* (1996) use a development of conventional mixture models (in which the means are constrained to lie on a spline) for handwritten digit recognition. Chen *et al.* (2011) use normal mixture models for recognising handwritten digits. Transformations are applied to the class-conditional mixture densities, and a gradient ascent method used to optimise an objective function, rather than standard EM maximum likelihood parameter estimation.
- Audio fingerprinting. Ramalingam and Krishnan (2006) use normal mixture models to model short-time Fourier transform features of audio clips. The models are used to match collected audio clips against a database of audio clips.
- Segmentation of vasculature in retinal images, with relevance to monitoring of diabetics. Soares *et al.* (2006) use normal mixture models to classify each pixel of a retinal image as vessel or nonvessel. Features for each pixel are obtained using the two-dimensional *Gabor wavelet transform*.
- Breast cancer prognosis. Falk *et al.* (2006) use a normal mixture model to model the joint distribution of cell nuclei features and cancer recurrence times. By conditioning on the cell nuclei features they predict breast cancer recurrence time. The normal mixture model approach outperformed decision tree approaches (see Chapter 7).
- Disease mapping. Rattanasiri *et al.* (2004) use a Poisson distribution mixture model to model incidence ratios of malaria. The resulting components are then used as classes, and the provinces of Thailand assigned to these classes using Bayes' theorem. A space–time mixture model was developed to model the dynamic nature of disease occurrences.
- Texture classification. Kim and Kang (2007) use normal mixture models within a scheme for texture classification and segmentation. The features modelled are obtained by applying wavelet transforms to the imagery.
- Modelling and synthesis of pedestrian trajectories. Johnson and Hogg (2002) use normal mixture models to model the joint distribution of latest location change and location history in pedestrian trajectories.

- Driver identification. Miyajima *et al.* (2007) use normal mixture models to identify drivers based on spectral features of pedal operation signals. The technology has relevance to customising intelligent driver assistance systems for individual drivers.
- Modelling of network data. Newman and Leicht (2007) model social network connections using a mixture distribution. Complex networks are considered in Chapter 12.

2.7 Summary and discussion

The approaches developed in this chapter towards discrimination have been based on estimation of the class-conditional density functions using parametric techniques (although mixture modelling is often referred to as *semiparametric*). It is certainly true that we cannot design a classifier that performs better than the actual Bayes' discriminant rule. No matter how sophisticated a classifier is, or how appealing it may be in terms of reflecting a model of human decision processes, it cannot achieve a lower error rate than the Bayes' classifier. Therefore a natural step is to estimate the components of the Bayes' rule from the data, namely the class-conditional probability density functions and the class priors.

In Sections 2.3 and 2.4 we considered discrimination based on normal models (i.e. modelling the measurements from each class as normal distributions). These linear and quadratic discriminants (or equivalently Gaussian classifiers) are widely used methods of supervised classification and are supported by many statistical packages. Their ease of specification and use is such that they are worth considering when tackling a classification problem. Problems occur when the class covariance matrices are close to singular. RDA is a regularisation scheme that includes a combination of covariance matrices and the addition of a penalty term to reduce problems with singularities in covariance matrices. RDA can improve classification performance when the sample size is too small for quadratic discriminant analysis to be viable. Non-normality of the data measurements can greatly reduce the performance of linear and quadratic discriminant functions. Transformation of variables to approximate normality prior to applying the discriminant functions can help.

In Section 2.5 we considered mixture models, and in particular the normal (Gaussian) mixture model. Modelling using normal mixtures is a simple and powerful way of developing the normal model to nonlinear discriminant functions. The parameters of the mixture model can be estimated using the iterative EM algorithm. Mixture models can be used for general distribution modelling, discrimination, and clustering (see Chapter 11).

2.8 Recommendations

An approach based on density estimation is not without its dangers of course. If incorrect assumptions are made about the form of the distribution in the parametric approach (and in many cases we will not have a physical model of the data generation process to use) or data points are sparse leading to poor estimates, then we cannot hope to achieve optimal performance.

However, the linear and quadratic (normal-based) rules are widely used, simple to implement and have been used with success in many applications. Therefore, it is worth applying such techniques to provide at least a baseline performance on which to build. It may prove to be sufficient.

Normal mixture models, optimised using the EM algorithm, provide an attractive and relatively easy procedure for general density modelling. We recommend their use, especially in situations where normal distributions are not believed to be appropriate (e.g. in the case of multiple modes, or long tails). Selection of the number of mixture components is the most difficult problem when using mixture models. However, for discrimination problems, a simple procedure like BIC minimisation, or validation set monitoring of classification rate, may be sufficient.

2.9 Notes and references

A comparison of the estimative and predictive (covered in Chapter 3) approaches is found in the articles by Aitchison *et al.* (1977) and Moran and Murphy (1979). McLachlan (1992a) gives a very thorough account of normal-based discriminant rules and is an excellent source of reference material. Simple procedures for correcting the bias of the discriminant rule are also given. Mkhadri *et al.* (1997) provide a review of regularisation in discriminant analysis.

Mixture distributions, and in particular the normal mixture model, are discussed in a number of texts. The book by Everitt and Hand (1981) provides a good introduction and a more detailed treatment is given by Titterington *et al.* (1985) (see also McLachlan and Basford, 1988). A thorough treatment, with recent methodological and computational developments, applications and software description is presented by McLachlan and Peel (2000). A review of mixture densities and the EM algorithm is given by Redner and Walker (1984). A thorough description of the EM algorithm and its extensions is provided in the book by McLachlan and Krishnan (1996). See also the review by Meng and van Dyk (1997), where the emphasis is on strategies for faster convergence. Software for the fitting of mixture models is publicly available, e.g. the NETLAB MATLAB™ routines documented in the book by Nabney (2002), and the PRTools MATLAB™ routines, documented in the book by van der Heiden *et al.* (2004).

Exercises

1. In the example application study of Section 2.3.3, is it appropriate to use a Gaussian classifier for the head injury data? Justify your answer.

2. Verify that the simple one-pass algorithm:

   (a) Initialise $\boldsymbol{S} = \boldsymbol{0}$, $\boldsymbol{m} = \boldsymbol{0}$.

   (b) For $r = 1$ to n do

      i. $\boldsymbol{d}_r = \boldsymbol{x}_r - \boldsymbol{m}$

      ii. $\boldsymbol{S} = \boldsymbol{S} + \left(1 - \frac{1}{r}\right) \boldsymbol{d}_r \boldsymbol{d}_r^T$

      iii. $\boldsymbol{m} = \boldsymbol{m} + \frac{\boldsymbol{d}_r}{r}$

   results in $\boldsymbol{m}$ as the sample mean and $\boldsymbol{S}$ as n times the sample covariance matrix.

3. Suppose that $\boldsymbol{B} = \boldsymbol{A} + \boldsymbol{u}\boldsymbol{u}^T$, where $\boldsymbol{A}$ is a nonsingular $(d \times d)$ matrix and $\boldsymbol{u}$ is a vector. Show that $\boldsymbol{B}^{-1} = \boldsymbol{A}^{-1} - k\boldsymbol{A}^{-1}\boldsymbol{u}\boldsymbol{u}^T\boldsymbol{A}^{-1}$, where $k = 1/(1 + \boldsymbol{u}^T\boldsymbol{A}^{-1}\boldsymbol{u})$ (Krzanowski and Marriott, 1996). Note how this can be used alongside the one-pass algorithm in Exercise 2, to update the estimate of an inverse covariance matrix when a new data vector is added to the data sample.
4. Show that the estimate of the covariance matrix given by

$$\hat{\boldsymbol{\Sigma}} = \frac{1}{n-1} \sum_{i=1}^{n} (\boldsymbol{x}_i - \boldsymbol{m})(\boldsymbol{x}_i - \boldsymbol{m})^T$$

where $\boldsymbol{m}$ is the sample mean, is unbiased.
5. For the Gaussian classifier using a common class covariance matrix (i.e. the normal-based linear discriminant) show that the maximum likelihood estimate for the common covariance matrix is the pooled within-class sample covariance matrix, $\boldsymbol{S}_W = \sum_{j=1}^{C} \frac{n_j}{n} \hat{\boldsymbol{\Sigma}}_j$, where n_j is the number of training data vectors for class ω_j, and $\hat{\boldsymbol{\Sigma}}_j$ is the maximum likelihood estimate of the covariance matrix using data from class ω_j only.
6. Suppose that the d-element vector $\boldsymbol{x}$ is normally distributed $N(\boldsymbol{\mu}, \boldsymbol{\Sigma}_j)$ in population j $(j = 1, 2)$, where $\boldsymbol{\Sigma}_j = \sigma_j^2[(1 - \rho_j)\boldsymbol{I} + \rho_j\boldsymbol{1}\boldsymbol{1}^T]$ and $\boldsymbol{1}$ denotes the d-vector all of whose elements are 1. Show that the optimal (i.e. Bayes) discriminant function is given, apart from an additive constant, by

$$-\frac{1}{2}(c_{11} - c_{12})Q_1 + \frac{1}{2}(c_{21} - c_{22})Q_2$$

where $Q_1 = (\boldsymbol{x} - \boldsymbol{\mu})(\boldsymbol{x} - \boldsymbol{\mu})^T$, $Q_2 = (\boldsymbol{1}^T(\boldsymbol{x} - \boldsymbol{\mu}))^2$, $c_{1j} = [\sigma_j^2(1 - \rho_j)]^{-1}$, and $c_{2j} = \rho_j[\sigma_j^2(1 - \rho_j)\{1 + (d-1)\rho_j\}]^{-1}$ (Krzanowski and Marriott, 1996).
7. Starting from Equation (2.36), verify the normal mixture model EM update formulae for the mean vectors and covariances matrices [i.e. Equations (2.37) and (2.38)].
8. Consider a gamma distribution of the form

$$p(x|\mu, m) = \frac{m}{\Gamma(m)\mu} \left(\frac{mx}{\mu}\right)^{m-1} \exp\left(-\frac{mx}{\mu}\right)$$

for mean μ and order parameter m. Derive the EM update equations for the π_j, μ_j and m_j for the gamma mixture (see Section 2.5.7)

$$p(x) = \sum_{j=1}^{g} \pi_j p(x|\mu_j, m_j)$$

9. Generate three datasets (train, validation and test sets) for the three-class, 21 variable, waveform data (Breiman *et al.*, 1984)

$$x_i = uh_1(i) + (1-u)h_2(i) + \epsilon_i \quad \text{(class 1)}$$
$$x_i = uh_1(i) + (1-u)h_3(i) + \epsilon_i \quad \text{(class 2)}$$
$$x_i = uh_2(i) + (1-u)h_3(i) + \epsilon_i \cdot \quad \text{(class 3)}$$

where $i = 1, \ldots, 21$; u is uniformly distributed on [0, 1]; ϵ_i are normally distributed with zero mean and unit variance; and the h_i are shifted triangular waveforms: $h_1(i) = \max(6-|i-11|, 0)$, $h_2(i) = h_1(i-4)$, $h_3(i) = h_1(i+4)$. Assume equal class priors. Construct a three-component mixture model for each class using a common covariance matrix across components and classes (Section 2.5.4). Investigate starting values for the means and covariance matrix and choose a model based on the validation set error rate. For this model, evaluate the classification error on the test set.

Compare the results with a linear discriminant classifier and a quadratic discriminant classifier constructed using the training set and evaluated on the test set.

3

Density estimation – Bayesian

Class-conditional density functions can be estimated with the aid of training data samples from the classes. These class-conditional density functions can be used within Bayes' rule to produce a discrimination rule. Bayesian approaches to parametric density estimation can be used to take into account parameter uncertainty due to the training data sampling. In this chapter, analytic, sampling and variational procedures for Bayesian estimation are all considered.

3.1 Introduction

In Chapter 2 we considered how parametric density estimates can be used within a Bayes' rule classifier. Classification decisions were based upon the probability of class membership $p(\omega_j|\boldsymbol{x})$, expressed as

$$p(\omega_j|\boldsymbol{x}) = p(\omega_j)\frac{p(\boldsymbol{x}|\hat{\boldsymbol{\theta}}_j)}{p(\boldsymbol{x})}$$

where $p(\boldsymbol{x}|\hat{\boldsymbol{\theta}}_j)$ is the estimate of the parametric class-conditional density for class ω_j, $p(\omega_j)$ is the prior class probability and $p(\boldsymbol{x})$ can be considered a normalisation constant. For each class, ω_j, the parameters $\hat{\boldsymbol{\theta}}_j$ were estimated using a sample of observations $\mathcal{D}_j = \{\boldsymbol{x}_1^j, \ldots, \boldsymbol{x}_{n_j}^j\}$ ($\boldsymbol{x}_i^j \in \mathbb{R}^d$) from class ω_j. Maximum likelihood estimation was used to obtain these parameter estimates. Potential variations in the estimates $\hat{\boldsymbol{\theta}}_j$ due to sampling variability of the training data were not taken into account.

As was noted in Chapter 2, a *predictive* or *Bayesian approach* can be used to take into account the sampling variability in the estimate of $\boldsymbol{\theta}_j$. Like the estimative approach, the dependence of the density at $\boldsymbol{x}$ on $\mathcal{D}_j$ is through the parameters, $\boldsymbol{\theta}_j$, used within the model assumed for the density. However, the Bayesian approach admits that we do not know the true value of these parameters $\boldsymbol{\theta}_j$. It does this by treating $\boldsymbol{\theta}_j$ as an unknown random variable,

Statistical Pattern Recognition, Third Edition. Andrew R. Webb and Keith D. Copsey.

represented by a posterior distribution taking into account prior information on the parameter, alongside the information contained in the data samples.

We replace our single estimate $p(\boldsymbol{x}|\hat{\boldsymbol{\theta}}_j)$ of the class-conditional probability density, $p(\boldsymbol{x}|\omega_j)$, with the predictive Bayesian estimate given by

$$\begin{aligned} p(\boldsymbol{x}|\omega_j) &= \int p(\boldsymbol{x}, \boldsymbol{\theta}_j|\mathcal{D}_j)d\boldsymbol{\theta}_j \\ &= \int p(\boldsymbol{x}|\boldsymbol{\theta}_j, \mathcal{D}_j)p(\boldsymbol{\theta}_j|\mathcal{D}_j)d\boldsymbol{\theta}_j \\ &= \int p(\boldsymbol{x}|\boldsymbol{\theta}_j)p(\boldsymbol{\theta}_j|\mathcal{D}_j)d\boldsymbol{\theta}_j \end{aligned} \tag{3.1}$$

where $p(\boldsymbol{\theta}_j|\mathcal{D}_j)$ is the Bayesian posterior density function for $\boldsymbol{\theta}_j$. The second line of (3.1) follows from the first via the definition of conditional probability densities, and the third line follows from the second by noting that conditioning on $\mathcal{D}_j$ provides no extra information on the density function $p(\boldsymbol{x}|\boldsymbol{\theta}_j)$.

By Bayes' theorem the posterior density of $\boldsymbol{\theta}_j$ may be expressed as

$$\begin{aligned} p(\boldsymbol{\theta}_j|\mathcal{D}_j) &= \frac{p(\mathcal{D}_j|\boldsymbol{\theta}_j)p(\boldsymbol{\theta}_j)}{p(\mathcal{D}_j)} \\ &= \frac{p(\mathcal{D}_j|\boldsymbol{\theta}_j)p(\boldsymbol{\theta}_j)}{\int p(\mathcal{D}_j|\boldsymbol{\theta}'_j)p(\boldsymbol{\theta}'_j)d\boldsymbol{\theta}'_j} \end{aligned} \tag{3.2}$$

where

- $p(\boldsymbol{\theta}_j)$ is the prior density for the parameters $\boldsymbol{\theta}_j$.
- $p(\mathcal{D}_j|\boldsymbol{\theta}_j)$ is the likelihood function, which is the probability density of the data sample $\mathcal{D}_j$ conditional on the value of the parameters $\boldsymbol{\theta}_j$.

In this chapter we introduce some of the Bayesian procedures (both analytic and numeric) that can be called upon in order to evaluate expressions of the form (3.1) and (3.2). Section 3.2 considers analytic approaches to Bayesian inference. Section 3.3 introduces sampling approaches to Bayesian inference, with Section 3.4 going on to consider Markov chain Monte Carlo (MCMC) sampling algorithms. Section 3.5 provides a worked discrimination example. Section 3.6 introduces the advanced topic of Sequential Monte Carlo (SMC) Samplers, another sampling approach to Bayesian inference. Section 3.7 considers Variational Bayes approximations, which provide an alternative means of approximating full Bayesian inference than sampling methods. Section 3.8 introduces Approximate Bayesian Computation, a recent development for use in problems where the likelihood function cannot be evaluated analytically.

For simplicity, unless specifically needed to distinguish between multiple classes, the index for the class ω_j is dropped in subsequent parts of this chapter. In keeping with many Bayesian texts, we adopt a relaxed approach to notation, often referring to a probability density function as a probability distribution. The reader should interpret this as being the probability distribution defined by the probability density function.

3.1.1 Basics

Essentially, Bayesian statistics concerns the updating of our prior beliefs on an unknown parameter in the light of observed data that has a relationship to that parameter. Our prior beliefs are represented by a prior distribution for the parameter. The distribution for the observed data given a value for the parameter is the likelihood distribution. The updated beliefs on the parameter are represented by its posterior distribution, which is related to the prior and likelihood distributions via Bayes' theorem.

Suppose that we have data, $\mathcal{D}$ (perhaps a set of noisy measurement vectors $\mathcal{D} = \{\boldsymbol{x}_1, \ldots, \boldsymbol{x}_n\}$), whose probability density function $p(\mathcal{D}|\boldsymbol{\theta})$, depends on the parameter vector $\boldsymbol{\theta}$. This density function is termed the likelihood function (when considered as a function of $\boldsymbol{\theta}$). In traditional statistics $\boldsymbol{\theta}$ is a fixed, but unknown, parameter and we want to determine a single estimate of $\boldsymbol{\theta}$ (or more pertinently make decisions depending on the value of this estimate). The Bayesian approach regards $\boldsymbol{\theta}$ as a realisation of a random variable, which has a prior probability density function $p(\boldsymbol{\theta})$ which represents our prior knowledge of $\boldsymbol{\theta}$. Our updated beliefs on $\boldsymbol{\theta}$ having observed the data $\mathcal{D}$ are represented by the posterior density of $\boldsymbol{\theta}$. This is given by Bayes' theorem as

$$p(\boldsymbol{\theta}|\mathcal{D}) = \frac{p(\mathcal{D}|\boldsymbol{\theta})p(\boldsymbol{\theta})}{\int_{\boldsymbol{\theta}'} p(\mathcal{D}|\boldsymbol{\theta}')p(\boldsymbol{\theta}')d\boldsymbol{\theta}'} \tag{3.3}$$

or in the discrete case[1]

$$p(\boldsymbol{\theta}|\mathcal{D}) = \frac{p(\mathcal{D}|\boldsymbol{\theta})p(\boldsymbol{\theta})}{\sum_{\boldsymbol{\theta}'} p(\mathcal{D}|\boldsymbol{\theta}')p(\boldsymbol{\theta}')}$$

The posterior density of $\boldsymbol{\theta}$ represents our revised views on the distribution of $\boldsymbol{\theta}$, given the information obtained from the observed data $\mathcal{D}$. This posterior distribution tells us all that we need to know about $\boldsymbol{\theta}$ and can be used to calculate summary statistics. Commonly, we calculate the posterior expectation of a function $h(\boldsymbol{\theta})$

$$E[h(\boldsymbol{\theta})|\mathcal{D}] = \frac{\int h(\boldsymbol{\theta})p(\mathcal{D}|\boldsymbol{\theta})p(\boldsymbol{\theta})d\boldsymbol{\theta}}{\int p(\mathcal{D}|\boldsymbol{\theta})p(\boldsymbol{\theta})d\boldsymbol{\theta}} \tag{3.4}$$

3.1.2 Recursive calculation

The posterior density can be calculated in a recursive manner. Suppose that the data $\mathcal{D}$ consists of n measurement vectors, so that $\mathcal{D} = \{\boldsymbol{x}_1, \ldots, \boldsymbol{x}_n\}$. If the measurements, $\boldsymbol{x}_i$, are obtained successively and are conditionally independent, we may write (3.3) as

$$p(\boldsymbol{\theta}|\boldsymbol{x}_1, \ldots, \boldsymbol{x}_n) = \frac{p(\boldsymbol{x}_n|\boldsymbol{\theta})p(\boldsymbol{\theta}|\boldsymbol{x}_1, \ldots, \boldsymbol{x}_{n-1})}{\int p(\boldsymbol{x}_n|\boldsymbol{\theta}')p(\boldsymbol{\theta}'|\boldsymbol{x}_1, \ldots, \boldsymbol{x}_{n-1})d\boldsymbol{\theta}'} \tag{3.5}$$

[1] We do not make this distinction in future discussions. When considering discrete distributions the reader should replace integration by summation, and probability density function by probability mass function.

This expresses the posterior distribution of $\boldsymbol{\theta}$ given n measurements in terms of the posterior distribution given $n-1$ measurements. Starting with $p(\boldsymbol{\theta})$, we may perform the operation (3.5) n times to obtain the posterior.

3.1.3 Proportionality

The denominator in the posterior density (3.3) is a normalising constant, independent of $\boldsymbol{\theta}$. We can therefore write

$$\begin{aligned} p(\boldsymbol{\theta}|\mathcal{D}) &\propto p(\mathcal{D}|\boldsymbol{\theta})p(\boldsymbol{\theta}) \\ &\propto \text{likelihood} \times \text{prior} \end{aligned} \tag{3.6}$$

where the proportionality is unambiguous due to the normalisation constraint of a probability density function

$$\int p(\boldsymbol{\theta}|\mathcal{D})d\boldsymbol{\theta} = 1$$

The proportionality is often further utilised by rearranging the terms in (3.6) so that

$$p(\mathcal{D}|\boldsymbol{\theta})p(\boldsymbol{\theta}) = g(\boldsymbol{\theta}, \mathcal{D})\psi(\mathcal{D})$$

where $\psi(\mathcal{D})$ is a function that depends on the data $\mathcal{D}$ only [i.e. a function that does not vary with $\boldsymbol{\theta}$], and $g(\boldsymbol{\theta}, \mathcal{D})$ is a function that depends on both $\boldsymbol{\theta}$ and $\mathcal{D}$. By separating out the multiplicative factors involving $\boldsymbol{\theta}$ we can then express the posterior density function as

$$p(\boldsymbol{\theta}|\mathcal{D}) \propto g(\boldsymbol{\theta}, \mathcal{D})$$

since the $\psi(\mathcal{D})$ factor can be incorporated into the normalisation constant

$$p(\boldsymbol{\theta}|\mathcal{D}) = \frac{g(\boldsymbol{\theta}, \mathcal{D})}{\int g(\boldsymbol{\theta}', \mathcal{D})d\boldsymbol{\theta}'}$$

Using such proportionality we can often simplify the process of developing Bayesian solutions, by ignoring irrelevant aspects until they are necessary. The concept is illustrated in the analytic solutions in the next section.

3.2 Analytic solutions

3.2.1 Conjugate priors

Bayes' theorem allows us to combine any prior, $p(\boldsymbol{\theta})$, with any likelihood, $p(\mathcal{D}|\boldsymbol{\theta})$, to give the posterior. However, it is convenient for particular likelihood functions to be combined with special forms of the prior that lead to simple, or at least tractable, solutions for the posterior. For a given likelihood, $p(\boldsymbol{x}|\boldsymbol{\theta})$, the family of prior distributions for which the posterior density,

$p(\boldsymbol{\theta}|\mathcal{D})$, is of the same functional form as the prior is called *conjugate* with respect to $p(\boldsymbol{x}|\boldsymbol{\theta})$. Some of the more common forms of conjugate priors are given by Bernardo and Smith (1994).

3.2.1.1 Poisson distribution example

As a simple example, suppose that we have n independent samples $\{x_1, \ldots, x_n\}$ from the Poisson distribution with unknown rate parameter $\lambda > 0$.

> **Poisson distribution**
>
> The probability mass function for the Poisson distribution with rate parameter $\lambda > 0$ is
>
> $$p(x|\lambda) = \frac{\lambda^x \exp(-\lambda)}{x!}$$
>
> where $x \geq 0$ is an integer. The mean and variance of this distribution are both λ.

The likelihood distribution is given by

$$p(x_1, \ldots, x_n|\lambda) = \prod_{i=1}^{n} p(x_i|\lambda) = \frac{\lambda^{\sum_{i=1}^n x_i} \exp(-\lambda n)}{\prod_{i=1}^n x_i!}$$

Now take as prior distribution for λ the gamma distribution with shape parameter α and inverse-scale parameter β.

> **Gamma distribution**
>
> The gamma distribution with shape parameter $\alpha > 0$ and inverse-scale parameter $\beta > 0$ has probability density function
>
> $$p(\gamma|\alpha, \beta) = \frac{\beta^\alpha}{\Gamma(\alpha)} \gamma^{\alpha-1} \exp(-\beta\gamma)$$
>
> The support of the distribution is $\gamma > 0$, the mean of the distribution is α/β and the variance is α/β^2.

The posterior density function for λ can be written

$$\begin{aligned}
p(\lambda|x_1, \ldots, x_n) &= \frac{p(x_1, \ldots, x_n|\lambda)p(\lambda)}{\int_{\lambda'} p(x_1, \ldots, x_n|\lambda')p(\lambda')d\lambda'} \\
&\propto p(x_1, \ldots, x_n|\lambda)p(\lambda) \\
&\propto \frac{\beta^\alpha}{\Gamma(\alpha)\prod_{i=1}^n x_i!} \lambda^{\alpha+\sum_{i=1}^n x_i - 1} \exp(-(\beta+n)\lambda)
\end{aligned}$$

Following Section 3.1.3 this can be simplified to

$$p(\lambda|x_1, \ldots, x_n) \propto \lambda^{\alpha+\sum_{i=1}^{n} x_i - 1} \exp(-(\beta + n)\lambda) \tag{3.7}$$

The key to expressing (3.7) as a distribution is to note that the probability density function of a gamma distribution with shape parameter $\alpha' = \alpha + \sum_{i=1}^{n} x_i$ and inverse scale parameter $\beta' = \beta + n$ can be expressed up to a normalisation constant as

$$p(\gamma) \propto \gamma^{\alpha'-1} \exp(-\beta'\gamma)$$

Hence, the posterior distribution of λ is a gamma distribution with parameters (α', β'), and therefore the gamma distribution is a conjugate prior for the parameter of a Poisson distribution. Arranging the relevant terms in a posterior density function until one 'spots' the underlying distribution is a key aspect of Bayesian analysis.

3.2.2 Estimating the mean of a normal distribution with known variance

Over the next two subsections we describe the analytic Bayesian learning approach for the problems of estimating the mean of a univariate normal distribution with known variance, and estimating unknown mean and covariance in a multivariate normal distribution, both using conjugate priors. Further details on estimating the parameters of normal models are given in the books by Fukunaga (1990) and Fu (1968).

3.2.2.1 Specification

Let the model for the density of a data measurement be normal with unknown mean μ and known variance σ^2. The likelihood, denoted $p(x|\mu)$, is then

$$p(x|\mu) = N(x; \mu, \sigma^2) = \frac{1}{\sqrt{2\pi}\sigma} \exp\left\{-\frac{1}{2}\left(\frac{x-\mu}{\sigma}\right)^2\right\}$$

We take a prior density for the mean μ, which is normal with mean μ_0 and variance σ_0^2 (μ_0 and σ_0^2 are termed hyper-parameters)

$$p(\mu) = N\left(\mu; \mu_0, \sigma_0^2\right) = \frac{1}{\sqrt{2\pi}\sigma_0} \exp\left\{-\frac{1}{2}\left(\frac{\mu-\mu_0}{\sigma_0}\right)^2\right\}$$

3.2.2.2 Posterior distribution

As will be shown later in this section, for independent measurements $\{x_1, \ldots, x_n\}$ our specification of prior and likelihood distributions gives rise to the posterior distribution

$$p(\mu|x_1, \ldots, x_n) = N\left(\mu; \mu_n, \sigma_n^2\right) = \frac{1}{\sqrt{2\pi}\sigma_n} \exp\left\{-\frac{1}{2}\left(\frac{\mu-\mu_n}{\sigma_n}\right)^2\right\} \tag{3.8}$$

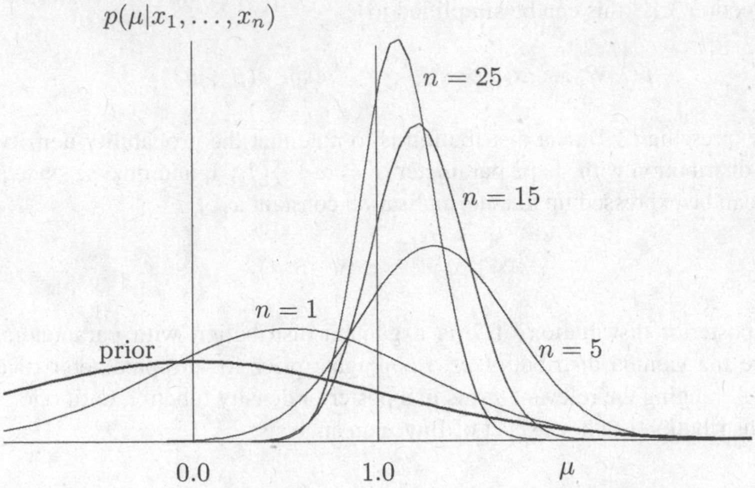

Figure 3.1 Bayesian learning of the mean of a normal distribution of known variance.

where

$$\frac{1}{\sigma_n^2} = \frac{1}{\sigma_0^2} + \frac{n}{\sigma^2}$$

$$\mu_n = \sigma_n^2 \left(\frac{\mu_0}{\sigma_0^2} + \frac{\sum_i x_i}{\sigma^2} \right) \tag{3.9}$$

Hence, the conjugate prior for the the mean of a normal distribution with known variance, is itself a normal distribution.

As $n \to \infty$, $\mu_n \to$ the sample mean, $m = \sum_i x_i / n$, and the posterior variance of μ, namely σ_n^2, tends to zero as $1/n$. Thus, as more samples are used to obtain the distribution (3.8), the contribution of the initial guesses σ_0 and μ_0 becomes smaller. This is illustrated in Figure 3.1. Samples, x_i, from a normal distribution with unit mean and unit variance have been generated. A normal model is assumed with mean μ and unit variance. Figure 3.1 plots [using (3.8)] the posterior density of the mean, μ, given different numbers of data samples and a prior distribution of the mean that is normal with $\mu_0 = 0$ and $\sigma_0^2 = 1$. As the number of samples increases, the posterior density narrows about the true mean.

3.2.2.3 Predictive density

As will be shown at the end of this section, substituting the posterior density into (3.1) gives the predictive density

$$p(x|x_1,\ldots,x_n) = \frac{1}{\sqrt{\sigma^2+\sigma_n^2}\sqrt{2\pi}} \exp\left\{ -\frac{1}{2}\frac{(x-\mu_n)^2}{\sigma^2+\sigma_n^2} \right\} \tag{3.10}$$

which is normal with mean μ_n and variance $\sigma^2 + \sigma_n^2$. This can be used as the class-conditional density estimate in the Bayesian classifier.

3.2.2.4 Derivation of the posterior distribution

To derive the posterior distribution we note that we have

$$\begin{aligned}
p(x_1, \ldots, x_n|\mu)p(\mu) &= p(\mu)\prod_{i=1}^{n} p(x_i|\mu) \\
&= \frac{1}{\sqrt{2\pi}\sigma_0}\exp\left\{-\frac{1}{2}\left(\frac{\mu-\mu_0}{\sigma_0}\right)^2\right\}\prod_{i=1}^{n}\left[\frac{1}{\sqrt{2\pi}\sigma}\exp\left\{-\frac{1}{2}\left(\frac{x_i-\mu}{\sigma}\right)^2\right\}\right]
\end{aligned}$$

which can be rearranged into the form

$$p(x_1, \ldots, x_n|\mu)p(\mu) = \frac{1}{\sigma^n\sigma_0}\frac{1}{(2\pi)^{(n+1)/2}}\exp\left\{-\frac{1}{2}\left(\frac{\mu-\mu_n}{\sigma_n}\right)^2\right\}\exp\left\{-\frac{k_n}{2}\right\} \tag{3.11}$$

where σ_n and μ_n are as in (3.9) and

$$k_n = \frac{\mu_0^2}{\sigma_0^2} - \frac{\mu_n^2}{\sigma_n^2} + \frac{\sum x_i^2}{\sigma^2}$$

Now

$$\int p(x_1, \ldots, x_n|\mu)p(\mu)d\mu = \frac{1}{\sigma^n\sigma_0}\frac{1}{(2\pi)^{(n+1)/2}}\exp\left\{-\frac{k_n}{2}\right\}\sqrt{2\pi}\sigma_n \tag{3.12}$$

since

$$\int \exp\left\{-\frac{1}{2}\left(\frac{\mu-\mu_n}{\sigma_n}\right)^2\right\}d\mu = \sqrt{2\pi}\sigma_n\int N\left(\mu;\mu_n,\sigma_n^2\right)d\mu = \sqrt{2\pi}\sigma_n$$

from the fact that the probability density function of the normal distribution $N(\mu; \mu_n, \sigma_n^2)$ integrates to unity over μ.

Evaluating (3.3) with the expressions in Equations (3.11) and (3.12) gives the posterior distribution as

$$\begin{aligned}
p(\mu|x_1, \ldots, x_n) &= \frac{p(x_1, \ldots, x_n|\mu)p(\mu)}{\int p(x_1, \ldots, x_n|\mu')p(\mu')d\mu'} \\
&= \frac{1}{\sqrt{2\pi}\sigma_n}\exp\left\{-\frac{1}{2}\left(\frac{\mu-\mu_n}{\sigma_n}\right)^2\right\}
\end{aligned}$$

which is the normal distribution with mean μ_n and variance σ_n^2.

3.2.2.5 Calculation utilising proportionality

A useful shortcut when deriving the posterior distribution arises from calculating the density function only up to proportionalities, as advocated in Section 3.1.3. Specifically, ignoring any multiplicative factors in the posterior density function that do not vary with μ, we write

$$\begin{aligned} p(\mu|x_1, \ldots, x_n) &\propto p(x_1, \ldots, x_n|\mu)p(\mu) \\ &\propto \exp\left\{-\frac{1}{2}\left[\left(\frac{\mu-\mu_0}{\sigma_0}\right)^2 + \sum_{i=1}^{n}\left(\frac{x_i-\mu}{\sigma}\right)^2\right]\right\} \end{aligned}$$

Completing the square for μ in the exponent, and ignoring additive terms in the exponent that do not vary with μ (since these terms can be incorporated into the normalisation factor), we have

$$p(\mu|x_1, \ldots, x_n) \propto \exp\left\{-\frac{1}{2}\left(\frac{\mu-\mu_n}{\sigma_n}\right)^2\right\} \tag{3.13}$$

where σ_n and μ_n are given in (3.9). By comparing (3.13) with the probability density function of a normal distribution, we can deduce that the missing normalisation constant must be $\frac{1}{\sqrt{2\pi}\sigma_n}$, and that the posterior distribution of μ is therefore $N(\mu_n, \sigma_n)$. As noted earlier, deducing distributions in such a manner is a useful tool when conducting Bayesian inference.

3.2.2.6 Derivation of the predictive density

The predictive density is given by

$$\begin{aligned} p(x|x_1, \ldots, x_n) &= \int p(x|\mu)p(\mu|x_1, \ldots, x_n)d\mu \\ &= \int \frac{1}{\sqrt{2\pi}\sigma}\exp\left\{-\frac{1}{2}\left(\frac{x-\mu}{\sigma}\right)^2\right\}\frac{1}{\sqrt{2\pi}\sigma_n}\exp\left\{-\frac{1}{2}\left(\frac{\mu-\mu_n}{\sigma_n}\right)^2\right\}d\mu \end{aligned}$$

Completing the square for μ in the exponent we have

$$p(x|x_1, \ldots, x_n) = \int \frac{1}{2\pi\sigma\sigma_n}\exp\left\{-\frac{\sigma^2+\sigma_n^2}{2\sigma^2\sigma_n^2}\left(\mu - \frac{x\sigma_n^2+\mu_n\sigma^2}{\sigma^2+\sigma_n^2}\right)^2 - \frac{1}{2}\frac{(x-\mu_n)^2}{\sigma^2+\sigma_n^2}\right\}d\mu$$

Then, we note from the properties of normal probability density functions that

$$\int \exp\left\{-\frac{\sigma^2+\sigma_n^2}{2\sigma^2\sigma_n^2}\left(\mu - \frac{x\sigma_n^2+\mu_n\sigma^2}{\sigma^2\sigma_n^2}\right)^2\right\}d\mu = \sqrt{\frac{2\pi\sigma^2\sigma_n^2}{\sigma^2+\sigma_n^2}}$$

and hence

$$p(x|x_1,\ldots,x_n) = \frac{1}{2\pi\sigma\sigma_n}\sqrt{\frac{2\pi\sigma^2\sigma_n^2}{\sigma^2+\sigma_n^2}}\exp\left\{-\frac{1}{2}\frac{(x-\mu_n)^2}{\sigma^2+\sigma_n^2}\right\}$$

which simplifies to (3.10).

3.2.3 Estimating the mean and the covariance matrix of a multivariate normal distribution

3.2.3.1 Specification

We now consider the multivariate problem in which the mean and the covariance matrix of a normal distribution are unknown. Let the model for the data be multivariate normal with mean $\boldsymbol{\mu}$ and covariance matrix $\boldsymbol{\Sigma}$

$$p(\boldsymbol{x}|\boldsymbol{\mu},\boldsymbol{\Sigma}) = N(\boldsymbol{x};\boldsymbol{\mu},\boldsymbol{\Sigma}) = \frac{1}{(2\pi)^{d/2}|\boldsymbol{\Sigma}|^{\frac{1}{2}}}\exp\left\{-\frac{1}{2}(\boldsymbol{x}-\boldsymbol{\mu})^T\boldsymbol{\Sigma}^{-1}(\boldsymbol{x}-\boldsymbol{\mu})\right\}$$

We wish to estimate the posterior distribution of $\boldsymbol{\mu}$ and $\boldsymbol{\Sigma}$ given independent measurements $\boldsymbol{x}_1,\ldots,\boldsymbol{x}_n \sim N(\boldsymbol{\mu},\boldsymbol{\Sigma})$.

When the mean $\boldsymbol{\mu}$ and the covariance matrix $\boldsymbol{\Sigma}$ of a normal distribution are to be estimated, the conjugate-prior choice of prior distribution is the Gauss-Wishart (also known as normal-Wishart) probability distribution in which:

- the mean is distributed according to a normal distribution with mean $\boldsymbol{\mu}_0$ and covariance matrix $\boldsymbol{K}^{-1}/\lambda$;
- $\boldsymbol{K}$ (the inverse of the covariance matrix $\boldsymbol{\Sigma}$) is distributed according to a Wishart distribution with parameters α and $\boldsymbol{\beta}$.

The Wishart distribution

The probability density function for a $d \times d$-dimensional Wishart distribution with parameters α and $\boldsymbol{\beta}$ is given by

$$\begin{aligned} p(\boldsymbol{W};\alpha,\boldsymbol{\beta}) &= \mathrm{Wi}_d(\boldsymbol{W};\alpha,\boldsymbol{\beta}) \\ &= c(d,\alpha)\,|\boldsymbol{\beta}|^{\alpha}\,|\boldsymbol{W}|^{(\alpha-(d+1)/2)}\exp\{-\mathrm{Tr}(\boldsymbol{\beta}\boldsymbol{W})\} \end{aligned} \quad (3.14)$$

where $\mathrm{Tr}(\boldsymbol{A}) = \sum_{i=1}^{d} A_{ii}$ is the trace of the $d \times d$ matrix $\boldsymbol{A}$, $2\alpha > d-1$, $\boldsymbol{\beta}$ is a symmetric non-singular $d \times d$ matrix, and

$$c(d,\alpha) = \left[\pi^{d(d-1)/4}\prod_{i=1}^{d}\Gamma\left(\frac{2\alpha+1-i}{2}\right)\right]^{-1} \quad (3.15)$$

The support of the Wishart distribution is symmetric positive definite matrices. The mean of the Wishart distribution parametrised above is $\mathrm{E}(\boldsymbol{W}) = \alpha\boldsymbol{\beta}^{-1}$, and the mean of the inverse matrix is $\mathrm{E}(\boldsymbol{W}^{-1}) = (\alpha - (d+1)/2)^{-1}\boldsymbol{\beta}$.

We note that despite its rather complicated probability density function, it is easy to draw samples from the Wishart distribution. Specifically, consider column vectors $\boldsymbol{x}_1, \ldots, \boldsymbol{x}_{2\alpha}$, $\boldsymbol{x}_i \in \mathbb{R}^d$, drawn independently from the d-variate normal distribution with zero mean and covariance matrix $\frac{1}{2}\boldsymbol{\beta}^{-1}$. Then if $2\alpha > d - 1$, $\sum_{i=1}^{2\alpha} \boldsymbol{x}_i\boldsymbol{x}_i^T$ has the Wishart distribution with parameters α and $\boldsymbol{\beta}$, i.e. $\sum_{i=1}^{2\alpha} \boldsymbol{x}_i\boldsymbol{x}_i^T \sim \mathrm{Wi}_d(\boldsymbol{W}; \alpha, \boldsymbol{\beta})$.

In the univariate case, the Wishart distribution is the gamma distribution with shape parameter $\alpha > 0$ and inverse-scale parameter $\beta > 0$.

Using the notation $N_d(\boldsymbol{\mu}|\boldsymbol{m}, \boldsymbol{A})$ for a d-variate normal distribution with mean $\boldsymbol{m}$ and inverse covariance matrix $\boldsymbol{A}$ (i.e. covariance matrix $\boldsymbol{A}^{-1}$) we have the following conjugate-prior distribution

$$\begin{aligned} p(\boldsymbol{\mu}, \boldsymbol{K}) &= \mathrm{N}_d(\boldsymbol{\mu}; \boldsymbol{\mu}_0, \lambda\boldsymbol{K})\mathrm{Wi}_d(\boldsymbol{K}; \alpha, \boldsymbol{\beta}) \\ &= \frac{|\lambda\boldsymbol{K}|^{1/2}}{(2\pi)^{d/2}}\exp\left\{-\frac{1}{2}\lambda(\boldsymbol{\mu} - \boldsymbol{\mu}_0)^T\boldsymbol{K}(\boldsymbol{\mu} - \boldsymbol{\mu}_0)\right\} \\ &\quad \times c(d, \alpha)\,|\boldsymbol{\beta}|^{\alpha}\,|\boldsymbol{K}|^{(\alpha-(d+1)/2)}\exp\{-\mathrm{Tr}(\boldsymbol{\beta}\boldsymbol{K})\} \end{aligned} \tag{3.16}$$

where $c(d, \alpha)$ is as defined in (3.15).

The term λ weights the confidence in $\boldsymbol{\mu}_0$ as the initial value of the mean, and α weights the initial confidence in the covariance matrix.

3.2.3.2 Posterior distribution

As will be shown later in this section, with the given choice of prior distribution for the parameters $\boldsymbol{\mu}$ and $\boldsymbol{K}$, the posterior distribution

$$p(\boldsymbol{\mu}, \boldsymbol{K}|\boldsymbol{x}_1, \ldots, \boldsymbol{x}_n) = \frac{p(\boldsymbol{x}_1, \ldots, \boldsymbol{x}_n|\boldsymbol{\mu}, \boldsymbol{K})p(\boldsymbol{\mu}, \boldsymbol{K})}{\int p(\boldsymbol{x}_1, \ldots, \boldsymbol{x}_n|\boldsymbol{\mu}', \boldsymbol{K}')p(\boldsymbol{\mu}', \boldsymbol{K}')d\boldsymbol{\mu}'d\boldsymbol{K}'} \tag{3.17}$$

is also Gauss-Wishart with the parameters $\boldsymbol{\mu}_0$, $\boldsymbol{\beta}$, λ and α replaced by (Fu, 1968):

$$\begin{aligned} \lambda_n &= \lambda + n \\ \alpha_n &= \alpha + n/2 \\ \boldsymbol{\mu}_n &= (\lambda\boldsymbol{\mu}_0 + n\boldsymbol{m})/\lambda_n \\ 2\boldsymbol{\beta}_n &= 2\boldsymbol{\beta} + (n-1)\boldsymbol{S} + \frac{n\lambda}{\lambda_n}(\boldsymbol{\mu}_0 - \boldsymbol{m})(\boldsymbol{\mu}_0 - \boldsymbol{m})^T \end{aligned} \tag{3.18}$$

where $\boldsymbol{m}$ is the sample mean and S is the unbiased estimate of the sample covariance matrix

$$\boldsymbol{S} = \frac{1}{n-1}\sum_{i=1}^{n}(\boldsymbol{x}_i - \boldsymbol{m})(\boldsymbol{x}_i - \boldsymbol{m})^T$$

That is

$$p(\boldsymbol{\mu}, \boldsymbol{K}|\boldsymbol{x}_1, \ldots, \boldsymbol{x}_n) = \mathrm{N}_d(\boldsymbol{\mu}; \boldsymbol{\mu}_n, \lambda_n \boldsymbol{K})\mathrm{Wi}_d(\boldsymbol{K}; \alpha_n, \boldsymbol{\beta}_n) \tag{3.19}$$

The posterior marginal for $\boldsymbol{K}$ (obtained by integrating with respect to $\boldsymbol{\mu}$) is Wishart $\mathrm{Wi}_d(\boldsymbol{K}|\alpha_n, \boldsymbol{\beta}_n)$, and the posterior conditional distribution of μ given $\boldsymbol{K}$ is normal $N_d(\boldsymbol{\mu}_n, \lambda_n K)$. Hence we can see that the Gauss-Wishart distribution is a conjugate prior for the mean and inverse covariance matrix (i.e. precision matrix) parameters of the multivariate normal distribution.

The posterior marginal for μ (obtained by integrating with respect to $\boldsymbol{K}$) is (see exercises at the end of the chapter)

$$p(\boldsymbol{\mu}|\boldsymbol{x}_1, \ldots, \boldsymbol{x}_n) = \mathrm{St}_d\left(\boldsymbol{\mu}; \boldsymbol{\mu}_n, \left(\alpha_n - \frac{d-1}{2}\right)\lambda_n \boldsymbol{\beta}_n^{-1}, 2\alpha_n - (d-1)\right)$$

which is the d-dimensional generalisation of the univariate Student distribution.

The multivariate Student distribution

The d-dimensional generalisation of the univariate Student's t-distribution has the following probability density function

$$\begin{aligned} p(\boldsymbol{x}; \boldsymbol{\mu}, \boldsymbol{\lambda}, \alpha) &= \mathrm{St}_d(\boldsymbol{x}; \boldsymbol{\mu}, \boldsymbol{\lambda}, \alpha) \\ &= \frac{\Gamma(\frac{1}{2}(\alpha + d))}{\Gamma(\frac{\alpha}{2})(\alpha\pi)^{d/2}} |\boldsymbol{\lambda}|^{\frac{1}{2}} \left[1 + \frac{1}{\alpha}(\boldsymbol{x} - \boldsymbol{\mu})^T \boldsymbol{\lambda}(\boldsymbol{x} - \boldsymbol{\mu})\right]^{-(\alpha+d)/2} \end{aligned} \tag{3.20}$$

were $\alpha > 0$ and $\boldsymbol{\lambda}$ is a $d \times d$ symmetric positive definite matrix. The support of the distribution is $\mathbb{R}^d$, and the mean and variance are $\mathrm{E}[\boldsymbol{x}] = \boldsymbol{\mu}$ and $\mathrm{Var}[\boldsymbol{x}] = \boldsymbol{\lambda}^{-1}(\alpha - 2)^{-1}\alpha$, respectively.

3.2.3.3 Predictive density and the resulting Bayes' decision rule

Substituting the posterior density into (3.1) gives the predictive density for the case of the normal distribution with unknown mean and covariance matrix

$$p(\boldsymbol{x}|\boldsymbol{x}_1, \ldots, \boldsymbol{x}_n) = \mathrm{St}_d\left(\boldsymbol{x}; \boldsymbol{\mu}_n, \frac{(2\alpha_n - (d-1))\lambda_n}{2(\lambda_n + 1)}\boldsymbol{\beta}_n^{-1}, 2\alpha_n - (d-1)\right) \tag{3.21}$$

which like the posterior marginal distribution of $\boldsymbol{\mu}$ is a Student distribution. The calculations behind this result are provided later in this section.

In the contexts of discrimination, the parameters α_n, λ_n, $\boldsymbol{\mu}_n$ and $\boldsymbol{\beta}_n$ may be calculated separately for each class using the training data for the class and the definitions in (3.18). The resulting conditional densities (3.21) may then be used as a basis for discrimination: assign $\boldsymbol{x}$ to class ω_i for which $g_i > g_j, j = 1, \ldots, C, j \neq i$, where

$$g_i = p(\boldsymbol{x}|\boldsymbol{x}_1, \ldots, \boldsymbol{x}_{n_i} \in \omega_i)p(\omega_i)$$

3.2.3.4 Derivation of the posterior distribution

The posterior distribution is derived using a similar procedure to that used when deriving the posterior distribution for the mean of the univariate normal distribution with known variance. Specifically:

1. Express the posterior density function only up to a constant factor (i.e. ignoring multiplicative factors that do not vary with $\boldsymbol{\mu}$ and $\boldsymbol{K}$).
2. Collect together terms based upon the forms of the distributions (e.g. completing the square within the exponents).
3. Deduce the results and any integrations by comparison of the densities with known forms.

We have

$$\begin{aligned} p(\boldsymbol{\mu}, \boldsymbol{K}|\boldsymbol{x}_1, \ldots, \boldsymbol{x}_n) &\propto p(\boldsymbol{x}_1, \ldots, \boldsymbol{x}_n|\boldsymbol{\mu}, \boldsymbol{K})p(\boldsymbol{\mu}, \boldsymbol{K}) \\ &\propto |\boldsymbol{K}|^{n/2}\exp\left\{-\frac{1}{2}\sum_{i=1}^{n}(\boldsymbol{x}_i - \boldsymbol{\mu})^T\boldsymbol{K}(\boldsymbol{x}_i - \boldsymbol{\mu})\right\} \\ &\quad \times |\boldsymbol{K}|^{1/2}\exp\left\{-\frac{1}{2}\lambda(\boldsymbol{\mu} - \boldsymbol{\mu}_0)^T\boldsymbol{K}(\boldsymbol{\mu} - \boldsymbol{\mu}_0)\right\} \\ &\quad \times |\boldsymbol{K}|^{(\alpha-(d+1)/2)}\exp\{-\mathrm{Tr}(\boldsymbol{\beta}\boldsymbol{K})\} \end{aligned}$$

Completing the square for $\boldsymbol{\mu}$ in the exponents, and rearranging the terms gives

$$\begin{aligned} p(\boldsymbol{\mu}, \boldsymbol{K}|\boldsymbol{x}_1, \ldots, \boldsymbol{x}_n) &\propto |\boldsymbol{K}|^{1/2}\exp\left\{-\frac{1}{2}\lambda_n(\boldsymbol{\mu} - \boldsymbol{\mu}_n)^T\boldsymbol{K}(\boldsymbol{\mu} - \boldsymbol{\mu}_n)\right\} \\ &\quad \times |\boldsymbol{K}|^{\alpha_n-(d+1)/2}\exp\{-f(K) - \mathrm{Tr}(\boldsymbol{\beta}\boldsymbol{K})\} \end{aligned} \tag{3.22}$$

where

$$f(\boldsymbol{K}) = \frac{1}{2}\left(\frac{n\lambda}{\lambda_n}(\boldsymbol{\mu}_0 - \boldsymbol{m})^T\boldsymbol{K}(\boldsymbol{\mu}_0 - \boldsymbol{m}) + \sum_{i=1}^{n}(\boldsymbol{x}_i - \boldsymbol{m})^T K(\boldsymbol{x}_i - \boldsymbol{m})\right) \tag{3.23}$$

and $\boldsymbol{\mu}_n$, α_n and λ_n are as defined in (3.18).

The first simplification is to note that the expression on the top line of (3.22) is proportional to a multivariate normal probability density function, so that

$$p(\boldsymbol{\mu}, \boldsymbol{K}|\boldsymbol{x}_1, \ldots, \boldsymbol{x}_n) \propto N_d(\boldsymbol{\mu}; \boldsymbol{\mu}_n, \lambda_n\boldsymbol{K})|\boldsymbol{K}|^{\alpha_n-(d+1)/2}\exp\{-f(K) - \mathrm{Tr}(\boldsymbol{\beta}\boldsymbol{K})\} \tag{3.24}$$

To proceed further we note that all terms making up (3.23) are of the form $\boldsymbol{u}^T\boldsymbol{K}\boldsymbol{u}$ (where $\boldsymbol{u}$ varies with the term under consideration, and is in each case a d-dimensional column vector). Since $\boldsymbol{u}^T\boldsymbol{K}\boldsymbol{u}$ is a scalar, it can be considered the trace of a 1×1 matrix $\boldsymbol{u}^T\boldsymbol{K}\boldsymbol{u}$. Then, noting that $\text{Tr}(\boldsymbol{AB}) = \text{Tr}(\boldsymbol{BA})$ for an $r \times s$ matrix $\boldsymbol{A}$ and an $s \times r$ matrix $\boldsymbol{B}$, we have

$$\boldsymbol{u}^T \boldsymbol{K} \boldsymbol{u} = \text{Tr}(\boldsymbol{u}^T \boldsymbol{K} \boldsymbol{u}) = \text{Tr}(\boldsymbol{u}\boldsymbol{u}^T \boldsymbol{K}) \tag{3.25}$$

Hence using the linear nature of the Trace operator

$$k_1 \text{Tr}(\boldsymbol{C}) + k_2 \text{Tr}(\boldsymbol{D}) = \text{Tr}(k_1 \boldsymbol{C} + k_2 \boldsymbol{D})$$

for two matrices $\boldsymbol{C}$ and $\boldsymbol{D}$ of the same size, we have

$$\begin{aligned} f(\boldsymbol{K}) &= \frac{1}{2}\text{Tr}\left(\frac{n\lambda}{\lambda_n}(\boldsymbol{\mu}_0 - \boldsymbol{m})(\boldsymbol{\mu}_0 - \boldsymbol{m})^T \boldsymbol{K} + \sum_{i=1}^{n} (\boldsymbol{x}_i - \boldsymbol{m})(\boldsymbol{x}_i - \boldsymbol{m})^T \boldsymbol{K}\right) \\ &= \frac{1}{2}\text{Tr}\left(\left(\frac{n\lambda}{\lambda_n}(\boldsymbol{\mu}_0 - \boldsymbol{m})(\boldsymbol{\mu}_0 - \boldsymbol{m})^T + (n-1)\boldsymbol{S}\right)\boldsymbol{K}\right) \end{aligned}$$

Hence using the linear nature of the Trace operator again we can write (3.24) as

$$p(\boldsymbol{\mu}, \boldsymbol{K}|\boldsymbol{x}_1, \ldots, \boldsymbol{x}_n) \propto N_d(\boldsymbol{\mu}; \boldsymbol{\mu}_n, \lambda_n \boldsymbol{K}) |\boldsymbol{K}|^{\alpha_n - (d+1)/2} \exp\left\{-\text{Tr}(\boldsymbol{\beta}_n \boldsymbol{K})\right\} \tag{3.26}$$

where $\boldsymbol{\beta}_n$ is as defined in (3.18).

By comparison with (3.16), and noting that any multiplicative factors that do not depend on $\boldsymbol{\mu}$ and $\boldsymbol{K}$ will be replaced by the unspecified normalising constant in (3.26), we deduce the desired Gauss-Wishart posterior distribution

$$p(\boldsymbol{\mu}, \boldsymbol{K}|\boldsymbol{x}_1, \ldots, \boldsymbol{x}_n) = N_d(\boldsymbol{\mu}; \boldsymbol{\mu}_n, \lambda_n \boldsymbol{K}) \text{Wi}_d(\boldsymbol{K}; \alpha_n, \boldsymbol{\beta}_n)$$

3.2.3.5 Derivation of the predictive density

The predictive density is given by

$$p(\boldsymbol{x}|\boldsymbol{x}_1, \ldots, \boldsymbol{x}_n) = \int p(\boldsymbol{x}|\boldsymbol{\mu}, \boldsymbol{K}) p(\boldsymbol{\mu}, \boldsymbol{K}|\boldsymbol{x}_1, \ldots, \boldsymbol{x}_n) d\boldsymbol{\mu} d\boldsymbol{K} \tag{3.27}$$

To derive the form quoted in (3.21), we note that because we have a conjugate-prior distribution, the expression being integrated in (3.27) is similar to that in the numerator of the posterior density function [see Equation (3.17)], the differences being that the Gauss-Wishart prior distribution with parameters $\{\lambda, \alpha, \boldsymbol{\mu}_0, \boldsymbol{\beta}_0\}$ is replaced by the Gauss-Wishart posterior distribution with parameters $\{\lambda_n, \alpha_n, \boldsymbol{\mu}_n, \boldsymbol{\beta}_n\}$, and that a single measurement $\boldsymbol{x}$ is being considered in the likelihood term, rather than the training data $\{\boldsymbol{x}_1, \ldots, \boldsymbol{x}_n\}$. We can therefore simplify the expression by following the procedure for deriving the joint posterior distribution for $\boldsymbol{\mu}$ and $\boldsymbol{\Sigma}$. Specifically, noting that no multiplicative factors involving $\boldsymbol{x}_i$ have been ignored in the calculations that led to (3.26), we can write immediately

$$\begin{aligned} p(\boldsymbol{x}|\boldsymbol{\mu}, \boldsymbol{K}) p(\boldsymbol{\mu}, \boldsymbol{K}|\boldsymbol{x}_1, \ldots, \boldsymbol{x}_n) &\propto N_d(\boldsymbol{\mu}; \boldsymbol{\mu}'_n, (\lambda_n + 1)\boldsymbol{K}) \\ &\quad \times |\boldsymbol{K}|^{\alpha_n + 1/2 - (d+1)/2} \exp\left\{-\text{Tr}(\boldsymbol{\beta}'_n \boldsymbol{K})\right\} \end{aligned} \tag{3.28}$$

where

$$\boldsymbol{\mu}'_n = (\lambda_n \boldsymbol{\mu}_n + \boldsymbol{x})/(\lambda_n + 1)$$
$$\boldsymbol{\beta}'_n = \boldsymbol{\beta}_n + \frac{\lambda_n}{2(\lambda_n + 1)}(\boldsymbol{\mu}_n - \boldsymbol{x})(\boldsymbol{\mu}_n - \boldsymbol{x})^T \tag{3.29}$$

The forms in (3.28) and (3.29) being derived by evaluating the expressions in (3.18) for a single data vector $\boldsymbol{x}$.

The integration with respect to $\boldsymbol{\mu}$ can then be conducted analytically (since $\boldsymbol{\mu}$ only occurs within the multivariate normal density term, which integrates to unity with respect to $\boldsymbol{\mu}$). Hence

$$p(\boldsymbol{x}|\boldsymbol{x}_1, \ldots, \boldsymbol{x}_n) \propto \int |\boldsymbol{K}|^{\alpha_n + 1/2 - (d+1)/2} \exp\left\{-\mathrm{Tr}(\boldsymbol{\beta}'_n \boldsymbol{K})\right\} d\boldsymbol{K} \tag{3.30}$$

Noting the form of the Wishart $\mathrm{Wi}_d(\alpha_n + 1/2, \boldsymbol{\beta}'_n)$ probability density function [see Equation (3.14) for details of the Wishart distribution], and that $\boldsymbol{\beta}'_n$ varies with $\boldsymbol{x}$ but α_n [and therefore $c(d, \alpha_n)$] does not, we can write (3.30) as

$$p(\boldsymbol{x}|\boldsymbol{x}_1, \ldots, \boldsymbol{x}_n) \propto |\boldsymbol{\beta}'_n|^{-(\alpha_n + 1/2)} \int \mathrm{Wi}_d(\boldsymbol{K}; \alpha_n + 1/2, \boldsymbol{\beta}'_n) d\boldsymbol{K}$$

Hence since all probability density functions integrate to unity, we obtain

$$p(\boldsymbol{x}|\boldsymbol{x}_1, \ldots, \boldsymbol{x}_n) \propto |\boldsymbol{\beta}'_n|^{-(\alpha_n + 1/2)}$$

Since $|\boldsymbol{AB}| = |\boldsymbol{A}||\boldsymbol{B}|$ for two $d \times d$ matrices $\boldsymbol{A}$ and $\boldsymbol{B}$ (the determinant of the product is the product of the determinants) we can use (3.29) to write

$$|\boldsymbol{\beta}'_n| = |\boldsymbol{\beta}_n| \left| \boldsymbol{I}_p\, d + \frac{\lambda_n}{2(\lambda_n + 1)} \boldsymbol{\beta}_n^{-1} (\boldsymbol{\mu}_n - \boldsymbol{x})(\boldsymbol{\mu}_n - \boldsymbol{x})^T \right|$$

To proceed we make use of Sylvester's determinant theorem, which states that for a $d \times s$ matrix $\boldsymbol{A}$ and an $s \times d$ matrix $\boldsymbol{B}$

$$|\boldsymbol{I}_d + \boldsymbol{AB}| = |\boldsymbol{I}_s + \boldsymbol{BA}| \tag{3.31}$$

and hence for two d-dimensional column vectors $\boldsymbol{u}$ and $\boldsymbol{v}$ we have

$$|\boldsymbol{I}_d + \boldsymbol{u}\boldsymbol{v}^T| = |\boldsymbol{I}_1 + \boldsymbol{v}^T \boldsymbol{u}|$$
$$= 1 + \boldsymbol{v}^T \boldsymbol{u}$$

Hence

$$|\boldsymbol{\beta}'_n| = |\boldsymbol{\beta}_n| \left(1 + \frac{\lambda_n}{2(\lambda_n + 1)} (\boldsymbol{\mu}_n - \boldsymbol{x})^T \boldsymbol{\beta}_n^{-1} (\boldsymbol{\mu}_n - \boldsymbol{x})\right)$$

Then defining

$$a = 2\alpha_n - (d-1)$$

we can write

$$p(\boldsymbol{x}|\boldsymbol{x}_1, \ldots, \boldsymbol{x}_n) \propto \left(1 + \frac{1}{a}(\boldsymbol{\mu}_n - \boldsymbol{x})^T \frac{a\lambda_n}{2(\lambda_n+1)} \boldsymbol{\beta}_n^{-1}(\boldsymbol{\mu}_n - \boldsymbol{x})\right)^{-(a+d)/2}$$

where the $|\boldsymbol{\beta}_n|^{-(a+d)/2}$ term has been incorporated into the as yet unspecified normalisation constant. By comparison with the probability density function of the multivariate student distribution [see Equation (3.20)] we obtain the desired result

$$p(\boldsymbol{x}|\boldsymbol{x}_1, \ldots, \boldsymbol{x}_n) = \mathrm{St}_d\left(\boldsymbol{x}; \boldsymbol{\mu}_n, \frac{(2\alpha_n - (d-1))\lambda_n}{2(\lambda_n+1)} \boldsymbol{\beta}_n^{-1}, 2\alpha_n - (d-1)\right)$$

3.2.4 Unknown prior class probabilities

To date in this section, we have considered estimation of the parameters of the class-conditional probability density functions. Recall that the predictive Bayesian classifier assigns $\boldsymbol{x}$ to class ω_i for which $g_i > g_j, j = 1, \ldots, C, j \neq i$, where

$$g_i = p(\boldsymbol{x}|\boldsymbol{x}_1, \ldots, \boldsymbol{x}_{n_i} \in \omega_i)p(\omega_i) \tag{3.32}$$

with $p(\boldsymbol{x}|\boldsymbol{x}_1, \ldots, \boldsymbol{x}_{n_i} \in \omega_i)$ the predictive density for class ω_i, and $p(\omega_i)$ the prior probability for class ω_i. It may happen that the prior class probabilities, $p(\omega_i)$, are unknown. In this case, we may treat them as parameters of the model that may be updated using the data.

Using $\boldsymbol{\pi} = (\pi_1, \ldots, \pi_C)$ to represent the prior class probabilities, we write

$$p(\omega_i|\boldsymbol{x}, \mathcal{D}) = \int p(\omega_i, \boldsymbol{\pi}|\boldsymbol{x}, \mathcal{D})d\boldsymbol{\pi} \tag{3.33}$$

(i.e. we marginalise over $\boldsymbol{\pi}$ in the joint distribution of $\{\omega_i, \boldsymbol{\pi}\}$ conditional on the test sample $\boldsymbol{x}$ and the training data $\mathcal{D}$). Using Bayes' theorem

$$p(\omega_i, \boldsymbol{\pi}|\boldsymbol{x}, \mathcal{D}) \propto p(\boldsymbol{x}|\omega_i, \boldsymbol{\pi}, \mathcal{D})p(\omega_i, \boldsymbol{\pi}|\mathcal{D}) \tag{3.34}$$

Since we are conditioning on the class ω_i we have

$$p(\boldsymbol{x}|\omega_i, \boldsymbol{\pi}, \mathcal{D}) = p(\boldsymbol{x}|\omega_i, \mathcal{D}) \tag{3.35}$$

Furthermore, the second term in (3.34) can (using the definition of conditional densities) be written

$$\begin{aligned} p(\omega_i, \boldsymbol{\pi}|D) &= p(\boldsymbol{\pi}|D)p(\omega_i|\boldsymbol{\pi}, D) \\ &= p(\boldsymbol{\pi}|D)\pi_i \end{aligned} \tag{3.36}$$

A suitable prior for $\pi_i = p(\omega_i)$ is a `Dirichlet distribution`, with parameters $\boldsymbol{a}_0 = (a_{01}, \ldots, a_{0C})$, $a_{i0} > 0$, $i = 1, \ldots, C$.

The Dirichlet distribution

The probability density function of the Dirichlet distribution with parameters $\boldsymbol{a}_0 = (a_{01}, \ldots, a_{0C})$, $a_{0i} > 0$ for $i = 1, \ldots, C$, is

$$p(\pi_1, \ldots, \pi_C) = \frac{\Gamma\left(\sum_i^C a_{0i}\right)}{\prod_i^C \Gamma(a_{0i})} \prod_{j=1}^{C} \pi_j^{a_{0j}-1}$$

where $\Gamma(u)$ is the gamma function. The support of the distribution is such that $0 < \pi_i < 1$, $i = 1, \ldots, C$, and $\sum_{i=1}^{c} \pi_i = 1$. It is therefore ideally suited to modelling the probabilities of categorical distributions. The component means of the Dirichlet distribution are given by $E[\pi_i] = \frac{a_i}{\sum_{j=1}^{C} a_j}$. Notationally we write the distribution as $\boldsymbol{\pi} \sim \text{Di}_C(\boldsymbol{\pi}; \boldsymbol{a}_0)$ for $\boldsymbol{\pi} = (\pi_1, \ldots, \pi_C)^T$.

The aspects of the training data measurements $\mathcal{D}$ that influence the distribution of the class probabilities, $\boldsymbol{\pi}$, are the numbers of patterns in each class, n_i, $i = 1, \ldots, C$. Modelling the distribution of the data (the n_i) given the prior class probabilities as being multinomial

$$p(\mathcal{D}|\boldsymbol{\pi}) = \frac{n!}{\prod_{l=1}^{C} n_l!} \prod_{l=1}^{C} \pi_l^{n_l}$$

then the posterior distribution

$$p(\boldsymbol{\pi}|\mathcal{D}) \propto p(\mathcal{D}|\boldsymbol{\pi})p(\boldsymbol{\pi})$$

is also distributed as $\text{Di}_C(\boldsymbol{\pi}; \boldsymbol{a})$, where $\boldsymbol{a} = \boldsymbol{a}_0 + \boldsymbol{n}$ and $\boldsymbol{n} = (n_1, \ldots, n_C)^T$, the vector of the numbers of patterns in each class.

Substituting the Dirichlet distribution for $p(\boldsymbol{\pi}|\mathcal{D})$ into (3.36), and using (3.34) and (3.35), we can express (3.33) as

$$p(\omega_i|\boldsymbol{x}, \mathcal{D}) \propto p(\boldsymbol{x}|\omega_i, \mathcal{D}) \int \pi_i \text{Di}_C(\boldsymbol{\pi}; \boldsymbol{a}) d\boldsymbol{\pi}$$

which replaces π_i in (3.32) by its expected value, $a_i/\sum_j a_j$. Thus, the posterior probability of class membership now becomes

$$p(\omega_i|\boldsymbol{x}, \mathcal{D}) = \frac{(n_i + a_{0i})p(\boldsymbol{x}|\omega_i, \mathcal{D})}{\sum_j (n_j + a_{0j})p(\boldsymbol{x}|\omega_j, \mathcal{D})} \tag{3.37}$$

There is, however, an important caveat. This treatment has assumed that the π_i, $i = 1, \ldots, C$ are unknown but can be estimated from the training data alongside prior information. The validity of these estimates depends on the sampling scheme used to gather the data.

Specifically, the procedure is relevant if the training data is obtained using random sampling from the true population. There are various modifications to (3.37) depending on the assumed knowledge concerning the π_i (Geisser, 1964).

3.2.5 Summary

The Bayesian approach described in this section involves two stages. The first stage is concerned with learning about the parameters of the distribution, $\boldsymbol{\theta}$, through the recursive calculation of the posterior density $p(\boldsymbol{\theta}|\boldsymbol{x}_1, \ldots, \boldsymbol{x}_n)$ for a specified prior. The recursive calculation is

$$p(\boldsymbol{\theta}|\boldsymbol{x}_1, \ldots, \boldsymbol{x}_n) \propto p(\boldsymbol{x}_n|\boldsymbol{\theta})p(\boldsymbol{\theta}|\boldsymbol{x}_1, \ldots, \boldsymbol{x}_{n-1})$$

For a suitable class-conditional density and choice of prior distribution, the posterior distribution for $\boldsymbol{\theta}$ is of the same form as the prior. This is called conjugacy. The second stage is the integration over $\boldsymbol{\theta}$ to obtain the predictive density $p(\boldsymbol{x}|\boldsymbol{x}_1, \ldots, \boldsymbol{x}_n)$ which may be viewed as making allowance for the variability in the estimate due to sampling. Although it is relatively straightforward to perform the integrations for the cases that have been considered here (i.e. when using normal distributions), it may be necessary to perform two multivariate numerical integrations for more complicated probability density functions. Typically, Bayesian sampling schemes, such as those to be introduced in Sections 3.3 and 3.4 are used for this. This is the case for the normal mixture model for which there exists no reproducing (conjugate) densities.

3.3 Bayesian sampling schemes

3.3.1 Introduction

In Section 3.2 we considered the Bayesian approach to density estimation in situations where the required integrals can be evaluated analytically. Over the next few sections we consider some of the computational machinery for practical implementation of Bayesian methods to problems for which the normalising integral cannot be evaluated analytically and numerical integration over possibly high-dimensional spaces is infeasible. The computational techniques described aim to draw samples from the posterior distribution. All inferences can then be made through consideration of these samples.

3.3.2 Summarisation

Suppose that we have drawn samples $\{\boldsymbol{\theta}^1, \ldots, \boldsymbol{\theta}^{N_s}\}$ from the posterior distribution $p(\boldsymbol{\theta}|\mathcal{D})$. Then we can approximate posterior expectations of functions $h(\boldsymbol{\theta})$ using sample averages as follows

$$\mathrm{E}[h(\boldsymbol{\theta})|\mathcal{D}] \approx \frac{1}{N_s}\sum_{t=1}^{N_s} h\left(\boldsymbol{\theta}^t\right) \tag{3.38}$$

without calculating the integrals in the numerator and denominator of (3.4).

Many quantities of interest can be expressed in terms of the posterior expectation of a function, including probability values. Suppose that $\boldsymbol{\theta}$ has posterior density function $p(\boldsymbol{\theta}|\mathcal{D})$. Let $h(\boldsymbol{\theta}) = I(\boldsymbol{\theta} \in \mathcal{A})$, for some region $\mathcal{A}$, where I is the indicator function (equal to 1 if the condition is true and zero otherwise). Then

$$\begin{aligned}\mathrm{E}[I(\boldsymbol{\theta} \in \mathcal{A})|\mathcal{D}] &= \int_{\boldsymbol{\theta}} I(\boldsymbol{\theta} \in \mathcal{A}) p(\boldsymbol{\theta}|\mathcal{D}) d\boldsymbol{\theta} \\ &= \int_{\boldsymbol{\theta} \in \mathcal{A}} p(\boldsymbol{\theta}|\mathcal{D}) d\boldsymbol{\theta}\end{aligned}$$

i.e. $\mathrm{E}[I(\boldsymbol{\theta} \in \mathcal{A})|\mathcal{D}]$ is the posterior probability of $\boldsymbol{\theta}$ being within the region $\mathcal{A}$. Our sampling approximation to this probability would be

$$\mathrm{E}[I(\boldsymbol{\theta} \in \mathcal{A})|\mathcal{D}] \approx \frac{1}{N_s} \sum_{t=1}^{N_s} I(\boldsymbol{\theta}^t \in \mathcal{A})$$

Other summaries could include plots of the marginal densities, obtained by using the samples within nonparametric methods of density estimation, such as *kernel methods* (see Chapter 4). The kernel density estimate of θ_i given samples $\{\theta_i^t, t = 1, \ldots, N_s\}$ is

$$p(\theta_i) = \frac{1}{N_s} \sum_{t=1}^{N_s} K\left(\theta_i, \theta_i^t\right)$$

where the *kernel function* $K(\theta, \theta^*)$ is centred at θ^*. Choices for kernels and their parameters are discussed in Chapter 4.

3.3.2.1 Rao-Blackwellisation

An alternative estimator for the marginal densities, due to Gelfand and Smith (1990) and termed the Rao-Blackwellised estimator, makes use of the conditional densities, $p(\theta_i|\boldsymbol{\theta}_{(i)}^t)$. The estimates are given by

$$p(\theta_i) = \frac{1}{N_s} \sum_{t=1}^{N_s} p\left(\theta_i|\boldsymbol{\theta}_{(i)}^t\right)$$

where $\boldsymbol{\theta}_{(i)}$ is the set of parameters with the ith parameter removed; that is, $\boldsymbol{\theta}_{(i)} = \{\theta_1, \ldots, \theta_{i-1}, \theta_{i+1}, \ldots, \theta_d\}$. The Rao-Blackwellised estimators estimate the tails of the distribution better than more general methods of density estimation (O'Hagan, 1994).

The Rao-Blackwellised estimator of $E[h(\theta_i)]$ is then

$$\mathrm{E}[h(\theta_i)] \approx \frac{1}{N_s} \sum_{t=1}^{N_s} \mathrm{E}\left[h(\theta_i)|\boldsymbol{\theta}_{(i)}^t\right] \tag{3.39}$$

The difference between (3.39) and (3.38) is that (3.39) requires an analytic expression for the conditional expectation so that it may be evaluated at each sample. For reasonably long runs, the improvement in using (3.39) over (3.38) is small.

Further discussion on Rao-Blackwellisation of sampling schemes is provided in Casella and Robert (1996), and in the context of Gibbs samplers (see Section 3.4.2) in J.S. Liu *et al.* (1994).

3.3.3 Sampling version of the Bayesian classifier

The class-conditional predictive density of (3.1) is given by

$$\begin{aligned} p(\boldsymbol{x}|\omega_j, \mathcal{D}_j) &= \int p(\boldsymbol{x}|\boldsymbol{\theta}_j) p(\boldsymbol{\theta}_j|\mathcal{D}_j) d\boldsymbol{\theta}_j \\ &= \mathrm{E}_{\boldsymbol{\theta}_j|\mathcal{D}_j}[p(\boldsymbol{x}|\boldsymbol{\theta}_j)] \end{aligned}$$

where the subscript $\boldsymbol{\theta}_j|\mathcal{D}_j$ has been used to emphasise that the expectation is with respect to the posterior distribution for $\boldsymbol{\theta}_j$.

Suppose that $\{\boldsymbol{\theta}_j^1, \ldots \boldsymbol{\theta}_j^{N_s}\}$ are samples from the posterior distribution $p(\boldsymbol{\theta}_j|\mathcal{D}_j)$. Then in line with (3.38) we can approximate the class-conditional density by

$$p(\boldsymbol{x}|\omega_j, \mathcal{D}_j) \approx \frac{1}{N_s} \sum_{t=1}^{N_s} p\left(\boldsymbol{x}|\boldsymbol{\theta}_j^t\right) \tag{3.40}$$

The Bayesian classifier is then: assign $\boldsymbol{x}$ to class ω_i for which $g_i > g_j, j = 1, \ldots, C, j \neq i$, where

$$g_i = \left(\sum_{t=1}^{N_s} p\left(\boldsymbol{x}|\boldsymbol{\theta}_i^t\right) \right) p(\omega_i)$$

3.3.4 Rejection sampling

A simple, although often inefficient, sampling scheme for general distributions is rejection sampling.

Let $f(\boldsymbol{\theta}) = g(\boldsymbol{\theta}) / \int g(\boldsymbol{\theta}') d\boldsymbol{\theta}'$ be the density from which we wish to sample. Rejection sampling uses a density $s(\boldsymbol{\theta})$ from which we can conveniently sample (cheaply) and requires that $g(\boldsymbol{\theta})/s(\boldsymbol{\theta})$ is bounded. The rejection sampling procedure is detailed in Algorithm 3.1. The probability density function of the accepted sample is $f(\boldsymbol{\theta})$ (see exercises), and multiple independent samples can be obtained by repeating the algorithm.

Depending on the choice of $s(\boldsymbol{\theta})$, many samples could be rejected before one is accepted, which can make the procedure inefficient. If $s(\boldsymbol{\theta})$ is close to the shape of $g(\boldsymbol{\theta})$, so that $g(\boldsymbol{\theta})/s(\boldsymbol{\theta}) \sim A$ for all $\boldsymbol{\theta}$, then the acceptance condition is almost always accepted.

Algorithm 3.1 Rejection sampling algorithm.

1. Specify a density $s(\boldsymbol{\theta})$ with the same support as $f(\boldsymbol{\theta}) = g(\boldsymbol{\theta})/\int g(\boldsymbol{\theta}')\,d\boldsymbol{\theta}'$ and for which $g(\boldsymbol{\theta})/s(\boldsymbol{\theta})$ is bounded.
2. Set A to be an upper bound of $g(\boldsymbol{\theta})/s(\boldsymbol{\theta})$.
3. Repeat the following until one $\boldsymbol{\theta}$ is accepted:
   - Sample a point $\boldsymbol{\theta}$ from the known distribution $s(\boldsymbol{\theta})$.
   - Sample u from the uniform distribution on [0, 1].
   - If $Au \leq g(\boldsymbol{\theta})/s(\boldsymbol{\theta})$ then accept $\boldsymbol{\theta}$.

3.3.5 Ratio of uniforms

The ratio of uniforms method (Kinderman and Monahan, 1977) can be used to obtain samples from univariate distributions.

Suppose that we require samples from the distribution with probability density function $f(\theta) = g(\theta)/\int g(\theta')d\theta'$. Let D denote the region in $\mathbb{R}^2$ such that

$$D = \{(u, v); 0 \leq u \leq \sqrt{g(v/u)}\}.$$

Then sampling a point uniformly from D and taking $\theta = v/u$ gives a sample from the density proportional to $g(\theta)$, i.e. from $f(\theta)$.

To make use of this procedure we need to be able to draw samples from the region D. This can often be done by first bounding the region D by a rectangle R, as shown in Figure 3.2.

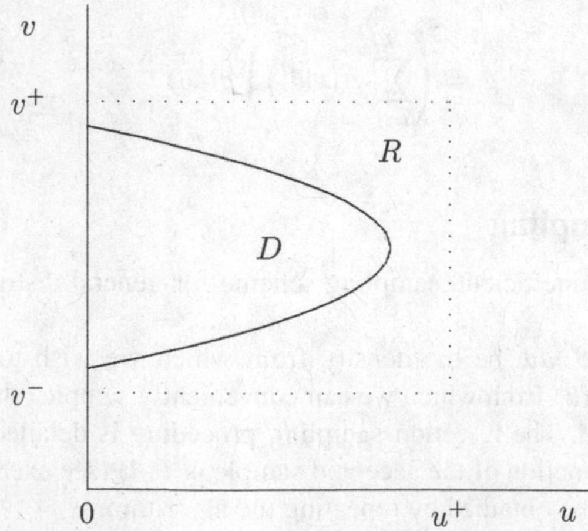

Figure 3.2 Envelope rectangle, R, for the region D defined by $D = \{(u, v); 0 \leq u \leq \sqrt{g(v/u)}\}$.

A sample from D can then be drawn by a simple application of rejection sampling: sample uniformly from the rectangle, R, bounding D and if (u, v) is in D then accept.

For $f(\theta)$ defined on $\theta \in \mathbb{R}$, the rectangle R can be specified by $[0, u^+] \times [v^-, v^+]$ where

$$u^+ = \sqrt{\max_\theta g(\theta)}$$

$$v^- = -\sqrt{\max_{\theta \leq 0}(\theta^2 g(\theta))}$$

$$v^+ = \sqrt{\max_{\theta \geq 0}(\theta^2 g(\theta))}$$

providing that all the quantities exist. The validity of the bound for u is clear, since $0 \leq u \leq \sqrt{g(v/u)} = \sqrt{g(\theta)} \Rightarrow 0 \leq u \leq u^+$. To see the validity of the bound for v, note that $v \in D$ implies that there is a u for which $0 \leq u \leq \sqrt{g(\theta)}$, and therefore for which $u^2 \leq g(\theta)$. Since $u = v/\theta$, the latter bound becomes $v^2 \leq \theta^2 g(\theta)$. An immediate bound for v is therefore $-\sqrt{\max_\theta(\theta^2 g(\theta))} \leq v \leq \sqrt{\max_\theta(\theta^2 g(\theta))}$. However, since $u \geq 0$, and $v = u\theta$, we know that θ and v must have the same sign. Hence, the tighter bound $v^- \leq v \leq v^+$ holds.

3.3.5.1 Proof of the validity of the ratio of uniforms method

To show the validity of the ratio of uniforms method, we need to make use of a result for the probability density function arising from a transformation of variables.

That result concerns the transformation of variables from $(X_1, \ldots, X_d)$ to $(Y_1, \ldots, Y_d)$, given by

$$\boldsymbol{Y} = \boldsymbol{g}(\boldsymbol{X})$$

where $\boldsymbol{g}(\boldsymbol{X}) = (g_1(\boldsymbol{X}), g_2(\boldsymbol{X}), \ldots, g_d(\boldsymbol{X}))^T$. Under such a transformation the probability density functions of $\boldsymbol{X}$ and $\boldsymbol{Y}$ are related by

$$p_Y(\boldsymbol{y}) = \frac{p_X(\boldsymbol{x})}{|J|}$$

where $|J|$ is the absolute value of the Jacobian determinant

$$J(x_1, \ldots, x_d) = \begin{vmatrix} \frac{\partial g_1}{\partial x_1} & \cdots & \frac{\partial g_1}{\partial x_d} \\ \vdots & \ddots & \vdots \\ \frac{\partial g_d}{\partial x_1} & \cdots & \frac{\partial g_d}{\partial x_d} \end{vmatrix}$$

We now apply the procedure to determine the joint probability density function of the transformed variables (u, θ), where $\theta = v/u$. The joint probability density function of (u, v) is given by

$$p(u, v) \propto I((u, v) \in D)$$

where $I(A)$ is the indicator function, equal to 1 if condition A holds, and zero otherwise. The condition $(u, v) \in D$ becomes $0 \le u \le \sqrt{g(\theta)}$, the Jacobian determinant is $J = 1/u$, and the probability density function of (u, θ) is therefore given by

$$p(u, \theta) \propto I(0 \le u \le \sqrt{g(\theta)})u$$

The normalising constant is

$$\int_\theta \int_u I(0 \le u \le \sqrt{g(\theta)})ud u d\theta = \int_\theta \int_{u=0}^{\sqrt{g(\theta)}} udu = \frac{1}{2}\int_\theta g(\theta)d\theta$$

The marginal distribution of θ is then given by

$$p(\theta) = \int_u p(u, \theta)du = \frac{1}{\frac{1}{2}\int_{\theta'} g(\theta')d\theta'} \int_{u=0}^{\sqrt{g(\theta)}} udu = \frac{g(\theta)}{\int g(\theta')d\theta'}$$

3.3.6 Importance sampling

Suppose that we have a set of N_s independent samples $\{\boldsymbol{\theta}^1, \ldots, \boldsymbol{\theta}^{N_s}\}$ from a distribution with probability density function proportional to $q(\boldsymbol{\theta})$. Importance sampling (Geweke, 1989; Smith and Gelfand, 1992) provides a means for using these samples to make inference on a second distribution with density function proportional to $f(\boldsymbol{\theta})$.

If the support of the two distributions is the same, the importance sampling result is that the expectation of a function $h(\boldsymbol{\theta})$ when $\boldsymbol{\theta}$ is distributed according to $f(\boldsymbol{\theta})$ can be approximated by

$$\mathrm{E}_f[h(\boldsymbol{\theta})] \approx \frac{1}{\sum_{t=1}^{N_s} w^t} \sum_{t=1}^{N_s} w^t h(\boldsymbol{\theta}^t) \tag{3.41}$$

where we have defined a set of unnormalised importance weights

$$w^s = f(\boldsymbol{\theta}^s)/q(\boldsymbol{\theta}^s), s = 1, \ldots, N_s \tag{3.42}$$

The distribution $q(\boldsymbol{\theta})$ is termed the importance sampling proposal distribution.

The power of the procedure arises if it is easy to draw samples form the distribution defined by $q(\boldsymbol{\theta})$, but not easy to draw samples directly from the distribution defined by $f(\boldsymbol{\theta})$. The closer $q(\boldsymbol{\theta})$ is to $f(\boldsymbol{\theta})$ the better the accuracy of the estimates, for a given sample size. An indicator that the estimates might be unreliable is when the importance sampling weights have a large variation. Degeneracy problems can arise if the weights vary too widely, since essentially this leads to the estimate being based on only the samples with the largest weights.

3.3.6.1 Posterior samples via the prior distribution

Consider the special case where the distribution defined by $f(\boldsymbol{\theta})$ is the posterior distribution arising from a likelihood function $p(\boldsymbol{x}|\boldsymbol{\theta})$ and prior distribution $p(\boldsymbol{\theta})$, and the distribution defined by $q(\boldsymbol{\theta})$ is the prior distribution $p(\boldsymbol{\theta})$. Then the importance weights defined in (3.42) become

$$w^t = \frac{p(\boldsymbol{x}|\boldsymbol{\theta}^t)p\left(\boldsymbol{\theta}^t\right)}{p\left(\boldsymbol{\theta}^t\right)} = p(\boldsymbol{x}|\boldsymbol{\theta}^t).$$

Thus, we have a mechanism for making inference on a posterior distribution by sampling from the prior, and weighting the samples by the likelihood function.

3.3.6.2 Particle filters

The approach described above forms the basis of particle filters (Gordon *et al.*, 1993) for sequential Bayesian parameter estimation. These are considered in detail in the book by Doucet *et al.* (2001), and also in the tutorial paper by Arulampalam *et al.* (2002). Particle filters provide a very powerful approach to problems where observations arrive sequentially and one is interested in performing inference online, e.g. tracking of manoeuvring objects such as aircraft from radar returns, and people in video imagery.

3.3.6.3 Use within a Bayesian classifier

As noted in Section 3.3.3, when considering classification problems, we have the following expression for the predictive density for class ω_j

$$p(\boldsymbol{x}|\omega_j, \mathcal{D}_j) = \mathrm{E}_{\boldsymbol{\theta}_j|\mathcal{D}_j}\left[p(\boldsymbol{x}|\boldsymbol{\theta}_j)\right]$$

Under the importance sampling approximation, if $f_j(\boldsymbol{\theta}_j)$ is proportional to the posterior density function $p(\boldsymbol{\theta}_j|\mathcal{D}_j)$, and $\{\boldsymbol{\theta}_j^1, \ldots, \boldsymbol{\theta}_j^{N_s}\}$ are samples from the probability distribution with probability density function proportional to $q_j(\boldsymbol{\theta}_j)$, then

$$p(\boldsymbol{x}|\omega_j, \mathcal{D}_j) \approx \frac{1}{\sum_{t=1}^{N_s} w_j^t} \sum_{t=1}^{N_s} w_j^t p\left(\boldsymbol{x}|\boldsymbol{\theta}_j^t\right)$$

where the weights w_j^t are as defined in (3.42)

$$w_j^t = \frac{f_j\left(\boldsymbol{\theta}_j^t\right)}{q_j\left(\boldsymbol{\theta}_j^t\right)}$$

A suitable choice for $f_j(\boldsymbol{\theta}_j)$ is the product of the likelihood function and the prior density function

$$f_j(\boldsymbol{\theta}_j) = p(\mathcal{D}_j|\boldsymbol{\theta}_j)p(\boldsymbol{\theta}_j)$$

which has proportionality to $p(\boldsymbol{\theta}_j^t|\mathcal{D}_j)$ via Bayes' theorem. This gives rise to the following weights

$$w_j^t = \frac{p\left(\mathcal{D}_j|\boldsymbol{\theta}_j^t\right)p\left(\boldsymbol{\theta}_j^t\right)}{q_j\left(\boldsymbol{\theta}_j^t\right)}$$

The Bayesian classifier is then: assign $\boldsymbol{x}$ to class ω_i for which $g_i > g_j, j = 1, \ldots, C, j \neq i$, where

$$g_i = \left(\frac{1}{\sum_{t=1}^{N_s} w_i^t} \sum_{t=1}^{N_s} w_i^t p\left(\boldsymbol{x}|\boldsymbol{\theta}_i^t\right) \right) p(\omega_i)$$

3.3.6.4 Derivation of the importance sampling approximations

We now prove the validity of the importance sampling approximations. Let the normalisation constant for the probability density function proportional to $f(\boldsymbol{\theta})$ be k_f [so that the probability density function is given by $k_f f(\boldsymbol{\theta})$], and that for the proposal distribution defined by $q(\boldsymbol{\theta})$ be k_q. Then we can express the expectation of a function $h(\boldsymbol{\theta})$ with respect to the distribution defined by $f(\boldsymbol{\theta})$ as follows

$$\begin{aligned} \mathrm{E}_f[h(\boldsymbol{\theta})] &= \int_{\boldsymbol{\theta}} k_f f(\boldsymbol{\theta}) h(\boldsymbol{\theta}) d\boldsymbol{\theta} \\ &= \int_{\boldsymbol{\theta}} k_f f(\boldsymbol{\theta}) h(\boldsymbol{\theta}) \frac{k_q q(\boldsymbol{\theta})}{k_q q(\boldsymbol{\theta})} d\boldsymbol{\theta} = \frac{k_f}{k_q} \int_{\boldsymbol{\theta}} k_q q(\boldsymbol{\theta}) \frac{f(\boldsymbol{\theta}) h(\boldsymbol{\theta})}{q(\boldsymbol{\theta})} d\boldsymbol{\theta} \\ &= \frac{k_f}{k_q} \mathrm{E}_q \left[\frac{f(\boldsymbol{\theta}) h(\boldsymbol{\theta})}{q(\boldsymbol{\theta})} \right] \end{aligned} \tag{3.43}$$

Now suppose that we define $h(\boldsymbol{\theta}) = 1$, for all $\boldsymbol{\theta}$. Then $\mathrm{E}_f[h(\boldsymbol{\theta})] = 1$ by definition, and (3.43) therefore gives

$$1 = \frac{k_f}{k_q} \mathrm{E}_q \left[\frac{f(\boldsymbol{\theta})}{q(\boldsymbol{\theta})} \right] \tag{3.44}$$

Returning to a general function $h(\boldsymbol{\theta})$ within (3.43), we can use the expression for k_f/k_q in (3.44) to obtain

$$\mathrm{E}_f[h(\boldsymbol{\theta})] = \frac{\mathrm{E}_q \left[\frac{f(\boldsymbol{\theta}) h(\boldsymbol{\theta})}{q(\boldsymbol{\theta})} \right]}{\mathrm{E}_q \left[\frac{f(\boldsymbol{\theta})}{q(\boldsymbol{\theta})} \right]} \tag{3.45}$$

Now let $\{\boldsymbol{\theta}^1, \ldots, \boldsymbol{\theta}^{N_s}\}$ be samples from the distribution defined by $q(\boldsymbol{\theta})$. We then have

$$\mathrm{E}_q \left[\frac{f(\boldsymbol{\theta}) h(\boldsymbol{\theta})}{q(\boldsymbol{\theta})} \right] \approx \frac{1}{N_s} \sum_{t=1}^{N_s} \frac{f\left(\boldsymbol{\theta}^t\right) h\left(\boldsymbol{\theta}^t\right)}{q\left(\boldsymbol{\theta}^t\right)} = \frac{1}{N_s} \sum_{t=1}^{N_s} w^t h\left(\boldsymbol{\theta}^t\right)$$

and

$$\mathrm{E}_q \left[\frac{f(\boldsymbol{\theta})}{q(\boldsymbol{\theta})} \right] \approx \frac{1}{N_s} \sum_{t=1}^{N_s} \frac{f\left(\boldsymbol{\theta}^t\right)}{q\left(\boldsymbol{\theta}^t\right)} = \frac{1}{N_s} \sum_{t=1}^{N_s} w^t$$

where the weights w^t are as defined in (3.42). Replacing the top and bottom expectations in (3.45) by these approximations we obtain the importance sampling result of (3.41). The bias introduced by this ratio of estimates decreases as the number of samples increases.

3.4 Markov chain Monte Carlo methods

3.4.1 Introduction

A group of methods known as Markov chain Monte Carlo (MCMC) methods has proven to be effective at generating samples asymptotically from the posterior distribution without knowing the normalising constant. Inference about model parameters can then may be made by forming sample averages, as in previous discussions. MCMC methodology may be used to analyse complex problems, no longer requiring users to force the problem into an oversimplified framework for which analytic treatment is possible.

3.4.2 The Gibbs sampler

We begin with a description of the Gibbs sampler, one of the most popular MCMC methods, and discuss some of the issues that must be addressed for practical implementation. Both the Gibbs sampler and a more general algorithm, the Metropolis–Hastings algorithm (see Section 3.4.3), have formed the basis for many variants. Good reviews of the Gibbs sampler and its origins are provided by Gelfand (2000) and Casella and George (1992).

Let $f(\boldsymbol{\theta})$ denote the posterior density function from which we wish to draw samples; $\boldsymbol{\theta}$ is a d-dimensional parameter vector, $(\theta_1, \ldots, \theta_d)^T$. We may not know $f(\boldsymbol{\theta})$ exactly, but we know a function $g(\boldsymbol{\theta})$, where $f(\boldsymbol{\theta}) = g(\boldsymbol{\theta})/\int g(\boldsymbol{\theta}')d\boldsymbol{\theta}'$. For example, $f(\boldsymbol{\theta})$ could be the posterior density function arising from a likelihood function $p(\mathcal{D}|\boldsymbol{\theta})$ and a prior density $p(\boldsymbol{\theta})$, and the normalisation constant $p(\mathcal{D}) = \int p(\mathcal{D}|\boldsymbol{\theta})p(\boldsymbol{\theta})$ might not be known analytically. In such cases we consider the function $g(\boldsymbol{\theta}) = p(\mathcal{D}|\boldsymbol{\theta})p(\boldsymbol{\theta})$.

Let $\boldsymbol{\theta}_{(i)}$ be the set of parameters with the ith parameter removed; that is, $\boldsymbol{\theta}_{(i)} = \{\theta_1, \ldots, \theta_{i-1}, \theta_{i+1}, \ldots, \theta_d\}$. We assume that we are able to draw samples from the one-dimensional conditional distributions, $f(\theta_i|\boldsymbol{\theta}_{(i)})$. These conditional distributions can be derived from the normalisation of $g(\theta_i|\boldsymbol{\theta}_{(i)})$, the function g regarded as a function of θ_i alone, all other parameters being fixed (this concept is discussed in more detail later in this section).

Gibbs sampling is a simple algorithm that consists of drawing samples from these distributions in a cyclical way as described in Algorithm 3.2. Figure 3.3 illustrates the Gibbs sampler for a bivariate distribution. The θ_1 and θ_2 components are updated alternately, producing moves in the horizontal and vertical directions.

In the Gibbs sampling algorithm, the distribution of $\boldsymbol{\theta}^t$ given all previous values $\boldsymbol{\theta}^0, \boldsymbol{\theta}^1, \ldots, \boldsymbol{\theta}^{t-1}$ depends only on $\boldsymbol{\theta}^{t-1}$. This is the Markov property and the sequence generated is termed a *Markov chain*.

3.4.2.1 Requirements

For the distribution of $\boldsymbol{\theta}^t$ to converge to a stationary distribution (a distribution that does not depend on $\boldsymbol{\theta}^0$ or t), the Markov chain must be *aperiodic*, *irreducible* and *positive recurrent*. A Markov chain is aperiodic if it does not oscillate between different subsets in a regular

Algorithm 3.2 Gibbs sampling algorithm.

- Choose an arbitrary starting value for $\boldsymbol{\theta}$, $\boldsymbol{\theta}^0 = (\theta_1^0, \ldots, \theta_d^0)^T$ from the support of the prior/posterior distribution.
- At stage $t+1$ of the iterative algorithm:
  - draw a sample, θ_1^{t+1} from $f(\theta_1|\theta_2^t, \ldots, \theta_d^t)$;
  - draw a sample, θ_2^{t+1} from $f(\theta_2|\theta_1^{t+1}, \theta_3^t, \ldots, \theta_d^t)$;
  - draw a sample, θ_3^{t+1} from $f(\theta_3|\theta_1^{t+1}, \theta_2^{t+1}, \theta_4^t, \ldots, \theta_d^t)$;
  - and continue through the variables, finally drawing a sample, θ_d^{t+1} from $f(\theta_d|\,\theta_1^{t+1}, \ldots, \theta_{d-1}^{t+1})$;
  - set $\boldsymbol{\theta}^{t+1} = (\theta_1^{t+1}, \ldots, \theta_d^{t+1})^T$.
- After a large number of iterations, provided that a number of requirements (detailed in the main text) are met, the vectors $\boldsymbol{\theta}^t$ behave like a random draw from the joint density $f(\boldsymbol{\theta})$.

periodic way. Recurrence is the property that all sets of $\boldsymbol{\theta}$ values will be reached infinitely often at least from almost all starting points. A Markov chain is irreducible if it can reach all possible $\boldsymbol{\theta}$ values from any starting point.

Figure 3.4 illustrates a distribution that is uniform on $([0, 1] \times [0, 1]) \bigcup ([1, 2] \times [1, 2])$, the union of two nonoverlapping unit squares. Consider the Gibbs sampler that uses the coordinate directions as sampling directions. For a point, $\theta_1 \in [0, 1]$, the conditional distribution of θ_2 given θ_1 is uniform over [0, 1]. Similarly, for a point, $\theta_2 \in [0, 1]$, the conditional distribution of θ_1 given θ_2 is uniform over [0, 1]. We can see that if the starting point for θ_1 is in [0, 1], then the Gibbs sampler will generate a point θ_2, also in [0, 1]. The next step of the algorithm will generate a value for θ_1, also in [0, 1], and so on. Therefore, successive

Figure 3.3 Gibbs sampler.

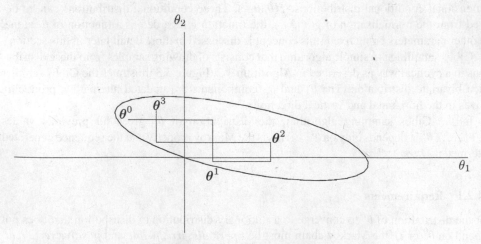

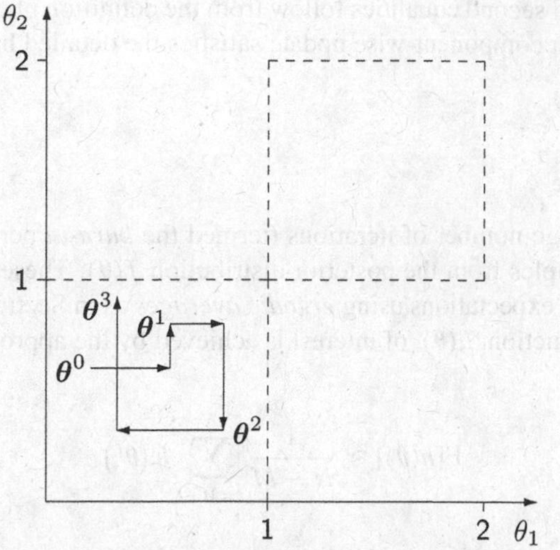

Figure 3.4 Illustration of a reducible (and therefore invalid) Gibbs sampler.

values of $\boldsymbol{\theta}^t$ will be uniformly distributed on the square ([0, 1] × [0, 1]). The square ([1, 2] × [1, 2]) will not be visited. Conversely, a starting point in ([1, 2] × [1, 2]) will yield a limiting distribution uniform on ([1, 2] × [1, 2]). Thus the limiting distribution depends on the starting value and therefore is not irreducible.

3.4.2.2 Detailed balance

By designing a *transition kernel* $\mathrm{K}(\boldsymbol{\theta}, \boldsymbol{\theta}')$ (the probability of moving from $\boldsymbol{\theta}$ to $\boldsymbol{\theta}'$) that satisfies *detailed balance* (time-reversibility)

$$f(\boldsymbol{\theta})\mathrm{K}(\boldsymbol{\theta}, \boldsymbol{\theta}') = f(\boldsymbol{\theta}')\mathrm{K}(\boldsymbol{\theta}', \boldsymbol{\theta}) \tag{3.46}$$

for all pairs of states $(\boldsymbol{\theta}, \boldsymbol{\theta}')$ in the support of f, then the stationary distribution is the target distribution of interest, $f(\boldsymbol{\theta})$.

For the Gibbs sampler the transition kernel can be written as a product of component-wise transition kernels that are the conditional probability densities

$$\mathrm{K}(\boldsymbol{\theta}, \boldsymbol{\theta}') = \prod_{i=1}^{d} f(\theta_i' | \theta_1', \ldots, \theta_{i-1}', \theta_{i+1}, \ldots, \theta_d)$$

The detailed balanced condition for the Gibbs sampler is met separately for each component-wise transition kernel. As an example, for the first component-wise update we have

$$\begin{aligned} f(\theta_1, \ldots, \theta_d) f(\theta_1' | \theta_2, \ldots, \theta_d) &= f(\theta_1 | \theta_2, \ldots, \theta_d) f(\theta_2, \ldots, \theta_d) f(\theta_1' | \theta_2, \ldots, \theta_d) \\ &= f(\theta_1', \theta_2, \ldots, \theta_d) f(\theta_1 | \theta_2, \ldots, \theta_d) \end{aligned}$$

where both the first and second equalities follow from the definition of conditional probability densities. Thus the first component-wise update satisfies the detailed balance equation, as do subsequent updates.

3.4.2.3 Convergence

After a sufficiently large number of iterations (termed the *burn-in* period), the samples $\{\boldsymbol{\theta}^t\}$ will be dependent samples from the posterior distribution $f(\boldsymbol{\theta})$. These samples may be used to obtain estimators of expectations using *ergodic averages* as in Section 3.3.2. Evaluation of the expectation of a function, $h(\boldsymbol{\theta})$, of interest is achieved by the approximation

$$\mathrm{E}[h(\boldsymbol{\theta})] \approx \frac{1}{N-M} \sum_{t=M+1}^{N} h\left(\boldsymbol{\theta}^t\right)$$

where N is the number of iterations of the Gibbs sampler and M is the number of initial samples discarded (the burn-in period).

In an implementation of Gibbs sampling, there are a number of practical considerations to be addressed. These include the length of the burn-in period, M; the length of the sequence, N; and any spacing between samples taken from the final sequence of iterations (the final sequence may be subsampled in an attempt to produce approximately independent samples and to reduce the amount of storage required).

The length of the chain should be long enough for it to 'forget' its starting value and such that all regions of the parameter space have been traversed by the chain. The limiting distribution should not depend on its starting value, $\boldsymbol{\theta}^0$, but the length of the sequence will depend on the correlation between the variables. Correlation between the θ_is will tend to slow convergence.

It can be difficult to know when the Gibbs sampler has converged as the Gibbs sampler can spend long periods in a relatively small region, thus giving the impression of convergence. The most commonly used method for determining the burn-in period is through visual inspection of plots of the output values, $\boldsymbol{\theta}^t$, and making a subjective judgement. More formal tools, *convergence diagnostics*, exist and we refer to Raftery and Lewis (1996), Gelman (1996), Cowles and Carlin (1996) and Mengersen *et al.* (1998) for further details of the most popular methods. However, convergence diagnostics do not tell when a chain has converged, but tell when it has not converged – sometimes.

There are various approaches for reducing correlation (and hence speeding up convergence), including *reparametrisation* and *grouping variables.*

Reparametrisation transforms the set $\boldsymbol{\theta}$ using a linear transformation to a new set $\boldsymbol{\phi}$ with zero correlation between the variables. The linear transformation is calculated using an estimate of the covariance matrix calculated using a short initial sequence (in pattern recognition, the process of deriving a set of uncorrelated variables that is a linear combination of the original variables is *principal components analysis*, which we describe in Chapter 10). The Gibbs sampler then proceeds using the variables $\boldsymbol{\phi}$, provided that it is straightforward to sample from the new conditionals $f(\phi_i|\boldsymbol{\phi}_{(i)})$. The process may be repeated until the correlation in the final sequence is small, hopefully leading to more rapid convergence.

Grouping variables means that at each step of the iteration a sample from a multivariate distribution $f(\boldsymbol{\theta}_i|\boldsymbol{\theta}_{(i)})$ is generated, where $\boldsymbol{\theta}_i$ is a subvector of $\boldsymbol{\theta}$ and $\boldsymbol{\theta}_{(i)}$ is the set of remaining variables. Provided correlations between variables are caused primarily by correlations between components of the subvectors, with low correlations between subvectors, we can hope for more rapid convergence. A method for sampling from the multivariate distribution, $f(\boldsymbol{\theta}_i|\boldsymbol{\theta}_{(i)})$, (which may be complex) is required.

3.4.2.4 Starting point

The starting point is any point you do not mind having in the sequence. Preliminary runs, started where the last one ended, will give you some feel for suitable starting values. There is some argument to say that since the starting point is a legitimate point from the sequence (although perhaps in the tail of the distribution), it would be visited anyway by the Markov chain, at some stage; hence there is no need for burn-in. However, using a burn-in period and removing initial samples will make estimators approximately unbiased.

3.4.2.5 Parallel runs

Instead of running one chain until convergence, it is possible to run multiple chains (with different starting values) as an approach to monitoring convergence, although more formal methods exist (Raftery and Lewis, 1996; Roberts, 1996), and for obtaining independent observations from $f(\boldsymbol{\theta})$. This is a somewhat controversial issue since independent samples are not required in many cases, and certainly not for ergodic averaging [Equation (3.38)]. Comparing several chains may help in identifying convergence. For example, are estimates of quantities of interest consistent between runs? In such cases, it is desirable to choose different starting values, $\boldsymbol{\theta}^0$, for each run, widely dispersed.

In practice, you will probably do several runs if computational resources permit, either to compare related probability models or to gain information about a chosen model such as burn-in length. Then, you would perform a long run in order to obtain samples for computing statistics.

3.4.2.6 Sampling from conditional distributions

The Gibbs sampler requires ways of sampling from the conditional distributions $f(\theta_i|\boldsymbol{\theta}_{(i)})$. If $f(\theta_i|\boldsymbol{\theta}_{(i)})$ is a standard distribution, then it is likely that algorithms exist for drawing samples from it. For algorithms for sampling from some of the more common distributions, see, for example, Devroye (1986) and Ripley (1987). If a conditional distribution is not recognised to be of a standard form for which a bespoke sampling algorithm exists, then other sampling schemes must be employed for drawing samples from the distribution. As we shall see in Section 3.4.3, a general purpose procedure known as Metropolis–Hastings sampling can be used given the probability density functions (up to a normalising constant) of the conditional probability distributions. Alternatively, use can be made of the sampling techniques described in Section 3.3.

We can obtain the probability density functions of the conditional distributions easily from the full distribution $f(\boldsymbol{\theta})$, or a function $g(\boldsymbol{\theta})$ such that $f(\boldsymbol{\theta}) = g(\boldsymbol{\theta})/\int g(\boldsymbol{\theta}')d\boldsymbol{\theta}'$. Initially, the conditional distribution $f(\theta_i|\boldsymbol{\theta}_{(i)})$ can be written

$$f\left(\theta_i|\boldsymbol{\theta}_{(i)}\right) = \frac{f(\boldsymbol{\theta})}{f(\boldsymbol{\theta}_{(i)})} \tag{3.47}$$

where $f(\boldsymbol{\theta}_{(i)})$ is the marginal distribution of $\boldsymbol{\theta}_{(i)}$ given by

$$f\left(\boldsymbol{\theta}_{(i)}\right) = \int_{\theta_i} f(\boldsymbol{\theta}) d\theta_i$$

Following the discussion in Section 3.1.3 we can write (3.47) as

$$f\left(\theta_i|\boldsymbol{\theta}_{(i)}\right) \propto f(\boldsymbol{\theta}) \propto g(\boldsymbol{\theta}) \propto g_i(\boldsymbol{\theta}_i, \boldsymbol{\theta}_{(i)})$$

where we have decomposed $g(\boldsymbol{\theta})$ into two factors as follows

$$g(\boldsymbol{\theta}) = g_i\left(\boldsymbol{\theta}_i, \boldsymbol{\theta}_{(i)}\right) g_{(i)}(\boldsymbol{\theta}_{(i)})$$

[note that such a factorisation for $g(\boldsymbol{\theta})$ always exists, since we have the default case of $g_i(\boldsymbol{\theta}_i, \boldsymbol{\theta}_{(i)}) = g(\boldsymbol{\theta})$ and $g_{(i)}(\boldsymbol{\theta}_{(i)}) = 1$]. Knowing the simplest form of a conditional probability density function is useful for determining whether it belongs to a known family (by spotting that it has a known form), and if not is useful in sampling algorithms (such as the Metropolis–Hastings method in Section 3.4.3).

As an example we show how conditional distributions can be derived in such a manner for the components of a multivariate normal distribution. Suppose that $\boldsymbol{X}$ is a d-dimensional vector from the multivariate normal distribution $N(\boldsymbol{X}|\boldsymbol{\mu}, \boldsymbol{\Sigma})$, and that we wish to determine the conditional distribution $\boldsymbol{X}_1|\boldsymbol{X}_2$ where $\boldsymbol{X}_1$ is the vector made from the first d_1 components of $\boldsymbol{X}$ and $\boldsymbol{X}_2$ is the vector made from the last $(d - d_1)$ components of $\boldsymbol{X}$.

We partition the mean vector and the covariance matrix as follows

$$\mu = \begin{pmatrix} \boldsymbol{\mu}_1 \\ \boldsymbol{\mu}_2 \end{pmatrix}, \quad \Sigma = \begin{pmatrix} \boldsymbol{\Sigma}_{1,1} & \boldsymbol{\Sigma}_{1,2} \\ \boldsymbol{\Sigma}_{2,1} & \boldsymbol{\Sigma}_{2,2} \end{pmatrix}$$

where $\boldsymbol{\mu}_1$ is a d_1-dimensional vector, $\boldsymbol{\mu}_2$ is a $(d - d_1)$-dimensional vector, $\boldsymbol{\Sigma}_{1,1}$ is a $d_1 \times d_1$-dimensional matrix, $\boldsymbol{\Sigma}_{1,2}$ is a $d_1 \times (d - d_1)$-dimensional matrix, $\boldsymbol{\Sigma}_{2,1} = \boldsymbol{\Sigma}_{1,2}^T$, and $\boldsymbol{\Sigma}_{2,2}$ is a $(d - d_1) \times (d - d_1)$-dimensional matrix.

A result for inverting a matrix divided into blocks is

$$\boldsymbol{\Sigma}^{-1} = \begin{pmatrix} \boldsymbol{\Sigma}_{1,1} & \boldsymbol{\Sigma}_{1,2} \\ \boldsymbol{\Sigma}_{2,1} & \boldsymbol{\Sigma}_{2,2} \end{pmatrix}^{-1} = \begin{pmatrix} \boldsymbol{\Sigma}_{c,1}^{-1} & -\boldsymbol{\Sigma}_{1,1}^{-1}\boldsymbol{\Sigma}_{1,2}\boldsymbol{\Sigma}_{c,2}^{-1} \\ -\boldsymbol{\Sigma}_{2,2}^{-1}\boldsymbol{\Sigma}_{2,1}\boldsymbol{\Sigma}_{c,1}^{-1} & \boldsymbol{\Sigma}_{c,2}^{-1} \end{pmatrix}$$

where

$$\boldsymbol{\Sigma}_{c,1} = \boldsymbol{\Sigma}_{1,1} - \boldsymbol{\Sigma}_{1,2}\boldsymbol{\Sigma}_{2,2}^{-1}\boldsymbol{\Sigma}_{2,1}$$
$$\boldsymbol{\Sigma}_{c,2} = \boldsymbol{\Sigma}_{2,2} - \boldsymbol{\Sigma}_{2,1}\boldsymbol{\Sigma}_{1,1}^{-1}\boldsymbol{\Sigma}_{1,2}$$

Now for a symmetric matrix Σ, we can simplify the expression further to

$$\mathbf{\Sigma}^{-1} = \begin{pmatrix} \mathbf{\Sigma}_{c,1}^{-1} & -\mathbf{\Sigma}_{c,1}^{-1}\mathbf{\Sigma}_{1,2}\mathbf{\Sigma}_{2,2}^{-1} \\ -(\mathbf{\Sigma}_{c,1}^{-1}\mathbf{\Sigma}_{1,2}\mathbf{\Sigma}_{2,2}^{-1})^T & \mathbf{\Sigma}_{c,2}^{-1} \end{pmatrix}$$

since the inverse of a symmetric matrix is also symmetric.

Using the above results we have

$$\begin{aligned} p(\boldsymbol{x}_1|\boldsymbol{x}_2) &\propto p(\boldsymbol{x}_1, \boldsymbol{x}_2) \\ &\propto \exp\left[-0.5(\boldsymbol{x}_1 - \boldsymbol{\mu}_1)^T \mathbf{\Sigma}_{c,1}^{-1}(\boldsymbol{x}_1 - \boldsymbol{\mu}_1) - 0.5(\boldsymbol{x}_2 - \boldsymbol{\mu}_2)^T \mathbf{\Sigma}_{c,2}^{-1}(\boldsymbol{x}_2 - \boldsymbol{\mu}_2)\right. \\ &\qquad \left. - (\boldsymbol{x}_1 - \boldsymbol{\mu}_1)^T \mathbf{\Sigma}_{c,1}^{-1}\mathbf{\Sigma}_{1,2}\mathbf{\Sigma}_{2,2}^{-1}(\boldsymbol{x}_2 - \boldsymbol{\mu}_2)\right] \end{aligned}$$

Ignoring terms that can be dealt with in the normalisation constant (i.e. ignoring multiplicative factors that do not vary with $\boldsymbol{x}_1$) we have

$$\begin{aligned} p(\boldsymbol{x}_1|\boldsymbol{x}_2) &\propto \exp\left[-0.5\boldsymbol{x}_1^T \mathbf{\Sigma}_{c,1}^{-1}\boldsymbol{x}_1 - \boldsymbol{x}_1^T \mathbf{\Sigma}_{c,1}^{-1}\left(\boldsymbol{\mu}_1 + \mathbf{\Sigma}_{1,2}\mathbf{\Sigma}_{2,2}^{-1}(\boldsymbol{x}_2 - \boldsymbol{\mu}_2)\right)\right] \\ &\propto \exp\left[-0.5(\boldsymbol{x}_1 - \boldsymbol{\mu}_{c,1})^T \mathbf{\Sigma}_{c,1}^{-1}(\boldsymbol{x}_1 - \boldsymbol{\mu}_{c,1})\right] \\ &\propto N(\boldsymbol{x}_1; \boldsymbol{\mu}_{c,1}, \mathbf{\Sigma}_{c,1}) \end{aligned}$$

where

$$\boldsymbol{\mu}_{c,1} = \boldsymbol{\mu}_1 + \mathbf{\Sigma}_{1,2}\mathbf{\Sigma}_{2,2}^{-1}(\boldsymbol{x}_2 - \boldsymbol{\mu}_2)$$

Hence the conditional distribution $\boldsymbol{X}_1|(\boldsymbol{X}_2 = \boldsymbol{x}_2)$ is also normal with mean vector $\boldsymbol{\mu}_{c,1}$ and covariance matrix $\mathbf{\Sigma}_{c,1}$.

3.4.2.7 Gibbs sampler example

To illustrate the Gibbs sampler we consider the univariate equivalent of the estimation problem considered in Section 3.2.3 (i.e. estimation of the mean and variance of a normal distribution). The data are normally distributed with mean μ and variance σ^2. The prior distribution for μ is $N(\mu_0, \sigma^2/\lambda_0)$ and the prior distribution for $\tau = 1/\sigma^2$ is the gamma distribution with shape parameter α_0 and inverse-scale parameter β_0 (σ^2 therefore has an inverse-gamma distribution with shape parameter α_0 and scale parameter β_0).

A Gibbs sampling approach is used for generating samples from the joint posterior density of μ and σ^2 given independent measurements $x_1, \ldots, x_n$ from $N(\mu, \sigma^2)$. The steps are:

1. Initialise σ^2. In this example, a sample is taken from the prior distribution. Note, however, that this would not necessarily be the best approach for a diffuse prior, since one might end up deep in the tails, requiring a long burn-in period.
2. Sample μ given the sample for σ^2.
3. Then for a given μ, sample σ^2, and return to step 2 to generate the next pair of samples.

The Gibbs sampler requires the conditional distributions. For ease of notation we work with the precision $\tau = 1/\sigma^2$, rather than the variance σ^2. We have the following joint probability density function

$$\begin{aligned} p(\mu, \tau|x_1, \ldots, x_n) &\propto p(\mu|\tau)p(\tau)p(x_1, \ldots, x_n|\mu, \tau) \\ &\propto \tau^{1/2} \exp\left(-\frac{\tau\lambda_0(\mu-\mu_0)^2}{2}\right)\tau^{\alpha_0-1}\exp[-\beta_0\tau] \\ &\quad \times \tau^{n/2}\exp\left(-\frac{\tau\sum_{i=1}^n (x_i-\mu)^2}{2}\right) \end{aligned} \tag{3.48}$$

The probability density function of $\tau = 1/\sigma^2$ given the mean μ is

$$\begin{aligned} p(\tau|\mu, x_1, \ldots, x_n) &\propto p(\mu, \tau|x_1, \ldots, x_n) \\ &\propto \tau^{\alpha_0+n/2+1/2-1}\exp\left[-\tau\left(\beta_0 + \frac{\sum_{i=1}^n (x_i-\mu)^2 + \lambda_0(\mu-\mu_0)^2}{2}\right)\right] \\ &\propto \tau^{\alpha_n+1/2-1}\exp[-\tau(\beta_n + \lambda_n(\mu-\mu_n)^2/2)] \\ &\propto \text{Ga}(\tau; \alpha_n + 1/2, \beta_n + \lambda_n(\mu-\mu_n)^2/2) \end{aligned}$$

where

$$\begin{aligned} \lambda_n &= \lambda_0 + n \\ \mu_n &= (\lambda_0\mu_0 + n\bar{x})/\lambda_n \\ \alpha_n &= \alpha_0 + n/2 \\ \beta_n &= \beta_0 + \frac{1}{2}\sum_{i=1}^n (x_i-\bar{x})^2 + \frac{n\lambda_0}{2\lambda_n}(\mu_0-\bar{x})^2 \end{aligned}$$

with $\bar{x}$ the mean of $\{x_1, \ldots, x_n\}$. Hence, given the mean, $1/\sigma^2$ has a gamma distribution with shape parameter $\alpha_n + 1/2$ and inverse-scale parameter $\beta_n + \lambda_n(\mu-\mu_n)^2/2$.

In a similar manner we find that given $\tau = 1/\sigma^2$, μ is normally distributed with mean μ_n (the prior updated by the data samples) and variance σ^2/λ_n.

To assess performance of the Gibbs sampler, we note that the required posterior distributions can be calculated analytically. From the univariate equivalent of (3.19) we know that the posterior distribution is given by

$$p(\mu, \tau|x_1, \ldots, x_n) = \text{N}_1(\mu; \mu_n, (\tau\lambda_n)^{-1})\text{Ga}(\tau; \alpha_n, \beta_n)$$

The true marginal posteriors of μ and σ^2 can therefore be calculated analytically; μ has a (Student) t-distribution with mean μ_n, degrees of freedom $2\alpha_n$ and scaling parameter parameter $\lambda_n\alpha_n/\beta_n$, such that

$$p(\mu|x_1, \ldots, x_n) \propto \left(1 + \frac{\lambda_n\alpha_n}{\beta_n}\frac{1}{2\alpha_n}(\mu-\mu_n)^2\right)^{-(2\alpha_n+1)/2}$$

Table 3.1 Summary statistics for μ and σ^2. The true values are calculated from the known marginal posterior distributions. The short run values are calculated from a Gibbs sampler run of 1000 samples less a burn-in period of 500 samples. The long run values are calculated after a run of 10000 samples, less a burn-in period of 500 samples.

	True	Short run	Long run
Mean μ	0.063	0.082	0.063
Var. μ	0.039	0.042	0.039
Mean σ^2	0.83	0.83	0.82
Var. σ^2	0.081	0.085	0.079

and the inverse of σ^2 has a gamma distribution with shape parameter α_n and inverse-scale parameter β_n.

In our example, the data $x_1, \ldots, x_n$, comprise 20 points from a normal distribution of zero mean and unit variance. The parameters of the prior distribution of μ and $\tau = 1/\sigma^2$ are taken to be $\lambda_0 = 1$, $\mu_0 = 1$, $\alpha_0 = 1/2$, $\beta_0 = 1/2$. As shown in Table 3.1, taking the first 500 samples as burn-in, and using the remaining 500 samples to calculate summary statistics, gives values for the mean and variance of μ and σ^2 that are reasonably close to the true values (calculated analytically), although there is some error in the mean of μ. A longer sequence gives values very close to the truth. Figure 3.5 shows the first 1000 samples in a sequence of μ and σ^2 samples.

Despite the good performance of the Gibbs sampler, one should note that since the analytic solution is available, one would never choose to use a Gibbs sampler for this problem.

3.4.3 Metropolis–Hastings algorithm

The Metropolis–Hastings algorithm is a widely used technique for sampling from distributions which are not amenable to direct sampling. This includes distributions for which the probability density function is only known up to proportionality (e.g. the normalising constant of a posterior distribution may be unknown). It uses a *proposal distribution*, from which

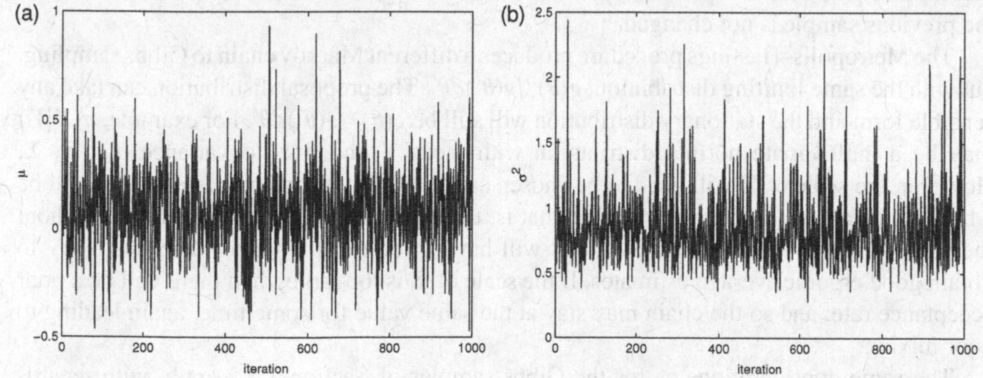

Figure 3.5 One thousand iterations from Gibbs sampler for a normal-inverse gamma posterior density. (a) Samples of μ; (b) samples of σ^2.

Algorithm 3.3 Metropolis–Hastings algorithm.

- Choose an arbitrary starting value for $\boldsymbol{\theta}$, $\boldsymbol{\theta}^0$ from the support of the prior/posterior distribution.
- At stage $t+1$ of the iterative algorithm, do the following:
  1. Draw a sample $\boldsymbol{\theta}$ from the proposal distribution $q(\boldsymbol{\theta}|\boldsymbol{\theta}^t)$.
  2. Draw a sample u from the uniform distribution on [0, 1].
  3. Evaluate:

$$a(\boldsymbol{\theta}^t, \boldsymbol{\theta}) = \min\left(1, \frac{g(\boldsymbol{\theta})q(\boldsymbol{\theta}^t|\boldsymbol{\theta})}{g(\boldsymbol{\theta}^t)\,q(\boldsymbol{\theta}|\boldsymbol{\theta}^t)}\right) \tag{3.49}$$

  4. If $u \leq a(\boldsymbol{\theta}^t, \boldsymbol{\theta})$ then accept $\boldsymbol{\theta}$ and set $\boldsymbol{\theta}^{t+1} = \boldsymbol{\theta}$, else reject $\boldsymbol{\theta}$ and set $\boldsymbol{\theta}^{t+1} = \boldsymbol{\theta}^t$.
- After a large number of iterations, the resulting samples $\boldsymbol{\theta}^t$ behave like a random draw from the density $f(\boldsymbol{\theta})$.

---

sampling is easy, and accepts a sample with a probability that depends on the proposal density and the (unnormalised) density from which we wish to sample.

Let $f(\boldsymbol{\theta})$ denote the posterior density function of the posterior distribution from which we wish to draw samples. As for previously discussed sampling techniques, we may not know $f(\boldsymbol{\theta})$ exactly, but we know a function $g(\boldsymbol{\theta})$, where $f(\boldsymbol{\theta}) = g(\boldsymbol{\theta})/\int g(\boldsymbol{\theta}')d\boldsymbol{\theta}'$. As an example, $g(\boldsymbol{\theta})$ could be the product of a prior density function and a likelihood function, so that $f(\boldsymbol{\theta})$ is a posterior density function.

Let $\boldsymbol{\theta}^t$ be the current sample. In the Metropolis–Hastings algorithm, a proposal distribution, that may depend on $\boldsymbol{\theta}^t$, is specified. We denote this $q(\boldsymbol{\theta}|\boldsymbol{\theta}^t)$. The Metropolis–Hastings algorithm is given in Algorithm 3.3. At each iteration, the sample drawn from the proposal distribution $q(\boldsymbol{\theta}|\boldsymbol{\theta}^t)$ is accepted with probability $a(\boldsymbol{\theta}^t, \boldsymbol{\theta})$ [given in Equation (3.49)], otherwise the previous sample is not changed.

The Metropolis–Hastings procedure produces a different Markov chain to Gibbs sampling, but with the same limiting distribution, $g(\boldsymbol{\theta})/\int g(\boldsymbol{\theta}')d\boldsymbol{\theta}'$. The proposal distribution can take any sensible form and the stationary distribution will still be $g(\boldsymbol{\theta})/\int g(\boldsymbol{\theta}')d\boldsymbol{\theta}'$. For example, $q(\boldsymbol{X}|\boldsymbol{Y})$ may be a multivariate normal distribution with mean $\boldsymbol{Y}$ and fixed covariance matrix, $\boldsymbol{\Sigma}$. However, the scale of $\boldsymbol{\Sigma}$ will need to be chosen carefully. If it is too small, then there will be a high acceptance rate, but poor *mixing*; that is, the chain may not move rapidly throughout the support of the target distribution and will have to be run for longer than necessary to obtain good ergodic average estimates. If the scale of $\boldsymbol{\Sigma}$ is too large, then there will be a poor acceptance rate, and so the chain may stay at the same value for some time, again leading to poor mixing.

The same considerations as for the Gibbs sampler of Section 3.4.2 apply with regards to a burn-in period, starting points, parallel runs and convergence diagnostics, and are not repeated here.

For symmetric proposal distributions, $q(\boldsymbol{X}|\boldsymbol{Y}) = q(\boldsymbol{Y}|\boldsymbol{X})$, and the acceptance probability reduces to

$$\min\left(1, \frac{g(\boldsymbol{\theta})}{g\left(\boldsymbol{\theta}^t\right)}\right)$$

For the symmetric proposal, $q(\boldsymbol{X}|\boldsymbol{Y}) = q(|\boldsymbol{X} - \boldsymbol{Y}|)$, so that q is a function of the difference between $\boldsymbol{X}$ and $\boldsymbol{Y}$ only. The algorithm is the *random-walk* Metropolis algorithm.

It is permissible to have the proposal distribution independent of the current location, $q(\boldsymbol{X}|\boldsymbol{Y}) = q(\boldsymbol{X})$. This is referred to as an *independence chain* (Tierney, 1994). It is important that such a proposal distribution has the same support as the desired target distribution (i.e. it has non-zero density at the same locations), and that the tails of the proposal density dominate those of the target density.

3.4.3.1 Detailed balance

The Markov-chain created by the Metropolis–Hastings sampler satisfies the detailed balance condition [Equation (3.46)] required for the stationary distribution to be the desired target distribution.

The Metropolis–Hastings kernel for continuous variables is given by

$$K(\boldsymbol{\theta}, \boldsymbol{\theta}') = \begin{cases} q(\boldsymbol{\theta}'|\boldsymbol{\theta})a(\boldsymbol{\theta}, \boldsymbol{\theta}') & \text{if} \quad \boldsymbol{\theta}' \neq \boldsymbol{\theta} \\ 1 - \int_{\boldsymbol{\theta}''} q(\boldsymbol{\theta}''|\boldsymbol{\theta})a(\boldsymbol{\theta}, \boldsymbol{\theta}'')d\boldsymbol{\theta}'' & \text{if} \quad \boldsymbol{\theta}' = \boldsymbol{\theta} \end{cases}$$

where the $\boldsymbol{\theta}' = \boldsymbol{\theta}$ expression reflects the fact that the Metropolis–Hastings sampler can reject proposed moves. Clearly, detailed balance is met for the rejected portion of the move, so it remains to show it for $\boldsymbol{\theta}' \neq \boldsymbol{\theta}$. We have

$$\begin{aligned} g(\boldsymbol{\theta})q(\boldsymbol{\theta}'|\boldsymbol{\theta})a(\boldsymbol{\theta}, \boldsymbol{\theta}') &= g(\boldsymbol{\theta})q(\boldsymbol{\theta}'|\boldsymbol{\theta})\min\left(1, \frac{g(\boldsymbol{\theta}')q(\boldsymbol{\theta}|\boldsymbol{\theta}')}{g(\boldsymbol{\theta})q(\boldsymbol{\theta}'|\boldsymbol{\theta})}\right) \\ &= \min(g(\boldsymbol{\theta})q(\boldsymbol{\theta}'|\boldsymbol{\theta}), g(\boldsymbol{\theta}')q(\boldsymbol{\theta}|\boldsymbol{\theta}')) \\ &= g(\boldsymbol{\theta}')q(\boldsymbol{\theta}|\boldsymbol{\theta}')\min\left(\frac{g(\boldsymbol{\theta})q(\boldsymbol{\theta}'|\boldsymbol{\theta})}{g(\boldsymbol{\theta}')q(\boldsymbol{\theta}|\boldsymbol{\theta}')}, 1\right) \\ &= g(\boldsymbol{\theta}')q(\boldsymbol{\theta}|\boldsymbol{\theta}')a(\boldsymbol{\theta}', \boldsymbol{\theta}) \end{aligned}$$

thus showing that detailed balance is met and that the stationary distribution is therefore the desired target distribution.

3.4.3.2 Single-component Metropolis–Hastings

The Metropolis–Hastings algorithm given above updates all components of the parameter vector, $\boldsymbol{\theta}$, in one step. An alternative approach is to update a single component at a time.

Algorithm 3.4 Single-component Metropolis–Hastings algorithm (Metropolis-within-Gibbs).

At stage t of the iterative algorithm, do the following:

- Draw a sample, Y from the proposal distribution $q_1(\theta_1|\theta_1^t, \ldots, \theta_d^t)$.
- Accept the sample with probability:

$$\alpha = \min\left(1, \frac{g\left(Y|\theta_2^t, \ldots, \theta_d^t\right) q_1\left(\theta_1^t|Y, \theta_2^t, \ldots, \theta_d^t\right)}{g\left(\theta_1^t|\theta_2^t, \ldots, \theta_d^t\right) q_1\left(Y|\theta_1^t, \theta_2^t, \ldots, \theta_d^t\right)}\right)$$

  – If Y is accepted, then set $\theta_1^{t+1} = Y$, else set $\theta_1^{t+1} = \theta_1^t$.
- Continue through the variables as in the Gibbs sampler (Algorithm 3.2), finally drawing a sample, Y from the proposal distribution $q_d(\theta_d|\theta_1^{t+1}, \ldots, \theta_{d-1}^{t+1}, \theta_d^t)$:
  – Accept the sample with probability:

$$\alpha = \min\left(1, \frac{g\left(Y|\theta_1^{t+1}, \ldots, \theta_{d-1}^{t+1}\right) q\left(\theta_d^t|\theta_1^{t+1}, \ldots, \theta_{d-1}^{t+1}, Y\right)}{g\left(\theta_d^t|\theta_1^{t+1}, \ldots, \theta_{d-1}^{t+1}\right) q\left(Y|\theta_1^{t+1}, \ldots, \theta_{d-1}^{t+1}, \theta_d^t\right)}\right)$$

  – If Y is accepted, then set $\theta_d^{t+1} = Y$, else set $\theta_d^{t+1} = \theta_d{}^t$.

This requires specification of functions $g(\theta_i|\boldsymbol{\theta}_{(i)})$ (where $\boldsymbol{\theta}_{(i)}$ is the set of parameters with the ith parameter removed), proportional to the posterior conditional probability density function $f(\theta_i|\boldsymbol{\theta}_{(i)})$ for θ_i. The iterative procedure is presented in Algorithm 3.4.

In this single-component update case, for proposal distributions that are the conditionals of the multivariate distribution that we wish to sample, i.e. $q(\theta_i|\boldsymbol{\theta}_{(i)}) = f(\theta_i|\boldsymbol{\theta}_{(i)})$, then the sample is always accepted. The algorithm is then identical to Gibbs sampling. The single-component update Metropolis–Hastings classifier is therefore often referred to as Metropolis-within-Gibbs.

3.4.3.3 Mixture and cycle kernels

The single-component Metropolis–Hastings algorithm above is a special case of the cycle strategy, in which a set of proposal distributions $q_1(\boldsymbol{\theta}'|\boldsymbol{\theta}), \ldots, q_k(\boldsymbol{\theta}'|\boldsymbol{\theta})$ is specified, and we cycle through each proposal in turn from one iteration to the next. Another powerful approach is to use mixtures of kernels. Here, selection probabilities $\pi_1, \ldots, \pi_k$ (satisfying $\pi_j \geq 0, j = 1, \ldots, k$ and $\sum_{j=1}^k \pi_j = 1$) are assigned alongside the proposal distributions $q_1(\boldsymbol{\theta}'|\boldsymbol{\theta}), \ldots, q_k(\boldsymbol{\theta}'|\boldsymbol{\theta})$, and at each iteration we select one of the proposal distributions according to its selection probability. As noted in Andrieu *et al.* (2003) the power of mixtures of kernels is that one global proposal can explore large regions of the state space, whilst another local proposal (such as a random walk) can explore local areas. This enables the Metropolis–Hastings algorithm to explore multimodal densities.

3.4.3.4 Choice of proposal distribution in Metropolis–Hastings

If the distribution that we wish to approximate, f, is unimodal, and is not heavy-tailed (loosely, heavy-tailed means that it tends to zero slower than the exponential distribution, but also the term is used to describe distributions with infinite variance), then an appropriate choice for the proposal distribution might be normal, with parameters chosen to be a best fit of $\log(q)$ to $\log(g)$ ($f = g/\int g$). For more complex distributions, the proposal could be a multivariate normal, or mixtures of multivariate normal, but for distributions with heavy tails, t-distributions might be used. For computational efficiency, q should be chosen so that it can be easily sampled and evaluated. Often a random walk algorithm is used (symmetric proposal distribution), and can give surprisingly good results. Chib and Greenberg (1995) discuss choice of the scale of a random walk proposal density (e.g. the standard deviation of a normal proposal centred on the current point). They note that the scale should be such that the acceptance rate of moves is approximately 0.45 in univariate cases and approximately 0.23 as the number of dimensions approaches infinity.

3.4.4 Data augmentation

Introducing auxiliary variables can often lead to more simple and efficient MCMC sampling methods, with improved mixing. If we require samples from a posterior $p(\boldsymbol{\theta}|\mathcal{D})$, then the basic idea is to notice that it may be easier to sample from $p(\boldsymbol{\theta}, \boldsymbol{\phi}|\mathcal{D})$, where $\boldsymbol{\phi}$ is a set of auxiliary variables. In some applications, the choice of $\boldsymbol{\phi}$ may be obvious, in others some experience is necessary to recognise suitable choices. The distribution $p(\boldsymbol{\theta}|\mathcal{D})$ is then simply a marginal of the augmented distribution, $p(\boldsymbol{\theta}, \boldsymbol{\phi}|\mathcal{D})$, and the method of sampling is termed the *data augmentation method*. Statistics concerning the distribution of $p(\boldsymbol{\theta}|\mathcal{D})$ can be obtained by using the $\boldsymbol{\theta}$ components of the samples of the augmented parameter vector $(\boldsymbol{\theta}, \boldsymbol{\phi})$ and ignoring the $\boldsymbol{\phi}$ components.

One type of problem where data augmentation is used is that involving missing data. Suppose that we have a dataset $\mathcal{D}$ and some 'missing values', $\boldsymbol{\phi}$. In a classification problem, where there are some unlabelled data available for training the classifier, $\boldsymbol{\phi}$ represents the class labels of these data (see Section 3.5). Alternatively, there may be incomplete pattern vectors; that is, for some patterns, measurements on some of the variables may be absent.

The posterior distribution of parameters $\boldsymbol{\theta}$ is given by

$$p(\boldsymbol{\theta}|\mathcal{D}) \propto \int p(\mathcal{D}, \boldsymbol{\phi}|\boldsymbol{\theta})d\boldsymbol{\phi}p(\boldsymbol{\theta})$$

However, it may be difficult to marginalise the joint density $p(\mathcal{D}, \boldsymbol{\phi}|\boldsymbol{\theta})$ and it is often simpler to obtain samples of the augmented vector $(\boldsymbol{\theta}, \boldsymbol{\phi})$. In this case:

$p(\boldsymbol{\theta}|\mathcal{D}, \boldsymbol{\phi})$ is the posterior based on the complete data, which is easy to sample, either directly or by use of MCMC methods (for example Metropolis–Hastings).

$p(\boldsymbol{\phi}|\boldsymbol{\theta}, \mathcal{D})$ is the sampling distribution for the missing data; again, typically easy to sample.

The Metropolis-within-Gibbs procedure is to draw samples from these two distributions alternately.

3.4.5 Reversible jump Markov chain Monte Carlo

MCMC methods can be extended to cases where the model order is unknown through the use of reversible jump Markov chain Monte Carlo (RJMCMC) procedures, first developed by Green (1995). As well as updating the parameters at a constant model dimension, as in standard MCMC techniques, RJMCMC allows changes in the model dimension, by proposing new parameters, or removing existing parameters.

We briefly outline the mechanics behind *birth* (adding a new parameter) and *death* (removing an existing parameter) dimension-changing moves. Birth and death moves are defined as a pair, and are proposed and then accepted or rejected according to a mechanism which ensures dimension-matching and detailed balance.

Suppose that we propose a move from model M_k, with state space of dimension k, to model $M_{k'}$, with state space of dimension k', where $k' = k + d$, for some integer $d \geq 1$. This dimension-changing proposal is made by drawing a d-dimensional vector of continuous random variables, $\boldsymbol{u}_d$, independently of the current state $\boldsymbol{\theta}_k$, and then defining the new state $\boldsymbol{\theta}_{k'}$ using a bijection, $\boldsymbol{\theta}_{k'} = \theta'(\boldsymbol{\theta}_k, \boldsymbol{u}_d)$. Since $\dim(\boldsymbol{\theta}_k) + \dim(\boldsymbol{u}_d) = \dim(\boldsymbol{\theta}_{k'})$, such a proposal satisfies the dimension-matching condition. Detailed balance is satisfied by specifying the acceptance probability for the move as

$$a(\boldsymbol{\theta}_k, \boldsymbol{\theta}_{k'}) = \min\left(1, \frac{p(\boldsymbol{\theta}_{k'}|\mathcal{D})q_{k',k}}{p(\boldsymbol{\theta}_k|\mathcal{D})q_{k,k'}p(\boldsymbol{u}_d)}\left|\frac{\partial \boldsymbol{\theta}_{k'}}{\partial(\boldsymbol{\theta}_k, \boldsymbol{u}_d)}\right|\right)$$

where $p(\boldsymbol{\theta}|\mathcal{D})$ denotes the posterior distribution of the parameters $\boldsymbol{\theta}$ given our data $\mathcal{D}$; $q_{k,k'}$ and $q_{k',k}$ are the probabilities of choosing a move from model M_k to $M_{k'}$, and from $M_{k'}$ to M_k, respectively; $p(\boldsymbol{u}_d)$ denotes the probability density function of the random variables $\boldsymbol{u}_d$; and $\frac{\partial \boldsymbol{\theta}_{k'}}{\partial(\boldsymbol{\theta}_k, \boldsymbol{u}_d)}$ is a Jacobian, arising from the change of variable from $(\boldsymbol{\theta}_k, \boldsymbol{u}_d)$ to $\boldsymbol{\theta}_{k'}$. In the event of any discrete random variables being generated as part of the birth move their probability mass is incorporated into the move selection probability, $q_{k,k'}$. Similarly, the probability mass of discrete random variables involved in the corresponding death move are incorporated into the death move selection probability $q_{k',k}$. An example in the death move case might be selecting a parameter at random to remove.

The acceptance probability can be written in a more interpretable manner as

$$a(\boldsymbol{\theta}_k, \boldsymbol{\theta}_{k'}) = \min(1, r(\boldsymbol{\theta}_k, \boldsymbol{\theta}_{k'}))$$

where

$$r(\boldsymbol{\theta}_k, \boldsymbol{\theta}_{k'}) = (\text{posterior ratio}) \times (\text{proposal ratio}) \times (\text{Jacobian})$$

The paired death move reducing the dimension of the state space, defined via the inverse of the bijection θ', has acceptance probability $a(\boldsymbol{\theta}_{k'}, \boldsymbol{\theta}_k) = \min(1, r(\boldsymbol{\theta}_k, \boldsymbol{\theta}_{k'})^{-1})$.

Further details of the RJMCMC methodology are available in the paper by Green (1995), and a number of papers developing RJMCMC samplers for specific problems. An RJMCMC methodology for univariate normal mixture models that jointly models the number of components and the mixture component parameters is presented by Richardson and Green

(1997).[2] Work of Andrieu and Doucet (1999) develops an RJMCMC algorithm for joint Bayesian model selection and parameter estimation for sinusoids in white Gaussian noise. The fixed-dimension aspects of their approach are considered in Section 3.4.7.

3.4.6 Slice sampling

The premise behind slice sampling (Neal, 2003) is that samples from a distribution can be obtained by sampling uniformly from the region under a plot of its probability density function (or more generally a plot of a function proportional to its probability density function). Slice sampling is a special case of the Gibbs sampler using data augmentation (Section 3.4.4).

Suppose that we wish to draw samples from a univariate random variable, $\theta \in \mathbb{R}$, whose probability density function, $f(\theta)$, is known up to a proportionality, $f(\theta) = g(\theta)/(\int g(\theta')d\theta')$. The slice sampler introduces an auxiliary variable, $u \in \mathbb{R}$, and defines the joint distribution of (θ, u) to be uniform within the region under the curve $u = g(\theta)$. The joint probability density function of (θ, u) is therefore

$$f(\theta, u) = \frac{1}{\int g(\theta')d\theta'} I(0 \leq u \leq g(\theta))$$

The marginal density of θ under the above joint distribution is

$$f(\theta) = \int_u f(\theta, u)du = \int_{u=0}^{g(\theta)} \frac{1}{\int g(\theta')d\theta'} du = \frac{g(\theta)}{\int g(\theta')d\theta'}$$

Hence the marginal distribution for θ is the desired distribution. Therefore, if we have samples $\{(\theta^{(s)}, u^{(s)}), s = 1, \ldots, N\}$ from $f(\theta, u)$ (i.e. drawn uniformly within the area under the curve $u = g(\theta)$), we can obtain samples $\{\theta^{(s)}, s = 1, \ldots, N\}$ from the distribution defined by $f(\theta)$ by discarding the samples for u.

A Gibbs sampling approach is adopted for obtaining samples from $f(\theta, u)$

- Initialise the algorithm with a sample θ^0 from the support of $f(\theta)$.
- At step t of the iterative algorithm:
  - Draw a sample, u^t from the distribution $f(u|\theta^{t-1})$, which by construction is the uniform distribution on $[0, g(\theta^{t-1})]$. This is the *vertical slice* of the algorithm.
  - Draw a sample, θ^t from the distribution $f(\theta|u^t)$, which by construction is the uniform distribution over the set S, where $S = \{\theta; u^t \leq g(\theta)\}$. This is the *horizontal slice* of the algorithm.

As with the standard Gibbs sampler, after an initial burn-in period, the samples will be (dependent) samples from the posterior distribution $f(\theta, u)$. Approximately independent samples can be obtained by subsampling the full set of samples.

[2] The associated correction is in Richardson and Green (1998).

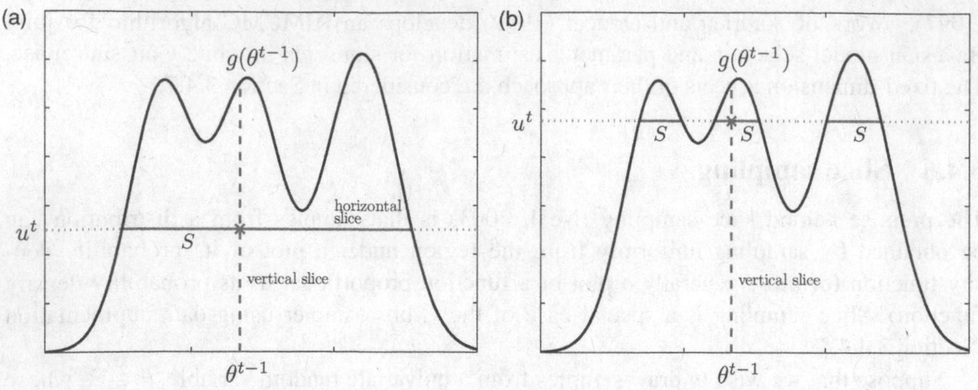

Figure 3.6 Single-variable slice sampling. In (a) the horizontal slice consists of a single interval S, whilst in (b) the horizontal slice consists of three disjoint intervals.

The core stage of the algorithm is illustrated in Figure 3.6. Whilst the sampling associated with the vertical slice is trivial, the horizontal slice can prove difficult, since the set S might be hard to determine. Specifically, as shown in Figure 3.6(b), the set S might consist of a number of disjoint intervals. Even if the set S is a single interval, determining the end-points might require numerical root-finding procedures (e.g. one of those in Press *et al.*, 1992).

Neal (2003) advocates replacing the sampling associated with the horizontal slice with a scheme that does not sample uniformly from the set S, but that does leave the desired distribution invariant. Suppose that we have drawn u^t in the uniform sampling associated with the vertical slice from $[0, g(\theta^{t-1})]$. Then, by construction, the point θ^{t-1} lies within the region $S = \{\theta; u^t \leq g(\theta)\}$. An initial interval of width w is specified to contain the point θ^{t-1}. The positioning of this interval is random, i.e. the lower limit, L_1, of the initial interval is drawn from a uniform distribution on $[\theta^{t-1} - w, \theta^{t-1}]$. The upper limit of the interval is $R_1 = L_1 + w$. This initial interval is then expanded via a *stepping out* procedure [an alternative *doubling* procedure is also presented by Neal (2003)].

A maximum expanded interval width of mw is pre-specified, where $m \geq 1$ is an integer (and can be infinite if desired). At each horizontal slice, a sample, J_L, is generated uniformly from the set $\{0, \ldots, m-1\}$, and is such that the subsequent stepping out interval growth procedure does not allow the lower interval limit to be below $L_1 - J_L w$, or the upper interval limit to be above $L_1 + (m - J_L)w$. This constraint, along with the random positioning of the initial interval, is required for the distribution to be left invariant under the new update scheme.

The *stepping out* procedure to create an interval $I = [L, R]$ is:

- Set $L = L_1 - i_L w$, where i_L is the smallest integer in the range $0 \leq i \leq J_L$ for which $u^t > g(L_1 - iw)$, i.e. is the first integer for which the lower limit of the expanding interval falls outside the set S, or is J_L if such a value does not exist.

- Set $R = R_1 + i_R w$, where i_R is the smallest integer in the range $0 \leq i \leq m - J_L - 1$ for which $u^t > g(R_1 + iw)$, i.e. is the first integer for which the upper limit of the expanding interval falls outside the set S, or is $m - J_L - 1$ if such a value does not exist.

Having specified a new interval $I = [L, R]$ at a given horizontal slice, the horizontal slice is completed by sampling from $S \cap I$. One of the following procedures can be used:

1. Sample uniformly from the interval $I = [L, R]$ until a sample θ is obtained within the set S, i.e. until $u^t \leq g(\theta)$. Then set $\theta^t = \theta$.
2. Sample θ uniformly from the interval $I = [L, R]$. If θ lies within the set S, then set $\theta^t = \theta$. If θ does not lie within the set S, shrink the interval I to $I = [\theta, R]$ if $\theta < \theta^{t-1}$ and $I = [L, \theta]$ if $\theta > \theta^{t-1}$. Repeat the procedure until a sample is accepted (which will occur eventually since $\theta^{t-1} \in S$).

The second procedure will be more efficient than the first if $S \cap I$ is only a small portion of the initial interval I.

The proof that the above scheme results in valid samples from the desired distribution is given by Neal (2003) and involves showing that the updates leave the desired distribution invariant. This is done by showing detailed balance. It is noted that caution is need if modifying the above procedure, because many 'seemingly reasonable' modifications result in the proof being invalidated.

Slice sampling can be used to obtain samples from multivariate distributions by applying the univariate procedure to each variable in turn, in the same manner as the Gibbs sampler algorithm (i.e. replacing each conditional sampling stage of the Gibbs sampler with a sample generated using an iteration of the univariate slice sampler, with target distribution the univariate conditional distribution). Neal (2003) also discusses more sophisticated versions of the multivariate slice sampler.

3.4.7 MCMC example – estimation of noisy sinusoids

In this example, we seek to model a time series as a sum of k sinusoids of unknown amplitude, frequency and phase. The approach and example is extracted from that of Andrieu and Doucet (1999). They treat the number of sinusoids as an unknown variable to be estimated, but for simplicity in this discussion we consider a fixed number. Denoting the amplitudes by $\boldsymbol{\psi} = (\psi_1, \ldots, \psi_k)$, the frequencies by $\boldsymbol{\omega} = (\omega_1, \ldots, \omega_k)$ and the phases by $\boldsymbol{\phi} = (\phi_1, \ldots, \phi_k)$, we model the data as

$$y = h(x; \boldsymbol{\xi}) + \epsilon = \sum_{j=1}^{k} \psi_j \cos(\omega_j x + \phi_j) + \epsilon \tag{3.50}$$

where $\epsilon \sim N(0, \sigma^2)$ and $\boldsymbol{\xi} = \{(\psi_j, \omega_j, \phi_j), j = 1, \ldots, k\}$. Thus

$$p(y|x; \boldsymbol{\theta}) = \frac{1}{\sqrt{2\pi\sigma^2}} \exp\left\{\frac{-(y - h(x; \boldsymbol{\xi}))^2}{2\sigma^2}\right\}$$

where the parameters of the density are $\boldsymbol{\theta} = (\boldsymbol{\xi}, \sigma^2)$. The training data, $\mathcal{D} = \{y_i, i = 1, \ldots, n\}$, comprise n measurements of y at regular intervals, $x_i = i$; $i = 0, 1, \ldots, n-1$. Assuming independent noise, we have

$$p(\mathcal{D}|\boldsymbol{\theta}) \propto \prod_{i=1}^{n} \frac{1}{\sigma} \exp\left\{\frac{-(y_i - h(x_i; \boldsymbol{\xi}))^2}{2\sigma^2}\right\}$$

The task is to ascertain information about the sinusoids given the training data and to provide predictions on y given a new sample x. A Bayesian approach is followed, with a prior distribution specified for the parameters $\boldsymbol{\theta}$, represented by the probability density function $p(\boldsymbol{\theta})$. Then using Bayes' theorem, we have the following posterior distribution

$$p(\boldsymbol{\theta}|\mathcal{D}) \propto p(\mathcal{D}|\boldsymbol{\theta})P(\boldsymbol{\theta})$$

The approach is to use an MCMC algorithm to draw samples from the posterior distribution. The resulting samples $\boldsymbol{\theta}^t, t = M, \ldots, N$, where M is the burn-in period and N is the length of the MCMC sequence, then describe the parameters of interest. For predicting a new sample, we take

$$p(y|x_n) \approx \frac{1}{N-M} \sum_{t=M+1}^{N} p(y|x_n; \boldsymbol{\theta}^t)$$

It is convenient to reparametrise the model as

$$y_i = \sum_{j=1}^{k} (g_j \cos(\omega_j x_i) + h_j \sin(\omega_j x_i)) + \epsilon_i$$

where $g_j = \psi_j \cos(\phi_j)$ and $h_j = -\psi_j \sin(\phi_j)$ represent the new amplitudes of the problem, which lie in the range $(-\infty, \infty)$. This may be written as

$$\boldsymbol{y} = \boldsymbol{D}\boldsymbol{a} + \boldsymbol{\epsilon}$$

where $\boldsymbol{y}^T = (y_1, \ldots, y_n)$; $\boldsymbol{a}^T = (g_1, h_1, \ldots, g_k, h_k)$ is the $2k$-dimensional vector of amplitudes, and $\boldsymbol{D}$ is an $n \times 2k$ matrix, defined by

$$D_{i,j} = \begin{cases} \cos(\omega_j x_i) & j \text{ odd;} \\ \sin(\omega_j x_i) & j \text{ even.} \end{cases}$$

Data

Data are generated according to the model given in (3.50), with $k = 3$, $n = 64$, $\boldsymbol{\omega} = 2\pi(0.2, 0.2 + 1/n, 0.2 + 2/n)$, $\boldsymbol{\psi} = (\sqrt{20}, \sqrt{2\pi}, \sqrt{20})$, $\sigma = 2.239$ and $\boldsymbol{\phi} = (0, \pi/4, \pi/3)$. The time series is shown in Figure 3.7.

Prior

The prior distribution for the random variables, $(\boldsymbol{\omega}, \sigma^2, \boldsymbol{a})$, is written as

$$p(\boldsymbol{\omega}, \sigma^2, \boldsymbol{a}) = p(\boldsymbol{\omega})p(\sigma^2)p(\boldsymbol{a}|\boldsymbol{\omega}, \sigma^2)$$

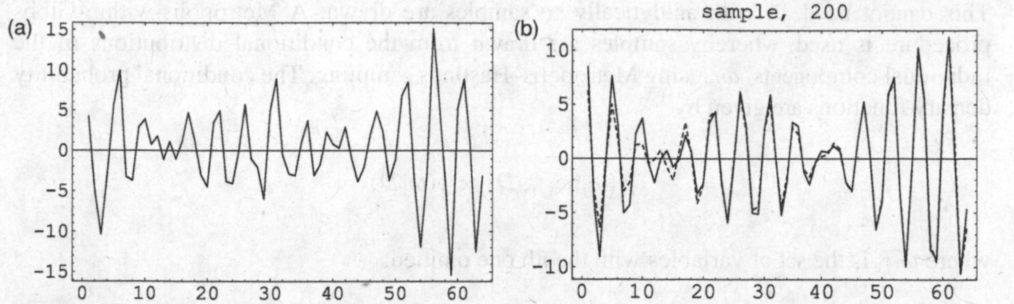

Figure 3.7 Data for sinusoids estimation problem (a); with reconstruction, based on the 200th set of MCMC samples [dashed line, (b)].

where

$$p(\boldsymbol{\omega}) = \frac{1}{\pi^k}\mathrm{I}[\boldsymbol{\omega} \in [0, \pi]^k]$$

$$p(\boldsymbol{a}|\boldsymbol{\omega}, \sigma^2) = N(\boldsymbol{a}; \boldsymbol{0}, \sigma^2\boldsymbol{\Sigma}), \quad \text{where} \quad \boldsymbol{\Sigma}^{-1} = \delta^{-2}\boldsymbol{D}^T\boldsymbol{D}$$

$$p(\sigma^2) = \mathrm{Ig}(\sigma^2; \nu_0/2, \gamma_0/2)$$

Here $\mathrm{Ig}(\sigma^2; \nu, \gamma)$ is the probability density function for σ^2 distributed according to the inverse-gamma distribution with shape parameter ν and scale parameter γ. This distribution is such that $1/\sigma^2$ has a gamma distribution with shape parameter ν and inverse-scale parameter γ. Hyperparameter values of $\delta^2 = 50$, $\nu_0 = 0.01$ and $\gamma_0 = 0.01$ have been used in the illustrated example.

Posterior

The posterior distribution can be re-arranged to

$$p\left(\boldsymbol{a}, \boldsymbol{\omega}, \sigma^2|\mathcal{D}\right) \propto \frac{1}{\sigma^{2\left(\frac{n+\nu_0}{2}+k+1\right)}} \exp\left[\frac{-(\gamma_0 + \boldsymbol{y}^T\boldsymbol{P}\boldsymbol{y})}{2\sigma^2}\right] I\left[\boldsymbol{\omega} \in [0, \pi]^k\right]$$
$$\times|\boldsymbol{\Sigma}|^{-1/2} \exp\left[\frac{-(\boldsymbol{a}-\boldsymbol{m})^T \boldsymbol{M}^{-1}(\boldsymbol{a}-\boldsymbol{m})}{2\sigma^2}\right] \tag{3.51}$$

where

$$\boldsymbol{M}^{-1} = \boldsymbol{D}^T\boldsymbol{D} + \boldsymbol{\Sigma}^{-1}, \quad \boldsymbol{m} = \boldsymbol{M}\boldsymbol{D}^T\boldsymbol{y}, \quad \text{and} \quad \boldsymbol{P} = \boldsymbol{I}_n - \boldsymbol{D}\boldsymbol{M}\boldsymbol{D}^T$$

The amplitude, $\boldsymbol{a}$, and variance, σ^2, can be integrated out analytically, leaving the following marginal posterior density for $\boldsymbol{\omega}$

$$p(\boldsymbol{\omega}|\mathcal{D}) \propto (\gamma_0 + \boldsymbol{y}^T\boldsymbol{P}\boldsymbol{y})^{-\frac{n+\nu_0}{2}}$$

This cannot be dealt with analytically so samples are drawn. A Metropolis-within-Gibbs procedure is used, whereby samples are drawn from the conditional distributions of the individual components, ω_j, using Metropolis–Hastings sampling. The conditional probability density functions are given by

$$p\left(\omega_j|\boldsymbol{\omega}_{(j)}, \mathcal{D}\right) \propto p(\boldsymbol{\omega}|\mathcal{D})$$

where $\boldsymbol{\omega}_{(j)}$ is the set of variables with the jth one omitted.

Metropolis–Hastings proposal distributions

We use a mixture Metropolis–Hastings kernel at each update step. Specifically, we choose randomly between two possible proposal distributions. The first, chosen with probability 0.2, is given by

$$q_j(\omega'_j|\boldsymbol{\omega}) \propto \sum_{l=0}^{n-1} p_l \mathrm{I}\left[\frac{l\pi}{n} < \omega'_j < \frac{(l+1)\pi}{n}\right]$$

where p_l is the squared modulus of the Fourier transform of the data at frequency $l\pi/n$. This proposal allows regions of interest of the posterior distribution to be reached quickly and also helps prevent the Markov chain getting stuck with one solution for ω_j. The second proposal distribution is the normally distributed random walk (mean zero, and standard deviation $\pi/(2n)$). This second proposal ensures irreducibility of the Markov chain. The algorithm is initialised with a sample from the prior for $\boldsymbol{\omega}$.

Sampling the remaining parameters

Having drawn samples for $\boldsymbol{\omega}$, the amplitudes can be sampled using

$$\boldsymbol{a}|(\boldsymbol{\omega}, \sigma^2, \mathcal{D}) \sim N(\boldsymbol{m}, \sigma^2\boldsymbol{M})$$

which comes from (3.51). The noise variance can be sampled using

$$\sigma^2|\boldsymbol{\omega}, \mathcal{D} \sim \mathrm{Ig}\left(\frac{n+\nu_0}{2}, \frac{\gamma_0 + \boldsymbol{y}^T\boldsymbol{P}\boldsymbol{y}}{2}\right)$$

which comes from (3.51) after analytical integration of $\boldsymbol{a}$.

Results

Convergence for the illustrated example was very quick, with a burn-in of less than 100 iterations required.

Figure 3.8 gives plots of samples of the parameter of the noise, σ (true value, 2.239), and frequencies, $\boldsymbol{\omega}$. Figure 3.7 shows the reconstruction of the data using the 200th set of MCMC samples.

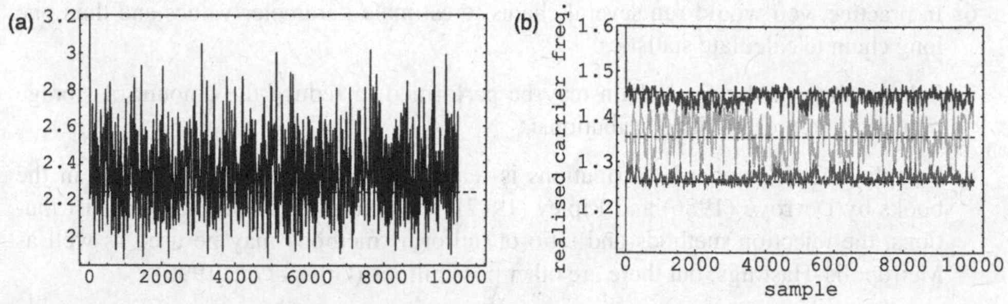

Figure 3.8 Ten thousand iterations from the MCMC algorithm: (a) samples of the noise standard deviation; (b) samples of the frequencies.

3.4.8 Summary

MCMC methods can provide effective approaches to inference problems in situations where analytic evaluation of posterior distributions is not feasible. Their main strength is in their flexibility. They enable Bayesian approaches to be used within real-world problems without having to make unnecessarily restrictive assumptions regarding prior distributions, or simplifications to the likelihoods, in order to may make the mathematics tractable. Originating in the statistical physics literature, the development in Bayesian statistics has been driven by the difficulty in performing numerical integration. The main disadvantages concern uncertainty over convergence, and hence over the accuracy of the estimates computed from the samples, and the computational expense of some procedures.

In many respects, the implementation of these methods is still something of an art, with several trial runs being performed in order to explore models and parameter values. Recent research has looked at adaptive MCMC procedures, whereby the parameters of the MCMC algorithm are adjusted to optimise the algorithm performance (Andrieu and Thoms, 2008). Techniques are required for reducing the amount of computation per iteration. Run times can be long, caused by poor mixing.

The main features of the MCMC method are as follows:

1. It performs iterative sampling from a proposal distribution. The samples may be univariate, multivariate or a subvector of the parameter vector. A special case is Gibbs sampling when samples from conditional probability distributions are drawn.
2. The samples provide a summary of the posterior probability distribution. They may be used to calculate summary statistics either by averaging functions of the samples [Equation (3.38)] or by averaging conditional expectations [Rao-Blackwellisation, Equation (3.39)].
3. Correlated variables lead to longer convergence.
4. The parameters of the method are N, the sequence length; M, the burn-in period; and $q(.|.)$, the proposal distribution.
5. Burn-in refers to the early part of the chain that is removed before function estimation to reduce bias.

6. In practice, you would run several chains to estimate parameter values and then one long chain to calculate statistics.
7. Subsampling of the final chain may be performed to reduce the amount of storage required to represent the distribution.
8. Sampling from standard distributions is readily performed using algorithms in the books by Devroye (1986) and Ripley (1987), for example. For nonstandard distributions, the rejection methods and ratio of uniforms methods may be used as well as Metropolis–Hastings, but there are other possibilities (Gilks *et al.*, 1996).

3.4.9 Notes and references

There are many excellent textbooks describing the theory, use and applications of MCMC methods, including those by Gilks *et al.* (1996), M.–H. Chen *et al.* (2000), Robert and Casella (2004) and Gamerman and Lopes (2006).

Texts by Robert and Casella (2009), and Albert (2009) discuss the use of MCMC methods within the R programming language, while the text by Ntzoufras (2009) covers the use of the WinBUGS (Windows Bayesian Inference using Gibbs Sampling) MCMC software. WinBUGS is a software package that enables MCMC models to be rapidly programmed and assessed. A discussion on the software is available in the paper by Lunn *et al.* (2000). In recent years the OpenBUGS project at the University of Helsinki, Finland, has been developing an open-source version of the WinBUGS software.

Good reference papers on MCMC methods include those by Brooks (1998), Kass *et al.* (1998), Besag (2000) and Andrieu *et al.* (2003). A detailed account of the theory behind Monte Carlo techniques for obtaining characteristics of posterior distributions is provided by Tierney (1994).

3.5 Bayesian approaches to discrimination

In this section we apply the Bayesian learning methods discussed so far in this chapter to the discrimination problem, making use of analytic solutions where we can, but using the numerical techniques where that is not possible.

3.5.1 Labelled training data

Let $\mathcal{D}$ denote the dataset used to train the classifier. In the first instance, let it comprise a set of labelled patterns $\{(\boldsymbol{x}_i, z_i), i = 1, \ldots, n\}$, where $z_i = j$ implies that the pattern $\boldsymbol{x}_i$ is from class ω_j. Given pattern $\boldsymbol{x}$ from an unknown class, we would like to predict its class membership; that is, we require

$$p(z = j|\mathcal{D}, \boldsymbol{x}) \quad j = 1, \ldots, C$$

where z is the class indicator variable corresponding to $\boldsymbol{x}$. The Bayes' decision rule for minimum error is to assign $\boldsymbol{x}$ to the class for which $p(z = j|\mathcal{D}, \boldsymbol{x})$ is the greatest. The above may be written

$$p(z = j|\mathcal{D}, \boldsymbol{x}) \propto p(\boldsymbol{x}|\mathcal{D}, z = j)p(z = j|\mathcal{D}) \tag{3.52}$$

where the constant of proportionality does not depend on class.

The first term in (3.52), $p(\boldsymbol{x}|\mathcal{D}, z = j)$, is the class-conditional probability density function for measurements $\boldsymbol{x}$ from class ω_j. If we assume a model for the density, with parameters $\boldsymbol{\Phi}_j$, then this may be written [see Equation (3.1)]

$$p(\boldsymbol{x}|\mathcal{D}, z = j) = \int p(\boldsymbol{x}|\boldsymbol{\Phi}_j)p(\boldsymbol{\Phi}_j|\mathcal{D}, z = j)d\boldsymbol{\Phi}_j$$

For certain special cases of the density model, $p(\boldsymbol{x}|\boldsymbol{\Phi}_j)$, we may evaluate this analytically. For example, as we have seen in Section 3.2.3, a normal model with parameters $\boldsymbol{\mu}$ and $\boldsymbol{\Sigma}$ with Gauss-Wishart priors leads to a posterior distribution of the parameters that is also Gauss-Wishart and a multivariate Student t-distribution for the density $p(\boldsymbol{x}|\mathcal{D}, z = j)$.

If we are unable to obtain an analytic solution, then a numerical approach will be required. For example, if we use one of the MCMC methods of Section 3.4 to draw samples from the posterior density of the parameters, $p(\boldsymbol{\Phi}_j|\mathcal{D})$, we may approximate $p(\boldsymbol{x}|\mathcal{D}, z = j)$ by

$$p(\boldsymbol{x}|\mathcal{D}, z = j) \approx \frac{1}{N - M} \sum_{t=M+1}^{N} p\left(\boldsymbol{x}|\boldsymbol{\Phi}_j^t\right)$$

where $\boldsymbol{\Phi}_j^t$ are samples generated from the MCMC process and M and N are the burn-in period and run length, respectively.

The second term in (3.52) is the probability of class ω_j given the dataset, $\mathcal{D}$. Thus, it is the prior probability updated by the data. We saw in Section 3.2.4 that if we assume a Dirichlet prior for $\boldsymbol{\pi}$, the vector of prior class probabilities, i.e.

$$p(\boldsymbol{\pi}) = \mathrm{Di}_C(\boldsymbol{\pi}; \boldsymbol{a}_0)$$

then

$$p(z = j|\mathcal{D}) = E[\pi_j|\mathcal{D}] = \frac{a_{0j} + n_j}{\sum_{j=1}^{C}(a_{0j} + n_j)} \tag{3.53}$$

where n_j is the number of training examples for class ω_j in $\mathcal{D}$.

3.5.2 Unlabelled training data

The case of classifier design using a normal model when the training data comprise both labelled and unlabelled patterns is considered by Lavine and West (1992). It provides an example of Gibbs sampling that uses some of the analytic results of Section 3.2.3. We summarise the approach here.

Let the dataset $\mathcal{D} = \{(\boldsymbol{x}_i, z_i), i = 1, \ldots, n; \boldsymbol{x}_i^u, i = 1, \ldots, n_u\}$, where $\boldsymbol{x}_i^u$ are the unlabelled patterns. Let $\boldsymbol{\mu} = \{\boldsymbol{\mu}_i, i = 1, \ldots, C\}$ and $\boldsymbol{\Sigma} = \{\boldsymbol{\Sigma}_i, i = 1, \ldots, C\}$ be the set of class means and covariance matrices and $\boldsymbol{\pi}$ the class priors. Denote by $z^u = \{z_i^u, i = 1, \ldots, n_u\}$, the set of unknown class labels.

The parameters of the model are $\theta = \{\boldsymbol{\mu}, \boldsymbol{\Sigma}, \boldsymbol{\pi}, z^u\}$. Taking a Gibbs sampling approach, we successively draw samples from three conditional distributions:

1. Sample from $p(\boldsymbol{\mu}, \boldsymbol{\Sigma}|\boldsymbol{\pi}, z^u, \mathcal{D})$. This density may be written

$$p(\boldsymbol{\mu}, \boldsymbol{\Sigma}|\boldsymbol{\pi}, z^u, \mathcal{D}) = \prod_{i=1}^{C} p(\boldsymbol{\mu}_i, \boldsymbol{\Sigma}_i|z^u, \mathcal{D})$$

   the product of C independent Gauss-Wishart distributions given by (3.19). Note that since we are conditioning on z^u all the training data is labelled.

2. Sample from $p(\boldsymbol{\pi}|\boldsymbol{\mu}, \boldsymbol{\Sigma}, z^u, \mathcal{D})$. This is a Dirichlet distribution, $\mathrm{Di}_C(\boldsymbol{\pi}; \boldsymbol{a})$, independent of $\boldsymbol{\mu}$ and $\boldsymbol{\Sigma}$. The parameters $\boldsymbol{a}$ are given by $\boldsymbol{a} = \boldsymbol{a}_0 + \boldsymbol{n}$, with $\boldsymbol{n} = (n_1, \ldots, n_C)$, where n_j is the number of patterns in class ω_j as determined by $\mathcal{D}$ and z^u.
3. Sample from $p(z^u|\boldsymbol{\mu}, \boldsymbol{\Sigma}, \boldsymbol{\pi}, \mathcal{D})$. Since the samples z_i^u are conditionally independent, using Bayes' theorem we require samples from

$$\begin{aligned} p\left(z_i^u = j|\boldsymbol{\mu}, \boldsymbol{\Sigma}, \boldsymbol{\pi}, \mathcal{D}\right) &\propto p\left(z_i^u = j|\boldsymbol{\pi}\right) p\left(\boldsymbol{x}_i^u|\boldsymbol{\mu}_j, \boldsymbol{\Sigma}_j, z_i^u = j\right) \\ &\propto \pi_j p\left(\boldsymbol{x}_i^u|\boldsymbol{\mu}_j, \boldsymbol{\Sigma}_j, z_i^u = j\right) \end{aligned} \tag{3.54}$$

   the product of the prior and the normal density of class ω_j at $\boldsymbol{x}_i^u$. The constant of proportionality is chosen so that the sum over classes is unity. Sampling a value for z_i^u is then trivial.

The Gibbs sampling procedure produces a set of samples $\{\boldsymbol{\mu}^t, \boldsymbol{\Sigma}^t, \boldsymbol{\pi}^t, (z^u)^t, t = 1, \ldots, N\}$, which may be used to classify the unlabelled patterns and future observations. We ignore an initial burn-in period of M samples.

To classify the unlabelled patterns in the training set, we use

$$p\left(z_i^u = j|\mathcal{D}\right) = \frac{1}{N-M} \sum_{t=M+1}^{N} p\left(z_i^u = j|\boldsymbol{\mu}_j^t, \boldsymbol{\Sigma}_j^t, \pi_j^t, \mathcal{D}\right)$$

where the terms in the summation are, by (3.54), products of the prior and class-conditional density, evaluated and normalised for each set of samples from the MCMC iterations. This is a Rao-Blackwellised estimate (see Section 3.3.2), where we estimate the quantity via conditional expectations rather than directly from the samples $(z^u)^t$.

To classify a new pattern $\boldsymbol{x}$, we require $p(z = j|\boldsymbol{x}, \mathcal{D})$, given by

$$p(z = j|\boldsymbol{x}, \mathcal{D}) \propto p(z = j|\mathcal{D}) p(\boldsymbol{x}|\mathcal{D}, z = j) \tag{3.55}$$

The first term in the product, $p(z = j|\mathcal{D})$, is

$$p(z = j|\mathcal{D}) = E[\pi_j|\mathcal{D}] = \frac{1}{N-M} \sum_{t=M+1}^{N} E[\pi_j|\mathcal{D}, (z^u)^t]$$

The expectation in the summation can be evaluated using (3.53), since the conditioning on $(z^u)^t$ means that all the training data are now labelled. This is again an example of Rao-Blackwellisation (see Section 3.3.2) in that we estimate the quantity via conditional expectations rather than directly from the samples π^t. The second term in (3.55) can be written as

$$p(\boldsymbol{x}|\mathcal{D}, z=j) \approx \frac{1}{N-M} \sum_{t=M+1}^{N} p(\boldsymbol{x}|\mathcal{D}, (z^u)^t, z=j)$$

the sum of multivariate Student probability density functions [each density function in the summation has the same form as (3.21) of Section 3.2.3, since the conditioning on $(z^u)^t$ means that all the training data is now labelled].

3.5.2.1 Illustration

The illustration given here is based on that of Lavine and West (1992). Two-dimensional data from three classes are generated from equally weighted non-normal distributions. Defining matrices

$$\boldsymbol{C}_1 = \begin{pmatrix} 5 & 1 \\ 3 & 5 \end{pmatrix}, \boldsymbol{C}_2 = \begin{pmatrix} 0 & 1 \\ 1 & 5 \end{pmatrix}, \boldsymbol{C}_3 = \begin{pmatrix} 5 & 0 \\ 3 & 1 \end{pmatrix}$$

then an observation $\boldsymbol{x}_i$ from class ω_j is generated according to

$$\boldsymbol{x}_i = \boldsymbol{C}_j \begin{pmatrix} u_i \\ 1-u_i \end{pmatrix} + \boldsymbol{\epsilon}_i$$

where u_i is uniform over [0, 1] and $\boldsymbol{\epsilon}_i$ is normally-distributed with zero mean and diagonal covariance matrix, $\boldsymbol{I}/2$.

The labelled training data are shown in Figure 3.9.

Using a normal model, with diagonal covariance matrix, for the density of each class, an MCMC approach using Gibbs sampling is taken. The training set consists of 1200 labelled and 300 unlabelled patterns. The priors for the mean and variances are normal and inverse gamma, respectively. The parameter values are initialised as samples from the prior. Figure 3.10 shows components of the mean and covariance matrix that are produced from the chain. The WinBUGS software package (Lunn *et al.* 2000) has been used to complete the MCMC sampling.

3.6 Sequential Monte Carlo samplers

3.6.1 Introduction

In Section 3.3.6 we introduced importance sampling and noted that it forms the basis of particle filters for sequential Bayesian parameter estimation. Sequential Monte Carlo (SMC) samplers (Del Moral *et al.*, 2006) are the next generation of particle filters, which allow the power of the particle filter methodology to be applied to general parameter estimation problems. They

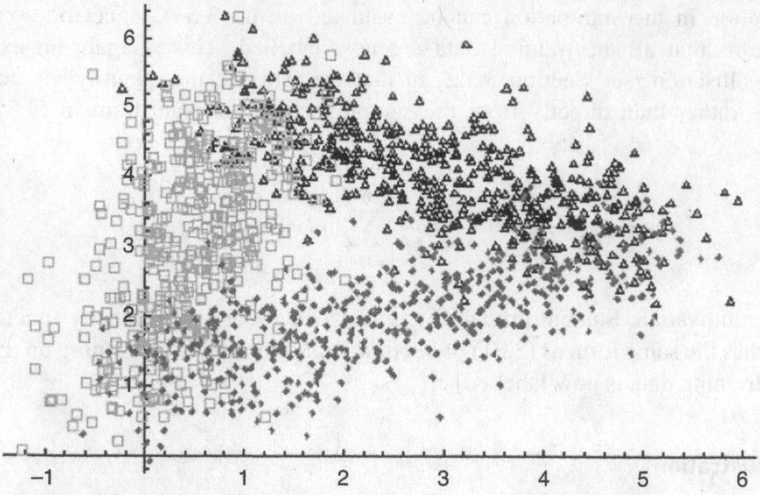

Figure 3.9 Three-class (triangles, class 1; squares, class 2; and diamonds, class 3), two-dimensional training data.

work by specifying an artificial sequence of distributions defined on the same fixed state space, with the sequence finishing at the distribution of interest. The motivation is that rather than attempting to sample from the desired distribution immediately, we instead draw samples from the distribution via a sequence of samples from intermediate distributions, which gradually approach the distribution of interest. SMC samplers can be used for generating samples from complicated probability distributions, and hence for estimating probability densities for use in Bayesian classifiers.

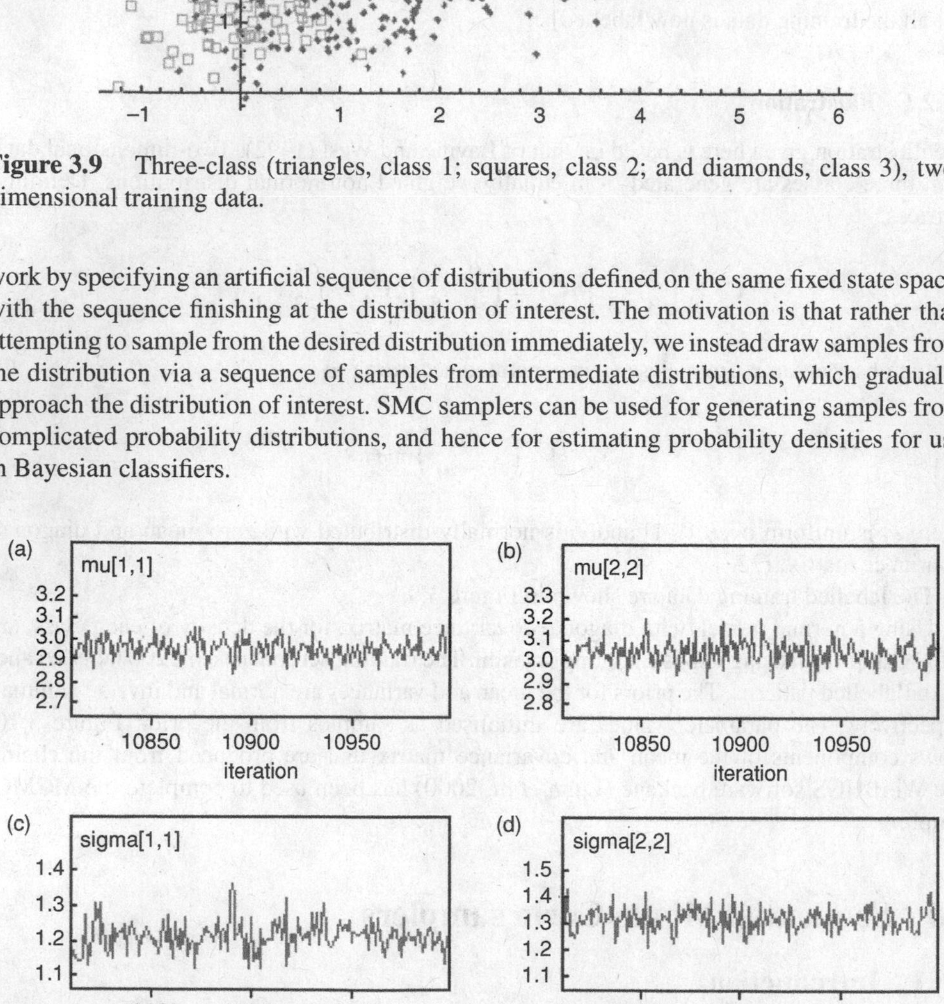

Figure 3.10 MCMC samples of μ and Σ. (a) μ_1 component of class 1; (b) μ_2 component of class 2; (c) $\Sigma_{1,1}$ component of class 1; (d) $\Sigma_{2,2}$ component of class 2.

This development of the particle filtering methodology to general parameter estimation problems uses an innovation of adding auxiliary variables to the state space under consideration. In the sequential parameter estimation problem addressed by particle filters, each time step brings a new variable (the latest state of the parameter under consideration), and therefore the dimension of the state space grows with time. As an example, suppose that we are tracking the location, $\boldsymbol{x}$, of a target at discrete time steps. At time step $t-1$ the state space is $(\boldsymbol{x}_1, \ldots, \boldsymbol{x}_{t-1})$, i.e. the time history of locations up to and including time state $t-1$. Moving to time state t, we increment the state space by the latest location $\boldsymbol{x}_t$, giving an enlarged state space $(\boldsymbol{x}_1, \ldots, \boldsymbol{x}_t)$. This expanding state space is critical to efficient application of particle filters. Indeed, the lack of an expanding state space prevents straightforward application of the technology to estimation of static parameters. SMC samplers remove this difficulty through the addition of auxiliary variables, that cause the state space to grow with time.

3.6.2 Basic methodology

We present a brief overview of the SMC sampler methodology. For detailed discussions and justifications the reader is referred to Del Moral *et al.* (2006) and Peters (2005).

Suppose that we wish to draw samples from the distribution $\pi(\boldsymbol{\theta})$ on the parameter space $\boldsymbol{\theta} \in \mathbb{R}^n$. The first step in defining an SMC sampler for sampling from this distribution is to define a sequence of target distributions $\pi_1(\boldsymbol{\theta}), \ldots, \pi_T(\boldsymbol{\theta})$, all on the same parameter space $\boldsymbol{\theta} \in \mathbb{R}^n$, with the following properties:

- $\pi_T(\boldsymbol{\theta}) = \pi(\boldsymbol{\theta})$, the desired distribution (i.e. the final distribution in the sequence is the desired distribution);
- $\pi_1(\boldsymbol{\theta})$ is a distribution from which it is easy to sample;
- the difference between distributions $\pi_t(\boldsymbol{\theta})$ and $\pi_{t+1}(\boldsymbol{\theta})$, $t = 1, \ldots, T-1$, is not too large.

The basic premise behind the SMC sampler is to draw samples from $\pi(\boldsymbol{\theta})$ by sampling first from the easy to sample from initial distribution $\pi_1(\boldsymbol{\theta})$, and then moving to the distribution $\pi(\boldsymbol{\theta})$, via a sequence of $T-1$ easy steps (the intermediate distributions).

3.6.2.1 Example sequences of target distributions

Suppose that $\pi(\boldsymbol{\theta})$ is the posterior distribution arising from a prior distribution $p(\boldsymbol{\theta})$, a likelihood function $p(\boldsymbol{x}|\boldsymbol{\theta})$, and a dataset $\mathcal{D} = \{\boldsymbol{x}_1, \ldots, \boldsymbol{x}_n\}$. Two potential sequences of SMC target distributions are then:

- $\pi_t(\boldsymbol{\theta}) = p(\boldsymbol{\theta}|\boldsymbol{x}_1, \ldots, \boldsymbol{x}_t)$, $t = 1, \ldots, n$, i.e. data tempering in which the tth target distribution is the posterior distribution based upon the first t data samples.
- $\pi_t(\boldsymbol{\theta}) \propto h(\boldsymbol{\theta})^{1-\alpha_t}\pi(\boldsymbol{\theta})^{\alpha_t}$, where $0 = \alpha_1 < \alpha_2 < \ldots < \alpha_T = 1$, and $h(\boldsymbol{\theta})$ is a distribution that is relatively easy to sample from. Often, $h(\boldsymbol{\theta})$ can be the prior distribution $p(\boldsymbol{\theta})$.

For determining the most likely parameters (i.e. the *maximum a posteriori* parameters that maximise the posterior distribution), a simulated annealing approach can be used. Here we set $\pi_t(\boldsymbol{\theta}) \propto \pi(\boldsymbol{\theta})^{\alpha_t}$, for $\alpha_t \geq 0$ an increasing sequence, tending to infinity as $t \to \infty$.

3.6.2.2 Underlying SMC sampler procedure

The underlying SMC sampler procedure is as follows:

- Step 1: Draw N samples $\{\boldsymbol{\theta}_1^{(1)}, \ldots, \boldsymbol{\theta}_1^{(N)}\}$ from the initial distribution $\pi_1(\boldsymbol{\theta})$.
- Steps $t = 2, \ldots, T$. Adjust the samples $\{\boldsymbol{\theta}_{t-1}^{(1)}, \ldots, \boldsymbol{\theta}_{t-1}^{(N)}\}$ from the distribution $\pi_{t-1}(\boldsymbol{\theta})$ to become samples $\{\boldsymbol{\theta}_t^{(1)}, \ldots, \boldsymbol{\theta}_t^{(N)}\}$ from the distribution $\pi_t(\boldsymbol{\theta})$.

Trying to apply standard importance sampling (Section 3.3.6) to adjust the samples in steps $t = 2, \ldots, T$ would typically require the evaluation of intractable integrals when calculating the importance weights (Del Moral *et al.*, 2006). This is what prevents the basic particle filter from being used for this problem.

SMC samplers use a trick of augmenting the state space with auxiliary variables to circumvent this problem. The auxiliary variables are the previous estimates of the state. From one distribution to the next, the state space therefore increases by $\mathbb{R}^d$ (the dimensionality of $\boldsymbol{\theta}$), such that at step t of the algorithm the state space is of size $\mathbb{R}^{td}$, and is given by $(\boldsymbol{\theta}_1, \ldots, \boldsymbol{\theta}_t)$. The target distribution at step t is augmented to take in the larger state space as follows

$$\tilde{\pi}_t(\boldsymbol{\theta}_1, \ldots, \boldsymbol{\theta}_t) = \pi_t(\boldsymbol{\theta}_t) \prod_{i=1}^{t-1} L_i(\boldsymbol{\theta}_{i+1}, \boldsymbol{\theta}_i)$$

where the $L_i(\boldsymbol{\theta}_{i+1}, \boldsymbol{\theta}_i)$ are backwards transition kernels, termed *L-kernels* in the SMC sampler literature. The original target distributions, $\pi_t(\boldsymbol{\theta}_t)$, can be obtained by marginalisation of the augmented distributions

$$\begin{aligned}
\int \tilde{\pi}_t(\boldsymbol{\theta}_1, \ldots, \boldsymbol{\theta}_t) d\boldsymbol{\theta}_1 \ldots d\boldsymbol{\theta}_{t-1} &= \pi_t(\boldsymbol{\theta}_t) \int \prod_{i=1}^{t-1} L_i(\boldsymbol{\theta}_{i+1}, \boldsymbol{\theta}_i) d\boldsymbol{\theta}_1 \ldots d\boldsymbol{\theta}_{t-1} \\
&= \pi_t(\boldsymbol{\theta}_t) \prod_{i=1}^{t-1} \left(\int L_i(\boldsymbol{\theta}_{i+1}, \boldsymbol{\theta}_i) d\boldsymbol{\theta}_i \right) \\
&= \pi_t(\boldsymbol{\theta}_t)
\end{aligned}$$

The significance of these augmentations is that particle filtering ideas can be used to draw samples from the distributions $\tilde{\pi}_t(\boldsymbol{\theta}_1, \ldots, \boldsymbol{\theta}_t)$, $t = 1, \ldots, T$, in sequence, starting from $\tilde{\pi}_1(\boldsymbol{\theta}_1) = \pi_1(\boldsymbol{\theta}_1)$ and ending at the distribution $\tilde{\pi}_T(\boldsymbol{\theta}_1, \ldots, \boldsymbol{\theta}_T)$. Samples from $\pi(\boldsymbol{\theta})$ can then be obtained by marginalisation of the samples from the final distribution, i.e. extracting the samples $\{\boldsymbol{\theta}_T{}^{(s)}, s = 1, \ldots, N\}$ from the final set of samples $\{(\boldsymbol{\theta}_1{}^{(s)}, \ldots, \boldsymbol{\theta}_T{}^{(s)}), s = 1, \ldots, N\}$. These samples can then be used in the manner described in Section 3.3.2 (and Section 3.3.3 if used within a Bayesian classifier).

Algorithm 3.5 Outline of an SMC sampler.

- Use importance sampling (Section 3.3.6) to draw samples $\{\boldsymbol{\theta}_1^{(s)}, s = 1, \ldots, N\}$ from the distribution $\pi_1(\boldsymbol{\theta}_1)$. The resulting samples have importance sampling weights $\{w_1^{(s)}, s = 1, \ldots, N\}$.
- For $t = 2, \ldots, T$:
  - Use a forward transition kernel $K_t(\boldsymbol{\theta}_{t-1}, \boldsymbol{\theta}_t)$ to move the samples $\{\boldsymbol{\theta}_{t-1}^{(1)}, \ldots, \boldsymbol{\theta}_{t-1}^{(N)}\}$ created in step $t-1$ to $\{\boldsymbol{\theta}_t^{(1)}, \ldots, \boldsymbol{\theta}_t^{(N)}\}$.
  - Reweight each sample according to (3.56), obtaining new weights $\{w_t^{(s)}, s = 1, \ldots, N\}$.
  - Normalise the weights as follows:

$$\tilde{w}_t^{(s)} = \frac{w_t^{(s)}}{\sum_{m=1}^{N} w_t^{(m)}}, \quad s = 1, \ldots, N$$

  - Conduct a resampling step (see Algorithm 3.6).
- Use the final set of resampled samples $\{\boldsymbol{\theta}_T^{(1)}, \ldots, \boldsymbol{\theta}_T^{(N)}\}$ as weighted samples (with weights specified in the resampling step) from the desired distribution $\pi(\boldsymbol{\theta})$.

3.6.2.3 SMC sampler algorithm

An outline of an SMC sampler is provided in Algorithm 3.5, and is now discussed in more detail.

At the start of step $t > 1$, we have a set of samples $\{(\boldsymbol{\theta}_1^{(s)}, \ldots, \boldsymbol{\theta}_{t-1}^{(s)}), s = 1, \ldots, N\}$ with weights $\{w_{t-1}^{(s)}, s = 1, \ldots, N\}$. These approximate the $(t-1)$th target distribution $\tilde{\pi}_{t-1}(\boldsymbol{\theta}_1, \ldots, \boldsymbol{\theta}_{t-1})$. A forward transition kernel $K_t(\boldsymbol{\theta}_{t-1}, \boldsymbol{\theta}_t)$ is used to move the samples $\{\boldsymbol{\theta}_{t-1}^{(1)}, \ldots, \boldsymbol{\theta}_{t-1}^{(N)}\}$ created in step $t-1$ to $\{\boldsymbol{\theta}_t^{(1)}, \ldots, \boldsymbol{\theta}_t^{(N)}\}$, giving updated samples $\{(\boldsymbol{\theta}_1^{(s)}, \ldots, \boldsymbol{\theta}_t^{(s)}), s = 1, \ldots, N\}$. Specifically, sample $\boldsymbol{\theta}_t^{(s)}$ is obtained by sampling from the kernel $K_t(\boldsymbol{\theta}_{t-1}^{(s)}, \boldsymbol{\theta}_t)$. As an example, the kernel could be a Metropolis–Hastings update step (see Section 3.4.3). With such an update step we first propose a sample $\boldsymbol{\theta}_t^*$ from a proposal distribution $q_t(\boldsymbol{\theta}_t|\boldsymbol{\theta}_{t-1}^{(s)})$. Then, with probability

$$a\left(\boldsymbol{\theta}_{t-1}^{(s)}, \boldsymbol{\theta}_t^*\right) = \min\left(1, \frac{\pi_t\left(\boldsymbol{\theta}_t^*\right) q\left(\boldsymbol{\theta}_{t-1}^{(s)}|\boldsymbol{\theta}_t^*\right)}{\pi_t\left(\boldsymbol{\theta}_{t-1}^{(s)}\right) q\left(\boldsymbol{\theta}_t^*|\boldsymbol{\theta}_{t-1}^{(s)}\right)}\right)$$

we accept the proposed sample, and set $\boldsymbol{\theta}_t^{(s)} = \boldsymbol{\theta}_t^*$. Otherwise we set $\boldsymbol{\theta}_t^{(s)} = \boldsymbol{\theta}_{t-1}^{(s)}$. The range of potential proposal distributions is as considered in Section 3.4.3, and could include random walks centred around the current point. It then remains to update the weights for these samples.

Using a sequential importance sample argument, weights $w_t^{(s)}$ at step t are obtained from the weights $w_{t-1}^{(s)}$ at step $t-1$ as follows

$$\begin{aligned} w_t^{(s)} &= w_{t-1}^{(s)} \frac{\tilde{\pi}_t\left(\boldsymbol{\theta}_1^{(s)}, \ldots, \boldsymbol{\theta}_t^{(s)}\right)}{\tilde{\pi}_{t-1}\left(\boldsymbol{\theta}_1^{(s)}, \ldots, \boldsymbol{\theta}_{t-1}^{(s)}\right) K_t\left(\boldsymbol{\theta}_{t-1}^{(s)}, \boldsymbol{\theta}_t^{(s)}\right)} \\ &= w_{t-1}^{(s)} \frac{\pi_t\left(\boldsymbol{\theta}_t^{(s)}\right) \prod_{i=1}^{t-1} L_i\left(\boldsymbol{\theta}_{i+1}^{(s)}, \boldsymbol{\theta}_i^{(s)}\right)}{\pi_{t-1}\left(\boldsymbol{\theta}_{t-1}^{(s)}\right)\left(\prod_{i=1}^{t-2} L_i\left(\boldsymbol{\theta}_{i+1}^{(s)}, \boldsymbol{\theta}_i^{(s)}\right)\right) K_t\left(\boldsymbol{\theta}_{t-1}^{(s)}, \boldsymbol{\theta}_t^{(s)}\right)} \\ &= w_{t-1}^{(s)} \frac{\pi_t\left(\boldsymbol{\theta}_t^{(s)}\right) L_{t-1}\left(\boldsymbol{\theta}_t^{(s)}, \boldsymbol{\theta}_{t-1}^{(s)}\right)}{\pi_{t-1}\left(\boldsymbol{\theta}_{t-1}^{(s)}\right) K_t\left(\boldsymbol{\theta}_{t-1}^{(s)}, \boldsymbol{\theta}_t^{(s)}\right)} \end{aligned} \tag{3.56}$$

Specification of the L-kernel, $L_{t-1}(\boldsymbol{\theta}_t^{(s)}, \boldsymbol{\theta}_{t-1}^{(s)})$, is discussed next.

3.6.2.4 L-kernels

The L-kernels, which are not present in the MCMC approaches of Section 3.4, affect the performance of the sampler, and require further discussion. Del Moral *et al.* (2006) and Peters (2005) discuss the choice of optimal and suboptimal L-kernels. A common suboptimal choice for use alongside MCMC forward transition kernels [e.g. when $K_t(\boldsymbol{\theta}_{t-1}, \boldsymbol{\theta}_t)$ is a Metropolis–Hastings kernel] is to choose the L-kernel so that is satisfies the detailed balanced equation [see Equation (3.46)]

$$\pi_t(\boldsymbol{\theta}_t) L_{t-1}(\boldsymbol{\theta}_t, \boldsymbol{\theta}_{t-1}) = \pi_t(\boldsymbol{\theta}_{t-1}) K_t(\boldsymbol{\theta}_{t-1}, \boldsymbol{\theta}_t)$$

Then the weights in (3.56) become

$$\begin{aligned} w_t^{(s)} &= w_{t-1}^{(s)} \frac{\pi_t\left(\boldsymbol{\theta}_t^{(s)}\right)\left(\frac{\pi_t\left(\boldsymbol{\theta}_{t-1}^{(s)}\right) K_t\left(\boldsymbol{\theta}_{t-1}^{(s)}, \boldsymbol{\theta}_t^{(s)}\right)}{\pi_t\left(\boldsymbol{\theta}_t^{(s)}\right)}\right)}{\pi_{t-1}\left(\boldsymbol{\theta}_{t-1}^{(s)}\right) K_t\left(\boldsymbol{\theta}_{t-1}^{(s)}, \boldsymbol{\theta}_t^{(s)}\right)} \\ &= w_{t-1}^{(s)} \frac{\pi_t\left(\boldsymbol{\theta}_{t-1}^{(s)}\right)}{\pi_{t-1}\left(\boldsymbol{\theta}_{t-1}^{(s)}\right)} \end{aligned} \tag{3.57}$$

which do not actually depend on the parameters drawn at step t of the algorithm. The performance (in terms of efficiency) of a sampler using such an L-kernel is better when the difference between successive distributions in the sequence is small, i.e. when $\pi_t(\boldsymbol{\theta}) \approx \pi_{t-1}(\boldsymbol{\theta})$. The weights provided in (3.57) are the same as those in Annealed Importance Sampling (Neal, 2001).

3.6.2.5 Resampling

A major problem with the procedure as described in the text is that the SMC sampler will suffer from degeneracy, in that after a few iterations, nearly all samples will have negligible

Algorithm 3.6 Resampling step within an SMC sampler.

Starting from a set of samples with normalised weights $\{\tilde{w}_t^{(s)}, s = 1, \ldots, N\}$:

- Calculate the *effective sample size* (ESS)

$$\text{ESS}_t = \frac{1}{\sum_{s=1}^{N} \left(\tilde{w}_t^{(s)}\right)^2} \tag{3.58}$$

- If $\text{ESS}_t < N/2$ conduct the following resampling step:
  - Sample N times from the current set of samples (with replacement), with the probability of selecting the ith sample at any stage being equal to the normalised weight $\tilde{w}_t^{(i)}$ for that sample.
  - Specify equal weights $w_t^{(s)} = 1/N$, $s = 1, \ldots, N$ for the new samples.
- Otherwise set $w_t^{(s)} = \tilde{w}_t^{(s)}$, $s = 1, \ldots, N$, and use the original samples.

weight, with possibly just one sample dominating. Untreated, this would have a severe impact on the performance of any technique relying on the samples, since the resulting sample estimates of the desired distribution would be unreliable. The solution is to use a resampling stage, as in the standard particle filter literature (Doucet *et al.*, 2001; Arulampalam *et al.*, 2002). Resampling results in samples being replicated or discarded in a way that preserves the underlying distribution.

The simplest form of resampling is multinomial resampling, in which the samples are resampled according to a multinomial distribution (with replacement). Specifically, we sample N times from the current set of samples, with the probability of selecting the ith sample at any stage being equal to the normalised weight $\tilde{w}_t^{(i)}$ for that sample. Improvements are possible (e.g. residual sampling and stratified sampling), and are discussed in the aforementioned particle filtering literature. After resampling, each new sample is set to have a weight of $w_t^{(m)} = 1/N$.

Over time, degeneracy is inevitable without the use of resampling, since the variance of the importance sampling weights increases over time. Hence resampling is necessary. However, since the resampling stage results in a further approximation, it is good practice to only conduct a resampling stage when the samples are showing signs of degeneracy. A simple procedure is described in Algorithm 3.6. The approach is based upon monitoring the *effective sample size* (ESS) calculated using the normalised weights [Equation (3.58)]. The ESS takes values between 1 (degenerate, with just a single sample having all the weight) and N (nondegenerate, with all samples having equal weight). A small effective sample size is therefore indicative of degeneracy, and resampling should take place if ESS_t falls below a pre-specified threshold.

3.6.3 Summary

SMC samplers are an advanced sampling methodology that can be used for generating samples from complicated probability distributions. They combine ideas from particle filters, a popular technique in the target tracking community, with MCMC methods.

Key factors that affect the performance of SMC samplers are:

- the form of the L-kernels, $L_i(\boldsymbol{\theta}_{i+1}, \boldsymbol{\theta}_i)$;
- the form of the forward transition kernels, $K_t(\boldsymbol{\theta}_{t-1}, \boldsymbol{\theta}_t)$;
- the sequence of target distributions, $\pi_1(\boldsymbol{\theta}), \ldots, \pi_T(\boldsymbol{\theta})$.

Although relatively easy to code up, the advanced nature of the topic is such that SMC samplers are not recommended for readers without experience of implementing MCMC and particle filter methods.

The combination of particle filters with MCMC methods for general purpose Bayesian inference is a fast moving research topic, with many interesting recent papers (Andrieu *et al.*, 2010).

3.7 Variational Bayes

3.7.1 Introduction

Variational Bayes methods provide an alternative means of approximating nontrivial Bayesian posterior distributions to Bayesian sampling schemes such as MCMC algorithms. Since the late 1990s they have been popular in the machine learning literature. They approximate the posterior distribution by a simpler *variational distribution*. This variational distribution is then used in lieu of the actual posterior distribution in subsequent calculations, such as when determining predictive densities using (3.1) for use in a Bayes' rule classifier.

In this section we provide a description and simple example for the use of Variational Bayes methods for Bayesian inference. Good accounts of Variational Bayes methods are provided in the text by Bishop (2007), and the PhD theses of Beal (2003) and Winn (2004). Tutorial papers are provided by Jaakkola (2000) and Tzikas *et al.* (2008).

3.7.2 Description

Suppose that we have set of measurements (observed data) $\mathcal{D}$, and a set of parameters $\boldsymbol{\theta}$. As in standard Bayesian methods, inference is conducted with regards to the posterior distribution represented by the probability density function $p(\boldsymbol{\theta}|\mathcal{D})$. Variational Bayes methods seek to approximate $p(\boldsymbol{\theta}|\mathcal{D})$ by the probability density function $q(\boldsymbol{\theta})$, known as the variational approximation. The Kullback–Leibler (KL) divergence (also referred to as the Kullback–Leibler cross-entropy measure) is used as the measure of closeness between $q(\boldsymbol{\theta})$ and $p(\boldsymbol{\theta}|\mathcal{D})$

$$KL(q(\boldsymbol{\theta})|p(\boldsymbol{\theta}|\mathcal{D})) = \int q(\boldsymbol{\theta})\log\left(\frac{q(\boldsymbol{\theta})}{p(\boldsymbol{\theta}|\mathcal{D})}\right)d\boldsymbol{\theta}$$

where the integrals are replaced by summations if the parameters $\boldsymbol{\theta}$ are discrete-valued rather than continuous. The KL divergence satisfies $KL(q(\boldsymbol{\theta})|p(\boldsymbol{\theta}|\mathcal{D})) \geq 0$ with equality if and only if $q(\boldsymbol{\theta}) = p(\boldsymbol{\theta}|\mathcal{D})$. Variational Bayes methods seek to minimise $KL(q(\boldsymbol{\theta})|p(\boldsymbol{\theta}|\mathcal{D}))$ subject to a constrained form for $q(\boldsymbol{\theta})$. A common constraint is a factorisation of $q(\boldsymbol{\theta})$ into a product of marginal densities for components of $\boldsymbol{\theta}$, e.g.

$$q(\boldsymbol{\theta}) = \prod_{i=1}^{k} q_i(\boldsymbol{\theta}_i) \tag{3.59}$$

where $\boldsymbol{\theta} = \boldsymbol{\theta}_1 \cup \boldsymbol{\theta}_2 \cup \ldots \cup \boldsymbol{\theta}_k$ and $\boldsymbol{\theta}_i \cap \boldsymbol{\theta}_j = \emptyset$, for $i \neq j$ (i.e. the sets are disjoint and cover $\boldsymbol{\theta}$). Such a factorisation is sometimes referred to as a mean field approximation, due to relations to similar procedures in statistical physics.

Rather than attempting to minimise the KL divergence directly we first define

$$L(q(\boldsymbol{\theta})) = \int q(\boldsymbol{\theta}) \log\left(\frac{p(\mathcal{D}, \boldsymbol{\theta})}{q(\boldsymbol{\theta})}\right) d\boldsymbol{\theta} \tag{3.60}$$

and note that

$$\log(p(\mathcal{D})) = L(q(\boldsymbol{\theta})) + KL(q(\boldsymbol{\theta})|p(\boldsymbol{\theta}|\mathcal{D})) \tag{3.61}$$

where $p(\mathcal{D})$ is the marginal density of the data, and $\log(p(\mathcal{D}))$ is often referred to as the log-evidence for the model. Then, since $p(\mathcal{D})$ is independent of the choice for $q(\boldsymbol{\theta})$, minimising $KL(q(\boldsymbol{\theta})|p(\boldsymbol{\theta}|\mathcal{D}))$ is equivalent to maximising $L(q(\boldsymbol{\theta}))$. Hence the Variational Bayes procedure seeks to maximise $L(q(\boldsymbol{\theta}))$ subject to the constrained form for $q(\boldsymbol{\theta})$.

A common alternative derivation for Variational Bayes methods comes from noting that $L(q(\boldsymbol{\theta}))$ is a lower bound on the logarithm of the marginal density $p(\mathcal{D})$. Variational Bayes methods are then motivated as attempting to maximise the lower bound on the marginal density. Such derivations are often made without directly considering the KL divergence, since we can write

$$\begin{aligned}
\log(p(\mathcal{D})) &= \log\left(\int p(\mathcal{D}, \boldsymbol{\theta}) d\boldsymbol{\theta}\right) \\
&= \log\left(\int q(\boldsymbol{\theta}) \frac{p(\mathcal{D}, \boldsymbol{\theta})}{q(\boldsymbol{\theta})} d\boldsymbol{\theta}\right) \\
&\geq \int q(\boldsymbol{\theta}) \log\left(\frac{p(\mathcal{D}, \boldsymbol{\theta})}{q(\boldsymbol{\theta})}\right) d\boldsymbol{\theta} = L(q(\boldsymbol{\theta}))
\end{aligned}$$

where the equality on the middle line applies for any distribution $q(\boldsymbol{\theta})$ defined over the same domain as $p(\mathcal{D}, \boldsymbol{\theta})$, and the inequality on the bottom line arises from a use of Jensen's inequality applied to the logarithmic function

$$\mathrm{E}[\log(T)] \leq \log(\mathrm{E}[T])$$

The term *variational* within the expression *Variational Bayes* relates to the form of the underlying optimisation procedure. A functional mapping has been defined that takes both the desired probability density function and the variational approximation and returns as output a real value (the KL divergence between the two). Subject to a number of constraints (typically factorisations), Variational Bayes seeks to determine the variational approximation that minimises the output value. Thus it seeks an input function that minimises a functional mapping. This general problem lies with the field of mathematics known as *calculus of variations*.

3.7.2.1 The KL divergence

At this point we note that the the KL divergence is not symmetric, since for general distributions

$$KL(q(\boldsymbol{\theta})|p(\boldsymbol{\theta}|\mathcal{D})) \neq KL(p(\boldsymbol{\theta}|\mathcal{D})|q(\boldsymbol{\theta}))$$

As is noted by Winn (2004), the Variational Bayes minimisation of $KL(q(\boldsymbol{\theta})|p(\boldsymbol{\theta}|\mathcal{D}))$, rather than the reverse option of minimising $KL(q(\boldsymbol{\theta})|p(\boldsymbol{\theta}|\mathcal{D}))$, can significantly affect the properties of the resultant approximating distribution. Minimising $KL(q(\boldsymbol{\theta})|p(\boldsymbol{\theta}|\mathcal{D}))$ favours approximations $q(\boldsymbol{\theta})$ whose probability mass lies entirely within high probability regions of $p(\boldsymbol{\theta}|\mathcal{D})$, without requiring that all high probability regions of $p(\boldsymbol{\theta}|\mathcal{D})$ are covered. In contrast, minimising $KL(p(\boldsymbol{\theta}|\mathcal{D})|q(\boldsymbol{\theta}))$ favours approximations $q(\boldsymbol{\theta})$ whose probability mass covers all the high probability regions of $p(\boldsymbol{\theta}|\mathcal{D})$, even if this results in regions where $q(\boldsymbol{\theta})$ is high but $p(\boldsymbol{\theta}|\mathcal{D})$ is low.

Figure 3.11 illustrates this with a simple example in which the posterior distribution $p(x)$ is a two-component normal mixture distribution, but the approximating distribution $q(x)$ is only a univariate normal distribution [and therefore cannot faithfully approximate $p(x)$]. The solid line shows the target probability density function $p(x)$. The dashed line shows

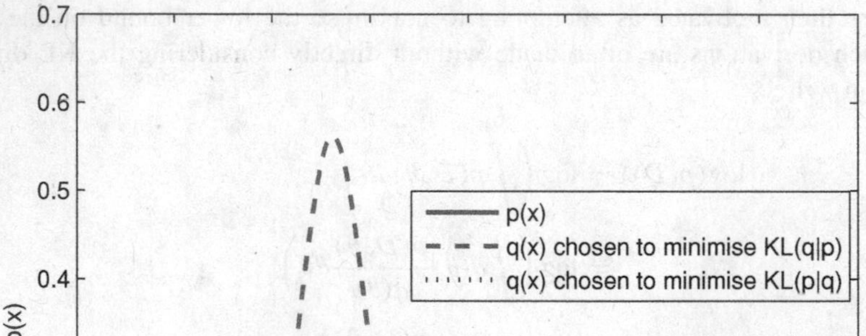

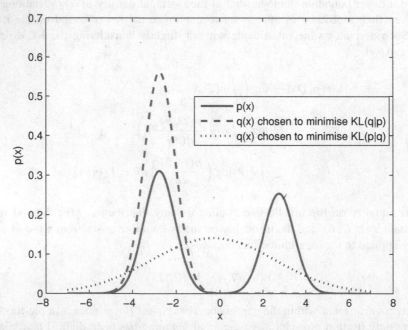

Figure 3.11 An illustration of the distributions favoured by the alternative forms of the Kullback–Leibler divergence, when the target density $p(x)$ is bimodal, but the approximation $q(x)$ is unimodal.

the (empirically determined) normal probability density function $q(x)$ that minimises the KL divergence $KL(q(x)|p(x))$ (the empirical determination relates to searching over a grid of values for the mean and standard deviation of the approximating normal distribution). This distribution has almost all its probability mass within a mode of $p(x)$. The dotted line shows the (empirically determined) normal probability density function $q(x)$ that minimises $KL(p(x)|q(x))$. This distribution covers all the high probability regions of $p(x)$, even though this results in assignment of high mass to areas of low probability of $p(x)$ (i.e. the area between the two modes).

If the effect of potentially missing high probability regions of $p(\boldsymbol{\theta}|\mathcal{D})$ is considered serious in a given application, then alternative approaches to Variational Bayes methods should be used, such as the sampling approaches considered earlier in this chapter.

3.7.3 Factorised variational approximation

Suppose that the factorised approximation given in (3.59) is used for the variational distribution. Then $L(q(\boldsymbol{\theta}))$ defined in (3.60) is given by

$$\begin{aligned} L(q(\boldsymbol{\theta})) &= \int \left(\prod_{i=1}^{k} q_i(\boldsymbol{\theta}_i) \right) \log \left(\frac{p(\mathcal{D}, \boldsymbol{\theta})}{\prod_{r=1}^{k} q_r(\boldsymbol{\theta}_r)} \right) d\boldsymbol{\theta} \\ &= \int \left(\prod_{i=1}^{k} q_i(\boldsymbol{\theta}_i) \right) \log(p(\mathcal{D}, \boldsymbol{\theta})) d\boldsymbol{\theta} - \int \left(\prod_{i=1}^{k} q_i(\boldsymbol{\theta}_i) \right) \left(\sum_{r=1}^{k} \log(q_r(\boldsymbol{\theta}_r)) \right) d\boldsymbol{\theta} \\ &= \int \left(\prod_{i=1}^{k} q_i(\boldsymbol{\theta}_i) \right) \log(p(\mathcal{D}, \boldsymbol{\theta})) d\boldsymbol{\theta} - \sum_{i=1}^{k} \int q_i(\boldsymbol{\theta}_i) \log(q_i(\boldsymbol{\theta}_i)) d\boldsymbol{\theta}_i \end{aligned} \tag{3.62}$$

where the last equality arises from noting that

$$\int \left(\prod_{i=1}^{k} q_i(\boldsymbol{\theta}_i) \right) \log(q_r(\boldsymbol{\theta}_r)) d\boldsymbol{\theta} = \int q_r(\boldsymbol{\theta}_r) \log(q_r(\boldsymbol{\theta}_r)) d\boldsymbol{\theta}_r$$

Variational calculus can be used to provide equations for the forms of the $q_i(\boldsymbol{\theta}_i)$ at local minima of $L(q(\boldsymbol{\theta}))$. However, a more accessible derivation of the equations is provided by Winn (2004), and is described here. By defining

$$\mathrm{E}_{(j)}[\log(p(\mathcal{D}, \boldsymbol{\theta}))] = \int \log(p(\mathcal{D}, \boldsymbol{\theta})) \left(\prod_{i \neq j} q_i(\boldsymbol{\theta}_i) d\boldsymbol{\theta}_i \right)$$

which is the expectation of $\log(p(\mathcal{D}, \boldsymbol{\theta}))$ with respect to all variational distribution factors apart from $q_j(\boldsymbol{\theta}_j)$, the dependence of (3.62) on a specific factor $q_j(\boldsymbol{\theta}_j)$ can be emphasised

$$L(q(\boldsymbol{\theta})) = \int q_j(\boldsymbol{\theta}_j) \mathrm{E}_{(j)}[\log(p(\mathcal{D}, \boldsymbol{\theta}))] d\boldsymbol{\theta}_j - \sum_{i=1}^{k} \int q_i(\boldsymbol{\theta}_i) \log(q_i(\boldsymbol{\theta}_i)) d\boldsymbol{\theta}_i \tag{3.63}$$

We then define a new marginal distribution

$$Q_j(\boldsymbol{\theta}_j) = \frac{1}{Z_j} \exp\left(\mathrm{E}_{(j)}[\log(p(\mathcal{D}, \boldsymbol{\theta}))]\right)$$

where Z_j is the normalisation factor that makes $Q_j(\boldsymbol{\theta}_j)$ a probability density

$$Z_j = \int \exp\left(\mathrm{E}_{(j)}[\log(p(\mathcal{D}, \boldsymbol{\theta}))]\right) d\boldsymbol{\theta}_j$$

This enables (3.63) to be rewritten as

$$L(q(\boldsymbol{\theta})) = \int q_j(\boldsymbol{\theta}_j) \log(Q_j(\boldsymbol{\theta}_j)) d\boldsymbol{\theta}_j + \log(Z_j) - \sum_{i=1}^{k} \int q_i(\boldsymbol{\theta}_i) \log(q_i(\boldsymbol{\theta}_i)) d\boldsymbol{\theta}_i$$

Now

$$KL(q_j(\boldsymbol{\theta}_j)|Q_j(\boldsymbol{\theta}_j)) = \int q_j(\boldsymbol{\theta}_j) \log\left(\frac{q_j(\boldsymbol{\theta}_j)}{Q_j(\boldsymbol{\theta}_j)}\right) d\boldsymbol{\theta}_j$$

so

$$L(q(\boldsymbol{\theta})) = -KL(q_j(\boldsymbol{\theta}_j)|Q_j(\boldsymbol{\theta}_j)) + \log(Z_j) - \sum_{i \neq j} \int q_i(\boldsymbol{\theta}_i) \log(q_i(\boldsymbol{\theta}_i)) d\boldsymbol{\theta}_i$$

where only the first term depends on $q_j(\boldsymbol{\theta}_j)$. Hence to maximise $L(q(\boldsymbol{\theta}))$ with respect to the factor $q_j(\boldsymbol{\theta}_j)$ we need only minimise the KL divergence $KL(q_j(\boldsymbol{\theta}_j)|Q_j(\boldsymbol{\theta}_j))$. This is done by setting

$$q_j(\boldsymbol{\theta}_j) = Q_j(\boldsymbol{\theta}_j) = \frac{1}{Z_j} \exp\left(\mathrm{E}_{(j)}[\log(p(\mathcal{D}, \boldsymbol{\theta}))]\right)$$

This can be expressed in terms of logarithms as

$$\log(q_j(\boldsymbol{\theta}_j)) = \mathrm{E}_{(j)}[\log(p(\mathcal{D}, \boldsymbol{\theta}))] + \text{constant} \tag{3.64}$$

where the constant is that required to normalise the distribution.

Evaluation of (3.64) requires the variational distribution factors $q_i(\boldsymbol{\theta}_i)$, $i \neq j$. Hence repeating the approach for each factor $q_i(\boldsymbol{\theta}_i)$ results in k coupled equations. A solution is found by adopting an iterative estimation procedure:

- Initialise the factors $q_j(\boldsymbol{\theta}_j)$, $j = 2, \ldots, k$. The precise form of initialisation will vary depending on the problem under consideration. As we will see in the example of Section 3.7.4, this initialisation does not necessarily require specification of full distributions – summary statistics are often sufficient.

- Update each $q_j(\boldsymbol{\theta}_j)$ in turn, for $j = 1, \ldots, k$, each time calculating (3.64) using the most recent estimates for $q_i(\boldsymbol{\theta}_i)$, $i \neq j$.
- Repeat the above step until convergence.

Since each update is carried out in turn, the above procedure increases (or keeps the same) the value of the lower bound $L(\boldsymbol{\theta})$ at each step. Convergence of the procedure can be monitored by calculating the values of $L(\boldsymbol{\theta})$ after each iteration. However, it is important to note that convergence may be only to a local maximum. Any reordering of the factors can be used within the updates.

3.7.4 Simple example

To illustrate the Variational Bayes procedure we consider the same univariate normal distribution problem as used to illustrate the Gibbs sampler in Section 3.4.2. The data are normally distributed with mean μ and variance σ^2. The prior distribution for μ is $N(\mu_0, \sigma^2/\lambda_0)$ and the prior distribution for $\tau = 1/\sigma^2$ is the gamma distribution with shape parameter α_0 and inverse-scale parameter β_0 (σ^2 therefore has an inverse-gamma distribution with shape parameter α_0 and scale parameter β_0).

We develop a Variational Bayes approximation to the posterior distribution of μ and σ^2 given independent measurements $x_1, \ldots, x_n$ from $N(\mu, \sigma^2)$. The variational distribution $q(\mu, \tau)$ is factorised into a product of marginal distributions

$$q(\mu, \tau) = q_\mu(\mu) q_\tau(\tau) \tag{3.65}$$

3.7.4.1 The $q_\mu(\mu)$ factor

Using (3.64) we have the following expression for $q_\mu(\mu)$

$$\log(q_\mu(\mu)) = \int_\tau q(\tau) \log(p(x_1, \ldots, x_n, \mu, \tau)) d\tau + \text{constant} \tag{3.66}$$

Now from (3.48) of Section 3.4.2, ignoring additive terms independent of μ and τ, we can express $\log(p(x_1, \ldots, x_n, \mu, \tau))$ as

$$\begin{aligned}\log(p(x_1, \ldots, x_n, \mu, \tau)) = &-\tau \left(\beta_0 + \frac{\sum_{i=1}^{n}(x_i - \mu)^2 + \lambda_0(\mu - \mu_0)^2}{2} \right) \\ &+ \left(\alpha_0 + \frac{n+1}{2} - 1 \right) \log(\tau) + \text{constant}\end{aligned} \tag{3.67}$$

Completing the square for μ, and noting that any additive terms independent of μ can be absorbed into the normalisation constant within Equation (3.66) we have

$$\log(q_\mu(\mu)) = \frac{-1}{2} \left(\int_\tau \tau q(\tau) d\tau \right) \lambda_n (\mu - \mu_n)^2 + \text{constant}$$

where

$$\lambda_n = \lambda_0 + n$$
$$\mu_n = (\lambda_0 \mu_0 + n\bar{x})/\lambda_n \tag{3.68}$$

with $\bar{x}$ the mean of $\{x_1, \ldots, x_n\}$. Hence

$$q_\mu(\mu) \propto \exp\left(\frac{-1}{2}\lambda_n E_{q_\tau}[\tau](\mu - \mu_n)^2\right) \tag{3.69}$$

where $E_{q_\tau}[\tau]$ is the expected value of τ under the variational distribution $q_\tau(\tau)$. By comparison of (3.69) with the probability density function of the normal distribution, we deduce that

$$q_\mu(\mu) = N\left(\mu; \mu_n, (\lambda_n E_{q_\tau}[\tau])^{-1}\right) \tag{3.70}$$

i.e. $q_\mu(\mu)$ has a normal probability density function with mean μ_n and variance $(\lambda_n E_{q_\tau}[\tau])^{-1}$. Conveniently, the only dependence on $q_\tau(\tau)$ is through $E_{q_\tau}[\tau]$, the mean value for τ under the $q_\tau(\tau)$ factor of the variational distribution. Note that the normal distribution arises entirely from the specification in (3.64) and has not been assumed during algorithm design.

3.7.4.2 The $q_\tau(\tau)$ factor

For $q_\tau(\tau)$ we have

$$\log(q_\tau(\tau)) = \int_\mu q(\mu)\log(p(x_1, \ldots, x_n, \mu, \tau))d\mu + \text{constant}$$

Starting from (3.67), $\log(p(x_1, \ldots, x_n, \mu, \tau))$ can be expanded and reorganised to

$$\begin{aligned}\log(p(x_1, \ldots, x_n, \mu, \tau)) = &-\tau\left(\beta_0 + \lambda_n\left(\mu^2/2 - \mu_n\mu\right) + \gamma_n\right)\\ &+(\alpha_n - 1)\log(\tau) + \text{constant}\end{aligned}$$

where

$$\alpha_n = \alpha_0 + (n+1)/2$$
$$\gamma_n = \frac{1}{2}\left(\lambda_0\mu_0^2 + \sum_{i=1}^n x_i^2\right)$$

Hence

$$\log(q_\tau(\tau)) = (\alpha_n - 1)\log(\tau) - \tau\beta_{\{\mathrm{E}_\mu[\mu],\mathrm{E}_\mu[\mu^2]\}} + \text{constant}$$

where

$$\beta_{\{\mathrm{E}_\mu[\mu],\mathrm{E}_\mu[\mu^2]\}} = \beta_0 + \lambda_n(\mathrm{E}_\mu[\mu^2]/2 - \mu_n\mathrm{E}_\mu[\mu]) + \gamma_n \tag{3.71}$$

This gives the following gamma distribution for the variational factor for τ

$$q_\tau(\tau) = \mathrm{Ga}(\tau; \alpha_n, \beta_n(\mathrm{E}_\mu[\mu], \mathrm{E}_\mu[\mu^2])) \tag{3.72}$$

The only dependence on $q_\mu(\mu)$ is through $E_{q_\mu}[\mu]$ and $E_{q_\mu}[\mu^2]$, the first two moments of the variational distribution for μ. As with the $q_\mu(\mu)$ factor, note that the form of $q_\tau(\tau)$ (in the case a gamma distribution) arises entirely from the specification in (3.64) and has not been assumed during algorithm design.

3.7.4.3 The optimisation process

The variational modelling procedure for this example is therefore:

- Initialise the mean of the variational distribution for τ, $E_{q_\tau}[\tau]$.
- Set $E_{q_\mu}[\mu] = \mu_n$, where μ_n is calculated according to (3.68). This value does not change throughout the procedure.
- Repeat the following steps until convergence:
  - Set $E_{q_\mu}[\mu^2] = (\lambda_n E_{q_\tau}[\tau])^{-1} + {\mu_n}^2$, the second moment of the normal distribution with mean μ_n and variance $(\lambda_n E_{q_\tau}[\tau])^{-1}$, using the most recent estimate of $E_{q_\tau}[\tau]$.
  - Calculate $\beta_{\{\mathrm{E}_\mu[\mu], \mathrm{E}_\mu[\mu^2]\}}$ according to (3.71) using the most recent estimate of $E_{q_\mu}[\mu^2]$, and the constant value for $E_{q_\mu}[\mu]$.
  - Set $E_{q_\tau}[\tau] = \alpha_n / \beta_{\{\mathrm{E}_\mu[\mu], \mathrm{E}_\mu[\mu^2]\}}$, the mean of the gamma distribution with shape parameter α_n and inverse-scale parameter $\beta_{\{\mathrm{E}_\mu[\mu], \mathrm{E}_\mu[\mu^2]\}}$.

Convergence of the procedure can be identified when the changes in the calculated statistics $E_{q_\tau}[\tau]$ and $E_{q_\mu}[\mu^2]$ are small. Alternatively the lower bound $L(q(\mu, \tau))$, defined in (3.62) can be calculated and monitored. Given the gamma and normal distributions, this is straightforward to calculate in this example.

On convergence of the procedure, the variational distribution factors are then as specified in (3.70) and (3.72), using the converged statistics. This variational distribution can then be used in lieu of the posterior distribution in all other calculations, e.g. when estimating predictive densities (Equation (3.1)) for use within a classifier based upon Bayes rule. However, at this point, it is important to note that the analytic solution is available for this problem (see Section 3.4.2), and therefore one does not need to use the described Variational Bayes procedure. The example is included because the steps taken are similar to those that would be adopted in more complicated problems, for which the analytic solution may not be available.

3.7.4.4 Example results

In our example, the data x_i, comprise 20 points from a normal distribution of zero mean and unit variance. The parameters of the prior distribution of μ and $\tau = 1/\sigma^2$ are taken to be $\lambda_0 = 1$, $\mu_0 = 1$, $\alpha_0 = 1/2$, $\beta_0 = 1/2$.

The Variational Bayes probability density function, evaluated over a grid, is displayed in Figure 3.12(a), alongside the actual probability density function Figure 3.12(b). A plot of the actual error between the two [i.e. $p(\mu, \tau|x_1, \ldots, x_{20}) - q(\mu, \tau)$, so that negative values

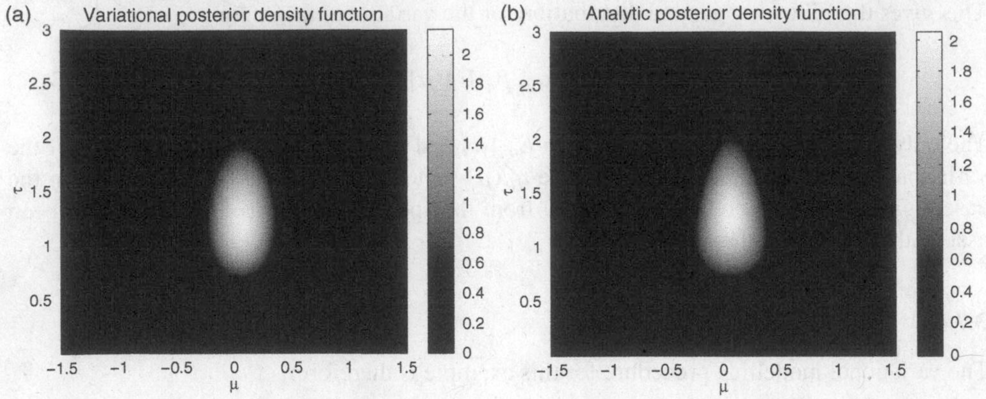

Figure 3.12 Variational Bayes estimate of the probability density function (a) and the actual probability density function (b).

indicate the variational density being larger than the actual density] is provided in Figure 3.13. The figure highlights that the mode of the variational distribution is more compact than that of the actual posterior distribution. A table of summary statistics for μ and σ^2 is provided in Table 3.2, which also shows the values for the MCMC example of Section 3.4.2, which used the same data. The variational values are close to the true values, with only the variance estimates showing larger errors than the MCMC values.

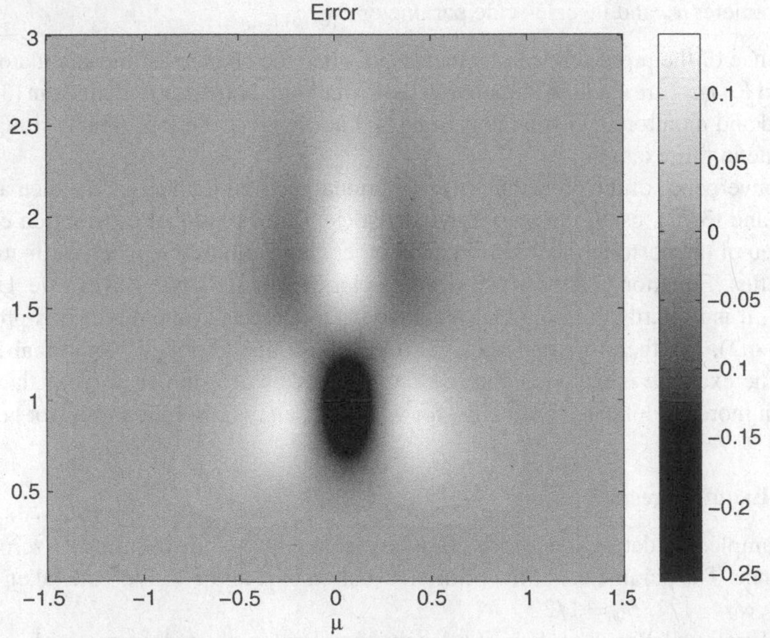

Figure 3.13 Error between the actual probability density function and the variational posterior density function, i.e. $p(\mu, \tau|x_1, \ldots, x_{20}) - q(\mu, \tau)$.

Table 3.2 Summary statistics for μ and σ^2. The true values are calculated from the known marginal posterior densities. The MCMC runs are as presented in Table 3.1. The variational values are calculated analytically from the variational distribution.

	True	MCMC short run	MCMC long run	Variational
Mean μ	0.063	0.082	0.063	0.063
Var. μ	0.039	0.042	0.039	0.035
Mean σ^2	0.83	0.83	0.82	0.82
Var. σ^2	0.081	0.085	0.079	0.075

3.7.5 Use of the procedure for model selection

Suppose that we are interested in performing model selection between different models (e.g. determining the most appropriate number of components in a mixture model), based upon the observed data $\mathcal{D}$. The posterior distribution over models can be written

$$p(m|\mathcal{D}) = \frac{p(m)p(\mathcal{D}|m)}{p(\mathcal{D})}$$

where $p(m)$ is the prior probability of the mth model structure, and $p(\mathcal{D}|m)$ is the marginal density of the data under the mth model structure. The most likely model is the one that maximises $p(m|\mathcal{D})$.[3]

Since

$$p(\mathcal{D}|m) = \int p(\mathcal{D}, \boldsymbol{\theta}_m|m)d\boldsymbol{\theta}_m = \int p(\mathcal{D}|\boldsymbol{\theta}_m, m)p(\boldsymbol{\theta}_m|m)d\boldsymbol{\theta}_m$$

there is a penalisation of overly large model structures, through a more diffuse prior $p(\boldsymbol{\theta}_m|m)$ for models with a large number of degrees of freedom than models with a small number of degrees of freedom. This effect, referred to as *Occam's razor*, can reduce overfitting of models. In addition, large models can be penalised through the model prior probabilities $p(m)$.

Variational Bayes methods can be used for model selection, through determination of the Variational Bayes approximation separately for each model structure. For the mth model structure, $L_m(q(\boldsymbol{\theta}))$ is optimised with respect to the model parameters $\boldsymbol{\theta}$ for that structure. As can be seen from an extension to (3.61), each $L_m(q(\boldsymbol{\theta}))$ provides a lower bound on the logarithm of the marginal density $p(\mathcal{D}|m)$. Hence the posterior model order probabilities can be approximated as

$$p(m|\mathcal{D}) \approx \frac{p(m)\exp(L_m(q(\boldsymbol{\theta})))}{\sum_{m'} p(m')\exp(L_{m'}(q(\boldsymbol{\theta})))}$$

Attias (2000) uses such an approach for mixture models.

[3] Note that a full Bayesian treatment would integrate over all model structures. However, this may not be practical computationally, and therefore a single model structure is generally selected.

An alternative approach for mixture models is proposed by Corduneanu and Bishop (2001). They show that by optimising the mixture component probabilities to maximise the marginal density of the data conditional on the mixture component probabilities [i.e. $p(\mathcal{D}|\boldsymbol{\pi})$ where $\boldsymbol{\pi}$ is the mixture component probability vector], unnecessary components are suppressed. Hence, provided that the initial number of components modelled is sufficiently large, the model order can be ascertained without running experiments at many different model orders. A two-step iterative procedure is adopted. Variational procedures are used to obtain lower bounds on $p(\mathcal{D}|\boldsymbol{\pi})$. These variational procedures are alternated with an optimisation step which maximises the lower bound with respect to $\boldsymbol{\pi}$. It is found that the component probabilities for unnecessary mixture components converge to zero.

3.7.6 Further developments and applications

Much work on Variational Bayes methods has considered problems in which there are hidden or latent variables in addition to the model parameters. In mixture models these hidden variables are the missing component labels z_i associated with each data measurement $\boldsymbol{x}_i$ (see Chapter 2). Beal (2003) derives the Variational Bayes equations whilst explicitly separating hidden variables from the model parameters. In this manner similarities between Variational Bayes methods and the EM algorithm (Chapter 2) are highlighted. Specifically, the EM algorithm is rederived using a variational approach in which there are no constraints on the form of the variational distributions. An approximated version of the EM algorithm can be obtained for problems in which the EM calculations are intractable, by constraining the variational distributions, e.g. through factorisations or by using parametrised families for the variational distributions, such as normal distributions.

In the simple example of Section 3.7.4 only the factorisation in (3.65) needed to be specified in order to provide the Variational Bayes solution. In more complicated problems it may not be possible to calculate the required expressions of (3.64) analytically. In such cases, as noted by Beal (2003), one needs to assume parametric forms for some of the variational distributions, e.g. normal distributions. Optimisation of $L(q(\boldsymbol{\theta}))$ is then with respect to the parameters of these distributions.

Bishop (1999) develops a Bayesian version of principal components analysis (PCA) (see Chapter 10). The approach chooses the number of principal components to retain through a procedure known as *automatic relevance determination*. The procedure uses a *hyperparameter* (hyperparameters are parameters on the prior distributions) for each weight vector defining a principal component, which can 'turn off' the effect of the weight vector, thereby effectively reducing the number of retained principal components (initially the maximum possible number of components is considered). A variational approach is used to determine the posterior distributions, with additional factorisations arising automatically, on top of the initial variational factorisation of variables.

Tzikas *et al.* (2008) demonstrate the application of Variational Bayes techniques to problems of linear regression and mixture models. An early introduction to variational methods, and their use in graphical models such as hidden Markov models is provided by Jordan (1999). McGrory and Titterington (2009) develop Variational Bayes methods for hidden Markov models where the number of hidden states is an unknown.

Attias (2000) applies Variational Bayes methods to blind source separation, in which one attempts to separate out linearly mixed signals. Here additional parametric constraints are placed on one of the factorised variational distributions, since for the considered problem calculation of (3.64) without such constraints is not tractable.

Winn and Bishop (2005) (see also Winn, 2004) develop a Variational Message Passing algorithm for applying variational inference to Bayesian networks. The approach is a variational version of the belief propagation algorithms used in Bayesian networks (Pearl, 1988), and applies to conjugate-exponential models. Conjugate-exponential models are models where the conditional and prior distributions are all conjugate and belong to the exponential family of distributions (Bernardo and Smith, 1994). A software package, termed VIBES (Variational Inference in Bayesian Networks), has been developed for the algorithm, with a view to simplifying use of Variational Bayes techniques.

Ghahramani and Beal (1999) calculate a variational distribution for use as the importance sampling proposal distribution within an importance sampler (see Section 3.3.6). The resulting weighted samples are then a more truthful representation of the actual posterior distribution than the variational distribution alone. The advantage of using the variational procedure is that it provides an intelligent proposal distribution that improves the efficiency of the importance sampler compared with using a naive proposal distribution.

3.7.7 Summary

Variational Bayes provides an alternative means of approximating full Bayesian inference than approaches based on drawing samples from the posterior distribution (such as MCMC sampling, see Section 3.4). The advantage over Bayesian sampling approaches is reduced computational cost. In its commonest form, Variational Bayes uses a factorised approximation to the posterior distribution, with the forms of the factors selected to minimise the KL divergence to the actual posterior distribution. The approach can be used for model selection, and practitioners have developed and applied it successfully to a number of real problems.

The simplifications (e.g. factorisations) made to the Variational Bayes distributions, may have unforeseen effects on the quality of the approximation of the variational Bayes distribution to the true posterior distribution. This may degrade performance when the variational distributions are then used with a classifier, compared with a full Bayesian sampling approach. Furthermore, the procedures often converge only to local optima. However, for some problems, this potential reduction in performance may be acceptable given the reduced computational cost of the procedure compared with Bayesian sampling schemes.

As was noted in the discussion on the use of the KL divergence, the procedure might not be appropriate if the true posterior distribution is expected to have multiple modes. This is because variational approximations often tend to represent only one of the modes. The selected mode may vary depending on the initialisation of the variational procedures, the order of updates in the iterative optimisation procedure, and the relative posterior probabilities of the modes.

3.8 Approximate Bayesian Computation

3.8.1 Introduction

At the heart of the use of Bayesian inference within statistical pattern recognition problems is the likelihood function $p(\mathcal{D}|\boldsymbol{\theta})$, i.e. the probability density function of the data measurements (or features) $\mathcal{D}$ given the (unknown) model parameters $\boldsymbol{\theta}$. The posterior distribution of the model parameters is calculated using the likelihood function $p(\mathcal{D}|\boldsymbol{\theta})$ and the prior distribution $p(\boldsymbol{\theta})$ via Bayes' theorem

$$p(\boldsymbol{\theta}|\mathcal{D}) \propto p(\mathcal{D}|\boldsymbol{\theta})p(\boldsymbol{\theta})$$

In all cases considered so far the functional form of the likelihood $p(\mathcal{D}|\boldsymbol{\theta})$ is known, and may be evaluated. However, in some complex scenarios (e.g. problems in statistical genetics and environmental studies) the likelihood will be analytically or computationally intractable (e.g. it might involve summing probabilities over lots of potential hidden states). This prevents the use of analytic Bayesian inference, and also use of the Bayesian sampling schemes and Variational Bayes techniques considered earlier in this chapter. This is because these techniques all require knowledge of, or evaluation of, the likelihood function.

Arising initially from the field of statistical genetics (Tavaré *et al.*, 1997; Pritchard *et al.*, 1999; Beaumont *et al.*, 2002) Approximate Bayesian Computation (ABC) techniques provide a means for drawing samples from a posterior distribution without evaluating the likelihood function. This is achieved through the use of computer simulation models, which simulate the measurements that could be received for a given set of parameters. Rather than evaluating the likelihood function, one draws samples from the likelihood distribution. As in standard Bayesian sampling algorithms, potential samples from the posterior distribution are proposed. However, when accepting, rejecting and weighting these samples, rather than calculating the likelihood function, the algorithms compare simulated data with the actual measured data.

Over the next few sections we introduce three of the most common forms of ABC samplers, namely ABC Rejection Sampling, ABC MCMC Sampling, and ABC Population Monte Carlo Sampling. Unlike the outputs of previously considered Bayesian sampling schemes, the resulting ABC samples cannot be used within a Bayesian classifier in the manner described in Section 3.3.3. This is because the classification rule requires evaluation of the likelihood function for each class [Equation (3.40)], which is the very thing that ABC sampling algorithms have been designed to avoid. One possibility is to build the classification decision into an ABC sampling algorithm. Some model selection ABC algorithms are therefore discussed in Section 3.8.5. Model selection in ABC algorithms is very much an active research area (Didelot *et al.*, 2011).

3.8.2 ABC Rejection Sampling

Suppose that we wish to draw samples from a posterior distribution $p(\boldsymbol{\theta}|\mathcal{D})$, arising from a prior distribution $p(\boldsymbol{\theta})$ and a likelihood distribution $p(\mathcal{D}|\boldsymbol{\theta})$, where the measurement data $\mathcal{D}$ is discrete. A rejection sampling procedure (see Section 3.3.4) for drawing samples from the posterior distribution is:

1. Sample a point $\boldsymbol{\theta}$ from $p(\boldsymbol{\theta})$.
2. Accept $\boldsymbol{\theta}$ with probability $p(\mathcal{D}|\boldsymbol{\theta})$.

This is continued until a sufficient number of samples have been accepted. To see that this is rejection sampling, note that in the notation of Section 3.3.4

$$g(\boldsymbol{\theta}) = p(\boldsymbol{\theta})p(\mathcal{D}|\boldsymbol{\theta}), \quad s(\boldsymbol{\theta}) = p(\boldsymbol{\theta}), \quad g(\boldsymbol{\theta})/s(\boldsymbol{\theta}) = p(\mathcal{D}|\boldsymbol{\theta}).$$

Therefore, since the data $\mathcal{D}$ are discrete, the upper bound A for $g(\boldsymbol{\theta})/s(\boldsymbol{\theta})$ is 1. Hence we accept the sample if the point drawn from the uniform distribution on $[0, 1]$ is less than or equal to $p(\mathcal{D}|\boldsymbol{\theta})$, i.e. with probability $p(\mathcal{D}|\boldsymbol{\theta})$.

Now suppose that we cannot calculate the likelihood function $p(\mathcal{D}|\boldsymbol{\theta})$, but can simulate measurements from it (e.g. via a computational model). In such circumstances the rejection sampling procedure for discrete measurement data can be replaced by the following likelihood-free procedure:

1. Sample a point $\boldsymbol{\theta}$ from $p(\boldsymbol{\theta})$.
2. Simulate $\mathcal{D}'$ from the measurement model using the sampled parameters $\boldsymbol{\theta}$.
3. Accept $\boldsymbol{\theta}$ if $\mathcal{D}' = \mathcal{D}$.

That the above procedure is equivalent to the rejection sampling algorithm is clear, since the probability of generating a discrete data sample $\mathcal{D}'$ equal to $\mathcal{D}$ is $p(\mathcal{D}|\boldsymbol{\theta})$, as required. However, the approach is not applicable if we have continuous data measurements, because the simulated data must be equal to the measured data for a sample to be accepted, which occurs with zero probability in the continuous case. In such cases an approximate procedure is needed, which leads to the following version of ABC Rejection Sampling (applicable to both discrete and continuous measurements):

1. Sample a point $\boldsymbol{\theta}$ from $p(\boldsymbol{\theta})$.
2. Simulate $\mathcal{D}'$ from the measurement model using the sampled parameters $\boldsymbol{\theta}$.
3. Accept $\boldsymbol{\theta}$ if $\rho(\mathcal{D}, \mathcal{D}') \leq \epsilon$.

Here $\rho(\mathcal{D}, \mathcal{D}')$ is a distance metric between the actual measurements $\mathcal{D}$ and the simulated measurements $\mathcal{D}'$ (e.g. the Euclidean distance), and $\epsilon > 0$ is the ABC distance tolerance. The samples drawn from Rejection ABC Sampling are then approximate samples from the posterior distribution, with the quality of the approximation increasing as $\epsilon > 0$ decreases. The choice of ϵ is a trade-off between accuracy of the approximation, and the computational effort. Clearly, using a very small value of ϵ with continuous data is likely lead to a large number of samples being rejected, making the procedure computationally expensive. If $\epsilon = 0$ we recover the original procedure for discrete measurement data.

To reduce the computational expense of the procedure for medium-to-high dimensional data $\mathcal{D}$ it is common to consider lower-dimensional summary statistics of the data, rather than the full data. Suppose that we calculate summary statistics $S(\mathcal{D})$ for the dataset $\mathcal{D}$. The ABC Rejection Sampling procedure becomes:

1. Sample a point $\boldsymbol{\theta}$ from $p(\boldsymbol{\theta})$.
2. Simulate $\mathcal{D}'$ from the measurement model using the sampled parameters $\boldsymbol{\theta}$.
3. Calculate summary statistics $S(\mathcal{D}')$ from the simulated data $\mathcal{D}'$.
4. Accept $\boldsymbol{\theta}$ if $\rho_S(S(\mathcal{D}), S(\mathcal{D}')) \leq \epsilon$.

Here the distance metric $\rho_S(S, S')$ operates on the summary statistics. If the summary statistics are *sufficient statistics*[4] for $\boldsymbol{\theta}$ then this approach introduces no further approximation over the previous version based upon the full measurements. If the summary statistics are not sufficient, then it represents a further approximation, the quality of which will vary depending on how close they are to being sufficient.

Whilst the ABC rejection algorithm is easy to implement, it suffers from all the disadvantages of standard rejection sampling, in particular a high rejection rate if the prior and posterior distributions are not similar. This problem is exacerbated by the extra approximations introduced by the ABC sampling procedure. When the computational cost of generating samples from the measurement distribution is high, this can encourage practitioners to increase the acceptance threshold ϵ, with a consequent degradation in the quality of the sample approximations.

3.8.3 ABC MCMC Sampling

The computational inefficiency of ABC Rejection Sampling motivates the development of ABC versions of more sophisticated Bayesian sampling schemes. One such algorithm is a likelihood-free version of Metropolis–Hastings MCMC sampling (see Section 3.4.3), first proposed by Marjoram *et al.* (2003).

As with the preceding discussion on ABC Rejection Sampling, suppose that we are seeking to draw samples from a posterior distribution $p(\boldsymbol{\theta}|\mathcal{D})$, where the measurement data $\mathcal{D}$ are continuous or discrete, the prior distribution is given by $p(\boldsymbol{\theta})$, and the likelihood function $p(\mathcal{D}|\boldsymbol{\theta})$ cannot be calculated analytically, but can have measurements simulated from it. We suppose that we can calculate summary statistics $S(\mathcal{D})$ from the data $\mathcal{D}$.

We consider the Metropolis–Hastings algorithm, where $\boldsymbol{\theta}^t$ is our current sample from the posterior, and our proposal distribution is given by $q(\boldsymbol{\theta}|\boldsymbol{\theta}^t)$. The next iteration of the ABC MCMC (Metropolis–Hastings) algorithm is as follows:

1. Draw a sample $\boldsymbol{\theta}$ from the proposal distribution $q(\boldsymbol{\theta}|\boldsymbol{\theta}^t)$.
2. Simulate $\mathcal{D}'$ from the measurement model using the sampled parameters $\boldsymbol{\theta}$.
3. Calculate summary statistics $S(\mathcal{D}')$ from the simulated data $\mathcal{D}'$.
4. If $\rho_S(S(\mathcal{D}), S(\mathcal{D}')) \leq \epsilon$ then proceed to step 5, otherwise set $\boldsymbol{\theta}^{t+1} = \boldsymbol{\theta}^t$ and jump to step 6.
5. Set $\boldsymbol{\theta}^{t+1} = \boldsymbol{\theta}$ with probability $a(\boldsymbol{\theta}^t, \boldsymbol{\theta})$ where
$$a(\boldsymbol{\theta}^t, \boldsymbol{\theta}) = \min\left(1, \frac{p(\boldsymbol{\theta})q(\boldsymbol{\theta}^t|\boldsymbol{\theta})}{p\left(\boldsymbol{\theta}^t\right)q(\boldsymbol{\theta}|\boldsymbol{\theta}^t)}\right)$$
Otherwise set $\boldsymbol{\theta}^{t+1} = \boldsymbol{\theta}^t$.
6. If more samples are required repeat the process starting from $\boldsymbol{\theta}^{t+1}$.

[4] Sufficient statistics are summary statistics of a dataset which are such that the distribution of the data is independent of the parameters of the underlying distribution when conditioned on the statistic (Bernardo and Smith, 1994).

Note that as in ABC Rejection Sampling we have used a distance metric $\rho_S(S, S')$ operating on the summary statistics, and an ABC distance tolerance $\epsilon > 0$.

For discrete data $\mathcal{D}$, the validity of a nonapproximated version of the above sampler can be proved by showing detailed balance [Equation (3.46)]. In the nonapproximated version, $\rho_S(S(\mathcal{D}), S(\mathcal{D}')) = ||\mathcal{D}' - \mathcal{D}||$ and $\epsilon = 0$ (i.e. to accept a sample we require $\mathcal{D}' = \mathcal{D}$ as in the first likelihood-free rejection sampling algorithm). The transition kernel from $\boldsymbol{\theta}$ to $\boldsymbol{\theta}'$ for a non-trivial move (i.e. an accepted move, $\boldsymbol{\theta}' \neq \boldsymbol{\theta}$) is therefore given by

$$K(\boldsymbol{\theta}, \boldsymbol{\theta}') = q(\boldsymbol{\theta}'|\boldsymbol{\theta})p(\mathcal{D}|\boldsymbol{\theta}')a(\boldsymbol{\theta}, \boldsymbol{\theta}')$$

and therefore

$$\begin{aligned}
p(\boldsymbol{\theta}|\mathcal{D})K(\boldsymbol{\theta}, \boldsymbol{\theta}') &= = \frac{p(\mathcal{D}|\boldsymbol{\theta})p(\boldsymbol{\theta})}{p(\mathcal{D})} q(\boldsymbol{\theta}'|\boldsymbol{\theta})p(\mathcal{D}|\boldsymbol{\theta}')a(\boldsymbol{\theta}, \boldsymbol{\theta}') \\
&= \frac{p(\mathcal{D}|\boldsymbol{\theta}')p(\boldsymbol{\theta}')}{p(\mathcal{D})} \frac{p(\boldsymbol{\theta})}{p(\boldsymbol{\theta}')} \frac{q(\boldsymbol{\theta}'|\boldsymbol{\theta})}{q(\boldsymbol{\theta}|\boldsymbol{\theta}')} q(\boldsymbol{\theta}|\boldsymbol{\theta}')p(\mathcal{D}|\boldsymbol{\theta})a(\boldsymbol{\theta}, \boldsymbol{\theta}') \\
&= p(\boldsymbol{\theta}'|\mathcal{D})q(\boldsymbol{\theta}|\boldsymbol{\theta}')p(\mathcal{D}|\boldsymbol{\theta}) \frac{q(\boldsymbol{\theta}'|\boldsymbol{\theta})p(\boldsymbol{\theta})}{q(\boldsymbol{\theta}|\boldsymbol{\theta}')p(\boldsymbol{\theta}')} \min\left(1, \frac{q(\boldsymbol{\theta}|\boldsymbol{\theta}')p(\boldsymbol{\theta}')}{q(\boldsymbol{\theta}'|\boldsymbol{\theta})p(\boldsymbol{\theta})}\right) \\
&= p(\boldsymbol{\theta}'|\mathcal{D})q(\boldsymbol{\theta}|\boldsymbol{\theta}')p(\mathcal{D}|\boldsymbol{\theta})a(\boldsymbol{\theta}', \boldsymbol{\theta}) \\
&= p(\boldsymbol{\theta}'|\mathcal{D})K(\boldsymbol{\theta}', \boldsymbol{\theta})
\end{aligned}$$

i.e. we have detailed balance.

A practical difficulty with the ABC MCMC Sampling scheme is that it can become stuck in the tails of the distribution for long periods of time, with the result that the algorithm needs to be run for long periods to obtain good mixing. The problem arises because in the tails the simulated data are likely to be a poor match to the measured data (which are likely to have been generated from parameters that are not in the tails of the distribution) and therefore the moves will be rejected.

3.8.4 ABC Population Monte Carlo Sampling

Some of the difficulties with ABC algorithms can be avoided if we gradually move to the distribution of interest, via a sequence of target distributions, as in SMC samplers (Section 3.6). The sequence of target distributions is the sequence of ABC approximations of the posterior distributions with different distance tolerances, ϵ. Specifically, the tth distribution in the sequence uses an ABC distance tolerance of ϵ_t, where the ϵ_t, $t = 1, \ldots, T$ are a strictly decreasing sequence, $\epsilon_1 > \epsilon_2 > \cdots > \epsilon_T > 0$. ϵ_T is the final tolerance, set sufficiently small that the ABC posterior approximation defined by it should be a good approximation to the actual posterior distribution. The motivation behind such a tempered sequence of tolerances is that for larger tolerances the samples will move around more, because the simulated data are more likely to be accepted. This helps initial mixing. As the distance tolerance decreases the samples then focus in on the actual posterior distribution.

An example of such an approach is the ABC Population Monte Carlo (PMC) algorithm (Beaumont *et al.*, 2009; Toni *et al.*, 2009). Using the same notation as earlier in this section, the ABC PMC algorithm is as follows:

- Initialisation: Perform ABC Rejection Sampling with (fairly large) ABC tolerances, ϵ_1. Continue until N samples $\{\boldsymbol{\theta}_1^{(1)}, \ldots, \boldsymbol{\theta}_1^{(N)}\}$ have been accepted (remembering that this will take more than N steps, since many proposed samples will be rejected). Assign equal weights, $w_1^{(s)} = 1/N, s = 1, \ldots, N$, to the samples.
- For iterations $t = 2, \ldots, T$:
  - For $i = 1, \ldots, N$:
    1. Select $\boldsymbol{\theta}'$ from the previous population $\{\boldsymbol{\theta}_{t-1}^{(1)}, \ldots, \boldsymbol{\theta}_{t-1}^{(N)}\}$ according to the probabilities $\{w_{t-1}^{(1)}, \ldots, w_{t-1}^{(n)}\}$, i.e. set $\boldsymbol{\theta}' = \boldsymbol{\theta}_{t-1}^{(j)}$ with probability $w_{t-1}^{(j)}$.
    2. Perturb the sample according to the perturbation kernel $\psi(\boldsymbol{\theta}', \boldsymbol{\theta})$, giving an updated sample $\boldsymbol{\theta}''$. An example perturbation kernel is a multivariate normal distribution, centred on the point $\boldsymbol{\theta}'$.
    3. Simulate $\mathcal{D}'$ from the measurement model using the sampled parameters $\boldsymbol{\theta}''$.
    4. Calculate summary statistics $S(\mathcal{D}')$ from the simulated data $\mathcal{D}'$.
    5. If $\rho_S(S(\mathcal{D}), S(\mathcal{D}')) \leq \epsilon_t$ then proceed to step 6, otherwise return to step 1.
    6. Set $\boldsymbol{\theta}_t^{(i)} = \boldsymbol{\theta}''$.
    7. Using a sequential importance sampling argument set:

$$W_t^{(i)} = \frac{p(\boldsymbol{\theta}_t^{(i)})}{\sum_{s=1}^{N} w_{t-1}^{(s)} \psi(\boldsymbol{\theta}_{t-1}^{(s)}, \boldsymbol{\theta}_t^{(i)})}$$

where $\psi(\boldsymbol{\theta}_{t-1}^{(s)}, \boldsymbol{\theta}_t^{(s)})$ is the probability density function of the perturbation kernel, e.g. $\psi(\boldsymbol{\theta}_{t-1}^{(s)}, \boldsymbol{\theta}_t^{(s)}) = N(\boldsymbol{\theta}_t^{(s)}; \boldsymbol{\theta}_{t-1}^{(s)}, \Omega)$ for a multivariate normal perturbation with covariance matrix Ω.

  - Normalise the weights as $w_t^{(i)} = W_t^{(i)}/(\sum_{s=1}^{n} W_t^{(s)}), i = 1, \ldots, N$.

3.8.5 Model selection

The samples output by ABC samplers can be used to estimate summary statistics on the parameters of interest. However, as has been noted in the initial discussion, they cannot be used within a Bayesian classifier as described in Section 3.3.3, since these classifiers require evaluation of the likelihood function for each class [Equation (3.40)]. Such an evaluation of the likelihood function is the very thing that ABC algorithms have been designed to avoid.

However, in some circumstances it is possible to use a model selection algorithm to determine the class (i.e. model) to which an object belongs. An ABC rejection model selection algorithm has been proposed by Grelaud *et al.* (2009), and is as follows:

- For $s = 1, \ldots, N$

  1. Sample a candidate model m' according to the prior model probabilities, $\{\pi_1, \ldots, \pi_M\}$, where M is the number of models.

  2. Sample a point $\boldsymbol{\theta}$ from $p(\boldsymbol{\theta}|m')$, the prior distribution for the parameters of model m'.

  3. Simulate $\mathcal{D}'$ from the measurement model for the m'th model using the sampled parameters $\boldsymbol{\theta}$.

  4. Calculate summary statistics $S(\mathcal{D}')$ from the simulated data $\mathcal{D}'$.

  5. If $\rho_S(S(\mathcal{D}), S(\mathcal{D}')) \leq \epsilon$ accept the sample and set $m^{(s)} = m'$, $\boldsymbol{\theta}^{(s)} = \boldsymbol{\theta}$. Otherwise reject and return to step 1.

Once N points have been accepted, the marginal model probabilities can be approximated as follows

$$P(m = m'|\mathcal{D}) \approx \frac{1}{N} \sum_{s=1}^{N} I(m^{(s)} = m')$$

The object can then be assigned to the class with the maximum marginal model probability.

The above approach is designed for situations where the class measurement distributions have different parametric models, rather than different parameter values of the same underlying parametric model. However, in a general discrimination problem, the approach could still be used if the prior parameter distributions, $p(\boldsymbol{\theta}|m)$, are infact posterior parameter distributions, $p(\boldsymbol{\theta}|\mathcal{D}_{(m)}, m)$, estimated using separate training datasets, $\mathcal{D}_{(m)}$, for each model type. Estimation of the posterior parameter distributions will need to have been through earlier ABC sampling procedures.

Model selection versions of the ABC PMC algorithm have been proposed by Toni *et al.* (2009) and Toni and Stumpf (2010). However, Didelot *et al.* (2011) add a note of caution, in that there is bias when ABC is used to approximate the *Bayes Factor*[5] between two models, with this bias arising from the use of sufficient statistics within the ABC procedures (Robert *et al.*, 2011).

3.8.6 Summary

ABC algorithms are a fast moving field of Bayesian statistics, developed for use in situations where the likelihood function is analytically or computationally intractable, but where it is possible to simulate data measurements from it. ABC sampling algorithms propose samples using similar procedures to standard Bayesian sampling, but accept, reject and weight

[5] The Bayes Factor is the posterior to prior odds ratio for two competing models, and can be used for model comparison (Bernardo and Smith, 1994).

those samples by comparing simulated data measurements with the actual measured data, rather than evaluating likelihood functions. The comparisons are typically based on summary statistics of the data, which may introduce unacceptable errors if they are far from sufficient statistics. Use of ABC for model selection, and therefore classification, is an active research area, and many improvements to the outlined ABC rejection model selection algorithm can be expected.

3.9 Example application study

The problem

This application concerns automatic restoration of damaged and contaminated images (Everitt and Glendinning, 2009).

Summary

The approach uses a semi-parametric model to recover an underlying image from damaged and contaminated data. A Bayesian approach is adopted with an MCMC sampling algorithm used to estimate both parametric-components describing known physical characteristics of the image (such as lighting effects) and also weights associated with a more flexible semi-parametric component describing the underlying image.

The data

The approach is used to restore a corruption of the standard *Lena* image (an image used commonly for demonstrating performance of image processing algorithms), with the corruption consisting of random missing pulses (ranging from 50 to 95% missing) and a global nonlinear lighting effect. The approach is also used to detect eye blink artifacts in electroencephalographic (EEG) imagery.

Model

The approach describes the value of an image $Y(x, z)$ at location (x, z) by

$$Y(x, z) = f(\beta, x, z) + h(x, z) + Z(x, z)$$

where $Z(x, z)$ represents independent identically distributed white noise, $f(\beta, x, z)$ is the parametric component, and $h(x, z)$ is a locally smooth component, describing the underlying image.

For the Lena image, a nonlinear lighting effect is modelled by the parametric component. For the EEG application, the effect of eye blink artifacts is incorporated via consideration of two alternative models for the parametric component, one containing an eye-blink effect, and the other a no-blink model. For each EEG image, the posterior model probability is used to select the best fit, and therefore to determine whether an eye blink artifact has occurred.

Algorithm

The nonparametric component is modelled using basis function expansions. Prior distributions are specified for the basis function weights, and for the parameters of the parametric components. A Metropolis-within-Gibbs MCMC procedure (see Section 3.4.3) is used to draw samples from the posterior distribution. Point estimates for the semi-parametric model, and therefore the underlying image, are obtained from the posterior sample means. An additional importance sampling step is used to estimate the posterior model probabilities in the EEG application.

Results

The approach recovers the underlying Lena image, with substantial distortions only appearing at extreme levels of missing pixels. Applied to EEG data, the approach consistently assigns the largest posterior model probability to the correct instance of with or without eye blink artifacts.

3.10 Application studies

A compilation of examples of applications of Bayesian methodology is given in the book by French and Smith (1997). This includes applications in clinical medicine, flood damage analysis, nuclear plant reliability and asset management. Lock and Gelman (2010) use analytic Bayesian inference to integrate previous election results and pre-election polls to make predictions on the 2008 US election.

As computational power continues to increase, use of Bayesian sampling techniques in real applications has increased dramatically:

- Copsey and Webb (2000) use an MCMC treatment of normal mixture models to classify mobile ground targets in inverse synthetic aperture radar (ISAR) images.

- Davy *et al.* (2002) use MCMC methods to classify chirp signals, with application to radar target identification, and knock detection in car engines.

- Bentow (1999) develops an MCMC algorithm to assess the stability of ecological ordinations. These are orderings of environmental species (or sites), such that similar species are placed close to one another, and dissimilar species are kept further apart. Ecological ordination is used by ecologists to study the effects of environmental factors on communities.

- Lyons *et al.* (2008) use Bayesian inference with an MCMC sampler to estimate posterior distributions for chloroform concentrations in tap water and ambient household air, given biomonitoring data collected at a later time.

- Everitt and Glendinning (2009) develop an MCMC algorithm for automatic restoration of damaged and contaminated images, and use the approach to identify blinking artifacts in EEG images.

- Lane (2010) considers a Bayesian approach to super-resolution. Both analytic and MCMC schemes are developed. The work has application to automatic target detection and recognition using radar imagery.

- Maskell (2008) considers the use of Bayesian models for information fusion, including an example of classification of objects from sequences of images using particle filters.

- Numerous applications of the particle filtering methodology for sequential parameter estimation are discussed in the book by Doucet *et al.* (2001). These applications include tracking of manoeuvring objects, monitoring of semiconductor growth using spectroscopic ellipsometry, and tracking of human motion with relevance to gesture recognition.

- Montemerlo *et al.* (2003) use a particle filter within their FastSLAM algorithms for simultaneous localisation and mapping (SLAM). SLAM is a means of navigation, in which an autonomous vehicle travelling through an unknown environment builds a map of the environment, while simultaneously estimating its own position within the map. Low accuracy motion estimates provided by inertial navigation systems are supplemented by relative measurements taken of a number of landmarks of opportunity, and the map of the environment is built up by estimating the positions of the landmarks.

- SMC samplers have been used for financial modelling. Jasra *et al.* (2011) discuss modelling of stochastic volatility in stock market data. Jasra and Del Moral (2011) discuss how SMC methods can be used for option pricing.

- Lane *et al.* (2009), develop an ABC version of an SMC sampler for the problem of source term estimation (STE) for chemical, biological, radiological and nuclear (CBRN) defence. STE relates to using sensor measurements to infer the locations, times, amounts and types of material in multiple CBRN releases.

- Toni and Stumpf (2010) use a model selection ABC PMC algorithm to ascertain whether different strains of the influenza virus share the same spread dynamics.

3.11 Summary and discussion

The approaches developed in this chapter towards discrimination have been based on estimation of the class-conditional density functions using Bayesian techniques. Unlike the maximum likelihood approaches of Chapter 2, these approaches take into account parameter variability due to data sampling. The procedures are more complicated than those in Chapter 2, since a Bayesian approach to density estimation can only be treated analytically for simple distributions. For problems in which the normalising integral in the denominator of the expression for a posterior density cannot be evaluated analytically, either Bayesian sampling approaches must be employed or approximations (such a variational approximations) must be used.

Bayesian sampling methods, including the Gibbs sampler, can be applied routinely allowing efficient practical application of Bayesian methods, at least for some problems. They work by drawing samples from the posterior distribution. There are many developments of the basic methodology that has been presented, particularly with respect to computational implementation of the Bayesian approach. These include strategies for improving MCMC; monitoring convergence; and adaptive MCMC methods. A good starting point is the book by Gilks *et al.* (1996), and other texts mentioned within this chapter. Developments of the MCMC methodology to problems when observations arrive sequentially and one is interested in performing inference online are described by Doucet *et al.* (2001). A key difficulty with developing MCMC solutions is that it can be hard to spot mistakes made when coding the algorithms, since these bugs can easily be confused with sampling variability and convergence difficulties. The WinBUGS software (Lunn *et al.*, 2000) addresses this issue by removing the need for the developer to explicitly calculate and code the posterior distributions.

Variational Bayes methods have proved popular in the machine learning community since the late 1990s, and provide an alternative approach to sampling methods for approximating Bayesian posterior distributions. A disadvantage with variational approaches is that it can be hard to ascertain the significance of the variational approximations.

A field known as Approximate Bayesian Computation addresses Bayesian inference in problems where the measurement likelihood is analytically or computationally intractable, but can be sampled from. The practical utility of such technology for discrimination problems is still an area requiring further research.

Approaches to discrimination can make use of unlabelled test samples to refine models. The procedure described in Section 3.5 implements an iterative procedure to classify test data. Although the procedure is attractive in that it uses the test data to refine knowledge about the parameters, its iterative nature may prevent its application in problems with real-time requirements.

3.12 Recommendations

As with the estimative approach to density estimation considered in Chapter 2, an approach based on Bayesian estimates of class-conditional densities is not without its dangers. If incorrect assumptions are made about the forms of the distributions in the Bayesian approach (in particular the likelihood distributions) then we cannot hope to achieve optimal performance, no matter how sophisticated the scheme for estimating the posterior distribution. Indeed, often pattern recognition and machine learning practitioners wrongly criticise the Bayesian methodology as being inadequate for a given problem, when it is in fact the underlying measurement model that is wrong. Bayesian models may considerably out-perform estimative approaches if the amount of training data is small, and are therefore to be recommended in such situations. Bayesian sampling schemes are recommended for dealing with complex posterior distributions.

3.13 Notes and references

Bayesian learning is discussed in many of the standard pattern recognition texts including Fu (1968), Fukunaga (1990), Young and Calvert (1974) and Hand (1981a). Geisser (1964) presents methods for Bayesian learning of means and covariance matrices under various assumptions on the parameters. Bayesian methods for discrimination are described by Lavine and West (1992) and West (1992).

More detailed treatments of Bayesian inference than provided in this text are given by O'Hagan (1994) and Bernardo and Smith (1994); see also Robert (2001), Lee (2004) and Gelman *et al.* (2004).

MCMC methods and sampling techniques are described in books by Gilks *et al.* (1996), M.-H. Chen *et al.* (2000), Robert and Casella (2004) and Gamerman and Lopes (2006). Bishop (2007) considers Variational Bayes methods in some detail.

Exercises

1. Consider the Gauss-Wishart posterior distribution of Section 3.2.3

$$p(\boldsymbol{\mu}, \boldsymbol{K}|\boldsymbol{x}_1, \ldots, \boldsymbol{x}_n) = \mathrm{N}_d(\boldsymbol{\mu}|\boldsymbol{\mu}_n, \lambda_n \boldsymbol{K})\mathrm{Wi}_d(\boldsymbol{K}|\alpha_n, \boldsymbol{\beta}_n)$$

Show that the marginal posterior distribution for $\boldsymbol{\mu}$ is

$$p(\boldsymbol{\mu}|\boldsymbol{x}_1, \ldots, \boldsymbol{x}_n) = \mathrm{St}_d\left(\boldsymbol{\mu}; \boldsymbol{\mu}_n, \left(\alpha_n - \frac{d-1}{2}\right)\lambda_n \boldsymbol{\beta}_n^{-1}, 2\alpha_n - (d-1)\right)$$

where St_d is the d-dimensional generalisation of the univariate Student distribution [see Equation (3.20) for the definition]. Hints: First express the desired distribution as the integral with respect to $\boldsymbol{K}$ of the full posterior distribution. Make use of (3.25) to enable the integral to be calculated via comparison with the probability density function of a Wishart distribution. Then, following a similar procedure to the derivation of the predictive density in Section 3.2.3, make use of Sylvester's determinant theorem [Equation (3.31)] to derive the required result.

2. Show that if γ has the gamma distribution with shape parameter α and inverse-scale parameter β, i.e. a probability density function of

$$p(\gamma) = \frac{\beta^\alpha}{\Gamma(\alpha)}\gamma^{\alpha-1}\exp(-\beta\gamma), \quad \gamma > 0$$

then $\theta = 1/\gamma$ has an inverse-gamma distribution with shape parameter α and scale parameter β, i.e. a probability density function of

$$p(\theta) = \frac{\beta^\alpha}{\Gamma(\alpha)}\theta^{-\alpha-1}\exp\left(\frac{-\beta}{\theta}\right), \quad \theta > 0$$

3. Show that rejection sampling (see Section 3.3.4) gives rise to exact samples from the required distribution. Hint: evaluate the cumulative distribution function $P(x \leq x_0)$ as $P(x \leq x_0|x \text{ is accepted})$, which can be expressed as $P(x \leq x_0, x \text{ is accepted})/P(x \text{ is accepted})$.

4. Consider the block decomposition of the $d \times d$ non-singular matrix $\boldsymbol{A}$

$$\boldsymbol{A} = \begin{pmatrix} \boldsymbol{A}_{1,1} & \boldsymbol{A}_{1,2} \\ \boldsymbol{A}_{2,1} & \boldsymbol{A}_{2,2} \end{pmatrix}$$

where $\boldsymbol{A}_{1,1}$ is a $d_1 \times d_1$-dimensional matrix, $\boldsymbol{A}_{1,2}$ is a $d_1 \times (d - d_1)$-dimensional matrix, $\boldsymbol{A}_{2,1}$ is a $(d - d_1) \times d_1$-dimensional matrix, and $\boldsymbol{A}_{2,2}$ is a $(d - d_1) \times (d - d_1)$-dimensional matrix. Show that the inverse of such a matrix is

$$\boldsymbol{A}^{-1} = \begin{pmatrix} \boldsymbol{A}_{c,1}^{-1} & -\boldsymbol{A}_{1,1}^{-1}\boldsymbol{A}_{1,2}\boldsymbol{A}_{c,2}^{-1} \\ -\boldsymbol{A}_{2,2}^{-1}\boldsymbol{A}_{2,1}\boldsymbol{A}_{c,1}^{-1} & \boldsymbol{A}_{c,2}^{-1} \end{pmatrix}$$

where:

$$\mathbf{A}_{c,1} = \mathbf{A}_{1,1} - \mathbf{A}_{1,2}\mathbf{A}_{2,2}^{-1}\mathbf{A}_{2,1}$$

$$\mathbf{A}_{c,2} = \mathbf{A}_{2,2} - \mathbf{A}_{2,1}\mathbf{A}_{1,1}^{-1}\mathbf{A}_{1,2}$$

5. The transition kernel in MCMC algorithms should be such that the stationary (limiting) distribution of the Markov chain is the target distribution. For continuous variables, we require

$$f(\boldsymbol{\theta}') = \int_{\boldsymbol{\theta}} f(\boldsymbol{\theta})K(\boldsymbol{\theta}, \boldsymbol{\theta}')d\boldsymbol{\theta} \tag{3.73}$$

Show that if a transition kernel $\mathrm{K}(\boldsymbol{\theta}, \boldsymbol{\theta}')$ satisfies detailed balance with respect to $f(\boldsymbol{\theta})$

$$f(\boldsymbol{\theta})\mathrm{K}(\boldsymbol{\theta}, \boldsymbol{\theta}') = f(\boldsymbol{\theta}')\mathrm{K}(\boldsymbol{\theta}', \boldsymbol{\theta})$$

then (3.73) is met. Show further that the Gibbs sampler with transition kernel

$$\mathrm{K}(\boldsymbol{\theta}, \boldsymbol{\theta}') = \prod_{i=1}^{d} f(\theta_i' | \theta_1', \ldots, \theta_{i-1}', \theta_{i+1}, \ldots, \theta_d)$$

satisfies (3.73).

6. For the distribution illustrated by Figure 3.4, show that a suitable linear transformation of the coordinate system, to new variables ϕ_1 and ϕ_2, will lead to an irreducible chain.

7. For the Variational Bayes example of Section 3.7.4, determine an exact expression for the upper bound $L(q(\mu, \tau))$, defined in (3.62). Implement the Variational Bayes optimisation procedure and monitor the convergence of $L(q(\mu, \tau))$.

8. Implement ABC Rejection Sampling and ABC MCMC algorithms for the univariate normal distribution example considered in Sections 3.4.2 and 3.7.4. Comment on the relative efficiencies of the approaches.

4

Density estimation – nonparametric

Nonparametric methods of density estimation can provide class conditional density estimates for use in Bayes' rule. k-nearest neighbour methods, histogram approaches, Bayesian networks, kernel methods of density estimation and copulas are introduced in this chapter. Methods for reducing the computational cost of k-nearest-neighbour classifiers are discussed in detail.

4.1 Introduction

Many of the classification methods discussed in this book require knowledge of the class-conditional probability density functions. Given these functions, we can apply Bayes rule or the likelihood ratio test (see Chapter 1) and make a decision as to which class a pattern $\boldsymbol{x}$ should be assigned to. In some cases we may be able to make simplifying assumptions regarding the form of the density function; for example, that it is normal or it is a normal mixture (see Chapter 2). In these cases we are left with the problem of estimating the parameters that describe the densities from available data samples.

In many cases, however, we cannot assume that the density is characterised by a set of parameters and we must resort to *nonparametric* methods of density estimation; that is, there is no formal structure for the density prescribed. There are many methods that have been used for nonparametric density estimation and in the following sections we shall consider five of them, namely k-nearest-neighbour, the histogram approach with generalisation to include Bayesian networks, kernel-based methods, expansion by basis functions and copulas.

4.1.1 Basic properties of density estimators

First, we shall consider some basic properties of density estimators, that will be referred to at various points in the text.

Statistical Pattern Recognition, Third Edition. Andrew R. Webb and Keith D. Copsey.

4.1.1.1 Unbiasedness

If $\boldsymbol{X}_1, \ldots, \boldsymbol{X}_n$ are independent and identically distributed d-dimensional random variables with continuous density $p(\boldsymbol{x})$

$$p(\boldsymbol{x}) \geq 0 \qquad \int_{\mathbb{R}^d} p(\boldsymbol{x}) d\boldsymbol{x} = 1 \tag{4.1}$$

the problem is to estimate $p(\boldsymbol{x})$ given measurements on these variables.

If the estimator $\hat{p}(\boldsymbol{x})$ also satisfies (4.1), then it is biased (Rosenblatt, 1956). That is, if we impose the condition that our estimator is itself a density [in that it satisfies (4.1)], it is biased

$$\mathrm{E}[\hat{p}(\boldsymbol{x})] \neq p(\boldsymbol{x})$$

where

$$\mathrm{E}[\hat{p}(\boldsymbol{x})] = \int \hat{p}(\boldsymbol{x}|\boldsymbol{x}_1 \ldots \boldsymbol{x}_n) p(\boldsymbol{x}_1) \ldots p(\boldsymbol{x}_n) d\boldsymbol{x}_1 \ldots d\boldsymbol{x}_n$$

the expectation over the random variables $\boldsymbol{X}_1, \ldots, \boldsymbol{X}_n$. Although estimators can be derived that are asymptotically unbiased, $\mathrm{E}[\hat{p}(\boldsymbol{x})] \to p(\boldsymbol{x})$ as $n \to \infty$, in practice we are limited by the number of samples that we have.

4.1.1.2 Consistency

There are other measures of discrepancy between the density and its estimate. The (pointwise) mean squared error (MSE) is defined by

$$\mathrm{MSE}_x(\hat{p}) = \mathrm{E}[(\hat{p}(\boldsymbol{x}) - p(\boldsymbol{x}))^2]$$

where the subscript x is used to denote that MSE is a function of $\boldsymbol{x}$. The above equation may be written

$$\mathrm{MSE}_x(\hat{p}) = \mathrm{var}(\hat{p}(\boldsymbol{x})) + \{\mathrm{bias}(\hat{p}(\boldsymbol{x}))\}^2$$

where

$$\begin{aligned} \mathrm{var}(\hat{p}(\boldsymbol{x})) &= \mathrm{E}[(\hat{p}(\boldsymbol{x}) - \mathrm{E}[\hat{p}(\boldsymbol{x})])^2] \\ \mathrm{bias}(\hat{p}(\boldsymbol{x})) &= \mathrm{E}[\hat{p}(\boldsymbol{x})] - p(\boldsymbol{x}) \end{aligned}$$

If $\mathrm{MSE}_x \to 0$ for all $\boldsymbol{x} \in \mathbb{R}^d$, then $\hat{p}$ is a *pointwise consistent estimator of p in the quadratic mean.*

Global measures of accuracy are the integrated squared error (ISE)

$$\mathrm{ISE} = \int [\hat{p}(\boldsymbol{x}) - p(\boldsymbol{x})]^2 d\boldsymbol{x}$$

and the mean integrated square error (MISE)

$$\mathrm{MISE} = \mathrm{E}\left[\int [\hat{p}(\boldsymbol{x}) - p(\boldsymbol{x})]^2 d\boldsymbol{x}\right]$$

with the MISE representing an average over all possible datasets. Since the order of the expectation and the integral may be reversed, the MISE is equivalent to the integral of the MSE, i.e. the sum of the integrated variance and the integrated squared bias

$$\mathrm{MISE} = \int \mathrm{var}(\hat{p}(\boldsymbol{x}))d\boldsymbol{x} + \int \{\mathrm{bias}(\hat{p}(\boldsymbol{x}))\}^2 d\boldsymbol{x}$$

4.1.1.3 Density estimates

Although one might naïvely expect that density estimates have to satisfy the property (4.1), this need not be the case. We shall want them to be pointwise consistent, so that we can get arbitrarily close to the true density given enough samples. Consideration has been given to density estimates that may be negative in parts in order to improve the convergence properties. Also, as we shall see in a later section, the integral constraint may be relaxed. The k-nearest-neighbour density estimate has an infinite integral.

4.2 *k*-nearest-neighbour method

4.2.1 *k*-nearest-neighbour classifier

The k-nearest-neighbour classifier is a popular classification technique. This popularity is primarily due to its simple, and intuitively appealing, specification. The k-nearest neighbour procedure for classifying a measurement $\boldsymbol{x}$ to one of C classes is as follows:

- Determine the k nearest training data vectors to the measurement $\boldsymbol{x}$, using an appropriate distance metric (see Section 4.2.3).
- Assign $\boldsymbol{x}$ to the class with the most representatives (votes) within the set of k nearest vectors.

The only aspects requiring pre-specification are the number of neighbours, k, the distance metric, and the training dataset.

4.2.1.1 *k*-nearest-neighbour decision rule

Suppose that our training dataset consists of pairs $(\boldsymbol{x}_i, z_i)$, $i = 1, \ldots, n$, where $\boldsymbol{x}_i$ is the ith training data vector, and z_i is the corresponding class indicator (such that $z_i = j$ if the ith training data vector is an example from class ω_j). Let the distance metric between two measurements $\boldsymbol{x}$ and $\boldsymbol{y}$ be denoted $d(\boldsymbol{x}, \boldsymbol{y})$. An example would be the Euclidean distance metric, $d(\boldsymbol{x}, \boldsymbol{y}) = |\boldsymbol{x} - \boldsymbol{y}|$, with other metrics discussed in Section 4.2.3. The Euclidean metric is only appropriate if the features (components or coordinates) making up a data vector have similar scales. If this is not the case, then the features that make up the data vectors can be rescaled (e.g. by a linear scaling so that each feature has unit standard deviation when considered across all available training vectors irrespective of class).

The basic implementation of the k-nearest-neighbour classifier applied to a test sample $\boldsymbol{x}$ first calculates $\delta_i = d(\boldsymbol{x}_i, \boldsymbol{x})$, for $i = 1, \ldots, n$. We then determine the indices $\{a_1, \ldots, a_k\}$ of the k smallest values of δ_i (such that $\delta_{a_i} \leq \delta_j$ for all $j \notin \{a_1, \ldots, a_k\}$, $i = 1, \ldots, k$, and $\delta_{a_1} \leq \cdots \leq \delta_{a_k}$). We define k_j to be the number of patterns (data vectors) within the nearest set of k whose class is ω_j

$$k_j = \sum_{i=1}^{k} I(z_{a_i} = j) \tag{4.2}$$

where $I(a = b)$ is the indicator function, equal to one if $a = b$ and 0 otherwise. The *k-nearest-neighbour decision rule* is then to assign $\boldsymbol{x}$ to ω_m if

$$k_m \geq k_j \quad \text{for all } j \tag{4.3}$$

Such a rule could produce more than one winning class, i.e. if $\max_{j=1,\ldots,C} k_j = k'$, there could be multiple $j \in \{1, \ldots, C\}$ for which $k_j = k'$. There are several ways of breaking such ties:

- Ties may be broken arbitrarily.
- $\boldsymbol{x}$ may be assigned to the class of the nearest neighbour.
- $\boldsymbol{x}$ may be assigned to the class, out of the classes with tying values of k', that has nearest mean vector to $\boldsymbol{x}$ (with the mean vector calculated over the k' samples).
- $\boldsymbol{x}$ may be assigned to the most compact class of the tying classes, i.e. to the one for which the distance to the k'th member is the smallest. This does not require any extra computation.

Dudani (1976) proposes a distance-weighted rule in which weights are assigned to the k nearest neighbours, with closest neighbours being weighted more heavily. A pattern is assigned to the class for which the weights of the representatives among the k neighbours sum to the greatest value. Here we replace (4.2) with

$$k'_j = \sum_{i=1}^{k} w_{a_i} I(z_{a_i} = j)$$

and use these weighted values within the decision rule (4.3). A specification for the weights is

$$w_{a_j} = \begin{cases} \dfrac{\delta_{a_k} - \delta_{a_j}}{\delta_{a_k} - \delta_{a_1}} & \text{if} \quad \delta_{a_k} \neq \delta_{a_1} \\ 1 & \text{if} \quad \delta_{a_k} = \delta_{a_1} \end{cases}$$

in which the weight varies from a minimum of zero for the most distant (i.e. kth neighbour) to a maximum of one for the nearest neighbour. Alternatively, the reciprocals of the distance between the test pattern and its k neighbours can be used

$$w_{a_j} = \frac{1}{\delta_{a_j}} \quad \text{if} \quad \delta_{a_j} \neq 0$$

For this second case, we would assign to the nearest neighbour if we have $\delta_{a_1} = 0$ (i.e. an exact match).

When assessing the k-nearest-neighbour classifier on a pattern recognition problem, it is important to have a realistic split between training and test data. If there is overlap between the training and test data, the resulting performance estimates may be severely over-optimistic. Indeed, if the test sample is actually in the training data, then the nearest neighbour will be the test sample sample being classified, which is clearly undesirable.

4.2.1.2 Choice of k

The number of neighbours k is selected as a trade-off between choosing a value large enough to reduce the sensitivity to noise, and choosing a value small enough that the neighbourhood does not extend to the domain of other classes. Typically a procedure such as cross-validation (see Chapter 13) is used to optimise the choice of k.

4.2.1.3 Nearest-neighbour classifier

If $k = 1$ we have the nearest-neighbour classifier, which assigns a test sample to the class of its nearest neighbour in the training data. Such a classifier is commonly used. At the very least, it is recommended as a baseline classifier, to which the performance of more complicated classifiers should be compared.

If the Euclidean distance metric is used

$$d(\boldsymbol{x}_i, \boldsymbol{x})^2 = |\boldsymbol{x}_i - \boldsymbol{x}|^2 = \boldsymbol{x}^T\boldsymbol{x} - 2\boldsymbol{x}^T\boldsymbol{x}_i + \boldsymbol{x}_i^T\boldsymbol{x}_i$$

and therefore the nearest-neighbour rule is to assign the test sample to the class of the training data vector $\boldsymbol{x}_m$ satisfying

$$\boldsymbol{x}^T\boldsymbol{x}_m - \frac{1}{2}\boldsymbol{x}_m^t\boldsymbol{x}_m > \boldsymbol{x}^T\boldsymbol{x}_i - \frac{1}{2}\boldsymbol{x}_i^t\boldsymbol{x}_i \quad \text{for all } i \neq m$$

Thus the rule has the form of a piecewise linear discriminant function, as considered in Section 1.6.3.

4.2.2 Derivation

4.2.2.1 Density estimates

The k-nearest-neighbour rule is in fact equivalent to applying Bayes' rule to class conditional densities estimated using a simple nonparametric method of density estimation.

The probability that a point x' falls within a volume V centred at a point $\boldsymbol{x}$ is given by

$$\theta = \int_{V(\boldsymbol{x})} p(\boldsymbol{x})d\boldsymbol{x}$$

where the integral is over the volume V. For a small volume

$$\theta \sim p(\boldsymbol{x})V \tag{4.4}$$

The probability, θ, may be approximated by the proportion of samples falling within V. If k is the number of samples, out of a total of n, falling within V (k is a function of $\boldsymbol{x}$) then

$$\theta \sim \frac{k}{n} \tag{4.5}$$

Equations (4.4) and (4.5) combine to give an approximation for the density

$$\hat{p}(\boldsymbol{x}) = \frac{k}{nV} \tag{4.6}$$

The k-nearest-neighbour approach to density estimation is to fix the probability, k/n (or, equivalently, for a given number of samples n, to fix k) and to determine the volume V which contains k samples centred on the point $\boldsymbol{x}$. For example, if $\boldsymbol{x}_k$ is the kth-nearest-neighbour point to $\boldsymbol{x}$, then V may be taken to be a sphere, centred at $\boldsymbol{x}$ of radius $|\boldsymbol{x} - \boldsymbol{x}_k|$ [the volume of a sphere of radius r in n dimensions is $2r^n\pi^{\frac{n}{2}}/(n\Gamma(n/2))$, where $\Gamma(x)$ is the gamma function]. The ratio of the probability to this volume gives the density estimate. As will be seen later, this is in contrast to the basic histogram approach which is to fix the cell size and then determine the number of points lying within it.

One of the parameters to choose is the value of k. If it is too large, then the estimate will be smoothed and fine detail averaged out. If it is too small, then the probability density estimate is likely to be spiky. This is illustrated in Figures 4.1 (peaks truncated) and 4.2, where 13 samples are plotted on the x-axis, and the k-nearest-neighbour density estimate shown for $k = 1$ and 2.

One thing to note about the density estimate is that it is not in fact a density. The integral under the curve is infinite. This is because for large enough $|\boldsymbol{x}|$, the estimate varies as $1/|\boldsymbol{x}|$. However, it can be shown that the density estimator is asymptotically unbiased and consistent if

$$\lim_{n\to\infty} k(n) = \infty$$

$$\lim_{n\to\infty} \frac{k(n)}{n} = 0$$

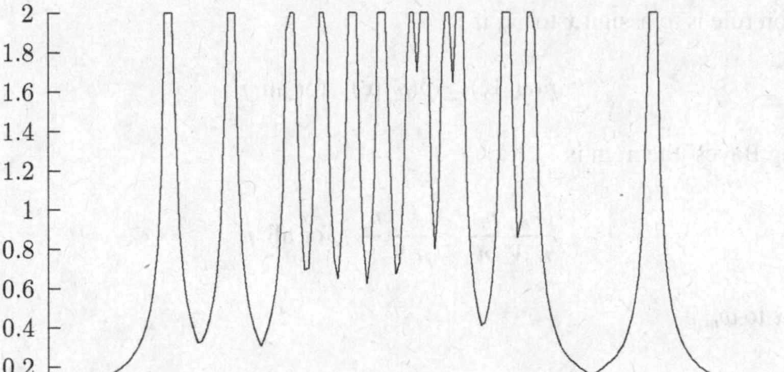

Figure 4.1 Nearest-neighbour density estimates for $k = 1$.

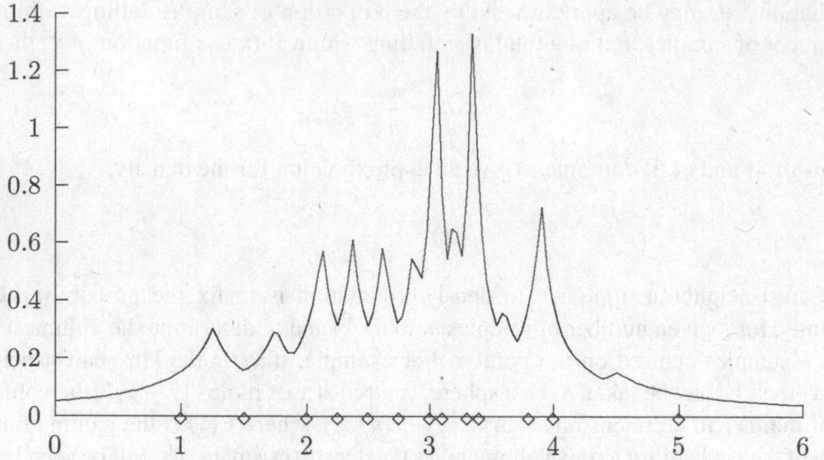

Figure 4.2 Nearest-neighbour density estimates for $k = 2$.

4.2.2.2 *k*-nearest-neighbour decision rule

Having obtained an expression for a density estimate, we can now use this in a decision rule. Suppose that in the first k samples there are k_m in class ω_m (so that $\sum_{m=1}^{C} k_m = k$). Let the total number of samples in class ω_m be n_m (so that $\sum_{m=1}^{C} n_m = n$). Then we may estimate the class-conditional density, $p(\boldsymbol{x}|\omega_m)$, as

$$\hat{p}(\boldsymbol{x}|\omega_m) = \frac{k_m}{n_m V} \tag{4.7}$$

and the prior probability, $p(\omega_m)$, as

$$\hat{p}(\omega_m) = \frac{n_m}{n}$$

The decision rule is to assign $\boldsymbol{x}$ to ω_m if

$$\hat{p}(\omega_m|\boldsymbol{x}) \geq \hat{p}(\omega_j|\boldsymbol{x}) \quad \text{for all } j$$

which using Bayes' theorem is

$$\frac{k_m}{n_m V}\frac{n_m}{n} \geq \frac{k_j}{n_j V}\frac{n_j}{n} \quad \text{for all } j$$

i.e. assign $\boldsymbol{x}$ to ω_m if

$$k_m \geq k_j \quad \text{for all } j$$

Thus, the decision rule is as defined in (4.3), i.e. to assign $\boldsymbol{x}$ to the class that receives the largest vote amongst the k-nearest-neighbours.

4.2.3 Choice of distance metric

The most commonly used metric in measuring the distance of a new sample from a training data vector is Euclidean distance. Therefore, since all variables are treated equally, the input variables must be scaled to ensure that the k-nearest-neighbour rule is independent of measurement units. Without such a scaling, if one input variable varies over a range of say 100–1000, and another in the range 1–10 (due perhaps to choice of measurement units), then the variable with the range 100–1000 will dominate the Euclidean distance calculations, and therefore the classification rule. This is undesirable, since the dominating variable may not be the variable containing the most discriminating information. Scaling may be as simple as applying a separate affine transformation to each variable, such that the mean and standard deviation of that variable over all the training data (pooled across classes) are zero and one respectively. Let $\boldsymbol{x}_1, \ldots, \boldsymbol{x}_n$ be the training data vectors combined across all classes. Then we can calculate

$$\mu_j = \frac{1}{n}\sum_{i=1}^{n} x_{ij}$$

$$\sigma_j = \sqrt{\frac{1}{n-1}\sum_{i=1}^{n}(x_{ij}-\mu_j)^2}$$

and normalise any training data vector or test sample $\boldsymbol{y}$ as follows

$$y'_j = \frac{y_j - \mu_j}{\sigma_j}$$

The generalisation of Euclidean distance with input variable scaling is

$$d(\boldsymbol{x},\boldsymbol{y}) = \left\{(\boldsymbol{x}-\boldsymbol{y})^T\boldsymbol{A}(\boldsymbol{x}-\boldsymbol{y})\right\}^{\frac{1}{2}} \tag{4.8}$$

for a matrix $\boldsymbol{A}$. Choices for $\boldsymbol{A}$ have been discussed by Fukunaga and Flick (1984).

Todeschini (1989) assesses six global metrics (Table 4.1) on 10 datasets after four ways of standardising the data (Table 4.2). The maximum scaling standardisation procedure performed well, and was found to be robust to the choice of distance metric.

Another development of the Euclidean rule is proposed by van der Heiden and Groen (1997)

$$d_p(\boldsymbol{x},\boldsymbol{y}) = \left\{(\boldsymbol{x}^{(p)}-\boldsymbol{y}^{(p)})^T(\boldsymbol{x}^{(p)}-\boldsymbol{y}^{(p)})\right\}^{\frac{1}{2}}$$

where $\boldsymbol{x}^{(p)}$ is a transformation of the vector $\boldsymbol{x}$ defined for each component, x_i, of the vector $\boldsymbol{x}$ by

$$x_i^{(p)} = \begin{cases} \left(x_i^p - 1\right)/p & \text{if } 0 < p \le 1 \\ \log(x_i) & \text{if } p = 0 \end{cases}$$

In experiments on radar range profiles of aircraft, van der Heiden and Groen (1997) evaluate the classification error as a function of p.

Table 4.1 Distance metrics assessed by Todeschini.

Cosine metric	$d(\boldsymbol{x}, \boldsymbol{y}) = \frac{\boldsymbol{x}.\boldsymbol{y}}{\lvert\boldsymbol{x}\rvert\lvert\boldsymbol{y}\rvert} = \frac{\sum_{i=1}^{d} x_i y_i}{\sqrt{\sum_{i=1}^{d} x_i^2}\sqrt{\sum_{i=1}^{d} y_i^2}}$
Canberra metric	$d(\boldsymbol{x}, \boldsymbol{y}) = \frac{1}{d}\sum_{i=1}^{d} \frac{\lvert x_i - y_i\rvert}{x_i + y_i}$
Euclidean metric	$d(\boldsymbol{x}, \boldsymbol{y}) = \lvert\boldsymbol{x} - \boldsymbol{y}\rvert = \sqrt{\sum_{i=1}^{d}(x_i - y_i)^2}$
Lagrange metric	$d(\boldsymbol{x}, \boldsymbol{y}) = \max_{i=1,\ldots,d} \lvert x_i - y_i\rvert$
Lance–Williams metric	$d(\boldsymbol{x}, \boldsymbol{y}) = \frac{\sum_{i=1}^{d}\lvert x_i - y_i\rvert}{\sum_{i=1}^{d}(x_i + y_i)}$
Manhattan metric	$d(\boldsymbol{x}, \boldsymbol{y}) = \sum_{i=1}^{d} \lvert x_i - y_i\rvert$

Friedman (1994) considers basic extensions to the k-nearest-neighbour method and presents a hybrid between a k-nearest-neighbour rule and a recursive partitioning method (see Chapter 7) in which the metric depends on position in the data space. In some classification problems (those in which there is unequal influence of the input variables on the classification performance), this can offer improved performance. Myles and Hand (1990) assess *local* metrics where the distance between $\boldsymbol{x}$ and $\boldsymbol{y}$ depends on local estimates of the posterior probability.

In the discriminant adaptive nearest-neighbour approach (Hastie and Tibshirani, 1996), a local metric is defined in which, loosely, the nearest-neighbour region is parallel to the decision boundary. It is in the region of the decision boundary where most misclassifications occur. As an illustration, consider Figure 4.3. The nearest neighbour to the point $\boldsymbol{x}$ is that labelled 1 and $\boldsymbol{x}$ is classified as $\diamond$. However, if we measure in a coordinate system orthogonal to the decision boundary, then the distance between two points is the difference between their distances from the decision boundary, and the point labelled 2 is the nearest neighbour. In this case, $\boldsymbol{x}$ is classified as $+$. The procedure of Hastie and Tibshirani uses a local definition of the matrix $\boldsymbol{A}$ in (4.8), based on local estimates of the *within* and

Table 4.2 Standardisation approaches assessed by Todeschini. μ_j and σ_j are the training data estimates of the mean and standard deviation of the jth variable. U_j and L_j are the upper and lower bounds of the jth variable. $y_{i,j}$ is the jth variable of the ith training data example.

Autoscaling	$x'_j = \frac{x_j - \mu_j}{\sigma_j}$
Maximum scaling	$x'_j = \frac{x_j}{U_j}$
Range scaling	$x'_j = \frac{x_j - L_j}{U_j - L_j}$
Profiles	$x'_j = \frac{x_j}{\sqrt{\sum_{i=1}^{n}\left(y_{i,j}^2\right)}\sqrt{\sum_{j=1}^{d}\left(x_j^2\right)}}$

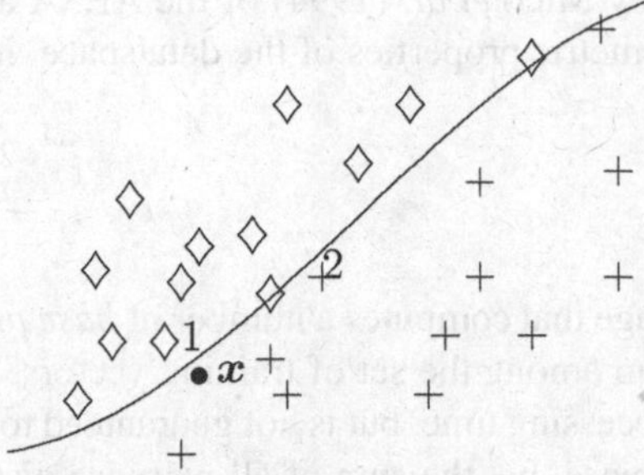

Figure 4.3 Discriminant adaptive nearest-neighbour illustration.

between-class scatter matrices. This procedure can offer substantial improvements in some problems.

4.2.4 Properties of the nearest-neighbour rule

The asymptotic misclassification rate of the nearest-neighbour rule, e, satisfies the condition (Cover and Hart, 1967)

$$e^* \leq e \leq e^* \left(2 - \frac{Ce^*}{C-1}\right)$$

where e^* is the Bayes probability of error and C is the number of classes. Thus in the large sample limit, the nearest-neighbour error rate is bounded above by twice the Bayes' error rate. The inequality may be inverted to give

$$\frac{C-1}{C} - \sqrt{\frac{C-1}{C}}\sqrt{\frac{C-1}{C} - e} \leq e^* \leq e$$

The left-most quantity is a lower bound on the Bayes' error rate. Therefore, any classifier must have an error rate greater than this value.

4.2.5 Linear approximating and eliminating search algorithm

Identifying the nearest neighbour of a given observation vector from among a set of training vectors is conceptually straightforward with n distance calculations to be performed. However, as the number n in the training set becomes large, this computational overhead may become excessive.

Many algorithms for reducing the nearest-neighbour search time involve significant computational overhead of preprocessing the prototype dataset (i.e. the set of training vectors) in order to form a distance matrix [see Dasarathy (1991) for a summary]. There is also the overhead of storing $n(n-1)/2$ distances. There are many approaches to this problem, of which the linear approximating and eliminating search algorithm (LAESA) is one.

LAESA is a development by Micó *et al.* (1994) of the AESA algorithm of Vidal (1986, 1994). The algorithm uses the metric properties of the data space, in the form of the triangle inequality.

4.2.5.1 Preprocessing

LAESA has a preprocessing stage that computes a number of *base prototypes* that are in some sense maximally separated from among the set of training vectors. This preprocessing stage can be achieved in linear preprocessing time, but is not guaranteed to provide the best possible set of base prototypes (in the sense that the sum of all pairwise distances between members of the set of base prototypes is a maximum). The LAESA algorithm requires the storage of a distance array of size n by n_b, the number of base prototypes. This will place an upper bound on the permitted number of base prototypes. Figure 4.4 illustrates a set of 21 training samples in two dimensions and four base prototypes chosen by the base prototype algorithm of Micó *et al.* (1994).

The preprocessing stage begins by first selecting a base prototype, $\boldsymbol{b}_1$, arbitrarily from the set of training data vectors (referred to as prototypes). The distance of $\boldsymbol{b}_1$ to every member of the remaining prototypes is calculated and stored in an array A. The second base prototype, $\boldsymbol{b}_2$, is the prototype that is furthest from $\boldsymbol{b}_1$ (i.e. the prototype corresponding to the largest value of A). The distance of this second base prototype to every remaining prototype is calculated. These distance values are added to those stored in the array A (i.e. the distance between base prototype $\boldsymbol{b}_2$ and the prototype $\boldsymbol{x}$ is added to that between base prototype $\boldsymbol{b}_1$ and $\boldsymbol{x}$, if $\boldsymbol{x}$ is not one of the first two base prototypes). Thus, A represents the accumulated distances of nonbase prototypes to base prototypes. The third base prototype is the one for which the accumulated distance is the greatest. This process of selecting a base prototype, calculating distances and accumulating the distances continues until the required number of base prototypes has been selected. As the selection process proceeds, the distances between base prototypes and the training set vectors are stored in an n by n_b array, D. The array, A, after addition of m prototypes is then the sum of the first m columns of D.

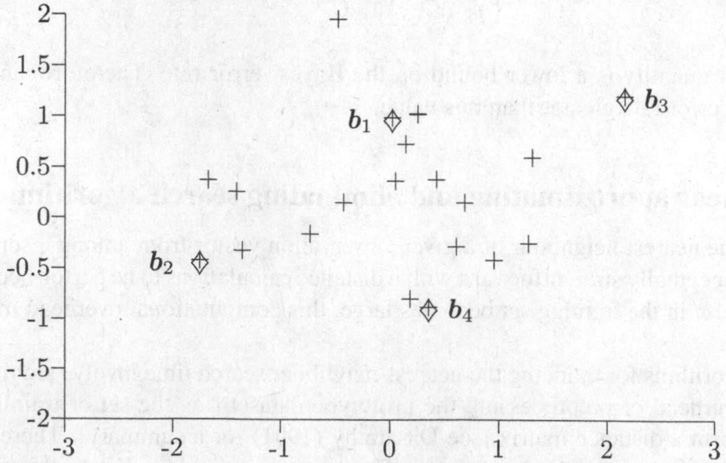

Figure 4.4 Selection of base prototypes (◇) from the dataset (+).

4.2.5.2 Searching algorithm

The LAESA searching algorithm uses the set of base prototypes and the interpoint distances between these vectors and those in the training set (i.e. the distances stored in the array D) as follows. Let $\boldsymbol{x}$ be the test sample (whose nearest neighbour from the set of prototypes we seek), $\boldsymbol{n}$ be the current nearest neighbour (i.e. the nearest neighbour within a subset of prototypes considered to date) at a distance $d(\boldsymbol{x}, \boldsymbol{n})$, and $\boldsymbol{q}$ be a base prototype whose distance to $\boldsymbol{x}$ has been computed at an earlier stage of the algorithm (Figure 4.5).

The condition for a prototype (training data vector) $\boldsymbol{p}$ to be rejected as a nearest-neighbour candidate is

$$d(\boldsymbol{x}, \boldsymbol{p}) \geq d(\boldsymbol{x}, \boldsymbol{n})$$

This requires the calculation of the distance $d(\boldsymbol{x}, \boldsymbol{p})$. However, using the triangle inequality, a lower bound on the distance $d(\boldsymbol{x}, \boldsymbol{p})$ is given by

$$d(\boldsymbol{x}, \boldsymbol{p}) \geq |d(\boldsymbol{p}, \boldsymbol{q}) - d(\boldsymbol{x}, \boldsymbol{q})|$$

which can be computed without calculating any additional distances, since $d(\boldsymbol{p}, \boldsymbol{q})$ can be read from the array D, and $d(\boldsymbol{x}, \boldsymbol{q})$ has been computed at an earlier stage. If this lower bound exceeds the current nearest-neighbour distance then clearly we may reject $\boldsymbol{p}$, without computing $d(\boldsymbol{x}, \boldsymbol{p})$. We may go further by stating

$$d(\boldsymbol{x}, \boldsymbol{p}) \geq G(p) \triangleq \max_{\boldsymbol{q}} |d(\boldsymbol{p}, \boldsymbol{q}) - d(\boldsymbol{x}, \boldsymbol{q})| \tag{4.9}$$

where the maximum is over all base prototypes considered so far in the iterative process. Therefore if $G(p) \geq d(\boldsymbol{x}, \boldsymbol{n})$, we may reject $\boldsymbol{p}$ without computing $d(\boldsymbol{x}, \boldsymbol{p})$.

To commence the procedure for a given $\boldsymbol{x}$, the algorithm selects a base prototype as an initial candidate $\boldsymbol{s}$ for a nearest neighbour and removes this from the set of prototypes. The distance $d(\boldsymbol{x}, \boldsymbol{s})$ is calculated, and stored as $d(\boldsymbol{x}, \boldsymbol{n})$, the current nearest-neighbour distance. The algorithm then searches through the remaining set of prototypes, and rejects all prototypes whose lower bound on the distance to $\boldsymbol{x}$, calculated using (4.9), exceeds the distance $d(\boldsymbol{x}, \boldsymbol{n})$. At this initial stage, each prototype's lower bound (4.9) is based on the selected base prototype, $\boldsymbol{s}$, only. A record of this lower bound is stored in an array, G.

The rejection procedure may lead to base prototypes being rejected. There is therefore an option to only eliminate base prototypes if an initial number of iterations has taken place. In

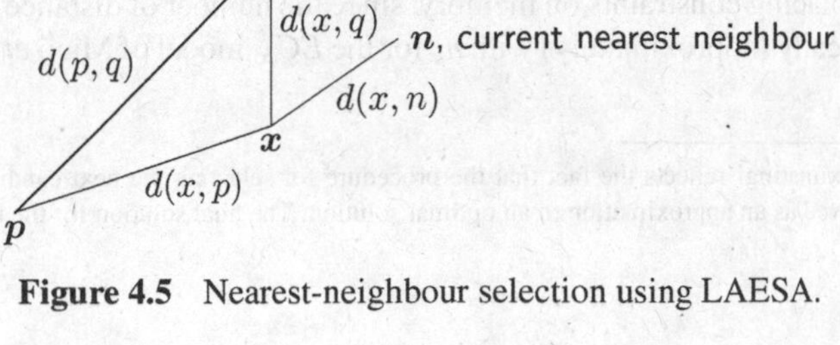

Figure 4.5 Nearest-neighbour selection using LAESA.

the EC_∞ version of LAESA no such option is used, and all base prototypes failing the lower bound test are rejected.

Out of the remaining prototypes, the algorithm selects as the next candidate s' for the nearest neighbour the base prototype for which the lower bound (stored in G) is a minimum (assuming the base prototypes have not been eliminated, otherwise a nonbase prototype sample is chosen). The distance $d(\boldsymbol{x}, \boldsymbol{s}')$ is calculated. The candidate vector $\boldsymbol{s}'$ need not necessarily be nearer than the previous choice, though if it is the nearest neighbour is updated. If $\boldsymbol{s}'$ is a base prototype the lower bounds stored in the array G are updated according to (4.9). The dataset is searched through again, rejecting prototypes (training vectors) whose lower bound distance is greater than the current nearest-neighbour distance (of course, if $\boldsymbol{s}'$ is neither a base prototype, nor a new nearest-neighbour candidate, then there is no need to run these comparisons again, because nothing will have changed).

This process is repeated and is summarised in the five following steps:

1. Distance computing – calculate the distance of $\boldsymbol{x}$ to the candidate $\boldsymbol{s}$ for a nearest neighbour.
2. Update the prototype (training data vector) nearest to $\boldsymbol{x}$ if necessary.
3. Update the lower bounds, $G(p)$ if $\boldsymbol{s}$ is a base prototype.
4. Eliminating – eliminate the prototypes with lower bounds greater than the current nearest-neighbour distance.
5. Approximating[1] – select the candidates for the next nearest neighbour.

Termination of the procedure occurs when all possible candidates for the nearest neighbour have been either eliminated, or selected as a candidate. At termination, the current nearest neighbour is the final nearest neighbour. Maximum-depth stopping rules can be applied, whereby after running the procedure for a pre-specified number of iterations, full distance calculations are performed for all remaining prototypes.

Further details are given by Micó *et al.* (1994). Figure 4.6 gives an illustration of the LAESA procedure (using the EC_∞ condition) for the dataset of Figure 4.4. The test sample $\boldsymbol{x}$ is at the origin (0, 0) and the first two choices for $\boldsymbol{s}$ are shown. The remaining samples are those left after two passes through the dataset (two distance calculations).

4.2.5.3 Discussion

There are two factors governing the choice of the number of base prototypes, n_b. One is the amount of storage available. An array of size $n \times n_b$ must be stored. This could become prohibitive if n and n_b are large. At the other extreme, too small a value of n_b will result in a large number of distance calculations if n is large. We suggest that you choose a value as large as possible without placing constraints on memory, since the number of distance calculations decreases monotonically (approximately) with n_b for the EC_∞ model of Micó *et al.* (1994) given

[1] The term 'approximating' reflects the fact that the procedure for selecting the next candidate for the nearest neighbour can be derived as an approximation to an optimal solution. The final solution for the nearest neighbour is exact.

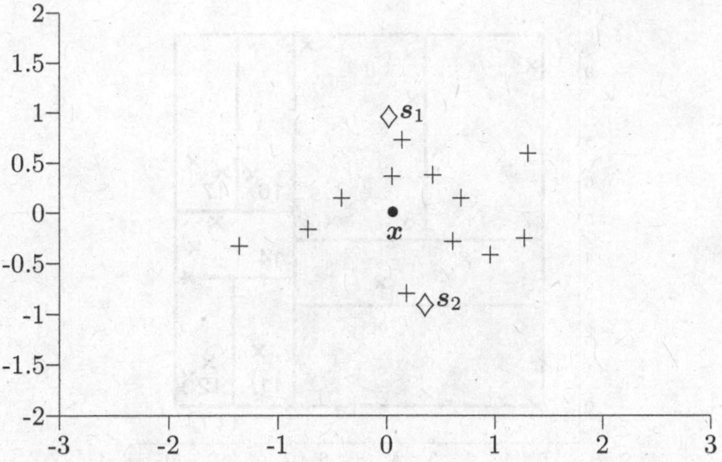

Figure 4.6 Nearest-neighbour selection using LAESA; $\boldsymbol{s}_1$ and $\boldsymbol{s}_2$ are the first two candidates for nearest neighbour.

above. However, it should be noted that if the distance calculations are computationally cheap, the additional overhead introduced by the procedure may remove its computational benefits.

As noted by Moreno-Seco *et al.* (2002), the LAESA procedure can be extended to determine k nearest neighbours, rather than just the nearest neighbour. The updated search procedure selects k initial base prototypes, to provide an initial estimate of the kth-nearest-neighbour distance. It then proceeds as standard LAESA, with alterations that prototypes are only eliminated if their lower bound distance is larger than the distance to the current kth nearest neighbour, and that the kth nearest neighbour rather than nearest neighbour is maintained.

4.2.6 Branch and bound search algorithms: kd-trees

4.2.6.1 Introduction

Branch and bound search algorithms (Lawler and Wood, 1966; see also Chapter 10) can be used to compute the k nearest neighbours of a test pattern more efficiently than a naïve linear search over all the training data vectors. Two popular, and related, branch and bound search algorithms are kd-trees (Friedman *et al.*, 1977) and ball-trees (Fukunaga and Narendra, 1975).

In both cases the approach is to:

1. Create a tree structure at the training stage, that assigns hierarchically the training vectors to subsets, each of which is a well-controlled geometrical region/structure.

2. Search for the k nearest neighbours of a test pattern, via a branch and bound method applied to the tree.

The k-dimensional tree (kd-tree) is a binary-tree partition of a k-dimensional data space into nonoverlapping hyperrectangles, aligned to the coordinate axes. Note here that k refers to the dimensionality of the input data (previously referred to as d), and not the number of nearest

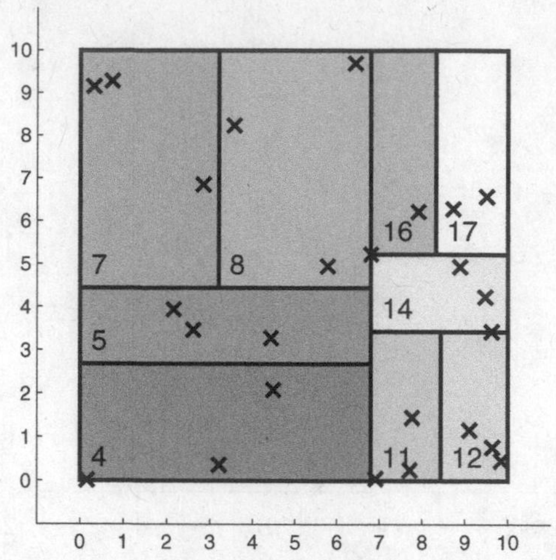

Figure 4.7 Rectangles partitioning the data space. Only the leaf (also known as terminal) node rectangles are displayed, labelled by node number. Each training vector is marked by a cross.

training vectors in the *k*-nearest-neighbour classifier. The use of kd-trees for improving the efficiency of *k*-nearest-neighbour search was proposed by Friedman *et al.* (1977).

The tree structures and procedures are similar to those in the classification trees that are discussed in Chapter 7. A two-dimensional example of a kd-tree partition of a data space is provided in Figure 4.7, with the tree structure illustrated in Figure 4.8.

The kd-tree structure can be understood by examining the intermediate levels of the tree in Figures 4.9 and 4.10. The first hyperrectangle [Figure 4.9(a)] corresponds to node 1

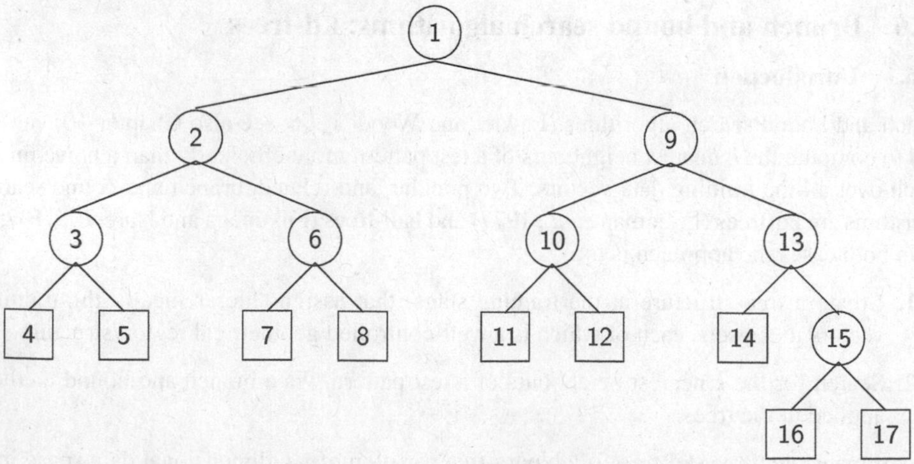

Figure 4.8 Example kd-tree. Leaf nodes are marked by squares, and non-leaf nodes by circles.

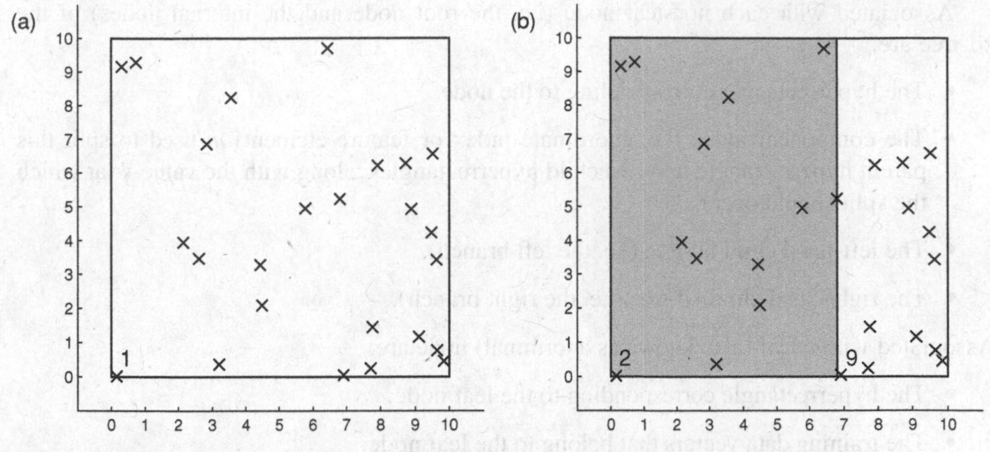

Figure 4.9 First two layers of rectangles partitioning the data space. Each training vector is marked by a cross, and the rectangles labelled by node number.

(the root node) and contains all the training data vectors. This top-level hyperrectangle is split into the hyperrectangles of nodes 2 and 9. These child hyperrectangles are shown in Figure 4.9(b). The split is along one coordinate axis of the parent hyperrectangle. Examining Figure 4.10 we can see that the child hyperrectangles are split successively as one moves down the tree. Each time, the parent hyperrectangle split is along one of the coordinate axes, and creates two nonoverlapping child hyperrectangles. This splitting process is such that each child hyperrectangle is a sub-portion of its parent hyperrectangle and the union of the child hyperrectangles is equal to the parent hyperrectangle.

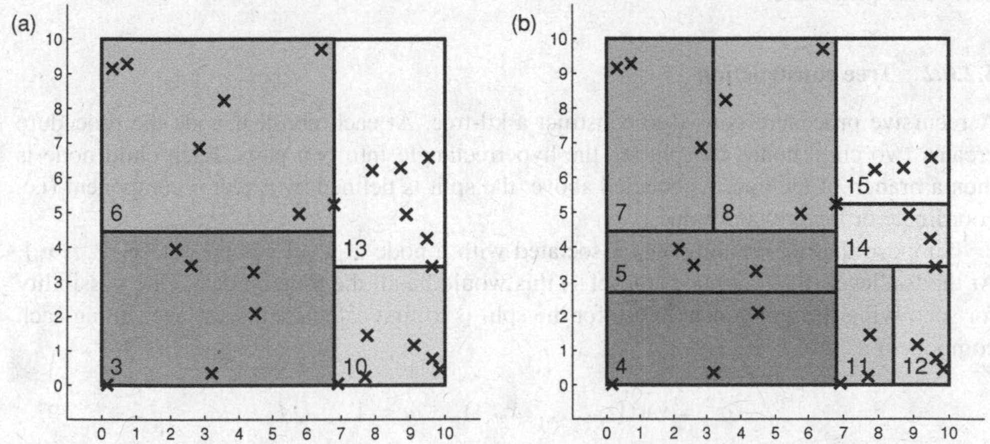

Figure 4.10 Layers three (a) and four (b) of the kd-tree, partitioning the data space. Each training vector is marked by a cross, and the rectangles labelled by node number.

Associated with each non-leaf node (i.e. the root node and the internal nodes) of the kd-tree are:

- The hyperrectangle corresponding to the node.
- The component index (i.e. coordinate index or feature element) m used to split this parent hyperrectangle into two child hyperrectangles, along with the value ψ at which the split should occur.
- The left-hand child kd-tree (i.e. the left branch).
- The right-hand child kd-tree (i.e. the right branch).

Associated with a leaf (also known as a terminal) node are:

- The hyperrectangle corresponding to the leaf node.
- The training data vectors that belong to the leaf node.

Training data vectors are assigned to leaf nodes according to which leaf node hyperrectangle they lie within. Rather than searching over all leaf nodes, we can determine an assignment by passing the data vector down through the tree. Suppose that we are at a node with split component index m, and split value ψ. The routing rule through the tree is to move to the left child if $x_m < \psi$ and the right child if $x_m \geq \psi$, where x_m is the value of the mth component (feature) of the data vector being passed down the tree. Such a procedure starts at the root node, and continues until we reach a leaf node.

Associated with any leaf node is therefore a hierarchy of hyperrectangles, one from each level, and each of which encloses the training samples at the leaf node. For example, the training data vector at coordinate (4.5, 2.1) belongs to leaf node 4, within the hierarchy of nodes $\{1, 2, 3, 4\}$. The key to using kd-trees for efficient k-nearest-neighbour search, is to note that if we can determine easily that none of the k nearest neighbours can lie within a hyperrectangle at a node high up the tree, then we do not need to search over the training vectors associated with the leaf nodes that lie beneath that node. This is the branch and bound part of the procedure.

4.2.6.2 Tree construction

A recursive procedure is used to construct a kd-tree. At each non-leaf node the procedure creates two child nodes by splitting the hyperrectangle into two parts. Each child node is then a branch of the tree. As detailed above, the split is defined by a vector component (i.e. coordinate or feature) and value.

Suppose that the training data associated with a node at level s is the set $\{\boldsymbol{x}_1, \ldots, \boldsymbol{x}_n\}$. At the top level (the root node at level 1) this would be all the training data. One possibility for specifying the component to use for the split is to first calculate the variance along each component

$$\sigma_m^2 = \text{var}(\{x_{1m}, \ldots, x_{nm}\}), \quad m = 1, \ldots, d$$

where x_{im} is the mth component value for the ith member of the training set for the node. The component to be used for the split is the the one with the largest variance, i.e. m for which $\sigma_m^2 \geq \sigma_{m'}$, for $m' = 1, \ldots, d$. An alternative to using the variance is to choose the component with the largest range. Constraints can be added if desired, so that we do not use the same component to split a child as was used to split the parent to create that child, or so that we cycle through all components as one progresses down the tree.

Having chosen the component (feature) to split, it is common to choose the split value ψ according to one of the following rules:

- $\psi = \text{median}(\{x_{1m}, \ldots, x_{nm}\})$, the median component value.
- $\psi = \text{mean}(\{x_{1m}, \ldots, x_{nm}\})$, the mean component value.

The median option should be used if one desires a perfectly balanced tree. However, if the data are not uniformly distributed, the combination of the median option with variance-based component selection is more prone than the mean option to creating narrow hyperrectangles. This is undesirable, since near hypercubic hyperrectangles have better theoretical properties.

The hyperrectangles for the two child nodes differ from the hyperrectangle for their parent node only in the ranges for the selected split component m. Specifically, if the parent hyperrectangle range for the split component is $[h_{m1}, h_{m2}]$, then

$$h_{m1}^{(L)} = h_{m1}, \quad h_{m2}^{(L)} = \psi, \quad h_{m1}^{(R)} = \psi, \quad h_{m2}^{(R)} = h_{m2}$$

where the superscripts L and R are used to denote left and right child nodes, respectively.

The training data $\{\boldsymbol{x}_1, \ldots, \boldsymbol{x}_n\}$ is split into a left-child set $\{\boldsymbol{x}_i, i \in L\}$ and a right-child set $\{\boldsymbol{x}_i, i \in R\}$, by applying the split rule to component m at value ψ.

Termination of the recursions occurs either at a maximum depth to the tree, or when the amount of training data associated with a node is sufficiently small (in the extreme case perhaps just a single training data example).

The computation required to build a tree is $O(dn\log(n))$ where d is the dimensionality of the data, and n the number of training samples (Friedman *et al.*, 1977).

4.2.6.3 Determination of nearest neighbours

We now describe the procedure for determining the nearest neighbour[2] of a test sample $\boldsymbol{x}$:

1. Descend the tree to find the leaf node whose hyperrectangle contains the test sample. The descent procedure is simple. Starting at the top node, we compare the test sample to the splitting value of the node, and then proceed down the winning branch to the appropriate child node, continuing in this manner until we reach the appropriate leaf node.

2. Determine an initial candidate for the nearest neighbour by searching over the training data vectors covered by the leaf node. The chosen training data vector, $\boldsymbol{n}$, is the current approximation to the nearest neighbour, with distance $d(\boldsymbol{x}, \boldsymbol{n}) = |\boldsymbol{x} - \boldsymbol{n}|$.

3. Construct a bounding hypersphere, centred at the test sample $\boldsymbol{x}$, and with radius equal to the current nearest-neighbour distance, $d(\boldsymbol{x}, \boldsymbol{n})$.

[2] Extension to k nearest neighbours requires a bit of extra work at the initialisation stage, but otherwise is straightforward.

4. Proceed up and down the tree, removing (termed *pruning*) branches where possible according to the procedure outlined in the next few paragraphs, and if not descending to leaf nodes. Once a branch has been pruned for a given test sample, we do not need to consider nodes that lie within that branch of the tree, and in particular do not need to search over the training vectors that belong to the leaf nodes of that branch.

5. Whenever a leaf node is reached, search over all training data vectors covered by it (i.e. calculate the distances to them from the test sample). If a lower distance than that of the current nearest neighbour is found, update the current nearest neighbour n to this new data vector, and the bounding hypersphere radius to the distance to this new nearest neighbour.

6. Continue with steps 4 and 5 until all leaf nodes have either been removed as part of the pruning procedure (hopefully at a high-level in the tree), or searched directly for nearest neighbours. The current nearest neighbour at this stage is the overall nearest neighbour.

We now consider the pruning process. From step 2, we have a current best estimate, $\boldsymbol{n}$, for the nearest neighbour of test sample $\boldsymbol{x}$. This gives a lower bound on the nearest neighbour distance of $d(\boldsymbol{x}, \boldsymbol{n})$. We construct a bounding hypersphere, centred at the test sample $\boldsymbol{x}$, and with radius equal to $d(\boldsymbol{x}, \boldsymbol{n})$. This is illustrated in Figure 4.11.

By construction, any closer training vector to the test sample than the current nearest neighbour must lie within this bounding hypersphere. Furthermore, if any training vector lies within the hypersphere, the hypersphere must intersect with each hyperrectangle in the

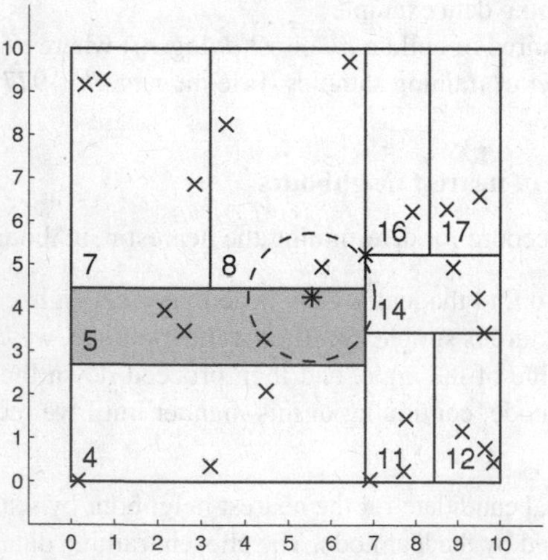

Figure 4.11 Bounding hypersphere around the test sample (dashed circle), with radius equal to the distance to the current estimate of the nearest neighbour. The test sample is marked by an asterisk, and the training vectors are marked by crosses. The initial nearest neighbour estimate belongs to the leaf node 5 within which the test sample lies. Training vectors within this leaf node are the only ones searched over to date (hence the different shading). Note that this initial nearest-neighbour estimate is not the overall nearest neighbour, which actually lies in leaf node 8.

hierarchy of hyperrectangles corresponding to that training vector. Therefore, if we find a node of the tree with a branch whose hyperrectangle does not intersect the hypersphere, we know that none of the training samples under that branch can be the nearest neighbour, and can therefore prune that branch of the tree.

The key test is therefore whether the hypersphere intersects with a hyperrectangle. A hypersphere and a hyperrectangle will intersect if and only if the distance from the closest point, $\boldsymbol{y}$, in the hyperrectangle to the centre, $\boldsymbol{x}$, of the hypersphere, is less than or equal to the radius $r = d(\boldsymbol{x}, \boldsymbol{n})$. Determining the closest point y of a hyperrectangle to the point $\boldsymbol{x}$ can be done on a component-by-component (coordinate-by-coordinate) basis

$$y_i = \begin{cases} h_{i1} & \text{if} \quad x_i < h_{i1} \\ x_i & \text{if} \quad x_i \in [h_{i1}, h_{i2}] \\ h_{i2} & \text{if} \quad x_i > h_{i2} \end{cases}$$

where $[h_{i1}, h_{i2}]$ is the range for the ith component of the hyperrectangle.

Now suppose that we consider a node of the tree with a branch whose top hyperrectangle does intersect with the bounding hypersphere. In that case we cannot prune that branch, and instead need to descend it. The descent involves pruning sub-branches where possible (if a hyperrectangle does not intersect the hypersphere), and if not proceeding to the leaf nodes. Whenever a leaf node is reached we search over all training data vectors covered by it (possibly using procedures analogous to the LAESA algorithm of Section 4.2.5). If a closer nearest neighbour is found then we update the current nearest neighbour. It is therefore more efficient to organise the descent through the tree in such a manner that if both branches at a node intersect with the bounding hypersphere, we proceed down the branch associated with the child node that is closest to the test sample first. This is because this branch is more likely to result in a new nearest neighbour than the other branch, and if the current nearest neighbour is updated, we are more likely to reject the sibling node than if we had used the original nearest-neighbour estimate.

We now outline this procedure for our test sample in Figure 4.11, making use of the tree and node specifications in Figures 4.8–4.10. After determining the nearest neighbour within the enclosing leaf node 5 (see Figure 4.11), we have leaf nodes $\{4, 7, 8, 11, 12, 14, 16, 17\}$ still to consider. Following Figure 4.8 we move up the tree from leaf node 5, to its parent, node 3, and then consider whether the hyperrectangle of node 5's sibling, node 4, intersects the bounding hypersphere. By inspection of Figure 4.11, the calculation will show that it does not intersect, and we can therefore reject leaf node 4, without searching over its training vectors. We then proceed up the tree to node 2 (Figure 4.8), and examine its other child, node 6. This hyperrectangle is that of nodes 7 and 8 combined, and does intersect with the hypersphere. So we proceed down the tree from node 6 and examine its children, nodes 7 and 8. Node 7 does not intersect the bounding hypersphere, and therefore by elimination node 8 must (since if it did not, we would not have had the intersection at its parent node 6). Node 8 is a leaf node so we search over its training vectors for candidate nearest neighbours. One is found, leading to the updated bounding hypersphere illustrated in Figure 4.12.

Having updated the hypersphere we note that all leaf nodes beneath node 2 have been considered, and therefore proceed up the tree to node 1. Here we need to consider whether the hyperrectangle of node 9 intersects with the bounding hypersphere. This hyperrectangle is the amalgam of the white hyperrectangles in Figure 4.12, and does not intersect with the bounding hypersphere. The entire branch can therefore be eliminated, and the algorithm completes, with

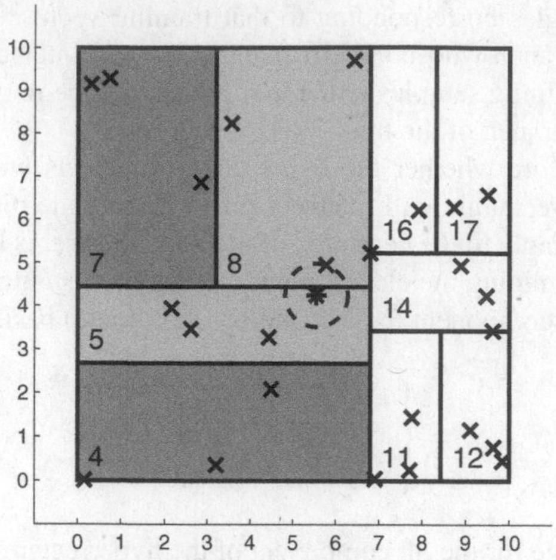

Figure 4.12 Bounding hypersphere around the test sample (dashed circle), with radius equal to the distance to the updated estimate of the nearest neighbour. The test sample is marked by an asterisk, and the training vectors are marked by crosses. Searched leaf nodes are light grey, eliminated leaf nodes dark grey, and leaf nodes yet to be searched or to be eliminated are white.

the nearest neighbour estimated correctly. Note that the hyperrectangle associated with node 9 does intersect with the original hypersphere of Figure 4.11. Its elimination here indicates the benefits of a structured search through the tree.

In this example there were 25 training vectors, of which six were searched over as candidates for the nearest neighbour (three within leaf node 5, and three within leaf node 8). In addition, as part of the pruning process, four distance calculations were made concerning the closest points of the hyperrectangles of nodes 4, 6, 7 and 9 to the bounding hypersphere. Proceeding down the tree to find the initial containing leaf node involved comparisons of one component at each of nodes 1, 2 and 3. Therefore the number of complete distance calculations is less than half that required for a linear search over all 25 training vectors.

The average computational cost of a search is $O(\log(n))$ where n is the number of training data samples (Friedman *et al.*, 1977). Friedman's analysis for this is based on an ideal case of the hyperrectangles being approximately hypercubical, each containing almost the same number of training samples. In worst case scenarios, the complexity may be similar to a linear search.

Unfortunately, it is noted (Moore, 1991) that the computational cost scales poorly with the dimensionality of the data. Indeed, for more than 10 dimensions, the benefits of kd-tree search over a linear search are often removed.

4.2.7 Branch and bound search algorithms: ball-trees

The use of ball-trees for k-nearest-neighbour search is similar in concept to the use of kd-trees, and was proposed by Fukunaga and Narendra (1975) (see also Uhlmann, 1991).

Ball-trees differ from kd-trees, in that rather than creating a hierarchy of hyperrectangles, a hierarchy of hyperspheres (i.e. balls) is created. This is advantageous, because the balls can be made to be tighter around the training data vectors than the hyperrectangles used in

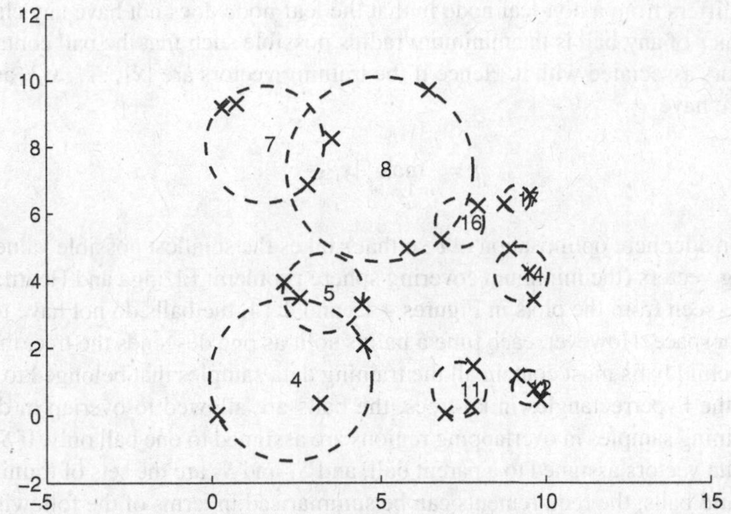

Figure 4.13 Ball-trees partitioning the data space. Only the leaf node balls are displayed, labelled by node number. Each training vector is marked by a cross.

kd-trees. This results in pruning being more commonplace in ball-tree searches than in kd-tree searches, with resultant improvements in computational efficiency.

A two-dimensional example of a ball-tree is provided in Figure 4.13. The tree structure of the nodes is the same as that for the kd-tree in Figure 4.8. Illustrations of the balls at the second and third layers of the network are provided in Figure 4.14.

Associated with each non-leaf node of the ball-tree are:

- The ball (hypersphere) corresponding to the node, characterised by its centre $\boldsymbol{c}$ and radius r.
- The left-hand and right-hand child balls (branches).
- The training data vectors associated with the ball (i.e. the training data vectors that lie inside the ball).

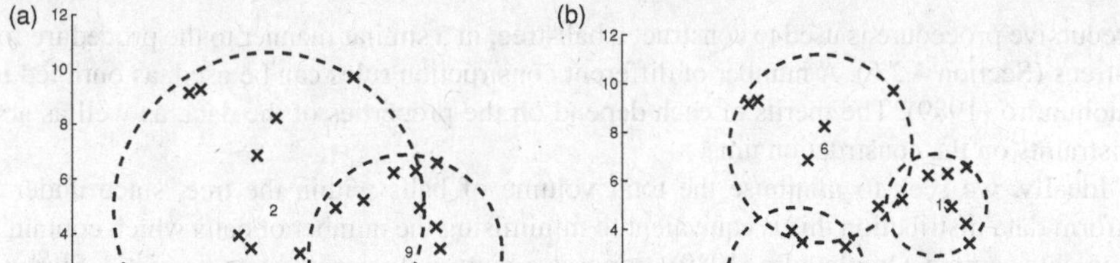

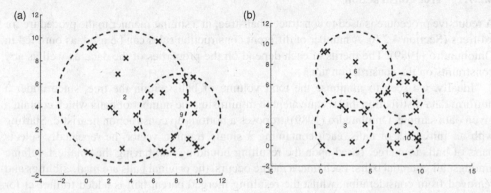

Figure 4.14 Layers two (a) and three (b) of the example ball-tree. Each training vector is marked by a cross, and the balls labelled by node number.

A leaf node differs from a non-leaf node in that the leaf node does not have any children.

The radius r of any ball is the minimum radius possible such that the ball contains all the training vectors associated with it. Hence if the training vectors are $\{\boldsymbol{x}_1, \ldots, \boldsymbol{x}_n\}$, and the ball centre is $\boldsymbol{c}$, we have

$$r = \max_{i=1,\ldots,n} |\boldsymbol{x}_i - \boldsymbol{c}|$$

We do not consider here optimisation of $\boldsymbol{c}$ so that r takes the smallest possible value given the set of training vectors (the minimum covering sphere problem; Elzinga and Hearn, 1972).

As can be seen from the plots in Figures 4.13 and 4.14, the balls do not have to partition the entire data space. However, each time a ball is split as one descends the tree, the union of the resulting child balls must contain all the training data samples that belonged to the parent ball. Unlike the hyperrectangles in kd-trees, the balls are allowed to overlap in data space. However, training samples in overlapping regions are assigned to one ball only. If S_0 is the set of training data vectors assigned to a parent ball, and S_1 and S_2 are the sets of training vectors for the the child balls, the requirements can be summarised in terms of the following rules:

$$S_1 \cup S_2 = S_0, \qquad S_1 \cap S_2 = \phi$$

Ideally, the centres of the child balls, $\boldsymbol{c}_1$ and $\boldsymbol{c}_2$, are selected in a manner so that

$$\begin{aligned} \boldsymbol{x} \in S_1 &\Rightarrow \|\boldsymbol{x} - \boldsymbol{c}_1\| \leq \|\boldsymbol{x} - \boldsymbol{c}_2\| \\ \boldsymbol{x} \in S_2 &\Rightarrow \|\boldsymbol{x} - \boldsymbol{c}_2\| \leq \|\boldsymbol{x} - \boldsymbol{c}_1\| \end{aligned} \tag{4.10}$$

However, the k-nearest-neighbour search procedure does not rely on the above property.

The routing rule through the tree is to start at the root node and then at each level proceed to the child node with the closest centre to the test vector under consideration. Note that if the ball-tree construction is such that (4.10) is not satisfied for a node, applying the routing procedure to the training data will not result in the correct assignment of training data to leaf nodes. In such an event, we keep the assignments used in tree construction.

4.2.7.1 Tree construction

A recursive procedure is used to construct a ball-tree, in a similar manner to the procedure for kd-trees (Section 4.2.6). A number of different construction rules can be used, as outlined in Omohundro (1989). The merits of each depend on the properties of the data, as well as any constraints on the construction time.

Ideally, we seek to minimise the total volume of balls within the tree, since under a uniform data distribution this is equivalent to minimising the number of balls which contain a given data sample. Omohundro (1989) proposes a bottom-up construction heuristic. Starting with an initial set of balls, each containing a single training vector, he repeatedly selects pairs of balls to merge, based upon the resulting bounding ball having the smallest volume amongst all potential pairs. Each time a merge occurs, the original balls are made siblings and removed from consideration, whilst the resulting merged parent ball is added to the set for consideration. Eventually, a single ball enclosing all the data is reached. Such ball-trees often have good behaviour, but the construction time may be prohibitive for some applications.

Analogous procedures to the top-down approach of kd-trees can also be considered. These start at a single ball covering all the data points and then recursively split the balls into two. Splitting rules can be similar to those for kd-trees. Specifically, choosing a component to split based upon maximum variance (or maximum range) and then choosing the median value as the value to split at. Training vectors with component values less than the split value are assigned to the left child ball, with the remainder being assigned to the right child ball.

Witten and Frank (2005) advocate the following rule for splitting the training data of a parent ball into two child sets:

- choose the training vector furthest from the centre of the parent ball;
- choose a second training vector that is furthest from the first selected vector;
- assign each remaining training vector to either the first set or the second set, depending on whether it is closer to the first or second selected vector.

Once the training vectors belonging to a child ball are specified, an enclosing ball is determined. Ideally, the minimum covering sphere (Elzinga and Hearn, 1972) would be calculated. However, in practice, approximations can be used, e.g. balls centred about the mean of the training vectors, or about the midpoint of the minimum bounding rectangle of the training vectors. Given the centre point, the radius is uniquely determined as the minimum possible to enclose all points.

4.2.7.2 Determination of nearest neighbours

The nearest-neighbour search process differs from that of the kd-tree (described in Section 4.2.6) only in the tests for pruning a branch (and trivially also the rule for descending the tree).

The rule for pruning branches when searching for the nearest neighbour to a test sample $\boldsymbol{x}$ can be developed in an analogous manner to that in kd-trees (in terms of intersecting balls), or can utilise the triangle inequality (as utilised in the LAESA search algorithm, see Section 4.2.5). Let $\boldsymbol{x}$ be the test sample, $\boldsymbol{n}$ be the current best estimate for the nearest neighbour, $\boldsymbol{c}$ be the centre of the ball under consideration, and r be the radius of the ball under consideration. Now consider a training vector $\boldsymbol{x}_i$ within the ball. The set-up is illustrated in Figure 4.15. Using the triangle inequality we have that

$$d(\boldsymbol{x}, \boldsymbol{x}_i) \geq d(\boldsymbol{x}, \boldsymbol{c}) - d(\boldsymbol{c}, \boldsymbol{x}_i)$$

Now since $\boldsymbol{x}_i$ lies inside a ball centred at $\boldsymbol{c}$ and of radius r, we know that $d(\boldsymbol{c}, \boldsymbol{x}_i) \leq r$, and hence

$$d(\boldsymbol{x}, \boldsymbol{x}_i) \geq d(\boldsymbol{x}, \boldsymbol{c}) - r$$

This gives a lower bound on the distance from the test sample to any training vector within the ball. Hence if

$$d(\boldsymbol{x}, \boldsymbol{c}) - r \geq d(\boldsymbol{x}, \boldsymbol{n})$$

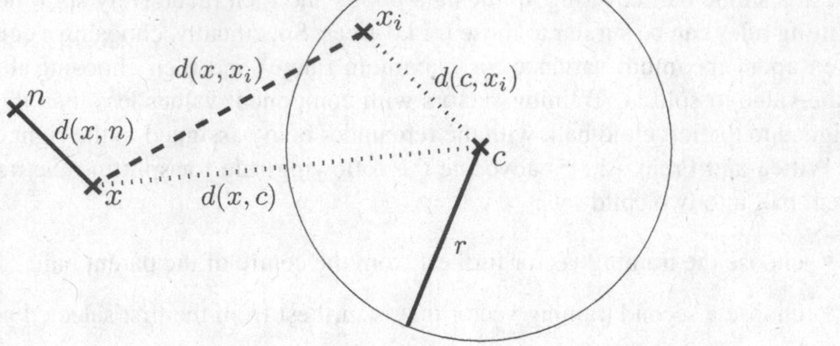

Figure 4.15 The ball-tree pruning rule. x is the test sample, n the current best estimate for the nearest neighbour, c the centre of the ball under consideration, r the radius of the ball, and x_i is any training vector within the ball. The triangle inequality is used to provide a lower bound on $d(x, x_i)$, the distance between the test sample and a training vector within the ball.

we know that the ball cannot contain a closer neighbour than the current estimate, and can therefore prune the ball (and all children of the ball).

An additional pruning test is advocated by Kamgar-Parsi and Kanal (1985). They also prune the ball if

$$d(\boldsymbol{x}, \boldsymbol{n}) + d(\boldsymbol{x}, \boldsymbol{c}) < r_{\min}$$

where $r_{\min}$ is the minimum distance from the centre of the ball to a training sample contained within the ball.

As with the kd-tree, training vectors at leaf nodes can either be searched exhaustively as potential nearest-neighbour candidates, or further bounds can be utilised, as advocated in Fukunaga and Narendra (1975).

4.2.8 Editing techniques

One of the disadvantages of the k-nearest-neighbour rule is that it requires the storage of all n data samples. If n is large, then this may mean an excessive amount of storage. Furthermore, a major disadvantage may be the computation time for obtaining the k nearest neighbours. The search algorithms of Sections 4.2.5, 4.2.6 and 4.2.7 address this latter disadvantage. As alternatives, there have been several studies aimed at reducing the number of class prototypes with the joint aims of increasing computational efficiency and increasing the generalisation error rate.

We shall consider algorithms for two procedures for reducing the number of prototypes (training vectors). The first of these belongs to the family of *editing* techniques. These techniques process the training set with the aim of removing prototypes that contribute to the misclassification rate. This is illustrated in Figure 4.16 for a two-class problem. Figure 4.16 plots samples from two overlapping distributions, together with the Bayes decision boundary. Each region, to the left and right of the boundary, contains prototypes that are misclassified by a Bayes classifier. Removing these to form Figure 4.17 gives two homogeneous sets

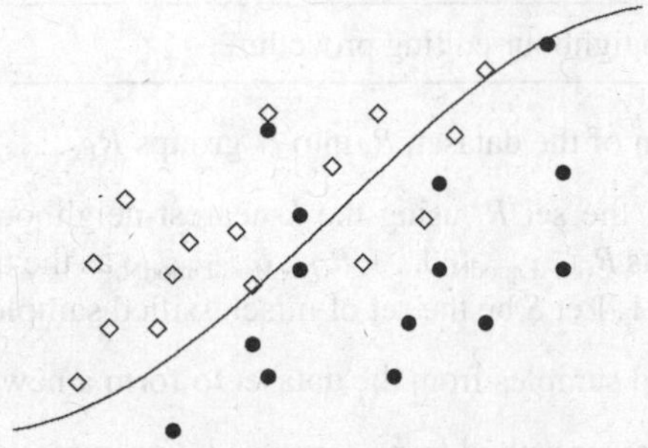

Figure 4.16 Editing illustration – samples and Bayes decision boundary.

of prototypes with a nearest-neighbour decision boundary that approximates the Bayes decision boundary.

The second technique that we consider is that of *condensing*. This aims to reduce the number of prototypes per class without changing the nearest-neighbour approximation to the Bayes decision boundary substantially.

4.2.8.1 Editing procedure

The basic editing procedure is as follows. Given a training set R (of known classification), together with a classification rule η, let S be the set of samples misclassified by the classification rule η. Remove these from the training set to form $R = R - S$ and repeat the procedure until a stopping criterion is met. Thus, on completion of the procedure, we have a set of samples correctly classified by the rule.

One implementation of applying this scheme to the k-nearest-neighbour rule is provided in Algorithm 4.1.

If $M = 1$, then we have a *modified holdout* method of error estimation and taking $k = 1$ gives the multiedit algorithm of Devijver and Kittler (1982). Taking $M = N - 1$ (so that all remaining sets are used) gives an N-fold cross-validation error estimate. If N is equal to the number of samples in the training set (and $M = N - 1$), then we have the leave-one-out error estimate which is Wilson's method of editing (Wilson, 1972). Note that after the first iteration,

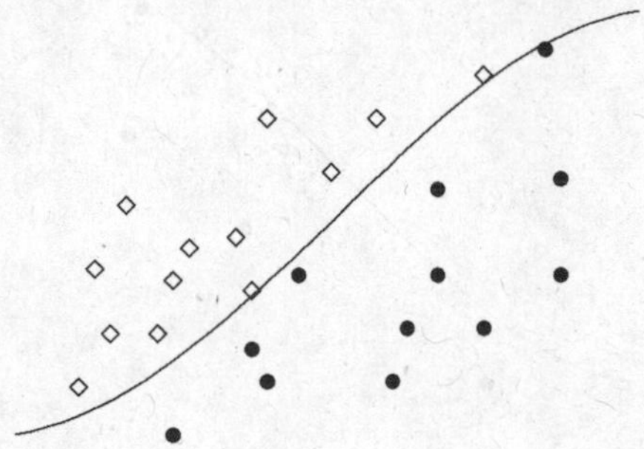

Figure 4.17 Editing illustration – edited dataset.

Algorithm 4.1 *k*-nearest-neighbour editing procedure.

1. Make a random partition of the dataset, R, into N groups $R_1, \ldots, R_N$.
2. Classify the samples in the set R_i using the k-nearest-neighbour decision rule with the union of the 'next' M sets $R_{(i+1)\text{modN}} \cup \ldots \cup R_{(i+M-1)\text{modN}}$ as the training set, for $i = 1, \ldots, N$ where $1 \leq M \leq N - 1$. Let S be the set of misclassified samples.
3. Remove all misclassified samples from the dataset to form a new dataset, $R = R - S$.
4. If the last I iterations have resulted in no samples being removed from the dataset, then terminate the algorithm, otherwise go back to step 1.

the number of training samples has reduced and the number of partitions cannot exceed the number of samples. For small datasets, an editing procedure using a cross-validation method of error estimation is preferred to multiedit (Ferri and Vidal, 1992a; Ferri *et al.*, 1999).

4.2.8.2 Condensing procedure

The editing algorithm above creates homogeneous sets of clusters of samples. The basic idea behind condensing is to remove those samples deeply embedded within each cluster that do not contribute significantly to the nearest-neighbour approximation to the Bayes decision region. The procedure that we describe is due to Hart (1968). We begin with two areas of store, labelled A and B. One sample is placed in A and the remaining samples in B. Each sample point in B is classified using the nearest-neighbour rule with the contents of A (initially a single vector) as prototypes. If a sample is classified correctly, it is returned to B, otherwise it is added to A. The procedure terminates when a complete pass through the set B fails to transfer any points.

The final contents of A constitute the condensed subset to be used with the nearest-neighbour rule.

The result of applying a condensing procedure to the data considered in Figure 4.16 is shown in Figure 4.18. There can be considerable reduction in the number of training samples.

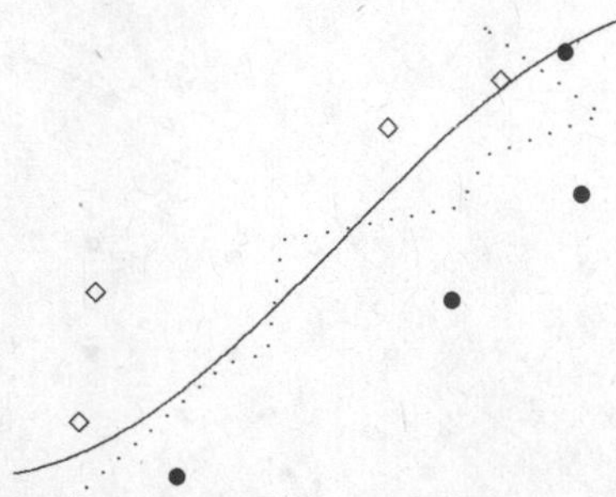

Figure 4.18 Condensing illustration. The solid line represents the Bayes decision boundary; the dotted line represents the approximation based on condensing.

Table 4.3 Results for multiedit–condensing for image segmentation (after Ferri and Vidal, 1992b).

		$I = 6$		$I = 10$	
	Original	Multiedit	Condense	Multiedit	Condense
Size	1513	1145	22	1139	28
Average errors over four test images	18.26	10.55	10.10	10.58	8.95

4.2.8.3 Application study

Ferri and Vidal (1992b) apply both editing and condensing techniques to image data gathered for a robotic harvesting application. The problem consists of detecting the location of pieces of fruit within a scene. The data comprise six images, captured into arrays of 128×128 pixels; two used for training, four for test. From the training images, a training set of 1513 10-dimensional feature vectors (each corresponding to a different pixel location) spanning three classes (fruit, leaves, sky) is constructed. The feature vectors are obtained from the colour information of the central pixel and that of its four surrounding neighbours. The chrominance (UV) components of the YUV colour space values are used (easily obtained from RGB values), giving 10 features per pixel location.

The editing algorithm is run with a value of N (the number of partitions of the dataset) chosen randomly from $\{3, 4, 5\}$ at each iteration of the editing process. Results are reported for values of I (the number of iterations in the editing procedure with no samples removed from the training set) of 6 and 10, and are summarised in Table 4.3. The condensing procedure is applied once to the outputs of the $I = 6$ editing procedure, and twice to the outputs of the $I = 10$ editing procedure (using different random initialisation of the procedure).

Condensing reduces the size of the dataset considerably. Editing reduces the error rate, with a further smaller reduction after condensing.

4.2.9 Example application study

The problem

This application is to develop credit scoring techniques for assessing the creditworthiness of consumer loan applicants (Henley and Hand, 1996).

Summary

The approach developed is one based on a *k*-nearest-neighbour method, with an adjusted version of the Euclidean distance metric that attempts to incorporate knowledge of class separation contained in the data.

The data

The data comprised measurements on 16 nominal or ordinal variables (resulting from a variable selection procedure), for a set of credit applicants, split into training and test sets of sizes 15054 and 4132, respectively. Making use of the categorical form of the variables, the data were preprocessed into a ratio form, so that the jth allowable value on the ith feature/variable was replaced by $\log(p_{ij}/q_{ij})$, where p_{ij} is the proportion of those classified good in value (attribute) j of feature (variable) i and q_{ij} is the proportion characterised as bad in value j of feature i.

The model

A k-nearest-neighbour classifier is used. The quadratic metric (4.8) is used, where the positive definite symmetric matrix, $\boldsymbol{A}$, is given by

$$\boldsymbol{A} = (\boldsymbol{I} + D\boldsymbol{w}\boldsymbol{w}^T)$$

Here D is a distance parameter and $\boldsymbol{w}$ is in the direction of the equiprobability contours of $p(g|\boldsymbol{x})$, the posterior probability of class g; that is, $\boldsymbol{w}$ is a vector parallel to the decision boundary. This is similar to the discriminant adaptive nearest-neighbour classifier mentioned in Section 4.2.3. $\boldsymbol{w}$ was estimated using a regression procedure.

Training procedure

The value of D was chosen to minimise the *bad risk rate*, the proportion of bad risk applicants accepted for a pre-specified proportion accepted, based on the training set. The value of k was also based on the training set. The performance assessment criterion used in this study is not the usual one of error rate used in most studies involving k-nearest-neighbour classifiers. In this investigation, the proportion to be accepted is pre-specified and the aim is to minimise the number of bad-risk applicants accepted.

Results

The main conclusion of the study was that the k-nearest-neighbour approach was fairly insensitive to the choice of parameters and it is a practical classification rule for credit scoring.

4.2.10 Further developments

There are many varieties of the k-nearest-neighbour method. The probability density estimator on which the k-nearest-neighbour decision rule is based has been studied by Buturović (1993), who proposes modifications to (4.7) for reducing the bias and the variance of the estimator.

There are other preprocessing schemes to reduce nearest-neighbour search time. Approximation–elimination algorithms for fast nearest-neighbour search are given by Vidal (1994) and reviewed and compared by Ramasubramanian and Paliwal (2000).

Variants of the branch and bound approaches used in kd-trees and ball-trees, are proposed by Niemann and Goppert (1988) and Jiang and Zhang (1993). An extension of kd-trees to consider decompositions into hyperrectangles that are not parallel to the coordinate axes is considered by Sproull (1991). Liu *et al.* (2006) show how improvements to ball-tree algorithms can be made in binary classification problems, by asking the question 'Are at least t of the k nearest neighbours from the first class?', rather than 'What are the k nearest neighbours of the test sample?'. The answer to the former question being the information required to classify the data according to the k-nearest neighbour classification rule.

Increased speed of finding neighbours is usually bought at increased preprocessing or storage (Djouadi and Bouktache, 1997). Dasarathy (1994a) proposes an algorithm for reducing the number of prototypes in nearest-neighbour classification using a scheme based on the concept of an optimal subset selection. This 'minimal consistent set' is derived using an iterative procedure and on the datasets tested gives improved performance over condensed nearest neighbour.

A review of the literature on computational procedures is given by Dasarathy (1991). More recently, a survey of some of the advantages and disadvantages of a range of different nearest-neighbour techniques is provided by Bhatia and Vandana (2010). Bajramovic *et al.* (2006) compare the performance of a range of nearest-neighbour procedures on an object recognition task.

Hamamoto *et al.* (1997) propose generating bootstrap samples by linearly combining local training samples. This increases the training set, rather than reducing it, but gives improved performance over conventional *k* nearest neighbour, particularly in high dimensions.

There are theoretical bounds on the Bayes' error rate for the *k*-nearest-neighbour method. We have given those for the nearest-neighbour rule. For small samples, the true error rate may be very different from the Bayes' error rate. The effects on the error rates of *k*-nearest-neighbour rules of finite sample size have been investigated by Fukunaga and Hummels (1987a, b) who demonstrate that the bias in the nearest-neighbour error decreases slowly with sample size, particularly when the dimensionality of the data is high. This indicates that increasing the sample size is not an effective means of reducing bias when dimensionality is high. However, a means of compensating for the bias is to obtain an expression for the rate of convergence of the error rate and to predict the asymptotic limit by evaluating the error rate on training sets of different sample size. This has been explored further by Psaltis *et al.* (1994) who characterise the error rate as an asymptotic series expansion. Leave-one-out procedures for error rate estimation are presented by Fukunaga and Hummels (1987b, 1989), who find that sensitivity of error rate to the choice of *k* (in a two-class problem) can be reduced by appropriate selection of a threshold for the likelihood function considered in the derivation of the *k*-nearest-neighbour decision rule (see Section 4.2.2).

4.2.11 Summary

Nearest-neighbour methods have received considerable attention over the years. The simplicity of the approach has made the nearest-neighbour method very popular with researchers and practitioners. It is perhaps conceptually the simplest of the classification rules that we present and the decision rule has been summed up as 'judge a person by the company he keeps' (Dasarathy, 1991). The approach requires a set of labelled prototypes that are used to classify a set of test patterns. The simple-minded implementation of the *k*-nearest-neighbour rule (calculating the distance of a test pattern from every member of the training set and retaining the class of the *k* nearest patterns for a decision) is likely to be computationally expensive for a large dataset, but for many applications it may well prove acceptable. If you get the answer in minutes, rather than seconds, you may probably not be too worried. The LAESA algorithm trades off storage requirements against computation. LAESA relies on the selection of base prototypes and computes distances of the stored prototypes from these base prototypes. Ball-trees and kd-trees arrange the prototypes into the form of a binary tree, so that branch and bound procedures can be used to determine the *k* nearest neighbours.

Additional means of reducing the search time for classifying a pattern using a nearest-neighbour method (and increasing generalisation) are given by the editing and condensing procedures. Both reduce the number of prototypes; editing with the purpose of increasing generalisation performance and condensing with the aim of reducing the number of prototypes without significant degradation of the performance. Experimental studies have been performed by Hand and Batchelor (1978).

Improvements may be obtained through the use of alternative distance metrics, either local metrics that use local measures of within-class and between-class distance or non-Euclidean distance.

One question that we have not addressed in detail is the choice of k. The larger the value of k, the more robust is the procedure. Yet k must be much smaller than the minimum of n_i, the number of samples in class i, otherwise the neighbourhood is no longer the local neighbourhood of the sample (Dasarathy, 1991). In a limited study, Enas and Choi (1986) give a rule $k \approx N^{2/8}$ or $k \approx N^{3/8}$, where N is the population size. The approach described by Dasarathy is to use a cross-validation procedure to classify each sample in the training set, using the remaining samples, for various values of k and to determine overall performance. Take the optimum value of k as the one giving the smallest error rate, though the lower the value of k the better from a computational point of view. Keep this value fixed in subsequent editing and condensing procedures if used.

4.3 Histogram method

The histogram method is perhaps the oldest method of density estimation. It is the classical method by which a probability density is constructed from a set of samples.

In one dimension, the real line is partitioned into a number of equal-sized cells (Figure 4.19) and the estimate of the density at a point x is taken to be

$$\hat{p}(x) = \frac{n_j}{\sum_j^N n_j dx}$$

where n_j is the number of samples in the cell of width dx that straddles the point x, N is the number of cells and dx is the size of the cell. This generalises to

$$\hat{p}(\boldsymbol{x}) = \frac{n_j}{\sum_j n_j dV}$$

for a multidimensional observation space, where dV is the volume of bin j.

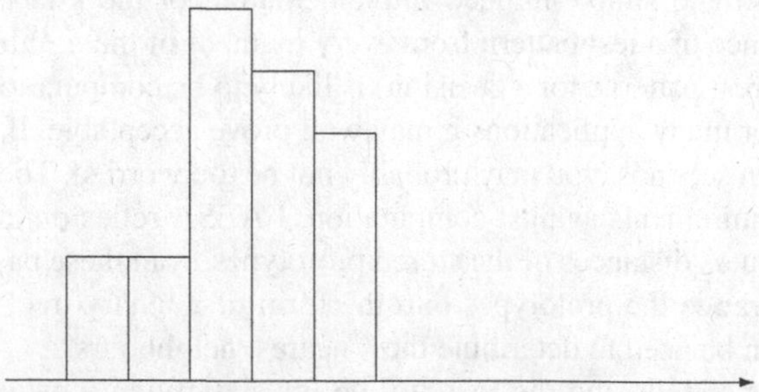

Figure 4.19 Histogram.

Although this is a very simple concept and easy to implement, and it has the advantage of not needing to retain the sample points, there are several problems with the basic histogram approach. First of all, it is seldom practical in high-dimensional spaces. In one dimension, there are N cells; in two dimensions, there are N^2 cells (assuming that each variable is partitioned into N cells). For data samples $\boldsymbol{x} \in \mathbb{R}^d$ (d-dimensional vector $\boldsymbol{x}$) there are N^d cells. This exponential growth in the number of cells means that in high dimensions a very large amount of data is required to estimate the density. For example, where the data samples are six-dimensional, then dividing each variable range into 10 cells (a not unreasonable figure) gives a million cells. In order to prevent the estimate being zero over a large region, many observations will be required. This is known as the *curse of dimensionality*. A second problem with the histogram approach is that the density estimate is discontinuous and falls abruptly to zero at the boundaries of the region. We shall now consider some approaches for overcoming these difficulties.

4.3.1 Data adaptive histograms

One approach to the problem of constructing approximations to probability density functions from a limited number of samples using d-dimensional histograms is to allow the histogram descriptors – location, shape and size – to adapt to the data. This is illustrated in Figure 4.20.

An early approach was by Sebestyen and Edie (1966) who described a sequential method to multivariate density estimation using cells that are hyperellipsoidal in shape.

4.3.2 Independence assumption (naïve Bayes)

Another approach for reducing the number of cells in high-dimensional problems is to make some simplifying assumptions regarding the form of the probability density function. One such simplification is to assume that the variables are independent so that $p(\boldsymbol{x})$ may be written in the form

$$p(\boldsymbol{x}) = \prod_{i=1}^{d} p(x_i)$$

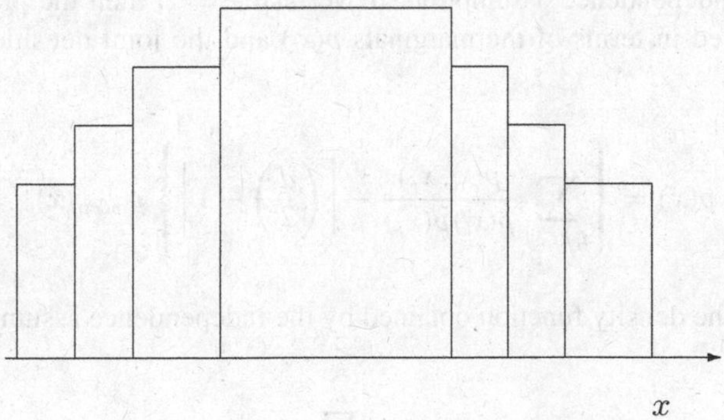

Figure 4.20 Variable cell size histogram.

where $p(x_i)$ are the individual (one-dimensional) densities of the components (features, elements, variables) of $\boldsymbol{x}$. Various names have been used to describe such a model including *naïve Bayes*, *idiot's Bayes* and *independence Bayes*. A histogram approach may be used for each density individually, giving Nd cells (assuming an equal number of cells, N, per variable), rather than N^d.

A particular implementation of the independence model is (Titterington *et al.*, 1981)

$$p(\boldsymbol{x}) \sim \left\{ \prod_{r=1}^{d} \frac{n(x_r) + \frac{1}{C_r}}{N(r) + 1} \right\}^B \tag{4.11}$$

where x_r is the rth variable (component) of $\boldsymbol{x}$, $n(x_r)$ is the number of samples with value x_r on variable r (i.e. the number of samples in the cell which covers x_r on variable r), $N(r)$ is the number of observations on variable r (this may vary due to missing data), C_r is the number of cells in variable r and B is an 'association factor' representing the 'proportion of nonredundant information' in the variables.

Note that the above expression takes account of missing data (which may be a problem in some categorical data problems). It has a non constant number of cells per variable. Furthermore, the $\frac{1}{C_r}$ offsets to the sample counts prevent a test sample density from being zero if for some reason no training data samples existed for the observed value of a given variable (in which case $n(x_r) = 0$). This is important, because if one component density is zero, the overall density will be zero, even if all the other components give large density values. Without such offsets, it would be possible for all the class conditional densities in a Bayes rule classifier to have zero weight.

Naïve Bayes models are popular due to their simplicity of implementation. The independence assumption is often used in parametric density estimates via normal distributions with diagonal covariance matrices (i.e. a product of univariate normal density functions), as considered in Section 2.4.

4.3.3 Lancaster models

Lancaster models are a means of representing the joint density function in terms of the marginal densities assuming all interactions higher than a certain order vanish. For example, if we assume that all interactions higher than order $s = 1$ vanish, then a Lancaster model is equivalent to the independence assumption. If we take $s = 2$, then the probability density function is expressed in terms of the marginals $p(x_i)$ and the joint densities $p(x_i, x_j)$, $i \neq j$ (Zentgraf, 1975) as

$$p(\boldsymbol{x}) = \left\{ \sum_{i,j,i<j} \frac{p(x_i, x_j)}{p(x_i)p(x_j)} - \left[\binom{d}{2} - 1 \right] \right\} p_{\text{indep}}(\boldsymbol{x})$$

where $p_{\text{indep}}(\boldsymbol{x})$ is the density function obtained by the independence assumption

$$p_{\text{indep}}(\boldsymbol{x}) = \prod_{k=1}^{d} p(x_k)$$

Lancaster models permit a range of models from the independence assumption to the full multinomial, but do have the disadvantage that some of the probability density estimates may be negative. Titterington *et al.* (1981) take the two-dimensional marginal estimates as

$$p(x_i, x_j) = \frac{n(x_i, x_j) + 1/(C_i C_j)}{N(i, j) + 1}$$

where the definitions of $n(x_i, x_j)$ and $N(i, j)$ are analogous to the definitions of $n(x_i)$ and $N(i)$ given previously for the independence model, and

$$p(x_i) = \left(\frac{n(x_i) + 1/C_i}{N(i) + 1} \right)^B$$

Titterington *et al.* (1981) adopt the independence model whenever the estimate of the joint probability density is negative.

4.3.4 Maximum weight dependence trees

Lancaster models are one way to capture dependencies between variables without making the sometimes unrealistic assumption of total independence on one hand, yet having a model that does not require an unrealistic amount of storage or number of observations. Chow and Liu (1968) propose a tree-dependent model in which the probability density function $p(\boldsymbol{x})$ is modelled as a tree-dependent distribution $p^t(\boldsymbol{x})$ that can be written as the product of $(d-1)$ pairwise conditional probability density functions:

$$p^t(\boldsymbol{x}) = \prod_{i=1}^{d} p(x_i | x_{j(i)}) \tag{4.12}$$

where $x_{j(i)}$ is the variable designated as the parent of x_i, with the root x_1 chosen arbitrarily and characterised by the prior probability $p(x_1|x_0) = p(x_1)$.

For example, the density

$$p^t(\boldsymbol{x}) = p(x_1)p(x_2|x_1)p(x_3|x_2)p(x_4|x_2)p(x_5|x_4)p(x_6|x_4)p(x_7|x_4)p(x_8|x_7) \tag{4.13}$$

has the tree depicted in Figure 4.21(a), with root x_1. An alternative tree representation of the same density is Figure 4.21(b), with root x_4, since (4.13) may be rearranged using Bayes' theorem as

$$p^t(\boldsymbol{x}) = p(x_4)p(x_1|x_2)p(x_3|x_2)p(x_2|x_4)p(x_5|x_4)p(x_6|x_4)p(x_7|x_4)p(x_8|x_7)$$

Indeed, any node may be taken as the root node. Applying the approach to discrete distributions (or discretised continuous distributions), if each variable can take N values, then the density (4.12) has $N(N-1)$ parameters for each of the conditional densities (recall that each conditional density must sum to one over the dependent variable) and $(N-1)$ parameters for the prior probability, giving a total of $N(N-1)(d-1) + N - 1$ parameters to estimate.

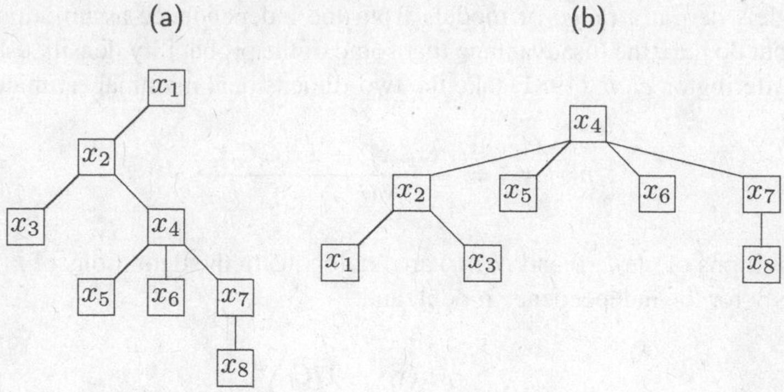

Figure 4.21 Tree representations.

The approach of Chow and Liu (1968) is to seek the tree-dependent distribution, $p^t(\boldsymbol{x})$ that best approximates the distribution $p(\boldsymbol{x})$. The KL divergence (also referred to as the Kullback–Leibler cross-entropy measure) is used as the measure of closeness in approximating $p(\boldsymbol{x})$ by $p^t(\boldsymbol{x})$

$$KL(p(\boldsymbol{x})|p^t(\boldsymbol{x})) = \int p(\boldsymbol{x})\log\left(\frac{p(\boldsymbol{x})}{p^t(\boldsymbol{x})}\right) d\boldsymbol{x}$$

or for discrete variables

$$KL(p(\boldsymbol{x})|p^t(\boldsymbol{x})) = \sum_{\boldsymbol{x}} p(\boldsymbol{x})\log\left(\frac{p(\boldsymbol{x})}{p^t(\boldsymbol{x})}\right)$$

where the sum is over all values that the variable $\boldsymbol{x}$ can take. They seek the tree-dependent distribution $p^\tau(\boldsymbol{x})$ such that $KL(p(\boldsymbol{x})|p^\tau(\boldsymbol{x})) \leq KL(p(\boldsymbol{x})|p^t(\boldsymbol{x}))$ over all t in the set of possible first-order dependence trees.

Using the mutual information between variables

$$I(X_i, X_j) = \sum_{x_i, x_j} p(x_i, x_j)\log\left(\frac{p(x_i, x_j)}{p(x_i)p(x_j)}\right)$$

to assign weights to every branch of the dependence tree, Chow and Liu (1968) show that (see the exercises at the end of the chapter) the tree-dependent distribution $p^t(\boldsymbol{x})$ that best approximates $p(\boldsymbol{x})$ is the one with maximum weight defined by

$$W = \sum_{i=1}^{d} I(X_i, X_{j(i)})$$

This is termed a *maximum weight dependence tree* (MWDT) or a *maximum weight spanning tree*. The steps in the algorithm to find an MWDT are provided in Algorithm 4.2.

Algorithm 4.2 Determination of a maximum weight dependence tree.

1. Compute the branch weights for all $d(d-1)/2$ variable pairs [i.e. $I(X_i, X_j)$ for all pairs $i < j$] and order them in decreasing magnitude.
2. Find the two branches with the largest weights, and take them as the first two branches of the tree.
3. Consider the next largest branch weight and add the corresponding branch to the tree if it does not form a cycle with the branches selected to date, otherwise discard it.
4. Repeat step 3 until $d-1$ branches have been selected in total (by construction these branches will cover all the variables).
5. The probability distribution may be computed by selecting an arbitrary root node and using the branch pairings in the tree to compute (4.12).

For discrete (or discretised continuous) data, the branch weights can be estimated from the marginal and pair sample frequencies within a training dataset. We approximate a branch weight $I(X_i, X_j)$ by

$$\hat{I}(X_i, X_j) = \sum_{u,v} f_{(x_i,x_j)}(u, v) \log\left(\frac{f_{(x_i,x_j)}(u, v)}{f_{x_i}(u) f_{x_j}(v)}\right)$$

where $f_{(x_i,x_j)}(u, v)$ is the proportion of training data examples for which $x_i = u$ and $x_j = v$, and $f_{x_i}(u)$ is the proportion of training data examples for which $x_i = u$.

The first stage in applying maximum weight dependence trees to a classification problem is to apply the algorithm to the dataset for each class individually to give C trees. These trees can then be used to specify the class conditional probability densities used in a Bayes' rule classifier.

There are several features that make maximum weight dependence trees attractive. The algorithm requires only second-order distributions but, unlike the second-order Lancaster model, it need store only $d-1$ of these. The tree is computed in $\mathcal{O}(d^2)$ steps (though additional computation is required in order to obtain the mutual information) and if $p(\boldsymbol{x})$ is indeed tree dependent then the approximation $p^t(\boldsymbol{x})$ is a consistent estimate in the sense that

$$\max_{\boldsymbol{x}} |p_n^{t(n)}(\boldsymbol{x}) - p(\boldsymbol{x})| \to 0 \text{ with probability 1 as } n \to \infty$$

where $p_n^{t(n)}$ is the tree-dependent distribution estimated from n independent samples of the distribution $p(\boldsymbol{x})$.

Example application

Applying the MWDT creation procedure to the six-dimensional head injury data of Titterington *et al.* (1981) produces (for class 1) the tree illustrated in Figure 4.22. The labels for the variables are described in Section 2.3.3 (x_1 – `Age`; x_2 – `EMV`; x_3 – `MRP`; x_4 – `Change`; x_5 – `Eye Indicant`; x_6 – `Pupils`).

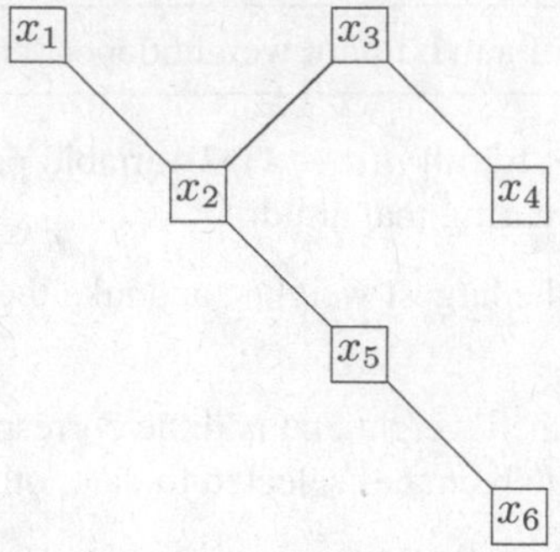

Figure 4.22 MWDT applied to head injury patient data (class 1).

To determine the tree-dependent distribution, select a root node (say node 1) and write the density using the figure as

$$p^t(\boldsymbol{x}) = p(x_1)p(x_2|x_1)p(x_3|x_2)p(x_5|x_2)p(x_4|x_3)p(x_6|x_5)$$

4.3.4.1 Further developments

The MWDT approximation can also be derived by minimising an upper bound on the Bayes' error rate under certain circumstances (Wong and Poon, 1989). Computational improvements in the procedure have been proposed by Valiveti and Oommen (1992, 1993), who suggest a chi-square metric in place of the expected mutual information measure.

A further development is to impose a common tree structure across all classes. The mutual information between variables is then written as

$$I(X_i, X_j) = \sum_{x_i, x_j, \omega} p(x_i, x_j, \omega) \log\left(\frac{p(x_i, x_j|\omega)}{p(x_i|\omega)p(x_j|\omega)}\right)$$

It is termed a tree-augmented naïve Bayesian network by Friedman *et al.* (1997), who provide a thorough evaluation of the model on 23 datasets from the UCI repository (Murphy and Aha, 1995) and two artificial datasets. Continuous attributes are discretised and patterns with missing values omitted from the analysis. Performance is measured in terms of classification accuracy with the holdout method used to estimate error rates (see Chapter 9). The model was compared with one in which separate trees were constructed for each class and the naïve Bayes model. Both tree models performed well in practice.

4.3.5 Bayesian networks

In Section 4.3.4 a development of the naïve Bayes model (which assumes independence between variables) was described. This was the MWDT model, which allows pairwise dependence between variables and is a compromise between approaches that specify all relationships between variables and the rather restrictive independence assumption. Bayesian networks also provide an intermediate model between these two extremes and have the tree-based models as a special case.

We introduce Bayesian networks by considering a graphical representation of a multivariate density. The chain rule allows a joint density, $p(x_1, \ldots, x_d)$, to be expressed in the form

$$p(x_1, \ldots, x_d) = p(x_d|x_1, \ldots, x_{d-1})p(x_{d-1}|x_1, \ldots, x_{d-2}) \ldots p(x_2|x_1)p(x_1) \quad (4.14)$$

The chain rule above is obtained by repeated applications of the product rule (i.e. the equation for conditional probabilities)

$$p(x_1, \ldots, x_i) = p(x_i|x_1, \ldots, x_{i-1})p(x_1, \ldots, x_{i-1})$$

for $i = 1, \ldots, d$.

We may depict the representation of the density in (4.14) graphically. This is illustrated in Figure 4.23 for $d = 6$ variables.

Each node in the graph represents a variable and the directed links denote the dependencies of a given variable. The *parents* of a given variable are those variables with directed links towards it. For example, the parents of x_5 are x_1, x_2, x_3 and x_4. The *root node* is the node without parents (the node corresponding to variable x_1). The probability density that such a figure depicts is the product of conditional densities

$$p(x_1, \ldots, x_d) = \prod_{i=1}^{d} p(x_i|\pi_i) \quad (4.15)$$

where π_i is the set of parents of x_i [cf. Equation (4.12), which is the same but with the constraint that each variable has at most one parent]. If π_i is empty, $p(x_i|\pi_i)$ is set to $p(x_i)$.

Figure 4.23 is the graphical part of the Bayesian network and Equation (4.15) is the density associated with the graphical representation. However, there is little to be gained in representing a full multivariate density $p(x_1, \ldots, x_6)$ as a product using the chain rule with the corresponding graph (Figure 4.23), unless we can make some simplifying assumptions concerning the dependence of variables. For example, suppose that

$$p(x_6|x_1, \ldots, x_5) = p(x_6|x_4, x_5)$$

that is, x_6 is independent of x_1, x_2, x_3 given x_4, x_5. Furthermore, suppose that

$$p(x_5|x_1, \ldots, x_4) = p(x_5|x_3)$$

$$p(x_4|x_1, x_2, x_3) = p(x_4|x_1, x_3)$$

$$p(x_3|x_1, x_2) = p(x_3)$$

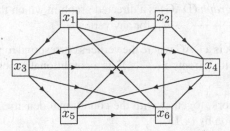

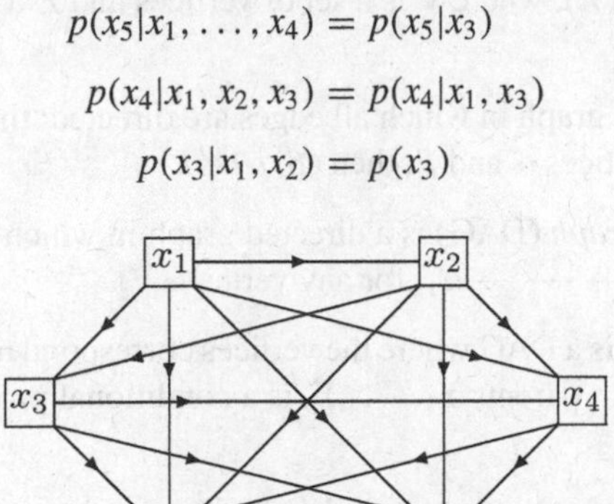

Figure 4.23 Graphical representation of the multivariate density $p(x_1, \ldots, x_6)$.

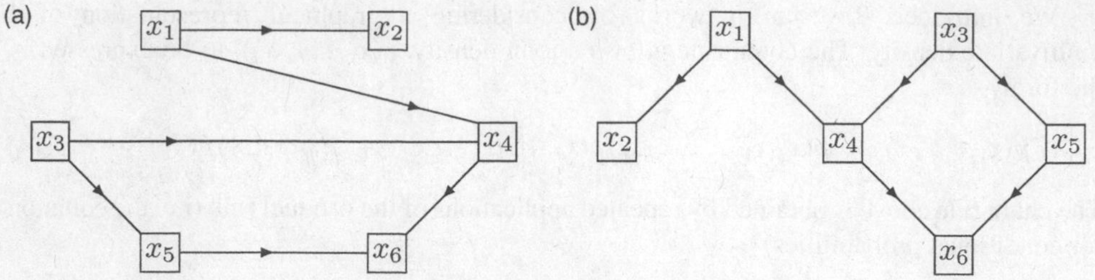

Figure 4.24 Graphical representations of the multivariate density $p(x_1, \ldots, x_6) = p(x_6|x_4, x_5)p(x_5|x_3)p(x_4|x_1, x_3)p(x_3)p(x_2|x_1)p(x_1)$.

Then the multivariate density may be expressed as the product

$$p(x_1, \ldots, x_6) = p(x_6|x_4, x_5)p(x_5|x_3)p(x_4|x_1, x_3)p(x_3)p(x_2|x_1)p(x_1) \tag{4.16}$$

This is depicted graphically in Figure 4.24; Figure 4.24(a) is obtained by removing links in Figure 4.23 and Figure 4.24(b) is an equivalent graph to make the 'parentage' more apparent. This figure, with the general probability interpretation (4.15), gives (4.16). Note that there are two root nodes.

The graphical representation of Bayesian networks is convenient for visualising dependencies. The factorisation of a multivariate density into a product of densities defined on perhaps only a few variables allows better nonparametric density estimates. For example, the product (4.16) requires densities defined on at most three variables.

4.3.5.1 Definitions

The following definitions are used in the literature on Bayesian networks and related techniques (see Chapters 11 and 12 for further discussions of graph theory, within the context of spectral clustering and analysis of complex networks).

- A *graph* is a pair (V, E), where V is a set of vertices and E a set of edges (connections between vertices).
- A directed graph is a graph in which all edges are directed: the edges are ordered pairs; if $(\alpha, \beta) \in E$, for vertices α and β, then $(\beta, \alpha) \notin E$.
- A *directed acyclic graph* (DAG) is a directed graph in which there are no cycles: there is no path $\alpha_1 \to \alpha_2 \to \cdots \to \alpha_1$, for any vertex α_1.
- A Bayesian network is a DAG where the vertices correspond to variables and associated with a variable X with parents $Y_1, \ldots, Y_p$ is a conditional probability density function, $p(X|Y_1, \ldots, Y_p)$.
- The Bayesian network, together with the conditional densities, specify a joint probability density function by (4.15).

4.3.5.2 Classification

As with the MWDT, we may construct a different Bayesian network to model the probability density function of each class $p(\boldsymbol{x}|\omega_i)$ separately. These densities may be substituted into Bayes' rule to obtain estimates of the posterior probabilities of class membership. Alternatively, we may construct a Bayesian network to model the joint density $p(\boldsymbol{x}, \omega)$, where ω is a class label. Again, we may evaluate this for each of the classes and use Bayes' rule to obtain $p(\omega|\boldsymbol{x})$. Separate networks for each class allow a more flexible model. Such a set of networks has been termed a *Bayesian multinet* (Friedman *et al.*, 1997).

4.3.5.3 Specifying the network

Specifying the structure of a Bayesian network consists of two parts: specifying the network topology and estimating the parameters of the conditional probability density functions. The topology may be specified by someone with an expert knowledge of the problem domain and who is able to make some statements about dependencies of variables. Hence, an alternative name for Bayesian networks is probabilistic expert systems: an *expert system* because the network encodes expert knowledge of a problem in its structure and *probabilistic* because the dependencies between variables are probabilistic. Acquiring expert knowledge can be a lengthy process. An alternative approach is to learn the graph from data if they are available, perhaps in a similar manner to the MWDT algorithm. In some applications, sufficient data may not be available.

In learning structure from data, the aim is to find the Bayesian network that best characterises the dependencies in the data. There are many approaches. Buntine (1996) reviews the literature on learning networks. Ideally, we would want to combine expert knowledge where available and statistical data. Heckerman *et al.* (1995; see also Cooper and Herskovits, 1992) discuss Bayesian approaches to network learning. Friedman and Koller (2003) consider an MCMC approach for estimating the posterior distributions of different network structures given a dataset. Their approach has two aspects; calculation of the marginal density of the data for networks with a known ordering of the features (an ordering of the features is such that if x_i precedes x_j in the ordering, then x_j cannot be a parent of x_i); and MCMC sampling over the space of possible feature orders. The book by Neapolitan (2003) provides a detailed treatment of learning of Bayesian networks from data, include both structure and parameter learning.

For nonparametric density estimation, the probability distributions are usually specified as a conditional probability table, with continuous variables discretised, perhaps as part of the structure learning process. For continuous density estimation, it is not essential to discretise the dependent variables and some nonparametric density estimates, perhaps based on *product kernels* (see Section 4.4), could be used for each set of discretised parent variables. More generally, the conditional densities can be prescribed parametric models, allowing some of the estimation procedures of Chapters 2 and 3 to be used.

4.3.5.4 Discussion

Bayesian networks provide a graphical representation of the variables in a problem and the relationship between them. This representation needs to be specified or learned from data. This structure, together with the conditional density functions, allows the multivariate density function to be specified through the chain rule (4.15). In a classification problem, a density may be estimated for each class and Bayes' rule used to obtain the posterior probabilities of class membership.

Bayesian networks have been used to model many complex problems, other than ones in classification, with the structure being used to calculate the conditional density of a variable given measurements made on some (or all) of the remaining variables. For example, suppose that we model the joint density $p(y, x_1, \ldots, x_d)$ as a Bayesian network, and that we are interested in the quantity $p(y|\boldsymbol{e})$, where $\boldsymbol{e}$ comprises measurements made on a subset of the variables, $x_1, \ldots, x_d$. Then, using Bayes' theorem, we may write

$$p(y|\boldsymbol{e}) = \frac{p(y, \boldsymbol{e})}{p(\boldsymbol{e})} = \frac{\int p(y, \boldsymbol{e}, \tilde{\boldsymbol{e}})d\tilde{\boldsymbol{e}}}{\int\int p(y, \boldsymbol{e}, \tilde{\boldsymbol{e}})d\tilde{\boldsymbol{e}}dy}$$

where $\tilde{\boldsymbol{e}}$ is the set of variables $x_1, \ldots, x_d$ not instantiated. Efficient algorithms that make use of the graphical structure have been developed for computing $p(y|\boldsymbol{e})$ in the case of discrete variables (Lauritzen and Spiegelhalter, 1988).

4.3.5.5 Further reading

Bayesian networks have become very popular over the past two decades, and are known by many other names such as belief networks, directed acyclic graphical models, probabilistic expert systems and probabilistic graphical models. A number of good textbooks are available for those interested in using Bayesian networks within statistical pattern recognition studies. These include texts by Pearl (1988), Jensen (1997, 2002), Jordan (1998), Neapolitan (2003), Darwiche (2009), Koller and Friedman (2009), Koski and Noble (2009), and Kjaerulff and Madsen (2010). A recent book edited by Pourret *et al.* (2008) considers a range of real-life applications of Bayesian networks, including medical diagnosis, crime risk evaluation and terrorism risk management. The book by Bishop (2007) is not a specific Bayesian network text but covers procedures for inference in Bayesian networks.

4.3.6 Example application study – naïve Bayes text classification

The problem

Rennie *et al.* (2003) consider the use of naïve Bayes classifiers in text classification. In text classification, one seeks to label a text document by the topic to which it refers, based on an analysis of the content of the document.

Summary

The authors discuss how naïve Bayes models have provided a popular approach to text classification over the years, primarily due to fast and easy means of implementation. However, they note that increasingly they have been performing poorly compared with other text classifiers in performance comparison experiments. To mitigate this they provide five heuristic alterations to the naïve Bayes classifier that increase its performance in such comparisons. These range from simple transformations of the features, to more radical approaches that remove the interpretation of the model as a Bayesian classifier.

The data

The authors consider three publicly available benchmark datasets in their experiments. The first, referred to as the 20 Newsgroups dataset, contains approximately 20 000 newsgroup documents, partitioned evenly across 20 different topics. The second is the Industry Sector dataset, made up of 9649 company web pages classified into 105 different industries. The third is the Reuters-21578 dataset containing documents from the Reuters newswire in 1987. The Reuters-21578 dataset has data classified to 90 topics. However, each document can be categorised into more than one topic (i.e. the data has multiple labels).

The naïve Bayes model

The 'bag of words' naïve Bayes model (McCallum and Nigam, 1998) is used as the baseline model in the experiments. A vocabulary of words is built for each dataset. The feature vector for a given document is then made up from the counts of the number of times each word in the vocabulary occurs in the document. Suppose that the vocabulary contains $|V|$ words. Then the feature vector associated with a document is the $|V|$-dimensional vector, f, such that f_i is the number of times that the ith word in the vocabulary occurs within the document. The class-conditional probability density (formally probability mass function) used for class ω_j in the classifier based on Bayes' rule is

$$p(f|\omega_j) = \left(\sum_{i=1}^{|V|} f_i\right)! \prod_{i=1}^{|V|} \frac{(\theta_{ji})^{f_i}}{f_i!}$$

where θ_{ji} is the independent probability that any given word in a document from class ω_j is the ith word in the vocabulary. This distribution is actually derived as a multinomial model for word frequency, and therefore represents a parametric rather than nonparametric density estimate.

The terms θ_{ji} are estimated from frequency counts in training data documents, along with offsets arising from incorporating prior information. The authors set

$$\hat{\theta}_{ji} = \frac{N_{ji} + \alpha_i}{N_j + \sum_{m=1}^{|V|} \alpha_m} \tag{4.17}$$

where N_{ji} is the number of occurrences of word i in training documents belonging to class ω_j, N_j is the total number of words in training documents belonging to class ω_j, and the α_i are constants relating to the prior distribution of Bayesian estimates for θ_{ji}. The authors set $\alpha_i = 1, i = 1, \ldots, V$.

Rennie *et al.* (2003) note that the form of the model is such that the decision rule is assign f to ω_i if $g_i > g_j$, for all $j \neq i$, where

$$\begin{aligned} g_j &= \log(p(\omega_j)) + \sum_{i=1}^{|V|} f_i \log(\hat{\theta}_{ji}) \\ &= \log(p(\omega_j)) + \sum_{i=1}^{|V|} f_i w_{ji} \end{aligned}$$

with $w_{ji} = \log(\hat{\theta}_{ji})$. This is a linear discriminant function.

Heuristic improvements

The following heuristic improvements are made to the classifier by the authors:

- Noting that empirical distributions of word counts within documents tend to have longer tails than would be suggested by the multinomial model, they replace the multinomial model of word counts by a power law distribution. This is implemented by transforming the word count features as follows

$$f_i' = \log(f_i + \kappa)$$

  where κ should be optimised to fit the data, but is set to $\kappa = 1$ in experiments.

- Noting that common words can adversely affect classification performance through random variations in occurrence across classes, they use an *inverse document frequency* transformation to reduce the effect of common words. The transformation is

$$f_i' = f_i \log\left(\frac{n}{n_i}\right)$$

  where n is the number of training documents, and n_i is the number of training documents which contain word i.

- Noting that the longer a document is, the more likely the naïve Bayes modelling assumption is to be violated through interdependencies between words, they apply the following transformation to reduce the influence of long documents:

$$f_i' = \frac{f_i}{\sqrt{\sum_{j=1}^{|V|}(f_j)^2}}$$

The above transformations are chained together in the order given.

Two more radical alterations to the naïve Bayes classifier are also advocated by the authors, which remove its interpretability as a Bayesian classifier. The first alteration is designed to improve the performance of the classifier when the amounts of available training data differ considerably from class to class (i.e. when the dataset is skewed). The approach is to replace estimation of the θ_{ji} occurrence probabilities for each word within class ω_j, with estimation of $\theta_{\tilde{j}i}$, the probability of the word for all classes except class ω_j (i.e. for the complement class). Equation (4.17) is replaced by

$$\hat{\theta}_{\tilde{j}i} = \frac{N_{\tilde{j}i} + \alpha_i}{N_{\tilde{j}} + \sum_{m=1}^{|V|}\alpha_m}$$

where $N_{\tilde{j}i}$ is the number of occurrences of word i in documents that do not belong to class ω_j, and $N_{\tilde{j}}$ is the total number of words in documents that do not belong to class ω_j. The adapted classification rule is then to assign $\boldsymbol{f}$ to ω_i if $g_i' > g_j'$ for all $j \neq i$, where

$$\begin{aligned} g_j' &= \log(p(\omega_j)) - \sum_{i=1}^{|V|} f_i \log(\hat{\theta}_{\tilde{j}i}) \\ &= \log(p(\omega_j)) - \sum_{i=1}^{|V|} f_i w_{\tilde{j}i} \end{aligned}$$

with $w_{\bar{j}i} = \log(\hat{\theta}_{\bar{j}i})$. A negative sign in front of the weights reflects the fact that documents that are a poor match to the complement class for ω_j should be assigned to class ω_j. It is argued by the authors that such an approach evens out the amount of training data used to estimate the discriminant function for each class, and therefore should reduce the bias in the estimates, and hence improve classification performance.

The final heuristic is to normalise the magnitudes of the weight vectors in the linear discriminant form of the classification rule. Specifically the weights w_{ji} are replaced by

$$\hat{w}'_{ji} = \frac{\log(\hat{\theta}_{ji})}{\sum_{k=1}^{|V|} |\log(\hat{\theta}_{ji})|}$$

It is argued by the authors that this can improve performance when there are correlations between word occurrences (which violates the assumption underlying the naïve Bayes model).

Results

On the datasets considered the authors find that their heuristic alterations lead to better performance than the original naïve Bayes model. The resulting classifier gave comparable performance with a support vector machine (SVM) classifier.[3]

4.3.7 Summary

In the development of the basic histogram approach described in this section, we have concentrated on methods for reducing the number of cells for high-dimensional data. The approaches described are most suited to data that are categorical with integer-labelled categories, though the categories themselves may or may not be ordered. The simplest models are the independence models, and on the head-injury data of Titterington *et al.* (1981) they gave consistently good performance over a range of values for B, the association factor (0.8 to 1.0) in Equation (4.11).

The independence assumption (naïve Bayes model) results in a very severe factorisation of the probability density function – clearly one that is unrealistic for many practical problems. Yet, as discussed by Hand and Yu (2001), it is a model that has had a long and successful history (see also, Domingos and Pazzani, 1997). Practical studies, particularly in medical areas, have shown it to perform surprisingly well. Hand (1992) provides some reasons why this may be so: its intrinsic simplicity means low variance in its estimates; although its probability density estimates are biased, this may not matter in supervised classification so long as $\hat{p}(\omega_1|\mathbf{x}) > \hat{p}(\omega_2|\mathbf{x})$ when $p(\omega_1|\mathbf{x}) > p(\omega_2|\mathbf{x})$; in many cases, variables have undergone a selection process to reduce dependencies. Naïve Bayes models have proven very popular in the field of text classification, with many variants of the basic model being proposed over the years (Lewis, 1998; Rennie *et al.*, 2003; Kim *et al.*, 2006; Chen *et al.*, 2009).

More complex interactions between variables may be represented using Lancaster models and MWDTs. Introduced by Chow and Liu (1968), MWDTs provide an efficient means of representing probability density functions using only second-order statistics.

---

[3] Support vector machines are discussed in Chapter 6, and are considered to be a state-of-the-art approach to text categorisation (Joachims, 1998).

Dependence trees are a special case of Bayesian networks which model a multivariate density as a product of conditional densities defined on a fewer number of variables. These networks may be specified by an expert, or learned from data. Bayesian networks are a powerful means of density estimation, used successfully in a wide range of real world problems. Many good dedicated text books exist which describe the procedures for learning and using Bayesian networks.

4.4 Kernel methods

One of the problems with the histogram approach, as discussed in the previous section, is that for a fixed cell dimension, the number of cells increases exponentially with dimension of the data vectors. This problem can be overcome somewhat by having a variable cell size. The k-nearest-neighbour method (in its simplest form) overcomes the problem by estimating the density using a cell in which the number of training samples is fixed and finds the cell volume that contains the nearest k. The kernel method (also known as the Parzen method of density estimation, after Parzen, 1962) fixes the cell volume and finds the number of samples within the cell and uses this to estimate the density.

Let us consider a one-dimensional example and let $\{x_1, \ldots, x_n\}$ be the set of observations or data samples that we shall use to estimate the density. We may easily write down an estimate of the cumulative distribution function as

$$\hat{P}(x) = \frac{\text{number of observations } \leq x}{n}$$

The density function, $p(x)$, is the derivative of the distribution, but the distribution is discontinuous (at observation values; Figure 4.25) and its derivative results in a set of spikes at the sample points, x_i, and a value zero elsewhere. However, we may define an estimate of the density as

$$\hat{p}(x) = \frac{\hat{P}(x+h) - \hat{P}(x-h)}{2h} \tag{4.18}$$

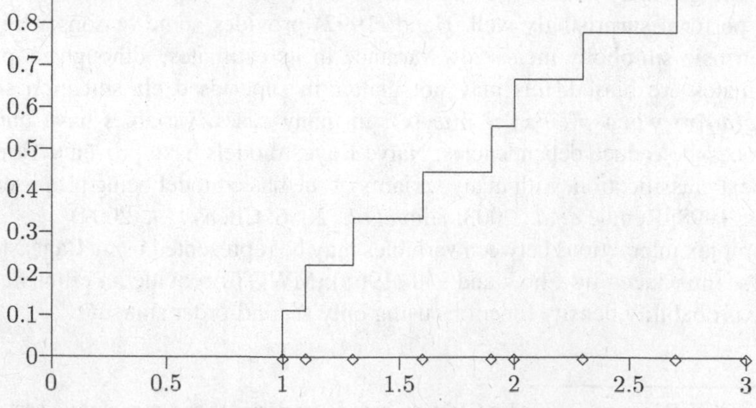

Figure 4.25 Cumulative distribution.

where h is a positive number. This is the proportion of observations falling within the interval $(x - h, x + h)$ divided by $2h$. This may be written as

$$\hat{p}(x) = \frac{1}{hn}\sum_{i=1}^{n} K\left(\frac{x - x_i}{h}\right) \tag{4.19}$$

where $K(z)$ is a *rectangular* (also referred to as *top hat*) kernel

$$K(z) = \begin{cases} 0 & |z| > 1 \\ \frac{1}{2} & |z| \leq 1 \end{cases} \tag{4.20}$$

Equations (4.19) and (4.20) give rise to (4.18), because for sample points, x_i, within h of x, the kernel function in (4.19) takes the value 1/2, and therefore the summation gives a value of $\frac{1}{2} \times$ (the number of observations within the interval). Figure 4.26 gives an estimate of the density using Equations (4.19) and (4.20) for the data used to form the cumulative distribution in Figure 4.25.

Figure 4.26 shows us that the density estimate is itself discontinuous. This arises from the fact that points within a distance h of x contribute a value $\frac{1}{2hn}$ to the density and points further away a value of zero. It is this jump from $\frac{1}{2hn}$ to zero that gives rise to the discontinuities. We can remove this, and generalise the estimator, by using a smoother weighting function than that given by (4.20). For example, we could have a weighting function $K_1(z)$ (also with the property that the integral over the real line is unity) that decreases as $|z|$ increases. Figure 4.27 plots the density estimate for a weighting given by the normal kernel

$$K_1(z) = \frac{1}{\sqrt{2\pi}}\exp\left\{-\frac{z^2}{2}\right\}$$

and a value of h of 0.2. This gives a smoother density estimate. Of course, it does not mean that this estimate is necessarily more 'correct' than that of Figure 4.26, but we might suppose that the underlying density is a smooth function and want a smooth estimate.

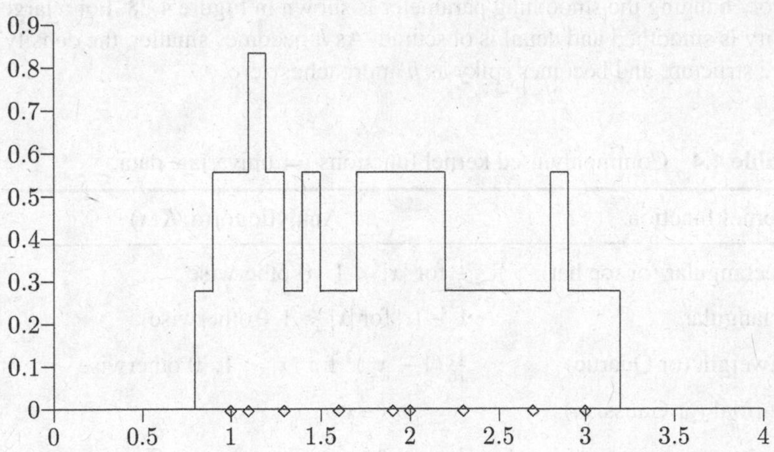

Figure 4.26 Probability density estimate with top hat kernel, $h = 0.2$.

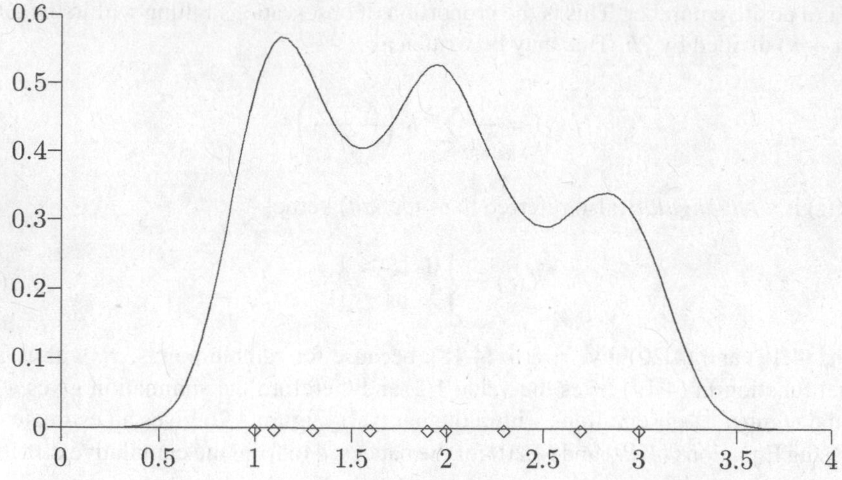

Figure 4.27 Probability density figure with Gaussian kernel and $h = 0.2$.

The above derivation, together with Figures 4.25–4.27, provide a motivation for the kernel method of density estimation, which we formulate as follows. Given a set of observations $\{x_1, \ldots, x_n\}$, an estimate of a density function, in one dimension, is taken to be

$$\hat{p}(x) = \frac{1}{nh} \sum_{i=1}^{n} K\left(\frac{x - x_i}{h}\right) \tag{4.21}$$

where $K(z)$ is termed the *kernel function* and h is the *spread* or *smoothing parameter* (sometimes termed the *bandwidth*). Examples of popular univariate kernel functions are given in Table 4.4.

Smoothing parameter

The effect of changing the smoothing parameter is shown in Figure 4.28. For a large value of h, the density is smoothed and detail is obscured. As h becomes smaller, the density estimate shows more structure and becomes spiky as h approaches zero.

Table 4.4 Commonly used kernel functions for univariate data.

Kernel function	Analytic form, $K(x)$
Rectangular (or top hat)	$\frac{1}{2}$ for $\lvert x\rvert < 1$, 0 otherwise
Triangular	$1 - \lvert x\rvert$ for $\lvert x\rvert < 1$, 0 otherwise
Biweight (or Quartic)	$\frac{15}{16}(1 - x^2)^2$ for $\lvert x\rvert < 1$, 0 otherwise
Normal (or Gaussian)	$\frac{1}{\sqrt{2\pi}}\exp(-x^2/2)$
Bartlett–Epanechnikov	$\frac{3}{4}(1 - x^2/5)/\sqrt{5}$ for $\lvert x\rvert < \sqrt{5}$, 0 otherwise

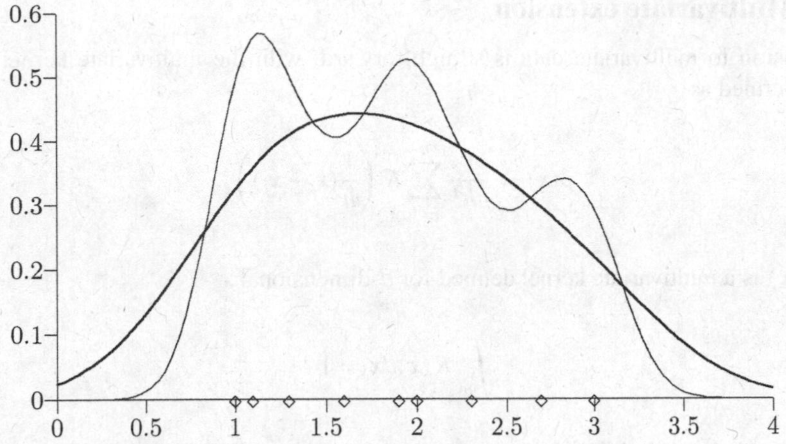

Figure 4.28 Probability density with different levels of smoothing ($h = 0.2$ and $h = 0.5$).

4.4.1 Biasedness

If we impose the conditions that the kernel $K(z) \geq 0$ and $\int K(z)dz = 1$, then the density estimate $\hat{p}(x)$ given by (4.21) also satisfies the necessary conditions for a probability density function, $p(x) \geq 0$ and $\int p(x)dx = 1$.

A theorem due to Rosenblatt (1956) implies that for positive kernels the density estimate will be biased for any finite sample size (see Section 4.1). That is, the estimate of the density averaged over an ensemble of datasets is a biased estimate of the true probability density function. In order to obtain an unbiased estimate, we would be required to relax the condition of positivity of the kernels. Thus the estimate of the density function would not necessarily be a density function itself since it may have negative values. To some people, that would not matter. After all, why should the properties of the *estimate* be the same as the *true* density? On the other hand, there are some who could not live with an estimate of a probability that had negative values, and so would readily accept a bias. They would also point out that asymptotically the estimate is unbiased (it is unbiased as the number of samples used in the estimate tends to infinity) and asymptotically consistent if certain conditions on the smoothing parameter and the kernel hold. These conditions on the kernel are

$$\int |K(z)|dz < \infty$$

$$\int K(z)dz = 1$$

$$\sup_z |K(z)| < \infty, \quad K(z) \text{ is finite everywhere}$$

$$\lim_{z \to \infty} |zK(z)| = 0$$

and the conditions on the smoothing parameter (see Section 4.4.3) are

$$\lim_{n \to \infty} h(n) = 0 \quad \text{for an asymptotic unbiased estimate}$$

$$\lim_{n \to \infty} nh(n) = \infty \quad \text{for an asymptotic consistent estimate}$$

4.4.2 Multivariate extension

The extension to multivariate data is straightforward, with the multivariate kernel density estimate defined as

$$\hat{p}(\boldsymbol{x}) = \frac{1}{nh^d} \sum_{i=1}^{n} K\left(\frac{1}{h}(\boldsymbol{x} - \boldsymbol{x}_i)\right)$$

where $K(\boldsymbol{x})$ is a multivariate kernel defined for d-dimensional $\boldsymbol{x}$

$$\int_{\mathbb{R}^d} K(\boldsymbol{x}) d\boldsymbol{x} = 1$$

and h is the window width.

Multivariate kernels (d variables) are usually radially symmetric unimodal densities such as the Gaussian kernel

$$K(\boldsymbol{x}) = (2\pi)^{-d/2} \exp(-\boldsymbol{x}^T \boldsymbol{x}/2)$$

and the Bartlett–Epanechnikov kernel

$$K(\boldsymbol{x}) = (1 - \boldsymbol{x}^T \boldsymbol{x})(d+2)/(2c_d) \quad \text{for } |\boldsymbol{x}| < 1 \quad (0 \text{ otherwise})$$

where $c_d = \pi^{d/2} / \Gamma((d/2) + 1)$ is the volume of the d-dimensional unit sphere.

Another form of the probability density function estimate commonly used is a sum of *product kernels* (note that this does not imply independence of the variables)

$$\hat{p}(\boldsymbol{x}) = \frac{1}{n} \frac{1}{h_1 \ldots h_d} \sum_{i=1}^{n} \prod_{j=1}^{d} K_j \left(\frac{[\boldsymbol{x} - \boldsymbol{x}_i]_j}{h_j}\right)$$

where there is a different smoothing parameter associated with each variable. The K_j can take any of the univariate forms in Table 4.4. Usually, the K_j are taken to be the same form.

More generally, we may take

$$\hat{p}(\boldsymbol{x}) = \hat{p}(\boldsymbol{x}; \boldsymbol{H}) = \frac{1}{n} \sum_{i=1}^{n} |\boldsymbol{H}|^{-1/2} K(\boldsymbol{H}^{-1/2}(\boldsymbol{x} - \boldsymbol{x}_i)) \tag{4.22}$$

where K is a d-variate spherically symmetric density function and $\boldsymbol{H}$ is a symmetric positive definite matrix. In a classification context, $\boldsymbol{H}$ is commonly taken to be $h_k^2 \hat{\boldsymbol{\Sigma}}_k$ for class ω_k, where h_k is a scaling for class ω_k and $\hat{\boldsymbol{\Sigma}}_k$ is the sample covariance matrix. Various approximations to the covariance are evaluated by Hamamoto *et al.* (1996) in situations where the sample size is small and the dimensionality is high.

4.4.3 Choice of smoothing parameter

One of the problems with this 'nonparametric' method is the choice of the smoothing parameter, h. If h is too small, the density estimator is a collection of n sharp peaks, positioned at the sample points. If h is too large, the density estimate is smoothed and structure in the probability density estimate is lost. The optimal choice of h depends on several factors. First, it depends on the data: the number of data points and their distribution. It also depends on the choice of the kernel function and on the optimality criterion used for its estimation. The maximum likelihood estimation of h that maximises the likelihood

$$p(\boldsymbol{x}_1, \ldots, \boldsymbol{x}_n | h)$$

is given by $h = 0$ – an estimate that consists of a spike at each data point and zero elsewhere. Therefore, some other technique must be used for estimating h. There are many possible methods. Surveys are given in the articles by Jones *et al.* (1996) and Marron (1988).

1. Find the average distance between samples to their kth nearest neighbour and use this for h. A value of $k = 10$ has been suggested (Hand, 1981a).

2. Find the value of h that minimises the mean integrated square error (MISE) between the density and its approximation. Since the actual density is unknown, an asymptotic expression is required for the MISE, calculated via a number of approximations (e.g. derivatives of the unknown density being approximated by the derivatives of a normal density). For a radially symmetric normal kernel, Silverman (1986) suggests

$$h = \sigma \left(\frac{4}{d+2} \right)^{\frac{1}{d+4}} n^{-\frac{1}{d+4}} \tag{4.23}$$

where a choice for σ is

$$\sigma^2 = \frac{1}{d} \sum_{i=1}^{d} s_{ii}$$

and s_{ii} are the diagonal elements of a sample covariance matrix, possibly a *robust* estimate (see Chapter 13). The above estimate will work well if the data come from a population that is normally distributed, but may over-smooth the density if the population is multimodal. A slightly smaller value may be appropriate. You could try several values and assess misclassification rate.

3. There are more sophisticated ways of choosing kernel widths based on least squares cross-validation (again requiring approximations since the actual density is unknown) or maximum likelihood cross-validation. In likelihood cross-validation the value of h is chosen to maximise (Duin, 1976)

$$\prod_{i=1}^{n} \hat{p}_i(\boldsymbol{x}_i)$$

where $\hat{p}_i(\boldsymbol{x}_i)$ is the density estimate based on $n - 1$ samples (all samples but the ith). The use of cross-validation estimates (i.e. specifying each kernel density without using the sample that the density will be evaluated at) removes the problem of the maximum likelihood estimate being given by $h = 0$ (i.e. by spikes at each data point). However, a major problem with this method is its reported poor performance in the heavy-tailed case.

4. Many bandwidth estimators have been considered for the univariate case. The basic idea behind the 'plug-in' estimate is to plug an estimate for the unknown curvature, $S \triangleq \int (p''(x))^2 dx$ (the integral of the square of the second derivative of the density), into the expression for h that minimises the asymptotic MISE

$$h = \left[\frac{c}{D^2 S n}\right]^{1/5}$$

where $c = \int K^2(t)dt$ and $D = \int t^2 K(t)dt$. Jones and Sheather (1991) propose a kernel estimate for the curvature, but this, in turn, requires an estimate of the bandwidth, which will be different from that used to estimate the density. Cao *et al.* (1994), in simulations, use

$$S = n^{-2} g^{-5} \sum_{i,j} K^{iv}\left(\frac{x_i - x_j}{g}\right)$$

where K^{iv} is the fourth derivative of K and the smoothing parameter g is given by

$$g = \left(\frac{2K^{iv}(0)}{D}\right)^{1/7} \hat{T}^{-1/7} n^{-1/7}$$

where $\hat{T}$ is a parametric estimator of $\int (p'''(t))^2 dt$. Development of the 'plug-in' ideas to bandwidth selectors for multivariate data have been considered by Wand and Jones (1994).

Cao *et al.* (1994) perform comparative studies on a range of smoothing methods for univariate densities. Although there is no uniformly best estimator, they find that the plug-in estimator of Sheather and Jones (1991) shows satisfactory performance over a range of problems.

The previous discussion has assumed that h is a fixed value over the whole of the space of data samples. The 'optimal' value of h may in fact be location dependent, giving a large value in regions where the data samples are sparse and a small value where the data samples are densely packed. There are two main approaches:

1. $h = h(x)$; that is, h depends on the location of the sample in the data space. Such approaches are often based on nearest-neighbour ideas.

2. $h = h(x_i)$; that is, h is fixed for each kernel and depends on the local density. These are termed *variable kernel* methods.

One particular choice for h is (Breiman *et al.*, 1977)

$$h_j = \alpha_k d_{jk} \tag{4.24}$$

where α_k is a constant multiplier and d_{jk} is the distance from $\boldsymbol{x}_j$ to its kth nearest neighbour in the training/design set. However, we still have the problem of parameter estimation – namely that of estimating α_k and k.

Breiman *et al.* (1977) find that good fits can be obtained over a wide range of k provided α_k satisfies

$$\beta_k \triangleq \frac{\alpha_k \overline{d_k}^2}{\sigma(d_k)} = \text{constant}$$

where $\overline{d_k}$ is the mean of the kth-nearest-neighbour distances ($\frac{1}{n}\sum_{j=1}^{n} d_{jk}$) and $\sigma(d_k)$ is their standard deviation. In their simulations, this constant was three to four times larger than the best value of h obtained for the fixed kernel estimator.

Other approaches that have a different bandwidth for the kernel associated with each data point employ a 'pilot' density estimate (i.e. an initial fixed kernel estimate of the density) to set the bandwidth. Abramson (1982) has proposed a bandwidth inversely proportional to the square root of the density, $hp^{-1/2}(x)$, which may lead to $\mathcal{O}(h^4)$ bias under certain conditions on the density (Terrell and Scott, 1992; Hall *et al.* 1995). Although a pilot estimate of $p(x)$ is required, the method is insensitive to the fine detail of the pilot (Silverman, 1986).

There have been several papers comparing the variable kernel approach with the fixed kernel method. Breiman *et al.* (1977) find superior performance compared with the fixed kernel approach. It seems that a variable kernel method is potentially advantageous when the underlying density is heavily skewed or long-tailed (Remme *et al.*, 1980; Bowman, 1985). Terrell and Scott (1992) report good performance of the Breiman *et al.* model for small-to-moderate sample sizes, but performance deteriorates as sample size grows compared with the fixed kernel approach.

Further investigations into variable kernel approaches include those of Krzyzak (1983) (who examines classification rules) and Terrell and Scott (1992). Terrell and Scott conclude that it is 'surprisingly difficult to do significantly better than the original fixed kernel scheme'. An alternative to the variable bandwidth kernel is the variable location kernel (Jones *et al.*, 1994), in which the location of the kernel is perturbed to reduce bias.

4.4.4 Choice of kernel

Another choice which we have to make in the form of our density estimate is the kernel function. Examples of different kernel functions were provided in Table 4.4. In practice, the most widely used kernel is the normal (Gaussian) form

$$K\left(\frac{x}{h}\right) = \frac{1}{h\sqrt{2\pi}}\exp\left\{-\frac{x^2}{2h^2}\right\}$$

with product kernels being used for multivariate density estimation. Alternatively, radially symmetric unimodal probability density functions such as the multivariate normal density function are used. There is evidence that the form is relatively unimportant, though the product form may not be ideal for the multivariate case. There are some arguments in favour of kernels that are not themselves densities and admit negative values (Silverman, 1986).

4.4.5 Example application study

The problem

Our practical problem is that addressed by Kraaijveld (1996). It relates to a problem in the oil industry: prediction of the type of subsurface (for example, sand, shale, coal – the lithofacies class) from physical properties such as electron density, velocity of sound and electrical resistivity obtained from a logging instrument lowered into a well.

Summary

The classifier is applied to data from wells which differ from those used to train the classifier. This is an example of a practical problem where the probability density function of data may vary with time (termed *population drift*; Hand, 1997). Thus, the test conditions differ from those used to define the training data. Kraaijveld (1996) showed how unlabelled test data can be used to improve the performance of a kernel approach to discriminant analysis where the training data are only approximately representative of the test conditions. This is done by adjusting the width of the kernel estimator using the observed (but unlabelled) test data. Such a use of the test set as part of the training procedure is feasible if the classification declarations are not needed online at the immediate time of measurement. This is because we can then collect together the test samples for batch processing at a later date.

The data

Twelve standard datasets were generated from data gathered from two fields and 18 different well sites. The datasets comprised different-sized feature sets (4–24 features) and 2, 3 and 4 classes.

The model

A Gaussian kernel is used, with the width estimated by maximising a modified likelihood function (solved using a numerical root-finding procedure and giving a width, s_1). An approximation, based on the assumption that the density at a point is determined by the nearest kernel only, is given by

$$s_2 = \sqrt{\frac{1}{dn}\sum_{i=1}^{n}|\boldsymbol{x}_i^* - \boldsymbol{x}_i|^2}$$

for a d-dimensional dataset $\{\boldsymbol{x}_i, i = 1, \ldots, n\}$, where $\boldsymbol{x}_i^*$ is the nearest sample to $\boldsymbol{x}_i$.

A robust estimate of kernel width is also derived using a test dataset, again using a modified likelihood criterion, to give s_3. The nearest-neighbour approximation is

$$s_4 = \sqrt{\frac{1}{dn_t}\sum_{i=1}^{n_t}|\boldsymbol{x}_i^* - \boldsymbol{x}_i^t|^2}$$

where the test set is $\{\boldsymbol{x}_i^t, i = 1, \ldots, n_t\}$, and $\boldsymbol{x}_i^*$ is now the nearest sample in the training set to $\boldsymbol{x}_i^t$. This provides a modification to the kernel width using the test set distribution (but not the test set labels). Thus, the test set is used as part of the training procedure.

Training procedure

Twelve experiments were defined by using different combinations of datasets as train and test. Five classifiers were assessed; namely Bayes' rule classifiers using kernel density estimates with bandwidth estimators s_1 to s_4, and also a baseline nearest-neighbour classifier.

Results

The robust methods led to an increase in performance (measured in terms of error rate) in all but one of the 12 experiments. The nearest-neighbour approximation to the bandwidth tended to underestimate the bandwidth by about 20%.

4.4.6 Further developments

For multivariate data, procedures for approximating the kernel density using a reduced number of kernels are described by Fukunaga and Hayes (1989a) and Babich and Camps (1996). Fukunaga and Hayes select a set from the given dataset by minimising an entropy expression. Babich and Camps use an agglomerative clustering procedure to find a set of prototypes and weight the kernels appropriately (cf. mixture modelling in Chapter 2). Jeon and Landgrebe (1994) use a clustering procedure, together with a branch and bound method, to eliminate data samples from the density calculation.

Zhang *et al.* (2006) consider the use of MCMC algorithms (see Chapter 3) to obtain optimal estimates of the bandwidth matrix used in the multivariate kernel defined in (4.22).

There are several approaches to density estimation that combine parametric and nonparametric approaches (*semiparametric* density estimators). Hjort and Glad (1995) describe an approach that multiplies an initial parametric start with a nonparametric kernel-type estimate for the necessary correction factor. Hjort and Jones (1996) find the best local parametric approximation to a density $p(x, \boldsymbol{\theta})$, where the parameter values, $\boldsymbol{\theta}$, depend on x.

Botev *et al.* (2010) relate the Gaussian kernel density estimator to the solution to a diffusion partial differential equation, and use this to motivate a kernel density estimator based on a linear diffusion partial differential equation.

The kernel methods described in this chapter apply to real-valued continuous quantities. We have not considered issues such as dealing with missing data (Titterington and Mill, 1983; Pawlak, 1993) or kernel methods for discrete data [see McLachlan (1992a) for a summary of kernel methods for other data types]. Karunamuni and Alberts (2005) consider how to reduce boundary effects when kernel density estimators are used on univariate data with bounded support.

Online approaches to obtaining kernel density estimates as the number of data points grows have been considered by Lambert *et al.* (1999), using a Taylor series expansion to Gaussian kernels, and by Kristan *et al.* (2011), using a compression and revitalization scheme to avoid the need to use all the data points.

4.4.7 Summary

Kernel methods, both for multivariate density estimation and regression, have been extensively studied. The idea behind kernel density estimation is very simple – put a 'bump' function over each data point and then add them up to form a density. One of the disadvantages of the kernel approach is the high computational requirements for large datasets – there is a kernel at every training data point that contributes to the density at a given point, x. Computation can be excessive and for large datasets it may be appropriate to use kernels other than the normal density in an effort to reduce computation. Also, since kernels are localised, only a small proportion will contribute to the density at a given point. Some preprocessing of the data will enable noncontributing kernels to be identified and omitted from a density calculation. For univariate density estimates, based on the normal kernel, Silverman (1982) proposes an efficient algorithm for computation based on the fact that the density estimate is a convolution of the data with the kernel and the Fourier transform is used to perform the convolution (Jones and Lotwick, 1984; Silverman, 1986). Speed improvements can be obtained in a similar manner to those used for k-nearest-neighbour methods – by reducing the number of prototypes (as in the edit and condense procedures).

k-nearest-neighbour methods may also be viewed as kernel approaches to density estimation in which the kernel has uniform density in the sphere centred at a point $\boldsymbol{x}$ and of radius equal to the distance to the kth nearest neighbour. The attractiveness of k-nearest-neighbour methods is that the kernel width varies according to the local density, but is discontinuous. The work of Breiman *et al.* (1977) on variable kernel methods is an attempt to combine the best features of k-nearest-neighbour methods with the fixed kernel approaches.

There may be difficulties in applying the kernel method in high dimensions. Regions of high density may contain few samples, even for moderate sample sizes. For example, in the 10-dimensional unit multivariate normal distribution[4] (Silverman, 1986), 99% of the mass of the distribution is at points at a distance greater than 1.6, whereas in one dimension, 90% of the distribution lies between ± 1.6. Thus, reliable estimates of the density can only be estimated for extremely large samples in high dimensions. As an indication of the sample sizes required to obtain density estimates, Silverman considers the special case of a unit multivariate normal distribution, and a kernel density estimate with normal kernels where the window width is chosen to minimise the mean squared error at the origin. In order that the relative mean square error, $\mathrm{E}[(\hat{p}(0) - p(0))^2/p^2(0)]$, is less than 0.1, a small number of samples is required in one and two dimensions (Table 4.5). However, for 10 dimensions, over 800 000 samples are necessary. Thus, in order to obtain accurate density estimates in high dimensions, an enormous sample size is necessary. Further, these results are likely to be optimistic and more samples would be required to estimate the density at other points in the distribution to the same accuracy.

Kernel methods are motivated by the asymptotic results and as such are only really relevant to low-dimensional spaces due to sample size considerations. However, as far as discrimination is concerned, we may not necessarily be interested in accurate estimates of the densities themselves, but rather the Bayes decision region for which approximate estimates of the densities may suffice. In practice, kernel methods do work well on multivariate data, in the sense that error rates similar to other classifiers can be achieved.

4.5 Expansion by basis functions

The method of density estimation based on an orthogonal expansion by basis functions was first introduced by Čencov (1962). The basic approach is to approximate a density function,

[4] The unit multivariate normal distribution is a spherical multivariate normal distribution, with the common variance set to one.

Table 4.5 Required sample size as a function of dimension for a relative mean square error at the origin of less than 0.1 when estimating a unit multivariate normal density using normal kernels with width chosen so that the mean square error at the origin is minimised (Silverman, 1986).

Dimensionality	Required sample size
1	4
2	19
3	67
4	223
5	768
6	2790
7	10 700
8	43 700
9	187 000
10	842 000

$p(x)$, by a weighted sum of orthogonal basis functions. We suppose that the density admits the expansion

$$p(x) = \sum_{i=1}^{\infty} a_i \phi_i(x) \tag{4.25}$$

where the $\{\phi_i\}$ form a complete orthonormal set of functions satisfying

$$\int k(x)\phi_i(x)\phi_j(x)dx = \lambda_i \delta_{ij} \tag{4.26}$$

for a *kernel* or *weighting function* $k(x)$, and $\delta_{ij} = 1$ if $i = j$ and zero otherwise. Thus, multiplying (4.25) by $k(x)\phi_i(x)$ and integrating gives

$$\lambda_i a_i = \int k(x)p(x)\phi_i(x)dx = \mathrm{E}[k(x)\phi_i(x)]$$

where the expectation is with respect to the distribution defined by $p(x)$. Given $\{x_1, x_2, \ldots, x_n\}$, a set of independently and identically distributed samples from $p(x)$, then the a_i can be estimated in an unbiased manner by

$$\lambda_i \hat{a}_i = \frac{1}{n}\sum_{j=1}^{n} k(x_j)\phi_i(x_j)$$

The orthogonal series estimator based on the sample $\{x_1, x_2, \ldots, x_n\}$ of $p(x)$ is then given by

$$\hat{p}_n(x) = \sum_{i=1}^{s} \frac{1}{n\lambda_i} \sum_{j=1}^{n} k(x_j)\phi_i(x_j)\phi_i(x) \tag{4.27}$$

where s is the number of terms retained in the expansion. The coefficients $\hat{a}_i$ may be computed sequentially

$$\lambda_i \hat{a}_i(r+1) = \frac{r}{r+1}\lambda_i \hat{a}_i(r) + \frac{1}{r+1}k(x_{r+1})\phi_i(x_{r+1})$$

where $\hat{a}_i(r+1)$ is the value obtained using $r+1$ data samples. This means that, given an extra sample point, it is a simple matter to update the coefficients. Also, a large number of data vectors could be used to calculate the coefficients without storing the data in memory.

A further advantage of the series estimator method is that the final estimate is easy to store. It is not in the form of a complicated analytic function, but a set of coefficients.

There are several disadvantages, however. First of all, the method is limited to low-dimensional data spaces. Although, in principle, the method may be extended to estimate multivariate probability density functions in a straightforward manner, the number of coefficients in the series increases exponentially with dimensionality. It is not an easy matter to calculate the coefficients.

Another disadvantage is that the density estimate is not necessarily a density. This may or may not be a problem depending on the application. Also, the density estimate is not necessarily nonnegative.

Many different functions have been used as basis functions. These include Fourier and trigonometric functions on [0, 1], and Legendre polynomials on [−1, 1]; and those with unbounded support such as Laguerre polynomials on [0, ∞] and Hermite functions on the real line. If we have no prior knowledge as to the form of $p(x)$, then the basis functions are chosen for their simplicity of implementation. The most popular orthogonal series estimator for densities with unbounded support is the Hermite series estimator. The normalised Hermite functions are given by

$$\phi_k(x) = \frac{\exp(-x^2/2)}{(2^k k! \pi^{\frac{1}{2}})^{\frac{1}{2}}} H_k(x)$$

where $H_k(x)$ is the kth Hermite polynomial

$$H_k(x) = (-1)^k \exp(x^2) \frac{d^k}{dx^k} \exp(-x^2)$$

The performance and the smoothness of the density estimator depends on the number of terms used in the expansion. Too few terms leads to over-smoothed densities. Different stopping rules [rules for choosing the number of terms, s, in the expansion (4.27)] have been proposed and are briefly reviewed by Izenman (1991). Kronmal and Tarter (1962) propose a stopping rule based on minimising a mean integrated square error. Termination occurs when the test

$$\hat{a}_j^2 > \frac{2}{n+1}\hat{b}_j^2$$

fails, where

$$\hat{b}_j^2 = \frac{1}{n}\sum_{k=1}^{n} \phi_j^2(x_k)$$

or, alternatively, when t or more successive terms fail the test. Practical and theoretical difficulties are encountered with this test, particularly with sharply peaked or multimodal densities and it could happen that an infinite number of terms pass the test. Alternative procedures for overcoming these difficulties have been proposed (Hart, 1985; Diggle and Hall, 1986).

4.6 Copulas

4.6.1 Introduction

An approach to modelling non-normal multivariate distributions that has been popular in financial communities is the use of copulas. Copulas are a means of combining marginal distributions to specify joint distributions with nontrivial dependencies between the variables. Instead of specifying a direct model for the joint density function, one models univariate marginal density functions and specifies a copula function to combine them into the joint density function.

In other chapters and sections of this book, we may refer to probability density functions and probability mass functions as being distributions. In this section we need to be more careful with notation. Here a distribution function refers to the cumulative distribution function. The distribution function, $F(x)$, for a random variable X, is given by $F(x) = P(X \leq x)$. In the continuous case the corresponding probability density function, $f(x)$, is given by $f(x) = \frac{\partial F(x)}{\partial x}$, so that $F(x) = \int_{-\infty}^{x} f(t)dt$.

4.6.2 Mathematical basis

Suppose that we are modelling a d-dimensional multivariate distribution, $\boldsymbol{u} = (u_1, \ldots, u_d)$, where the marginal distribution of each constituent variable is a uniform distribution over the range [0, 1], i.e. $U_i \sim U(0, 1)$, $i = 1, \ldots, n$. The copula function $C(u_1, \ldots, u_d)$ is the function that specifies the joint distribution

$$C(u_1, \ldots, u_d) = P(U_1 \leq u_1, \ldots, U_d \leq u_d)$$

The use of copulas for modelling multivariate distributions arises from a theorem by Sklar (1973) (French language original, Sklar, 1959). Sklar's theorem states that for a multivariate distribution function $F(x_1, \ldots, x_d)$, with marginal distribution functions $F_1(x_1), \ldots, F_d(x_d)$, there exists a d-dimensional copula function C such that

$$C(F_1(x_1), \ldots, F_d(x_d)) = F(x_1, \ldots, x_d) \tag{4.28}$$

Furthermore, if the marginal distributions are continuous, this copula function is unique. The above makes use of the property that if $F(x)$ is the distribution function of a random variable X [i.e. $F(x) = P(X \leq x)$], then $F(X) \sim U(0, 1)$, a uniform distribution over the range [0, 1]. The significance of Sklar's theorem is that there is a unique function that maps the univariate marginal distributions to the joint distribution. Thus, instead of specifying a direct model for the joint density function, we can model univariate density functions and specify a copula function to obtain the joint density function.

Through differentiation of (4.28), the probability density function for a distribution specified via a copula function is then given by

$$f(x_1, \ldots, x_n) = c(F_1(x_1), \ldots, F_d(x_d)) \prod_{i=1}^{d} f_i(x_i) \tag{4.29}$$

where $f_i(x_i)$ is the marginal probability density function for the ith variable, and:

$$c(u_1, \ldots, u_d) = \frac{\partial^d C(u_1, \ldots, u_d)}{\partial u_1 \ldots \partial u_d} \tag{4.30}$$

Such density functions can then be used within a discrimination rule based upon Bayes' theorem.

The significance of the above is that rather than specifying a multivariate density directly, we can:

1. model the marginal distributions of the variables;
2. specify a function, C, on the unit hypercube, that takes into account dependencies between the variables, enabling us to combine the marginal distributions into the joint distribution.

We may use both parametric (see Chapter 2) and nonparametric estimates (using the density estimation techniques considered earlier in this chapter) for the marginal density functions. The task is then to find a suitable form for the copula function (including specification of any parameters). This procedure may be simpler than estimating a full multivariate density. However, there is the risk that the chosen copula function does not adequately represent the dependencies between the variables.

4.6.3 Copula functions

In practice, when using copulas to model multivariate distributions, we utilise one of a family of parametrised copula functions. The parameters of the function are then optimised with the aid of a set of training data.

A commonly used family of copula functions is the Archimedean copula family. An Archimedean copula is of the form

$$C_\phi(u_1, \ldots, u_d) = \phi^{-1}\left(\sum_{i=1}^{d} \phi(u_i)\right)$$

where $\phi(x)$ is the generator of the copula. For C to be a valid copula function, $\phi(x)$ needs to be a strictly decreasing convex function with input range $(0, 1]$, and output range $[0, \infty)$, such that $\phi(1) = 0$, and typically $\lim_{x \to 0} \phi(x) = \infty$. Examples are (Frees and Valdez, 1998):

- Clayton, $\phi(x) = x^{-\alpha} - 1, \alpha > 0$,
- Gumbel, $\phi(x) = (-\log(x))^{\alpha}, \alpha \geq 1$.
- Frank, $\phi(x) = -\log\left(\frac{\exp(-\alpha x)-1}{\exp(-\alpha)-1}\right), \alpha \in (-\infty, \infty), \alpha \neq 0$.

We note that if the Gumbel copula is used with $\alpha = 1$, we actually have an independence model

$$C(u_1, \ldots, u_d) = \prod_{i=1}^{d} u_i$$

so that

$$F(x_1, \ldots, x_d) = C(F_1(x_1), \ldots, F_d(x_d)) = \prod_{i=1}^{d} F_i(x_i)$$

i.e. independence.

Alternative families of copulas can be obtained by specifying the 'correct' copula for a given multivariate distribution. To do this, we note that substituting $u_i = F_i(x_i)$ into (4.28) gives

$$C(u_1, \ldots, u_d) = F\left(F_1^{-1}(u_1), \ldots, F_d^{-1}(u_d)\right)$$

where $F_i^{-1}(u)$ is the inverse of the marginal distribution function for the ith variable. An example would be the Gaussian copula, in which C is chosen to be the copula for a zero-mean multivariate Gaussian distribution (parametrised by the covariance matrix Σ). Such a copula is easy to specify, since the marginal distributions of a multivariate Gaussian distribution are themselves Gaussian distributions. A discussion on the correlation structure to use within high-dimensional Gaussian copulas is given by Zezula (2009). Another example is a copula based upon the multivariate t-distribution (Demarta and McNeil, 2005).

A danger of using such families of copulas to model complicated multivariate distributions, is that the range of dependencies that the chosen family can model, might not be appropriate for the data being modelled. Widespread banking use of an approach to modelling credit risks (Li, 2000) that utilised Gaussian copula functions has been linked to considerable financial losses in recent years.

Venter (2001) considers the properties and features of a range of different copulas, with a view to selecting the most appropriate copula for a given dataset and problem. A consideration of how to select the most appropriate Archimedean copula to model data is given by Melchiori (2003).

4.6.4 Estimating copula probability density functions

Given a parametrised copula function, and parametrised forms for the marginal densities, we can use a maximum likelihood estimation procedure to optimise the likelihood given a set of training data. The probability density function underlying the likelihood function is that defined in (4.29). For a dataset $\{\boldsymbol{x}_1, \ldots, \boldsymbol{x}_n\}$, the log-likelihood under an assumption of independent data vectors is therefore given by

$$L(\alpha, \theta_1, \ldots, \theta_d) = \sum_{j=1}^{n} \log(c_\alpha(F_1(x_{j1}), \ldots, F_d(x_{jd})) + \sum_{j=1}^{n} \sum_{i=1}^{d} \log(f_i(x_{ji}))$$

where α is the parameter vector associated with the (derivative of) the copula function; $\theta_1, \ldots, \theta_d$ are the parameters of the marginal densities; and x_{ji} is the ith variable of the jth training data example $\boldsymbol{x}_j$. The parameters $\{\alpha, \theta_1, \ldots, \theta_d\}$ are the ones to be optimised. In practice, if the dimensionality of these parameters is large, optimising this likelihood function with respect to the free parameters can be difficult.

The Inference for Margins (IFM) method, described in Trivedi and Zimmer (2005) and Joe (2005), uses a two-stage parametric estimation procedure:

- Estimate the parameters of the marginal distributions in isolation, using maximum likelihood estimates for a given set of training data.
- Plug the parameters of the marginal distributions into the joint density given in (4.29), and optimise the parameters of the copula function to maximise the resulting likelihood. This is equivalent to maximising

$$L(\alpha) = \sum_{j=1}^{n} \log(c_\alpha(\hat{F}_1(x_{j1}), \ldots, \hat{F}_d(x_{jd})))$$

  with respect to the parameters α of the (derivative of) the copula function, where $\hat{F}_i(x_i)$ is the optimised parametric form for the marginal distribution of the ith variable, evaluated at x_i.

Such a procedure is computationally simpler than estimating all the parameters together in a single maximum likelihood estimation step. However, this separation into two phases will not necessarily produce the optimal solution.

A semiparametric version of the IFM method can be used to remove the requirement for parametric forms for the marginal densities:

- In the first stage, the marginal distributions are estimated as straightforward empirical distributions

$$\tilde{F}_i(y) = \tilde{P}(X_i \leq y) \approx \frac{1}{n}\sum_{j=1}^{n} I(x_{ji} \leq y)$$

- These empirical distributions are then used to approximate the marginal distributions in (4.29). The parameters of the copula function are selected to maximise the resulting semiparametric likelihood. This is equivalent to maximising

$$L(\alpha) = \sum_{j=1}^{n} \log(c_\alpha(\tilde{F}_1(x_{j1}), \ldots, \tilde{F}_d(x_{jd}))) \tag{4.31}$$

  with respect to the parameters α of the (derivative of) the copula function.

Kim *et al.* (2007), demonstrate that this semiparametric method can out-perform the parametric estimation methods, in cases where the parametric forms for the marginal densities are not known. An alternative to estimating empirical marginal distributions directly from the training data in the first stage is to obtain them via nonparametric density estimates, such as the kernel density estimates considered earlier in this chapter.

4.6.5 Simple example

We now illustrate the process of using copulas to estimate a probability density function. We consider a bivariate example, using the Frank Archimedean copula. The density to be estimated is a bivariate normal distribution, with mean vector $\mu = (0, 0)$, variances $(1.25, 1.75)$ and covariance cross-term of 0.43. 1000 training data samples are available for optimising the copula.

The bivariate Frank Archimedean copula defined in Section 4.6.3 can be expressed as (see Exercises)

$$C_\alpha(u_1, u_2) = \frac{-1}{\alpha} \log\left(1 + \frac{(\exp[-\alpha u_1] - 1)(\exp[-\alpha u_2] - 1)}{\exp[-\alpha] - 1}\right) \tag{4.32}$$

The derivative defined in (4.30), and required in (4.29), can then be expressed in the following form

$$c_\alpha(u_1, u_2) = \frac{-\alpha \exp[-\alpha(u_1 + u_2)]}{\exp[-\alpha] - 1} \times \left(1 + \frac{(\exp[-\alpha u_1] - 1)(\exp[-\alpha u_2] - 1)}{\exp[-\alpha] - 1}\right)^{-2} \tag{4.33}$$

We consider the semiparametric version of the IFM procedure described in Section 4.6.4. A Nelder–Meader (Press *et al.*, 1992) procedure is used to optimise (4.31) using the form of c_α in (4.33), alongside the empirical estimates of the marginal distributions. The overall probability density function given in (4.29) is estimated using the optimised copula, along with Gaussian kernel density estimates for the marginal probability density functions (see Section 4.4). The width of each Gaussian kernel density estimate is chosen in line with the expression in Equation (4.23).

The estimated probability density function, evaluated over a grid, is displayed in Figure 4.29(a), alongside the actual probability density function 4.29(b). The squared errors over the grid are displayed in Figure 4.30. Much of the error is due to errors in estimating the

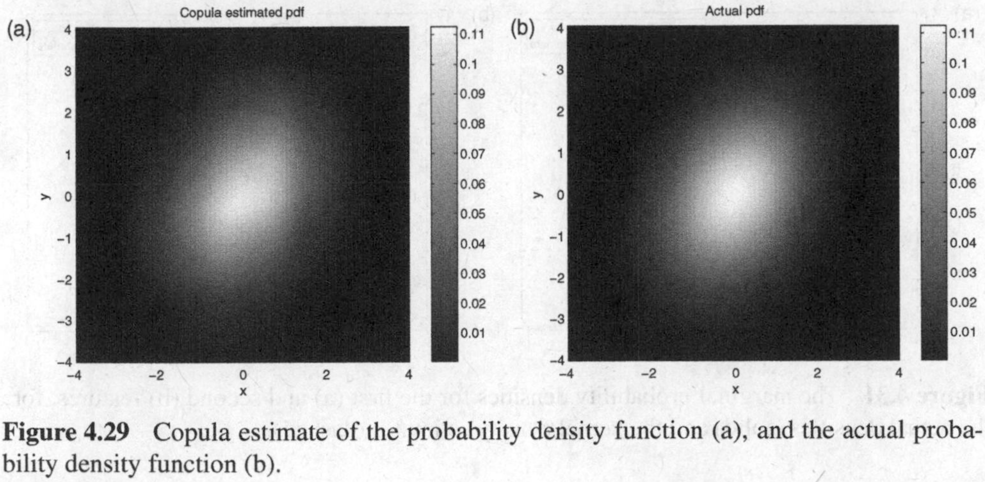

Figure 4.29 Copula estimate of the probability density function (a), and the actual probability density function (b).

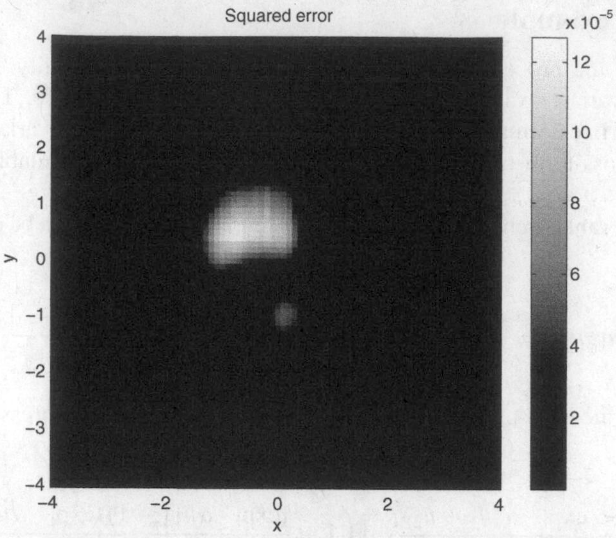

Figure 4.30 Squared errors between the copula estimate of the probability density function and the actual probability density function.

marginal densities, as is shown in the plots of the kernel density estimates of the marginal probability densities in Figure 4.31. However, for more complicated distributions the copula function itself is likely to be a major source of error.

4.6.6 Summary

Copulas are a means of combining marginal distributions to specify joint distributions with nontrivial dependencies between the variables. Joint density functions are obtained by modelling univariate density functions and specifying a copula function on a hypercube in order

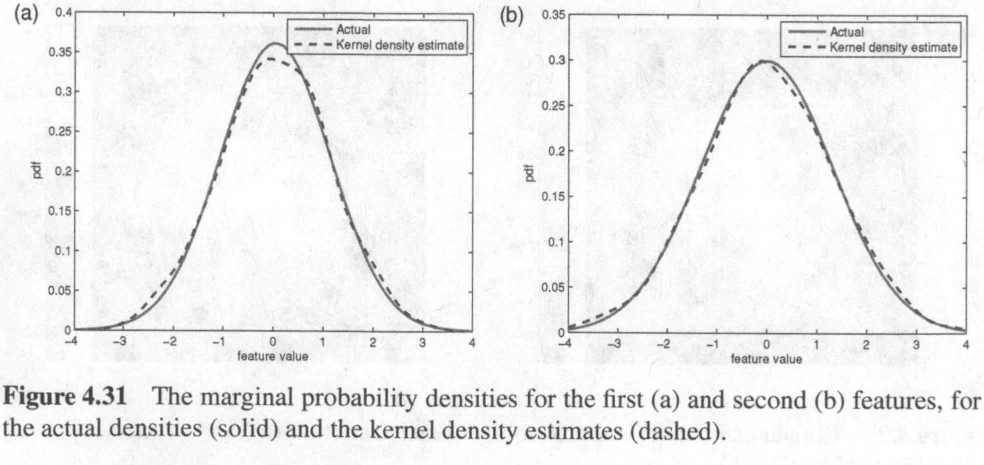

Figure 4.31 The marginal probability densities for the first (a) and second (b) features, for the actual densities (solid) and the kernel density estimates (dashed).

to obtain the joint density function. In some sense, little has been gained here. We are still required to define a multivariate function (the copula function). However, this function is defined on a hypercube, rather than over the domain of the variables (the latter being what is required in full multivariate density estimation). The potential gain comes from using a range of parametrised forms for the copula function. However, there is a risk that these parametrised forms do not adequately represent the correlations present in the variables, i.e. that the chosen copula function does not adequately represent the dependencies between the variables. In such cases, the resulting joint densities will be poor reflections of the true joint densities. Copulas should therefore be used with caution. A critical assessment of the utility of copulas is provided by Mikosch (2006).

4.7 Application studies

The nonparametric methods of density estimation described in this chapter have been applied to a wide range of problems. Applications of Bayesian networks include the following:

- Drug safety. Cowell *et al.* (1991) develop a Bayesian network for analysing a specific adverse drug reaction problem (drug-induced pseudomembranous colitis). The algorithm of Lauritzen and Spiegelhalter (1988) was used for manipulating the probability density functions in Bayesian networks [see also Spiegelhalter *et al.* (1991) for a clear account of the application of probabilistic expert systems].

- Endoscope navigation. In a study on the use of computer vision techniques for automatic guidance and advice in colon endoscopy, Kwoh and Gillies (1996) construct a Bayesian network (using subjective knowledge from an expert) and compare performance with a maximum weight dependence tree learned from data. The latter gave better performance.

- Geographic information processing. Stassopoulou *et al.* (1996) compare Bayesian networks and neural networks (see Chapter 6) in a study to combine remote sensing and other data for assessing the risk of desertification of burned forest areas in the Mediterranean region. A Bayesian network is constructed using information provided by experts. An equivalent neural network is trained and its parameters used to set the conditional probability tables in the Bayesian network.

- Image segmentation. Williams and Feng (1998) use a tree-structured network, in conjunction with a neural network, as part of an image labelling scheme. The conditional probability tables are estimated from training data using a maximum likelihood procedure based on the EM algorithm (see Chapter 2).

- Automated situation-assessment for naval anti-air warfare. Bladon *et al.* (2006) discuss the use of Bayesian networks within automated situation-assessment tools, which can be used as decision aids in a command and control system. They develop a Combat-ID and Threat Assessment tool for naval anti-air warfare, making probabilistic estimates of hostility by taking into account such factors as speed, Identification of Friend or Foe (IFF) system responses and airlane obedience. The networks are constructed using information provided by experts.

- Modelling operational risk to information technology infrastructure. Neil *et al.* (2008) show how Bayesian networks can be used to model the financial losses that could result from information technology infrastructure risk scenarios. Both discrete and non-Gaussian distributions are used in the model, with approximate inference taking place with the aid of a dynamic discretisation algorithm (Neil *et al.*, 2007), and an inference procedure known as the Junction Tree algorithm (Jensen, 2002).

- Oil and gas exploration. Martinelli *et al.* (2010) use a Bayesian network to model the dependencies between prospects in oil and gas exploration. The network is created using expert knowledge of the local geology. They use the network to answer the two questions of which locations should be drilled in an area where hydrocarbons have been discovered, and what drilling locations should be used when considering abandoning an area which to date has only had dry wells drilled?

Examples of applications of k-nearest-neighbour methods include the following:

- Target classification. Chen and Walton (1986) apply nearest-neighbour methods to ship and aircraft classification using radar cross-section measurements. Drake *et al.* (1994) compare several methods, including k nearest neighbour on multispectral imagery.

- Handwritten character recognition. There have been many studies in which nearest-neighbour methods have been applied to handwritten characters. Smith *et al.* (1994) use three distance metrics for a k-nearest-neighbour application to handwritten digits. Yan (1994) uses nearest neighbour with a multilayer perceptron to refine prototypes.

- Text categorisation. Han *et al.* (2001) use a k-nearest-neighbour classifier to assign documents to pre-specified categories. Feature vectors are obtained by specifying a vocabulary of words, and then creating a feature vector from the number of times each word occurs in the document. To remove the effect of document size, feature vectors are normalised so that their elements sum to one. Different weights are used for different features (word counts) when calculating the distance between two documents, resulting in a weight adjusted k-nearest-neighbour classifier. The weights are optimised using an iterative training procedure. Good performance is obtained on a range of benchmark text categorisation problems.

- Prostate cancer identification. Tahir *et al.* (2007) consider classification of prostate cancer on the basis of multispectral images of prostate biopsies. The aim is to classify the tissue into four groups (two benign, one a precursor to cancer, and one cancer). 128-dimensional feature vectors are extracted from multispectral images of prostate needle biopsies. A Tabu Search heuristic (Glover, 1989, 1990) is used to perform feature selection and feature weighting for a k-nearest-neighbour classification rule. An improvement in performance is obtained over previously published results.

- Speech recognition. Deselaers *et al.* (2007) use a nearest-neighbour classifier within a hidden Markov model (HMM) approach to speech recognition. The exponential of the negative within-class nearest-neighbour distance is used to estimate each state-conditional feature probability density in the HMM. A kd-tree is used to find each nearest neighbour. When the amount of available training data is small, the performance is better than a baseline of Gaussian mixture models within an HMM.

Example applications of kernel methods are:

- Chest pain prognosis. Scott *et al.* (1978) use a quartic kernel to estimate the density of plasma lipids in two groups (diseased and normal). The aim of the investigation was to ascertain the dependence of the risk of coronary artery disease on the joint variation of plasma lipid concentrations.
- Fruit fly classification. Sutton and Steck (1994) use Epanechnikov kernels in a two-class fruit fly discrimination problem.
- Visual surveillance. Elgammal *et al.* (2002) build estimates for the scene background in surveillance imagery, using kernel density estimation. Motion is then detected through poor matches to the learnt background distributions, a process known as *background subtraction*. Kernel density estimates are also used for modelling foreground objects.
- Crime hotspot mapping. Eck *et al.* (2005) consider crime mapping, in which high-crime-density areas are identified as hot spots, in order to assist policing. They recommend a quartic kernel density estimation approach to estimate a surface map of crime density, which can then be visualised. Chainey *et al.* (2008) extend the concept of crime hotspot mapping to using the maps to predict where crimes will occur in the future. They find that maps produced by kernel density estimation are better at predicting future street crime locations than less data adaptive approaches.
- Maritime surveillance. Ristic *et al.* (2008) consider the detection of vessels with anomalous trajectories at Port Adelaide in Australia. The motivation for identifying such anomalies is that abnormal routes and speeds might be indicative of customs evasion, drug smuggling, piracy or terrorism. The kinematic state of a vessel is given by its location and velocity. A kernel density estimate is built for these kinematic states using training data. Subsequent vessels are identified as anomalous if their kinematic probability density value estimated using the trained kernel density estimator is below a threshold. Real data are obtained from Automatic Identification System (AIS) broadcasts.
- Understanding tropical cyclones. Haikun *et al.* (2009) use kernel density estimation to model the genesis locations of tropical cyclones in the Northwestern Pacific. A product of normal kernels is used. The motivation is that understanding cyclone genesis could help in predicting future occurrence of tropical cyclones.
- Safety at military firing ranges. Glonek *et al.* (2010) use kernel density estimation to estimate ground impact distributions in guided missile firings. This is used to define the safety exclusion area during trials of guided air missiles.

Applications of copulas outside of the financial and insurance industries include:

- Climate research. Schölzel and Friederichs (2008) demonstrate the use of copulas for modelling meteorological variables at different locations, and for modelling multiple meteorological variables (e.g. precipitation averages and temperature minima) at a single location.
- Multimodal signal processing. Iyengar *et al.* (2009) show how copulas can be used for signal detection, with a view to processing multimodel data such as audio-video streams.

4.7.1 Comparative studies

A comparative study on 25 datasets of the naïve Bayes classifier and a tree-structured classifier is performed by Friedman *et al.* (1997). The tree-structured classifiers outperform the naïve Bayes model while maintaining simplicity.

Numerous studies have used naïve Bayes classifiers as a baseline text categorisation procedure. For example, Joachims (1998), Yang and Liu (1999) and Zhang and Oles (2001), all report naïve Bayes text classifiers as being outperformed by SVMs (SVMs are introduced in Chapter 6). The experiments of Yang and Liu (1999) show the naïve Bayes approach significantly underperforming an SVM and also a k-nearest-neighbour classifier. These results have led to a number of variants of naïve Bayes classifiers being proposed, such as the approaches of Rennie *et al.* (2003) and Kim *et al.* (2006), both of which result in a modified naïve Bayes performance that is competitive with that from an SVM.

A comparative study of different kernel methods applied to multivariate data is reported by Breiman *et al.* (1977) and Bowman (1985). Jones and Signorini (1997) perform a comparison of 'improved' univariate kernel density estimators [see also Cao *et al.* (1994) and Titterington (1980) for a comparison of kernels for categorical data].

The *Statlog* project (Michie *et al.*, 1994) provides a thorough comparison of a wide range of classification methods, including k nearest neighbour and kernel discriminant analysis (algorithm ALLOC80). ALLOC80 has a slightly lower error rate than k nearest neighbour (for the special case of $k = 1$ at any rate), but had longer training and test times. Further comparisons include those in Liu and White (1995).

4.8 Summary and discussion

The approaches developed in this chapter towards discrimination have been based on estimation of the class-conditional density functions using nonparametric techniques. It is certainly true that we cannot design a classifier that performs better than the Bayes discriminant rule. No matter how sophisticated a classifier is, or how appealing it may be in terms of reflecting a model of human decision processes, it cannot outperform the Bayes classifier for any proper performance criterion; in particular, achieve a lower error rate. Therefore a natural step is to estimate the components of the Bayes' rule from the data, namely the class-conditional probability density functions and the class priors. We shall see in later chapters that we do not need to model the density explicitly to get good estimates of the posterior probabilities of class membership.

We have described five nonparametric methods of density estimation: the k-nearest-neighbour method leading to the k-nearest-neighbour classifier; the histogram approach and developments to reduce the number of parameters (naïve Bayes, tree-structured density estimators and Bayesian networks); kernel methods of density estimation; series methods; and finally copula methods. With advances in computing in recent years, these methods have now become viable and nonparametric methods of density estimation have had an impact on nonparametric approaches to discrimination and classification. Of the methods described, for discrete data the developments of the histogram – the independence model, the Lancaster models and maximum weight dependence trees – are easy to implement. Learning algorithms for Bayesian networks can be computationally demanding. For continuous data, the kernel method is probably the most popular, with normal kernels with the same window width for each dimension. However, it is reported by Terrell and Scott (1992) that nearest-neighbour methods are superior to fixed kernel approaches to density estimation beyond four dimensions. The kernel method has also been applied to discrete data.

k-nearest-neighbour classifiers are easy to both code and understand, and are a popular classification technique. However, the naïve search algorithm is computationally demanding for large training datasets. A variety of techniques exist for reducing the search time of nearest-neighbour procedures.

In conclusion, an approach based on density estimation is not without its dangers of course. If incorrect assumptions are made about the form of the distribution in the parametric approach (and in many cases we will not have a physical model of the data generation process to use) or data points are sparse leading to poor kernel density estimates in the nonparametric approach, then we cannot hope to achieve good density estimates. However, the performance of a classifier, in terms of error rate, may not deteriorate too dramatically. Thus, it is a strategy worth trying.

4.9 Recommendations

1. Nearest-neighbour methods are easy to implement and are recommended as a starting point for a nonparametric approach. In the *Statlog* project (Michie *et al.*, 1994), the k-nearest-neighbour method came out best on the image datasets (top in four and runner-up in two of the six image datasets) and did very well on the whole.
2. For large datasets, some form of data reduction in the form of condensing/editing is advised when using nearest-neighbour methods.
3. If computational time is an issue, improved nearest-neighbour search procedures should be used, such as offered by the LAESA algorithm, or the ball-tree branch and bound search algorithm. However, it might not be worth implementing these procedures until after initial experiments have confirmed that the k-nearest-neighbour classifier gives satisfactory performance in the considered problem.
4. As density estimators, kernel methods are not appropriate for high-dimensional data, but if smooth estimates of the density are required they are to be preferred over k-nearest neighbour. Even poor estimates of the density may still give good classification performance.
5. For multivariate datasets, it is worth trying a simple independence model as a baseline. It is simple to implement, handles missing values easily and can give good performance.
6. Domain-specific and expert knowledge should be used where available. Bayesian networks are an attractive scheme for encoding such knowledge.

4.10 Notes and references

There is a large literature on nonparametric methods of density estimation. A good starting point is the book by Silverman (1986), placing emphasis on the practical aspects of density estimation. The book by Scott (1992) provides a blend of theory and applications, placing some emphasis on the visualisation of multivariate density estimates. Klemelä (2009) also considers multivariate density estimation and visualisation. The article by Izenman (1991) is to be recommended. Other texts are by Devroye and Györfi (1985) and Nadaraya (1989). The book by Wand and Jones (1995) presents a thorough treatment of kernel smoothing.

The literature on kernel methods for regression and density estimation is vast. A treatment of kernel density estimation can be found in most textbooks on density estimation. Silverman (1986) gives a particularly lucid account and the book by Hand (1982) provides a very good introduction and considers the use of kernel methods for discriminant analysis. Other treatments, more detailed than that presented here, may be found in the books by Scott (1992), McLachlan (1992a) and Nadaraya (1989).

The popularity of Bayesian networks in recent years is evidenced by the large number of good textbooks describing them. Detailed descriptions of Bayesian networks are available in texts by Pearl (1988), Jensen (1997, 2002), Jordan (1998), Neapolitan (2003), Darwiche (2009), Koller and Friedman (2009), Koski and Noble (2009) and Kjaerulff and Madsen (2010), amongst others.

A good overview of kd-trees for efficient k-nearest-neighbour search is provided by Moore (1991), and also in the course notes of Renals (2007). An implementation is available in the WEKA data mining software package, documented in Witten and Frank (2005), and also within the 2010 release of the Statistics toolbox of MATLAB®.

A good introduction to copulas is provided by Trivedi and Zimmer (2005), and from the viewpoint of actuarial statisticians by Frees and Valdez (1998). Further background and details on the use of copulas can be found in papers by Clemen and Reilly (1999), Embrechts *et al.* (2001), Dorey and Joubert (2007) and Chiou and Tsay (2008). A critical assessment of the utility of copulas is provided by Mikosch (2006).

Exercises

Dataset 1: Generate d-dimensional multivariate data (500 samples in train and test sets, equal priors) for two classes: for $\omega_1, \boldsymbol{x} \sim N(\boldsymbol{\mu}_1, \boldsymbol{\Sigma}_1)$ and for $\omega_2, \boldsymbol{x} \sim 0.5N(\boldsymbol{\mu}_2, \boldsymbol{\Sigma}_1) + 0.5N(\boldsymbol{\mu}_3, \boldsymbol{\Sigma}_3)$ where $\boldsymbol{\mu}_1 = (0, \dots, 0)^T$, $\boldsymbol{\mu}_2 = (2, \dots, 2)^T$, $\boldsymbol{\mu}_3 = (-2, \dots, -2)^T$ and $\boldsymbol{\Sigma}_1 = \boldsymbol{\Sigma}_2 = \boldsymbol{\Sigma}_3 = \boldsymbol{I}$, the identity matrix.

Dataset 2: Generate data from d-dimensional multivariate data (500 samples in train and test sets, equal priors) for three normally distributed classes with $\boldsymbol{\mu}_1 = (0, \dots, 0)^T$, $\boldsymbol{\mu}_2 = (2, \dots, 2)^T$, $\boldsymbol{\mu}_3 = (-2, \dots, -2)^T$ and $\boldsymbol{\Sigma}_1 = \boldsymbol{\Sigma}_2 = \boldsymbol{\Sigma}_3 = \boldsymbol{I}$, the identity matrix.

1. Compare and contrast the k-nearest-neighbour classifier with the Gaussian classifier. What assumptions do the models make? Also, consider such issues as training requirements, computation time, and storage and performance in high dimensions.

2. For three variables, X_1, X_2 and X_3 taking one of two values, 1 or 2, denote by P^{ij}_{ab} the probability that $X_i = a$ and $X_j = b$. Specifying the density as

$$P^{12}_{12} = P^{13}_{11} = P^{23}_{11} = P^{23}_{12} = P^{12}_{22} = P^{13}_{21} = P^{23}_{21} = P^{23}_{22} = \tfrac{1}{4}$$
$$P^{12}_{11} = P^{13}_{12} = \tfrac{7}{16};\, P^{12}_{21} = P^{13}_{22} = \tfrac{1}{16}$$

show that the Lancaster density estimate of the probability $p(X_1 = 2, X_2 = 1, X_3 = 2)$ is negative $(= -1/64)$.

3. Show that the tree-dependent distribution specified in (4.13) is the same as that given in (4.12).

4. For a tree-dependent distribution [Equation (4.12)]

$$p^t(\mathbf{x}) = \prod_{i=1}^{d} p(x_i|x_{j(i)})$$

show that minimising the Kullback–Leibler divergence

$$KL(p(\mathbf{x})|p^t(\mathbf{x})) = \sum_{\mathbf{x}} p(\mathbf{x})\log\left(\frac{p(\mathbf{x})}{p^t(\mathbf{x})}\right)$$

is equivalent to finding the tree that maximises

$$\sum_{i=1}^{d} \sum_{x_i, x_{j(i)}} p(x_i, x_{j(i)})\log\left(\frac{p(x_i, x_{j(i)})}{p(x_i)p(x_{j(i)})}\right)$$

Hint: note that $p(\mathbf{x})\log(p(\mathbf{x}))$ does not depend on the tree structure and also that

$$p(\mathbf{x}) = p(x_i, x_{j(i)})p(x_{-i,-j(i)}|x_i, x_{j(i)})$$

where $x_{-i,\,-j(i)}$ refers to the variables $x_1, \ldots, x_d$ but excluding x_i and $x_{j(i)}$.

5. Verify that the Bartlett–Epanechnikov kernel (see Table 4.4) satisfies the properties

$$\int K(t)dt = 1$$

$$\int tK(t)dt = 0$$

$$\int t^2 K(t)dt = k_2 \neq 0$$

6. Consider a sample of n observations $(x_1, \ldots, x_n)$ from a density $p(x)$. An estimate $\hat{p}$ is calculated using a kernel density estimate with Gaussian kernels for various bandwidths h. How would you expect the number of relative maxima of $\hat{p}$ to vary as h increases? Suppose that the x_i's are drawn from the Cauchy density $p(x) = 1/(\pi(1 + x^2))$. Show that the variance of X is infinite. Does this mean that the variance of the Gaussian kernel density estimate $\hat{p}$ is infinite?

7. Consider the multivariate kernel density estimate ($\mathbf{x} \in \mathbb{R}^d$)

$$\hat{p}(\mathbf{x}) = \frac{1}{nh^d}\sum_{i=1}^{n} K\left(\frac{1}{h}(\mathbf{x} - \mathbf{x}_i)\right)$$

Show that the k-nearest-neighbour density estimate given by (4.6) is a special case of the above for suitable choice of K and h (which varies with position, $\mathbf{x}$).

8. Show that if $F(x)$ is the distribution function of a random variable X [i.e. $F(x) = P(X \leq x)$], then $F(X) \sim U(0, 1)$, a uniform distribution over the range [0, 1].

9. Starting from the definition of the Frank copula as an Archimedean family kernel in Section 4.6.3, verify the bivariate expression given in (4.32), and hence the derivative in (4.33).

10. Implement a k-nearest-neighbour classifier using dataset 1 and investigate its performance as a function of dimensionality $d = 1, 3, 5, 10$ and k. Comment on the results.

11. For the data of dataset 1, implement a Gaussian kernel classifier. Construct a separate validation set to obtain a value of the kernel bandwidth (initialise at the value given by (4.23) and vary from this). Describe the results.

12. Implement a base prototype selection algorithm to select n_b base prototypes using dataset 2. Implement the LAESA procedure and plot the number of distance calculations in classifying the test data as a function of the number of base prototypes, n_b, for $d = 2, 4, 6, 8$ and 10.

13. Using dataset 2, implement a nearest-neighbour classifier with editing and condensing. Calculate the nearest-neighbour error rate, the error rate after editing, and the error rate after editing and condensing.

14. Using dataset 2, implement a kd-tree nearest-neighbour classifier, and a ball-tree nearest-neighbour classifier. Assess performance in terms of the distribution for the number of distance calculations required to find the nearest neighbour. Investigate the effects of using different depths of tree (i.e. different numbers of training data samples at the leaf nodes).

15. Using dataset 1, investigate the performance of a Bayesian classifier utilising copula density estimates. Compare the performance to that from a nearest-neighbour classifier. Investigate the effects of using different copula functions.

16. Using dataset 2, investigate procedures for choosing k in the k-nearest-neighbour method.

17. Consider nearest neighbour with edit and condense. Suggest ways of reducing the final number of prototypes by careful initialisation of the condensing algorithm. Plan a procedure to test your hypotheses. Implement it and describe your results.

5

Linear discriminant analysis

Discriminant functions that are linear in the features are constructed, resulting in (piecewise) linear decision boundaries. Different optimisation schemes give rise to different methods including the perceptron, Fisher's linear discriminant function and support vector machines. The relationship between these methods is discussed.

5.1 Introduction

This chapter deals with the problem of finding the weights of a linear discriminant function. Techniques for performing this task have sometimes been referred to as *learning algorithms* and we retain some of the terminology here even though the methods are ones of optimisation or training rather than learning. A linear discriminant function has already appeared in Chapter 2. In that chapter, it arose as a result of a normal distribution assumption for the class densities in which the class covariance matrices were equal. In this chapter, we make no distributional assumptions, but start from the assumption that the decision boundaries are linear. The algorithms have been extensively treated in the literature, but they are included here as an introduction to the nonlinear models discussed in the following chapter, since a stepping stone to the nonlinear models is the generalised linear model in which the discriminant functions are linear combinations of nonlinear functions.

The initial treatment is divided into two parts: the binary classification problem and the multiclass problem. Although the two-class problem is clearly a special case of the multiclass situation (and in fact all the algorithms in the multiclass section can be applied to two classes), the two-class case is of sufficient interest in its own right to warrant a special treatment. It has received a considerable amount of attention and many different algorithms have been proposed. Linear support vector machines (SVMs) are an example of a two-class linear discriminant function. However, due to their high prevalence in machine learning applications they are considered in a separate section.

Statistical Pattern Recognition, Third Edition. Andrew R. Webb and Keith D. Copsey.

5.2 Two-class algorithms

5.2.1 General ideas

In Chapter 1 we introduced the discriminant function approach to supervised classification; here we briefly restate that approach for linear discriminant functions.

Suppose we have a set of training patterns $\boldsymbol{x}_1, \ldots, \boldsymbol{x}_n$, each of which is assigned to one of two classes, ω_1 or ω_2. Using this design set, we seek a weight vector $\boldsymbol{w}$ and a threshold w_0 such that

$$\boldsymbol{w}^T\boldsymbol{x} + w_0 \begin{cases} > & 0 \\ < & 0 \end{cases} \Rightarrow \boldsymbol{x} \in \begin{cases} \omega_1 \\ \omega_2 \end{cases} \tag{5.1}$$

5.2.1.1 Decision surface

The decision surface is the hyperplane represented by the equation

$$g(\boldsymbol{x}) = \boldsymbol{w}^T\boldsymbol{x} + w_0 = 0$$

which has unit normal in the direction of $\boldsymbol{w}$, and a perpendicular distance $|w_0|/|\boldsymbol{w}|$ from the origin.

The distance of a pattern $\boldsymbol{x}$ to the decision hyperplane is given by $|r|$, where

$$r = g(\boldsymbol{x})/|\boldsymbol{w}| = (\boldsymbol{w}^T\boldsymbol{x} + w_0)/|\boldsymbol{w}| \tag{5.2}$$

with the sign of r indicating on which side of the decision hyperplane the pattern lies, and therefore to which class the pattern should be assigned.

5.2.1.2 Simplification

The linear discriminant rule given by Equation (5.1) can be written as:

$$\boldsymbol{v}^T\boldsymbol{z} \begin{cases} > & 0 \\ < & 0 \end{cases} \Rightarrow \boldsymbol{x} \in \begin{cases} \omega_1 \\ \omega_2 \end{cases}$$

where $\boldsymbol{z} = (1, x_1, \ldots, x_p)^T$ is the *augmented pattern vector* and $\boldsymbol{v}$ is a $(p+1)$-dimensional vector $(w_0, w_1, \ldots, w_p)^T$. In what follows, $\boldsymbol{z}$ could also be $(1, \phi_1(\boldsymbol{x}), \ldots, \phi_D(\boldsymbol{x}))^T$, with $\boldsymbol{v}$ a $(D+1)$-dimensional vector of weights, where $\{\Phi_i, i = 1, \ldots, D\}$ is a set of D functions of the original variables. Thus, we may apply these algorithms in a transformed feature space.

A sample in class ω_2 is classified correctly if $\boldsymbol{v}^T\boldsymbol{z} < 0$, i.e. if $\boldsymbol{v}^T(-\boldsymbol{z}) > 0$. If we were to redefine all samples in class ω_2 in the design set by their negative values and denote these redefined samples by $\boldsymbol{y}$, then we seek a value for $\boldsymbol{v}$ which satisfies:

$$\begin{aligned} &\boldsymbol{v}^T\boldsymbol{y} > 0 \qquad \text{for all } \boldsymbol{y}_i \text{ corresponding to } \boldsymbol{x}_i \text{ in the design set} \\ &[\boldsymbol{y}_i^T = (1, \boldsymbol{x}_i^T), \boldsymbol{x}_i \in \omega_1; \boldsymbol{y}_i^T = (-1, -\boldsymbol{x}_i^T), \boldsymbol{x}_i \in \omega_2] \end{aligned} \tag{5.3}$$

Ideally, we would like a solution for $\mathbf{v}$ that makes $\mathbf{v}^T\mathbf{y}$ positive for as many samples in the design set as possible. This minimises the misclassification error on the design set. If $\mathbf{v}^T\mathbf{y}_i > 0$ for all members of the design set then the data are said to be *linearly separable*.

However, it is difficult to minimise the number of misclassifications. Usually some other criterion is employed. The sections that follow introduce a range of criteria adopted for discrimination between two classes. Some are suitable if the classes are separable, others for overlapping classes. Some lead to algorithms that are deterministic, others can be implemented using stochastic algorithms.

5.2.2 Perceptron criterion

Perhaps the simplest criterion to minimise is the perceptron criterion function

$$J_P(\mathbf{v}) = \sum_{\mathbf{y}_i \in \mathcal{Y}} (-\mathbf{v}^T\mathbf{y}_i)$$

where $\mathcal{Y} = \{\mathbf{y}_i | \mathbf{v}^T\mathbf{y}_i < 0\}$ (the set of misclassified samples). J_P is proportional to the sum of the distances of the misclassified samples to the decision boundary [see Equation (5.2)].

5.2.2.1 Error-correction procedure

Since the criterion function J_P is continuous, we can use a gradient-based procedure, such as the method of steepest descent (Press *et al.*, 1992), to determine its minimum:

$$\frac{\partial J_P}{\partial \mathbf{v}} = \sum_{\mathbf{y}_i \in \mathcal{Y}} (-\mathbf{y}_i)$$

which is the sum of the misclassified patterns, and the method of steepest descent gives a movement along the negative of the gradient with update rule

$$\mathbf{v}_{k+1} = \mathbf{v}_k + \rho_k \sum_{\mathbf{y}_i \in \mathcal{Y}} \mathbf{y}_i \tag{5.4}$$

where ρ_k is a scale parameter that determines the step size. If the sample sets are separable, then this procedure is guaranteed to converge to a solution that separates the sets. Algorithms of the type (5.4) are sometimes referred to as *many-pattern adaptation* or *batch update* since all given pattern samples are used in the update of $\mathbf{v}$. The corresponding single-pattern adaptation scheme is

$$\mathbf{v}_{k+1} = \mathbf{v}_k + \rho_k \mathbf{y}_i \tag{5.5}$$

where $\mathbf{y}_i$ is a training sample that has been misclassified by $\mathbf{v}_k$. This procedure cycles through the training set, modifying the weight vector whenever a sample is misclassified. There are several types of *error correction* procedures of the form of (5.5). The *fixed increment* rule takes $\rho_k = \rho$, a constant and is the simplest algorithm for solving systems of linear inequalities.

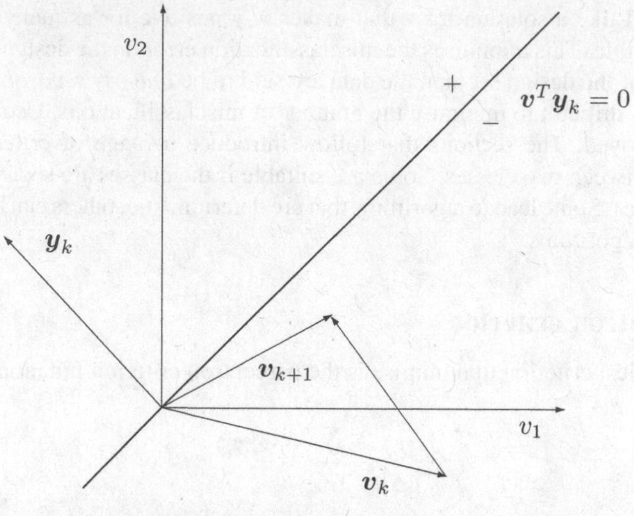

Figure 5.1 Perceptron training.

The error-correction procedure is illustrated geometrically in weight space in Figures 5.1 and 5.2. In Figure 5.1, the plane is partitioned by the line $v^T y_k = 0$. Since the current estimate of the weight vector v_k has $v_k y_k < 0$, the weight vector is updated by (for $\rho = 1$) adding on the pattern vector y_k. This moves the weight vector towards and possibly into (although not in Figure 5.1) the region $v^T y_k > 0$. In Figure 5.2, the lines $v^T y_k = 0$ are plotted for four separate training patterns.

If the classes are separable, the solution for v must lie in the shaded region (the *solution region* for which $v^T y_k > 0$ for all patterns y_k). A solution path is also shown starting from an initial estimate v_0. By presenting the patterns y_1, y_2, y_3, y_4 cyclically, a solution for v is obtained in five steps: the first change to v_0 occurs when y_2 is presented (v_0 is already on the

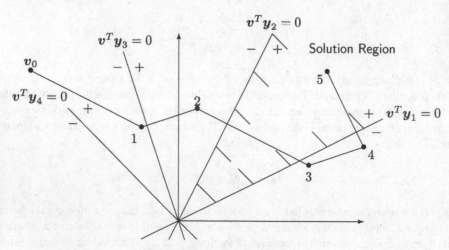

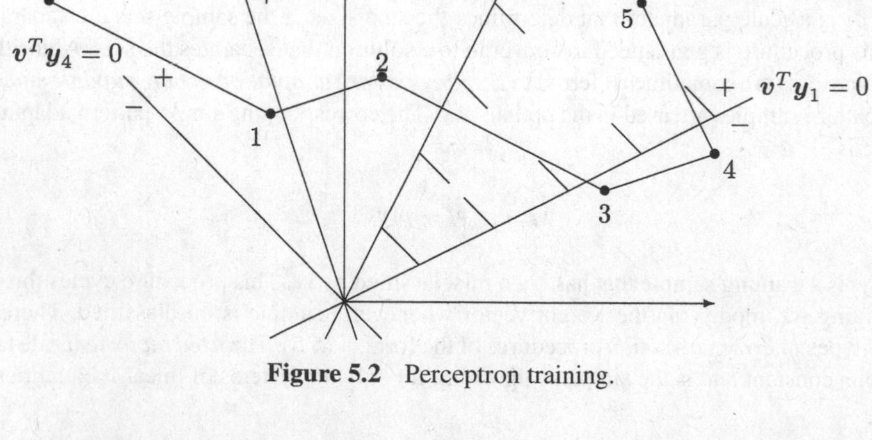

Figure 5.2 Perceptron training.

positive side of $\boldsymbol{v}^T\boldsymbol{y}_1 = 0$, so presenting $\boldsymbol{y}_1$ does not update $\boldsymbol{v}$). The second step adds on $\boldsymbol{y}_3$; the third $\boldsymbol{y}_2$ ($\boldsymbol{y}_4$ and $\boldsymbol{y}_1$ are not used because $\boldsymbol{v}$ is on the positive sides of both hyperplanes $\boldsymbol{v}^T\boldsymbol{y}_4 = 0$ and $\boldsymbol{v}^T\boldsymbol{y}_1 = 0$); the final two stages are the addition of $\boldsymbol{y}_3$, then $\boldsymbol{y}_1$. Thus, from the sequence

$$\boldsymbol{y}_1, \hat{\boldsymbol{y}}_2, \hat{\boldsymbol{y}}_3, \boldsymbol{y}_4, \boldsymbol{y}_1, \hat{\boldsymbol{y}}_2, \hat{\boldsymbol{y}}_3, \boldsymbol{y}_4, \hat{\boldsymbol{y}}_1$$

only those vectors with a caret are used. Note that it is possible for an adjustment to undo a correction previously made. In this example, although the iteration started on the right (positive) side of $\boldsymbol{v}^T\boldsymbol{y}_1 = 0$, successive iterations of $\boldsymbol{v}$ gave an estimate with $\boldsymbol{v}^T\boldsymbol{y}_1 < 0$ (at stages 3 and 4). Eventually a solution with $J_P(\boldsymbol{v}) = 0$ will be obtained for separable patterns.

5.2.2.2 Variants

There are many variants to the fixed-increment rule given in the previous section. We consider just a few of them here.

Absolute correction rule
Choose the value of ρ so that the value of $\boldsymbol{v}_{k+1}^T\boldsymbol{y}_i$ is positive. Hence:

$$\rho > |\boldsymbol{v}_k^T\boldsymbol{y}_i|/|\boldsymbol{y}_i|^2$$

where $\boldsymbol{y}_i$ is the misclassified pattern presented at the kth step. This follows from noting that for a misclassified pattern at the kth step:

$$\boldsymbol{v}_{k+1}^T\boldsymbol{y}_i = \boldsymbol{v}_k^T\boldsymbol{y}_i + \rho\boldsymbol{y}_i^T\boldsymbol{y}_i = -|\boldsymbol{v}_k^T\boldsymbol{y}_i| + \rho|\boldsymbol{y}_i|^2$$

The rule means that the iteration corrects for each misclassified pattern as it is presented. For example, ρ may be taken to be the smallest integer greater than $|\boldsymbol{v}_k^T\boldsymbol{y}_i|/|\boldsymbol{y}_i|^2$.

Fractional correction rule
This sets ρ to be a function of the distance to the hyperplane $\boldsymbol{v}^T\boldsymbol{y}_i = 0$, i.e.

$$\rho = \lambda|\boldsymbol{v}_k^T\boldsymbol{y}_i|/|\boldsymbol{y}_i|^2$$

where λ is the fraction of the distance to the hyperplane $\boldsymbol{v}^T\boldsymbol{y}_i = 0$, traversed in going from $\boldsymbol{v}_k$ to $\boldsymbol{v}_{k+1}$. If $\lambda > 1$, then pattern $\boldsymbol{y}_i$ will be classified correctly after the adjustment to $\boldsymbol{v}$.

Introduction of a margin, *b*
A margin, $b > 0$, is introduced (Figure 5.3) and the weight vector is updated whenever $\boldsymbol{v}^T\boldsymbol{y}_i \leq b$. Thus, the solution vector $\boldsymbol{v}$, must lie at a distance greater than $b/|\boldsymbol{y}_i|$ from each hyperplane $\boldsymbol{v}^T\boldsymbol{y}_i = 0$. The training procedures given above are still guaranteed to produce a solution when the classes are separable. One of the reasons often given for the introduction of a threshold is to aid generalisation. Without the threshold, some of the points in the data space may lie close to the separating boundary. Viewed in data space, all points $\boldsymbol{x}_i$ lie at a distance greater than $b/\boldsymbol{w}|$ from the separating hyperplane. Clearly, the solution is not unique and in Section 5.4 we address the problem of seeking a 'maximal margin' classifier.

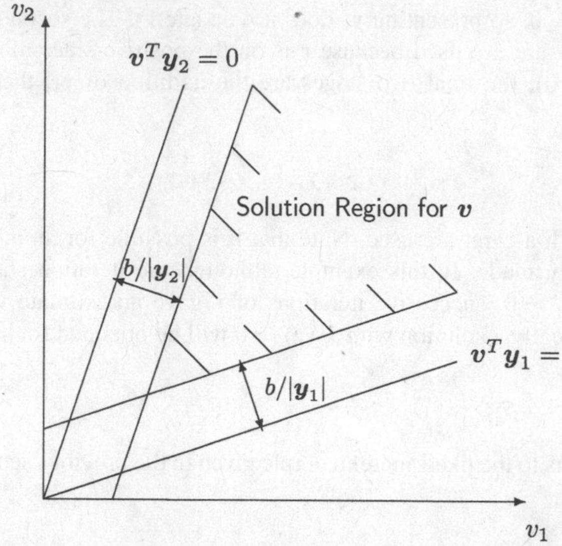

Figure 5.3 Solution region for a margin.

Variable increment ρ

One of the problems with the above procedures is that, although they will converge if the classes are separable, the solution for $\boldsymbol{v}$ will oscillate if the classes overlap. The error-correction procedure also converges (for linearly separable classes) if ρ_k satisfies the following conditions

$$\rho_k \geq 0$$
$$\sum_{k=1}^{\infty} \rho_k = \infty$$

and

$$\lim_{m \to \infty} \frac{\sum_{k=1}^{m} \rho_k^2}{\left(\sum_{k=1}^{m} \rho_k\right)^2} = 0$$

In many problems, we do not know *a priori* whether the samples are separable or not. If they are separable, then we would like a procedure that yields a solution that separates the classes. On the other hand, if they are not separable then a method with $\rho_k \to 0$ will decrease the effects of misclassified samples as the iteration proceeds. One possible choice is $\rho_k = 1/k$.

Relaxation algorithm

The *relaxation* or *Agmon–Mays algorithm* minimises the criterion

$$J_r = \frac{1}{2} \sum_{\boldsymbol{y}_i \in \mathcal{Y}} (\boldsymbol{v}^T \boldsymbol{y}_i - b)^2 / |\boldsymbol{y}_i|^2$$

where $\mathcal{Y}$ is $\{y_i | y_i^T v \leq b\}$. Thus not only do the misclassified samples contribute to J_r, but also do those correctly classified samples lying closer than $b/|v|$ to the boundary $v^T y = 0$. The basic algorithm is

$$v_{k+1} = v_k + \rho_k \sum_{y_i \in \mathcal{Y}_k} \frac{b - v_k^T y_i}{|y_i|^2} y_i$$

where $\mathcal{Y}_k$ is $\{y_i | y_i^T v_k \leq b\}$. This has a single-pattern scheme

$$v_{k+1} = v_k + \rho_k \frac{b - v_k^T y_i}{|y_i|^2} y_i$$

where $v_k^T y_i \leq b$ (that is, the patterns y_i that cause the vector v to be corrected). This is the same as the fractional correction rule with a margin.

5.2.3 Fisher's criterion

The approach adopted by Fisher was to find a linear combination of the variables that separates the two classes as much as possible. That is, we seek the direction along which the two classes are best separated in some sense. The criterion proposed by Fisher is the ratio of between-class to within-class variances. Formally, we seek a direction w such that

$$J_F = \frac{|w^T (m_1 - m_2)|^2}{w^T S_W w} \tag{5.6}$$

is a maximum, where m_1 and m_2 are the class means and S_W is the pooled within-class sample covariance matrix, in its bias-corrected form given by

$$\frac{1}{n-2}\left(n_1 \hat{\Sigma}_1 + n_2 \hat{\Sigma}_2\right)$$

where $\hat{\Sigma}_1$ and $\hat{\Sigma}_2$ are the maximum likelihood estimates of the covariance matrices of classes ω_1 and ω_2, respectively, and there are n_i samples in class ω_i ($n_1 + n_2 = n$). Maximising the above criterion gives a solution for the direction w. The threshold weight w_0 is determined by an allocation rule. The solution for w that maximises J_F can be obtained by differentiating J_F with respect to w and equating to zero. This yields

$$\frac{2w^T (m_1 - m_2)}{w^T S_W w} \left\{ (m_1 - m_2) - \left(\frac{w^T (m_1 - m_2)}{w^T S_W w} \right) S_W w \right\} = 0$$

Since we are interested in the direction of w (and noting that $w^T (m_1 - m_2)/w^T S_W w$ is a scalar), we must have

$$w \propto S_W^{-1} (m_1 - m_2) \tag{5.7}$$

We may take equality without loss of generality. The solution for $\boldsymbol{w}$ is a special case of the more general feature extraction criteria described in Chapter 10 that result in transformations that maximise a ratio of between-class to within-class variance. Therefore, it should be noted that Fisher's criterion does not provide us with an allocation rule, merely a mapping to a reduced dimension (actually one dimension in the two-class situation) in which discrimination is in some sense easiest. If we wish to determine an allocation rule, we must specify a threshold, w_0, so that we may assign $\boldsymbol{x}$ to class ω_1 if

$$\boldsymbol{w}^T\boldsymbol{x} + w_0 > 0$$

In Chapter 2 we have seen that if the data were normally distributed with equal covariance matrices, then the optimal decision rule is linear: assign $\boldsymbol{x}$ to ω_1 if $\boldsymbol{w}^T\boldsymbol{x} + w_0 > 0$ where [Equations (2.16) and (2.17)]

$$\begin{aligned} \boldsymbol{w} &= \boldsymbol{S}_W^{-1}(\boldsymbol{m}_1 - \boldsymbol{m}_2) \\ w_0 &= -\frac{1}{2}(\boldsymbol{m}_1 + \boldsymbol{m}_2)^T\boldsymbol{S}_W^{-1}(\boldsymbol{m}_1 - \boldsymbol{m}_2) - \log\left(\frac{p(\omega_2)}{p(\omega_1)}\right) \end{aligned}$$

Thus, the direction on to which $\boldsymbol{x}$ is projected in the Bayes classifier rule for normally distributed data with equal covariance matrices, is the same as that obtained through maximisation of (5.6) and given by (5.7). This suggests that if we take $\boldsymbol{w} = \boldsymbol{S}_W^{-1}(\boldsymbol{m}_1 - \boldsymbol{m}_2)$ (unit constant of proportionality giving equality in (5.7)), then we may choose a threshold to be given by w_0 above, although we note that it is optimal only for normally distributed classes with equal covariance matrices.

Note however that the discriminant direction (5.7) has been derived without any assumptions of normality. We have used normal assumptions to set a threshold for discrimination. In non-normal situations, a different threshold may be more appropriate. Nevertheless, we may still use the above rule in the more general non-normal case, giving: assign $\boldsymbol{x}$ to ω_1 if

$$\left\{\boldsymbol{x} - \frac{1}{2}(\boldsymbol{m}_1 + \boldsymbol{m}_2)\right\}^T \boldsymbol{w} > \log\left(\frac{p(\omega_2)}{p(\omega_1)}\right) \tag{5.8}$$

but it will not necessarily be optimal. Note that the above rule is not guaranteed to give a separable solution even if the two groups are separable.

5.2.4 Least mean-squared-error procedures

The perceptron and related criteria in Section 5.2.2 are all defined in terms of misclassified samples. In this section, all the data samples are used and we attempt to find a solution vector for which the *equality* constraints

$$\boldsymbol{v}^T\boldsymbol{y}_i = t_i$$

are satisfied for *positive* constants t_i. [Recall that the vectors $\boldsymbol{y}_i$ are defined by $\boldsymbol{y}_i^T = (1, \boldsymbol{x}_i^T)$ for $\boldsymbol{x}_i \in \omega_1$ and $\boldsymbol{y}_i^T = (-1, -\boldsymbol{x}_i^T)$ for $\boldsymbol{x}_i \in \omega_2$, or the $(D+1)$-dimensional vectors $(1, \boldsymbol{\phi}_i^T)$ and $(-1, -\boldsymbol{\phi}_i^T)$ for transformations $\boldsymbol{\phi}$ of the data $\boldsymbol{\phi}_i = \boldsymbol{\phi}(\boldsymbol{x}_i)$]. In general, it will not be possible to satisfy these constraints exactly and we seek a solution for $\boldsymbol{v}$ that minimises a cost function of the difference between $\boldsymbol{v}^T\boldsymbol{y}_i$ and t_i. The particular cost function we shall consider is the mean square error.

5.2.4.1 Solution

Let $\boldsymbol{Y}$ be the $n \times (p+1)$ matrix of sample vectors, with the ith row $\boldsymbol{y}_i$, and $\boldsymbol{t} = (t_1, \ldots, t_n)^T$. Then the sum-squared error criterion is

$$J_S = |\boldsymbol{Y}\boldsymbol{v} - \boldsymbol{t}|^2 \tag{5.9}$$

The solution for $\boldsymbol{v}$ minimising J_S is

$$\boldsymbol{v} = \boldsymbol{Y}^{\dagger}\boldsymbol{t} \tag{5.10}$$

where $\boldsymbol{Y}^{\dagger}$ is the pseudo-inverse of $\boldsymbol{Y}$ (Figure 5.4).

The general linear least squares problem may be solved using the singular value decomposition of a matrix. An $m \times n$ matrix $\boldsymbol{A}$ may be written in the form:

$$\boldsymbol{A} = \boldsymbol{U\Sigma V}^T = \sum_{i=1}^{r} \sigma_i \boldsymbol{u}_i \boldsymbol{v}_i^T$$

where:

- r is the rank of $\boldsymbol{A}$;
- $\boldsymbol{U}$ is an $m \times r$ matrix with columns $\boldsymbol{u}_1, \ldots, \boldsymbol{u}_r$, the *left singular vectors* and $\boldsymbol{U}^T\boldsymbol{U} = \boldsymbol{I}_r$, the $r \times r$ identity matrix;
- $\boldsymbol{V}$ is an $n \times r$ matrix with columns $\boldsymbol{v}_1, \ldots, \boldsymbol{v}_r$, the *right singular vectors* and $\boldsymbol{V}^T\boldsymbol{V} = \boldsymbol{I}_r$ also;
- $\boldsymbol{\Sigma} = \text{diag}(\sigma_1, \ldots, \sigma_r)$, the diagonal matrix of *singular values*, σ_i, $i = 1, \ldots, r$.

The singular values of $\boldsymbol{A}$ are the square roots of the nonzero eigenvalues of $\boldsymbol{AA}^T$ or $\boldsymbol{A}^T\boldsymbol{A}$. The *pseudo-inverse* or *generalised inverse* is the $n \times m$ matrix $\boldsymbol{A}^{\dagger}$

$$\boldsymbol{A}^{\dagger} = \boldsymbol{V\Sigma}^{-1}\boldsymbol{U}^T = \sum_{i=1}^{r} \frac{1}{\sigma_i}\boldsymbol{v}_i\boldsymbol{u}_i^T \tag{5.11}$$

and the solution for $\boldsymbol{x}$ that minimises the square error

$$|\boldsymbol{Ax} - \boldsymbol{b}|^2$$

is given by

$$\boldsymbol{x} = \boldsymbol{A}^{\dagger}\boldsymbol{b}$$

If the rank of $\boldsymbol{A}$ is less than n, then there is not a unique solution for $\boldsymbol{x}$ and singular value decomposition delivers the solution with minimum norm.

Figure 5.4 The pseudo-inverse and its properties. (*continued*)

The pseudo-inverse has the following properties:

$$\begin{aligned} AA^{\dagger}A &= A \\ A^{\dagger}AA^{\dagger} &= A^{\dagger} \\ (AA^{\dagger})^T &= AA^{\dagger} \\ (A^{\dagger}A)^T &= A^{\dagger}A \end{aligned}$$

Figure 5.4 *(continued)*

If $\boldsymbol{Y}^T\boldsymbol{Y}$ is nonsingular, then another form for Equation (5.10) is:

$$\boldsymbol{v} = (\boldsymbol{Y}^T\boldsymbol{Y})^{-1}\boldsymbol{Y}^T\boldsymbol{t} \tag{5.12}$$

This follows from writing Equation (5.9) as:

$$J_S = (\boldsymbol{Y}\boldsymbol{v} - \boldsymbol{t})^T(\boldsymbol{Y}\boldsymbol{v} - \boldsymbol{t})$$

and differentiating to obtain to obtain the following relationship at the minimum:

$$2\boldsymbol{Y}^T(\boldsymbol{Y}\boldsymbol{v} - \boldsymbol{t}) = 0$$

For the given solution for $\boldsymbol{v}$, the approximation to $\boldsymbol{t}$ is

$$\begin{aligned} \hat{\boldsymbol{t}} &= \boldsymbol{Y}\boldsymbol{v} \\ &= \boldsymbol{Y}\left(\boldsymbol{Y}^T\boldsymbol{Y}\right)^{-1}\boldsymbol{Y}^T\boldsymbol{t} \end{aligned}$$

A measure of how well the linear approximation fits the data is provided by the absolute error in the approximation, or *error sum of squares*, which is

$$|\hat{\boldsymbol{t}} - \boldsymbol{t}|^2 = |\{\boldsymbol{Y}(\boldsymbol{Y}^T\boldsymbol{Y})^{-1}\boldsymbol{Y}^T - \boldsymbol{I}\}\boldsymbol{t}|^2$$

and we define the *normalised error* as

$$\epsilon = \left(\frac{|\hat{\boldsymbol{t}} - \boldsymbol{t}|^2}{|\bar{\boldsymbol{t}} - \boldsymbol{t}|^2}\right)^{\frac{1}{2}}$$

where $\bar{\boldsymbol{t}} = \bar{t}\boldsymbol{1}$ and

$$\bar{t} = \frac{1}{n}\sum_{i=1}^{n} t_i \quad \text{is the mean of the values } t_i$$

and $\boldsymbol{1}$ is a vector of 1s. The denominator, $|\bar{\boldsymbol{t}} - \boldsymbol{t}|^2$, is the *total sums of squares* or *total variation*.

Thus, a normalised error close to zero represents a good fit to the data and a normalised error close to one means that the model predicts the data in the mean and represents a poor fit. The normalised error can be expressed in terms of the *multiple coefficient of determination*, R^2, used in ordinary least-squares regression as (Dillon and Goldstein, 1984)

$$R^2 = 1 - \epsilon^2$$

5.2.4.2 Relationship to Fisher's linear discriminant

We have still not said anything about the choice of the t_i. In this section we consider a specific choice which we write

$$t_i = \begin{cases} t_1 & \text{for all } \boldsymbol{x}_i \in \omega_1 \\ t_2 & \text{for all } \boldsymbol{x}_i \in \omega_2 \end{cases}$$

Order the rows of $\boldsymbol{Y}$ so that the first n_1 samples correspond to class ω_1 and the remaining n_2 samples correspond to class ω_2. Write the matrix $\boldsymbol{Y}$ as

$$\boldsymbol{Y} = \begin{bmatrix} \boldsymbol{u}_1 & \boldsymbol{X}_1 \\ -\boldsymbol{u}_2 & -\boldsymbol{X}_2 \end{bmatrix} \tag{5.13}$$

where $\boldsymbol{u}_i (i = 1, 2)$ is a vector of n_i 1s and there are n_i samples in class ω_i $(i = 1, 2)$. The matrix $\boldsymbol{X}_i$ has n_i rows containing the training set patterns for class ω_i and p columns. Then (5.12) may be written

$$\boldsymbol{Y}^T\boldsymbol{Y}\boldsymbol{v} = \boldsymbol{Y}^T\boldsymbol{t}$$

and on substitution for $\boldsymbol{Y}$ from (5.13) and $\boldsymbol{v}$ as $(w_0, \boldsymbol{w})^T$ this may be rearranged to give

$$\begin{bmatrix} n & n_1\boldsymbol{m}_1^T + n_2\boldsymbol{m}_2^T \\ n_1\boldsymbol{m}_1 + n_2\boldsymbol{m}_2 & \boldsymbol{X}_1^T\boldsymbol{X}_1 + \boldsymbol{X}_2^T\boldsymbol{X}_2 \end{bmatrix} \begin{bmatrix} w_0 \\ \boldsymbol{w} \end{bmatrix} = \begin{bmatrix} n_1t_1 - n_2t_2 \\ t_1n_1\boldsymbol{m}_1 - t_2n_2\boldsymbol{m}_2 \end{bmatrix}$$

where $\boldsymbol{m}_i$ is the mean of the rows of $\boldsymbol{X}_i$. The top row of the matrix gives a solution for w_0 in terms of $\boldsymbol{w}$ as

$$w_0 = \frac{-1}{n}(n_1\boldsymbol{m}_1^T + n_2\boldsymbol{m}_2^T)\boldsymbol{w} + \frac{n_1}{n}t_1 - \frac{n_2}{n}t_2 \tag{5.14}$$

and the second row gives

$$(n_1\boldsymbol{m}_1 + n_2\boldsymbol{m}_2)w_0 + (\boldsymbol{X}_1^T\boldsymbol{X}_1 + \boldsymbol{X}_2^T\boldsymbol{X}_2)\boldsymbol{w} = t_1\boldsymbol{m}_1n_1 - t_2\boldsymbol{m}_2n_2$$

Substituting for w_0 from (5.14) and rearranging gives

$$\left\{n\boldsymbol{S}_W + \frac{n_1n_2}{n}(\boldsymbol{m}_1 - \boldsymbol{m}_2)(\boldsymbol{m}_1 - \boldsymbol{m}_2)^T\right\}\boldsymbol{w} = (\boldsymbol{m}_1 - \boldsymbol{m}_2)\frac{n_1n_2}{n}(t_1 + t_2) \tag{5.15}$$

where $\boldsymbol{S}_W$ is the estimate of the assumed common covariance matrix, written in terms of $\boldsymbol{m}_i$ and $\boldsymbol{X}_i$, $i = 1, 2$, as

$$\boldsymbol{S}_W = \frac{1}{n}\left\{\boldsymbol{X}_1^T\boldsymbol{X}_1 + \boldsymbol{X}_2^T\boldsymbol{X}_2 - n_1\boldsymbol{m}_1\boldsymbol{m}_1^T - n_2\boldsymbol{m}_2\boldsymbol{m}_2^T\right\}$$

Whatever the solution for $\boldsymbol{w}$, the term

$$\frac{n_1n_2}{n}(\boldsymbol{m}_1 - \boldsymbol{m}_2)(\boldsymbol{m}_1 - \boldsymbol{m}_2)^T\boldsymbol{w}$$

in (5.15) is in the direction of $\boldsymbol{m}_1 - \boldsymbol{m}_2$. Thus, (5.15) may be written

$$nS_W\boldsymbol{w} = \alpha(\boldsymbol{m}_1 - \boldsymbol{m}_2)$$

for some constant of proportionality, α, with solution

$$\boldsymbol{w} = \frac{\alpha}{n}\boldsymbol{S}_W^{-1}(\boldsymbol{m}_1 - \boldsymbol{m}_2)$$

the same solution obtained for Fisher's linear discriminant (5.7). Thus, provided that the value of t_i is the same for all members of the same class, we recover Fisher's linear discriminant. We require that $t_1 + t_2 \neq 0$ to prevent a trivial solution for $\boldsymbol{w}$.

In the spirit of this approach, discrimination may be performed according to whether $w_0 + \boldsymbol{w}^T\boldsymbol{x}$ is closer in the least squares sense to t_1 than $-w_0 - \boldsymbol{w}^T\boldsymbol{x}$ is closer to t_2. That is, assign $\boldsymbol{x}$ to ω_1 if $|t_1 - (w_0 + \boldsymbol{w}^T\boldsymbol{x})|^2 < |t_2 + (w_0 + \boldsymbol{w}^T\boldsymbol{x})|^2$. Substituting for w_0 and $\boldsymbol{w}$, this simplifies to (assuming $\alpha(t_1 + t_2) > 0$): assign $\boldsymbol{x}$ to ω_1 if

$$\left(\boldsymbol{S}_W^{-1}(\boldsymbol{m}_1 - \boldsymbol{m}_2)\right)^T(\boldsymbol{x} - \boldsymbol{m}) > \frac{t_1 + t_2}{2}\frac{n_2 - n_1}{\alpha} \tag{5.16}$$

where $\boldsymbol{m}$ is the sample mean, $(n_1\boldsymbol{m}_1 + n_2\boldsymbol{m}_2)/n$. The threshold on the right-hand side of the inequality above is independent of t_1 and t_2 – see the exercises at the end of the chapter.

Of course, other discrimination rules may be used, particularly in view of the fact that the least squares solution gives Fisher's linear discriminant, which we know is the optimal discriminant for two normally distributed classes with equal covariance matrices. Compare the one above with (5.8) that incorporates the numbers in each class in a different way.

5.2.4.3 Optimal discriminant

Another important property of the minimum squared error solution is that it approaches the minimum mean-squared-error approximation to the Bayes discriminant function, $g(\boldsymbol{x})$

$$g(\boldsymbol{x}) = p(\omega_1|\boldsymbol{x}) - p(\omega_2|\boldsymbol{x})$$

in the limit as the number of samples tends to infinity.

In order to understand what this statement means, consider J_S given by (5.9) where $t_i = 1$ for all $\boldsymbol{y}_i$, so that

$$J_S = \sum_{\boldsymbol{x}\in\omega_1}(w_0 + \boldsymbol{w}^T\boldsymbol{x} - 1)^2 + \sum_{\boldsymbol{x}\in\omega_2}(w_0 + \boldsymbol{w}^T\boldsymbol{x} + 1)^2 \tag{5.17}$$

where we have assumed linear dependence of the $\boldsymbol{y}_i$ on the $\boldsymbol{x}_i$. Figure 5.5 illustrates the minimisation process taking place.

For illustration, five samples are drawn from each of two univariate normal distributions of unit variance and means 0.0 and 2.0 and plotted on the x-axis of Figure 5.5, $\diamond$ for class ω_1 and $\bullet$ for class ω_2. Minimising J_S means that the sum of the squares of the distances from the straight line in Figure 5.5 to either $+1$ for class ω_1 or -1 for class ω_2 is minimised. Also plotted in Figure 5.5 is the optimal Bayes discriminant, $g(x)$, for the two normal distributions.

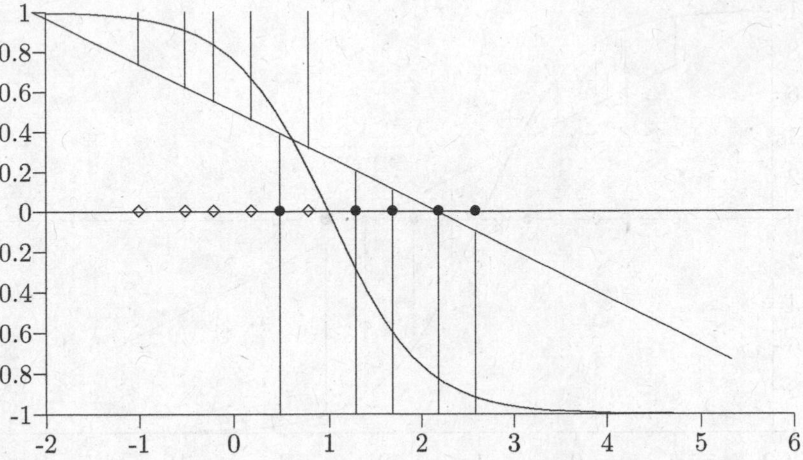

Figure 5.5 Optimality of least mean-squared-error rule – illustration of Equation (5.17).

As the number of samples, n, becomes large, the expression J_S/n tends to

$$\frac{J_S}{n} = p(\omega_1)\int (w_0 + \boldsymbol{w}^T\boldsymbol{x} - 1)^2 p(\boldsymbol{x}|\omega_1)d\boldsymbol{x} + p(\omega_2)\int (w_0 + \boldsymbol{w}^T\boldsymbol{x} + 1)^2 p(\boldsymbol{x}|\omega_2)d\boldsymbol{x}$$

Expanding and simplifying, this gives

$$\begin{aligned}\frac{J_S}{n} &= \int \left(w_0 + \boldsymbol{w}^T\boldsymbol{x}\right)^2 p(\boldsymbol{x})d\boldsymbol{x} + 1 - 2\int \left(w_0 + \boldsymbol{w}^T\boldsymbol{x}\right) g(\boldsymbol{x})p(\boldsymbol{x})d\boldsymbol{x} \\ &= \int \left(w_0 + \boldsymbol{w}^T\boldsymbol{x} - g(\boldsymbol{x})\right)^2 p(\boldsymbol{x})d\boldsymbol{x} + 1 - \int g^2(\boldsymbol{x})p(\boldsymbol{x})d\boldsymbol{x}\end{aligned}$$

Since only the first integral in the above expression depends on w_0 and $\boldsymbol{w}$, we have the result that minimising (5.17) is equivalent, as the number of samples becomes large, to minimising

$$\int (w_0 + \boldsymbol{w}^T\boldsymbol{x} - g(\boldsymbol{x}))^2 p(\boldsymbol{x})d\boldsymbol{x} \tag{5.18}$$

which is the minimum mean-squared-error approximation to the Bayes discriminant function. This is illustrated in Figure 5.6.

The expression (5.18) above is the square difference between the optimal Bayes discriminant and the straight line, integrated over the distribution, $p(\boldsymbol{x})$.

Note that if we were to choose a suitable basis $\phi_1, \ldots, \phi_D$, transform the feature vector $\boldsymbol{x}$ to $(\phi_1(\boldsymbol{x}), \ldots, \phi_D(\boldsymbol{x}))^T$ and then construct the linear discriminant function, we might get a closer approximation to the optimal discriminant, and the decision boundary would not necessarily be a straight line (or plane) in the original space of the variables, $\boldsymbol{x}$. Also, although asymptotically the solution gives the best approximation (in the least squares sense) to the Bayes discriminant function, it is influenced by regions of high density rather than

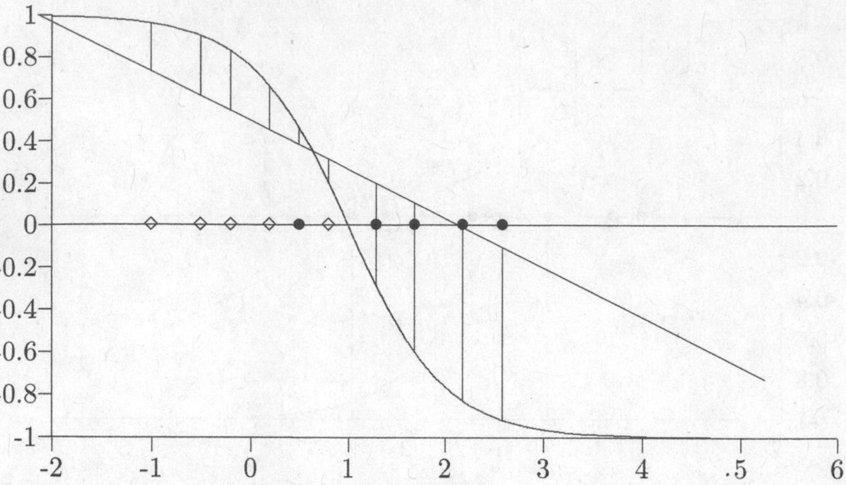

Figure 5.6 Least mean-squared-error approximation to the Bayes discriminant rule – illustration of Equation (5.18).

samples close to the decision boundary. Although Bayesian heuristics motivate the use of a linear discriminant trained by least squares, it can give poor decision boundaries in some circumstances (Hastie *et al.*, 1994).

5.2.4.4 Bounds on the error rate

The least mean-squared-error criterion possesses some attractive theoretical properties that we now quote without proof. Let E_1 denote the nearest-neighbour error rate, E_{mse} the least mean-squared-error rate and let $\boldsymbol{v}$ be the minimum error solution. Then (Devijver and Kittler, 1982)

$$\frac{J_S(\boldsymbol{v})/n}{1 - J_S(\boldsymbol{v})/n} \geq E_{mse}$$

$$J_S(\boldsymbol{v})/n \geq 2E_1 \geq 2E^*$$

$$J_S(\boldsymbol{v})/n = 2E_1 \Rightarrow E_{mse} = E^*$$

where E^* is the optimal Bayes' error rate.

The first condition gives an upper bound on the error rate [and may easily be computed from the values of $J_S(\boldsymbol{v})$ delivered by the algorithm above]. It seems sensible that if we have two possible sets of discriminant functions, ϕ_i and ν_i, then if $J_S^{\phi} < J_S^{\nu}$, then the set ϕ should be preferred since it gives a smaller upper bound for the error rate, E_{mse}. Of course, this is not sufficient but gives us a reasonable guideline.

The second condition shows that the value of the criterion function J_S/n is bounded below by twice the nearest-neighbour error rate E_1.

5.2.5 Further developments

The main developments of the two-class linear algorithms are as follows:

1. Multiclass algorithms. These are discussed in Section 5.3.
2. Support vector machines. These are discussed in Section 5.4.
3. Nonlinear methods. Many classification methods that produce nonlinear decision boundaries are essentially *linear models*: they are linear combinations of nonlinear functions of the variables. Radial basis function networks are one example. Thus the machinery developed in this chapter is important. This is examined further in Chapter 6.
4. Regularisation – introducing a parameter that controls the sensitivity of the technique to small changes in the data or training procedure and improves generalisation (see Section 5.3.5). This includes combining multiple versions of the same classifier, trained under different conditions (see Chapter 8).

5.2.6 Summary

In this section we have considered a range of techniques for performing linear discrimination in the two-class case. These are summarised in Table 5.1, where for completeness we have also included SVMs, to be considered in Section 5.4. They fall broadly into two groups: those techniques that minimise a criterion based on misclassified samples and those that use all samples, correctly classified or not. The former group includes the perceptron, relaxation and SVM algorithms. The latter group includes Fisher's criterion and criteria based on a least squares error measure, including the pseudo-inverse method.

A perceptron is a trainable threshold logic unit. During training, weights are adjusted to minimise a specific criterion. For two separable classes, the basic error-correction procedure converges to a solution in which the classes are separated by a linear decision boundary. If

Table 5.1 Summary of linear techniques

Procedure name	Criterion	Algorithm
Perceptron	$J_P(\boldsymbol{v}) = \sum_{\boldsymbol{y}_i \in \mathcal{Y}} (-\boldsymbol{v}^T\boldsymbol{y}_i)$	$\boldsymbol{v}_{k+1} = \boldsymbol{v}_k + \rho_k \sum_{\boldsymbol{y}_i \in \mathcal{Y}} \boldsymbol{y}_i$
Relaxation	$J_r = \frac{1}{2} \sum_{\boldsymbol{y}_i \in \mathcal{Y}} \frac{(\boldsymbol{v}^T\boldsymbol{y}_i - b)^2}{\lvert\boldsymbol{y}_i\rvert^2}$	$\boldsymbol{v}_{k+1} = \boldsymbol{v}_k + \rho_k \frac{b - \boldsymbol{v}_k^T\boldsymbol{y}_i}{\lvert\boldsymbol{y}_i\rvert^2}\boldsymbol{y}_i$
Fisher	$J_F = \frac{\lvert\boldsymbol{w}^T(\boldsymbol{m}_1 - \boldsymbol{m}_2)\rvert^2}{\boldsymbol{w}^T\boldsymbol{S}_W\boldsymbol{w}}$	$\boldsymbol{w} \propto \boldsymbol{S}_W^{-1}(\boldsymbol{m}_1 - \boldsymbol{m}_2)$
lms – pseudo-inverse	$J_S = \lvert\boldsymbol{Y}\boldsymbol{v} - \boldsymbol{t}\rvert^2$	$\hat{\boldsymbol{v}} = \boldsymbol{Y}^\dagger\boldsymbol{t}$
SVM	$\boldsymbol{w}^T\boldsymbol{w} + C\sum_i \xi_i$ subject to constraints (5.52)	quadratic programming

the classes are not separable, the training procedure must be modified to ensure convergence. More complex decision surfaces can be implemented by using combinations and layers of perceptrons (Nilsson, 1965; Minsky and Papert, 1988). This we shall discuss further in Chapter 6.

Some of the techniques will find a solution that separates two classes if they are separable, others do not. A further dichotomy is: the algorithms that converge for nonseparable classes and those that do not.

The least mean-squared-error design criterion is widely used in pattern recognition. It can be readily implemented in many of the standard computer programs for regression and we have shown how the discrimination problem may be viewed as an exercise in regression. The linear discriminant obtained by the procedure is optimal in the sense of providing a minimum mean-squared-error approximation to the Bayes discriminant function. The analysis of this section applies also to generalised linear discriminant functions [the variables $\boldsymbol{x}$ are replaced by $\boldsymbol{\phi}(\boldsymbol{x})$]. Therefore, choosing a suitable basis for the $\phi_j(\boldsymbol{x})$ is important since a good set will lead to a good approximation to the Bayes discriminant function.

One problem with the least mean-squared-error procedure is that it is sensitive to outliers and does not necessarily produce a separable solution, even when the classes are separable by a linear discriminant. Modifications of the least mean-squared rule to ensure a solution for separable sets have been proposed [the Ho–Kashyap procedure (Ho and Kashyap, 1965) which adjusts both the weight vector, $\boldsymbol{v}$, and the target vector $\boldsymbol{t}$], but the optimal approximation to the Bayes discriminant function when the sets overlap is no longer achieved.

5.3 Multiclass algorithms

There are several ways of extending two-class procedures to the multiclass case. We start with an outline of generic procedures for converting binary classifiers into multiclass classifiers. The structure of the remainder of this section then follows that of Section 5.2 with error-correction procedures, generalisations of Fisher's discriminant and minimum squared-error procedures.

5.3.1 General ideas

5.3.1.1 One-against-all

For C classes, construct C binary classifiers. The kth classifier is trained to discriminate patterns in class ω_k from those in the remaining classes. Thus, determine the weight vector, $\boldsymbol{w}^k$, and the threshold, ${w^k}_0$, such that

$$(\boldsymbol{w}^k)^T\boldsymbol{x} + w_0^k \begin{cases} > & 0 \\ < & 0 \end{cases} \Rightarrow \boldsymbol{x} \in \begin{cases} \omega_k \\ \omega_1, \ldots, \omega_{k-1}, \omega_{k+1}, \ldots, \omega_C \end{cases}$$

Ideally, for a given pattern $\boldsymbol{x}$, the quantity $g_k(\boldsymbol{x}) = (\boldsymbol{w}^k)^T\boldsymbol{x} + w_0^k$ will be positive for one value of k and negative for the remainder giving a clear indication of class. However, this procedure may result in a pattern $\boldsymbol{x}$ belonging to more than one class, or belonging to none.

If there is more than one class for which the quantity $g_k(\boldsymbol{x})$ is positive, $\boldsymbol{x}$ may be assigned to the class for which $((\boldsymbol{w}^k)^T\boldsymbol{x} + w_0^k)/|\boldsymbol{w}^k|$ (the distance to the hyperplane) is the largest. If all values of $g_k(\boldsymbol{x})$ are negative, then assign $\boldsymbol{x}$ to the class with smallest value of $|((\boldsymbol{w}^k)^T\boldsymbol{x} + w_0^k)|/|\boldsymbol{w}^k|$.

5.3.1.2 One-against-one

Construct $C(C-1)/2$ classifiers. Each classifier discriminates between two classes. A pattern $\boldsymbol{x}$ is assigned using each classifier in turn and a majority vote taken. This can lead to ambiguity, with no clear decision for some patterns.

5.3.1.3 Discriminant functions

A third approach is, for C classes, to define C linear discriminant functions $g_1(\boldsymbol{x}), \ldots, g_C(\boldsymbol{x})$ and assign $\boldsymbol{x}$ to class ω_i if

$$g_i(\boldsymbol{x}) = \max_j g_j(\boldsymbol{x})$$

that is, $\boldsymbol{x}$ is assigned to the class whose discriminant function is the largest value at $\boldsymbol{x}$. If

$$g_i(\boldsymbol{x}) = \max_j g_j(\boldsymbol{x}) \quad \Leftrightarrow \quad p(\omega_i|\boldsymbol{x}) = \max_j p(\omega_j|\boldsymbol{x})$$

then the decision boundaries obtained will be optimal in the sense of the Bayes' minimum error.

5.3.2 Error-correction procedure

A generalisation of the two-class error-correction procedure for $C > 2$ classes is to define C linear discriminants

$$g_i(\boldsymbol{x}) = \boldsymbol{v}_i^T \boldsymbol{z}$$

where $\boldsymbol{z}$ is the augmented data vector, $\boldsymbol{z}^T = (1, \boldsymbol{x}^T)$. The *generalised error-correction procedure* is used to train the classifier. Arbitrary initial values are assigned to the $\boldsymbol{v}_i$ and each pattern in the training set is considered one at a time. If a pattern belonging to class ω_i is presented and the maximum value of the discriminant functions is for the jth discriminant function (i.e. a pattern in class ω_i is classified as class ω_j) then the weight vectors $\boldsymbol{v}_i$ and $\boldsymbol{v}_j$ are modified according to

$$\begin{aligned} \boldsymbol{v}_i' &= \boldsymbol{v}_i + c\boldsymbol{z} \\ \boldsymbol{v}_j' &= \boldsymbol{v}_j - c\boldsymbol{z} \end{aligned}$$

where c is a positive correction increment. That is, the value of the ith discriminant function is increased for pattern $\boldsymbol{z}$ and the value of the jth discriminant is decreased. This procedure will converge in a finite number of steps if the classes are separable (Nilsson, 1965). Convergence may require the dataset to be cycled through several times (as in the two-class case).

Choosing c according to

$$c = (\boldsymbol{v}_j - \boldsymbol{v}_i)^T \frac{\boldsymbol{z}}{|\boldsymbol{z}|^2}$$

will ensure that after adjustment of the weight vectors, $\boldsymbol{z}$ will be correctly classified.

5.3.3 Fisher's criterion – linear discriminant analysis

The term *linear discriminant analysis* (LDA), although generically referring to techniques that produce discriminant functions that are linear in the input variables (and thus applies to the perceptron and all of the techniques of this chapter), is also used in a specific sense to refer to the technique of this subsection in which a linear transformation is sought that, in some sense, maximises between-class separability and minimises within-class variability. The characteristics of the method are:

1. A transformation is produced to a space of dimension at most $C - 1$, where C is the number of classes.
2. The transformation is distribution-free – for example, no assumption is made regarding normality of the data.
3. The axes of the transformed coordinate system can be ordered in terms of 'importance for discrimination'. Those most important can be used to obtain a graphical representation of the data by plotting the data in this coordinate system (usually, two or three dimensions).
4. Discrimination may be performed in this reduced-dimensional space using any convenient classifier. Often improved performance is achieved over the application of the rule in the original data space. If a nearest-class-mean type rule is employed, the decision boundaries are linear (and equal to those obtained by a Gaussian classifier under the assumption of equal covariance matrices for the classes).
5. Linear discriminant analysis may be used as a post-processor for more complex, nonlinear classifiers.

There are several ways of generalising the criterion J_F (5.6) to the multiclass case. Optimisation of these criteria yields transformations that reduce to Fisher's linear discriminant in the two-class case and that, in some sense, maximise the between-class scatter and minimise the within-class scatter. We present one approach here.

We consider the criterion

$$J_F(\boldsymbol{a}) = \frac{\boldsymbol{a}^T \boldsymbol{S}_B \boldsymbol{a}}{\boldsymbol{a}^T \boldsymbol{S}_W \boldsymbol{a}} \tag{5.19}$$

where the sample-based estimates of $\boldsymbol{S}_B$ and $\boldsymbol{S}_W$ are given by

$$\boldsymbol{S}_B = \sum_{i=1}^{C} \frac{n_i}{n} (\boldsymbol{m}_i - \boldsymbol{m})(\boldsymbol{m}_i - \boldsymbol{m})^T$$

and

$$S_W = \sum_{i=1}^{C} \frac{n_i}{n} \hat{\Sigma}_i$$

where $\boldsymbol{m}_i$ and $\hat{\boldsymbol{\Sigma}}_i$, $i = 1, \dots, C$ are the sample means and covariance matrices of each class (with n_i samples) and $\boldsymbol{m}$ is the sample mean. We seek a set of feature vectors $\boldsymbol{a}_i$ that maximise (5.19) subject to the normalisation constraint $\boldsymbol{a}_i^T \boldsymbol{S}_W \boldsymbol{a}_j = \delta_{ij}$ (class-centralised vectors in the transformed space are uncorrelated). This leads to the generalised symmetric eigenvector equation (Press *et al.*, 1992)

$$\boldsymbol{S}_B \boldsymbol{A} = \boldsymbol{S}_W \boldsymbol{A} \boldsymbol{\Lambda} \tag{5.20}$$

where $\boldsymbol{A}$ is the matrix whose columns are the $\boldsymbol{a}_i$ and $\boldsymbol{\Lambda}$ is the diagonal matrix of eigenvalues. If $\boldsymbol{S}_W^{-1}$ exists, this may be written

$$\boldsymbol{S}_W^{-1} \boldsymbol{S}_B \boldsymbol{A} = \boldsymbol{A} \boldsymbol{\Lambda} \tag{5.21}$$

The eigenvectors corresponding to the largest of the eigenvalues are used for feature extraction. The rank of $\boldsymbol{S}_B$ is at most $C - 1$; therefore the projection will be on to a space of dimension at most $C - 1$. The solution for $\boldsymbol{A}$ satisfying (5.20) and the constraint also diagonalises the between-class covariance matrix, $\boldsymbol{A}^T \boldsymbol{S}_B \boldsymbol{A} = \boldsymbol{\Lambda}$, the diagonal matrix of eigenvalues.

5.3.3.1 Solving the LDA generalised symmetric eigenvector equation

When the matrix $\boldsymbol{S}_W$ is not ill-conditioned with respect to inversion, the eigenvectors of the generalised symmetric eigenvector equation can be determined by solving the equivalent equation

$$\boldsymbol{S}_W^{-1} \boldsymbol{S}_B \boldsymbol{a} = \lambda \boldsymbol{a} \tag{5.22}$$

though note that the matrix $\boldsymbol{S}_W^{-1} \boldsymbol{S}_B$ is not symmetric. However, the system may be reduced to a symmetric eigenvector problem using the Cholesky decomposition (Press *et al.*, 1992) of $\boldsymbol{S}_W$, which allows $\boldsymbol{S}_W$ to be written as the product $\boldsymbol{S}_W = \boldsymbol{L}\boldsymbol{L}^T$, for a lower triangular matrix $\boldsymbol{L}$. Then, (5.22) is equivalent to

$$\boldsymbol{L}^{-1} \boldsymbol{S}_B (\boldsymbol{L}^{-1})^T \boldsymbol{y} = \lambda \boldsymbol{y}$$

where $\boldsymbol{y} = \boldsymbol{L}^T \boldsymbol{a}$. Efficient routines based on the QR algorithm (Stewart, 1973; Press *et al.*, 1992) may be used to solve the above eigenvector equation.

If $\boldsymbol{S}_W$ is close to singular, then $\boldsymbol{S}_W^{-1} \boldsymbol{S}_B$ cannot be computed accurately. One approach is to use the QZ (Stewart, 1973) algorithm, which reduces $\boldsymbol{S}_B$ and $\boldsymbol{S}_W$ to upper triangular form (with diagonal elements b_i and w_i, respectively) and the eigenvalues are given by the ratios $\lambda_i = b_i/w_i$. If $\boldsymbol{S}_W$ is singular, the system will have 'infinite' eigenvalues, and the ratio cannot be formed. These 'infinite' eigenvalues correspond to eigenvectors in the null space of

S_W. L.-F. Chen *et al.* (2000) propose using these eigenvectors, ordered according to b_i, for the LDA feature space.

There are other approaches. Instead of solving (5.20) or (5.21), we may determine $\boldsymbol{A}$ by solving two symmetric eigenvector equations successively. The solution is given by:

$$\boldsymbol{A} = \boldsymbol{U}_r \boldsymbol{\Lambda}_r^{-\frac{1}{2}} \boldsymbol{V}_\nu \tag{5.23}$$

where $\boldsymbol{U}_r = [\boldsymbol{u}_1| \ldots |\boldsymbol{u}_r]$ is the matrix whose columns are the eigenvectors of $\boldsymbol{S}_W$ with nonzero eigenvalues $\lambda_1, \ldots, \lambda_r$; $\boldsymbol{\Lambda}_r = \text{Diag}\{\lambda_1 \ldots \lambda_r\}$ and $\boldsymbol{V}_\nu$ is the matrix of eigenvectors of $\boldsymbol{S}'_B = \boldsymbol{\Lambda}_r^{-\frac{1}{2}} \boldsymbol{U}_r^T \boldsymbol{S}_B \boldsymbol{U}_r \boldsymbol{\Lambda}_r^{-\frac{1}{2}}$, which has the same eigenvalues as (5.20). This is the Karhunen–Loève transformation proposed by Kittler and Young (1973), and covered in Chapter 10. To see the validity of this process we note that the eigen-decomposition of the symmetric matrix $\boldsymbol{S}_W$ is:

$$\boldsymbol{U}_r \boldsymbol{\Lambda}_r \boldsymbol{U}_r^T = \boldsymbol{S}_W$$

and hence Equation (5.20) can be written:

$$\boldsymbol{S}_B \boldsymbol{A} = \boldsymbol{U}_r \boldsymbol{\Lambda}_r \boldsymbol{U}_r^T \boldsymbol{A} \boldsymbol{\Lambda} \tag{5.24}$$

Then substituting (5.23) into (5.24) and noting that $\boldsymbol{U}_r^T \boldsymbol{U}_r = \boldsymbol{I}_r$, the $r \times r$ identity matrix, gives:

$$\boldsymbol{S}_B \boldsymbol{U}_r \boldsymbol{\Lambda}_r^{-\frac{1}{2}} \boldsymbol{V}_\nu = \boldsymbol{U}_r \boldsymbol{\Lambda}_r^{\frac{1}{2}} \boldsymbol{V}_\nu \boldsymbol{\Lambda}$$

Pre-multiplying both sides of the above equation by $\boldsymbol{\Lambda}_r^{-\frac{1}{2}} \boldsymbol{U}_r^T$, and simplifying, gives:

$$\boldsymbol{S}'_B \boldsymbol{V}_\nu = \boldsymbol{V}_\nu \boldsymbol{\Lambda} \tag{5.25}$$

i.e. as desired, $\boldsymbol{V}_\nu$ are the eigenvectors of $\boldsymbol{S}'_B$ with eigenvalues as (5.20).

Cheng *et al.* (1992) describe several methods for determining optimal discriminant transformations when $\boldsymbol{S}_W$ is ill-conditioned. These include

1. *The pseudo-inverse method.* Replace $\boldsymbol{S}_W^{-1}$ by the pseudo-inverse, $\boldsymbol{S}_W^{\dagger}$ (Tian *et al.*, 1988).
2. *The perturbation method.* Stabilise the matrix $\boldsymbol{S}_W$ by adding a small perturbation matrix, $\boldsymbol{\Delta}$ (Hong and Yang, 1991). This amounts to replacing the singular values of $\boldsymbol{S}_W$, λ_r, by a small fixed positive value, δ if $\lambda_r < \delta$.
3. *The rank decomposition method.* This is a two-stage process, similar to the one given above (5.23), with successive eigendecompositions of the total scatter matrix, and between-class scatter matrix.

5.3.3.2 Discrimination

As in the two-class case, the transformation in itself does not provide us with a discrimination rule. The transformation is independent of the distributions of the classes and is defined in terms of matrices S_B and S_W. However, if we were to assume that the data were normally distributed, with equal covariance matrices (equal to the within-class covariance matrix, S_W) in each class and means m_i, then the discrimination rule is: assign x to class ω_i if $g_i \geq g_j$ for all $j \neq i, j = 1, \ldots, C$, where

$$g_i = \log(p(\omega_i)) - \frac{1}{2}(x - m_i)^T S_W^{-1}(x - m_i) \tag{5.26}$$

or, neglecting the quadratic terms in x since they do not depend on class,

$$g_i = \log(p(\omega_i)) - \frac{1}{2}m_i^T S_W^{-1} m_i + x^T S_W^{-1} m_i \tag{5.27}$$

the normal-based linear discriminant function (see Chapter 2).

If A is the linear discriminant transformation, then S_W^{-1} may be written (see the exercises at the end of the chapter)

$$S_W^{-1} = AA^T + A_\perp A_\perp^T$$

where $A_\perp^T(m_j - m) = 0$ for all j. Using the above expression for S_W^{-1} in (5.26) gives a discriminant function

$$g_i = \log(p(\omega_i)) - \frac{1}{2}(y(x) - y_i)^T(y(x) - y_i) - \frac{1}{2}x - m^T A_\perp A_\perp^T x - m \tag{5.28}$$

and ignoring terms that are constant across classes, discrimination is based on:

$$g_i = \log(p(\omega_i)) - \frac{1}{2}y_i^T y_i + y^T(x) y_i$$

a nearest class mean classifier in the transformed space, where $y_i = A^T m_i$ and $y(x) = A^T x$.

This is simply the Gaussian classifier of Chapter 2 applied in the transformed space.

5.3.4 Least mean-squared-error procedures

5.3.4.1 Introduction

As in Section 5.2.4, we seek a linear transformation of the data x [or the transformed data $\phi(x)$] that we can use to make a decision and which is obtained by minimising a squared-error measure. Specifically, let the data be denoted by the $n \times p$ matrix $X = [x_1|\ldots|x_n]^T$ and consider the minimisation of the quantity E,

$$\begin{aligned} E &= \left\| W X^T + w_0 \mathbf{1}^T - T^T \right\|^2 \\ &= \sum_{i=1}^{n} (W x_i + w_0 - t_i)^T (W x_i + w_0 - t_i) \end{aligned}$$

where $\boldsymbol{W}$ is a $C \times p$ matrix of weights; $\boldsymbol{w}_0$ is a C-dimensional vector of biases and $\mathbf{1}$ is an n-dimensional vector with each component equal to unity. The $n \times C$ matrix of constants $\boldsymbol{T}$, sometimes termed the *target* matrix, is defined so that the ith row is

$$\boldsymbol{t}_i = \boldsymbol{\lambda}_j = \begin{pmatrix} \lambda_{j1} \\ \vdots \\ \lambda_{jC} \end{pmatrix} \text{ for } \boldsymbol{x}_i \text{ in class } \omega_j; \tag{5.29}$$

that is, $\boldsymbol{t}_i$ has the same value for all patterns in the same class. Choice of the target values is considered later in this subsection. Minimising (5.29) with respect to $\boldsymbol{w}_0$ gives

$$\boldsymbol{w}_0 = \bar{\boldsymbol{t}} - \boldsymbol{W}\boldsymbol{m} \tag{5.30}$$

where

$$\bar{\boldsymbol{t}} = \frac{1}{n}\sum_{j=1}^{C} n_j \boldsymbol{\lambda}_j, \quad \text{the mean 'target' vector}$$

with n_j the number of data patterns from class ω_j, and

$$\boldsymbol{m} = \frac{1}{n}\sum_{i=1}^{n} \boldsymbol{x}_i, \quad \text{the mean data vector}$$

Substituting for $\boldsymbol{w}_0$ from (5.30) into (5.29) allows us to express the error, E, as

$$E = \|\boldsymbol{W}\hat{\boldsymbol{X}}^T - \hat{\boldsymbol{T}}^T\|^2 \tag{5.31}$$

where $\hat{\boldsymbol{X}}$ and $\hat{\boldsymbol{T}}$ are defined as

$$\hat{\boldsymbol{X}} \triangleq \boldsymbol{X} - \mathbf{1}\boldsymbol{m}^T$$
$$\hat{\boldsymbol{T}} \triangleq \boldsymbol{T} - \mathbf{1}\bar{\boldsymbol{t}}^T$$

(data and target matrices with zero mean rows); $\mathbf{1}$ is an n-dimensional vector of 1s. The minimum (Frobenius) norm solution for $\boldsymbol{W}$ that minimises E is

$$\boldsymbol{W} = \hat{\boldsymbol{T}}^T(\hat{\boldsymbol{X}}^T)^\dagger \tag{5.32}$$

where $(\hat{\boldsymbol{X}}^T)^\dagger$ is the Moore–Penrose pseudo-inverse of $\hat{\boldsymbol{X}}^T$ (see Section 5.2.4), and $\boldsymbol{X}^\dagger = (\boldsymbol{X}^T\boldsymbol{X})^{-1}\boldsymbol{X}^T$ if the inverse exists. The matrix of fitted values is given by

$$\tilde{\boldsymbol{T}} = \hat{\boldsymbol{X}}\hat{\boldsymbol{X}}^\dagger\hat{\boldsymbol{T}} + \mathbf{1}\bar{\boldsymbol{t}}^T \tag{5.33}$$

Thus, we can obtain a solution for the weights in terms of the data and the 'target matrix' $\boldsymbol{T}$, as yet unspecified.

5.3.4.2 Properties

Before we consider particular forms for $\boldsymbol{T}$, let us note one or two properties of the least mean-squared-error approximation. The large sample limit of (5.29) is

$$E/n \;\rightarrow\; E_\infty = \sum_{j=1}^{C} p(\omega_j)\mathrm{E}[|\boldsymbol{W}\boldsymbol{x} + \boldsymbol{w}_0 - \boldsymbol{\lambda}_j|^2]_j \tag{5.34}$$

where $p(\omega_j)$ is the prior probability (the limit of n_j/n) and the expectation, $\mathrm{E}[.]_j$, is with respect to the conditional distribution of $\boldsymbol{x}$ on class ω_j, i.e. for any function z of $\boldsymbol{x}$

$$\mathrm{E}[z(\boldsymbol{x})]_j = \int z(\boldsymbol{x})p(\boldsymbol{x}|\omega_j)d\boldsymbol{x}$$

The solution for $\boldsymbol{W}$ and $\boldsymbol{w}_0$ that minimises (5.34) also minimises (Wee, 1968; Devijver, 1973; also see the exercises at the end of the chapter)

$$E' = \mathrm{E}[|\boldsymbol{W}\boldsymbol{x} + \boldsymbol{w}_0 - \boldsymbol{\rho}(\boldsymbol{x})|^2] \tag{5.35}$$

where the expectation is with respect to the unconditional distribution $p(\boldsymbol{x})$ of $\boldsymbol{x}$ and $\boldsymbol{\rho}(\boldsymbol{x})$ is defined as

$$\boldsymbol{\rho}(\boldsymbol{x}) = \sum_{j=1}^{C} \boldsymbol{\lambda}_j p(\omega_j|\boldsymbol{x}) \tag{5.36}$$

Thus, $\boldsymbol{\rho}(\boldsymbol{x})$ may be viewed as a 'conditional target' vector; it is the expected target vector given a pattern $\boldsymbol{x}$, with the property that

$$\mathrm{E}[\boldsymbol{\rho}(\boldsymbol{x})] = \int \boldsymbol{\rho}(\boldsymbol{x})p(\boldsymbol{x})d\boldsymbol{x} = \sum_{j=1}^{C} p(\omega_j)\boldsymbol{\lambda}_j$$

the mean target vector. From (5.34) and (5.35), the discriminant vector that minimises E_∞ has minimum variance from the discriminant vector $\boldsymbol{\rho}$.

5.3.4.3 Choice of targets

The particular interpretation of $\boldsymbol{\rho}$ depends on the choice we make for the target vectors for each class, $\boldsymbol{\lambda}_j$. If we specify the prototype target matrix as

$$\lambda_{ji} = \text{loss in deciding } \omega_i \text{ when the true class is } \omega_j \tag{5.37}$$

then $\boldsymbol{\rho}(\boldsymbol{x})$ is the *conditional risk vector* (Devijver, 1973), where the conditional risk is the expected loss in making a decision, with the ith component of $\boldsymbol{\rho}(\boldsymbol{x})$ being the conditional risk of deciding in favour of ω_i. The Bayes' decision rule for minimum conditional risk is

$$\text{assign } \boldsymbol{x} \text{ to } \omega_i \text{ if } \rho_i(\boldsymbol{x}) \leq \rho_j(\boldsymbol{x}), \quad j = 1, \ldots, C$$

From (5.34) and (5.35), the discriminant rule that minimises the mean square error E has minimum variance from the optimum Bayes discriminant function $\boldsymbol{\rho}$ as the number of samples tends to infinity.

For a coding scheme in which

$$\lambda_{ij} = \begin{cases} 1 & i = j \\ 0 & \text{otherwise} \end{cases} \tag{5.38}$$

the vector $\boldsymbol{\rho}(\boldsymbol{x})$ is equal to the vector of posterior probabilities, with jth component $p(\omega_j|\boldsymbol{x})$. The Bayes discriminant rule for minimum error is

$$\text{assign } \boldsymbol{x} \text{ to } \omega_i \text{ if } \rho_i(\boldsymbol{x}) \geq \rho_j(\boldsymbol{x}), \quad j = 1, \ldots, C$$

The change in the direction of the inequality results from the fact that the terms λ_{ij} are now *gains* rather than losses. For this coding scheme, the least mean-squared-error solution for $\boldsymbol{W}$ and $\boldsymbol{w}_0$ gives a vector discriminant function that asymptotically has minimum variance from the vector of *a posteriori* probabilities, shown for the two-class case in Section 5.2.4.

5.3.4.4 Decision rule

The above asymptotic results suggest that we use the same decision rules to assign a pattern $\boldsymbol{x}$ to a class assuming that the linear transformation $\boldsymbol{W}\boldsymbol{x} + \boldsymbol{w}_0$ had produced $\boldsymbol{\rho}(\boldsymbol{x})$. For example, with the coding scheme (5.38) for $\boldsymbol{\Lambda}$, we would assign $\boldsymbol{x}$ to the class corresponding to the largest component of the discriminant function $\boldsymbol{W}\boldsymbol{x} + \boldsymbol{w}_0$. Alternatively, in the spirit of the least-squares approach, assign $\boldsymbol{x}$ to ω_i if

$$|\boldsymbol{W}\boldsymbol{x} + \boldsymbol{w}_0 - \boldsymbol{\lambda}_i|^2 < |\boldsymbol{W}\boldsymbol{x} + \boldsymbol{w}_0 - \boldsymbol{\lambda}_j|^2 \quad \text{for all } j \neq i \tag{5.39}$$

which leads to the linear discrimination rule; assign $\boldsymbol{x}$ to class ω_i if

$$\boldsymbol{d}_i^T \boldsymbol{x} + d_{0i} > \boldsymbol{d}_j^T \boldsymbol{x} + d_{0j} \quad \forall j \neq i$$

where

$$\begin{aligned} \boldsymbol{d}_i &= \boldsymbol{W}^T \boldsymbol{\lambda}_i \\ d_{0i} &= -|\boldsymbol{\lambda}_i|^2/2 + \boldsymbol{w}_0^T \boldsymbol{\lambda}_i \end{aligned}$$

For $\boldsymbol{\lambda}_i$ given by (5.38), this decision rule is identical to the one that treats the linear discriminant function $\boldsymbol{W}\boldsymbol{x} + \boldsymbol{w}_0$ as the vector of posterior probabilities, but it is not in general (Lowe and Webb, 1991).

We add a word of caution here. The result given above is an asymptotic result only. Even if we had a very large number of samples and a flexible set of basis functions $\phi(\boldsymbol{x})$ (replacing the measurements $\boldsymbol{x}$), then we do not necessarily achieve the Bayes optimal discriminant function. Our approximation may indeed become closer in the least-squares sense, but this is weighted in favour of higher-density regions, not necessarily at class boundaries.

A final result, which we shall quote without proof, is that for the 1-from-C coding scheme (5.38), the values of the vector $\boldsymbol{Wx} + \boldsymbol{w}_0$ do indeed sum to unity (Lowe and Webb, 1991). That is, if we denote the linear discriminant vector $\boldsymbol{z}$ as

$$\boldsymbol{z} = \boldsymbol{Wx} + \boldsymbol{w}_0$$

where $\boldsymbol{W}$ and $\boldsymbol{w}_0$ have been determined using the mean-squared-error procedure [Equations (5.30) and (5.32)] with the columns of $\boldsymbol{\Lambda} = [\boldsymbol{\lambda}_1, \ldots, \boldsymbol{\lambda}_C]$ being the columns of the identity matrix, then

$$\sum_{i=1}^{C} z_i = 1$$

that is, the sum of the discriminant function values is unity. This does not mean that the components of $\boldsymbol{z}$ can necessarily be treated as probabilities since some may be negative.

5.3.4.5 Incorporating priors

A more general form of (5.29) is a weighted error function

$$\begin{aligned} E &= \sum_{i=1}^{n} d_i(\boldsymbol{Wx}_i + \boldsymbol{w}_0 - \boldsymbol{t}_i)^T(\boldsymbol{Wx}_i + \boldsymbol{w}_0 - \boldsymbol{t}_i) \\ &= \|(\boldsymbol{WX}^T + \boldsymbol{w}_0\boldsymbol{1}^T - \boldsymbol{T}^T)\boldsymbol{D}\|^2 \end{aligned} \tag{5.40}$$

where the ith pattern is weighted by the real factor d_i, and $\boldsymbol{D}$ is diagonal with $\boldsymbol{D}_{ii} = \sqrt{d_i}$. Two different codings for the weighting matrix $\boldsymbol{D}$ are described.

Prior-weighted patterns

In this case, each pattern in the training set is weighted according to the *a priori* probabilities of class membership and the number in that class as

$$d_i = \frac{P_k}{n_k/n} \quad \text{for pattern } i \text{ in class } \omega_k$$

where P_k is the assumed known class probability (derived from knowledge regarding the relative expected class importance, or frequency of occurrence in operation) and n_k is the number of patterns in class ω_k in the training set.

The weighting above would be used in situations where the expected test conditions differ from the conditions described by the training data by the expected proportions in the classes. This may be a result of population drift (see Chapter 1), or when there are limited data available for training. The above weighting has been used by Munro *et al.* (1996) for learning low-probability events in order to reduce the size of the training set for a neural network.

Cluster-weighted patterns

The computational time for many training schemes increases with the training set size (especially if nonlinear neural network routines are being used). *Clustering* (see Chapter 11) is one means of finding a reduced set of prototypes that characterises the training dataset. There are many ways in which clustering may be used to preprocess the data – it could be applied to classes separately, or to the whole training set. For example, when applied to each class separately, the patterns for that class are replaced by the cluster means, with d_i for the new patterns set proportional to the number in the cluster. When applied to the whole dataset, if a cluster contains all members of the same class, those patterns are replaced in the dataset by the cluster mean and d_i is set to the number of patterns in the cluster. If a cluster contains members of different classes, all patterns are retained.

5.3.5 Regularisation

If the matrix $(\boldsymbol{X}^T\boldsymbol{X})$ is close to singular, an alternative to the pseudo-inverse approach is to use a *regularised* estimator. The error, E (5.31), is modified by the addition of a regularisation term to give

$$E = \|\boldsymbol{W}\hat{\boldsymbol{X}}^T - \hat{\boldsymbol{T}}^T\|^2 + \alpha\|\boldsymbol{W}\|^2$$

where α is a *regularisation* parameter or *ridge parameter*. The solution for $\boldsymbol{W}$ that minimises E is

$$\boldsymbol{W} = \hat{\boldsymbol{T}}^T\hat{\boldsymbol{X}}(\hat{\boldsymbol{X}}^T\hat{\boldsymbol{X}} + \alpha\boldsymbol{I}_p)^{-1}$$

We still have to choose the ridge parameter, α, which may be different for each output dimension (corresponding to class in a discrimination problem). There are several possible choices (Brown, 1993). The procedure of Golub *et al.* (1979) is to use a cross-validation estimate. The estimate $\hat{\alpha}$ for α is the value that minimises

$$\frac{|(\boldsymbol{I} - \boldsymbol{A}(\alpha))\hat{\boldsymbol{t}}|^2}{[\text{Tr}(\boldsymbol{I} - \boldsymbol{A}(\alpha))]^2}$$

where

$$A(\alpha) = \hat{\boldsymbol{X}}(\hat{\boldsymbol{X}}^T\hat{\boldsymbol{X}} + \alpha\boldsymbol{I})^{-1}\hat{\boldsymbol{X}}^T$$

and $\hat{\boldsymbol{t}}$ is one of the columns of $\hat{\boldsymbol{T}}$; i.e. measurements on one of the output variables, that is being predicted.

5.3.6 Example application study

The problem
Face recognition using linear discriminant analysis (LDA) (L.-F. Chen *et al.*, 2000).

Summary
A technique, based on LDA, that is appropriate for the small-sample problem (when the within-class matrix, S_W, is singular) is assessed in terms of error rate and computational requirements.

The data

The data comprise 10 different facial images of 128 people (classes). The raw images are 155 × 175 pixels. These images are reduced to 60 × 60 and after further processing and alignment, further dimension reduction, based on k-means clustering (see Chapter 11), to 32, 64, 128 and 256 values is performed.

The model

The classifier is simple. The basic method is to project the data to a lower dimension and perform classification using a nearest-neighbour rule.

Training procedure

The projection to the lower dimension is determined using the training data. Problems occur with the standard LDA approach when the $p \times p$ within-class scatter matrix, S_W is singular (of rank $s < p$). In this case, let $Q = [q_{s+1}, \ldots, q_p]$ be the $p \times (p - s)$ matrix of eigenvectors of S_W with zero eigenvalue; that is, those that map into the null space of S_W. The eigenvectors of $\tilde{S}_B$, defined by $\tilde{S}_B \stackrel{\triangle}{=} QQ^T S_B (QQ^T)^T$, are used to form the the feature vectors.

Results

Recognition rates were calculated using a leave-one-out cross validation strategy. Limited results on small sample size datasets show good performance compared with previously published techniques.

5.3.7 Further developments

The standard generalisation of Fisher's linear discriminant to the multiclass situation chooses as the columns of the feature extraction matrix A, those vectors a_i that maximise

$$\frac{a_i^T S_B a_i}{a_i^T S_W a_i} \tag{5.41}$$

subject to the orthogonality constraint

$$a_i^T S_W a_j = \delta_{ij}$$

i.e. the within-class covariance matrix in the transformed space is the identity matrix. For the two-class case ($C = 2$), only one discriminant vector is calculated. This is Fisher's linear discriminant.

An alternative approach, proposed by Foley and Sammon (1975) for the two-class case and generalised to the multiclass case by Okada and Tomita (1985), is to seek the vector a_i that maximises (5.41) subject to the constraints

$$a_i a_j = \delta_{ij}$$

The first vector, a_1, is Fisher's linear discriminant. The second vector, a_2, maximises (5.41) subject to being orthogonal to a_1 ($a_2 a_1 = 0$), and so on. A direct analytic solution for the problem is given by Duchene and Leclercq (1988), and involves determining an eigenvector of a nonsymmetric matrix. Okada and Tomita (1985) propose an iterative procedure for determining the eigenvectors. The transformation derived is not limited by the number of classes.

A development of this approach that uses error probability to select the vectors $\boldsymbol{a}_i$ is described by Hamamoto *et al.* (1991) and a comparison with the Fisher criterion on the basis of the discriminant score J is given by Hamamoto *et al.* (1993).

Another extension of LDA is to a transformation that is not limited by the number of classes. The set of discriminant vectors is augmented by an orthogonal set that maximises the projected variance. Thus the linear discriminant transformation is composed of two parts: a set of vectors determined, say, by the usual multiclass approach and a set orthogonal to the first set for which the projected variance is maximised. This combines LDA with a principal components analysis (Duchene and Leclercq, 1988).

There have been many developments of Fisher's linear discriminant both in the two-class case (Aladjem, 1991) and in the multiclass situation (Aladjem and Dinstein, 1992; Liu *et al.*, 1993). The aims have been either to determine transformations that aid in exploratory data analysis or interactive pattern recognition or that may be used prior to a linear or quadratic discriminant rule (Schott, 1993). Several methods have been proposed for a small number of samples (Tian *et al.*, 1988; Hong and Yang, 1991; Cheng *et al.*, 1992) when the matrix $\boldsymbol{S}_W$ may be singular. These are compared by Liu *et al.* (1992). Loog and Duin (2004) propose an extension that better deals with *heteroscedastic* data, i.e. data which is not well suited to modelling the class-conditional distributions as having equal covariance matrices. Their approach makes use of the Chernoff dissimilarity measure between probability distributions (see Chapter 10).

A development of the least mean-squared-error approach which weights the error in favour of patterns near the class boundary has been proposed by Al-Alaoui (1977). Al-Alaoui describes a procedure, starting with the generalised inverse solution, that progressively replicates misclassified samples in the training set. The linear discriminant is iteratively updated and the procedure is shown to converge for the two-class separable case. Repeating the misclassified samples in the dataset is equivalent to increasing the cost of misclassification of the misclassified samples or, alternatively, weighting the distribution of the training data in favour of samples near the class boundaries (see the description of boosting in Chapter 9).

5.3.8 Summary

In Chapter 2, we developed a multiclass linear discriminant rule based on normality assumptions for the class-conditional densities (with common covariance matrices). In this section, we have presented a different approach. We have started with the requirement of a linear discriminant rule, and sought ways of determining the weight vectors for each class. Several approaches have been considered including perceptron schemes, Fisher's discriminant rule and least-mean-square procedures. Linear SVMs (see Section 5.4) can also be derived in such a manner.

The least mean-squared-error approach has the property that asymptotically, the procedure produces discriminant functions that provide a minimum squared error approximation to the Bayes discriminant function. One of the problems of the least mean-squared-error approach is that it places emphasis on regions of high density, which are not necessarily at class boundaries.

In many practical studies, when a linear discriminant rule is used, it is often the normal-based linear discriminant rule of Chapter 2. However, the least mean-squared rule, with binary-coded target vectors, is important since it forms a part of some nonlinear regression models.

5.4 Support vector machines

5.4.1 Introduction

As we stated in the introduction to Section 5.2, algorithms for linear discriminant functions may be applied to the original variables or in a transformed feature space defined by nonlinear transformations of the original variables. Support vector machines are no exception. They implement a very simple idea – they map pattern vectors to a high-dimensional feature space where a 'best' separating hyperplane (the *maximal margin* hyperplane) is constructed (Figure 5.7).

In this section we introduce the basic ideas behind the support vector model and in Chapter 6 we develop the model further in the context of neural network classifiers. Much of the work on support vector classifiers relates to the binary classification problem, with the multiclass classifier constructed by combining several binary classifiers (see Section 5.4.4).

5.4.2 Linearly separable two-class data

Consider the binary classification task in which we have a set of training patterns $\{\boldsymbol{x}_i, i = 1, \ldots, n\}$ assigned to one of two classes, ω_1 and ω_2, with corresponding labels $y_i = \pm 1$. Denote the linear discriminant function $g(\boldsymbol{x})$

$$g(\boldsymbol{x}) = \boldsymbol{w}^T \boldsymbol{x} + w_0$$

with decision rule

$$\boldsymbol{w}^T \boldsymbol{x} + w_0 \begin{cases} > & 0 \\ < & 0 \end{cases} \Rightarrow \boldsymbol{x} \in \begin{cases} \omega_1 \text{ with corresponding numeric value, } y_i = +1 \\ \omega_2 \text{ with corresponding numeric value, } y_i = -1 \end{cases}$$

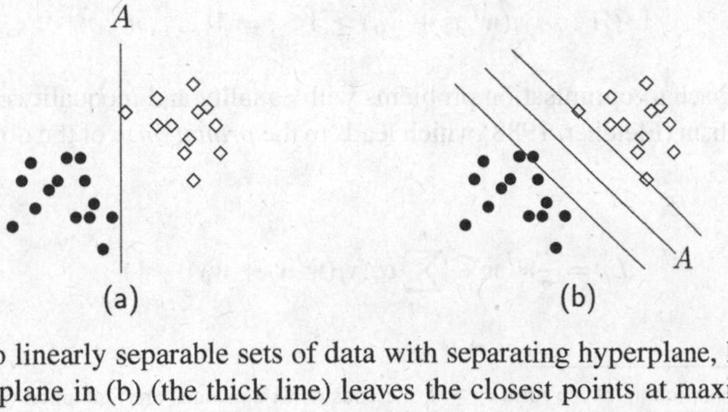

Figure 5.7 Two linearly separable sets of data with separating hyperplane, labelled *A*. The separating hyperplane in (b) (the thick line) leaves the closest points at maximum distance. The thin lines on either side of the hyperplane *A* identify the margin.

Thus, all training points are correctly classified if

$$y_i(\boldsymbol{w}^T\boldsymbol{x}_i + w_0) > 0 \text{ for all } i$$

This is an alternative way of writing (5.3).

Figure 5.7(a) shows two separable sets of points with a separating hyperplane, A. Clearly, there are many possible separating hyperplanes. The maximal margin classifier determines the hyperplane for which the *margin* – the sum of the distances from the separating hyperplane to the closest example from class ω_1 and the closest example from class ω_2 – is the largest. This is illustrated in Figure 5.7(b), where the margin is shown as the distance between two parallel hyperplanes on each side of the hyperplane A that separates the data, with each parallel hyperplane passing through the closest points (on a given side) to the separating hyperplane. The assumption is that the larger the margin, the better the generalisation error of the linear classifier defined by the separating hyperplane.

In Section 5.2.2, we saw that a variant of the perceptron rule was to introduce a margin, $b > 0$, and seek a solution so that

$$y_i(\boldsymbol{w}^T\boldsymbol{x}_i + w_0) \geq b \tag{5.42}$$

The perceptron algorithm yields a solution for which all points, $\boldsymbol{x}_i$ are at a distance greater than $b/|\boldsymbol{w}|$ from the separating hyperplane. A scaling of b and $\boldsymbol{w}$, w_0 leaves this distance unaltered and the condition (5.42) still satisfied. Therefore, without loss of generality, a value $b = 1$ may be taken, defining what are termed the *canonical hyperplanes*, $H_1 : \boldsymbol{w}^T\boldsymbol{x} + w_0 = +1$ and $H_2 : \boldsymbol{w}^T\boldsymbol{x} + w_0 = -1$, and we have

$$\begin{array}{ll} \boldsymbol{w}^T\boldsymbol{x}_i + w_0 \geq +1 & \text{for } y_i = +1 \\ \boldsymbol{w}^T\boldsymbol{x}_i + w_0 \leq -1 & \text{for } y_i = -1 \end{array} \tag{5.43}$$

The distance between each of these two hyperplanes and the separating hyperplane, $g(\boldsymbol{x}) = 0$, is $1/|\boldsymbol{w}|$, and the *margin* is $2/|\boldsymbol{w}|$. Figure 5.8 shows the separating hyperplane and the canonical hyperplanes for two separable datasets. The points that lie on the canonical hyperplanes are called *support vectors* (circled in Figure 5.8).

Therefore, maximising the margin means that we seek a solution that minimises $|\boldsymbol{w}|$ subject to the constraints

$$\mathcal{C}1: \quad y_i(\boldsymbol{w}^T\boldsymbol{x}_i + w_0) \geq 1 \quad i = 1, \ldots, n \tag{5.44}$$

A standard approach to optimisation problems with equality and inequality constraints is the Lagrange formalism (Fletcher, 1988) which leads to the *primal form* of the objective function, L_p, given by[1]

$$L_p = \frac{1}{2}\boldsymbol{w}^T\boldsymbol{w} - \sum_{i=1}^{n} \alpha_i(y_i(\boldsymbol{w}^T\boldsymbol{x}_i + w_0) - 1) \tag{5.45}$$

[1] For inequality constraints of the form $c_i \geq 0$, the constraints equations are multiplied by positive Lagrange multipliers and subtracted from the objective function.

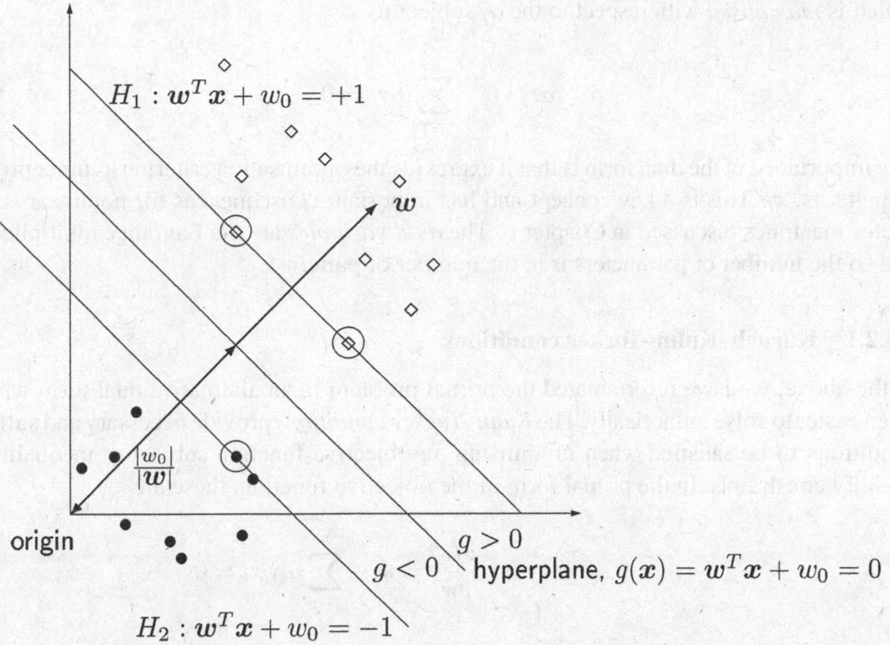

Figure 5.8 H_1 and H_2 are the canonical hyperplanes. The margin is the perpendicular distance between the separating hyperplane $[g(\boldsymbol{x}) = 0]$ and a hyperplane through the closest points (marked by a ring around the data points). These are termed the *support vectors*.

where $\{\alpha_i, i = 1, \ldots, n; \alpha_i \geq 0\}$ are the Lagrange multipliers. The *primal parameters* are $\boldsymbol{w}$ and w_0 and therefore the number of parameters is $p + 1$, where p is the dimensionality of the feature space.

The solution to the problem of minimising $\boldsymbol{w}^T\boldsymbol{w}$ subject to constraints (5.44) is equivalent to determining the saddlepoint of the function L_p, at which L_p is minimised with respect to $\boldsymbol{w}$ and w_0 and maximised with respect to the α_i. Differentiating L_p with respect to w_0 and $\boldsymbol{w}$ and equating to zero yields

$$\sum_{i=1}^{n} \alpha_i y_i = 0 \tag{5.46}$$

$$\boldsymbol{w} = \sum_{i=1}^{n} \alpha_i y_i \boldsymbol{x}_i \tag{5.47}$$

Substituting into (5.45) gives the *dual form* of the Lagrangian

$$L_D = \sum_{i=1}^{n} \alpha_i - \frac{1}{2} \sum_{i=1}^{n} \sum_{j=1}^{n} \alpha_i \alpha_j y_i y_j \boldsymbol{x}_i^T \boldsymbol{x}_j \tag{5.48}$$

which is *maximised* with respect to the α_i subject to

$$\alpha_i \geq 0 \quad \sum_{i=1}^{n} \alpha_i y_i = 0 \tag{5.49}$$

The importance of the dual form is that it expresses the optimisation criterion as inner products of patterns, $\boldsymbol{x}_i$. This is a key concept and has important consequences for nonlinear support vector machines discussed in Chapter 6. The *dual variables* are the Lagrange multipliers, α_i, and so the number of parameters is n, the number of patterns.

5.4.2.1 Karush–Kuhn–Tucker conditions

In the above, we have reformulated the primal problem in an alternative dual form which is often easier to solve numerically. The *Kuhn–Tucker conditions* provide necessary and sufficient conditions to be satisfied when minimising an objective function subject to inequality and equality constraints. In the primal form of the objective function, these are

$$\frac{\partial L_p}{\partial \boldsymbol{w}} = \boldsymbol{w} - \sum_{i=1}^{n} \alpha_i y_i \boldsymbol{x}_i = 0$$

$$\frac{\partial L_p}{\partial w_0} = -\sum_{i=1}^{n} \alpha_i y_i = 0$$

$$y_i(\boldsymbol{x}_i^T \boldsymbol{w} + w_0) - 1 \geq 0$$

$$\alpha_i \geq 0$$

$$\alpha_i(y_i(\boldsymbol{x}_i^T \boldsymbol{w} + w_0) - 1) = 0$$

In particular, the condition $\alpha_i(y_i(\boldsymbol{x}_i^T \boldsymbol{w} + w_0) - 1) = 0$ (known as the Karush–Kuhn–Tucker complementarity condition – product of the Lagrange multiplier and the inequality constraint) implies that for *active* constraints [the solution satisfies $y_i(\boldsymbol{x}_i^T \boldsymbol{w} + w_0) - 1) = 0$] then $\alpha_i \geq 0$; otherwise, for inactive constraints $\alpha_i = 0$. For active constraints, the Lagrange multiplier represents the sensitivity of the optimal value of L_p to the particular constraint (Cristianini and Shawe-Taylor, 2000). These data points with nonzero Lagrange multiplier lie on the canonical hyperplanes. They are termed the *support vectors* and are the most informative points in the dataset. If any of the other patterns (with $\alpha_i = 0$) were to be moved around (provided that they do not cross one of the outer – canonical – hyperplanes), they would not affect the solution for the separating hyperplane.

5.4.2.2 Classification

Re-casting the constrained optimisation problem in its dual form enables numerical quadratic programming solvers to be employed (see Section 5.4.6 for further discussion). Once the Lagrange multipliers, α_i, have been obtained, the value of w_0 may be found from one of the Karush–Kuhn–Tucker complementarity conditions

$$\alpha_i(y_i(\boldsymbol{x}_i^T \boldsymbol{w} + w_0) - 1) = 0$$

using any of the support vectors (patterns for which $\alpha_i \neq 0$). For reasons of numerical stability an average over all support vectors is preferred

$$n_{\mathcal{SV}} w_0 + \boldsymbol{w}^T \sum_{i \in \mathcal{SV}} \boldsymbol{x}_i = \sum_{i \in \mathcal{SV}} y_i \tag{5.50}$$

where $n_{\mathcal{SV}}$ is the number of support vectors, the summations are over the set of support vectors, $\mathcal{SV}$, and we have made use of $y_i^2 = 1$, for any i. The solution for $\boldsymbol{w}$ used in the above is given by (5.47):

$$\boldsymbol{w} = \sum_{i \in \mathcal{SV}} \alpha_i y_i \boldsymbol{x}_i \tag{5.51}$$

since $\alpha_i = 0$ for other patterns. Thus, the support vectors define the separating hyperplane.

A new pattern, $\boldsymbol{x}$, is classified according to the sign of

$$\boldsymbol{w}^T \boldsymbol{x} + w_0$$

Substituting for $\boldsymbol{w}$ and w_0 gives the linear discriminant: assign $\boldsymbol{x}$ to ω_1 if

$$\sum_{i \in \mathcal{SV}} \alpha_i y_i \boldsymbol{x}_i^T \boldsymbol{x} - \frac{1}{n_{\mathcal{SV}}} \sum_{i \in \mathcal{SV}} \sum_{j \in \mathcal{SV}} \alpha_i y_i \boldsymbol{x}_i^T \boldsymbol{x}_j + \frac{1}{n_{\mathcal{SV}}} \sum_{i \in \mathcal{SV}} y_i > 0$$

5.4.3 Linearly nonseparable two-class data

In many real-world practical problems there will be no linear boundary separating the classes and the problem of searching for an optimal separating hyperplane is meaningless. Even if we were to use sophisticated feature vectors, $\boldsymbol{\phi}(\boldsymbol{x})$, to transform the data to a high-dimensional feature space in which classes are linearly separable, this would lead to an overfitting of the data and hence poor generalisation ability. We shall return to nonlinear SVMs in Chapter 6.

However, we can extend the above ideas to handle nonseparable data by relaxing the constraints (5.43). We do this by introducing 'slack' variables ξ_i, $i = 1, \ldots, n$ into the constraints to give

$$\begin{aligned} &\boldsymbol{w}^T \boldsymbol{x}_i + w_0 \geq +1 - \xi_i && \text{for } y_i = +1 \\ &\boldsymbol{w}^T \boldsymbol{x}_i + w_0 \leq -1 + \xi_i && \text{for } y_i = -1 \\ &\xi_i \geq 0 && i = 1, \ldots, n \end{aligned} \tag{5.52}$$

For a point to be misclassified by the separating hyperplane, we must have $\xi_i > 1$ (the concept is illustrated in Figure 5.9).

Analogously to Equation (5.44) for the separable case, we can rewrite the constraints as:

$$\begin{aligned} &y_i(\boldsymbol{w}^T \boldsymbol{x}_i + w_0) \geq 1 - \xi_i && i = 1, \ldots, n \\ &\xi_i \geq 0 && i = 1, \ldots, n \end{aligned} \tag{5.53}$$

A convenient way to incorporate the additional cost due to nonseparability is to introduce an extra cost term to the cost function by replacing $\boldsymbol{w}^T \boldsymbol{w}/2$ by $\boldsymbol{w}^T \boldsymbol{w}/2 + C \sum_i \xi_i$ where C is a

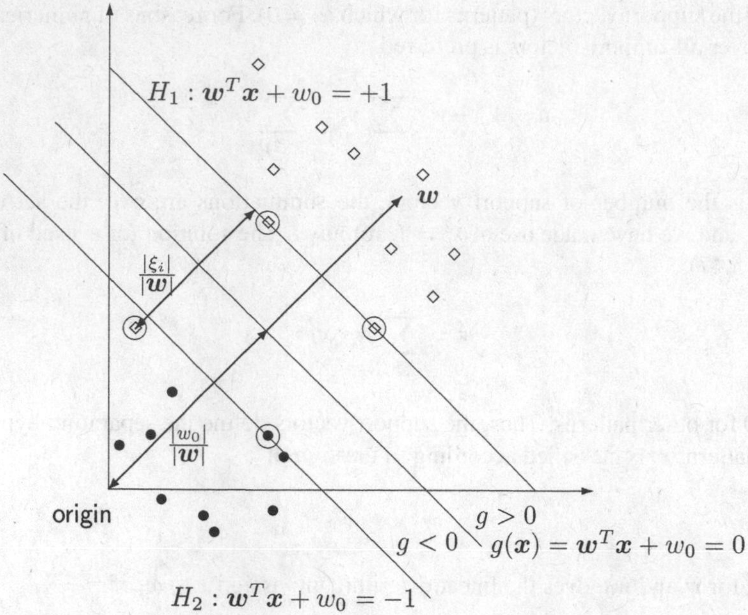

Figure 5.9 Linear separating hyperplane for nonseparable data.

'regularisation' parameter. The term $C\sum_i \xi_i$ can be thought of as measuring some amount of misclassification – the lower the value of C, the smaller the penalty for 'outliers' and a 'softer' margin. Other penalty terms are possible, for example, $C\sum_i \xi_i^2$ (Vapnik, 1998; Cristianini and Shawe-Taylor, 2000).

Thus, we minimise

$$\frac{1}{2}\boldsymbol{w}^T\boldsymbol{w} + C\sum_i \xi_i \tag{5.54}$$

subject to the constraints (5.53). The primal form of the Lagrangian (5.45) now becomes

$$L_p = \frac{1}{2}\boldsymbol{w}^T\boldsymbol{w} + C\sum_i \xi_i - \sum_{i=1}^{n} \alpha_i(y_i(\boldsymbol{w}^T\boldsymbol{x}_i + w_0) - 1 + \xi_i) - \sum_{i=1}^{n} r_i\xi_i \tag{5.55}$$

where $\alpha_i \geq 0$ and $r_i \geq 0$ are Lagrange multipliers; r_i are introduced to ensure positivity of ξ_i.

Differentiating with respect to $\boldsymbol{w}$ and w_0 still results in (5.46) and (5.47)

$$\sum_{i=1}^{n} \alpha_i y_i = 0 \tag{5.56}$$

$$\boldsymbol{w} = \sum_{i=1}^{n} \alpha_i y_i \boldsymbol{x}_i \tag{5.57}$$

and differentiating with respect to ξ_i yields

$$C - \alpha_i - r_i = 0 \tag{5.58}$$

Substituting the results (5.56) and (5.57) above into the primal form (5.55) and using (5.58) gives the dual form of the Lagrangian

$$L_D = \sum_{i=1}^{n} \alpha_i - \frac{1}{2} \sum_{i=1}^{n} \sum_{j=1}^{n} \alpha_i \alpha_j y_i y_j \boldsymbol{x}_i^T \boldsymbol{x}_j \tag{5.59}$$

which is the same form as the maximal margin classifier (5.48). This is maximised with respect to the α_i subject to

$$\sum_{i=1}^{n} \alpha_i y_i = 0 \tag{5.60}$$

$$0 \leq \alpha_i \leq C$$

The latter condition follows from (5.58) and $r_i \geq 0$. Thus, the only change to the maximisation problem is the upper bound on the α_i. Again, numerical quadratic programming solvers can be employed.

The Karush–Kuhn–Tucker complementarity conditions are

$$\alpha_i (y_i (\boldsymbol{x}_i^T \boldsymbol{w} + w_0) - 1 + \xi_i) = 0 \tag{5.61}$$

$$r_i \xi_i = (C - \alpha_i) \xi_i = 0$$

Patterns for which $\alpha_i > 0$ are termed the support vectors, and satisfy $y_i(\boldsymbol{x}_i^T \boldsymbol{w} + w_0) - 1 + \xi_i = 0$. Those satisfying $0 < \alpha_i < C$ must have $\xi_i = 0$ – that is, they lie on one of the canonical hyperplanes at a distance of $1/|\boldsymbol{w}|$ from the separating hyperplane (these support vectors are sometimes termed *margin vectors*). Nonzero slack variables can only occur when $\alpha_i = C$. In this case, the points $\boldsymbol{x}_i$ are misclassified if $\xi_i > 1$. If $\xi_i < 1$, they are classified correctly, but lie closer to the separating hyperplane than $1/|\boldsymbol{w}|$. As in the separable case, the value of w_0 is determined using the first Karush–Kuhn–Tucker complementarity condition [Equation (5.61)] for any support vector for which $0 < \alpha_i < C$ (for which $\xi_i = 0$). For reasons of numerical stability it is usual to take the average over samples for which $0 < \alpha_i < C$. Equations (5.50) and (5.51) become

$$n_{\widetilde{\mathcal{SV}}} w_0 + \boldsymbol{w}^T \sum_{i \in \widetilde{\mathcal{SV}}} \boldsymbol{x}_i = \sum_{i \in \widetilde{\mathcal{SV}}} y_i \tag{5.62}$$

$$\boldsymbol{w} = \sum_{i \in \mathcal{SV}} \alpha_i y_i \boldsymbol{x}_i \tag{5.63}$$

where $\mathcal{SV}$ is the set of support vectors with associated values of α_i satisfying $0 < \alpha_i \leq C$ and $\widetilde{\mathcal{SV}}$ is the set of $n_{\widetilde{\mathcal{SV}}}$ support vectors satisfying $0 < \alpha_i < C$ (those at the target

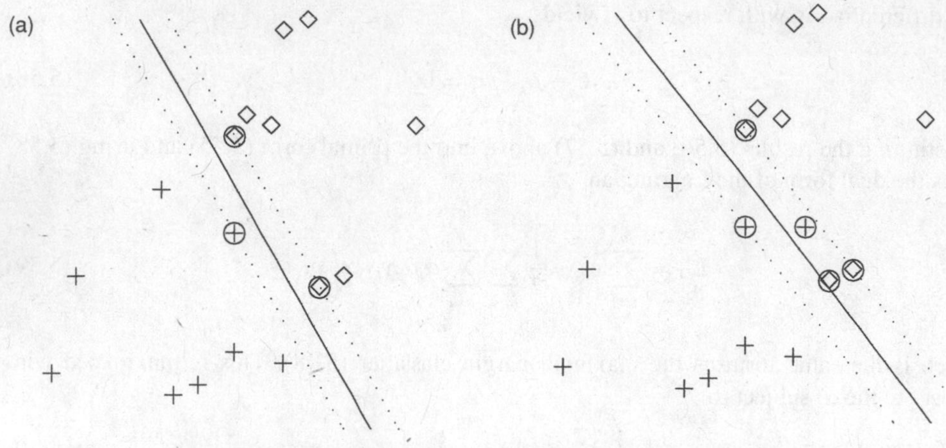

Figure 5.10 Linearly separable data (a) and nonseparable data [(b), $C = 20$].

distance of $1/|\boldsymbol{w}|$ from the separating hyperplane). The resulting linear discriminant is: assign $\boldsymbol{x}$ to ω_1 if:

$$\sum_{i \in \mathcal{SV}} \alpha_i y_i \boldsymbol{x}_i^T \boldsymbol{x} + \frac{1}{n_{\widetilde{\mathcal{SV}}}} \left\{ \sum_{j \in \widetilde{\mathcal{SV}}} y_j - \sum_{i \in \mathcal{SV}, j \in \widetilde{\mathcal{SV}}} \alpha_i y_i \boldsymbol{x}_i^T \boldsymbol{x}_j \right\} > 0$$

Figure 5.10 shows the optimal separating hyperplane for linearly separable and nonseparable data. The support vectors ($\alpha_i > 0$) are circled. All points that are not support vectors lie outside the margin strip ($\alpha_i = 0$, $\xi_i = 0$). In Figure 5.10(b), one of the support vectors (from class +) is incorrectly classified ($\xi_i > 1$).

The only free parameter is the regularisation parameter, C. A value may be chosen by varying C through a range of values and monitoring the performance of the classifier on a separate validation set, or by using cross-validation (see Chapter 13).

5.4.4 Multiclass SVMs

Support vector machines are most commonly applied in multiclass problems either by using the binary classifier in a one-against-all, or one-against-one situation, as described in Section 5.3.1

Alternatively, one can construct C linear discriminant functions simultaneously (Vapnik, 1998). Consider the linear discriminant functions

$$g_k(\boldsymbol{x}) = (\boldsymbol{w}^k)^T \boldsymbol{x} + w_0^k \quad k = 1, \ldots, C$$

We seek a solution for $\{(\boldsymbol{w}^k, w_0^k), k = 1, \ldots, C\}$ such that the decision rule: assign $\boldsymbol{x}$ to class ω_i if

$$g_i(\boldsymbol{x}) = \max_j g_j(\boldsymbol{x})$$

separates the training data without error. That is, there are solutions for $\{(w^k, w_0^k), k = 1, \ldots, C\}$ such that for all $k = 1, \ldots, C$,

$$(\boldsymbol{w}^k)^T \boldsymbol{x} + w_0^k - ((\boldsymbol{w}^j)^T \boldsymbol{x} + w_0^j) \geq 1$$

for all $\boldsymbol{x} \in \omega_k$ and for all $j \neq k$. This means that every pair of classes is separable. If a solution is possible, we seek a solution for which

$$\sum_{k=1}^{C} (\boldsymbol{w}^k)^T \boldsymbol{w}^k$$

is minimal. If the training data cannot be separated, slack variables are introduced and we minimise

$$L = \sum_{k=1}^{C} (\boldsymbol{w}^k)^T \boldsymbol{w}^k + C \sum_{i=1}^{n} \xi_i$$

subject to the constraints

$$(\boldsymbol{w}^k)^T \boldsymbol{x}_i + w_0^k - ((\boldsymbol{w}^j)^T \boldsymbol{x}_i + w_0^j) \geq 1 - \xi_i$$

for all $\boldsymbol{x}_i$ (where $\boldsymbol{x}_i \in \omega_k$), and for all $j \neq k$. The procedure for minimising the quantity L subject to inequality constraints is the same as that developed in the two-class case (Abe, 2010).

5.4.5 SVMs for regression

Support vector machines may also be used for problems in regression. Suppose that we have a dataset $\{(\boldsymbol{x}_i, y_i), i = 1, \ldots, n\}$ of measurements $\boldsymbol{x}_i$ on the independent variables and y_i on the response variables. Instead of the constraints (5.52) we have

$$\begin{array}{ll} (\boldsymbol{w}^T \boldsymbol{x}_i + w_0) - y_i \leq \epsilon + \xi_i & i = 1, \ldots, n \\ y_i - (\boldsymbol{w}^T \boldsymbol{x}_i + w_0) \leq \epsilon + \hat{\xi}_i & i = 1, \ldots, n \\ \xi_i, \hat{\xi} \geq 0 & i = 1, \ldots, n \end{array} \tag{5.64}$$

This allows a deviation between the target values y_i and the function f,

$$f(\boldsymbol{x}) = \boldsymbol{w}^T \boldsymbol{x} + w_0.$$

Two slack variables are introduced: one (ξ) for exceeding the target value by more than ϵ and the other ($\hat{\xi}$) for being more than ϵ below the target value (Figure 5.11).

As in the classification case, a loss function is minimised subject to the constraints (5.64). For a linear ϵ-insensitive loss, we minimise

$$\frac{1}{2} \boldsymbol{w}^T \boldsymbol{w} + C \sum_{i=1}^{n} (\xi_i + \hat{\xi}_i)$$

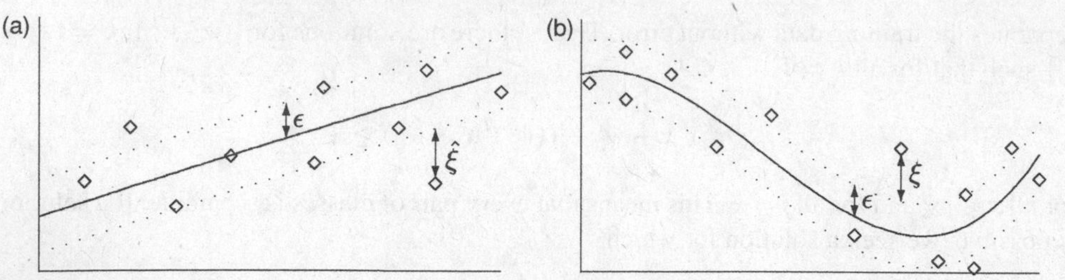

Figure 5.11 Linear (a) and nonlinear (b) SVM regression. The variables ξ and $\hat{\xi}$ measure the cost of lying outside the 'ϵ-insensitive band' around the regression function.

The primal form of the Lagrangian is

$$L_p = \frac{1}{2}\boldsymbol{w}^T\boldsymbol{w} + C\sum_{i=1}^{n}(\xi_i + \hat{\xi}_i) - \sum_{i=1}^{n}\alpha_i(\xi_i + \epsilon - (\boldsymbol{w}^T\boldsymbol{x}_i + w_0 - y_i)) - \sum_{i=1}^{n} r_i\xi_i$$

$$- \sum_{i=1}^{n}\hat{\alpha}_i(\hat{\xi}_i + \epsilon - (y_i - \boldsymbol{w}^T\boldsymbol{x}_i - w_0)) - \sum_{i=1}^{n}\hat{r}_i\hat{\xi}_i \tag{5.65}$$

where $\alpha_i, \hat{\alpha}_i \geq 0$ and $r_i, \hat{r}_i \geq 0$ are Lagrange multipliers. Differentiating with respect to $\boldsymbol{w}$, w_0, ξ_i and $\hat{\xi}_i$ gives

$$\boldsymbol{w} + \sum_{i=1}^{n}(\alpha_i - \hat{\alpha}_i)\boldsymbol{x}_i = 0$$

$$\sum_{i=1}^{n}(\alpha_i - \hat{\alpha}_i) = 0 \tag{5.66}$$

$$C - \alpha_i - r_i = 0$$

$$C - \hat{\alpha}_i - \hat{r}_i = 0$$

Substituting for $\boldsymbol{w}$ into Equation (5.65) and using the relations above gives the dual form

$$L_D = \sum_{i=1}^{n}(\hat{\alpha}_i - \alpha_i)y_i - \epsilon\sum_{i=1}^{n}(\hat{\alpha}_i + \alpha_i) - \frac{1}{2}\sum_{i=1}^{n}\sum_{j=1}^{n}(\hat{\alpha}_i - \alpha_i)(\hat{\alpha}_j - \alpha_j)\boldsymbol{x}_i^T\boldsymbol{x}_j \tag{5.67}$$

This is maximised subject to

$$\sum_{i=1}^{n}(\alpha_i - \hat{\alpha}_i) = 0 \tag{5.68}$$

$$0 \leq \alpha_i, \hat{\alpha}_i \leq C$$

which follows from (5.66) and $r_i, \hat{r}_i \geq 0$. The Karush–Kuhn–Tucker complementarity conditions are

$$\begin{aligned} \alpha_i(\xi_i + \epsilon - (\boldsymbol{w}^T \boldsymbol{x}_i + w_0 - y_i)) &= 0 \\ \hat{\alpha}_i(\hat{\xi}_i + \epsilon - (y_i - \boldsymbol{w}^T \boldsymbol{x}_i - w_0)) &= 0 \\ r_i\xi_i = (C - \alpha_i)\xi_i &= 0 \\ \hat{r}_i\hat{\xi}_i = (C - \hat{\alpha}_i)\hat{\xi}_i &= 0 \end{aligned} \tag{5.69}$$

These imply $\alpha_i\hat{\alpha}_i = 0$ and $\xi_i\hat{\xi}_i = 0$. To see that $\alpha_i\hat{\alpha}_i = 0$, note that if both $\alpha_i \neq 0$ and $\hat{\alpha}_i \neq 0$, the first two conditions in Equation (5.69) imply that $\xi_i + \epsilon - (\boldsymbol{w}^T\boldsymbol{x}_i + w_0 - y_i) = 0$ and $\hat{\xi}_i + \epsilon - (y_i - \boldsymbol{w}^T\boldsymbol{x}_i - w_0) = 0$. However, adding these two equations would give $\xi_i + \hat{\xi}_i + 2\epsilon = 0$, which is not possible, since $\epsilon > 0$ and $\xi_i, \hat{\xi}_i \geq 0$. Hence either $\alpha_i = 0$ or $\hat{\alpha}_i = 0$. That $\xi_i\hat{\xi}_i = 0$ follows from the bottom two conditions in Equation (5.69) together with $\alpha_i\hat{\alpha}_i = 0$.

Those patterns $\boldsymbol{x}_i$ with $\alpha_i > 0$ or $\hat{\alpha}_i > 0$ are support vectors. If $0 < \alpha_i < C$ or $0 < \hat{\alpha}_i < C$ then $(\boldsymbol{x}_i, y_i)$ lies on the boundary of the tube surrounding the regression function at distance ϵ. In the $0 < \alpha_i < C$ case this can be seen by first noting that the third condition in Equation (5.69) is such that $0 < \alpha_i < C$ implies that $\xi_0 = 0$. Using $\alpha_i \neq 0$ and $\xi_0 = 0$ in the first condition then implies that $\epsilon + y_i = \boldsymbol{w}^T\boldsymbol{x}_i + w_0$, as stated. Similarly for the second case. If $\alpha_i = C$ or $\hat{\alpha}_i = C$, then the point lies outside the tube. If $\alpha_i = 0$ and $\hat{\alpha}_i = 0$ then the point lies inside the tube (and is not a support vector).

Having optimised the parameters α_i and $\hat{\alpha}_i$ using a suitable numerical quadratic programming solver, the solution for $f(\boldsymbol{x})$ is then

$$f(\boldsymbol{x}) = \sum_{i=1}^{n} (\hat{\alpha}_i - \alpha_i)\boldsymbol{x}_i^T\boldsymbol{x} + w_0 \tag{5.70}$$

using the expression for $\boldsymbol{w}$ in (5.66). The parameter w_0 is chosen so that

$$\begin{aligned} f(\boldsymbol{x}_i) - y_i &= \epsilon \quad \text{for any } i \text{ with } 0 < \alpha_i < C \\ \text{or} \quad f(\boldsymbol{x}_i) - y_i &= -\epsilon \quad \text{for any } i \text{ with } 0 < \hat{\alpha}_i < C \end{aligned}$$

by the Karush–Kuhn–Tucker complementarity conditions above. As with the classification SVMs, for numerical reasons it is usual to average the expression for w_0 over all patterns for which $0 < \alpha_i < C$ or $0 < \hat{\alpha}_i < C$.

5.4.6 Implementation

There are many freely available and commercial software packages for solving quadratic programming optimisation problems, often based on standard numerical methods of nonlinear optimisation that iteratively hill-climb to find the maximum of the objective function. For very large datasets, however, they become impractical. Traditional quadratic programming algorithms require that the kernel be computed and stored in memory and can involve expensive matrix operations. There are many developments to handle large datasets. *Decomposition* methods (see Algorithm 5.1) apply the standard optimisation package to a fixed subset of the

Algorithm 5.1 The decomposition algorithm.

1. Set b, the size of the subset ($b < n$, the total number of patterns). Set $\alpha_i = 0$ for all patterns.
2. Choose b patterns from the training dataset to form a subset $\mathcal{B}$.
3. Solve the quadratic programming problem defined by the subset $\mathcal{B}$ using a standard routine.
4. Apply the model to all patterns in the training set.
5. If there are any patterns that do not satisfy the Karush–Kuhn–Tucker conditions, replace any patterns in $\mathcal{B}$ and the corresponding α_i with these patterns and their α_i values.
6. If not converged, go to step 3.

data and revise the subset in the light of applying the classification/regression model learned to the training data not in the subset.

Sequential minimal optimisation

A special development of the decomposition algorithm is the sequential minimal optimisation (SMO) algorithm (Platt, 1998), which optimises a subset of two parameters (each corresponding to a different pattern) at a time, for which the optimisation admits an analytic solution.

Each optimisation step consists of optimising two of the parameters (without loss of generality α_1 and α_2 in the following discussion), whilst keeping the remaining parameters $\alpha_3, \ldots, \alpha_n$ unchanged. The linear equality constraint [Equation (5.60)] is preserved in the optimisation. For this to happen the following must hold:

$$\alpha_1' y_1 + \alpha_2' y_2 = -\sum_{i=3}^{n} \alpha_i y_i = \alpha_1 y_1 + \alpha_2 y_2 \tag{5.71}$$

where α_1' and α_2' are the updated parameters.

Now for the binary classification problem, $y_1, y_2 \in \{-1, 1\}$. Hence (5.71) and $0 \leq \alpha_1'$, $\alpha_2' \leq C$ give the following bounds for α_2':

$$L \leq \alpha_2' \leq H$$

where:

$$L = \begin{cases} \max(0, \alpha_2 - \alpha_1) & \text{if} \quad y_1 \neq y_2 \\ \max(0, \alpha_1 + \alpha_2 - C) & \text{if} \quad y_1 = y_2 \end{cases}$$

and:

$$H = \begin{cases} \min(C, C + \alpha_2 - \alpha_1) & \text{if} \quad y_1 \neq y_2 \\ \min(C, \alpha_1 + \alpha_2) & \text{if} \quad y_1 = y_2 \end{cases}$$

As is derived in Cristianini and Shawe-Taylor (2000) (see Exercises) the optimal value for α'_2 is

$$\alpha'_2 = \max\left(\min\left(\alpha_2 + \frac{y_2(E_1 - E_2)}{\boldsymbol{x}_1^T\boldsymbol{x}_1 + \boldsymbol{x}_2^T\boldsymbol{x}_2 - 2\boldsymbol{x}_1^T\boldsymbol{x}_2}, H\right), L\right) \tag{5.72}$$

where for $j = 1, 2$

$$E_j = g(\boldsymbol{x}_j) - y_j = (\boldsymbol{w}^T\boldsymbol{x}_j + w_0) - y_j = \left(\sum_{i=1}^{n} \alpha_i y_i \boldsymbol{x}_i^T\boldsymbol{x}_j + w_0\right) - y_j \tag{5.73}$$

Having determined α'_2, Equation (5.71) can be used to obtain the following optimal value of α'_1

$$\alpha'_1 = \alpha_1 + y_1 y_2(\alpha_2 - \alpha'_2)$$

(here we have made use of the fact that $y^2{}_i = 1$, for any i).

Using Equation (5.63) the weight vector $\boldsymbol{w}$ is updated to $\boldsymbol{w}'$ according to

$$\boldsymbol{w}' = \boldsymbol{w} + (\alpha'_1 - \alpha_1)y_1\boldsymbol{x}_1 + (\alpha'_2 - \alpha_2)y_2\boldsymbol{x}_2$$

The bias parameter w_0 is updated so that the Karush–Kuhn–Tucker complementarity conditions of Equation (5.61) are satisfied for the updated parameters. If $0 < \alpha'_1 < C$ then w_0 is updated as follows:

$$w'_0 = -E_1 - (\alpha'_1 - \alpha_1)y_1\boldsymbol{x}_1^T\boldsymbol{x}_1 - (\alpha'_2 - \alpha_2)y_2\boldsymbol{x}_2^T\boldsymbol{x}_1 + w_0 \tag{5.74}$$

If $\alpha'_1 \in \{0, C\}$, whilst $0 < \alpha'_2 < C$, then the update rule becomes:

$$w'_0 = -E_2 - (\alpha'_1 - \alpha_1)y_1\boldsymbol{x}_1^T\boldsymbol{x}_2 - (\alpha'_2 - \alpha_2)y_2\boldsymbol{x}_2^T\boldsymbol{x}_2 + w_0 \tag{5.75}$$

If both $0 < \alpha'_1 < C$ and $0 < \alpha'_2 < C$ then the expressions in Equations (5.74) and (5.75) will be equal. If neither $0 < \alpha'_1 < C$ nor $0 < \alpha'_2 < C$ (i.e. $\alpha'_1, \alpha'_2 \in \{0, C\}$) then the bias is set to be the average of the values given by Equations (5.74) and (5.75).

The SVM outputs can then be updated for all patterns using the following:

$$g'(\boldsymbol{x}_i) = g(\boldsymbol{x}_i) + (\alpha'_1 - \alpha_1)y_1\boldsymbol{x}_1^T\boldsymbol{x}_i + (\alpha'_2 - \alpha_2)y_2\boldsymbol{x}_2^T\boldsymbol{x}_i + (w'_0 - w_0)$$

This allows the errors E_i to be updated.

The SMO algorithm successively updates pairs of parameters until convergence is reached. Heuristics are used to select these pairs. At each iteration the first pattern in the pair is set to be one that does not satisfy the Karush–Kuhn–Tucker complementarity conditions of Equation (5.61).[2] Initially, all patterns violating the Karush–Kuhn–Tucker conditions are selected in

[2] In most applications it is only necessary to have the conditions satisfied to within a small tolerance.

turn. However, once a complete pass through the dataset is completed, violating patterns that also belong to the set $\widetilde{\mathcal{SV}}$ (i.e. which satisfy $0 < \alpha_i < C$) are selected preferentially.

Suppose that the first selected parameter is α_{i_1} (i.e. the i_1th pattern). A second heuristic is used to determine the second parameter, α_{i_2}, to optimise. Specifically, i_2 is selected to maximise $|E_{i_1} - E_{i_2}|$, where the E_i are as specified in Equation (5.73) evaluated using the most recent weight vector and bias parameter expressions. In practice the procedure chooses i_2 such that E_{i_2} is the minimum in the set $\{E_i, i \neq i_1, i \in \widetilde{\mathcal{SV}}\}$ if $E_{i,1}$ positive, and such that E_{i_2} is the maximum in the set $\{E_i, i \neq i_1, i \in \widetilde{\mathcal{SV}}\}$ if $E_{i,1}$ is negative. Further details, including a number of special cases, are available in the pseudocode within Platt (1998) and Cristianini and Shawe-Taylor (2000).

Initialisation of the procedure consists of setting $\alpha_i = 0$ for all patterns (giving $\boldsymbol{w} = 0$ and $w_0 = 0$). Termination occurs when the Karush–Kuhn–Tucker complementarity conditions are met to within a tolerance for all patterns. Cristianini and Shawe-Taylor (2000) discuss a number of different convergence criteria, applicable to a range of SVM optimisation procedures.

Extension of the SMO algorithm to SVM regression problems is discussed in Cristianini and Shawe-Taylor (2000). Improvements to the SMO bias update procedure have been made by Keerthi *et al.* (2001). Fan *et al.* (2005) (see also Chang and Lin, 2011) use second-order information to improve the convergence time of an SMO algorithm.

5.4.7 Example application study

The problem

This application involves classification of land cover using remote sensing satellite imagery (Brown *et al.*, 2000).

Summary

A conventional classification technique developed in the remote sensing community (linear spectral mixture models) is compared, both theoretically and practically, with SVMs. Under certain circumstances the methods are equivalent.

The data

The data comprise measurements of two classes of land cover: developed and other (including slate, tarmac, concrete) and undeveloped and vegetation (including sand, water, soil, grass, shrubs, etc). The measurements are Landsat images of the Leicester, UK, suburbs in two frequency bands. Each image is 33×33 pixels in size and a pattern is a two-dimensional pair of measurements of corresponding pixels in each of the two bands. Training and test sets were constructed, the training set consisting of 'pure' pixels (those that relate to a region for which there is a single class).

The model

An SVM model was adopted. Linear separable and linear nonseparable SVMs were trained [by maximising (5.45) and (5.55)] using patterns selected from the training set.

Training procedure

The value of the regularisation parameter, C, was chosen to minimise a sum-squared-error criterion evaluated on the test set.

5.4.8 Summary

Support vector machines have been receiving increasing research interest in recent years. They provide an optimally separating hyperplane in the sense that the margin between two groups is maximised. Generalisation of this idea to the nonlinear classifier, discussed in Chapter 6, has led to classifiers with remarkable good generalisation ability.

5.5 Logistic discrimination

In the previous sections, discrimination is performed using values of a linear function of the data sample $\boldsymbol{x}$ [or the transformed data samples $\boldsymbol{\phi}(\boldsymbol{x})$]. We continue this theme here. We shall introduce logistic discrimination for the two-class case first and then consider the multigroup situation.

5.5.1 Two-class case

The basic assumption is that the difference between the logarithms of the class-conditional density functions is linear in the variables $\boldsymbol{x}$:

$$\log\left(\frac{p(\boldsymbol{x}|\omega_1)}{p(\boldsymbol{x}|\omega_2)}\right) = \beta_0 + \boldsymbol{\beta}^T\boldsymbol{x} \tag{5.76}$$

This model is an exact description in a wide variety of situations including (Anderson, 1982):

1. when the class-conditional densities are multivariate normal with equal covariance matrices;
2. multivariate discrete distributions following a loglinear model with equal interaction terms between classes;
3. when situations 1 and 2 are combined: both continuous and categorical variables describe each sample.

Hence, the assumption is satisfied by many families of distributions and has been found to be applicable to a wide range of real datasets that depart from normality.

It is a simple matter to show from (5.76) that the assumption is equivalent to

$$\begin{aligned} p(\omega_2|\boldsymbol{x}) &= \frac{1}{1+\exp\left(\beta_0' + \boldsymbol{\beta}^T\boldsymbol{x}\right)} \\ p(\omega_1|\boldsymbol{x}) &= \frac{\exp\left(\beta_0' + \boldsymbol{\beta}^T\boldsymbol{x}\right)}{1+\exp\left(\beta_0' + \boldsymbol{\beta}^T\boldsymbol{x}\right)} \end{aligned} \tag{5.77}$$

where $\beta_0' = \beta_0 + \log(p(\omega_1)/p(\omega_2))$.

Discrimination between two classes depends on the ratio $p(\omega_1|\boldsymbol{x})/p(\omega_2|\boldsymbol{x})$,

$$\text{assign } \boldsymbol{x} \text{ to } \begin{cases} \omega_1 \\ \omega_2 \end{cases} \text{ if } \frac{p(\omega_1|\boldsymbol{x})}{p(\omega_2|\boldsymbol{x})} \begin{cases} > \\ < \end{cases} 1$$

and substituting the expressions (5.77), we see that the decision about discrimination is determined solely by the linear function $\beta_0' + \beta^T x$ and is given by

$$\text{assign } x \text{ to } \begin{cases} \omega_1 \\ \omega_2 \end{cases} \text{if } \beta_0' + \beta^T x \begin{cases} > \\ < \end{cases} 0$$

This is an identical rule to that given in Section 5.2 on linear discrimination, and we gave several procedures for estimating the parameters. The only difference here is that we are assuming a specific model for the ratio of the class-conditional densities that leads to this discrimination rule, rather than specifying the rule *a priori*. Another difference is that we may use the models for the densities (5.77) to obtain maximum likelihood estimates for the parameters.

5.5.2 Maximum likelihood estimation

The parameters of the logistic discrimination model may be estimated using a maximum likelihood approach (Day and Kerridge, 1967; Anderson, 1982). An iterative nonlinear optimisation scheme may be employed using the likelihood function and its derivatives.

The estimation procedure depends on the sampling scheme used to generate the labelled training data (Anderson, 1982; McLachlan, 1992a) and three common sampling designs are considered by Anderson. These are (i) sampling from the mixture distribution (i.e. randomly selected data from the full distribution over all classes), (ii) sampling conditional on x in which x is fixed and one or more samples of the observation are taken (which may be ω_1 or ω_2 in a two-class discrimination problem) and (iii) separate sampling for each class in which the conditional distributions, $p(x|\omega_i)$, $i = 1, 2$, are sampled. Maximum likelihood estimates of $\boldsymbol{\beta}$ are independent of the sampling scheme though one of the sampling designs considered (separate sampling from each class) derives estimates for β_0 rather than β_0' (which is the term required for discrimination). We assume in our derivation below a mixture sampling scheme, which arises when a random sample is drawn from a mixture of the classes. Each of the sampling schemes above is discussed in detail by McLachlan (1992a).

The likelihood of the observations is

$$L = \prod_{r=1}^{n_1} p(x_{1r}|\omega_1) \prod_{r=1}^{n_2} p(x_{2r}|\omega_2) \quad r = 1, \ldots, n_s;\ s = 1, 2.$$

where $x_{sr} (s = 1, 2; r = 1, \ldots, n_s)$ are the observations in class ω_s.

This may be rewritten as

$$\begin{aligned} L &= \prod_{r=1}^{n_1} p(\omega_1|x_{1r}) \frac{p(x_{1r})}{p(\omega_1)} \prod_{r=1}^{n_2} p(\omega_2|x_{2r}) \frac{p(x_{2r})}{p(\omega_2)} \\ &= \frac{1}{p(\omega_1)^{n_1} p(\omega_2)^{n_2}} \prod_{\text{all } x} p(x) \prod_{r=1}^{n_1} p(\omega_1|x_{1r}) \prod_{r=1}^{n_2} p(\omega_2|x_{2r}) \end{aligned}$$

The factor

$$\frac{1}{p(\omega_1)^{n_1}p(\omega_2)^{n_2}}\prod_{\text{all } \boldsymbol{x}} p(\boldsymbol{x})$$

is independent of the parameters of the model [the assumption in Anderson (1982) and Day and Kerridge (1967) is that we are free to choose $p(\boldsymbol{x})$; the only assumption we have made is on the log-likelihood ratio]. Therefore maximising the likelihood L is equivalent to maximising

$$L' = \prod_{r=1}^{n_1} p(\omega_1|\boldsymbol{x}_{1r}) \prod_{r=1}^{n_2} p(\omega_2|\boldsymbol{x}_{2r})$$

or

$$\log(L') = \sum_{r=1}^{n_1} \log(p(\omega_1|\boldsymbol{x}_{1r})) + \sum_{r=1}^{n_2} \log(p(\omega_2|\boldsymbol{x}_{2r}))$$

and using the functional forms (5.77)

$$\log(L') = \sum_{r=1}^{n_1} (\beta_0' + \boldsymbol{\beta}^T\boldsymbol{x}_{1r}) - \sum_{\text{all } \boldsymbol{x}} \log\{1 + \exp(\beta_0' + \boldsymbol{\beta}^T\boldsymbol{x})\}$$

The gradient of $\log(L')$ with respect to the parameters β_j is

$$\frac{\partial \log L'}{\partial \beta_0'} = n_1 - \sum_{\text{all } \boldsymbol{x}} p(\omega_1|\boldsymbol{x})$$

$$\frac{\partial \log L'}{\partial \beta_j} = \sum_{r=1}^{n_1} (\boldsymbol{x}_{1r})_j - \sum_{\text{all } \boldsymbol{x}} p(\omega_1|\boldsymbol{x})x_j \quad j = 1, \ldots, p$$

Having written down an expression for the likelihood and its derivative, we may now use a nonlinear optimisation procedure to obtain a set of parameter values for which the function $\log(L')$ attains a local maximum. First of all, we need to specify initial starting values for the parameters. Anderson recommends taking zero as a starting value for all $p + 1$ parameters, $\beta_0', \beta_1, \ldots, \beta_p$. Except in two special cases (see below), the likelihood has a unique maximum attained for finite $\boldsymbol{\beta}$ (Anderson, 1982; Albert and Lesaffre, 1986). Hence, the starting point is in fact immaterial.

If the two classes are separable, then there are nonunique maxima at infinity. At each stage of the optimisation procedure, it is easy to check whether $\beta_0' + \boldsymbol{\beta}^T\boldsymbol{x}$ gives complete separation. If it does, then the algorithm may be terminated. The second situation when L does not have a unique maximum at a finite value of $\boldsymbol{\beta}$ occurs with discrete data when the proportions for one of the variables are zero for one of the values. In this case, the maximum value of L is at infinity. Anderson (1974) suggests a procedure for overcoming this difficulty, based on the assumption that the variable is conditionally independent of the remaining variables in each class.

5.5.3 Multiclass logistic discrimination

In the multiclass discrimination problem, the basic assumption is that, for C classes

$$\log\left(\frac{p(\boldsymbol{x}|\omega_s)}{p(\boldsymbol{x}|\omega_C)}\right) = \beta_{s0} + \boldsymbol{\beta}_s^T\boldsymbol{x}, \quad s = 1, \ldots, C-1$$

that is, the log-likelihood ratio is linear for any pair of likelihoods. Again, we may show that the posterior probabilities are of the form

$$p(\omega_s|\boldsymbol{x}) = \frac{\exp\left(\beta'_{s0} + \boldsymbol{\beta}_s^T\boldsymbol{x}\right)}{1 + \sum_{s=1}^{C-1}\exp\left(\beta'_{s0} + \boldsymbol{\beta}_s^T\boldsymbol{x}\right)} \quad s = 1, \ldots, C-1$$

$$p(\omega_C|\boldsymbol{x}) = \frac{1}{1 + \sum_{s=1}^{C-1}\exp\left(\beta'_{s0} + \boldsymbol{\beta}_s^T\boldsymbol{x}\right)}$$

where $\beta'_{s0} = \beta_{s0} + \log(p(\omega_s)/p(\omega_C))$. Also, the decision rule about discrimination depends solely on the linear functions $\beta'_{s0} + \boldsymbol{\beta}_j^T\boldsymbol{x}$ and the rule is: assign $\boldsymbol{x}$ to class ω_j if

$$\max\,\{\beta'_{s0} + \boldsymbol{\beta}_s^T\boldsymbol{x}\} = \beta'_{j0} + \boldsymbol{\beta}_j^T\boldsymbol{x} > 0, \quad s = 1, \ldots, C-1$$

otherwise assign $\boldsymbol{x}$ to class ω_C.

The likelihood of the observations is given by

$$L = \prod_{i=1}^{C}\prod_{r=1}^{n_i} p(\boldsymbol{x}_{ir}|\omega_i) \tag{5.78}$$

using the notation given previously. As in the two-class case, maximising L is equivalent to maximising

$$\log(L') = \sum_{s=1}^{C}\sum_{r=1}^{n_s}\log(p(\omega_s|\boldsymbol{x}_{sr})) \tag{5.79}$$

with derivatives

$$\frac{\partial \log L'}{\partial \beta'_{j0}} = n_j - \sum_{\text{all } \boldsymbol{x}} p(\omega_j|\boldsymbol{x})$$

$$\frac{\partial \log L'}{\partial (\beta_j)_l} = \sum_{r=1}^{n_j}(\boldsymbol{x}_{jr})_l - \sum_{\text{all } \boldsymbol{x}} p(\omega_j|\boldsymbol{x})x_l$$

Again, for separable classes, the maximum of the likelihood is achieved at a point at infinity in the parameter space, but the algorithm may be terminated when complete separation occurs. Also, zero marginal sample proportions cause maxima at infinity and the procedure of Anderson may be employed.

5.5.4 Example application study

The problem
This application is to predict in the early stages of pregnancy the feeding method (bottle or breast) a woman would use after giving birth (Cox and Pearce, 1997).

Summary
A 'robust' two-group logistic discriminant rule was developed and compared with the ordinary logistic discriminant. Both methods gave similar (good) performance.

The data
Data were collected on 1200 pregnant women from two district general hospitals. Eight variables were identified as being important to the feeding method: presence of children under 16 years of age in the household, housing tenure, taught at school about feeding babies, feeding intention, frequency of seeing own mother, feeding advice from relatives, how woman was fed, previous experience of breast feeding.

Some patterns were excluded from the analysis for various reasons: incomplete information, miscarriage, termination, refusal and delivery elsewhere. This left 937 cases for parameter estimation.

The model
Two models were assessed: the two-group ordinary logistic discrimination model [Equation (5.76)] and a robust logistic discrimination model, designed to reduce the effect of outliers on the discriminant rule,

$$\frac{p(\boldsymbol{x}|\omega_1)}{p(\boldsymbol{x}|\omega_2)} = \frac{c_1 + c_2 \exp[\beta_0 + \boldsymbol{\beta}^T \boldsymbol{x}]}{1 + \exp[\beta_0 + \boldsymbol{\beta}^T \boldsymbol{x}]}$$

where c_1 and c_2 are fixed positive constants.

Training procedure
The prior probabilities $p(\omega_1)$ and $p(\omega_2)$ are estimated from the data and c_1 and c_2 specified. This is required for the robust model, although it can be incorporated with β_0 is the standard model (see Section 5.5.1). Two set of experiments were performed, both determining maximum likelihood estimates for the parameters: training on the full 937 cases and testing on the same data; training on 424 cases from one hospital and testing on 513 cases from the second hospital.

Results
Both model gave similar performance, classifying around 85% cases correctly.

5.5.5 Further developments

Further developments of the basic logistic discriminant model have been to robust (i.e. reducing the effect of outliers) procedures (Cox and Ferry, 1993; Cox and Pearce, 1997) and to more general models. Several other discrimination methods are based on models of discriminant functions that are nonlinear functions of linear projections. These include the multilayer perceptron (Chapter 6), and *projection pursuit* (Friedman and Tukey, 1974; Friedman and Stuetzle, 1981; Friedman *et al.*, 1984; Huber, 1985; Friedman, 1987; Jones and Sibson, 1987) in which the linear projection and the form of the nonlinear function are simultaneously determined.

5.5.6 Summary

Logistic discrimination makes assumptions about the log-likelihood ratios of one population relative to a reference population. As with the methods discussed in the previous sections, discrimination is made by considering a set of values formed from linear transformations of the explanatory variables. These linear transformations are determined by a maximum likelihood procedure. It is a technique which lies between the linear techniques of Chapters 1 and 5 and the nonlinear methods of Chapter 6 in that it requires a nonlinear optimisation scheme to estimate the parameters (and the posterior probabilities are nonlinear functions of the explanatory variables), but discrimination is made using a linear transformation.

One of the advantages of the maximum likelihood approach is that asymptotic results regarding the properties of the estimators may readily be derived. Logistic discrimination has further advantages as itemised by Anderson (1982):

1. It is appropriate for both continuous and discrete-valued variables.
2. It is easy to use.
3. It is applicable over a wide range of distributions.
4. It has a relatively small number of parameters (unlike some of the nonlinear models discussed in Chapter 6).

5.6 Application studies

There have been several studies comparing logistic discrimination with LDA (Press and Wilson, 1978; Bull and Donner, 1987). Logistic discrimination has been found to work well in practice, particularly for data that depart significantly from normality – something that occurs often in practice. The *Statlog* project (Michie *et al.*, 1994) compared a range of classification methods on various datasets. It reports that there is little practical difference between linear and logistic discrimination. Both methods were in the top five algorithms.

There are many applications of SVMs, primarily concerned with the nonlinear variant, reported in Chapter 6. However, in a communications example, SVMs have been used successfully to implement a decision feedback equaliser (to combat distortion and interference), giving superior performance to the conventional minimum mean-squared-error approach (S. Chen *et al.*, 2000).

5.7 Summary and discussion

The discriminant functions discussed in the previous section are all linear in the components of $\boldsymbol{x}$ or the transformed variables $\phi_i(\boldsymbol{x})$ (generalised linear discriminant functions). We have described several approaches for determining the parameters of the model (error-correction schemes, least-squares optimisation, logistic model), but we have regarded the initial transformation, $\boldsymbol{\phi}$, as being prescribed. Some possible choices for the functions $\phi_i(\boldsymbol{x})$ were given in Chapter 1 and there are many other parametric and nonparametric forms that may be used, as we shall see in the next chapter. But how should we choose the functions Φ_i? A good choice for the Φ_i will lead to better classification performance for a subsequent linear classifier than simply applying a linear classifier to the variables x_i. On the other hand, if we were to use a complex nonlinear classifier after the initial transformation of the variables, then the choice for the Φ_i may not be too critical. Any inadequacies in a poor choice may be compensated for by the subsequent classification process. However, in general, if we have knowledge about which variables or transformations are useful for discrimination, then we should use it, rather than hope that our classifier will 'learn' the important relationships.

Many sets of basis functions have been proposed, and the more basis functions we use in our classifier, the better we might expect our classifier to perform. This is not necessarily the case, since increasing the dimension of the vector $\boldsymbol{\phi}$, by the inclusion of more basis functions, leads to more parameters to estimate in the subsequent classification stage. Although error rate on the training set may in fact decrease, the true error rate may increase as generalisation performance deteriorates.

For polynomial basis functions, the number of terms increases rapidly with the order of the polynomial, restricting such an approach to polynomials of low order. However, an important development is that of SVMs that replaces the need to calculate D-dimensional feature vectors $\boldsymbol{\phi}(\boldsymbol{x})$ with the evaluation of a kernel $K(\boldsymbol{x}, \boldsymbol{y})$ at points $\boldsymbol{x}$ and $\boldsymbol{y}$ in the training set.

In the next chapter, we turn to discriminant functions that are linear combinations of nonlinear functions, $\boldsymbol{\phi}$ (i.e. generalised linear discriminant functions). The radial basis function network defines the nonlinear function *explicitly*. It is usually of a prescribed form with parameters set as part of the optimisation process. They are usually determined through a separate procedure and the linear parameters are obtained using the procedures of this chapter. The SVM defines the nonlinear function *implicitly*, through the specification of a kernel function.

5.8 Recommendations

1. Linear schemes provide a baseline from which more sophisticated methods may be judged. They are easy to implement and should be considered before more complex methods.

2. The error correction procedure, or an SVM, can be used to test for separability. This might be important for high-dimensional datasets where classes may be separable due to the finite training set size. A classifier that achieves linear separability on a training set does not necessarily mean good generalisation performance.

3. Linear discriminant analysis, which is the multiclass extension of Fisher's linear discriminant, is recommended.

5.9 Notes and references

The theory of algorithms for linear discrimination is well developed. The books by Nilsson (1965) and Duda *et al.* (2001) provide descriptions of the most commonly used algorithms; see also Ho and Agrawala (1968) and Kashyap (1970). A treatment of the perceptron can be found in the book by Minsky and Papert (1988).

Logistic discrimination is described in the survey article by Anderson (1982) and the book by McLachlan (1992a).

Support vector machines were introduced by Vapnik and co-workers. The book by Vapnik (1998) provides a very good description with historical perspective. Cristianini and Shawe-Taylor (2000) present an introduction to SVMs aimed at students and practitioners. Burges (1998) provides a very good tutorial on SVMs for pattern recognition; see also Abe (2010).

Exercises

1. By considering two arbitrary points lying on the decision surface

$$\boldsymbol{w}^T\boldsymbol{x} + w_0 = 0$$

show that $\boldsymbol{w}$ is perpendicular to the surface. By considering the closest point in the surface to the origin, show that the perpendicular distance of the hyperplane from the origin is $|w_0|/|\boldsymbol{w}|$. Show that the distance of a pattern, $\boldsymbol{x}$, from the hyperplane is given by $|r|$, where

$$r = \frac{g(\boldsymbol{x})}{|\boldsymbol{w}|} = \left(\boldsymbol{w}^T\boldsymbol{x} + w_0\right)/|\boldsymbol{w}|$$

Hint: write $\boldsymbol{x}$ as a sum of two components: its projection onto the hyperplane, $\boldsymbol{x}_h$, and a component normal to the hyperplane.

2. Linear programming or linear optimisation techniques are procedures for maximising linear functions subject to equality and inequality constraints. Specifically, we find the vector $\boldsymbol{x}$ such that

$$z = \boldsymbol{a}_0^T\boldsymbol{x}$$

is minimised subject to the constraints

$$x_i \geq 0 \quad i = 1, \ldots, n \tag{5.80}$$

and the additional constraints

$$\begin{array}{lll} \boldsymbol{a}_i^T\boldsymbol{x} & \leq b_i & i = 1, \ldots, m_1 \\ \boldsymbol{a}_j^T\boldsymbol{x} & \geq b_j \geq 0 & j = m_1 + 1, \ldots, m_1 + m_2 \\ \boldsymbol{a}_k^T\boldsymbol{x} & = b_k \geq 0 & k = m_1 + m_2 + 1, \ldots, m_1 + m_2 + m_3 \end{array}$$

for given m_1, m_2 and m_3. Consider optimising the perceptron criterion function as a problem in linear programming. The perceptron criterion function, with a positive margin vector $\boldsymbol{b}$, is given by

$$J_P = \sum_{\boldsymbol{y}_i \in \mathcal{Y}} (b_i - \boldsymbol{v}^T \boldsymbol{y}_i)$$

where now $\boldsymbol{y}_i$ are the vectors satisfying $\boldsymbol{v}^T \boldsymbol{y}_i \leq b_i$. A margin is introduced to prevent the trivial solution $\boldsymbol{v} = 0$. This can be reformulated as a linear programming problem as follows. We introduce the *artificial variables* a_i and consider the problem of minimising

$$\mathcal{Z} = \sum_{i=1}^{n} a_i$$

subject to

$$a_i \geq 0$$
$$a_i \geq b_i - \boldsymbol{v}^T \boldsymbol{y}_i$$

Show that minimising $\mathcal{Z}$ with respect to $\boldsymbol{a}$ and $\boldsymbol{v}$ will minimise the perceptron criterion (observe that for fixed $\boldsymbol{v}$, minimising with respect to the a_i gives the perceptron objective function).

3. Evaluate the constant of proportionality, α, in (5.16) and hence show that the offset on the right-hand side of (5.16) does not depend on the choice for t_1 and t_2 and is given by

$$\frac{p_2 - p_1}{2} \left(\frac{1 + p_1 p_2 d^2}{p_1 p_2} \right)$$

where $p_i = n_i/n$ and d^2 is given by

$$d^2 = (\boldsymbol{m}_1 - \boldsymbol{m}_2)^T \boldsymbol{S}_W^{-1} (\boldsymbol{m}_1 - \boldsymbol{m}_2)^T$$

the Mahalanobis distance between two normal distributions of equal covariance matrices. Compare this with the optimal value for normal distributions, given by (5.8).

4. Show that the maximisation of

$$\mathrm{Tr}\{\boldsymbol{S}_W^{-1} \boldsymbol{S}_B\}$$

in the transformed space leads to the same feature space as the linear discriminant solution. Hint: write $\boldsymbol{S}_W$ in the transformed space as $\boldsymbol{W}^T \boldsymbol{S}_W \boldsymbol{W}$ and $\boldsymbol{S}_B$ in the transformed space as $\boldsymbol{W}^T \boldsymbol{S}_B \boldsymbol{W}$; then differentiate with respect to $\boldsymbol{W}$ and note that $(\boldsymbol{W}^T \boldsymbol{S}_W \boldsymbol{W})^{-1} \boldsymbol{W}^T \boldsymbol{S}_B \boldsymbol{W}$ may be written as $\boldsymbol{U} \boldsymbol{\Lambda} \boldsymbol{U}^{-1}$ for eigenvectors $\boldsymbol{U}$ and a diagonal matrix of eigenvectors, $\boldsymbol{\Lambda}$.

5. Show that the criterion

$$J_4 = \frac{\mathrm{Tr}\{A^T S_B A\}}{\mathrm{Tr}\{A^T S_W A\}}$$

is invariant to an orthogonal transformation of the matrix $\boldsymbol{A}$.

6. Consider the squared error,

$$\sum_{j=1}^{C} p(\omega_j)\mathrm{E}[\|\boldsymbol{W}\boldsymbol{x} + \boldsymbol{w}_0 - \boldsymbol{\lambda}_j\|^2]_j$$

where $\mathrm{E}[.]_j$ is with respect to the conditional distribution of $\boldsymbol{x}$ on class ω_j. By writing

$$\mathrm{E}[\|\boldsymbol{W}\boldsymbol{x} + \boldsymbol{w}_0 - \boldsymbol{\lambda}_j\|^2]_j = \mathrm{E}[\|(\boldsymbol{W}\boldsymbol{x} + \boldsymbol{w}_0 - \boldsymbol{\rho}(\boldsymbol{x})) + (\boldsymbol{\rho}(\boldsymbol{x}) - \boldsymbol{\lambda}_j)\|^2]_j$$

and expanding, show that the linear discriminant rule also minimises

$$E' = \mathrm{E}[\|\boldsymbol{W}\boldsymbol{x} + \boldsymbol{w}_0 - \boldsymbol{\rho}(\boldsymbol{x})\|^2]$$

where the expectation is with respect to the unconditional distribution $p(\boldsymbol{x})$ of $\boldsymbol{x}$ and $\boldsymbol{\rho}(\boldsymbol{x})$ is defined as

$$\boldsymbol{\rho}(\boldsymbol{x}) = \sum_{j=1}^{C} \boldsymbol{\lambda}_j p(\omega_j|\boldsymbol{x})$$

7. Show that if $\boldsymbol{A}$ is the linear discriminant transformation (transforming the within-class covariance matrix to the identity and diagonalising the between-class covariance matrix), then the inverse of the within-class covariance matrix may be written

$$S_W^{-1} = \boldsymbol{A}\boldsymbol{A}^T + \boldsymbol{A}_{\perp}\boldsymbol{A}_{\perp}^T$$

where $\boldsymbol{A}_{\perp}^T(\boldsymbol{m}_j - \boldsymbol{m}) = 0$ for all j (the columns of $\boldsymbol{A}$ are orthogonal to the space spanned by the vectors $\boldsymbol{m}_j - \boldsymbol{m}$). Hint: note that for an $n \times n$ orthogonal matrix, $\boldsymbol{V} = [\boldsymbol{v}_1, \ldots, \boldsymbol{v}_n]$, $\boldsymbol{V}\boldsymbol{V}^T = \sum_{i=1}^n \boldsymbol{v}_i\boldsymbol{v}_i^T = \sum_{i=1}^r \boldsymbol{v}_i\boldsymbol{v}_i^T + \sum_{i=r+1}^n \boldsymbol{v}_i\boldsymbol{v}_i^T = \boldsymbol{V}_r\boldsymbol{V}_r^T + \boldsymbol{V}_{\bar{r}}\boldsymbol{V}_{\bar{r}}^T$, where $\boldsymbol{V}_r = [\boldsymbol{v}_1, \ldots, \boldsymbol{v}_r]$ and $\boldsymbol{V}_{\bar{r}} = [\boldsymbol{v}_{r+1}, \ldots, \boldsymbol{v}_n]$.

8. Show that the outputs of a linear discriminant function, trained using a least-squares approach with 0–1 targets, sum to unity. Hint: using the solution $\boldsymbol{W}\boldsymbol{x} + \boldsymbol{w}_0$ of 5.3.4, show that $\boldsymbol{1}^T\boldsymbol{W}\boldsymbol{x} + \boldsymbol{w}_0 = 1$.

9. Generate data from three bivariate normal distributions, with means (– 4, –4), (0, 0), (4, 4) and identity covariance matrices; 300 samples in train and test sets; equal priors. Train a least-squares classifier (Section 5.3.4) and a linear discriminant function (Section 5.3.3) classifier. For each classifier, obtain the classification error and plot the data and the decision boundaries. What do you conclude from the results?

10. Derive the dual form of the Lagrangian for SVMs applied to the multiclass case.

11. For SVM regression with a quadratic (rather than linear) ϵ-insensitive loss, the primal form of the Lagrangian is written

$$L_p = \boldsymbol{w}^T\boldsymbol{w} + C\sum_{i=1}^{n}(\xi_i^2 + \hat{\xi}_i^2)$$

and is minimised subject to the constraints (5.64). Derive the dual form of the Lagrangian and state the constraints on the Lagrange multipliers.

6

Nonlinear discriminant analysis – kernel and projection methods

Developed primarily in the neural networks and machine learning literature, the radial basis function (RBF) network, the support vector machine (SVM) and the multilayer perceptron (MLP) are flexible models for nonlinear discriminant analysis that give good performance on a wide range of problems. RBFs are sums of radially symmetric functions; SVMs define the basis functions implicitly through the specification of a kernel; MLPs are nonlinear functions of linear projections of the data.

6.1 Introduction

In the previous chapter, classification of an object is achieved as the result of a linear transformation whose parameters were determined through some optimisation procedure. The linear transformation may be applied to the observed data, or some prescribed features of that data. Various optimisation schemes for the parameters were considered, including simple error-correction schemes (as in the case of the perceptron), least squares error minimisation and in the logistic discrimination model, the parameters were obtained through a maximum likelihood approach using a nonlinear optimisation procedure.

In this chapter, we generalise the discriminant model still further by assuming parametric forms for the discriminant functions g_j. Specifically, we assume a discriminant function of the form

$$g_j(\boldsymbol{x}) = \sum_{i=1}^{m} w_{ji}\phi_i(\boldsymbol{x}; \boldsymbol{\mu}_i) + w_{j0}, \quad j = 1, \ldots, C \tag{6.1}$$

Statistical Pattern Recognition, Third Edition. Andrew R. Webb and Keith D. Copsey.

where there are m nonlinear 'basis' functions, ϕ_i, each of which has n_m parameters $\boldsymbol{\mu}_i = \{\mu_{ik}, k = 1, \ldots, n_m\}$ (the number of parameters may differ between the ϕ_i, but here we shall assume an equal number), and use the discriminant rule:

$$\text{assign } \boldsymbol{x} \text{ to class } \omega_i \text{ if } g_i(\boldsymbol{x}) = \max_j g_j(\boldsymbol{x})$$

that is, $\boldsymbol{x}$ is assigned to the class whose discriminant function is the largest. In (6.1) the parameters of the model are the values w_{ji} and μ_{ik} and the number of basis functions, m.

Equation (6.1) is exactly of the form of a generalised linear discriminant function, but we allow some flexibility in the nonlinear functions ϕ_i. There are several special cases of (6.1); these include:

$\phi_i(\boldsymbol{x}; \boldsymbol{\mu}_i) \equiv x_i$	linear discriminant function $m = d$, dimension of $\boldsymbol{x}$.
$\phi_i(\boldsymbol{x}; \boldsymbol{\mu}_i) \equiv \phi_i(\boldsymbol{x})$	generalised linear discriminant function with fixed transformation; it can take any of the forms in Table 1.1 for example.

Equation (6.1) may be written

$$\boldsymbol{g}(\boldsymbol{x}) = \boldsymbol{W}\boldsymbol{\phi}(\boldsymbol{x}) + \boldsymbol{w}_0 \tag{6.2}$$

where $\boldsymbol{W}$ is the $C \times m$ matrix with (ji) component w_{ji}, $\boldsymbol{\phi}(\boldsymbol{x})$ is the m-dimensional vector with ith component $\phi_i(\boldsymbol{x}, \boldsymbol{\mu}_i)$ and $\boldsymbol{w}_0$ is the vector $(w_{10}, \ldots, w_{C0})^T$. Equation (6.2) may be regarded as providing a transformation of a data sample $\boldsymbol{x} \in \mathbb{R}^d$ to $\mathbb{R}^C$ through an intermediate space $\mathbb{R}^m$ defined by the nonlinear functions ϕ_i. This is a model of the *feed-forward type*. As we shall discuss later, models of this form have been widely used for functional approximation and (as with the linear and logistic models), they are not confined to problems in discrimination.

There are two problems to solve with the model (6.2). The first is to determine the complexity of the model or the model order. How many functions ϕ_i do we use (specify the value of m)? How complex should each function be (how many parameters do we allow)? The answers to these questions are data dependent. There is an interplay between model order, training set size and the dimensionality of the data. Unfortunately there is no simple equation relating the three quantities – it is very much dependent on the data distribution. The problem of model order selection is nontrivial and is very much an active area of research (see Chapter 13). The second problem is to determine the remaining parameters of the model ($\boldsymbol{W}$ and the $\boldsymbol{\mu}_i$), for a given model order. This is simpler and will involve some nonlinear optimisation procedure for minimising a cost function. We shall discuss several of the most commonly used forms.

In this chapter we introduce models that have been developed primarily in the neural network and machine learning literatures. The types of neural network model that we consider in this chapter are of the *feed-forward type* and there is a very strong overlap between these models and those developed in the statistical literature, particularly kernel discrimination and logistic regression. The radial basis function network and the multilayer perceptron may be thought of as a natural progression of the generalised linear discriminant models described in the previous chapter.

Therefore, for the purpose of the chapter, we consider neural network models to provide models of discriminant functions that are linear combinations of simple basis functions, usually of the same parametric form. The parameters of the basis functions, as well as the linear weights, are determined by a training procedure. Other models that we have described in earlier chapters could be termed neural network models (for example, linear discriminant analysis). Also, the development of classification and regression models in the neural network literature is no longer confined to such simple models.

Neural network optimisation criteria assume that we have a set of data samples $\{(\boldsymbol{x}_i, \boldsymbol{t}_i), i = 1, \ldots, n\}$ that we use to 'train' the model. In a regression problem, the $\boldsymbol{x}_i$ are measurements on the regressors (also known as predictor variables or input variables) and $\boldsymbol{t}_i$ are measurements on the dependent or response variables. In a classification problem, $\boldsymbol{t}_i$ are the class labels. In both cases, we wish to obtain an estimate of $\boldsymbol{t}$ given $\boldsymbol{x}$, a measurement. In the neural network literature, $\boldsymbol{t}_i$ are referred to as *targets*, the desired response of the model for measurements or *inputs*, $\boldsymbol{x}_i$; the actual responses of the model, $\boldsymbol{g}(\boldsymbol{x}_i)$, are referred to as *outputs*. In all the algorithms considered, standardisation of the inputs (e.g. linear scaling so that each input variable has zero mean and unit variance over the training data, or linear scaling so that all input variables range between -1 and 1) may improve algorithm performance (not least through ensuring that variables with large numeric range do not dominate those with small numeric range).

The issues in neural network development are the common ones of pattern recognition: model specification, training and model selection for good generalisation performance. Sections 6.2, 6.4 and 6.3 introduce three popular models, namely the radial basis function network, the nonlinear support vector machine (SVM) and the multilayer perceptron (MLP).

6.2 Radial basis functions

6.2.1 Introduction

Radial basis functions (RBFs) were originally proposed in the functional interpolation literature (see the review by Powell, 1987; Lowe, 1995a) and first used for discrimination by Broomhead and Lowe (1988). However, radial basis functions have been around in one form or another for a very long time. They are very closely related to kernel methods for density estimation and regression developed in the statistics literature (see Chapter 4) and to normal mixture models (Chapter 2).

The RBF may be described mathematically as a linear combination of radially symmetric nonlinear basis functions. The RBF provides a transformation of a pattern $\boldsymbol{x} \in \mathbb{R}^d$ to an n'-dimensional output space according to[1]

$$g_j(\boldsymbol{x}) = \sum_{i=1}^{m} w_{ji}\phi_i(|\boldsymbol{x} - \boldsymbol{\mu}_i|) + w_{j0}, \quad j = 1, \ldots, n' \tag{6.3}$$

The parameters w_{ji} are often referred to as the *weights*; w_{j0}, the *bias* and the vectors $\boldsymbol{\mu}_i$ the *centres*. The model (6.3) is very similar to the kernel density model described in Chapter 4 in which $n' = 1$, $w_{10} = 0$, and the number of centres m is taken to be equal to the number of data samples n, with $\boldsymbol{\mu}_i = \boldsymbol{x}_i$ (a centre at each data sample); $w_{ji} = 1/n$ and ϕ is one of the kernels

[1] Here we use n' to denote the dimensionality of the output space. In a classification problem, we usually have $n' = C$, the number of classes.

given in Chapter 4, sometimes referred to as the *activation function* in the neural network literature.

In the case of exact interpolation, a basis function is also positioned at each data point ($m = n$). Suppose we seek a mapping g from $\mathbb{R}^d$ to $\mathbb{R}$ (taking $n' = 1$) through the points $(\boldsymbol{x}_i, t_i)$ which satisfies the condition that $g(\boldsymbol{x}_i) = t_i$; that is, under the assumed model (6.3) (and ignoring the bias), we seek a solution for $\boldsymbol{w} = (w_1, \ldots, w_n)^T$ that satisfies

$$\boldsymbol{t} = \boldsymbol{\Phi} \boldsymbol{w}$$

where $\boldsymbol{t} = (t_1, \ldots, t_n)^T$ and $\boldsymbol{\Phi}$ is the $n \times n$ matrix with (i, j) element $\phi(|\boldsymbol{x}_i - \boldsymbol{x}_j|)$, for a nonlinear function ϕ. For a large class of functions Micchelli (1986) has shown that the inverse of $\boldsymbol{\Phi}$ exists and the solution for $\boldsymbol{w}$ is given by

$$\boldsymbol{w} = \boldsymbol{\Phi}^{-1} \boldsymbol{t}$$

However, exact interpolation is not a good thing to do in general. It leads to poor generalisation performance in many pattern recognition problems (see Chapter 1). The fitting function can be highly oscillatory. Therefore, we usually take $m < n$.

An often-cited advantage of the RBF model is its simplicity. Once the forms of the nonlinearity have been specified and the centres determined, we have a linear model whose parameters can be easily obtained by a least squares procedure, or indeed any appropriate optimisation procedure such as those described in Chapter 5.

For supervised classification, the RBF is used to construct a discriminant function for each class. Figure 6.1 illustrates a one-dimensional example.

Data drawn from two univariate normal distributions of unit variance and means of 0.0 and 2.0 are plotted on the axis of Figure 6.1. Normal kernels are positioned over centres selected from these data. The weights w_{ji} are determined using a least squares procedure and the discriminant functions $g_\bullet$ and $g_\diamond$ plotted in Figure 6.1. Thus, a linear combination

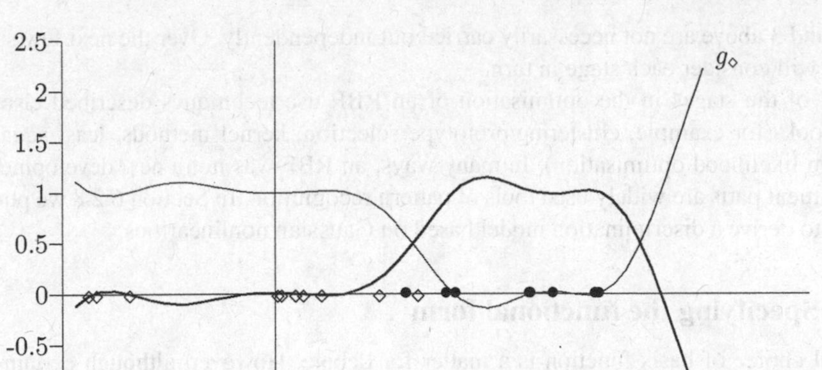

Figure 6.1 Discriminant functions constructed using a radial basis function network of normal kernels.

of 'blob'-shaped functions is used to produce two functions that can be used as the basis for discrimination.

6.2.2 Specifying the model

The basic RBF model is of the form

$$g_j(\boldsymbol{x}) = \sum_{i=1}^{m} w_{ji}\phi\left(\frac{|\boldsymbol{x} - \boldsymbol{\mu}_i|}{h}\right) + w_{j0}, \quad j = 1, \ldots, n'$$

that is, all the basis functions are of the same functional form ($\phi_i = \phi$) and a scaling parameter h has been introduced. In this model, there are five quantities to prescribe or to determine from the data:

- the form of the basis function, ϕ;
- the positions of the centres, $\boldsymbol{\mu}_i$;
- the smoothing parameter, h;
- the weights, w_{ji}, and bias w_{j0};
- the number of basis functions, m.

There are three main stages in constructing an RBF model:

1. Specify the nonlinear functions, ϕ. To a large extent, this is independent of the data and the problem (though the parameters of these functions are not).
2. Determine the number and positions of the centres, and the smoothing parameters. These values should be data dependent.
3. Determine the weights of the RBF. These values are data dependent.

Items 2 and 3 above are not necessarily carried out independently. Over the next few subsections we will consider each stage in turn.

Most of the stages in the optimisation of an RBF use techniques described elsewhere in this book (for example, clustering/prototype selection, kernel methods, least squares or maximum likelihood optimisation). In many ways, an RBF was not a new development; all its constituent parts are widely used tools of pattern recognition. In Section 6.2.8 we put them together to derive a discrimination model based on Gaussian nonlinearities.

6.2.3 Specifying the functional form

The ideal choice of basis function is a matter for debate. However, although certain types of problem may be matched inappropriately to certain forms of nonlinearity, the actual form of the nonlinearity is relatively unimportant (as in kernel density estimation) compared with the number and the positions of the centres. Typical forms of nonlinearity are given in Table 6.1. Note that some RBF nonlinearities produce smooth approximations, in that the fitting function and its derivatives are continuous. Others [for example, $z\log(z)$ and $\exp(-z)$] have discontinuous gradients with respect to the measurements $\boldsymbol{x}$.

Table 6.1 Radial basis function nonlinearities.

Nonlinearity	Mathematical form
Gaussian	$\exp(-z^2)$
Exponential	$\exp(-z)$
Quadratic	$z^2 + \alpha z + \beta$
Inverse quadratic	$1/[1 + z^2]$
Thin-plate spline	$z^\alpha \log(z)$
Trigonometric	$\sin(z)$

The two most popular forms are the thin-plate spline, $\phi(z) = z^2\log(z)$, and the normal or Gaussian form, $\phi(z) = \exp(-z^2)$. Each of these functions may be motivated from different perspectives (see Section 6.2.9): the normal form from a kernel regression and kernel density estimation point of view and the thin-plate spline from curve fitting (Lowe, 1995a). Indeed, each may be shown to be optimal under certain conditions: in fitting data in which there is normally distributed noise on the inputs, the normal form is the optimal basis function in a least-squares sense (Webb, 1994); in fitting a surface through a set of points and using a roughness penalty, the *natural* thin-plate spline is the solution (Duchon, 1976; Meinguet, 1979).

These functions are very different: one is compact and positive, the second diverges at infinity and is negative over a region. However, in practice, this difference is to some extent superficial since, for training purposes, the function ϕ need only be defined in the feature space over the range $[s_{\min}/h, s_{\max}/h]$, where

$$s_{\max} = \max_{ij} |\boldsymbol{x}_i - \boldsymbol{\mu}_j|$$

$$s_{\min} = \min_{ij} |\boldsymbol{x}_i - \boldsymbol{\mu}_j| \tag{6.4}$$

and therefore ϕ may be redefined over this region as $\hat{\phi}(s)$ given by

$$\hat{\phi} \leftarrow \frac{\phi(s/h) - \phi_{\min}}{\phi_{\max} - \phi_{\min}} \tag{6.5}$$

where $\phi_{\max}$ and $\phi_{\min}$ are the maximum and minimum values of ϕ over $[0, s_{\max}/h]$ (taking $s_{\min} = 0$); respectively and $s = |\boldsymbol{x} - \boldsymbol{\mu}_j|$ is the distance in the feature space. The redefined function satisfies $0 \leq \hat{\phi} \leq 1$. Scaling of ϕ may simply be compensated for by adjustment of weights $\{w_{ji}, i = 1, \ldots, m\}$ and bias $w_{j0}, j = 1, \ldots, n'$. The fitting function is unaltered.

Choice of the smoothing parameters is discussed in more detail in Section 6.2.5. As in kernel density estimation it is more important to choose appropriate values for the smoothing parameters than the functional form. There is some limited empirical evidence to suggest that thin-plate splines fit data better in high-dimensional settings (Lowe, 1995b).

6.2.4 The positions of the centres

Suppose that the number of basis functions, m, has been specified. The values of centres and weights may be found by minimising a suitable criterion (for example, least squares) using a nonlinear optimisation scheme. However, it is more usual to position the centres first, and then to calculate the weights using one of the optimisation schemes appropriate for linear models. Of course, this means that the optimisation criterion will not be at an extremum with respect to the positions of the centres, but in practice this does not matter.

Positioning of the centres can have a major effect on the performance of an RBF for discrimination and interpolation. In an interpolation problem, more centres should be positioned in regions of high curvature. In a discrimination problem, more centres should be positioned near class boundaries. There are several schemes commonly employed.

1. Select from dataset – random selection

Select randomly from the dataset. We would expect there to be more centres in regions where the density is greater. A consequence of this is that sparse regions may not be 'covered' by the RBF unless the smoothing parameter is adjusted. Random selection is an approach that is commonly used. An advantage is that it is fast. A disadvantage is that the fitting function, or the class labels in a supervised classification problem, is not taken into account; that is, it is an unsupervised placement scheme and may not provide the best solution for a mapping problem.

2. Clustering approach

The values for the centres obtained by the previous approach could be used as seeds for a *k*-means clustering algorithm, thus giving centres for RBFs as cluster centres. The *k*-means algorithm seeks to partition the data into *k* groups or *clusters* so that the within-group sum of squares is minimised; that is, it seeks the cluster centres, $\{\boldsymbol{\mu}_j, j = 1, \ldots, k\}$ that minimise $\sum_{j=1}^{k} S_j$ where the within-group sum of squares for group j is

$$S_j = \sum_{i=1}^{n} z_{ji} |\boldsymbol{x}_i - \boldsymbol{\mu}_j|^2$$

where $z_{ji} = 1$ if $\boldsymbol{x}_i$ is in group j (of size $n_j = \sum_{i=1}^{n} z_{ji}$) and zero otherwise; $\boldsymbol{\mu}_j$ is the mean of group j,

$$\boldsymbol{\mu}_j = \frac{1}{n_j} \sum_{i=1}^{n} z_{ji} \boldsymbol{x}_i$$

Algorithms for computing the cluster centres, $\boldsymbol{\mu}_j$, using *k*-means are described in Chapter 11.

Alternatively, any other clustering approach could be used: either pattern based, or dissimilarity matrix based by first forming a dissimilarity matrix (see Chapter 11).

3. Normal mixture model

If we are using Gaussian nonlinearities, then it seems sensible to use as centres (and indeed widths) the parameters resulting from a normal mixture model of the underlying distribution $p(\boldsymbol{x})$ (see Chapter 2). We model the distribution as a mixture of normal models

$$p(\boldsymbol{x}) = \sum_{j=1}^{g} \pi_j p(\boldsymbol{x}|\boldsymbol{\mu}_j, h)$$

where

$$p(\boldsymbol{x}|\boldsymbol{\mu}_j, h) = \frac{1}{(2\pi)^{d/2}h^d}\exp\left\{-\frac{1}{2h^2}(\boldsymbol{x}-\boldsymbol{\mu}_j)^T(\boldsymbol{x}-\boldsymbol{\mu}_j)\right\}$$

The values of h, π_j and $\boldsymbol{\mu}_j$ may be determined using the EM algorithm to maximise the likelihood (see Chapter 2). The weights π_j are ignored and the resulting normal basis functions, defined by h and $\boldsymbol{\mu}_i$, are used in the RBF model.

4. *k*-nearest neighbour initialisation

The approaches described above use the input data only to define the centres. Class labels, or the values of the dependent variables in a regression problem, are not used. Thus, unlabelled data from the same distribution as the training data may be used in the centre initialisation process. We now consider some supervised techniques. In Chapter 4 we found that in the k-nearest neighbour classifier not all data samples are required to define the decision boundary. We may use an *editing* procedure to remove those prototypes that contribute to the error (with the hope of improving the generalisation performance) and a *condensing* procedure to reduce the number of samples needed to define the decision boundary. The prototypes remaining after editing and condensing may be retained as centres for an RBF classifier.

5. Orthogonal least squares

The choice of RBF centres can be viewed as a problem in variable selection (see Chapter 10). Chen *et al.* (1991, 1992) consider the complete set of data samples to be candidates for centres and construct a set of centres incrementally. Suppose that we have a set of $k-1$ centres, positioned over $k-1$ different data points. At the kth stage, the centre selected from the remaining $n-(k-1)$ data samples that reduces the prediction error the most is added to the set of centres. Centres are added until a criterion that balances the fitting error against model complexity is minimised. The approach therefore performs model selection alongside centre position optimisation.

A naïve implementation of this method would solve for the network weights $n-(k-1)$ times at the kth stage, each time evaluating Equation (6.7) of Section 6.2.6. Chen *et al.* (1991) propose a scheme that reduces the computation based on an orthogonal least squares algorithm. An efficient procedure is proposed for calculating the error when each additional centre is introduced.

6.2.5 Smoothing parameters

Having obtained centres, we need to specify the smoothing parameters. It is important to choose an 'appropriate' value for the smoothing parameters in order to fit the structure in the data. This is a compromise between the one extreme of fitting noise in the data and the other of being unable to model the structure. The choice depends on the particular form we adopt for the nonlinearity. If it is Gaussian, then the normal mixture model approach to centre selection will lead to values for the widths of the distributions naturally. The other approaches will necessitate a separate estimation procedure.

There are several heuristics which were discussed in Chapter 4 on kernel density estimation. Although they are suboptimal, they are fast to calculate. An alternative is to choose the smoothing parameter that minimises a cross-validation estimate of the sum-square error.

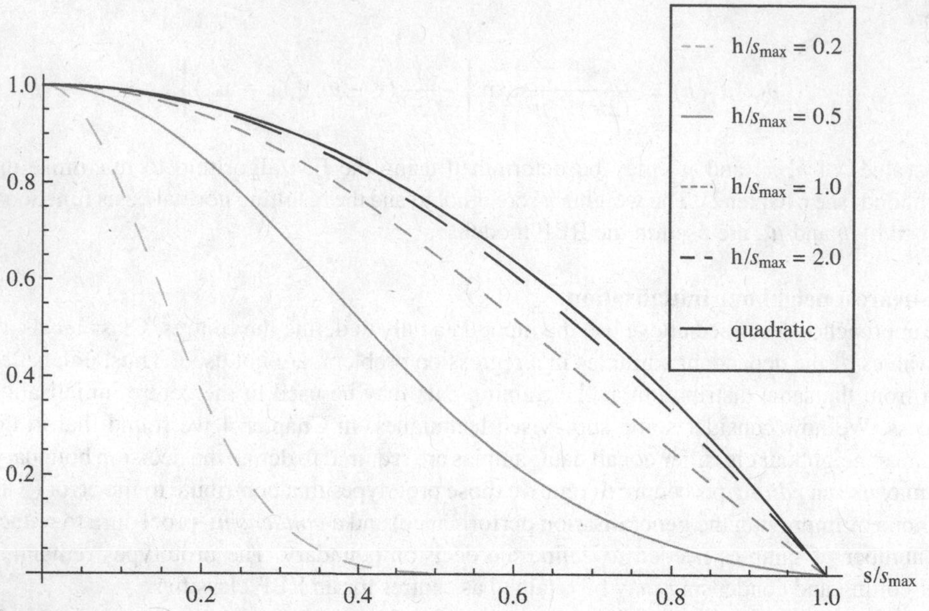

Figure 6.2 Normalised Gaussian basis functions, $\hat{\phi}(s)$, for $h/s_{max} = 0.2, 0.5, 1.0, 2.0$ and the limiting quadratic.

Figure 6.2 illustrates the normalised form [see the discussion surrounding Equations (6.4) and (6.5)] of the nonlinearity for the Gaussian basis function for several values of the smoothing parameter, h. For these Gaussian basis functions we see that there is little change in the shape of the normalised form for h/s_{max} greater than about 2. As $h \to \infty$, the normalised form for the nonlinearity tends to the quadratic

$$\hat{\phi}_\infty(s) \triangleq 1 - \frac{s^2}{s_{max}^2}$$

Thus, asymptotically, the normalised basis function is independent of h.

For large h/s_{max} (greater than about 2), changes in the value of h may be compensated for by adjustments of the weights, w_{ji}, and the RBF is a quadratic function of the input variables. For smaller values of h, the normalised function tends to the Gaussian form, thus allowing small-scale variation in the fitting function.

6.2.6 Calculation of the weights

The final stage of determining the RBF for a given model order is the calculation of the weights. The most popular technique for calculating the weights is based on the least squares error measure. In certain special cases, alternative techniques such as maximum likelihood procedures have been proposed. Examples are the generalised logistic model (see Section 6.4.5 for a description) and an EM algorithm for use with Gaussian basis functions (Lázaro *et al.*, 2003).

Least squares error measure

As in the linear case (Chapter 5), we seek to minimise a square error measure

$$E = \sum_{i=1}^{n} |\boldsymbol{t}_i - \boldsymbol{g}(\boldsymbol{x}_i)|^2$$
$$= \| -\boldsymbol{T}^T + \boldsymbol{W}\boldsymbol{\Phi}^T + \boldsymbol{w}_0\boldsymbol{1}^T \|^2 \tag{6.6}$$

with respect to the parameters w_{ij}, where $\boldsymbol{T} = [\boldsymbol{t}_1|\dots|\boldsymbol{t}_n]^T$ is an $n \times C$ target matrix whose ith row is the target for input $\boldsymbol{x}_i$; $\boldsymbol{\Phi} = [\boldsymbol{\phi}(\boldsymbol{x}_1)|\dots|\boldsymbol{\phi}(\boldsymbol{x}_n)]^T$ is an $n \times m$ matrix whose ith row is the set of basis functions values evaluated at $\boldsymbol{x}_i$ (and dependent on the centres $\boldsymbol{\mu}_j$); $\|\boldsymbol{A}\|^2 = \text{Tr}\{\boldsymbol{A}\boldsymbol{A}^T\} = \sum_{ij} A_{ij}^2$; and $\boldsymbol{1}$ is a $n \times 1$ vector of 1s.

Suppose that the parameters $\{\boldsymbol{\mu}\}$ of the nonlinear functions ϕ have been determined. The solution for $\boldsymbol{W}$ with minimum norm that minimises (6.6) is

$$\boldsymbol{W} = \hat{\boldsymbol{T}}^T(\hat{\boldsymbol{\Phi}}^T)^\dagger \tag{6.7}$$

where $\dagger$ denotes the pseudo-inverse of a matrix (Figure 5.4). The matrices $\hat{\boldsymbol{T}}$ and $\hat{\boldsymbol{\Phi}}$ have zero mean rows ($\hat{\boldsymbol{T}}^T\boldsymbol{1} = 0$; $\hat{\boldsymbol{\Phi}}^T\boldsymbol{1} = 0$) and are defined as

$$\hat{\boldsymbol{T}} \triangleq \boldsymbol{T} - \boldsymbol{1}\bar{\boldsymbol{t}}^T$$
$$\hat{\boldsymbol{\Phi}} \triangleq \boldsymbol{\Phi} - \boldsymbol{1}\bar{\boldsymbol{\phi}}^T$$

where

$$\bar{\boldsymbol{t}} = \frac{1}{n}\sum_{i=1}^{n} \boldsymbol{t}_i = \frac{1}{n}\boldsymbol{T}^T\boldsymbol{1}$$

$$\bar{\boldsymbol{\phi}} = \frac{1}{n}\sum_{i=1}^{n} \boldsymbol{\phi}(\boldsymbol{x}_i) = \frac{1}{n}\boldsymbol{\Phi}^T\boldsymbol{1}$$

are the mean values of the targets and the basis function outputs, respectively. The solution for $\boldsymbol{w}_0$ is

$$\boldsymbol{w}_0 = \bar{\boldsymbol{t}} - \boldsymbol{W}\bar{\boldsymbol{\phi}}$$

Therefore, we may solve for the weights $\boldsymbol{W}$ using a linear method such as a singular value decomposition.

In a classification problem, in which we have the 1-from-C coding for the class labels (that is, if $\boldsymbol{x}_i \in \omega_j$ then we have the target $\boldsymbol{t}_i = (0, 0, \dots, 0, 1, 0, \dots, 0)^T$, where the 1 is in the jth position), then using the pseudo-inverse solution (6.7) for the final layer weights gives the property that the components of $\boldsymbol{g}$ sum to unity for any data sample $\boldsymbol{x}$, i.e.

$$\sum_i g_i = 1$$

where

$$g = \hat{T}^T (\hat{\Phi}^T)^{\dagger} \Phi^T + w_0 \tag{6.8}$$

The values of g are not constrained to be positive. One way to ensure positivity is to transform the values g_i by replacing g_i by $\exp(-g_i)/\sum_j \exp(-g_j)$. This forms the basis of the generalised logistic model (sometimes called *softmax*).

One can incorporate priors into the error function through pattern weights, as in Equation (5.40).

Regularisation of the least squares error measure
Too many parameters in a model may lead to *over-fitting* of the data by the model and poor generalisation performance (see Chapter 1). One means of smoothing the model fit is to penalise the sum square error. Thus, we modify the square error measure (6.6) and minimise

$$E = \sum_{i=1}^{n} |t_i - g(x_i)|^2 + \alpha \int F(g(x))dx \tag{6.9}$$

where α is a *regularisation* parameter and F is a function of the complexity of the model. For example, in a univariate curve-fitting problem, a popular choice for $F(g)$ is $\partial^2 g/\partial x^2$, the second derivative of the fitting function g. In this case, the solution for g that minimises (6.9) is a *cubic spline* (Green and Silverman, 1994).

In the neural network literature, a penalising term of the form

$$\alpha \sum_i \tilde{w}_i^2$$

is often used, where the summation is over all adjustable network parameters, $\tilde{w}$. This procedure is termed *weight decay*.

For the generalised linear model (6.2) we have seen that the penalised error is taken to be (see Chapter 5)

$$E = \| -\hat{T}^T + W\hat{\Phi}^T \|^2 + \alpha \|W\|^2 \tag{6.10}$$

expressed in terms of the zero-mean matrices $\hat{T}$ and $\hat{\Phi}$; and α is termed the *ridge parameter* (see Chapter 5). The solution for W that minimises E is

$$W = \hat{T}^T \hat{\Phi}(\hat{\Phi}^T \hat{\Phi} + \alpha I_m)^{-1}$$

6.2.7 Model order selection

Finally, we address the question of how to choose the *number* of centres.[2] The model order selection problem is very similar to many of the problems of model complexity discussed

[2] Note that the orthogonal least squares centre selection procedure has a model selection criterion incorporated into the centre selection process.

elsewhere in this book (see Chapter 13); for example, how many clusters are best, how many components in a normal mixture, how do we determine intrinsic dimensionality, etc.? It is not easy to answer. The number depends on several factors including the amount and distribution of the data, the dimension and the form adopted for the nonlinearity. It is probably better to have many centres with limited complexity (single smoothing parameter) than an RBF with few centres and a complex form for the nonlinearity. There are several approaches to determining the number of centres. These include:

1. Using cross-validation. The cross-validation error (minimised over the smoothing parameter) is plotted as a function of the number of centres. The number of centres is chosen as that above which there is no appreciable decrease, or an increase in the cross-validation error [as used by Orr (1995) in forward selection of centres].

2. Monitoring the performance on a separate test set. This is similar to the cross-validation procedure above, except that the error is evaluated on a separate test set.

3. Using information complexity criteria. The sum-square error is augmented by an additional term that penalises complexity (Chen *et al.*, 1991).

4. If we are using a normal mixture model to set the positions and widths of centres, we may use one of the methods for estimating the number of components in a normal mixture (see Chapter 2).

6.2.8 Simple RBF

We have now set up the machinery for implementing a simple RBF. The stages in a simple implementation are as follows:

1. Specify the functional form for the nonlinearity.

2. Prescribe the number of centres m.

3. Determine the positions of the centres (for example, random selection, or the k-means algorithm).

4. Determine the smoothing parameters (for example, simple heuristic or cross-validation).

5. Map the data to the space spanned by the outputs of the nonlinear functions; i.e. for a given dataset $\boldsymbol{x}_i$, $i = 1, \ldots, n$, form the vectors $\boldsymbol{\phi}_i = \boldsymbol{\phi}(\boldsymbol{x}_i)$, $i = 1, \ldots, n$.

6. Solve for the weights and the biases using the minimum least squared error.

7. Calculate the final output on the train and test sets; classify the data if required.

The above is a simple prescription for an RBF network that uses unsupervised techniques for centre placement and width selection (and therefore is suboptimal). One of the often-quoted advantages of the RBF network is its simplicity – there is no nonlinear optimisation scheme required, in contrast to the multilayer perceptron classifier discussed later in this chapter. However, many of the sophisticated techniques for centre placement and width determination are more involved and increase the computational complexity of the model substantially. Nevertheless, the simple RBF can give acceptable performance for many applications.

6.2.9 Motivation

The RBF model may be motivated from several perspectives. We present the first two in a discrimination context and then two from a regression perspective.

6.2.9.1 Kernel discriminant analysis

Consider the multivariate kernel density estimate (see Chapter 4)

$$p(\boldsymbol{x}) = \frac{1}{nh^d} \sum_{i=1}^{n} K\left(\frac{1}{h}(\boldsymbol{x} - \boldsymbol{x}_i)\right)$$

where $K(\boldsymbol{x})$ is defined for d-dimensional $\boldsymbol{x}$ satisfying $\int_{\mathbb{R}^d} K(\boldsymbol{x})d\boldsymbol{x} = 1$. Suppose we have a set of samples $\boldsymbol{x}_i$ $(i = 1, \ldots, n)$ with n_j samples in class ω_j $(j = 1, \ldots, C)$. If we construct a density estimate for each class then the posterior probability of class membership can be written

$$p(\omega_j|\boldsymbol{x}) = \frac{p(\omega_j)}{p(\boldsymbol{x})} \frac{1}{n_j h^d} \sum_{i=1}^{n} z_{ji} K\left(\frac{1}{h}(\boldsymbol{x} - \boldsymbol{x}_i)\right) \tag{6.11}$$

where $p(\omega_j)$ is the prior probability of class ω_j and $z_{ji} = 1$ if $\boldsymbol{x}_i \in$ class ω_j, 0 otherwise. Thus, discrimination is based on a model of the form

$$\sum_{i=1}^{n} w_{ji} \phi_i(\boldsymbol{x} - \boldsymbol{x}_i) \tag{6.12}$$

where $\phi_i(\boldsymbol{x} - \boldsymbol{x}_i) = K((\boldsymbol{x} - \boldsymbol{x}_i)/h)$ and

$$w_{ji} = \frac{p(\omega_j)}{n_j} z_{ji} \tag{6.13}$$

[We neglect the term $p(\boldsymbol{x})h^d$ in the denominator of (6.11) since it is independent of j.] Equation (6.12) is of the form of an RBF with a centre at each data point and weights determined by class priors [Equation (6.13)].

6.2.9.2 Mixture models

In discriminant analysis by Gaussian mixtures (Chapter 2), the class-conditional density for class ω_j is expressed as

$$p(\boldsymbol{x}|\omega_j) = \sum_{r=1}^{R_j} \pi_{jr} p(\boldsymbol{x}|\boldsymbol{\theta}_{jr})$$

where class ω_j has R_j mixture components, π_{jr} are the mixing proportions ($\sum_{r=1}^{R_j} \pi_{jr} = 1$) and $p(\boldsymbol{x}|\boldsymbol{\theta}_{jr})$ is the density of the rth mixture component of class ω_j evaluated at $\boldsymbol{x}$ ($\boldsymbol{\theta}_{jr}$ denote the mixture component parameters: for normal mixtures, the mean and covariance matrix). The posterior probabilities of class membership are

$$p(\omega_j|\boldsymbol{x}) = \frac{p(\omega_j)}{p(\boldsymbol{x})} \sum_{r=1}^{R_j} \pi_{jr} p(\boldsymbol{x}|\boldsymbol{\theta}_{jr})$$

where $p(\omega_j)$ is the prior probability of class ω_j.

For normal mixture components, $p(\boldsymbol{x}|\boldsymbol{\theta}_{jr})$, with means $\boldsymbol{\mu}_{jr}$, $j = 1, \ldots, C; r = 1, \ldots, R_j$ and common spherical covariance matrices, $\sigma^2 \boldsymbol{I}$, we have discriminant functions of the form (6.3). There are $m = \sum_{j=1}^{C} R_j$ basis functions, centred at the $\boldsymbol{\mu}_{jr}$, with the weights set by the mixing proportions, class priors and class indicator variables.

6.2.9.3 Regularisation

Suppose that we have a dataset $\{(\boldsymbol{x}_i, t_i), i = 1, \ldots, n\}$ where $\boldsymbol{x}_i \in \mathbb{R}^d$ and we seek a smooth surface g,

$$t_i = g(\boldsymbol{x}_i) + \text{error}$$

One such approach is to minimise the penalised sum of squares

$$S = \sum_{i=1}^{n} (t_i - g(\boldsymbol{x}_i))^2 + \alpha J(g)$$

where α is a regularisation or *roughness* (Green and Silverman, 1994) parameter and $J(g)$ is a penalty term that measures how 'rough' the fitted surface is and has the effect of penalising 'wiggly' surfaces (see Chapter 5 and Section 6.2.6).

A popular choice for J is one based on mth derivatives. Taking J as

$$J(g) = \int_{\mathbb{R}^d} \sum \frac{m!}{\nu_1! \ldots \nu_d!} \left(\frac{\partial^m g}{\partial x_1^{\nu_1} \ldots \partial x_d^{\nu_d}} \right)^2 dx_1 \ldots dx_d$$

where the summation is over all nonnegative integers $\nu_1, \nu_2, \ldots, \nu_d$ such that $\nu_1 + \nu_2 + \cdots + \nu_d = m$, results in a penalty invariant under translations and rotations of the coordinate system (Green and Silverman, 1994).

Defining $\eta_{md}(r)$ by

$$\eta_{md}(r) = \begin{cases} \theta r^{2m-d} \log(r) & \text{if } d \text{ is even} \\ \theta r^{2m-d} & \text{if } d \text{ is odd} \end{cases}$$

and the constant of proportionality, θ, by

$$\theta = \begin{cases} (-1)^{m+1+d/2} 2^{1-2m} \pi^{-d/2} \frac{1}{(m-1)!} \frac{1}{(m-d/2)!} & \text{if } d \text{ is even} \\ \Gamma(d/2 - m) 2^{-2m} \pi^{-d/2} \frac{1}{(m-1)!} & \text{if } d \text{ is odd} \end{cases}$$

then (under certain conditions on the points $\boldsymbol{x}_i$ and m) the function g minimising $J(g)$ is a *natural thin-plate spline*. This is a function of the form

$$g(\boldsymbol{x}) = \sum_{i=1}^{n} b_i \eta_{md}(|\boldsymbol{x} - \boldsymbol{x}_i|) + \sum_{j=1}^{M} a_j \gamma_j(\boldsymbol{x})$$

where

$$M = \binom{m+d-1}{d}$$

and $\{\gamma_j, j = 1, \ldots, M\}$ is a set of linearly independent polynomials spanning the M-dimensional space of polynomials in $\mathbb{R}^d$ of degree less than m. The coefficients $\{a_j, j = 1, \ldots, M\}$, $\{b_i, i = 1, \ldots, n\}$ satisfy certain constraints [see Green and Silverman (1994) for further details]. Thus, the minimising function contains radially symmetric terms, η, and polynomials.

An alternative derivation based on a different form for the penalty terms leading to Gaussian RBFs is provided by Bishop (1995).

6.2.10 RBF properties

One of the properties of an RBF that has motivated its use in a wide range of applications both in functional approximation and in discrimination is that it is a universal approximator: it is possible (given certain conditions on the kernel function) to construct an RBF that approximates a given (integrable, bounded and continuous) function arbitrarily accurately (Park and Sandberg, 1993; Chen and Chen, 1995). This may require a very large number of centres. In most, if not all, practical applications the mapping we wish to approximate is defined by a finite set of data samples providing class-labelled data or, in an approximation problem, data samples and associated function values and implemented in finite precision arithmetic. Clearly, this limits the complexity of the model.

6.2.11 Example application study

The problem

The problem of source position estimation using measurements made on a radar focal-plane array using RBFs was treated by Webb and Garner (1999). This particular approach was motivated by a requirement for a compact integrated (hardware) implementation of a bearing estimator in a sensor focal plane (a solution was required that could readily be implemented in silicon on the same substrate as the focal-plane array).

Summary

The problem is one of prediction rather than discrimination: given a set of training samples $\{(\boldsymbol{x}_i, \theta_i); i = 1, \ldots, n\}$, where $\boldsymbol{x}_i$ is a vector of measurements on the independent variables (array calibration measurements in this problem) and θ_i is the response variable (position), a predictor, f, was sought such that given a new measurement, z, then $f(z)$ is a good estimate of the position of the source that gave rise to the measurements, z. However, the problem differs from a standard regression problem in that there is noise on the measurement vector, z, and this is similar to errors-in-variables models in statistics. Therefore, we seek a predictor that is robust to noise on the inputs.

The data

The training data comprised detector outputs of an array of 12 detectors, positioned in the focal plane of a lens. Measurements were made on the detectors as the lens scanned across a microwave point source (thus providing measurements of the *point-spread function* of the lens). There were 3721 training samples measured at a signal-to-noise ratio (SNR) of about 45 dB. The test data were recorded for the source at specific positions over a range of lower SNR.

The model

The model adopted was a standard RBF network with a Gaussian kernel with centres defined in the 12-dimensional space. The approach adopted to model parameter estimation was a standard least squares one. The problem may be regarded as one example from a wider class in discrimination and regression in which the expected operating conditions (the test conditions) differ from the training conditions in a known way. For example, in a discrimination problem, the class priors may differ considerably from the values estimated from the training data. In a least squares approach, this may be compensated for by modifying the sum-squared-error criterion appropriately (see Section 5.3.4). Also, allowance for expected population drift may be made by modifying the error criterion. In the source position estimation problem, the training conditions are considered 'noiseless' (obtained through a calibration procedure) and the test conditions differ in that there is noise (of known variance) on the data. Again, this can be taken into account by modifying the sum-squared-error criterion.

Training procedure

An RBF predictor was designed with centres chosen using a k-means procedure. A ridge regression-type solution was obtained for the weights (see Section 6.2.6 on regularisation), with the ridge parameter inversely proportional to a SNR term. Thus, there was no need to perform any search procedure to determine the ridge parameter. It can be set by measuring the SNR of the radar system. The theoretical development was validated by experimental results on a 12-element microwave focal-plane array in which two angle coordinates were estimated.

Results

It was shown that it was possible to compensate for noisy test conditions by using a regularisation solution for the parameters, with regularisation parameter proportional to the inverse of the SNR.

6.2.12 Further developments

There have been many developments of the basic RBF model in the areas of RBF design, learning algorithms and Bayesian treatments.

Developments of the k-means approach to RBF training to take account of the class labels of the data samples are described by Musavi *et al.* (1992), in which clusters contain samples from the same class, and Karayiannis and Mi (1997), in which a localised class-conditional variance is minimised as part of a network growing process.

Chang and Lippmann (1993) propose a supervised approach for allocating RBF centres near class boundaries. An alternative approach that chooses centres for a classification task in a supervised way based on the ideas of *support vectors* is described by Schölkopf *et al.* (1997) (see Chapter 5 and Section 6.3). In support vector learning of RBF networks with Gaussian basis functions, the separating surface obtained by an SVM approach (that is, the decision boundary) is a linear combination of Gaussian functions centred at selected training points (the support vectors). The number and location of centres is automatically determined. In a comparative study reviewing several approaches to RBF training, Schwenker *et al* (2001) find that the SVM learning approach is often superior on a classification task to the standard two-stage learning of RBFs (selecting or adapting the centres followed by calculating the weights).

The orthogonal least squares forward selection procedure has been developed to use a regularised error criterion by Chen *et al.* (1996) and Orr (1995), who uses a generalised cross-validation criterion as a stopping condition.

In the basic approach, all patterns are used at once in the calculation for the weights ('batch learning'). Online learning methods have been developed (Marinaro and Scarpetta, 2000). These enable the weights of the network to be updated sequentially according to the error computed on the last selected new example. This allows for possible temporal changes in the task being learned.

A Bayesian treatment that considers the number of basis functions and weights to be unknown has been developed by Holmes and Mallick (1998). A joint probability density function is defined over model dimension and model parameters. Using reversible jump Markov chain Monte Carlo methods (see Chapter 3), inference is made by integrating over the model dimension and parameters. A hierarchical Bayesian model for RBF networks has been developed by Andrieu *et al.* (2001), again making use of reversible jump Markov chain Monte Carlo.

6.2.13 Summary

RBFs are simple to construct, easy to train and find a solution for the weights rapidly. They provide a very flexible model and give very good performance over a wide range of problems, both for discrimination and for functional approximation. The RBF model uses many of the standard pattern recognition building blocks (clustering, least squares optimisation, for example). There are many variants of the RBF model (due to the choice of centre selection procedure, form of the nonlinear functions, procedure for determining the weights, and model selection method). This can make it difficult to draw meaningful conclusions about RBF performance over a range of studies on different applications, since the form of an RBF may vary from study to study.

The disadvantages of RBFs also apply to many, if not all, of the discrimination models covered in this book. That is, care must be taken not to construct a classifier that models noise in the data, or models the training set too well, which may give poor generalisation performance. Choosing a model of the appropriate complexity is very important as we emphasise repeatedly in this book. Regularising the solution for the weights can improve generalisation performance. Model selection requirements add to the computational requirements of the model, and thus the often-stated claim for RBF networks of simplicity is perhaps overstated. Yet it should be said that a simple scheme can give good performance and, not unusually, performance exceeds many of the more 'traditional' statistical classifiers.

6.3 Nonlinear support vector machines

6.3.1 Introduction

In Chapter 5 we introduced the SVM as a tool for finding the optimal separating hyperplane for linearly separable data and considered developments of the approach for situations when the data are not linearly separable. As we remarked in that chapter, the support vector algorithm may be applied in a transformed feature space, $\boldsymbol{\phi}(\boldsymbol{x})$, for some nonlinear function $\boldsymbol{\phi}$. Indeed, this is the principle behind many methods of pattern classification: transform the input features nonlinearly to a space in which linear methods may be applied (see also Chapter 1). We discuss this approach further in the context of nonlinear SVMs.

6.3.2 Binary classification

For the binary classification problem, we seek a discriminant function of the form

$$g(\boldsymbol{x}) = \boldsymbol{w}^T\boldsymbol{\phi}(\boldsymbol{x}) + w_0$$

with decision rule

$$\boldsymbol{w}^T\boldsymbol{\phi}(\boldsymbol{x}) + w_0 \begin{cases} > & 0 \\ < & 0 \end{cases} \Rightarrow \boldsymbol{x} \in \begin{cases} \omega_1 \text{ with corresponding numeric value } y_i = +1 \\ \omega_2 \text{ with corresponding numeric value } y_i = -1 \end{cases}$$

The SVM procedure determines the maximum margin solution through determination of the saddle point of a Lagrangian. The dual form of the Lagrangian [Equation (5.59)] becomes

$$L_D' = \sum_{i=1}^{n} \alpha_i - \frac{1}{2}\sum_{i=1}^{n}\sum_{j=1}^{n} \alpha_i\alpha_j y_i y_j \boldsymbol{\phi}^T(\boldsymbol{x}_i)\boldsymbol{\phi}(\boldsymbol{x}_j) \tag{6.14}$$

where $y_i = \pm 1$, $i = 1, \ldots, n$ are class indicator values and α_i, $i = 1, \ldots, n$ are Lagrange multipliers satisfying

$$\sum_{i=1}^{n} \alpha_i y_i = 0$$

$$0 \leq \alpha_i \leq C \tag{6.15}$$

for a 'regularisation' parameter, C. Maximising (6.14) subject to the constraints (6.15) leads to support vectors identified by nonzero values of α_i.

The solution for $\boldsymbol{w}$ [Equation (5.63)] is

$$\boldsymbol{w} = \sum_{i \in \mathcal{SV}} \alpha_i y_i \boldsymbol{\phi}(\boldsymbol{x}_i)$$

and classification of a new data sample $\boldsymbol{x}$ is performed according to the sign of

$$g(\boldsymbol{x}) = \sum_{i \in \mathcal{SV}} \alpha_i y_i \boldsymbol{\phi}^T(\boldsymbol{x}_i)\boldsymbol{\phi}(\boldsymbol{x}) + w_0 \tag{6.16}$$

where

$$w_0 = \frac{1}{n_{\widetilde{\mathcal{SV}}}} \left\{ \sum_{i \in \widetilde{\mathcal{SV}}} y_i - \sum_{i \in \mathcal{SV}, j \in \widetilde{\mathcal{SV}}} \alpha_i y_i \boldsymbol{\phi}^T(\boldsymbol{x}_i) \boldsymbol{\phi}(\boldsymbol{x}_j) \right\} \tag{6.17}$$

where $\mathcal{SV}$ is the set of support vectors with associated values of α_i satisfying $0 < \alpha_i \leq C$ and $\widetilde{\mathcal{SV}}$ is the set of $n_{\widetilde{\mathcal{SV}}}$ support vectors satisfying $0 < \alpha_i < C$ (those at the target distance of $1/|\boldsymbol{w}|$ from the separating hyperplane).

Optimisation of L_D (6.14) and the subsequent classification of a sample [(6.16) and (6.17)] relies only on scalar products between transformed feature vectors, which can be replaced by a kernel function

$$K(\boldsymbol{x}, \boldsymbol{y}) = \boldsymbol{\phi}^T(\boldsymbol{x}) \boldsymbol{\phi}(\boldsymbol{y})$$

Thus, we can avoid computing the transformation $\boldsymbol{\phi}(\boldsymbol{x})$ explicitly and replace the scalar product with $K(\boldsymbol{x}, \boldsymbol{y})$ instead. This is referred to as the *kernel trick*. The discriminant function (6.16) becomes

$$g(\boldsymbol{x}) = \sum_{i \in \mathcal{SV}} \alpha_i y_i K(\boldsymbol{x}_i, \boldsymbol{x}) + w_0 \tag{6.18}$$

The advantage of the kernel representation is that we need only use K in the training algorithm and even do not need to know $\boldsymbol{\phi}$ explicitly, provided that the kernel can be written as an inner product. In some cases (for example, the exponential kernel), the feature space is infinite-dimensional and so it is more efficient to use a kernel.

Solution of the quadratic programming optimisation problem to obtain the α_i (and therefore the support vectors) is as in the linear case (see Section 5.4.6), but with the scalar products between vectors replaced by kernel evaluations. For example, one can use the decomposition algorithm of Osuna *et al.* (1997), or the SMO algorithm proposed by Platt (1998).

6.3.3 Types of kernel

There are many types of kernel that may be used in an SVM. Table 6.2 lists some commonly used forms.

Table 6.2 Support vector machine kernels.

Nonlinearity	Mathematical form
Simple polynomial	$(1 + \boldsymbol{x}^T \boldsymbol{y})^p$
Polynomial	$(r + \gamma \boldsymbol{x}^T \boldsymbol{y})^p$, $\gamma > 0$
Gaussian	$\exp(-\lvert\boldsymbol{x} - \boldsymbol{y}\rvert^2/\sigma^2)$
Sigmoid	$\tanh(k\boldsymbol{x}^T \boldsymbol{y} - \delta)$

As an example, consider the kernel $K(\boldsymbol{x}, \boldsymbol{y}) = (1 + \boldsymbol{x}^T\boldsymbol{y})^p$ for $p = 2$ and $\boldsymbol{x}, \boldsymbol{y} \in \mathbb{R}^2$. This may be expanded as

$$\begin{aligned}(1 + x_1y_1 + x_2y_2)^2 &= 1 + 2x_1y_1 + 2x_2y_2 + 2x_1x_2y_1y_2 + x_1^2y_1^2 + x_2^2y_2^2 \\ &= \boldsymbol{\phi}^T(\boldsymbol{x})\boldsymbol{\phi}(\boldsymbol{y})\end{aligned}$$

where $\boldsymbol{\phi}(\boldsymbol{x}) = (1, \sqrt{2}x_1, \sqrt{2}x_2, \sqrt{2}x_1x_2, x_1^2, x_2^2)$.

Acceptable kernels must be expressible as an inner product in a feature space, which means that they must satisfy *Mercer's condition* (Courant and Hilbert, 1959; Vapnik, 1998): a kernel $K(\boldsymbol{x}, \boldsymbol{y}), \boldsymbol{x}, \boldsymbol{y} \in \mathbb{R}^d$ is an inner product in some feature space, or $K(\boldsymbol{x}, \boldsymbol{y}) = \boldsymbol{\phi}^T(\boldsymbol{x})\boldsymbol{\phi}(\boldsymbol{y})$, if and only if $K(\boldsymbol{x}, \boldsymbol{y}) = K(\boldsymbol{y}, \boldsymbol{x})$ and

$$\int K(\boldsymbol{x}, \boldsymbol{z}) f(\boldsymbol{x}) f(\boldsymbol{z}) d\boldsymbol{x} d\boldsymbol{z} \geq 0$$

for all functions f satisfying

$$\int f^2(\boldsymbol{x}) d\boldsymbol{x} < \infty$$

That is, $K(\boldsymbol{x}, \boldsymbol{y})$ may be expanded as

$$K(\boldsymbol{x}, \boldsymbol{y}) = \sum_{j=1}^{\infty} \lambda_j \hat{\phi}_j(\boldsymbol{x}) \hat{\phi}_j(\boldsymbol{y})$$

where λ_j and $\phi_j(\boldsymbol{x})$ are the eigenvalues and eigenfunctions satisfying

$$\int K(\boldsymbol{x}, \boldsymbol{y}) \phi_j(\boldsymbol{x}) d\boldsymbol{x} = \lambda_j \phi_j(\boldsymbol{x})$$

and $\hat{\phi}_j$ is normalised so that $\int \hat{\phi}_j^2(\boldsymbol{x}) d\boldsymbol{x} = 1$.

6.3.4 Model selection

The degrees of freedom of the SVM model are the choice of kernel. the parameters of the kernel and the choice of the regularisation parameter, C, which penalises the training errors. For most types of kernel, it is generally possible to find values for the kernel parameters for which the classes are separable. However, this is not a sensible strategy and leads to overfitting of the training data and poor generalisation to unseen data.

The simplest approach to model selection is to reserve a validation set that is used to monitor performance as the model parameters are varied. More expensive alternatives that make better use of the data are data re-sampling methods such as cross-validation and bootstrapping (see Chapter 9, in the context of classifier error rate estimation, and Chapter 13). Appropriate model selection is a critical aspect when training SVMs. Use of default parameters may well lead to disappointing performance. However, as with the smoothing parameters of

RBF networks (see Section 6.2.5), heuristics can be used to select initial values of SVM kernel parameters. For the Gaussian kernel, these could be based on those used in kernel density estimation (see Chapter 4).

Often a grid-based search approach is used (with performance estimated using cross-validation), as advocated by Chang and Lin (2011) when describing their LIBSVM C++ software for implementing SVMs (see also Hsu *et al.*, 2003). As noted by Hsu *et al.* (2003), one can reduce search time by using a hierarchical search strategy, starting with a coarse grid, and then performing finer searches within the best grids. They recommend parameter search covering many orders of magnitude for each parameter (e.g. exponential sequences). Staelin (2002) advocates iterative reductions in the search range for each parameter, each time centring the search around the previous best estimate.

Fröhlich and Zell (2005) propose learning an *online Gaussian process* model of the error surface in parameter space, and then training SVMs only at points for which the expected improvement in the fitted error is highest. Huang and Wang (2006) develop a *genetic algorithm* for SVM parameter selection and also selection of a subset of features to use as input data. Genetic algorithms (Goldberg, 1989) seek to use ideas from Darwinian natural selection to move a population of potential solutions towards an optimal solution. Lin *et al.* (2008) propose the use of *particle swarm optimisation* for joint SVM parameter and feature selection. Particle swarm optimisation (Kennedy and Eberhart, 1995) is a population-based search heuristic motivated by observance of birds flocking to a promising position.

Hastie *et al.* (2004) demonstrate how an alternative formulation of the SVM optimisation problem allows a path of SVM solutions for different values of the regularisation parameter C to be calculated efficiently. This removes the need to include the C parameter alongside the kernel parameters in a grid search.

6.3.5 Multiclass SVMs

Multiclass SVMs may be developed along the lines discussed in Chapter 5: whether by combining binary classifiers in a one-against-one or a one-against-all method, or by solving a single multiclass optimisation problem (the 'all-together' method). In an assessment of these methods, Hsu and Lin (2002) find that the all-together method yields fewer support vectors, but one-against-all is more suitable for practical use.

6.3.6 Probability estimates

A drawback with SVMs is that they do not produce estimates of the class membership probabilities. To address this for binary classification problems, Platt (2000) proposes training a binary SVM classifier in the usual manner and then learning a separate sigmoid function to map the SVM outputs [Equation (6.18)] into probabilities. The probabilities are modelled as

$$\hat{p}(y = +1|g(\boldsymbol{x})) = \frac{1}{1 + \exp(Ag(\boldsymbol{x}) + B)} \tag{6.19}$$

and

$$\hat{p}(y = -1|g(\boldsymbol{x})) = 1 - \hat{p}(y = +1|g(\boldsymbol{x}))$$

where A and B are parameters to be optimised. A maximum likelihood optimisation procedure is used to estimate A and B. The procedure requires the outputs of the trained SVM on labelled validation data (perhaps from a cross-validation procedure). The validation set target values $\{y_i, i = 1, \ldots, n\}$ are redefined as

$$t_i = \begin{cases} 0 & \text{if} \quad y_i = -1 \\ 1 & \text{if} \quad y_i = +1 \end{cases}$$

or in a regularised version as

$$t_i = \begin{cases} \dfrac{1}{\sum_{i=1}^n I(y_i = -1) + 2} & \text{if} \quad y_i = -1 \\ \\ \dfrac{\sum_{i=1}^n I(y_i = +1) + 1}{\sum_{i=1}^n I(y_i = +1) + 2} & \text{if} \quad y_i = 1 \end{cases}$$

where I is the indicator function, taking value one if its argument is true, and 0 otherwise. The maximum likelihood procedure is to select A and B to maximise the log-likelihood

$$l(A, B) = \sum_{i=1}^{n} (t_i \log p_i + (1 - t_i) \log(1 - p_i))$$

where p_i is as defined in (6.19) for $g(\boldsymbol{x}_i)$ the output for the ith pattern in the validation set. Numerical optimisation routines are used to solve the maximisation problem (Press *et al.*, 1992). Lin *et al.* (2007) discuss the optimisation step in more detail, including reduction of numerical stability issues.

An unfortunate property of the procedure is that the following relationship does not hold

$$g(\boldsymbol{x} > 0) \Leftrightarrow \hat{p}(y = +1) > 0.5$$

Hence the class with the largest probability might not be the same as the class predicted by the SVM.

An extension of the approach in order to provide multiclass probability estimates is provided by Wu *et al.* (2004). The approach is based upon the one-against-one multiclass procedure (see Section 5.3.1), in each case obtaining binary classification probability estimates using the procedure of Platt (2000). For a test pattern $\boldsymbol{x}$, denote by r_{ij} the probability estimate for class ω_i estimated by the classifier for class ω_i against class ω_j. The class probability vector $\boldsymbol{p} = (p_1, \ldots, p_C)$ for the pattern is optimised to be the vector which minimises

$$\sum_{i=1}^{C} \sum_{j \neq i} (r_{ji} p_i - r_{ij} p_j)^2$$

subject to $\sum_{i=1}^{C} p_i = 1$, $p_i \geq 0$, $i = 1, \ldots, C$. Further procedures for combining the one-against-one probability estimates are discussed by Wu *et al.* (2004) and also by Milgram *et al.* (2006).

6.3.7 Nonlinear regression

The linear SVMs for regression (see Section 5.4.5) may also be generalised to a nonlinear regression function in a similar manner to the SVMs for binary classification. If the nonlinear function is given by

$$f(\boldsymbol{x}) = \boldsymbol{w}^T \boldsymbol{\phi}(\boldsymbol{x}) + w_0$$

then Equation (5.67) is replaced by

$$L_D = \sum_{i=1}^{n} (\hat{\alpha}_i - \alpha_i) y_i - \epsilon \sum_{i=1}^{n} (\hat{\alpha}_i + \alpha_i) - \frac{1}{2} \sum_{i=1}^{n} \sum_{j=1}^{n} (\hat{\alpha}_i - \alpha_i)(\hat{\alpha}_j - \alpha_j) K(\boldsymbol{x}_i, \boldsymbol{x}_j)$$

where $K(\boldsymbol{x}, \boldsymbol{y})$ is a kernel satisfying Mercer's conditions. This is maximised subject to the constraints in Equation (5.68).

The solution for $f(\boldsymbol{x})$ is then [compare with (5.70)]

$$f(\boldsymbol{x}) = \sum_{i=1}^{n} (\hat{\alpha}_i - \alpha_i) K(\boldsymbol{x}, \boldsymbol{x}_i) + w_0$$

The parameter w_0 is chosen so that

$$\begin{aligned} & f(\boldsymbol{x}_i) - y_i = \epsilon \quad \text{for any } i \text{ with } 0 < \alpha_i < C \\ \text{or} \quad & f(\boldsymbol{x}_i) - y_i = -\epsilon \quad \text{for any } i \text{ with } 0 < \hat{\alpha}_i < C \end{aligned}$$

by the Karush–Kuhn–Tucker complementarity conditions [Equation (5.69)] with the scalar products replaced by kernel evaluations. As with the linear case, for numerical reasons it is usual to average the expression for w_0 over all patterns for which $0 < \alpha_i < C$ or $0 < \hat{\alpha}_i < C$.

6.3.8 Example application study

The problem

The application is to determine land-cover types (e.g. crop type) using airborne hyperspectral imagery of the land (Melgani and Bruzzone, 2004).

Summary

Determination of broad land-cover classes (e.g. forestry, crops and urban areas) can be addressed using imagery from multispectral sensors. Classification of different types or conditions of the same species (such as types of forest and types of crop) is more problematic, because of the reduced variation between these classes. Classification based upon imagery from hyperspectral sensors, with hundreds of observation channels, is advocated for such problems. The performance of SVM classifiers on hyperspectral data was compared with an RBF classifier and a k-nearest-neighbour classifier (see Chapter 4). Comparisons were made both before and after the use of feature selection criteria to reduce the dimensionality of the data.

The data

Hyperspectral imagery collected by the airborne visible/infrared imaging spectroradiometer (AVIRIS) sensor in trials in June 1992 in Indiana, USA, was used to test the performance of the algorithm. Two hundred spectral channels were used from the data. A nine-class problem was constructed from the data, with 4757 training data examples (ranging from 236 to 1245 examples for a class) and 4588 test examples.

Training procedure

Four different approaches for combining binary SVM classifiers to obtain a multiclass classifier (see Section 6.3.5) were considered: one-against-all; one-against-one and two hierarchical tree-based strategies (differing in the strategy used to define the pairs of combined classes to consider at each level of the tree). Two SVM kernels were considered: linear and Gaussian. SVM parameters (C and σ for the Gaussian kernel) were optimised by examining performance on the training data. The feature selection criteria were based on optimisation of a distance measure using a steepest ascent search procedure. The same features were therefore used for each classifier.

Results

The Gaussian kernel SVM provided the best performance in terms of classification accuracy, both with and without feature selection. The computational cost of applying the linear SVM classifier was an order of magnitude larger than that for the other classifiers. The one-against-one combination rule provided the best performance for the Gaussian kernel SVM, although the hierarchical tree combiners were more efficient computationally.

6.3.9 Further developments

There are many developments of the basic SVM model for discrimination and regression presented in this chapter.

Incorporation of priors and costs into the SVM model (to allow for test conditions that differ from training conditions, as often happens in practice) is addressed by Lin *et al.* (2002). Lauer and Bloch (2008) review techniques for incorporation of a range of additional knowledge on the data, including invariances to transformations of the data features and imbalances of the training set. Two types of approach are used to incorporate such information: (i) alterations of the kernel function to provide transformation-invariance; and (ii) extension of the training sets to reflect the prior knowledge (termed *sample methods*).

The basic regression model has been extended to take account of different ϵ-insensitive loss functions and ridge regression solutions (Vapnik, 1998; Cristianini and Shawe-Taylor, 2000). The ν-support vector classification algorithm (Schölkopf *et al.*, 2000) introduces a parameter to control the number of support vectors and training errors. The support vector method has also been applied to density estimation (Vapnik, 1998).

The relationship of the support vector method to other methods of classification is discussed by Guyon and Stork (1999) and Schölkopf *et al.* (2000). Sollich (2002) and Van Gestel *et al.* (2002) develop Bayesian frameworks for SVMs. Joachims (2006) considers an SVM approach to learning structured outputs, such as trees and sequences, which has relevance to natural language parsing.

6.3.10 Summary

Support vector machines comprise a class of algorithms that represent the decision boundary in a pattern recognition problem in terms of a small subset of the training samples, typically. This generalises to problems in regression through the ϵ-insensitive loss function that does not penalise errors below $\epsilon > 0$.

The loss function that is minimised comprises two terms, a term $\boldsymbol{w}^T \boldsymbol{w}$ that characterises model complexity and a second term that measures training error. A single parameter, C, controls the tradeoff between these two terms. The optimisation problem can be re-cast as a quadratic programming problem for both the classification and regression cases.

Several approaches for solving the multiclass classification problem have been proposed: combining binary classifiers and a one-step multiclass SVM approach.

SVMs have been applied to a wide range of applications, and demonstrated to be valuable for real-world problems. The generalisation performance often either matches or is significantly better that competing methods.

Once the kernel is fixed, SVMs have only one free parameter – the regularisation parameter that controls the balance between model complexity and training error. However, there are often parameters of the kernel that must be set and a poor choice can lead to poor generalisation. The choice of best kernel for a given problem is not resolved and special kernels have been derived for particular problems, for example document classification (Lodhi *et al.*, 2002).

6.4 The multilayer perceptron

6.4.1 Introduction

The MLP is yet another tool in the pattern recognition toolbox, and one that has received considerable use over the years. It was presented in a paper by Rumelhart *et al.* (1986) as an enhancement to the perceptron of Chapter 5 (in particularly providing an ability to classify data that is not linearly separable). Since that time it has been extensively employed in many pattern recognition tasks.

In order to introduce some terminology, let us consider a simple MLP model. We shall then consider some generalisations. The basic MLP produces a transformation of a pattern $\boldsymbol{x} \in \mathbb{R}^d$ to an n'-dimensional space according to

$$g_j(\boldsymbol{x}) = \sum_{i=1}^{m} w_{ji}\phi_i(\boldsymbol{\alpha}_i^T \boldsymbol{x} + \alpha_{i0}) + w_{j0}, \quad j = 1, \ldots, n' \tag{6.20}$$

The functions ϕ_i are fixed nonlinearities, usually identical and taken to be of the *logistic* form representing historically the mean firing rate of a neuron as a function of the input current

$$\phi_i(y) = \phi(y) = \frac{1}{1 + \exp(-y)} \tag{6.21}$$

Thus, the transformation (6.20) consists of projecting the data onto each of m directions described by the vectors $\boldsymbol{\alpha}_i = (\alpha_{i1}, \ldots, \alpha_{id})$; then transforming the projected data (offset by a bias α_{i0}) by the nonlinear functions $\phi_i(y)$; and finally, forming a linear combination using the weights w_{ji} (offset by a bias w_{j0}).

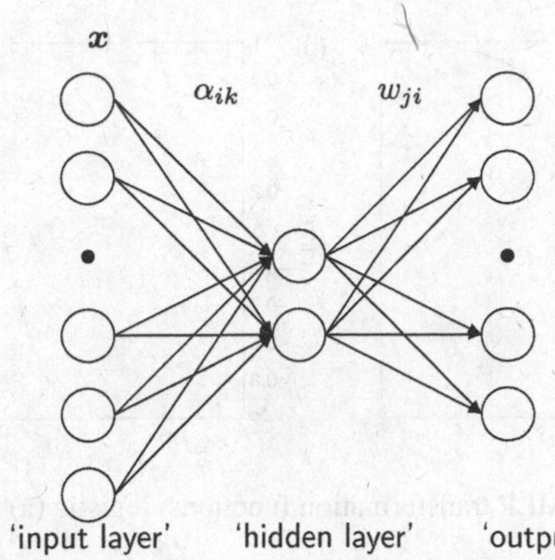

Figure 6.3 Single hidden layer multilayer perceptron.

The MLP is often presented in diagrammatic form (Figure 6.3).

The *input nodes* accept the data vector or pattern. There are weights associated with the links between the input nodes and the *hidden nodes* that accept the weighted combination $y = \boldsymbol{\alpha}_i^T \boldsymbol{x} + \alpha_{i0}$, and perform the nonlinear transformation $\phi(y)$. The *output nodes* take a linear combination of the outputs of the hidden nodes and deliver these as *outputs*. In principle, there may be many *hidden layers*, each one performing a transformation of a scalar product of the outputs of the previous layer and the vector of weights linking the previous layer to a given node. Also, there may be nonlinearities associated with the output nodes that, instead of producing a linear combination of weights w_{ji} and hidden unit outputs, perform nonlinear transformations of this value.

The emphasis of the treatment of MLPs in this section is on MLPs with a single hidden layer. Further, it is assumed either that the 'outputs' are a linear combination of the functions ϕ_i or at least that, in a discrimination problem, discrimination may be performed by taking a linear combination of the functions ϕ_i (a logistic discrimination model is of this type). There is some justification for using a *single* hidden layer model in that it has been shown that an MLP with a single hidden layer can approximate an arbitrary (continuous bounded integrable) function arbitrarily accurately (Hornik, 1993; see Section 6.4.4). Also, practical experience has shown that very good results can be achieved with a single hidden layer on many problems. There may be practical reasons for considering more than one hidden layer and it is conceptually straightforward to extend the analysis presented here to do so.

The MLP is a nonlinear model: the output is a nonlinear function of its parameters and the inputs, and a nonlinear optimisation scheme must be employed to minimise the selected optimisation criterion. Therefore, all that can be hoped for is a local extremum of the criterion function that is being optimised. This may give satisfactory performance but several solutions may have to be sought before an acceptable one is found.

6.4.2 Specifying the MLP structure

To specify the network structure we must prescribe the number of hidden layers, the number of nonlinear functions within each layer and the form of the nonlinear functions. Most of

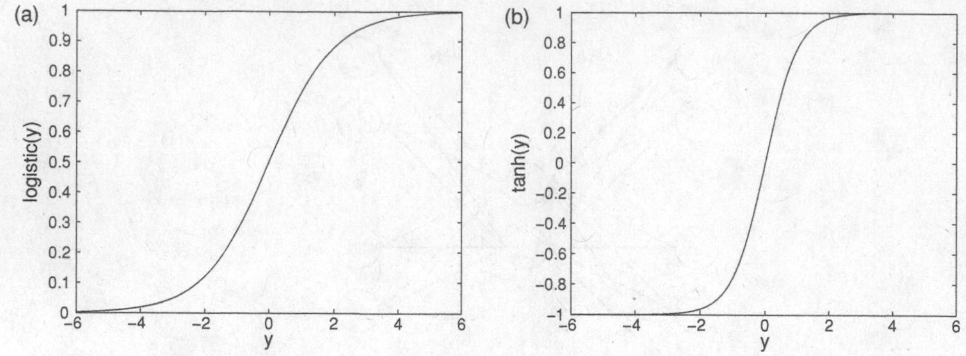

Figure 6.4 Standard MLP transformation functions: logistic (a) and tanh (b).

the MLP networks to be found in the literature consist of layers of logistic processing units (6.21) with each unit connected to every unit in the previous layer (*fully connected*) and no connections between units in nonadjacent layers.

A common alternative to the logistic processing unit is the hyperbolic tangent (tanh) function, given by

$$\phi_i(y) = \phi(y) = \tanh(y) = \frac{\exp(y) - \exp(-y)}{\exp(y) + \exp(-y)} \tag{6.22}$$

Both the logistic and tanh functions are illustrated in Figure 6.4.

Although fully connected MLPs are most common in the literature, it should be noted that many examples exist of MLP networks that are not fully connected. There has been some success with networks in which each processing unit is connected to only a small subset of the units in the previous layer. The units chosen are often part of a neighbourhood, especially if the input is some kind of image. In even more complex implementations the weights connected to units that examine similar neighbourhoods at different locations may be forced to be identical (shared weights), e.g. as used in Le Cun *et al.* (1989) for recognition of handwritten zip codes. Such advanced MLP networks, although of practical importance (particularly for image processing), are not universally applicable and are not considered further here.

6.4.3 Determining the MLP weights

There are two stages to optimisation of the MLP parameters. The first is the initialisation of the parameter values; the second is the implementation of a nonlinear optimisation scheme.

6.4.3.1 Weight initialisation

There are several schemes for initialising the weights. The weights of the MLP are often started at small random values. In addition, some work has been carried out to investigate the benefits of starting the weights at values obtained from simple heuristics. Given good starting values, training time is reduced, albeit at the expense of increased initialisation times.

For random initialisation, the weights are initialised with values drawn from a uniform distribution[3] over $[-\Delta, \Delta]$, where Δ depends on the scale of the values taken by the data. If all the variables are equally important and the sample variance is σ^2, then $\Delta = 1/(d\sigma)$ is a reasonable choice (d variables). Hush *et al.* (1992) assess several weight initialisation schemes and support initialisation to small random values.

There have been studies that use various pattern recognition schemes for initialisation. For example, Weymaere and Martens (1994) propose a network initialisation procedure that is based on k-means clustering (see Chapter 11) and nearest-neighbour classification. Brent (1991) uses decision trees (see Chapter 7) for initialisation. An approach that initialises an MLP to deliver class conditional probabilities under the assumption of independence of variables is described in Lowe and Webb (1990). This really only applies to categorical variables in which the data can be represented as binary patterns.

6.4.3.2 Optimisation

Many different optimisation criteria and many nonlinear optimisation schemes have been considered for the MLP. Over the years, the MLP has been explored in depth, particularly in the engineering literature. This has sometimes been by researchers who do not have a real application, but whose imagination has been stimulated by the 'neural network' aspect of the MLP. It would be impossible to offer anything but a brief introduction to optimisation schemes for the MLP.

Most optimisation schemes involve the evaluation of a function and its derivatives with respect to a set of parameters. Here, the parameters are the multilayer perceptron weights and the function is a chosen error criterion. In the following discussion we consider least-squares minimisation as the error criterion. In Section 6.4.5 we consider a logistic discrimination model.

Least squares error minimisation

The error to be minimised is the average squared distance between the approximation given by the model and the 'desired' value

$$E = \sum_{k=1}^{n} |\boldsymbol{t}_k - \boldsymbol{g}(\boldsymbol{x}_k)|^2 \tag{6.23}$$

where $\boldsymbol{g}(\boldsymbol{x}_k)$ is the vector of 'outputs' and $\boldsymbol{t}_k$ is the desired pattern (sometimes termed the *target*) for training data sample $\boldsymbol{x}_k$. In a regression problem, $\boldsymbol{t}_k$ are the dependent variables; in a discrimination problem, $\boldsymbol{t}_k = (t_{k1}, \ldots, t_{kC})^T$ are the class labels usually coded as

$$t_{kj} = \begin{cases} 1 & \text{if } \boldsymbol{x}_k \text{ is in class } \omega_j \\ 0 & \text{otherwise} \end{cases}$$

Modifications of the error criterion to take into account the effect of priors and alternative target codings to incorporate costs of misclassification are described in Section 5.3.4.

Most of the nonlinear optimisation schemes used for the MLP require the error and its derivative to be evaluated for a given set of weight values.

[3] Initialisation using samples from normal distributions is also common.

The derivative of the error with respect to a weight v (which at the moment can represent either a weight α, between inputs and hidden units, or a weight w, between the hidden units and the outputs – see Figure 6.3) can be expressed as[4]

$$\frac{\partial E}{\partial v} = -2 \sum_{k=1}^{n} \sum_{l=1}^{n'} (\boldsymbol{t}_k - \boldsymbol{g}(\boldsymbol{x}_k))_l \frac{\partial g_l(\boldsymbol{x}_k)}{\partial v} \tag{6.24}$$

The derivative of g_l with respect to v, for v one of the weights w, $v = w_{ji}$ say, is

$$\frac{\partial g_l}{\partial w_{ji}} = \begin{cases} \delta_{lj} & i = 0 \\ \delta_{lj}\phi_i(\boldsymbol{\alpha}_i^T \boldsymbol{x}_k + \alpha_{i0}) & i \neq 0 \end{cases} \tag{6.25}$$

and for the weights α, $v = \alpha_{ji}$ say, is

$$\frac{\partial g_l}{\partial \alpha_{ji}} = w_{lj} \frac{\partial \phi_j}{\partial \alpha_{ji}} = \begin{cases} w_{lj} \dfrac{\partial \phi_j}{\partial y} & i = 0 \\ \cdot & \\ w_{lj} \dfrac{\partial \phi_j}{\partial y} x_{ki} & i \neq 0 \end{cases} \tag{6.26}$$

where x_{ki} is the ith element of input $\boldsymbol{x}_k$, and $\partial\phi_j/\partial y$ is the derivative of ϕ with respect to its argument, given by

$$\frac{\partial \phi}{\partial y} = \frac{\exp(-y)}{(1 + \exp(-y))^2} = \phi(y)(1 - \phi(y)) \tag{6.27}$$

for the logistic form (6.21) and evaluated at $y = \boldsymbol{\alpha}_j^T \boldsymbol{x}_k + \alpha_{j0}$. Equations (6.24), (6.25), (6.26) and (6.27) can be combined to give expressions for the derivatives of the square error with respect to the weights w and α.

Back-propagation

In the above example, we explicitly calculated the derivatives for a single hidden layer network. Here we consider a more general treatment. *Back-propagation* is the term given to efficient calculation of the derivative of an error function in multilayer networks. We write the error, E, as

$$E = \sum_{k=1}^{n} E^k$$

where E^k is the contribution to the error from pattern k. For example, in (6.23) we have

$$E^k = |\boldsymbol{t}_k - \boldsymbol{g}(\boldsymbol{x}_k)|^2$$

We refer to Figure 6.5 for general notation: let the weights between layer $o - 1$ and layer o be w_{ji}^o (the weight connecting the ith node of layer $o - 1$ to the jth node of layer o);

[4] Again, we use n' to denote the output dimensionality; in a classification problem, $n' = C$, the number of classes.

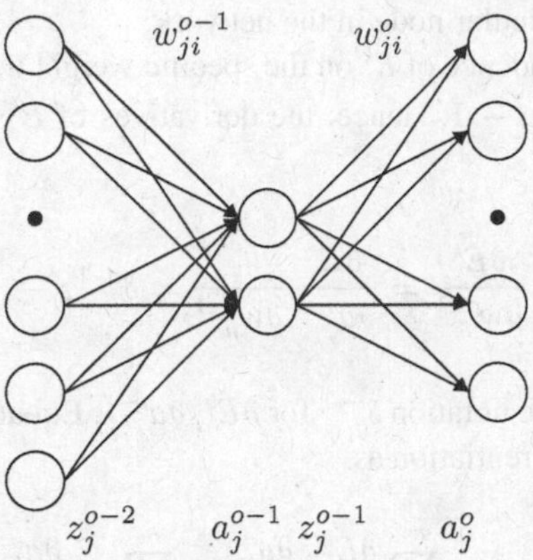

Figure 6.5 Notation used for back-propagation.

let a_j^o be the inputs to the nonlinearity at layer o and z_j^o be the outputs of the nonlinearity, so that

$$a_j^o = \sum_{i=1}^{n_o-1} w_{ji}^o z_i^{o-1} \tag{6.28}$$

$$z_i^{o-1} = \phi\left(a_i^{o-1}\right)$$

where n_{o-1} is the number of nodes at layer $o-1$ and ϕ is the (assumed common) nonlinearity associated with a node.[5] These quantities are calculated during the process termed *forward propagation*: namely the calculation of the error, E, from its constituent parts, E^k.

Now let layer o be the final layer of the network. The term E^k is a function of the inputs to the final layer,

$$E^k = E^k\left(a_1^o, \ldots, a_{n_o}^o\right)$$

For example, with a network with linear output units and using a square error criterion

$$E^k = \sum_{j=1}^{n_o} \left(t_{kj} - a_j^o\right)^2$$

where t_{kj} is the jth component of the kth target vector.

The derivatives of the error E are simply the sums of the derivatives of E^k, the contribution to the error by the kth pattern. Since the only dependence of E^k on the specific weight w_{ji}^o is through the input a_j^o at the final layer, the derivatives of E^k with respect to the final layer of weights, w_{ji}^o, are given by

$$\frac{\partial E^k}{\partial w_{ji}^o} = \frac{\partial E^k}{\partial a_j^o}\frac{\partial a_j^o}{\partial w_{ji}^o} = \delta_j^o z_i^{o-1} \tag{6.29}$$

[5] Strictly, the terms a_j^o and z_j^{o-1} should be subscripted by k since they depend on the kth pattern.

where we use the convenient shorthand notation δ_j^o to denote $\partial E^k / \partial a_j^o$, the derivative with respect to the input of a particular node in the network.

Similarly, the only dependence of E^k on the specific weight w_{ji}^{o-1} is through the input a_j^{o-1} to the nonlinearity at layer $o-1$. Hence, the derivatives of E^k with respect to the weights w_{ji}^{o-1} are given by

$$\frac{\partial E^k}{\partial w_{ji}^{o-1}} = \frac{\partial E^k}{\partial a_j^{o-1}} \frac{\partial a_j^{o-1}}{\partial w_{ji}^{o-1}} = \delta_j^{o-1} z_i^{o-2} \tag{6.30}$$

where we have again used the notation δ_j^{o-1} for $\partial E^k / \partial a_j^{o-1}$. Equation (6.30) may be expanded using the chain rule for differentiation as

$$\delta_j^{o-1} = \sum_{l=1}^{n_o} \frac{\partial E^k}{\partial a_l^o} \frac{\partial a_l^o}{\partial a_j^{o-1}} = \sum_{l=1}^{n_o} \delta_l^o \frac{\partial a_l^o}{\partial a_j^{o-1}}$$

Using the relationships (6.28), we have

$$\frac{\partial a_l^o}{\partial a_j^{o-1}} = \frac{\partial}{\partial a_j^{o-1}} \left(\sum_{i=1}^{n_{o-1}} w_{li}^o \phi\left(a_i^{o-1}\right) \right) = w_{lj}^o \phi'\left(a_j^{o-1}\right)$$

giving the *back-propagation formula*

$$\delta_j^{o-1} = \phi'\left(a_j^{o-1}\right) \sum_{l=1}^{n_o} \delta_l^o w_{lj}^o \tag{6.31}$$

The above result allows the derivatives of E^k with respect to the inputs to a particular node to be expressed in terms of derivatives with respect to the inputs to nodes higher up the network; that is, nodes in layers closer to the output layer. Once calculated, these are combined with node outputs in equations of the form (6.29) and (6.30) to give derivatives with respect to the weights. It is is easy to extend this formulation to consider more than one hidden layer.

Equation (6.31) requires the derivatives with respect to the nonlinearities ϕ to be specified. For ϕ given by (6.21)

$$\phi'(a) = \phi(a)(1 - \phi(a))$$

For ϕ given by (6.22)

$$\phi'(a) = (1 - \phi(a)^2)$$

The derivative calculation also requires the initialisation of δ_j^o, the derivative of the kth component of the error with respect to a_j^o. For the sum-square error criterion with linear output units, this is

$$\delta_j^o = -2\left(t_{kj} - a_j^o\right) \tag{6.32}$$

The above scheme is applied recursively to each layer to calculate derivatives. The term *back-propagation networks* is often used to describe multilayer perceptrons employing such a calculation, though strictly *back-propagation* refers to the method of derivative calculation, rather than the type of network.

Bias terms

Each layer in the network, up to the output layer, may have biases: the terms α_{i0} and w_{j0} in the single hidden layer model of Equation (6.20). The bias node at any given layer may be regarded as a node with a fixed output of one [i.e. $\phi(.) \equiv 1$], which does not therefore require any connections to the preceding layer. We typically assign an index of 0 to a bias node, and extend the expression for a_j^o in (6.28) to

$$a_j^o = \sum_{i=1}^{n_{o-1}} w_{ji}^o z_i^{o-1} + w_{j0}^o$$

Hence (6.29) becomes

$$\frac{\partial E^k}{\partial w_{j0}^o} = \frac{\partial E^k}{\partial a_j^o} \frac{\partial a_j^o}{\partial w_{j0}^o} = \delta_j^o$$

and (6.30) becomes

$$\frac{\partial E^k}{\partial w_{j0}^{o-1}} = \frac{\partial E^k}{\partial a_j^{o-1}} \frac{\partial a_j^{o-1}}{\partial w_{j0}^{o-1}} = \delta_j^{o-1} \tag{6.33}$$

Note that under the convention (rule) $z_0^{o'} \equiv 1$ for the bias output from any layer o', these expressions for the bias term derivatives are the same as those for the full weight derivatives.

Optimisation using the derivatives

Now that we can evaluate the error and its derivatives, we can, in principle, use one of many nonlinear optimisation schemes available (Press *et al.*, 1992).

A simple, although inefficient, approach is to use the technique of *gradient descent* (also known as *steepest descent*). In gradient descent each weight is updated by an amount depending on the gradient of the error with respect to that weight

$$w_{ji}^o \rightarrow w_{ji}^o - \lambda \sum_{k=1}^{n} \frac{\partial E^k}{\partial w_{ji}^o}$$

where $\lambda > 0$ is the update rate. Such an approach can require a large number of steps before reaching a (local) minimum (Press *et al.*, 1992).

We have found that the *conjugate gradients algorithm* with the Polak–Ribière update scheme works well on many practical problems. For problems with a small number of parameters (less than about 250), the Broyden–Fletcher–Goldfarb–Shanno (BFGS) optimisation scheme (which is a quasi-Newton nonlinear optimisation method) is recommended (Webb *et al.*, 1988; Webb and Lowe, 1988). Storage of the inverse Hessian matrix ($n_p \times n_p$ where

n_p is the number of parameters) limits its use in practice for large networks. Both conjugate gradients and BFGS are described in Press *et al.* (1992).

For further details of optimisation algorithms see, for example, van der Smagt (1994) and Karayiannis and Venetsanopolous (1993).

Stopping criterion

The most common stopping criterion used in the nonlinear optimisation schemes is to terminate the algorithm when the relative change in the error is less than a specified amount, or the maximum number of allowed iterations has been exceeded.

An alternative that has been employed for classification problems is to cease training when the classification error (either on the training set or preferably a separate validation set) stops decreasing (see Chapter 9 for model selection). Another strategy is based on growing and pruning in which a network larger than necessary is trained and parts that are not needed are removed. See Reed (1993) for a survey of pruning algorithms.

Training strategies

There are several training strategies that may be used to minimise the error. The one that we have described above uses the complete set of patterns to calculate the error and its derivatives with respect to the weights, w. These may then be used in a nonlinear optimisation scheme to find the values for which the error is a minimum. This is known as *batch learning*.

A different approach that has been commonly used is to update the parameters using individual gradients $\partial E^k/\partial w$. This is a form of *online learning*. At each iteration of the chosen nonlinear optimisation procedure, rather than calculating the gradient using all the patterns, the gradient for just a single pattern is used. The pattern considered varies (randomly or in sequence) with each iteration of the optimisation scheme. Such an approach is commonly used with the gradient descent technique. In practice, although the total error will decrease, it will not converge to a minimum, but tend to fluctuate around the local minimum. An improvement is a *stochastic update* scheme which also uses single patterns at a time to calculate the gradient, but decreases the influence of each gradient calculation (i.e. decreases the update rate) as the iteration number increases. This allows the approach to converge to a minimum. The stochastic update scheme can reduce the computational cost of MLP training using a large dataset, and also have beneficial effects through allowing greater exploration of the parameter space than full batch updating.

Finally, incremental training is a heuristic technique for speeding up the overall learning time of neural networks whilst simultaneously improving the final classification performance. The method is simple: first the network is partially trained on a subset of the training data (there is no need to train to completion) and then the resulting network is further tuned using the entire training database. The motivation is that the subset training will perform a coarse search of weight space to find a region that ‘solves’ the initial problem. The hope is that this region will be a useful place to start for training on the full dataset. The technique is often used with a small subset used for initial training and progressing through larger and larger training databases as the performance increases. The number of patterns used in each subset will vary according to the task although one should ensure that sufficient patterns representative of the data distribution are present to prevent the network over-fitting the subset data. In a discrimination problem, there should, of course, be samples from all classes in the initial training subset.

6.4.3.3 Variations to the optimisation scheme

Most algorithms for determining the weights of an MLP (including that described above) do not make direct use of the fact that the discriminant function is (usually specified to be) linear in the final layer of weights w. Specifically, they solve for all the weights using a nonlinear optimisation scheme. Another approach is to alternate between solving for the final layer weights $\boldsymbol{w}$ (using a linear pseudo-inverse method) and adjusting the parameters of the nonlinear functions ϕ, namely $\boldsymbol{\alpha}$ [see Equation (6.20)]. This is equivalent to regarding the final layer weights as a function of the parameters $\boldsymbol{\alpha}$ (Webb and Lowe, 1988; Stäger and Agarwal, 1997).

6.4.4 Modelling capacity of the MLP

If we are free to choose the weights and nonlinear functions, then a single-layer MLP can approximate any continuous function to any degree of accuracy if and only if the nonlinear function is not a polynomial (Leshno *et al.*, 1993). Since such networks are simulated on a computer, the nonlinearity of the hidden nodes are finite polynomials. Thus, they are not capable of universal approximation (Wray and Green, 1995). However, for most practical purposes, the lack of a universal approximation property is irrelevant.

Classification properties of MLPs are addressed by Faragó and Lugosi (1993). Let L^* be the Bayes' error rate (see Chapter 1); let g_{kn} be an MLP with one hidden layer of k nodes (with step function nonlinearities) trained on n samples to minimise the number of errors on the training data [with error probability $L(g_{kn})$]; then, provided k is chosen so that

$$k \to \infty$$

$$k\frac{\log(n)}{n} \to 0$$

as $n \to \infty$, then

$$\lim_{n\to\infty} L(g_{kn}) = L^*$$

with probability 1. Thus, the classification error approaches the Bayes' error as the number of training samples becomes large, provided k is chosen to satisfy the conditions above. However, although this result is attractive, the problem of choosing the parameters of g_{kn} to give minimum errors on a training set is computationally difficult.

6.4.5 Logistic classification

We now consider an alternative error criterion to the least squares error used within Section 6.4.3. The criterion is based upon the generalised logistic model.

6.4.5.1 Generalised logistic model

In the multiclass case, the basic assumption for the *generalised logistic discrimination* is

$$\log\left(\frac{p(\boldsymbol{x}|\omega_s)}{p(\boldsymbol{x}|\omega_C)}\right) = \beta_{s0} + \boldsymbol{\beta}_s^T\boldsymbol{\phi}(\boldsymbol{x}), \quad s = 1, \ldots, C-1$$

where $\boldsymbol{\phi}(\boldsymbol{x})$ is a nonlinear function of the variables $\boldsymbol{x}$, with parameters $\{\boldsymbol{\mu}\}$; that is, the log-likelihood ratio is a linear combination of the nonlinear functions, ϕ. The posterior probabilities are of the form (see Chapter 5)

$$p(\omega_s|\boldsymbol{x}) = \frac{\exp(\beta'_{s0} + \boldsymbol{\beta}_s^T \boldsymbol{\phi}(\boldsymbol{x}))}{1 + \sum_{j=1}^{C-1} \exp\left(\beta'_{j0} + \boldsymbol{\beta}_j^T \boldsymbol{\phi}(\boldsymbol{x})\right)}, \quad s = 1, \ldots, C-1$$

$$p(\omega_C|\boldsymbol{x}) = \frac{1}{1 + \sum_{j=1}^{C-1} \exp\left(\beta'_{j0} + \boldsymbol{\beta}_j^T \boldsymbol{\phi}(\boldsymbol{x})\right)} \tag{6.34}$$

where $\beta'_{s0} = \beta_{s0} + \log(p(\omega_s)/p(\omega_C))$. Discrimination depends solely on the $C-1$ functions $\beta'_{s0} + \boldsymbol{\beta}_s^T \phi(\boldsymbol{x})(s = 1, \ldots, C-1)$ with the decision:

$$\text{assign } \boldsymbol{x} \text{ to class } \omega_j \text{ if } \max_{s=1,\ldots,C-1} \beta'_{s0} + \boldsymbol{\beta}_s^T \boldsymbol{\phi}(\boldsymbol{x}) = \beta'_{j0} + \boldsymbol{\beta}_j^T \boldsymbol{\phi}(\boldsymbol{x}) > 0$$

else assign $\boldsymbol{x}$ to class ω_C.

The parameters of the model (in this case, the $\boldsymbol{\beta}$ terms and $\{\boldsymbol{\mu}\}$, the set of parameters on which ϕ depends) may be determined using a maximum likelihood procedure. Following the same arguments as for logistic discrimination in Chapter 5, for a dataset $\{\boldsymbol{x}_1, \ldots, \boldsymbol{x}_n\}$ we seek the parameters that maximise

$$\log(L') = \sum_{i=1}^{C} \sum_{\boldsymbol{x}_r \in \omega_i} \log(p(\omega_i|\boldsymbol{x}_r)) \tag{6.35}$$

These parameters can be found through the use of some form of numerical optimisation scheme, such as conjugate gradient methods or quasi-Newton procedures (Press *et al.*, 1992).

The parameters of the generalised logistic model may also be found using a least square error procedure by minimising

$$\sum_{i=1}^{n} |\boldsymbol{t}_i - \boldsymbol{p}(\omega_{j(i)}|\boldsymbol{x}_i)|^2$$

where $\boldsymbol{t}_i = (0, 0, \ldots, 0, 1, 0, \ldots, 0)^T$ (a one in the jth position) for $\boldsymbol{x}_i \in \omega_j$ and $\boldsymbol{p}$ is the model for the vector of *a posterior* probabilities given by (6.34).

6.4.5.2 MLP model

We now consider the generalised logistic discrimination for an MLP model. The basic assumption is

$$\log\left(\frac{p(\boldsymbol{x}|\omega_j)}{p(\boldsymbol{x}|\omega_C)}\right) = \sum_{i=1}^{m} w_{ji}^o \phi_i\left(a_i^{o-1}\right) + w_{j0}^o, \quad j = 1, \ldots, C-1$$

where ϕ_i is the ith nonlinearity (a logistic function) and a_i^{o-1} represents the input to the ith nonlinearity. The terms a_i^{o-1} may be linear combinations of the outputs of a previous layer in a layered network system and depend on the parameters of a network. The generalised logistic discrimination posterior probabilities of (6.34) become

$$p(\omega_j|\mathbf{x}) = \frac{\exp\left(\sum_{i=1}^{m} w_{ji}^o \phi_i\left(a_i^{o-1}\right) + w'_{j0}\right)}{1+\sum_{s=1}^{C-1}\exp\left(\sum_{i=1}^{m} w_{si}^o \phi_i\left(a_i^{o-1}\right) + w'_{s0}\right)}, \quad j = 1,\ldots,C-1$$

$$p(\omega_C|\mathbf{x}) = \frac{1}{1+\sum_{s=1}^{C-1}\exp\left(\sum_{i=1}^{m} w_{si}^o \phi_i\left(a_i^{o-1}\right) + w'_{s0}\right)} \quad (6.36)$$

where $w'_{j0} = w_{j0}^o + \log(p(\omega_j)/p(\omega_C))$. The discrimination rule is to assign to the class with the largest posterior probability.

In terms of the MLP model (Figure 6.5), Equation (6.36) can be thought of as a final normalising layer in which w_{ji}^o are the final layer weights and the set of $C-1$ inputs to the final layer (terms of the form $a_j^o = \sum_i w_{ji}^o \phi_i(a_i^{o-1}) + w_{j0}^o$, $j = 1,\ldots, C-1$) is offset by prior terms $\log(p(\omega_j)/p(\omega_C))$ and normalised through (6.36) to give C outputs $p(\omega_j|\mathbf{x})$, $j = 1,\ldots,C$ (the posterior class probability for class ω_C being specified so that the sum of the posterior probabilities is one).

Estimates of the parameters of the model may be achieved by maximising (6.35) using the posterior probability estimates given in (6.36). This is equivalent to minimising

$$E = -\sum_{i=1}^{n}\sum_{k=1}^{C} t_{ik}\log(p(\omega_k|\mathbf{x}_i)) \quad (6.37)$$

where t_{ik} is the kth element of the vector $\mathbf{t}_i$, which is the target vector for $\mathbf{x}_i$: a vector of zeros, except that there is a 1 in the position corresponding to the class of $\mathbf{x}_i$. The above criterion is of the form $\sum_i E^i$, where

$$E^i = -\sum_{k=1}^{C} t_{ik}\log(p(\omega_k|\mathbf{x}_i)) \quad (6.38)$$

Given this expression for the error, and noting that

$$p(\omega_j|\mathbf{x}) = \frac{\exp\left(a_j^o + \log(p(\omega_j)/p(\omega_C))\right)}{1+\sum_{s=1}^{C-1}\exp\left(a_s^o + \log(p(\omega_{js})/p(\omega_C))\right)}, \quad j = 1,\ldots,C\ 1$$

$$p(\omega_C|\mathbf{x}) = \frac{1}{1+\sum_{s=1}^{C-1}\exp\left(a_s^o + \log(p(\omega_{js})/p(\omega_C))\right)} \quad (6.39)$$

we can use the back-propagation algorithm considered previously. This requires $\delta_j^o = \partial E^i/\partial a_j^o$, $j = 1,\ldots, C-1$, which is given by

$$\delta_j^o = \frac{\partial E^i}{\partial a_j^o} = -t_{ij} + p(\omega_j|\mathbf{x}_i)$$

the difference between the jth component of the target vector and the jth output for pattern i [compare with (6.32); see Exercises].

6.4.6 Example application study

The problem

This application involves prediction of the power that is anticipated to be produced by each turbine in a wind farm, for use as a maintenance indicator (Shuhui *et al.*, 2001).

Summary

Lower-than-expected power outputs from a wind turbine may be an indicator of the need for maintenance. The study therefore predicts the power outputs for each turbine in a wind farm, which are then compared against the actual outputs. Power prediction is complicated because the power generated by wind turbines changes rapidly due to variations in wind speed and direction. Moreover, wind conditions vary at different locations in a wind farm, and usually are not measured at each turbine. The study uses a separate MLP for each turbine to predict power generation, based upon wind measurements (velocity and direction) at two locations in the wind farm.

The data

The study used real data collected at Central and South West Services Fort Davis wind farm in Texas, USA. In 1996 the wind farm consisted of 12 turbines, and two meteorological towers providing wind measurements (speed and direction) collected as 10 min averages. For each turbine, 1500 sets of 10 min average data measurements from March 1996 were used to train (optimise) the MLP neural network for the turbine under consideration. Performance was assessed using data from April 1996.

The model

Single hidden layer MLPs were used, with 4 input nodes (wind speed and direction measurements from the two meteorological towers), 8 hidden nodes, and 1 output node. Bias nodes were added to the input and hidden layers. Hyperbolic tangent hidden layer nonlinearities were used. A separate MLP was trained for each wind turbine, with target outputs the power output of the turbine.

Training procedure

Data preprocessing took place before the wind speed and direction measurements were used as inputs to the MLP networks. Higher wind speeds were compressed to reflect limitations on power output when winds are strong. Additionally, since power output is less influenced by wind direction than wind speed, the wind direction measurement inputs were compressed into a more limited range than the wind speed. This was done in such a manner as to emphasise wind direction in certain directions with environmental significance. The precise forms of these preprocessing functions were determined by trial and error. The preprocessing was found to both reduce MLP training time, and improve prediction performance.

Randomly generated initial weights were used in the training process, with back-propagation calculations of mean-squared-error function derivatives. Weights were reinitialised if performance was unsatisfactory.

Results

Prediction performance from the MLPs was better than that for a model based upon the turbine manufacturer's performance curve for the turbines (which amongst other things assumes that the wind speed and direction at each turbine is the same as at the measurement locations). The percentage differences between estimated and measured power generation for all the turbines were around 2% in the April 1996 test data.

6.4.7 Bayesian MLP networks

An interesting development of MLPs is the application of Bayesian inference techniques to the fitting of neural network models, which has been shown to lead to practical benefits on real problems.

In the *predictive approach* (Ripley, 1996), the posterior distribution of the observed target value (t) given an input vector (x) and the dataset ($\mathcal{D}$), $p(t|x, \mathcal{D})$, is obtained by integrating over the posterior distribution of network weights, w

$$p(t|x, \mathcal{D}) = \int p(t|x, w)p(w|\mathcal{D})dw \tag{6.40}$$

where $p(w|\mathcal{D})$ is the *posterior distribution* for w and $p(t|x, w)$ is the distribution of outputs for the given model, w, and input, x. For a dataset $\mathcal{D} = \{(x_i, t_i), i = 1, \ldots, n\}$ of measurement vectors x_i with associated targets t_i, the posterior of the weights w may be written

$$p(w|\mathcal{D}) = \frac{p(w, \mathcal{D})}{p(\mathcal{D})} = \frac{1}{p(\mathcal{D})}p(w)\prod_{i=1}^{n} p(t_i|x_i, w)p(x_i|w)$$

assuming independent samples. Further, if we assume that $p(x_i|w)$ does not depend on w, then

$$p(w|\mathcal{D}) \propto p(w)\prod_{i=1}^{n} p(t_i|x_i, w)$$

The *maximum a posteriori* (MAP) estimate of w is that for which $p(w|\mathcal{D})$ is a maximum. Specifying a spherical Gaussian prior distribution for the weights

$$p(w) \propto \exp\left(-\frac{\alpha}{2}\|w\|^2\right)$$

where the parameter α defines the diagonal covariance matrix $(1/\alpha)I$; and a zero-mean spherical Gaussian noise model so that

$$p(t|x, w) \propto \exp\left(-\frac{\beta}{2}|t - g(x; w)|^2\right)$$

for the diagonal covariance matrix, $(1/\beta)I$ and network output, $g(x; w)$; then

$$p(w|\mathcal{D}) \propto \exp\left(-\frac{\beta}{2}\sum_{i} |t_i - g(x_i; w)|^2 - \frac{\alpha}{2}\|w\|^2\right)$$

The MAP estimate is that for which

$$S(\boldsymbol{w}) \triangleq \frac{1}{2}\sum_i |\boldsymbol{t}_i - \boldsymbol{g}(\boldsymbol{x}_i; \boldsymbol{w})|^2 + \frac{\alpha}{2\beta}\|\boldsymbol{w}\|^2 \tag{6.41}$$

is a minimum. This is the regularised solution [Equation (6.10)] derived as a solution for the MAP estimate for the parameters $\boldsymbol{w}$.

Equation (6.41) is not a simple function of $\boldsymbol{w}$ and may have many local minima (many local peaks of the posterior density), and the integral in (6.40) is computationally difficult to evaluate. Bishop (1995) approximates $S(\boldsymbol{w})$ using a Taylor expansion around its minimum value, $\boldsymbol{w}_{\text{MAP}}$ (although, as we have noted above, there may be many local minima) to write

$$S(\boldsymbol{w}) = S(\boldsymbol{w}_{\text{MAP}}) + \frac{1}{2}(\boldsymbol{w} - \boldsymbol{w}_{\text{MAP}})^T \boldsymbol{A}(\boldsymbol{w} - \boldsymbol{w}_{\text{MAP}})$$

where $\boldsymbol{A}$ is the *Hessian* matrix

$$A_{ij} = \left.\frac{\partial}{\partial w_i}\frac{\partial}{\partial w_j} S(\boldsymbol{w})\right|_{\boldsymbol{w}=\boldsymbol{w}_{\text{MAP}}}$$

to give

$$p(\boldsymbol{w}|\mathcal{D}) \propto \exp\left\{-\frac{\beta}{2}(\boldsymbol{w} - \boldsymbol{w}_{\text{MAP}})^T \boldsymbol{A}(\boldsymbol{w} - \boldsymbol{w}_{\text{MAP}})\right\}$$

Also, expanding $g(\boldsymbol{x}; \boldsymbol{w})$ around $\boldsymbol{w}_{\text{MAP}}$ (assuming for simplicity a scalar quantity)

$$g(\boldsymbol{x}; \boldsymbol{w}) = g(\boldsymbol{x}; \boldsymbol{w}_{\text{MAP}}) + (\boldsymbol{w} - \boldsymbol{w}_{\text{MAP}})^T \boldsymbol{h}$$

where $\boldsymbol{h}$ is the gradient vector, evaluated at $\boldsymbol{w}_{\text{MAP}}$, gives

$$p(t|\boldsymbol{x}, \mathcal{D}) \propto \int \exp\left\{-\frac{\beta}{2}[t - g(\boldsymbol{x}; \boldsymbol{w}_{\text{MAP}}) - \Delta\boldsymbol{w}^T\boldsymbol{h}]^2 - \frac{\beta}{2}\Delta\boldsymbol{w}^T\boldsymbol{A}\Delta\boldsymbol{w}\right\} d\boldsymbol{w} \tag{6.42}$$

where $\Delta\boldsymbol{w} = (\boldsymbol{w} - \boldsymbol{w}_{\text{MAP}})$. This may be evaluated to give (Bishop, 1995)

$$p(t|\boldsymbol{x}, \mathcal{D}) = \frac{1}{(2\pi\sigma_t^2)^{\frac{1}{2}}}\exp\left\{-\frac{1}{2\sigma_t^2}(t - g(\boldsymbol{x}; \boldsymbol{w}_{\text{MAP}}))^2\right\} \tag{6.43}$$

where the variance σ_t^2 is given by

$$\sigma_t^2 = \frac{1}{\beta}(1 + \boldsymbol{h}^T\boldsymbol{A}^{-1}\boldsymbol{h}) \tag{6.44}$$

Equation (6.43) describes the distribution of output values for a given input value, $\boldsymbol{x}$, given the dataset. The expression can be used to provide error bars on the estimate. Further discussion of the Bayesian approach is given by Bishop (1993) and Ripley (1996).

The approach has been extended by Cawley and Talbot (2005) to use a diagonal (rather than spherical) Gaussian prior distribution for the weights. This allows redundant weights to be removed from the network, hence providing a means to reduce over-fitting of the training data.

6.4.8 Projection pursuit

An alternative projection-based nonlinear-modelling approach to the MLP is *projection pursuit*. Projection pursuit has been used for exploratory data analysis, density estimation and in multiple regression problems (Friedman and Tukey, 1974; Friedman and Stuetzle, 1981; Friedman *et al.*, 1984; Huber, 1985; Friedman, 1987; Jones and Sibson, 1987).

Projection pursuit models the regression surface as a sum of nonlinear functions of linear combinations of the variables (as in the basic MLP model)

$$y = \sum_{j=1}^{m} \phi_j \left(\boldsymbol{\beta}_j^T \boldsymbol{x} \right)$$

where the parameters $\boldsymbol{\beta}_j$, $j = 1, \ldots, m$, and the function ϕ_j are determined from the data and y is the response variable. However, unlike the MLP model, the nonlinear functions are also determined from the data. Friedman and Stuetzle (1981), and Jones and Sibson (1987) describe an optimisation procedure. Zhao and Atkeson (1996) use an RBF smoother for the nonlinear functions, and solve for the weights as part of the optimisation process (Kwok and Yeung, 1996).

Projection pursuit may produce projections of the data that reveal structure that is not apparent by using the coordinates axes or simple projections such as principal components (see Chapter 10). However, a disadvantage is that 'projection pursuit will uncover not only true but also spurious structure' (Huber, 1985). In a comparative study between projection pursuit and the MLP on simulated data, Hwang *et al.* (1994) report similar accuracy and similar computation times, but projection pursuit requires fewer functions.

6.4.9 Summary

The MLP is a model that, in its simplest form, can be regarded as a generalised linear discriminant function in which the nonlinear functions ϕ are flexible and adapt to the data. This is the way that we have chosen to introduce it in this chapter as it forms a natural progression from linear discriminant analysis, through generalised linear discriminant functions with fixed nonlinearities to the MLP. The parameters of the model are determined through a nonlinear optimisation scheme, which is one of the drawbacks of the model. Computation time may be excessive.

There have been many assessments of gradient-based optimisation algorithms, variants and alternatives. Webb *et al.* (1988; see also Webb and Lowe, 1988) compared various gradient-based schemes on several problems. They found that the Levenberg–Marquardt optimisation scheme gave best overall performance for networks with a small number of parameters, but favoured conjugate gradient methods for networks with a large number of parameters. Further comparative studies include those of Karayiannis and Venetsanopolous (1993), van der Smagt (1994), Stäger and Agarwal (1997) and Alsmadi *et al.* (2009). The addition of extra terms to the error involving derivatives of the error with respect to the input has been considered by Drucker and Le Cun (1992) as a means of improving generalisation (Bishop, 1993; Webb, 1994). The latter approach of Webb was motivated from an *error-in-variables* perspective (noise on the inputs). The addition of noise to the inputs as a means of improving generalisation has also been assessed by Holmström and Koistinen (1992) and Matsuoka (1992).

The MLP is a very flexible model, giving good performance on a wide range of problems in discrimination and regression. We have presented only a very basic model. There are many variants, some adapted to particular types of problem such as time series (for example, time-delay neural networks). Growing and pruning algorithms for MLP construction have also been considered in the literature, as well as the introduction of regularisation in the optimisation criteria in order to prevent over-fitting of the data. The implementation in hardware for some applications has received attention. There are several commercial products available for MLP design and implementation.

6.5 Application studies

The term 'neural networks' is used to describe models that are combinations of many simple processing units – for example, the RBF is a linear combination of kernel basis functions and the multilayer perceptron is a weighted combination of logistic (or other nonlinearity) units – but the boundary between what does and what does not constitute a neural network is rather vague. In many respects, the distinction is irrelevant, but it still persists, largely through historical reasons relating to application domain, with neural network methods tending to be developed more in the engineering and computer science literature.

Applications of neural networks are widespread and many classification studies mentioned in this book will include an assessment of an MLP model in particular. Zhang (2000) surveys the area of classification from a neural network perspective. Some review articles in specific application domains include:

- Face processing. Valentin *et al.* (1994) review connectionist models (combining many simple processing units) for face recognition (see also Samal and Iyengar, 1992).
- Speech recognition. Morgan and Bourlard (1995) review the use of artificial neural networks (ANNs) in automatic speech recognition, and describe hybrid hidden Markov models and ANN models.
- Image compression. A summary of the use of neural network models as signal processing tools for image compression is provided by Dony and Haykin (1995) and Jiang (1999).
- Fault diagnosis. Sorsa *et al.* (1991) consider several neural architectures, including the MLP, for process fault diagnosis.
- Chemical science. Sumpter *et al.* (1994) discuss neural computing in chemical science.
- Target recognition. Reviews of neural networks for automatic target recognition are provided by Roth (1990) and Rogers *et al.* (1995).
- Financial engineering. Refenes *et al.* (1997) and Burrell and Folarin (1997) consider neural networks in financial engineering.

There have been many special issues of journals focusing on different aspects of neural networks, including everyday applications (Dillon *et al.*, 1997), industrial electronics

(Chow, 1993), general applications (Lowe, 1994), signal processing (Constantinides *et al.*, 1997; Unbehauen and Luo, 1998), target recognition, image processing (Chellappa *et al.*, 1998), machine vision (Dracopoulos and Rosin, 1998), oceanic engineering (Simpson, 1992), process engineering (Fernändez de Cañete and Bulsari, 2000), human–computer interaction (Yasdi, 2000), financial engineering (Abu-Mostafa *et al.*, 2001) and data mining and knowledge discovery (Bengio *et al.*, 2000). This latter area is one that is currently receiving considerable attention. Over the last twenty years there has been a huge increase in the amount of information available from the Internet, business and other sources. One of the challenges in the area of *data mining and knowledge discovery* is to develop models that can handle large datasets, with a large number of variables (high dimensional). Many standard data analysis techniques may not scale suitably.

There have been many comparative studies assessing the performance of neural networks in terms of speed of training, memory requirements, and classification performance (or prediction error in regression problems), comparing the results with statistical classifiers. Probably the most comprehensive comparative study is that provided by the *Statlog* project (Michie *et al.*, 1994). A neural network method (an RBF) gave best performance on only one out of 22 datasets, but provided close to best performance in nearly all cases. Other comparative studies include, for example, assessments on character recognition (Logar *et al.*, 1994), fingerprint classification (Blue *et al.*, 1994) and remote sensing (Serpico *et al.*, 1996).

Applications of MLPs, trained using a Bayesian approach, to image analysis are given by Vivarelli and Williams (2001) and Lampinen and Vehtari (2001). Skabar (2005) uses Bayesian MLP techniques to predict which geographical areas might contain certain types of mineral deposits.

There is an increasing amount of application and comparative studies involving SVMs. These include:

- Financial time series prediction. Cao and Tay (2001) investigate the feasibility of using SVMs in financial forecasting [see also Van Gestel *et al.* (2001) and Cao *et al.* (2009)]. An SVM with Gaussian kernel is applied to multivariate data (5 or 8 variables) relating to the closing price of the S&P Daily Index in the Chicago Mercantile.

- Drug design. This is an application in structure–activity relationship analysis, a technique used to reduce the search for new drugs. Combinatorial chemistry enables the synthesis of millions of new molecular compounds at a time. Statistical techniques that direct the search for new drugs are required to provide an alternative to testing every molecular combination. Burbidge *et al.* (2001) compare SVMs with an RBF network and a classification tree (see Chapter 7). The training time for the classification tree was much smaller than for the other methods, but significantly better performance (measured in terms of error rate) was obtained with the SVM.

- Cancer diagnosis. There have been several applications of SVMs to disease diagnosis. Furey *et al.* (2000) address the problem of tissue sample labelling using measurements from DNA microarray experiments. The datasets comprise measurements on ovarian tissue; human tumour and normal colon tissues; and bone marrow and blood samples from patients with leukaemia. Similar performance to a linear perceptron was achieved. Further cancer studies are reported by Guyon *et al.* (2002) and Ramaswamy *et al.* (2001).

- Radar image analysis. Zhao *et al.* (2000) compare three classifiers, including an SVM, on an automatic target recognition task using synthetic aperture radar (SAR) data.

Experimental results show that the SVM and an MLP gave similar performance, but superior to nearest neighbour.

- Prediction of protein secondary structure. As a step towards the goal of predicting three-dimensional protein structures directly from protein sequences Hua and Sun (2001) use an SVM to classify secondary structure (as helix, sheets, or coil) based on sequence features.
- Vehicle driver assistance systems. Schaack *et al.* (2009) use SVMs within a system for detecting pedestrians and vehicles using video imagery. Coordinate transformations were made using MLPs.
- Handwriting recognition. Bahlmann *et al.* (2002) use an SVM for online handwriting recognition. The SVM utilises an application specific kernel incorporating ideas from *dynamic time warping* to deal with varying lengths and temporal distortion of online handwriting.
- Computer forensics. De Vel *et al.* (2001) use SVMs to analyse e-mail messages. In the growing field of *computer forensics*, of particular interest to investigators is the misuse of e-mail for the distribution of messages and documents that may be unsolicited, inappropriate, unauthorised or offensive. The objective of the study was to classify e-mails as belonging to a particular author.
- Text categorisation. Joachims (1998) and Lodhi *et al.* (2002) consider the use of SVMs for categorising text documents. Joachims (1998) uses features based on word frequencies within each document (a *bag of words* model), along with polynomial or Gaussian kernels. Lodhi *et al.* (2002) adopt a different approach of representing documents as symbol sequences, requiring the development of a string subsequence kernel, specifically designed for text sequences.
- Plagiarism detection. With the advent of the Internet, accusations of plagiarism are becoming more common, particularly with regard to student essays. Diederich (2006) advocates the use of SVMs for attributing authorship based upon an analysis of the words present in documents.

6.6 Summary and discussion

In this chapter we have developed the basic linear discriminant model to one in which the model is essentially linear, but the decision boundaries are nonlinear. The RBF model is implemented in a straightforward manner and, in its simplest form, it requires little more than a matrix pseudo-inverse operation to determine the weights of the network. This hides the fact that optimum selection of the numbers and positions of centres is a more complicated process. Nevertheless, simple rules of thumb can result in acceptable values giving good performance.

The MLP takes a sum of univariate nonlinear functions ϕ of linear projections of the data, $\boldsymbol{x}$, onto a weight vector $\boldsymbol{\alpha}$, and uses this for discrimination (in a classification problem). The nonlinear functions are of a prescribed form (usually logistic) and a weighted sum is formed. Optimisation of the objective function is performed with respect to the projection directions and the weights in the sum.

There have been many links between neural networks, such as the MLP, and established statistical techniques. The basic idea behind neural network methods is the combination of simple nonlinear processing in a hierarchical manner, together with efficient algorithms for their optimisation. In this chapter we have confined our attention primarily to single hidden layer networks, where the links to statistical techniques are more apparent. However, more complex networks can be built for specific tasks. Also, neural networks are not confined to the feed-forward types for supervised pattern classification or regression as presented here. Networks with feedback from one layer to the previous layer, and unsupervised networks have been developed for a range of applications.

The SVM defines the basis functions implicitly through the definition of a kernel function in the data space and has been found to give very good performance on many problems. There are few parameters to set: the kernel parameters and the regularisation parameter. These can be varied to give optimum performance on a validation set. The SVM focuses on the decision boundary and the standard model is not suitable for the nonstandard situation where the operating conditions differ from the training conditions due to drifts in values for costs and priors.

6.7 Recommendations

The nonlinear discriminant methods described in this chapter are easy to implement and there are many sources of software for applying these methods to a dataset. Before applying these techniques you should consider the reasons for doing so. Do you believe that the decision boundary is nonlinear? Is the performance provided by linear techniques below that desired or believed to be achievable? Moving to neural network techniques or SVMs may be one way to achieve improved performance. This is not guaranteed. If the classes are not separable, a more complex model will not make them so. It may be necessary to make measurements on additional variables.

The following recommendations are made:

1. A simple pattern recognition technique (e.g. Gaussian classifier, Chapter 2, k nearest neighbour, Chapter 4, linear discriminant analysis, Chapter 5) should be implemented as a baseline before considering neural network methods.
2. A simple RBF (unsupervised selection of centres, weights optimised using a squared error criterion) should be tried to get a feel for whether nonlinear methods provide some gain for your problem.
3. Data preprocessing based on the specific application under consideration should be used (this may help reduce the risk of MLP training becoming stuck in local minima).
4. Data should be standardised (e.g. linear rescaling to have zero mean and unit variance for each input; or linear rescaling so that each element lies within the interval $[-1, 1]$) before application of the algorithms. Be sure to apply the same linear rescalings to test data as are derived and applied for training data.
5. Take care in model selection so that over-training does not result. Use cross-validation or a separate validation set for model selection. For RBFs use a regularised solution for the weights.
6. An MLP should be trained using a batch training method unless the dataset is very large (in this case, divide the dataset into subsets, randomly selected from the original dataset).

7. For a model that provides approximations to the posterior probabilities that enable changes of priors and costs to be incorporated into a trained model, an RBF should be used.

8. For classification problems in high dimensional spaces where training data are representative of test conditions and misclassification rate is an acceptable measure of classifier performance, SVMs should be implemented.

6.8 Notes and references

There are many developments of the techniques described in this chapter, in addition to other neural network methods. A description of these is beyond the scope of this book, but the use of neural techniques in pattern recognition and the relationship to statistical and structural pattern recognition can be found in the book by Schalkoff (1992). Views of neural networks from statistical perspectives, and relationships to other pattern classification approaches, are provided by Barron and Barron (1988), Ripley (1994, 1996), Cheng and Titterington (1994) and Holmström *et al.* (1997).

Bishop (1995, 2007) and Theodoridis and Koutroumbas (2008) both provide thorough accounts of feed-forward neural networks, whilst Duda *et al.* (2001) provide a good account of the MLP and related procedures. A more engineering perspective is provided by Haykin (1994). Relationships to other statistical methods and other insights are described by Ripley (1996). Tarassenko (1998) provides a basic introduction to neural network methods (including RBFs, MLPs, recurrent networks and unsupervised networks) with an emphasis on applications.

A good summary of Bayesian perspectives is given in the book by Ripley (1996); see also Bishop (1995, 2007), MacKay (1995), Buntine and Weigend (1991) and Thodberg (1996).

Although many of the features of SVMs can be found in the literature of the 1960s (large margin classifiers, optimisation techniques and sparseness, slack variables), the basic SVM for nonseparable data was not introduced until 1995 (Cortes and Vapnik, 1995). The book by Vapnik (1998) provides an excellent description of SVMs. A very good self-contained introduction is provided by Cristianini and Shawe-Taylor (2000). The tutorial by Burges (1998) is an excellent concise account. A comprehensive treatment of this field is provided in the book by Schölkopf and Smola (2001).

MATLAB™ routines for RBFs and MLPs are documented in the book by Nabney (2002). MLPs, SVMs and RBFs are available in the open source WEKA Data Mining Software, described in Witten and Frank (2005). Many packages are available for implementing SVMs, such as the freely available LIBSVM C++ software package (Chang and Lin, 2011; Hsu *et al.*, 2003).

Exercises

Dataset 1: Five hundred samples in training, validation and test sets; d-dimensional; three classes; class $\omega_1 \sim N(\boldsymbol{\mu}_1, \boldsymbol{\Sigma}_1)$; class $\omega_2 \sim 0.5N(\boldsymbol{\mu}_2, \boldsymbol{\Sigma}_2) + 0.5N(\boldsymbol{\mu}_3, \boldsymbol{\Sigma}_3)$ (a mixture model, see Chapter 2); class $\omega_3 \sim 0.2N(\boldsymbol{\mu}_4, \boldsymbol{\Sigma}_4) + 0.8N(\boldsymbol{\mu}_5, \boldsymbol{\Sigma}_5)$; $\boldsymbol{\mu}_1 = (-2, 2, \ldots, 2)^T$, $\boldsymbol{\mu}_2 = (-4, -4, \ldots, -4)^T$, $\boldsymbol{\mu}_3 = (4, 4, \ldots, 4)^T$, $\boldsymbol{\mu}_4 = (0, 0, \ldots, 0)^T$, $\boldsymbol{\mu}_5 = (-4, 4, \ldots, 4)^T$ and $\boldsymbol{\Sigma}_i$ as the identity matrix; equal class priors.

Dataset 2: Generate time series data according to the iterative scheme

$$u_{t+1} = \frac{4\left(1 - \frac{\Delta^2}{2}\right) u_t - \left(2 + \mu\Delta\left(1 - u_t^2\right)\right) u_{t-1}}{2 - \mu\Delta\left(1 - u_t^2\right)}$$

initialised with $u_{-1} = u_0 = 2$. Plot the generated time series. Construct training and test sets of 500 patterns $(\boldsymbol{x}_i, \boldsymbol{t}_i)$ where $\boldsymbol{x}_i = (u_i, u_{i+1})^T$ and $\boldsymbol{t}_i = u_{i+2}$. Thus the problem is one of time series prediction: predict the next sample in the series given the previous two samples. Take $\mu = 4$, $\Delta = \pi/50$.

1. Compare an RBF classifier with a *k*-nearest-neighbour classifier. Take into account the type of classifier, computational requirements on training and test operations and the properties of the classifiers.
2. Compare and contrast an RBF classifier and a classifier based on kernel density estimation.
3. Implement a simple RBF classifier: *m* Gaussian basis functions of width *h* and centres selected randomly from the data. Using data from dataset 1, evaluate performance as a function of dimensionality, *d*, and number of basis functions, where *h* is chosen based on the validation set.
4. For dataset 2, train an RBF for a $\exp(-z^2)$ and a $z^2\log(z)$ nonlinearity for varying numbers of centres. Once trained, use the RBF in a generative mode: initialise u_{t-1}, u_t (as a sample from the training set), predict u_{t+1}; then predict u_{t+2} from u_t and u_{t+1} using the RBF. Continue for 500 samples. Plot the generated time series. Investigate the sensitivity of the final generated time series to starting values, number of basis functions and the form of the nonlinearity.
5. Consider the optimisation criterion [see Equations (5.40) and (6.6) for notation]

$$E = \|(-\boldsymbol{T}^T + \boldsymbol{W}\boldsymbol{\Phi}^T + \boldsymbol{w}_0\boldsymbol{1}^T)\boldsymbol{D}\|^2$$

where $\boldsymbol{D}$ is a diagonal matrix with $\boldsymbol{D}_{ii} = \sqrt{d_i}$ and the d_i are weights for the *i*th pattern. By minimising with respect to $\boldsymbol{w}_0$ and substituting back the solutions for $\boldsymbol{w}_0$, show that this may be written

$$E = \|(-\hat{\boldsymbol{T}}^T + \boldsymbol{W}\hat{\boldsymbol{\Phi}}^T)\boldsymbol{D}\|^2$$

where $\hat{\boldsymbol{T}}$ and $\hat{\boldsymbol{\Phi}}$ are zero-mean matrices. By minimising with respect to $\boldsymbol{W}$ and substituting into the expression for E, show that minimising E with respect to the parameters of the nonlinear functions, $\boldsymbol{\phi}$ is equivalent to maximising $\mathrm{Tr}(\boldsymbol{S}_B\boldsymbol{S}_T^{\dagger})$, where

$$\boldsymbol{S}_T = \frac{1}{n}\hat{\boldsymbol{\Phi}}^T\boldsymbol{D}^2\hat{\boldsymbol{\Phi}}$$

$$\boldsymbol{S}_B = \frac{1}{n^2}\hat{\boldsymbol{\Phi}}^T\boldsymbol{D}^2\hat{\boldsymbol{T}}\hat{\boldsymbol{T}}^T\boldsymbol{D}^2\hat{\boldsymbol{\Phi}}$$

6. For $\boldsymbol{D}$ equal to the identity matrix in the above and a target coding scheme

$$t_{ik} = \begin{cases} a_k & \boldsymbol{x}_i \in \omega_k \\ 0 & \text{otherwise} \end{cases}$$

determine the value of a_k for which $\boldsymbol{S}_B$ is the conventional between-class covariance matrix of the hidden output patterns.

7. Given a normal mixture model for each of C classes (see Chapter 2) with a common spherical covariance matrix across classes, together with class priors $p(\omega_i)$, $i = 1, \ldots, C$, construct an RBF network whose outputs are proportional to the posterior probabilities of class membership. Write down the forms for the centres and weights.

8. Implement an SVM classifier and assess performance using dataset 1. Use a Gaussian kernel and choose kernel width and regularisation parameter using a validation set. Investigate performance as a function of dimensionality, d.

9. Let K_1 and K_2 be kernels defined on $\mathbb{R}^d \times \mathbb{R}^d$. Show that the following are also kernels.

$$\begin{aligned} K(\boldsymbol{x}, \boldsymbol{z}) &= K_1(\boldsymbol{x}, \boldsymbol{z}) + K_2(\boldsymbol{x}, \boldsymbol{z}) & \\ K(\boldsymbol{x}, \boldsymbol{z}) &= aK_1(\boldsymbol{x}, \boldsymbol{z}) & \text{where } a \text{ is a positive real number} \\ K(\boldsymbol{x}, \boldsymbol{z}) &= g(\boldsymbol{x})g(\boldsymbol{z}) & \text{where } g(.) \text{ is a real-valued function on } \boldsymbol{x} \end{aligned}$$

10. Show that a support vector machine with a spherically-symmetric kernel function satisfying Mercer's conditions implements an RBF classifier with numbers of centres and positions chosen automatically by the SVM algorithm.

11. Discuss the differences and similarities between an MLP and an RBF classifier with reference to network structure, geometric interpretation, initialisation of parameters and algorithms for parameter optimisation.

12. For the logistic form of the nonlinearity in an MLP

$$\phi(z) = \frac{1}{1 + \exp(-z)}$$

show that

$$\frac{\partial \phi}{\partial z} = \phi(z)(1 - \phi(z))$$

13. Verify that error derivatives calculated using back-propagation [Equations (6.29)–(6.33)] are the same as those obtained from the direct expressions for the single layer model [Equations (6.24)–(6.27)].

14. Consider an MLP with a final normalising layer ('softmax')

$$z_j^o = \frac{\exp(a_j^o)}{\sum_{k=1}^{C} \exp(a_k^o)}$$

where a_j^o are the inputs to the final layer. How would you initialise the MLP to give a nearest class mean classifier?

15. Describe two ways of estimating the posterior probabilities of class membership without modelling the class densities explicitly and using Bayes' theorem; state the assumptions and conditions of validity.

16. Logistic classification. Given (6.38) and (6.39) derive the expression for the derivative

$$\frac{\partial E^i}{\partial a_j^0} = -t_{ij} + o_{ij}$$

where $\boldsymbol{o}_i$ is the output for input pattern $\boldsymbol{x}_i$.

17. Using the results that for a nonsingular matrix $\boldsymbol{A}$ and vector $\boldsymbol{u}$

$$(\boldsymbol{A} + \boldsymbol{u}\boldsymbol{u}^T)^{-1} = \boldsymbol{A}^{-1} - \boldsymbol{A}^{-1}\boldsymbol{u}\boldsymbol{u}^T\boldsymbol{A}^{-1}/(1 + \boldsymbol{u}^T\boldsymbol{A}^{-1}\boldsymbol{u})$$

derive the result (6.43) from (6.42) with σ_t^2 given by (6.44).

7

Rule and decision tree induction

Classification or decision trees lie at the intersection of the areas of statistical pattern recognition and machine learning. On one hand they are an example of a nonparametric approach that models the classification/regression function as a weighted sum of basis functions. On the other hand, they can be used to generate interpretable rules, which can be very important in many applications. The basic model is described and then developed in two directions: rule induction and the modelling of continuous functions.

7.1 Introduction

Classification trees, or decision trees, are conceptually very simple models that differ in one important aspect from the approaches to classification discussed in earlier chapters: the representation of a classification rule as a tree, with nodes labelled as features, edges labelled as values (or a set of values) of those features and leaves as class labels enables the interpretation of the classifier. This is essential in some applications where a degree of explanation of the classification decision is required.

Decision trees are capable of modelling complex nonlinear decision boundaries. Decision tree rules automatically determine those features important for classification as part of the construction process. Thus, feature selection (see Chapter 10) and classification are integrated. In common with many methods of discrimination that are based on an expansion of the classification rule into sums of basis functions – the radial basis function uses a weighted sum of univariate functions of radial distances; the multilayer perceptron uses a weighted sum of sigmoidal functions of linear projections – a decision tree model uses an expansion into indicator functions of multidimensional rectangles. It achieves this through a process known as recursive partitioning: the data space is recursively partitioned into smaller and smaller hyperrectangles. The MARS (multivariate adaptive regressions splines) model is a development of the recursive partitioning approach that permits continuous and smooth basis functions.

Statistical Pattern Recognition, Third Edition. Andrew R. Webb and Keith D. Copsey.

The *top-down induction of decision trees* (TDIDT) is another term used to describe tree construction. It is one of many methods of rule induction, particularly appropriate for categorical data and the study of approaches that integrate probabilistic methods and rule-based methods has received considerable attention in recent years.

There are three main sections to this chapter. The chapter begins with a description of the basic decision tree model and methods for its construction and then develops the idea in two different directions. Section 7.3 introduces rule-based approaches and presents two basic paradigms:

- rule extraction from decision trees (indirect methods);
- rule induction using a *sequential covering* approach (direct methods).

Section 7.4 considers the development of the recursive partitioning approach employed for decision tree construction to the modelling of continuous functions through the MARS procedure.

7.2 Decision trees

7.2.1 Introduction

A classification tree or a *decision tree* is an example of a multistage decision process. Instead of using the complete set of features jointly to make a decision, different subsets of features are used at different levels of the tree. Let us illustrate with an example. Consider the problem of deciding whether to make a loan to an applicant based on the measurements of three variables: applicant's income, marital status and previous credit rating. We approach the decision-making problem by asking a series of questions about the applicant. The first question we ask concerns the applicant's income: is it high, medium or low? If it is high, we decide that a loan may be made. If the income is medium, we ask a further question: What is the applicant's marital status? If the client is married, we make the loan; if not, then a loan is not made. If the income is low, we ask a question concerning the credit rating of the applicant. If the credit rating is good, we make the loan, if not then a loan is not made.

The series of questions above and their answers may be represented as a hierarchical decision tree (Figure 7.1) of nodes and directed edges. There are three types of node:

- **Root node**: this is the node that has no incoming edges. It is usually at the top of the tree (trees are commonly drawn 'upside-down').
- **Internal nodes**: nodes with one incoming edge and two or more outgoing edges.
- **Leaf** or **terminal nodes**: nodes with one incoming edge.

Associated with each nonterminal node (a root or internal node) is one of the features and the edges emanating from a node represent a value, or set of values, that that feature may take. Associated with each leaf or terminal node is a class label. In Figure 7.1, the root node is the feature 'income', the internal nodes are 'marital status' and 'credit rating', and the leaf nodes are the class labels 'loan' and 'no loan'. The series of questions, represented as a tree, is a description of the decision boundary. Classifying a test pattern proceeds by starting at the top of the tree and evaluating the feature for the test pattern and then taking the appropriate path down the tree.

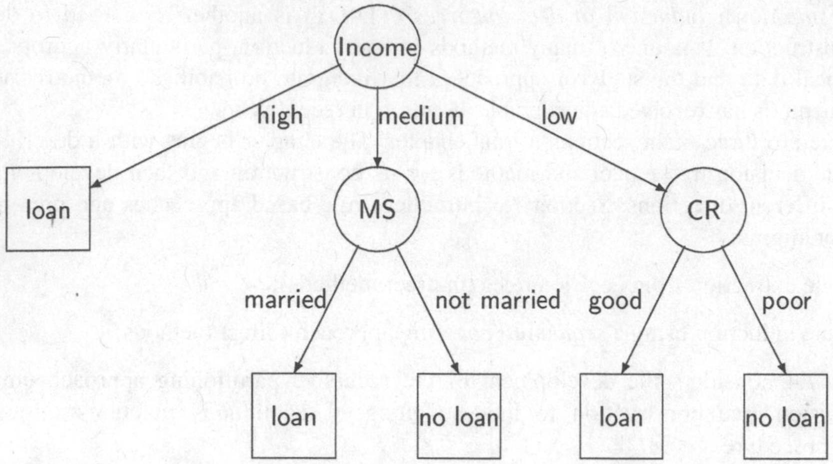

Figure 7.1 A decision tree for a loan example. CR, credit rating; MS, marital status.

Classifier construction is the process of designing a decision tree given a set of data. This involves deciding whether the node at the end of an edge is a terminal node; if not, which feature should be associated with that node and how that feature should be partitioned (how many edges emanate from the node and the range of values of the features that those edges correspond to).

A second example is given by Figure 7.2, which shows a classification tree solution to the head injury data problem of Chapter 2. Figure 7.2 is an example of a *binary decision tree*, in that there are only two edges leading away from a node. Thus, at each node, the values of the feature are partitioned into two sets. For a continuous variable, this is a threshold on the feature. Now suppose we wish to classify the pattern $\boldsymbol{x} = (5, 4, 6, 2, 2, 3)$. Beginning at the root, we compare the value of x_6 with the threshold 2. Since the threshold is exceeded, we proceed to the *right child*. We then compare the value of the variable x_5 with the threshold 5. This is not exceeded, so we proceed to the *left child*. The decision at this node leads us to the terminal node with classification label ω_3. Thus, we assign the pattern $\boldsymbol{x}$ to class ω_3. Note that the tree constructed from the training data does not use all the features – the training procedure has selected those features important for classification.

The decision tree of Figure 7.2 is a conceptually simple approximation to a complex decision that breaks up the decision into a series of simpler decisions at each node. The number of decisions required to classify a pattern depends on the pattern.

Binary trees successfully partition the feature space at each node into two parts. In the example above, the partitions are hyperplanes parallel to the coordinate axes. Figure 7.3 illustrates this in two dimensions for a two-class problem and Figure 7.4 shows the corresponding binary tree. The tree gives 100% classification performance on the design set. The tree is not unique for the given partition. Other trees also giving 100% performance are possible.

The decisions at each node need not be thresholds on a single variable (giving hyperplanes parallel to coordinate axes as decision boundaries), but could involve a linear or nonlinear combination of variables. In fact, the data in Figure 7.3 are linearly separable by a line

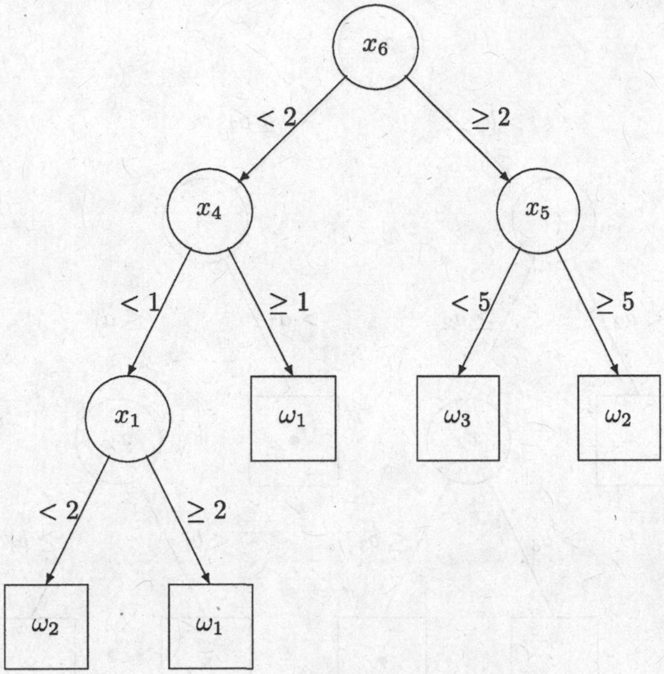

Figure 7.2 Decision tree for the head injury patient data: the variables $x_1, \ldots, x_6$ are `age`, `EMV score (eye, motor, verbal response score)`, `MRP (a summary of the motor responses in all four limbs)`, `change`, `eye indicant` and `pupils` and the classes ω_1, ω_2 and ω_3 are dead/vegetative, severely disabled and moderate or good recovery, respectively.

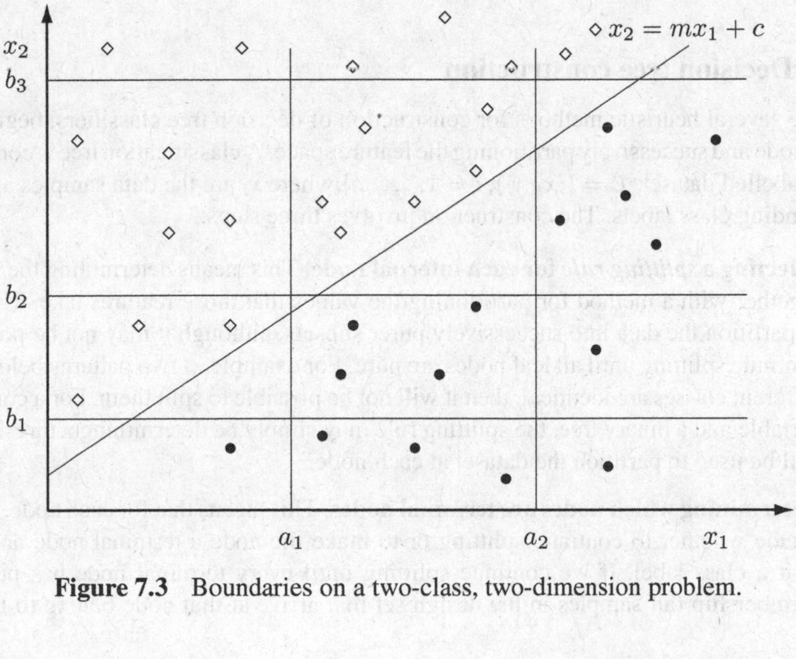

Figure 7.3 Boundaries on a two-class, two-dimension problem.

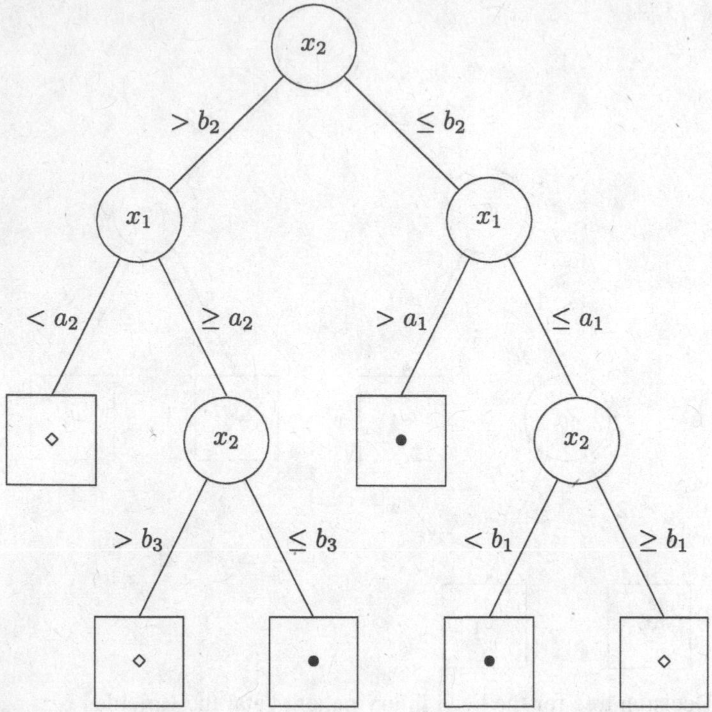

Figure 7.4 Binary decision tree for the two-class, two-dimension data of Figure 7.3.

that is not parallel to the coordinate axes and the decision rule: assign $\boldsymbol{x}$ to class $\diamond$ if $x_2 - mx_1 - c > 0$ classifies all samples correctly.

7.2.2 Decision tree construction

There are several heuristic methods for construction of decision tree classifiers, beginning at the root node and successively partitioning the feature space. A classification tree is constructed using a labelled dataset, $\mathcal{L} = \{(\boldsymbol{x}_i, y_i), i = 1, \ldots, n\}$ where $\boldsymbol{x}_i$ are the data samples and y_i the corresponding class labels. The construction involves three steps:

- **Selecting a *splitting rule* for each internal node.** This means determining the features, together with a method for partitioning the values that those features take. The aim is to partition the data into successively purer subsets, although it may not be possible to continue splitting until all leaf nodes are pure. For example, if two patterns belonging to different classes are identical, then it will not be possible to split them. For a continuous variable and a binary tree, the splitting rule may simply be determining a threshold that will be used to partition the dataset at each node.
- **Determining which nodes are terminal nodes.** This means that for each node, we must decide whether to continue splitting or to make the node a terminal node and assign to it a class label. If we continue splitting until every terminal node has pure class membership (all samples in the design set that arrive at that node belong to the same

class), then we are likely to end up with a large tree that overfits the data and gives a poor error rate on an unseen test set. Alternatively, relatively impure terminal nodes (nodes for which the corresponding subset of the design set has mixed class membership) lead to small trees that may underfit the data (see Chapter 1). Several *stopping rules* have been proposed in the literature, but the approach suggested by Breiman *et al.* (1984) is successively to grow and selectively prune the tree, using *cross-validation* to choose the subtree with the lowest estimated misclassification rate.

- **Assigning class labels to terminal nodes.** This is relatively straightforward and labels can be assigned by minimising the estimated misclassification rate.

We shall now consider each of these stages in turn.

7.2.3 Selection of the splitting rule

A splitting rule is a prescription for deciding which variable, or combination of variables, should be used at a node to divide the data samples into subgroups, and for deciding how the values that the variable takes should be partitioned.

7.2.3.1 Defining the split

The way in which a split is expressed depends on the nature of the feature.

- **Binary feature.** There are two possible outcomes, one for each feature value.
- **Nominal feature.** Since a nominal feature can take many distinct values, there could be as many child nodes of a nominal feature node as there are values that the nominal feature can take. Thus, there is a different path in the tree for each feature value. Alternatively, values may be grouped, reducing the number of child nodes.
- **Ordinal feature.** As with nominal features, there could be a single path from the node for each feature value or values could be grouped. However, grouping should maintain the ordering of the feature values.
- **Continuous feature.** A split consists of a condition on the coordinates of a vector $\boldsymbol{x} \in \mathbb{R}^p$. For example, we may define a split s_p to be

$$s_p = \{\boldsymbol{x} \in \mathbb{R}^p; x_4 \leq 8.2\}$$

a threshold on an individual feature, or

$$s_p = \{\boldsymbol{x} \in \mathbb{R}^p; x_2 + x_7 \leq 2.0\}$$

a threshold on a linear combination of features. Nonlinear functions have also been considered (Gelfand and Delp, 1991). The above examples are binary splits at a node, but multiway splits may be considered by partitioning a variable into more than two segments.

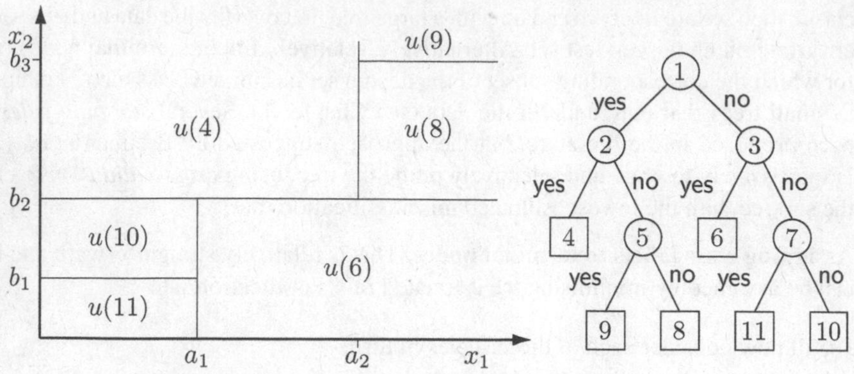

Figure 7.5 Decision tree and decision regions for the decision tree of Figure 7.4.

7.2.3.2 Selecting the best split

The question we now have to address is how to split the data (denoted by $\mathcal{L} = \{(\boldsymbol{x}_i, y_i); i = 1, \ldots, n\}$, the set of patterns and their labels) that lie in the subspace $u(t)$ at node t [Figure 7.5 shows the regions $u(t)$ associated with the terminal nodes for the binary decision tree of Figure 7.4]. This is achieved through the definition of node impurity measures, which are expressed in terms of estimates of $p\left(\omega_j|\boldsymbol{x} \in u(t)\right)$, the probability that the class of a pattern, $\boldsymbol{x}$, is ω_j, given that it falls into node t, given by

$$p\left(\omega_j|\boldsymbol{x} \in u(t)\right) = p(\omega_j|t) \sim \frac{N_j(t)}{N(t)}$$

the proportion of patterns belonging to class ω_j at node t, where $N(t)$ denotes the number of samples for which $\boldsymbol{x}_i \in u(t)$ and $N_j(t)$ the number of samples for which $\boldsymbol{x}_i \in u(t)$ and $y_i = \omega_j$ $(\sum_j N_j(t) = N(t))$.

Following Breiman *et al.* (1984) we define the node impurity function, $\mathcal{I}(t)$, to be

$$\mathcal{I}(t) = \phi(p(\omega_1|t), \ldots, p(\omega_C|t))$$

where ϕ is a function defined on all C-tuples $(q_1, \ldots, q_C)$ satisfying $q_j \geq 0$ and $\sum_j q_j = 1$. It has the properties

1. ϕ is a maximum only when $q_j = 1/C$ for all j.
2. It is a minimum when for some j, $q_j = 1$ and $q_i = 0$ for all $i \neq j$.
3. It is a symmetric function of $q_1, \ldots, q_C$.

Several different forms for $\mathcal{I}(t)$ have been used. Examples include:

- ***Gini criterion***

$$\mathcal{I}(t) = \sum_{i \neq j} p(\omega_i|t)p(\omega_j|t) = 1 - \sum_i \left[p\left(\omega_i|t\right)\right]^2$$

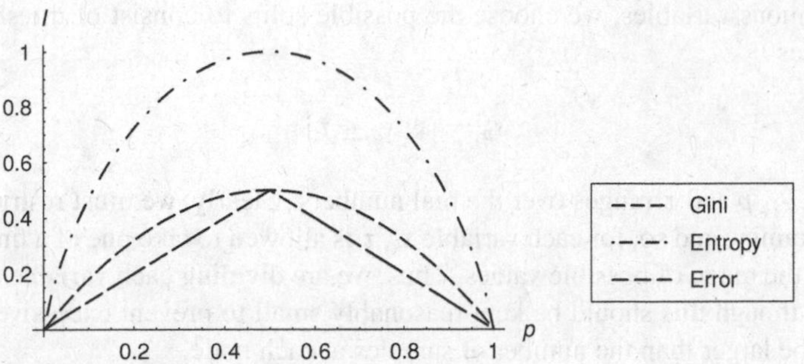

Figure 7.6 Examples of node impurity functions for a binary classification problem as a function of $p = p(\omega_1|t)$.

- **Entropy**

$$\mathcal{I}(t) = -\sum_i p(\omega_i|t)\log_2(p(\omega_i|t))$$

- **Classification error**

$$\mathcal{I}(t) = 1 - \max_i\,[p(\omega_i|t)]$$

These are easily computed for a given split. Figure 7.6 plots these measures as a function of $p = p(\omega_1|t)$ $(p(\omega_2|t) = 1 - p)$ for a binary classification problem ($C = 2$). The functions have their peak at $p = 1/C = 1/2$ (the class distribution is uniform) and their minimum when $p = 0$ or $p = 1$ (when the patterns all belong to the same class).

One measure of the goodness of a split is the difference between the impurity function of a parent node (before splitting) and the impurity of the child nodes (after splitting). For k child nodes we define

$$\Delta\mathcal{I}(s_p, t) \triangleq \mathcal{I}(t) - \sum_{i=1}^{k}\mathcal{I}(t_j)\frac{N(t_j)}{N(t)}$$

where t_j is a child of node t and $N(t_j)$ is the number of patterns that fall into that child node, and we seek a split that maximises $\Delta\mathcal{I}$. When $\mathcal{I}$ is the entropy measure, $\Delta\mathcal{I}$ is known as *Information Gain*.

Since, for a given node t, $\mathcal{I}(t)$ is the same for all splits, maximising $\Delta\mathcal{I}$ over all splits is equivalent to minimising the weighted average impurity of the child nodes; i.e. finding the split that makes the child nodes most pure.

At each stage of the classification tree procedure, we must decide which variable to split and how to make that split. The variable and split combination that results in the greatest reduction in impurity is chosen. For a binary variable, there is only one way to split, so no search over splits is required. Nominal or ordinal variables may produce a binary or multi-way split. For example, the variable 'income' in Figure 7.1 could be split in a binary manner – {low, medium} and {high} or {low} and {medium, high} – or a three-way split – {low}, {medium} and {high} – as shown in the figure.

For continuous variables, we choose the possible splits to consist of thresholds on individual variables

$$s_p = \{\boldsymbol{x}; x_k \leq \tau\}$$

where $k = 1, \ldots, p$ and τ ranges over the real numbers. Clearly, we must restrict the number of splits we examine and so, for each variable x_k, τ is allowed to take one of a finite number of values within the range of possible values. Thus, we are dividing each variable into a number of categories, though this should be kept reasonably small to prevent excessive computation and need not be larger than the number of samples at each node.

There are many other approaches to splitting. Some are given in the survey by Safavian and Landgrebe (1991). The approach described above assumes ordinal variables. For a categorical variable with N unordered outcomes, partitioning on that variable means considering 2^N partitions. By exhaustive search, it is possible to find the optimal one, but this could result in excessive computation time if N is large. Techniques for nominal variables can be found in Breiman *et al.* (1984) and Chou (1991).

7.2.4 Terminating the splitting procedure

So now we can grow our tree by successively splitting nodes, but how do we stop? We could continue until each terminal node contained one observation only. This would lead to a very large tree, for a large dataset, that would over-fit the data. Thus, we could achieve very good classification performance on the dataset used to train the model (perhaps even an error rate of zero), but the model's generalisation performance (the performance on an independent dataset representative of the true operating conditions) is likely to be poor. There are two basic approaches to avoiding overfitting in classification tree design.

- **Implement a stopping rule:** we do not split the node if the change in the impurity function is less than a pre-specified threshold. The difficulty with this approach is the specification of the threshold. In addition, even if the current splits results in a small decrease in impurity measure, further splits, if allowed to happen, may lead to a larger decrease.

- **Pruning:** we grow a tree until the terminal nodes have pure (or nearly pure) class membership and then prune it, replacing a subtree with a terminal node whose class label is determined from the patterns that lie in the region of data space corresponding to that node. This can lead to better performance than a stopping rule. We now discuss one pruning algorithm.

These approaches lead to a simpler tree and more easily interpreted decision rules.

7.2.4.1 Pruning algorithms

Many algorithms prune the tree to minimise an error that incorporates a term that depends on model complexity, and thus penalising complex models to prevent overfitting. There are several approaches. The CART[1] (classification and regression trees; Breiman *et al.*, 1984) algorithm employs one such approach.

[1] CART is a registered trademark of California Statistical Software, Inc.

Let $R(t)$ be real numbers associated with each node t of a given tree T. If t is a terminal node, i.e. $t \in \tilde{T}$, where $\tilde{T}$ denotes the set of terminal nodes, then $R(t)$ could represent the proportion of misclassified samples – the number of samples in $u(t)$ that do not belong to the class associated with the terminal node, defined to be $M(t)$, divided by the total number of data points, n

$$R(t) = \frac{M(t)}{n} \qquad t \in \tilde{T}$$

More generally, in a regression problem, $R(t)$ could represent a mean square error between the predicted value and the response variable values of the patterns.

Let $R_\alpha(t) = R(t) + \alpha$ for a real number α. Set[2]

$$R(T) = \sum_{t \in \tilde{T}} R(t)$$

$$R_\alpha(T) = \sum_{t \in \tilde{T}} R_\alpha\,(t) = R(T) + \alpha \left|\tilde{T}\right|$$

In a classification problem, $R(T)$ is the *estimated misclassification rate*, $|\tilde{T}|$ denotes the cardinality of the set $\tilde{T}$, $R_\alpha(T)$ is the *estimated complexity–misclassification rate* of a classification tree and α is a constant that can be thought of as the complexity cost per terminal node. If α is small, then there is a small penalty for having a large number of nodes. As α increases, the *minimising subtree* [the subtree $T' \leq T$ that minimises $R_\alpha(T')$] has fewer terminal nodes.

Let the quantity $R(t)$ be written as

$$R(t) = r(t)p(t)$$

where $r(t)$ is the resubstitution estimate of the probability of misclassification (or *apparent error rate*, see Chapter 9) given that a case falls into node t,

$$r(t) = 1 - \max_{\omega_j} p(\omega_j|t)$$

and $p(t)$ and $p(\omega_j|t)$ are defined as

$$p(t) = \frac{N(t)}{n} \tag{7.1}$$

an estimate of $p(\boldsymbol{x} \in u(t))$ based on the dataset $\mathcal{L}$;

$$p(\omega_j|t) = \frac{N_j(t)}{N(t)} \tag{7.2}$$

an estimate of $p(y = \omega_j|\boldsymbol{x} \in u(t))$ based on the dataset $\mathcal{L}$. Thus, if t is taken to be a terminal node, $R(t)$ is the contribution of that node to the total error.

---

[2] The argument of R may be a tree or a node; capital letters denote a tree.

Let T_t be the subtree with root t (the part of the tree that grows out from the node t as a root node). If $R_\alpha(T_t) < R_\alpha(t)$, then the cost complexity of the subtree is less than that for the node t. This occurs for small α. As α increases, equality is achieved when

$$\alpha = \frac{R(t) - R(T_t)}{N_d(t) - 1}$$

where $N_d(t)$ is the number of terminal nodes in T_t, i.e. $N_d(t) = |\tilde{T}_t|$ and termination of the tree at t is preferred.

7.2.5 Assigning class labels to terminal nodes

Let $\lambda(\omega_j, \omega_i)$ be the cost of assigning a pattern $\boldsymbol{x}$ to ω_i when $\boldsymbol{x} \in \omega_j$. Denote the patterns that lie in $u(t)$, the part of the data space corresponding to node t, by $\{\boldsymbol{x}_i\}$, with corresponding class labels $\{y_i\}$, where $y_i \in \{\omega_1, \ldots, \omega_C\}$. The terminal node is assigned the class label ω_k for which

$$\sum_{\boldsymbol{x}_j \in u(t)} \lambda\left(y_j, \omega_k\right)$$

is a minimum. For the equal cost loss matrix,

$$\lambda(\omega_j, \omega_i) = \begin{cases} 1 & i \neq j \\ 0 & i = j \end{cases}$$

the terminal node is assigned to the class that gives the smallest number of misclassifications.

7.2.6 Decision tree pruning – worked example

7.2.6.1 Terminology

The approach we present is based on the CART description of Breiman *et al.* (1984). A *binary decision tree* is defined to be a set T of positive integers together with two functions $l(.)$ and $r(.)$ from T to $T \cup \{0\}$. Each member of T corresponds to a node in the tree. Figure 7.7 shows a tree and the corresponding values of $l(t)$ and $r(t)$ (denoting the left and right nodes, respectively).

1. For each $t \in T$, either $l(t) = 0$ and $r(t) = 0$ (a terminal node) or $l(t) > 0$ and $r(t) > 0$ (a nonterminal node).
2. Apart from the root node (the smallest integer, $t = 1$ in the above) there is a unique *parent* $s \in T$ of each node; that is, for $t \neq 1$, there is an s such that either $t = l(s)$ or $t = r(s)$.

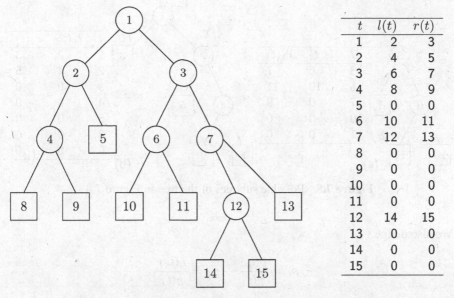

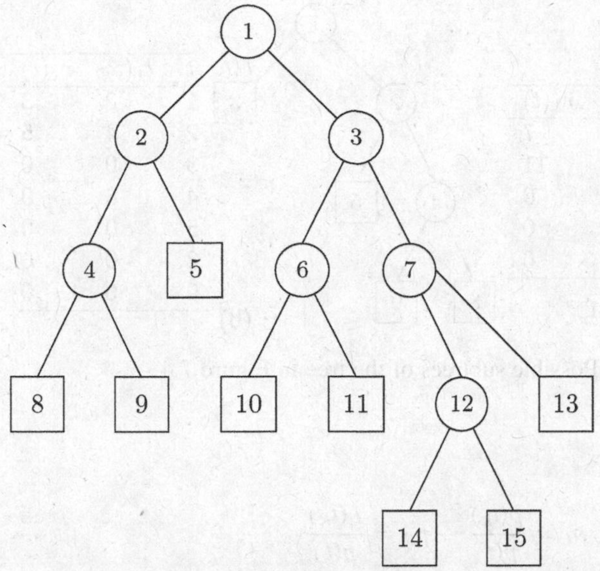

t	$l(t)$	$r(t)$
1	2	3
2	4	5
3	6	7
4	8	9
5	0	0
6	10	11
7	12	13
8	0	0
9	0	0
10	0	0
11	0	0
12	14	15
13	0	0
14	0	0
15	0	0

Figure 7.7 Classification tree and the values of $l(t)$ and $r(t)$.

A *subtree* is a nonempty subset T_1 of T together with two functions l_1 and r_1 such that

$$l_1(t) = \begin{cases} l(t) & \text{if } l(t) \in T_1 \\ 0 & \text{otherwise} \end{cases}$$

$$r_1(t) = \begin{cases} r(t) & \text{if } r(t) \in T_1 \\ 0 & \text{otherwise} \end{cases} \tag{7.3}$$

and provided that T_1, $l_1(.)$ and $r_1(.)$ form a tree. For example, the set $\{3, 6, 7, 10, 11\}$ together with (7.3) forms a subtree, but the sets $\{2, 4, 5, 3, 6, 7\}$ and $\{1, 2, 4, 3, 6, 7\}$ do not; in the former case because there is no parent for both 2 and 3 and in the latter case because $l_1(2) > 0$ and $r_1(2) = 0$. A *pruned subtree* T_1 of T is a subtree of T that has the same root. This is denoted by $T_1 \leq T$. Thus, example (b) in Figure 7.8 is a *pruned* subtree, but example (a) is not (though it is a subtree).

Let $\tilde{T}$ denote the set of terminal nodes (the set $\{5, 8, 9, 10, 11, 13, 14, 15\}$ in Figure 7.7). Let $\{u(t), t \in \tilde{T}\}$ be a partition of the data space $\mathbb{R}^p$ (that is, $u(t)$ is a subspace of $\mathbb{R}^p$ associated with a terminal node such that $u(t) \cap u(s) = \emptyset$ for $t \neq s, t, s \in \tilde{T}$; and $\cup_{t\in\tilde{T}} u(t) = \mathbb{R}^p$). Let $\omega_{j(t)} \in \{\omega_1, \ldots, \omega_C\}$ denote one of the class labels. Then, a *classification tree* consists of the tree T, together with the class labels $\{\omega_{j(t)}, t \in \tilde{T}\}$ and the partition $\{u(t), t \in \tilde{T}\}$. All we are saying here is that associated with each terminal node is a region of the data space that we label as belonging to a particular class. There is also a subspace of $\mathbb{R}^p$, $u(t)$ associated with each *nonterminal* node, being the union of the subspaces of the terminal nodes that are its descendants.

A classification tree is constructed using the labelled dataset, $\mathcal{L} = \{(\boldsymbol{x}_i, y_i), i = 1, \ldots, n\}$ of data samples $\boldsymbol{x}_i$ and the corresponding class labels, y_i.

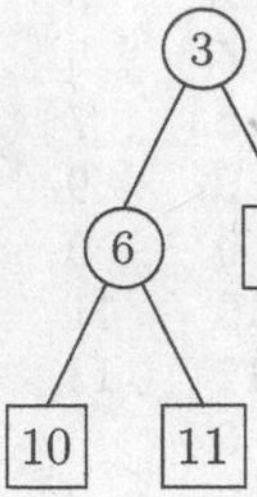

t	$l_1(t)$	$r_1(t)$
3	6	7
6	10	11
7	0	0
10	0	0
11	0	0

(a)

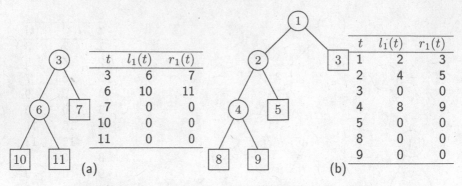

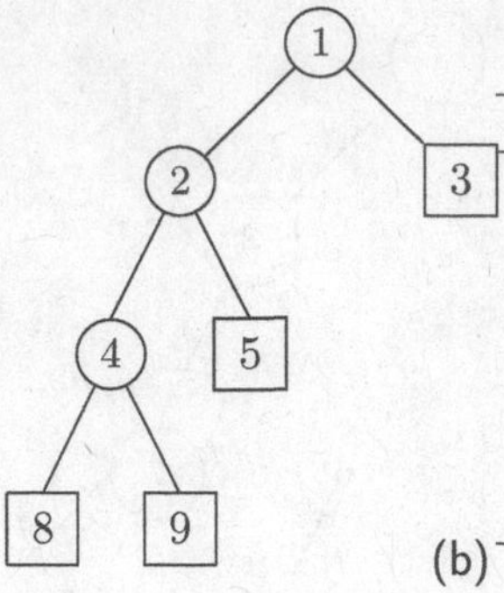

t	$l_1(t)$	$r_1(t)$
1	2	3
2	4	5
3	0	0
4	8	9
5	0	0
8	0	0
9	0	0

(b)

Figure 7.8 Possible subtrees of the tree in Figure 7.4.

We also define

$$p_L = \frac{p(t_L)}{p(t)} \qquad p_R = \frac{p(t_R)}{p(t)}$$

where $t_L = l(t)$, $t_R = r(t)$, as estimates of $p(\boldsymbol{x} \in u(t_L)|\boldsymbol{x} \in u(t))$ and $p(\boldsymbol{x} \in u(t_R)|\boldsymbol{x} \in u(t))$ based on $\mathcal{L}$, respectively.

We may assign a label to each node, t, according to the proportions of samples from each class in $u(t)$: assign label ω_j to node t if

$$p(\omega_j|t) = \max_i p(\omega_i|t)$$

We have now covered most of the terminology that we shall use. Table 7.1 illustrates some of these concepts using the data of Figure 7.3 and the tree of Figure 7.4.

7.2.6.2 Algorithm description

We describe the CART pruning algorithm using the tree of Figure 7.9. In Figure 7.9, each terminal node has been labelled with a single number $R(t)$, the amount by which that node contributes to the error rate. Each nonterminal node has been labelled by two numbers. The number to the left of the node is the value of $R(t)$, the contribution to the error rate if that node were a terminal node. The number to the right is $g(t)$, defined by

$$g(t) = \frac{R(t) - R(T_t)}{N_d(t) - 1} \tag{7.4}$$

Thus, the value of $g(t)$ for node $t = 2$, say, is $0.03 = [0.2 - (0.01 + 0.01 + 0.03 + 0.02 + 0.01)]/4$.

The first stage of the algorithm searches for the node with the smallest value of $g(t)$. This is node 12, with a value of 0.0075. This is now made a terminal node and the value of $g(t)$ recalculated for all its ancestors. This is shown in Figure 7.10.

The values of the nodes in the subtree beginning at node 2 are unaltered. The process is now repeated. The new tree is searched to find the node with the smallest value of $g(t)$. In this

Table 7.1 Tree table: class ω_1, $\diamond$; class ω_2, $\bullet$.

t	node	$l(t)$	$r(t)$	$N(t)$	$N_1(t)$	$N_2(t)$	$p(t)$	$p(1\|t)$	$p(2\|t)$	p_L	p_R
1	$x_2 > b_2$	2	3	35	20	15	1	$\frac{20}{35}$	$\frac{15}{35}$	$\frac{22}{35}$	$\frac{13}{35}$
2	$x_1 < a_2$	4	5	22	17	5	$\frac{22}{35}$	$\frac{17}{22}$	$\frac{5}{22}$	$\frac{15}{22}$	$\frac{7}{22}$
3	$x_1 > a_1$	6	7	13	4	9	$\frac{13}{35}$	$\frac{4}{13}$	$\frac{9}{13}$	$\frac{9}{13}$	$\frac{4}{13}$
4	$\diamond$	0	0	15	15	0	$\frac{15}{35}$	1	0		
5	$x_2 > b_3$	9	8	7	2	5	$\frac{7}{35}$	$\frac{2}{7}$	$\frac{5}{7}$	$\frac{5}{7}$	$\frac{2}{7}$
6	$\bullet$	0	0	9	0	9	$\frac{9}{35}$	0	1		
7	$x_2 < b_1$	11	10	4	3	1	$\frac{4}{35}$	$\frac{3}{4}$	$\frac{1}{4}$	$\frac{3}{4}$	$\frac{1}{4}$
8	$\bullet$	0	0	5	0	5	$\frac{5}{35}$	0	1		
9	$\diamond$	0	0	2	2	0	$\frac{2}{35}$	1	0		
10	$\diamond$	0	0	3	3	0	$\frac{3}{35}$	1	0		
11	$\bullet$	0	0	1	0	1	$\frac{1}{35}$	0	1		

case, there are two nodes, 6 and 9, each with a value of 0.01. Both are made terminal nodes and again the values of $g(t)$ for the ancestors recalculated. Figure 7.11 gives the new tree (T^3). Node 4 now becomes a new terminal node. This continues until all we are left with is the root node. Thus, the pruning algorithm generates a succession of trees. We denote the tree at the kth stage by T^k. Table 7.2 gives the value of the error rate for each successive tree, together

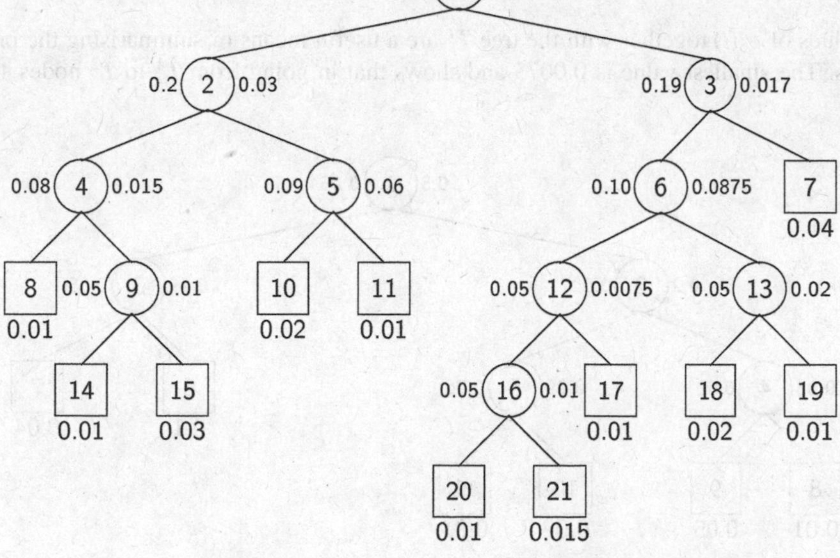

Figure 7.9 Pruning example; original tree.

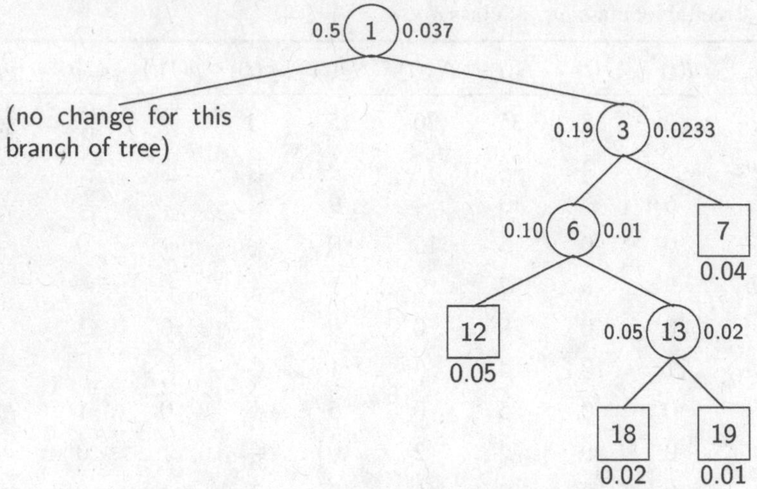

Figure 7.10 Pruning example; pruning at node 12.

with the number of terminal nodes in each tree and the value of $g(t)$ at each stage (denoted by α_k) that is used in the pruning to generate tree T^k. The tree T^k has all internal nodes with a value of $g(t) > \alpha_k$.

The results of the pruning algorithm are summarised in Figure 7.12.

In this figure is shown the original tree together with the values $g_6(t)$ for the internal nodes, where g_k is defined recursively ($0 \leq k \leq K - 1$, where K is the number of pruning stages):

$$g_k = \begin{cases} g(t) & t \in T^k - \tilde{T}^k \quad (t \text{ an internal node of } T^k) \\ g_{k-1}(t) & \text{otherwise} \end{cases}$$

The values of $g_6(t)$ together with the tree T^1 are a useful means of summarising the pruning process. The smallest value is 0.0075 and shows that in going from T^1 to T^2 nodes 16, 17,

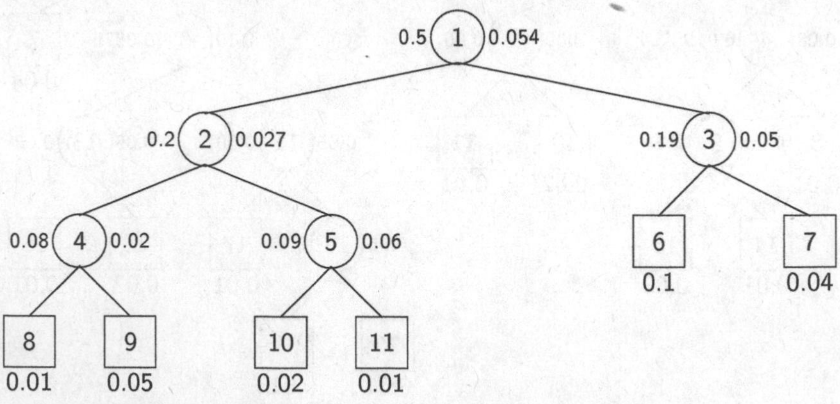

Figure 7.11 Pruning example; pruning at nodes 6 and 9.

Table 7.2 Tree results.

	α_k	$\lvert\tilde{T}^k\rvert$	$R(T^k)$
1	0	11	0.185
2	0.0075	9	0.2
3	0.01	6	0.22
4	0.02	5	0.25
5	0.045	3	0.34
6	0.05	2	0.39
7	0.11	1	0.5

20 and 21 are removed. The smallest value of the pruned tree is 0.01 and therefore in going from T^2 to T^3 nodes 12, 13, 14, 15, 18 and 19 are removed; from T^3 to T^4 nodes 8 and 9 are removed; from T^4 to T^5 nodes 4, 5, 10 and 11; from T^5 to T^6 nodes 6 and 7 and finally nodes 2 and 3 are removed to obtain T^7 the tree with a single node.

An explicit algorithm for the above process is given by Breiman *et al.* (1984).

7.2.7 Decision tree construction methods

Having considered approaches to growing and pruning decision trees and estimating their error rate, we are now in a position to present methods for tree construction using these features.

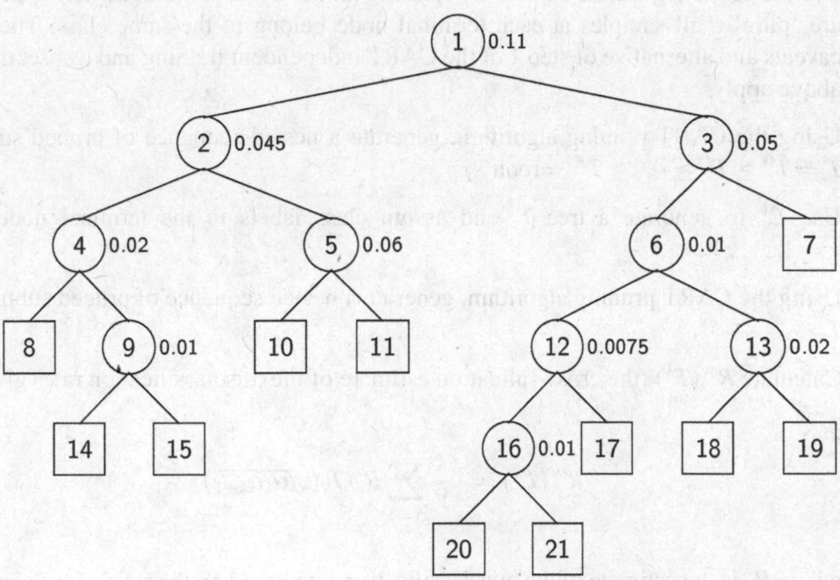

Figure 7.12 Pruning example; summary of pruning process.

7.2.7.1 CART independent training and test set method

The stages in the CART independent training and test set method are as follows. Assume that we are given a training set $\mathcal{L}_r$ and a test set (or more correctly a validation set) $\mathcal{L}_s$ of data samples. These are generated by partitioning the design set approximately into two groups.

The CART independent training and test set method is as follows:

1. Use the set $\mathcal{L}_r$ to generate a tree T by splitting all the nodes until all the terminal nodes are 'pure' – all samples at each terminal node belong to the same class. It may not be possible to achieve this with overlapping distributions, therefore an alternative is to stop when the number in each terminal node is less than a given threshold or the split of a node t results in a left child node t_L or a right child node t_R with $\min(N(t_L), N(t_R)) = 0$.
2. Use the CART pruning algorithm to generate a nested sequence of subtrees T^k using the set $\mathcal{L}_s$.
3. Select the smallest subtree for which $R(T^k)$ is a minimum.

7.2.7.2 CART cross-validation method

For the cross-validation method, the training set $\mathcal{L}$ is divided into V subsets $\mathcal{L}_1, \ldots, \mathcal{L}_V$, with approximately equal numbers in each class. Let $\mathcal{L}^v = \mathcal{L} - \mathcal{L}_v$, $v = 1, \ldots, V$. We shall denote by $T(\alpha)$ the pruned subtree with all internal nodes having a value of $g(t) > \alpha$. Thus, $T(\alpha) = T^k$, the pruned subtree at the kth stage where k is chosen so that $\alpha_k \leq \alpha \leq \alpha_{k+1}$ $(\alpha_{K+1} = \infty)$.

The CART cross-validation method is as follows.

1. Use the set $\mathcal{L}$ to generate a tree T by splitting all the nodes until all the terminal nodes are 'pure' – all samples at each terminal node belong to the same class. The same caveats and alternative of step 1 of the CART independent training and test set method above apply.
2. Using the CART pruning algorithm, generate a nested sequence of pruned subtrees $T = T^0 \geq T^1 \geq \cdots \geq T^K = \text{root}(T)$.
3. Use $\mathcal{L}^v$ to generate a tree T_v and assign class labels to the terminal nodes, for $v = 1, \ldots, V$.
4. Using the CART pruning algorithm, generate a nested sequence of pruned subtrees of T_v.
5. Calculate $R^{\text{cv}}(T^k)$ (the cross-validation estimate of the misclassification rate) given by

$$R^{\text{cv}}(T^k) = \frac{1}{V}\sum_{v=1}^{V} R_v(T_v(\sqrt{\alpha_k \alpha_{k+1}}))$$

where R_v is the estimate of the misclassification rate based on the set $\mathcal{L}_v$ for the pruned subtree $T_v(\sqrt{\alpha_k \alpha_{k+1}})$.

6. Select the smallest $T^* \in \{T^0, \ldots, T^K\}$ such that

$$R^{cv}(T^*) = \min_k R^{cv}(T^k)$$

7. Estimate the misclassification rate by

$$\hat{R}(T^*) = R^{cv}(T^*)$$

The procedure presented in this section implements one of many approaches to recursive tree design which use a growing and pruning approach and that have emerged as reliable techniques for determining right-sized trees. The CART approach is appropriate for datasets of continuous or discrete variables with ordinal or nominal significance, including datasets of mixed variable types.

There have been many different growing and pruning strategies proposed in the literature. Quinlan (1987) describes and assesses four pruning approaches, the motivation behind the work being to simplify decision trees in order to use the knowledge in expert systems. The use of information theoretic criteria in tree construction has been considered by several authors. Quinlan and Rivest (1989) describe an approach based on the *minimum description length principle* and Goodman and Smyth (1990) present a top-down mutual information algorithm for tree design. A comparative study of pruning methods for decision trees is provided by Esposito *et al.* (1997) [see also comments by Kay (1997) and Mingers (1989)]. Averaging, as an alternative to pruning, is discussed by Oliver and Hand (1996).

In addition to the top-down approach to decision tree construction, other strategies include the bottom-up approach of Landeweerd *et al.* (1983). See the review by Murthy (1998).

7.2.8 Other issues

7.2.8.1 Missing data

The procedure given above contains no mechanism for handling missing data. The CART algorithm deals with this problem through the use of *surrogate splits*. If the best split of a node is s on the variable x_m say, then the split s^* that predicts s most accurately, on a variable x_j other than x_m, is termed the *best surrogate for* s. Similarly, a second best surrogate on a variable other than x_m and x_j can be found, and so on.

A tree is constructed in the usual manner, but at each node t, the best split s on a variable x_m is found by considering only those samples for which a value for x_m is available. Objects are assigned to the groups corresponding to t_L and t_R according to the value on x_m. If this is missing for a given test pattern, the split is made using the best surrogate for s (that is, the split is made on a different variable). If this value is also missing, the second best surrogate is used, and so on until the sample is split. Alternatively, we may use procedures that have been developed for use by conventional classifiers for dealing with missing data (see Chapter 13).

7.2.8.2 Priors and costs

The definitions (7.1) and (7.2) assume that the prior probabilities for each class, denoted by $\pi(j)$, are equal to N_j/n. If the distribution on the design set is not proportional to the expected occurrence of the classes, then the resubstitution estimates of the probabilities that a sample falls into node t, $p(t)$, and that it is in class ω_j given that it falls into node t, $p(j|t)$, are defined as

$$\begin{aligned} p(t) &= \sum_{j=1}^{C} \pi(j) \frac{N_j(t)}{N_j} \\ p(j|t) &= \frac{\pi(j) N_j(t) / N_j}{\sum_{j=1}^{C} \pi(j) N_j(t) / N_j} \end{aligned} \tag{7.5}$$

In the absence of costs (or assuming an equal cost loss matrix – see Chapter 1), the misclassification rate is given by

$$R(T) = \sum_j \sum_{i \neq j} q(i|j)\pi(j) \tag{7.6}$$

where $q(i|j)$ is the proportion of samples of class ω_j defined as class ω_i by the tree. If λ_{ji} is the cost of assigning an object of class ω_j to ω_i, then the misclassification cost is

$$R(T) = \sum_j \sum_{i \neq j} \lambda_{ji} q(i|j)\pi(j)$$

This may be written in the same form as (7.6), with redefined priors, provided that λ_{ji} ($j \neq i$) is independent of i (that is, the cost of misclassifying an object of class ω_j does not depend on the class to which it is assigned); thus, $\lambda_{ji} = \lambda_j$ and the priors are redefined as

$$\pi'(j) = \frac{\lambda_j \pi(j)}{\sum_j \lambda_j \pi(j)}$$

In general, with nonconstant costs of misclassification for each class, then the costs cannot be incorporated into modified priors.

7.2.9 Example application study

The problem

In drug development research, it is a challenging task to examine high-throughput screening data (often requiring the testing of millions of compounds generated using combinatorial chemistry in a matter of days) for potential interest. Han *et al.* (2008) developed decision tree approaches for screening the data.

Summary

A decision tree model was developed to discriminate compound bioactivities by using their chemical structure fingerprints.

The data
A model was developed for each of four protein targets selected from the PubChem BioAssay database. For each protein target, there were between about 61 000 and 99 000 compounds tested, with a small percentage (but hundreds) identified as active compounds.

The model
The decision tree model developed was based on the C4.5 algorithm (Quinlan, 1993).

Training procedure
A 10-fold cross-validation procedure was used as part of the assessment procedure together with a tree pruning procedure.

Results
The results suggested that biologically interesting compounds could be identified using decision tree models developed for specific protein targets. These models could then be used as part of a filtering process to select active compounds for that target.

7.2.10 Further developments

The splitting rules described have considered only a single variable. Some datasets may be naturally separated by hyperplanes that are not parallel to the coordinate axes, and Chapter 5 concentrated on means for finding linear discriminants. The basic CART algorithm attempts to approximate these surfaces by multidimensional rectangular regions, and this can result in very large trees. Extensions to this procedure that allow splits not orthogonal to axes (termed *oblique decision trees*) are described by Loh and Vanichsetakul (1988), Wu and Zhang (1991), Murthy *et al.* (1994), Shah and Sastry (1999) and Heath *et al.* (1993) (and indeed can be found in the CART book of Breiman *et al.*, 1984). Sankar and Mammone (1991) use a neural network to recursively partition the data space and allow general hyperplane partitions. Pruning of this *neural tree classifier* is also addressed [see also Sethi and Yoo (1994) for multifeature splits methods using perceptron learning].

Tree-based methods for vector quantisation are described by Gersho and Gray (1992). A tree-structured approach reduces the search time in the encoding process. Pruning the tree results in a variable-rate vector quantiser and the CART pruning algorithm may be used. Other approaches for growing a variable-length tree (in the vector quantisation context) without first growing a complete tree are described by Riskin and Gray (1991). Crawford (1989) describes some extensions to CART that improve upon the cross-validation estimate of the error rate and allow for incremental learning – updating an existing tree in the light of new data. Such *incremental decision trees* (Kalles and Morris, 1996) may be required if the properties of the data are time-dependent (and so the tree is required to adapt) or if the dataset is too large to process all at one time.

The kd approach for searching for nearest neighbours (see Chapter 4) exploits a tree approach similar to that presented in this chapter.

One of the problems with nominal variables with a large number of categories is that there may be many possible partitions to consider. Chou (1991) presents a clustering approach to finding a locally optimum partition without exhaustive search. Buntine (1992) develops a Bayesian statistics approach to tree construction that uses Bayesian techniques for node splitting, pruning and averaging of multiple trees (also assessed as part of the *Statlog* project). A Bayesian CART algorithm is described by Denison *et al.* (1998a).

7.2.11 Summary

Decision trees are nonparametric approaches for constructing classification models. Their advantages are that they can be compactly stored; they efficiently classify new samples and have demonstrated good generalisation performance on a wide variety of problems. Possible disadvantages are the difficulty in designing an optimal tree, leading perhaps to large trees with poor error rates on certain problems, particularly if the separating boundary is complicated and a binary decision tree, with decision boundaries parallel to the coordinate axes is used. Also, most approaches are nonadaptive – they use a fixed training set and additional data may require redesign of the tree.

One of the main attractions of decision trees is their simplicity: they employ a top-down recursive partitioning approach and, in particular, CART performs binary splits on single variables in a recursive manner. Although finding an optimal tree can be time consuming, there are efficient tree-construction algorithms available. Classification of a test sample is fast – it may require only a few simple tests. Yet despite its simplicity, a decision tree is able to give performance superior to many traditional methods on complex nonlinear datasets of many variables. Of course there are possible generalisations of the model to multiway splitting on linear (or even nonlinear) combinations of variables (oblique decision trees), but there is no strong evidence that these will lead to improved performance. In fact, the contrary has been reported by Breiman and Friedman (1988). Also, univariate splitting has the advantage that the models can be interpreted more easily, particularly on small datasets. Lack of interpretability by many, if not most, of the other methods of discrimination described in this book can be a serious shortcoming in many applications. Although different node impurity functions may lead to a different choice of variable to split, there is much consistency in the performance of the different functions.

Another advantage of decision trees is that the approach has been extensively evaluated and tested, both by the workers who have developed them and numerous researchers who have implemented the software. In addition, the tree-structured approach may be used for regression using the same impurity function (for example, a least-squares error measure) to grow the tree as to prune it.

Decision trees are a nonparametric approach to classification. An alternative to decision trees is to use a parametric approach with the inherent underlying assumptions. The use of nonparametric methods in discrimination, regression and density estimation has been increasing with the continuing development of computing power. In most applications it is the cost of data collection that far exceeds any other cost and the nonparametric approach is appealing because it does not make the (often gross) assumptions regarding the underlying population distributions that other discrimination methods do.

7.3 Rule induction

7.3.1 Introduction

In the previous section, we presented an approach to classification based on the construction of decision trees using a recursive partitioning algorithm. Such decision trees may be used to generate a set of IF-THEN rules that represents the learned model. Rule-based approaches produce descriptive models that can be easier to interpret than the decision tree classifier. In this section, we consider two major approaches to rule induction:

- rule extraction from decision trees;
- rule induction using a sequential covering approach.

Much of the work in this area has been developed in the machine learning literature and can be found in books on data mining. Nevertheless, it is included here since increasingly many application domains require some degree of classification rule interpretability: it is necessary for a classifier to 'explain' its decision and approaches that provide a description of the decision boundary in terms of rules are of growing interest. In addition, the area of research that bridges rule-based and statistical approaches is an active research area.

7.3.1.1 Notation and terminology

A rule-based classifier represents the model as a set of IF-THEN rules. An IF-THEN rule, r, is an expression of the form

r: IF *condition* THEN *conclusion*

For example, in the loan applicant decision tree example (Figure 7.1), we have the rule $r1$

$r1$: IF income = medium AND marital status = married THEN loan

The left-hand side of the rule (the IF part) is the **rule antecedent** or **precondition** and contains a **conjunction** of tests on features:

$$condition = (A_1 \text{ op } v_1) \wedge (A_2 \text{ op } v_2) \wedge \cdots \wedge (A_k \text{ op } v_k)$$

where $\wedge$ is the conjunction operator (AND); op is a logical operator from the set $\{=, \neq, <, >, \leq, \geq\}$; and (A_j, v_j) are variable-value pairs. In the above condition, $k = 2$; $(A_1, v_1) =$ (income, low); $(A_2, v_2) =$ (marital status, married); and op is = for all tests.

The right-hand side of the rule (the THEN part) is the **rule consequent**, which for classification problems is the predicted class.

7.3.1.2 Coverage and accuracy

A rule *covers* a pattern $\boldsymbol{x}$ if the antecedent is satisfied (the rule is said to be **triggered**). For example, a loan applicant with $\boldsymbol{x} =$ (medium, married, good) satisfies the antecedent of the rule $r1$ above, or $r1$ covers $\boldsymbol{x}$. The **coverage** of a rule r is the proportion of patterns covered by the rule:

$$\text{Coverage } (r) = \frac{n_{covers}\,(r)}{N}$$

where $n_{covers}(r)$ is the number of patterns covered by the rule r and N is the total number of patterns in the dataset.

The **accuracy** of a rule r is the proportion of patterns covered that the rule correctly classifies,

$$\text{Accuracy}\,(r) = \frac{n_{correct}\,(r)}{n_{covers}\,(r)}$$

where $n_{correct}(r)$ is the number of patterns correctly classified by rule r.

7.3.1.3 Using a rule-based classifier

Suppose that our rule-based classifier consists of a set of rules $(r_1, \ldots, r_n)$. A rule-based classifier predicts the class of a given pattern $\boldsymbol{x}$ using the predictions of the rules r_i that are triggered by $\boldsymbol{x}$. If no two rules are triggered by the same pattern $\boldsymbol{x}$, the rules are said to be **mutually exclusive**. If there is one rule for every combination of feature values, the rules are said to be **exhaustive**. In this case, every potential pattern triggers a rule, or there is a rule for every part of the feature space.

7.3.1.4 Handling conflict and indecision

A set of mutually exclusive and exhaustive rules ensures that every pattern is covered by exactly one rule. However, not all rule-based classifiers have this property:

- a rule set may not be exhaustive – in this case there may be patterns for which no rule triggers;
- a rule set may not be mutually exclusive – several rules may be triggered by the same record. If each of the rules triggered produces the same class prediction, then there is no conflict to resolve. However, if a conflict arises, a strategy for resolution must be provided.

No rule triggered

If no rule is triggered, then we need a strategy for assigning a class to the pattern $\boldsymbol{x}$. We add a default rule to the rule set which triggers when other rules fail. This rule specifies a default class based on the training data and may typically be the majority class of the patterns in the training set not covered by a rule.

Multiple rules triggered

There are many approaches to conflict resolution when multiple rules are triggered. One approach is to use all the predictions and to make a class decision based on the number of votes cast for a particular class by the set of rules. This may simply be a majority vote, or the votes weighted by the rule's accuracy. An alternative approach is to order the rules and to proceed through the rule list in order until the first one is triggered and to use that rule's class prediction.

CLASS-BASED ORDERING With this scheme, the rules are sorted according to their class prediction, so that rules belonging to the same class appear together in the list. Classes are ranked according to the 'importance' of the class (for example, the most common class is placed first in the list).

RULE-BASED ORDERING The rules are ordered according to some measure of rule quality [for example, accuracy, coverage, size (the number of features tested in the condition)]. A disadvantage of this approach is that a given rule in the list implies the negation of rules preceding it. This can make interpretation difficult, particularly for lower-ranked rules.

7.3.2 Generating rules from a decision tree

To extract a set of rules from a decision tree, every path from the root node to a leaf node is expressed as a classification rule. The rule antecedent is the conjunction of the test conditions along the path from the root to the leaf. The rule consequent (the class prediction) is the class label assigned to the leaf node. The rules extracted from Figure 7.1 are:

IF income = high THEN loan

IF income = medium AND marital status = married THEN loan

IF income = medium AND marital status = not married THEN no loan

IF income = low AND credit rating = good THEN loan

IF income = low AND credit rating = poor THEN no loan

The rules extracted from the tree are mutually exclusive and exhaustive. In some cases, the set may be large or difficult to interpret and it may be possible to simplify some of the rules. However, the rules obtained after simplification may no longer be mutually exclusive.

7.3.2.1 Rule pruning in C4.5

The decision tree algorithm C4.5 has a mechanism to express the learnt decision tree as an ordered list of IF-THEN rules. The number of rules is usually substantially fewer than the number of leaf nodes in the tree since a rule pruning procedure is employed. This removes one of the conjuncts in turn and assesses the classification error rate. Often it leads to rules that are more easily interpreted.

7.3.2.2 Conflict resolution in C4.5

C4.5 adopts a simple conflict resolution scheme. It uses a class-based ordering: rules that predict the same class are grouped together and then the classes are ordered using the minimum description length principle (Quinlan, 1993; see Chapter 13). The first rule in the list that covers a pattern is then used to predict the class of the pattern.

7.3.3 Rule induction using a sequential covering algorithm

Extracting rules directly from a dataset, without generating a decision tree first, can be achieved using the sequential covering algorithm. Rules are learned sequentially, extracted for one class at a time, with the order of the classes depending on several factors such as class priors and misclassification costs.

Algorithm 7.1 Sequential covering algorithm.

Let $\mathcal{D}(\mathcal{R})$ be the set of training patterns and their class labels not covered by the rule set $\mathcal{R}$. Initially, $\mathcal{D} = \{(\boldsymbol{x}_i, z_i), i = 1, \ldots, n\}$, the dataset of patterns $\boldsymbol{x}_i$ and their corresponding class labels, $z_i \in \{\omega_1, \ldots, \omega_C\}$.

Let $\mathcal{A}$ be the set of features and corresponding values the features take: $\mathcal{A} = \{(X_j, v_j)\}$. So for the loan example, $\mathcal{A}$ = {(income, high), (income, medium), (income, low), (MS, married), (MS, not married), (CR, good), (CR, poor)}.

Initialise the order of the classes, $Y = \{y_1, \ldots, y_C\}$ (where $\{y_1, \ldots, y_C\}$ is a permutation of $\{\omega_1, \ldots, \omega_C\}$).

1. Initialise $\mathcal{R} = \{\ \}$, the initial rule list.
2. For each class $y \in Y$
   Repeat
   (a) Rule, r = Learn-One-Rule $(\mathcal{D}(\mathcal{R}), \mathcal{A}, y)$
   (b) Add r to the current list of rules, $\mathcal{R}$: $\mathcal{R} = \mathcal{R} \vee r$
   (c) Remove patterns from $\mathcal{D}$ covered by the rule, r
   until a terminating condition is satisfied.
3. Return the rule set, $\mathcal{R}$.

7.3.3.1 Sequential covering algorithm

A description of the approach is provided by Algorithm 7.1. It uses the Learn-One-Rule function described below.

Suppose that we wish to extract a set of rules for class ω_j. The patterns for class ω_j are labelled as positive and all remaining patterns are labelled as negative. We seek a rule that covers as many of the patterns of class ω_j as possible, and ideally none (or few) of the patterns of the remaining classes. Denote this rule by r_1. All the training patterns covered by r_1 are removed from the dataset and the algorithm finds the next best rule to classify correctly the patterns from class ω_j. The process continues until a terminating condition is satisfied (when the added rule fails to improve the accuracy of the set of rules on a validation set). Once this is achieved, rules are generated for the next class. The outcome is a set of classification rules, ordered by class.

The approach is depicted pictorially in Figure 7.13. Figure 7.13(a) shows the distribution of three classes, labelled as ▲, ■ and ♦. First, we seek rules for the class ▲, which is labelled as the positive class and the remaining classes as the negative class [Figure 7.13(b)]. The rule R1, whose coverage is shown in Figure 7.13(c) is extracted first since it is the rule covering the greatest proportion of positive patterns. The patterns that it covers are removed from the dataset [Figure 7.13(d) and a second rule, R2, is sought. The coverage is shown in Figure 7.13(e) and the patterns removed from the training dataset. The algorithm then proceeds to find the rules for the next class [Figure 7.13(f)].

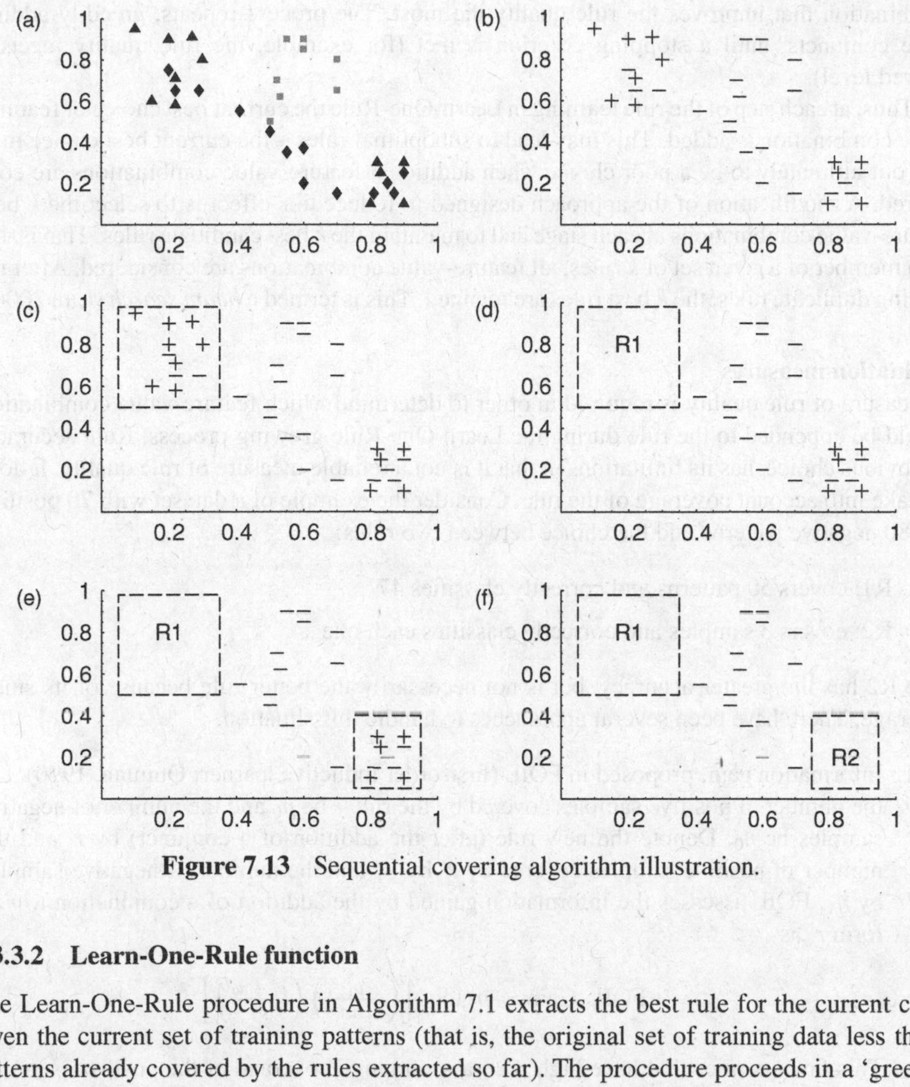

Figure 7.13 Sequential covering algorithm illustration.

7.3.3.2 Learn-One-Rule function

The Learn-One-Rule procedure in Algorithm 7.1 extracts the best rule for the current class given the current set of training patterns (that is, the original set of training data less those patterns already covered by the rules extracted so far). The procedure proceeds in a 'greedy' fashion – making a locally optimal decision at each stage of the algorithm. It seeks an optimal rule by refining the current best rule until a stopping criterion is reached.

Rule growing

Rules are generally grown under a general-to-specific or specific-to-general strategy. Under the general-to specific approach, the initial rule is of the form

IF <empty> THEN class=y

that is, the condition is empty and the rule consequent is the class under consideration. The rule covers all patterns in the current set of patterns and would have poor quality, as assessed according to some rule evaluation measure (discussed below). The rule is grown by adding conjuncts (tests on the feature values) to improve the rule's quality. There may be many features, each one taking many possible values, and one approach is to choose the feature-value combination that improves the rule quality the most. The process repeats, greedily adding more conjuncts, until a stopping criterion is met (for example, the rule quality meets a desired level).

Thus, at each step of the rule learning in Learn-One-Rule the current best choice of feature-value combination is added. This may lead to suboptimal rules – the current best choice may turn out ultimately to be a poor choice when additional feature-value combinations are considered. A modification of the approach designed to reduce this effect is to select the k best feature-value combinations at each stage and to maintain the k best candidate rules. That is, for each member of a given set of k rules, all feature-value combinations are considered. After removing duplicate rules, the k best rules are retained. This is termed a *beam search* (Liu, 2006).

Evaluation measures

A measure of rule quality is required in order to determine which feature-value combination should be appended to the rule during the Learn-One-Rule growing process. Rule accuracy, an obvious choice, has its limitations in that it is not a reliable measure of rule quality. It does not take into account coverage of the rule. Consider the example of a dataset with 70 positive and 80 negative patterns and the choice between two rules:

R1: covers 50 patterns and correctly classifies 47

R2: covers 3 samples and correctly classifies each one

Rule R2 has the greater accuracy, but is not necessarily the better rule because of its small coverage. There have been several approaches to handle this situation.

1. Information gain, proposed in FOIL (first-order inductive learner; Quinlan, 1990). Let the number of positive samples covered by the rule r be p_0 and the number of negative samples be n_0. Denote the new rule (after the addition of a conjunct) by r' and the number of positive samples covered by r' by p_1 and the number of negative samples by n_1. FOIL assesses the information gained by the addition of a conjunction to r to form r' as

$$\text{FOIL\_Gain} = p_1 \log_2 \left[\left(\frac{p_1}{p_1 + n_1} \right) \left(\frac{p_0 + n_0}{p_0} \right) \right]$$

This favours rules that have high accuracy and high coverage of the positive patterns. For the example above, $p_0 = 70$, $n_0 = 80$, $p_1 = 47$ and $n_1 = 3$ for rule R1, giving an information gain of 46.5; for rule R2, $p_1 = 3$ and $n_1 = 0$ and the information gain is 2.73.

2. Likelihood ratio. A statistical test is used to prune rules that have poor coverage. It is used to assess whether the observed distribution of the patterns among classes is attributed to chance. We compute the likelihood ratio test statistic

$$L = 2 \sum_{i=1}^{C} f_i \log \left(\frac{f_i}{e_i} \right)$$

where C is the number of classes, f_i is the observed frequency of each of the classes among the patterns covered by the rule and e_i is the expected frequency of the rule if the rule made predictions at random (it depends on the class priors and the coverage of the rule). The statistic has a χ^2 distribution with $C-1$ degrees of freedom. A large value of L suggests there is a significant difference in the number of correct predictions made by the rule than random guessing. The performance of the rule is not due to chance. The rule giving the larger value of L is the preferred rule. For example, using the illustration above, the expected frequency for the positive class, $e_+ = 50 \times 70/150 = 23.3$ for Rule R1 and $e_- = 50 \times 80/150 = 26.7$, giving

$$L(\mathrm{R1}) = 2\left[47\log_2\left(\tfrac{47}{23.3}\right) + 3\log_2\left(\tfrac{3}{26.7}\right)\right] = 90.95$$

For rule R2, $e_+ = 3 \times 70/150 = 1.4$ and $e_- = 3 \times 80/150 = 1.6$, giving

$$L(\mathrm{R2}) = 2\left[3\log_2\left(\tfrac{3}{1.4}\right) + 0 \times \log_2\left(\tfrac{0}{1.6}\right)\right] = 5.46$$

Rule pruning

Note that the Learn-One-Rule algorithm does not employ a separate validation set; rule quality is assessed on the basis of the training data alone. Therefore, the resulting rule is likely to overfit the data and provide an optimistic estimate of performance. One approach to improve performance is to remove conjuncts, one at a time, while monitoring the performance on a separate validation set. If the error decreases, on the removal of a conjunct, the pruned rule is preferred.

7.3.3.3 Illustration

RIPPER (repeated incremental pruning to produce error reduction) is an approach proposed by Cohen (1995) to address some of the computational issues arising from the application of existing rule induction systems applied to large noisy datasets. For multiple classes, the classes are ordered $\omega_1, \ldots, \omega_C$, where ω_1 is the least prevalent and ω_C the most prevalent. Then, starting with ω_1 as the positive class and the remaining classes as the negative class, a rule set is built to separate the positive class from the negative class, one rule at a time. After growing a rule, it is immediately pruned using the separate pruning set to monitor performance. The rule is then added to the rule set provided that it does not violate a stopping condition.

Rule growing and pruning

Beginning with an empty set of conjunctions, a conjunction is added that maximises FOIL's information gain criterion until the rule covers no examples from the negative class. The rule is then pruned by deleting the conjunct that maximises the function

$$v^* = \frac{p-n}{p+n}$$

where p (n) is the number of positive (negative) examples in the pruning set covered by the rule, subject to performance on the pruning set not deteriorating. The rule is then added to the rule set (if it does not violate the stopping condition). The positive and negative examples covered by the rule are removed from the training set and, if there are still positive examples remaining, a new rule is grown.

Stopping condition

The stopping condition employed by RIPPER is based on the *description length* of the rule set and the examples. If the new rule increases the total description length by d bits [Cohen (1995) takes $d = 64$], then no further rules are added to the rule set. The rule set is then examined and rules are deleted that reduce the total description length.

Optimisation of the rule set

The final stage of RIPPER is a rule optimisation process that examines each rule in turn and generates a *revision rule* and a *replacement rule*. The MDL principle is then used to decide whether the final set includes the original rule, its revision rule or its replacement rule (for further details see Cohen, 1995).

7.3.4 Example application study

The problem

Discovery of classification rules for Egyptian rice diseases (El-Telbany et al., 2006). The main aim of the research is the discovery and control of diseases and an automatic rule discovery approach was taken to help farmers identify diseases from collected evidence. Formalising knowledge that can benefit agriculture required an approach that had interpretability. Disease identification is important as it is estimated that rice diseases cause losses that reduce the potential yield by 15%.

Summary

The C4.5 decision tree algorithm was used to discover classification rules and performance suitable for accurate classification of the disease demonstrated.

The data

The data consisted of measurements on six features [variety, age, part (e.g. leaves, grains), appearance, colour, temperature] for each of the five most important diseases (blight, brown-spot, false smut, white tip nematode and stem rot). The features were a mixture of continuous and categorical variables. A total of 206 samples were collected.

The model

A decision tree was constructed using the C4.5 algorithm (Quinlan, 1993).

Training procedure

The algorithm was trained using the default parameters for C4.5 in WEKA (Witten and Frank, 2005). A cross-validation procedure was used to assess performance.

Results

Rules were extracted from the decision tree. Two of the resulting rule set are as follows:

IF appearance=spot AND colour=olive THEN disease=blight

IF appearance=spot AND colour=brown AND age$\leq$55 THEN disease=brown-spot.

A good classification accuracy was achieved.

7.3.5 Further developments

Rule induction approaches induce a set of rules so that (in the case of two classes) a pattern is classified as belonging to the positive class if it satisfies one of the rules. We have described approaches for handling multiclass problems. These approaches provide a description of the classification rule in terms of the features and values of the patterns.

On the other hand, many of the techniques presented earlier in this book classify an object by defining a probability distribution over the feature space and providing a probability of class membership.

Integrating probabilistic and rule induction approaches (in areas such as probabilistic inductive logic programming) has been addressed by several authors and is an active area of research (Getoor and Taskar, 2007; De Raedt *et al.*, 2008).

7.3.6 Summary

Rule-based classifiers offer descriptive models that are interpretable. Rules may be extracted directly from a decision tree (leading to a set of mutually exclusive and exhaustive rules) or using a rule induction approach such as sequential covering. These methods may be regarded as providing a basis function model for the discriminant function, where the basis function is a hyperrectangle in the feature space and associated with each basis function is a class label. In the case of the rule induction methods, the basis functions overlap, permitting complex decision boundaries to be modelled more easily, but requiring a conflict resolution strategy since multiple rules giving conflicting class predictions may be triggered by a pattern.

7.4 Multivariate adaptive regression splines

7.4.1 Introduction

The multivariate adaptive regression splines (MARS)[3] (Friedman, 1991) approach may be considered as a continuous generalisation of the regression tree methodology treated separately in Section 7.2 and the presentation here follows Friedman's approach.

Suppose that we have a dataset of n measurements on p variables, $\{\boldsymbol{x}_i, i = 1, \ldots, n\}$, $\boldsymbol{x}_i \in \mathbb{R}^p$, and corresponding measurements on the response variable $\{y_i, i = 1, \ldots, n\}$. We shall assume that the data have been generated by a model

$$y_i = f(\boldsymbol{x}) + \epsilon$$

where ϵ denotes a residual term. Our aim is to construct an approximation, $\hat{f}$, to the function f.

7.4.2 Recursive partitioning model

The recursive partitioning model takes the form

$$\hat{f}(\boldsymbol{x}) = \sum_{m=1}^{M} a_m B_m(\boldsymbol{x})$$

[3] MARS is a trademark of Salford Systems.

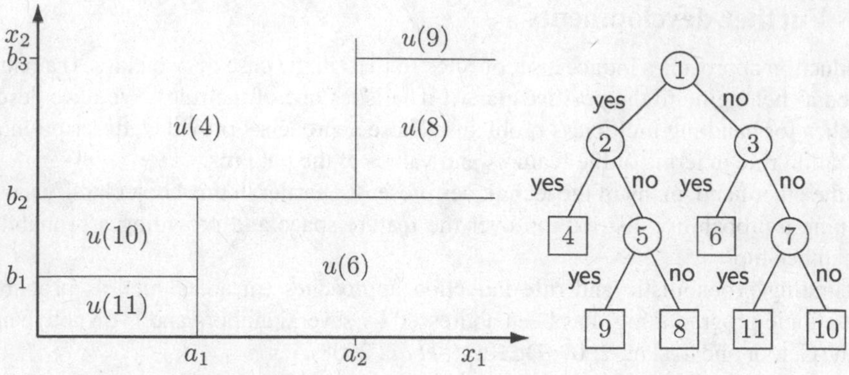

Figure 7.14 Classification tree and decision regions; the decision nodes (circular) are characterised as follows: node 1, $x_2 > b_2$; node 2, $x_1 < a_2$; node 5, $x_2 > b_3$; node 3, $x_1 > a_1$; node 7, $x_2 < b_1$. The square nodes correspond to regions in the feature space.

The basis functions B_m are produced recursively and take the form

$$B_m(\boldsymbol{x}) = I\{\boldsymbol{x} \in u_m\}$$

where I is an indicator function with value unity if the argument is true and zero otherwise. We have used the notation that $\{u_m, m = 1, \ldots, M\}$ is a partition of the data space $\mathbb{R}^p$ (that is, u_i is a subspace of $\mathbb{R}^p$ such that $u_i \cap u_j = \emptyset$ for $i \neq j$, and $\cup_i u_i = \mathbb{R}^p$). The set $\{a_m, m = 1, \ldots, M\}$ are the coefficients in the expansion whose values are determined (often) by a least-squares minimisation procedure for fitting the approximation to the data. In a classification problem, a regression for each class separately may be performed (using binary dependent variables, with value equal to 1 for $\boldsymbol{x}_i \in$ class ω_j, and zero otherwise) giving C functions $\hat{f}_j$, on which discrimination is based.

The basis functions are produced by a recursive partitioning algorithm (Friedman, 1991) and can be represented as a product of step functions. Consider the partition produced by the tree given in Figure 7.14. The first partition is on the variable x_2 and divides the plane into two regions, giving basis functions

$$H[(x_2 - b_2)] \quad \text{and} \quad H[-(x_2 - b_2)]$$

where

$$H(x) = \begin{cases} 1 & x \geq 0 \\ 0 & \text{otherwise} \end{cases}$$

The region $x_2 < b_2$ is partitioned again, on variable x_1 with threshold a_1, giving the basis functions

$$H[-(x_2 - b_2)]H[+(x_1 - a_1)] \quad \text{and} \quad H[-(x_2 - b_2)]H[-(x_1 - a_1)]$$

The final basis functions for the tree comprise the products

$$\begin{array}{l} H[-(x_2 - b_2)]H[-(x_1 - a_1)]H[+(x_2 - b_1)] \\ H[-(x_2 - b_2)]H[-(x_1 - a_1)]H[-(x_2 - b_1)] \\ H[-(x_2 - b_2)]H[+(x_1 - a_1)] \\ H[+(x_2 - b_2)]H[-(x_1 - a_2)] \\ H[+(x_2 - b_2)]H[+(x_1 - a_2)]H[+(x_2 - b_3)] \\ H[+(x_2 - b_2)]H[+(x_1 - a_2)]H[-(x_2 - b_3)] \end{array}$$

Thus, each basis function is a product of step functions, H.

In general, the basis functions of the recursive partitioning algorithm have the form

$$B_m(\boldsymbol{x}) = \prod_{k=1}^{K_m} H[s_{km}(x_{v(k,m)} - t_{km})] \tag{7.7}$$

where s_{km} take the values ± 1 and K_m is the number of splits that give rise to $B_m(\boldsymbol{x})$; $x_{v(k,\,m)}$ is the variable split and t_{km} are the thresholds on the variables.

The MARS procedure is a generalisation of this recursive partitioning procedure in the following ways.

7.4.2.1 Continuity

The recursive partitioning model described above is discontinuous at region boundaries. This is due to the use of the step function H. The MARS procedure replaces these step functions by spline functions. The two-sided truncated power basis functions for qth-order splines are

$$b_q^{\pm}(x - t) = [\pm(x - t)]_+^q$$

where $[.]_+$ denotes that the positive part of the argument is considered. The basis functions $b_q^+(x)$ are illustrated in Figure 7.15.

The step function, H, is the special case of $q = 0$. The MARS algorithm employs $q = 1$. This leads to a continuous function approximation, but discontinuous first derivatives.

The basis functions now take the form

$$B_m^q(\boldsymbol{x}) = \prod_{k=1}^{K_m} [s_{km}(x_{v(k,m)} - t_{km})]_+^q \tag{7.8}$$

where the t_{km} are referred to as the *knot locations*.

7.4.2.2 Retention of parent functions

The basic recursive partitioning algorithm replaces an existing parent basis function by its product with a step function and its product with the reflected step function. The number of basis functions therefore increases by one on each split. The MARS procedure retains the parent basis function. Thus the number of basis functions increases by two at each split. This provides a much more flexible model that is capable of modelling classes of functions

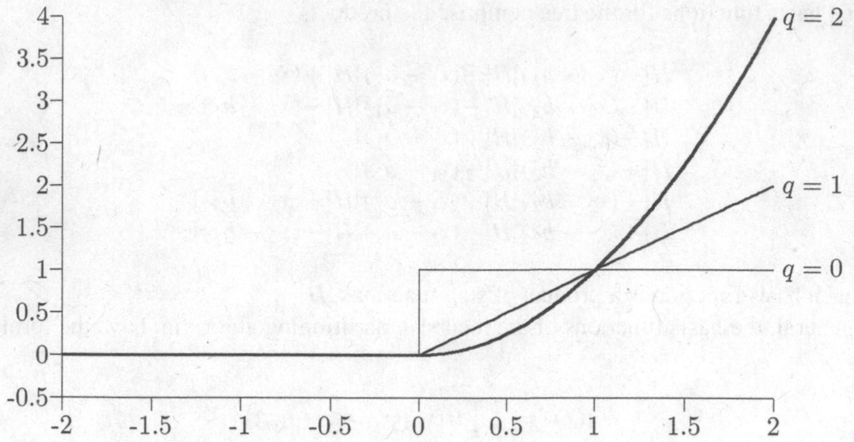

Figure 7.15 Spline functions, $b_q^+(x)$, for $q = 0, 1, 2$.

that have no strong interaction effects, or strong interactions involving at most a few of the variables, as well as higher-order interactions. A consequence of the retention of the parent functions is that the regions corresponding to the basis functions may overlap.

7.4.2.3 Multiple splits

The basic recursive partitioning algorithm allows multiple splits on a single variable: as part of the modelling procedure, a given variable may be selected repeatedly and partitioned. Thus basis functions comprise products of repeated splits on a given variable (the basis functions for the tree in Figure 7.14 contain repeated splits). In the continuous generalisation, this leads to functional dependencies of higher order than q [see Equation (7.8)] on individual variables. To take advantage of the properties of *tensor product spline basis functions*, whose factors involve a different variable, MARS restricts the basis functions to products involving single splits on a given variable. By reselecting the same parent for splitting (on the same variable), the MARS procedure is able to retain the flexibility of the repeated split model (it trades depth of tree for breadth of tree).

The second stage in the MARS strategy is a pruning procedure; basis functions are deleted one at a time – the basis function being removed being that which results in the fit being improved the most (or degrades it the least). A separate validation set could be chosen to estimate the goodness of fit of the model. The lack-of-fit criterion proposed for the MARS algorithms is a modified form of the generalised cross-validation criterion of Craven and Wahba (1979). In a discrimination problem, a model could be chosen that minimises error rate on the validation set.

In applying MARS to discrimination problems, one approach is to use the usual binary coding for the response function, $f_j, j = 1, \ldots, C$, with $f_j(\boldsymbol{x}) = 1$ if $\boldsymbol{x}$ is in class ω_j and zero otherwise. Each class may have a separate MARS model; the more parsimonious model of having common basis functions for all classes will reduce the computation. The weights a_i in the MARS algorithms are replaced by vectors $\boldsymbol{a}_i$, determined to minimise a generalised cross-validation measure.

7.4.3 Example application study

The problem

This application is to construct reliable network intrusion detection systems for protecting information systems security against the increasing threat of cyber attacks (Mukkamala *et al.*, 2004).

Summary

The detection and classification of network traffic behaviour types based on an examination of patterns of user activity is a problem requiring the analysis of often huge amounts of data. Three nonparametric models (MARS, support vector machines and neural networks) were trained to predict traffic patterns in a network.

The data

The data, supplied by MIT Lincoln Laboratory for a DARPA intrusion detection evaluation, comprised records in five classes of behaviour: normal, probing (surveillance), denial of service, user to super-user (unauthorised access to super-user), and remote to local (unauthorised access from a remote machine). Separate training and test sets were generated.

The model

Three types of model were assessed. The MARS model used cubic splines to fit the data with 5 basis functions.

Training procedure

Five MARS models were employed for the classification (one developed for each of the classes). A model was generated that overfitted the data and a backwards pruning procedure employed to remove unwanted basis functions.

Results

For this study, MARS gave superior classification performance for the most important classes (user to super-user and remote to local). The development of the MARS approach for dynamic environments, essential in applications such as intrusion detection, was identified as an important way forward.

7.4.4 Further developments

A development of the basic model to include procedures for mixed ordinal and categorical variables is provided by Friedman (1993). POLYMARS (Stone *et al.*, 1997) is a development to handle a categorical response variable, with application to classification problems.

Time series versions of MARS (TSMARS) have been developed for forecasting applications (De Gooijer *et al.*, 1998). A Bayesian approach to MARS (BMARS) fitting, which averages over possible models (leading to a consequent loss interpretability of the final model), is described by Denison *et al.* (1998b) [see also Holmes and Denison (2003) for an application in a classification context].

7.4.5 Summary

MARS is a method for modelling high-dimensional data in which the basis functions are a product of spline functions. MARS searches over threshold positions on the variables, or *knots*, across all variables and interactions. Once it has done this, it uses a least squares regression to provide estimates of the coefficients of the model.

MARS can be used to model data comprising measurements on variables of different type – both continuous and categorical. The optimal model is achieved by growing a large model and then pruning the model by removing basis functions until a lack-of-fit criterion is minimised. In common with decision trees, it selects the variables important for classification, thus performing automatic variable selection.

7.5 Application studies

Applications of decision tree and rule induction methodology are varied and include the following:

- Intrusion detection. Komviriyavut *et al.* (2009) evaluate C4.5 and RIPPER on an online dataset in a study to detect attacks on computer networks. The network data collected were processed to produce 13 features. Rules were developed to classify the data into Normal, Denial of Service attack and Probe attack. The training data comprised 7200 records equally distributed between the three classes. The test data comprised 4800 records, also equally distributed. RIPPER obtained 17 rules. Both approaches gave similar (very good) results (classification rates greater than 98%). [see also Saravanan (2009)].
- Power systems security. Swarnkar and Niazi (2005) [see also Swarup *et al.* (2005) and Wehenkel and Pavella (1993)] develop a decision tree approach based on CART for online transient security evaluation of a power system to predict system security and provide a preventative control strategy. Such an approach is an alternative to neural networks, which although showing promise do not offer the explanation of a decision tree.
- Predicting stroke inpatient rehabilitation outcome. A classification tree model (CART) was developed to predict rehabilitation outcomes for stroke patients (Falconer *et al.*, 1994). The data comprised measurements on 51 ordinal variables on 225 patients. A classification tree was used to identify those variables most informative for predicting favourable and unfavourable outcomes. The resulting tree used only 4 of the 51 variables measured on admission to a university-affiliated rehabilitation institute, improving the ability to predict rehabilitation outcomes, and correctly classifying 88% of the sample.
- Gait events. In a study into the classification of phases of the gait cycle (as part of the development of a control system for functional electrical stimulation of the lower limbs), Kirkwood *et al.* (1989) develop a decision tree approach that enable redundant combinations of variables to be identified. Efficient rules are derived with high performance accuracy.

- Thyroid diseases. In a study to use a decision tree to synthesise medical knowledge, Quinlan (1986) employs C4, a descendant of ID3 that implements a pruning algorithm, to generate a set of high-performance rules. The data consist of input from a referring doctor (patient's age, sex and 11 true–false indicators); a clinical laboratory (up to six assay results) and a diagnostician. Thus, variables are of mixed type, with missing values and some misclassified samples. The pruning algorithm leads to an improved simplicity and intelligibility of derived rules.

Other applications have been in the areas of telecommunications, marketing and industrial applications. See Langley and Simon (1995) for a review of the applications of machine learning and rule induction.

There have been several comparisons with neural networks and other discrimination methods; for example,

- Digit recognition. Brown *et al.* (1993) compare a classification tree with a multilayer perceptron (MLP) on a digit recognition (extracted from license plate images) problem (with application to highway monitoring and tolling). All features were binary and the classification tree performance was poorer than the MLP, but performance improved when features that were a combination of the original variables were included.

- Various datasets. Curram and Mingers (1994) compared an MLP with a decision tree on several datasets. The decision tree was susceptible to noisy data, but had the advantage of providing insight. Shavlik *et al.* (1991) compared ID3 with an MLP on four datasets, finding that the MLP handled noisy data and missing features slightly better than ID3, but took considerably longer to train.

Other comparative studies include speaker-independent vowel classification and load forecasting (Atlas *et al.*, 1989), finding the MLP superior to a classification tree; disk drive manufacture quality control and the prediction of chronic problems in large-scale communication networks (Apté *et al.*, 1994).

The MARS methodology has been applied to problems in classification and regression including the following:

- Intrusion detection systems. The development of efficient and reliable intrusion detection systems is essential for protecting information systems security from the increasing threat of cyber attack. This is a great challenge. Mukkamala *et al.* (2004) compare MARS with neural networks and support vector machines on a problem to classify attack into four main categories. MARS was found to be superior to the other methods in its ability to classify the most important classes in terms of attack severity.

- Ecology. Leathwick *et al.* (2005) use MARS for predicting the distributions of freshwater fish. MARS is used to model the nonlinear relationship between a set of environmental variables and the occurrence of fifteen species of fish. The R implementation of MARS was used to obtain a set of basis functions that were used in further modelling.

- Economic time series. Sephton (1994) uses MARS to model three economic time series: annual US output, capital and labour inputs; interest rates and exchange rates using a generalised cross-validation score for model selection.

- Telecommunications. Duffy *et al.* (1994) compare neural networks with CART and MARS on two telecommunications problems: modelling switch processor memory (a regression problem) and characterising traffic data (speech and modem data at three different baud rates) (a four-class discrimination problem).

- Particle detection. In a comparative study of four methods of discrimination, Holmström and Sain (1997) compare MARS with a quadratic classifier, a neural network and kernel discriminant analysis on a problem to detect a weak signal against a dominant background. Each event (in the two-class problem) was described by 14 variables and the training set comprised 5000 events. MARS appeared to give best performance.

7.6 Summary and discussion

Recursive partitioning methods have a long history and have been developed in many different fields of endeavour. Complex decision regions can be approximated by the union of simpler decision regions. A major step in this research was the development of CART, a simple nonparametric method of partitioning data. The approach described in this chapter for constructing a classification tree is based on that work. The basic idea has then been developed in two different directions: rule induction and the modelling of continuous functions.

Decision trees may be used to generate a set of IF-THEN rules that provide an interpretable classification rule important in some applications. There are other approaches to rule induction and one based on a sequential covering algorithm has been described.

There have been many comparative studies between decision trees and neural networks, especially MLPs. Both approaches are capable of modelling complex data. The MLP usually takes longer to train and does not provide the insight of a tree, but performance has often been better than a tree on the datasets used for the evaluation (which may favour an MLP anyway since they variables are continuous). Further work is required.

The multivariate adaptive regression spline approach is a recursive partitioning method that utilises products of spline functions as the basis functions. Like CART, it is also well-suited to model mixed variable (discrete and continuous) data.

7.7 Recommendations

Although it cannot be said that classification trees perform substantially better than other methods (and for a given problem, there may be a parametric method that will work better – but you do not know which one to choose), their simplicity and consistently good performance on a wide range of datasets have led to their widespread use in many disciplines. We recommend that you try them for yourselves.

Specifically, classification tree approaches are recommended:

1. For complex datasets in which you believe decision boundaries are nonlinear and decision regions can be approximated by the sum of simpler regions.

2. For problems where it is important to gain an insight into the data structure and the classification rule, the explanatory power of trees may lead to results that are easier to communicate than other techniques.

3. For problems with data consisting of measurements on variables of mixed type (continuous, ordinal, nominal).

4. Where ease of implementation is required.
5. Where speed of classification performance is important – the classifier performs simple tests on (single) variables.

The above recommendations apply equally to methods of rule induction, particularly the interpretability aspect.

MARS is simple to use and is recommended for high-dimensional regression problems, for problems involving variables of mixed type and for problems where some degree of interpretability of the final solution is required.

7.8 Notes and references

There is a very large literature on classification trees in the areas of pattern recognition, artificial intelligence, statistics and the engineering sciences, but by no means exclusively within these disciplines. Many developments have taken place by researchers working independently and there are many extensions of the approach described in this chapter as well as many alternatives. A survey is provided by Safavian and Landgrebe (1991); see also Feng and Michie (1994). Several tree-based approaches were assessed as part of the *Statlog* project (Michie *et al.*, 1994) including CART, which proved to be one of the better ones because it can incorporate costs into the decision.

Quinlan developed C4.5, building on his earlier algorithm ID3. The decision tree algorithm implemented in the machine learning software WEKA (Witten and Frank, 2005), J48, is an implementation of C4.5. A later development, with added features, is C5.0.

Examples of rule induction algorithms that employ sequential covering are CN2 (Clark and Niblett, 1989) and RIPPER (Cohen, 1995).

MARS was introduced by Friedman (1991). Software for CART, other decision tree software and MARS is publicly available (from Salford Systems and in the WEKA and R software packages).

Exercises

1. In Figure 7.5, if regions $u(4)$ and $u(6)$ correspond to class ω_1 and $u(8)$, $u(9)$, $u(10)$ and $u(11)$ to class ω_2, construct a multilayer perceptron with the same decision boundaries.
2. A standard classification tree produces binary splits on a single variable at each node. For a two-class problem, using the results of Chapter 5, describe how to construct a tree that splits on a linear combination of variables at a given node.
3. The predictability index (relative decrease in the proportion of incorrect predictions for split s on node t) is written (using the notation of Section 7.2.6)

$$\tau(\omega|s) = \frac{\sum_{j=1}^{C} p^2(\omega_j|t_L)p_L + \sum_{j=1}^{C} p^2(\omega_j|t_R)p_R - \sum_{j=1}^{C} p^2(\omega_j|t)}{1 - \sum_{j=1}^{C} p^2(\omega_j|t)}$$

Show that the decrease in impurity when passing from one group to two subgroups for the Gini criterion can be written

$$\Delta\mathcal{I}(s,t) = \sum_{j=1}^{C} p^2(\omega_j|t_L)p_L + \sum_{j=1}^{C} p^2(\omega_j|t_R)p_R - \sum_{j=1}^{C} p^2(\omega_j|t)$$

and hence that maximising the predictability also maximises the decrease in impurity.

4. Consider the two-class problem with bivariate distributions characterised by x_1 and x_2 which can take three values. The training data are given by:

	Class ω_1				Class ω_2		
	x_1				x_1		
	1	2	3		1	2	3
1	3	0	0	1	0	1	4
x_2 2	1	6	0	x_2 2	0	5	7
3	4	1	2	3	1	0	1

where, for example, there are four training samples in class ω_1 with $x_1 = 1$ and $x_2 = 3$. Using the Gini criterion, determine the split (variable and value) for the root node of a tree.

5. Construct a classification tree using dataset 1 from the exercises in Chapter 6. Initially allow 10 splits per variable. Monitor the performance on the validation set as the tree is grown. Prune using the validation set to monitor performance. Investigate the performance of the approach for $p = 2$, 5 and 10. Describe the results and compare with a linear discriminant analysis.

6. Show that MARS can be recast in the form

$$a_0 + \sum_i f_i(x_i) + \sum_{ij} f_{ij}(x_i, x_j) + \sum_{ijk} f_{ijk}(x_i, x_j, x_k) + \cdots$$

8

Ensemble methods

The classification techniques considered so far in this book predict the class membership of a pattern based on a single classifier induced from the dataset. We now ask the question whether we can gain improvements in performance by combining the outputs of several classifiers. *Ensemble methods* (or *classifier combination techniques*) has been an area of growing research in recent years and is related to developments in the *data fusion* literature where, in particular, the problem of decision fusion (combining decisions from multiple target detectors) is being addressed extensively.

8.1 Introduction

The approach to classifier design commonly taken is to identify a candidate set of plausible models; to train the classifiers using a training set of labelled patterns and to adopt the classifier that gives the best generalisation performance, estimated using an independent test set assumed representative of the true operating conditions. This results in a single 'best' classifier that may then be applied throughout the feature space. In the following chapter we shall address the question of how we measure classifier performance to select a 'best' classifier.

We now consider the potential of combining classifiers for datasets with complex decision boundaries. It may happen that, out of our set of classifiers, no single classifier is clearly best (using some suitable performance measure, such as error rate). However, the set of misclassified samples may differ from one classifier to another. Thus, the classifiers may give complementary information and combination could prove useful.

A simple example is illustrated in Figure 8.1. Two linear classifiers, Cl_1 and Cl_2, are defined on a univariate data space with the rule

$$\begin{aligned} Cl_i(\boldsymbol{x}) > 0 &\Rightarrow x \in \text{class } \bullet \\ Cl_i(\boldsymbol{x}) < 0 &\Rightarrow x \in \text{class } \diamond \end{aligned} \tag{8.1}$$

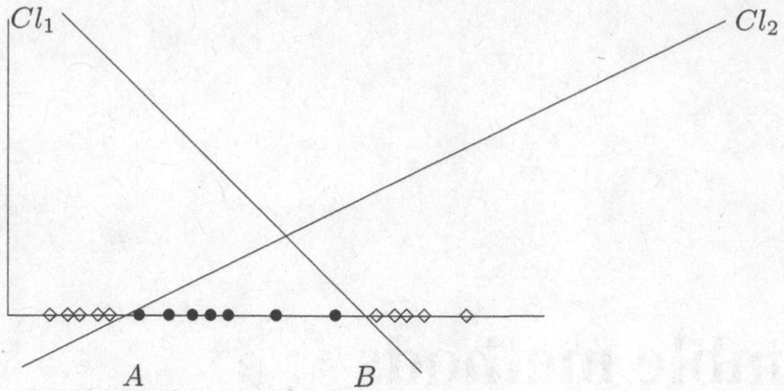

Figure 8.1 Two linear discriminants defined on a univariate data space.

Thus, classifier Cl_1 predicts class • for data points to the left of B and class ◇ for points to the right of B. Classifier Cl_2 predicts class • for points to the right of A and class ◇ for points to the left. Neither classifier obtains 100% performance on the dataset. However, combining them with the rule:

IF Cl_1 AND Cl_2 predict • THEN assign x to •
ELSE assign x to ◇

gives 100% performance.

The idea of combining classifiers is not a new one. Early work on multiclass discrimination developed techniques for combining the results of two-class discrimination rules (Devijver and Kittler, 1982) defining different rules for different parts of a feature space. The terms 'dynamic classifier selection' (Woods *et al.*, 1997) and 'classifier choice system' (Hand *et al.*, 2001) have been used for classifier systems that attempt to predict the best classifier for a given region of feature space. The term 'classifier fusion' or 'multiple classifier system' usually refers to the combination of predictions from multiple classifiers to yield a single class prediction.

8.2 Characterising a classifier combination scheme

Combination schemes may themselves be classified according to several characteristics including their structure, input data type, form of component classifiers and training requirements. In this section, we summarise some of the main features of a combiner according to the following taxonomy:

- feature space of classifiers – same or different;
- level of combiner – data, feature or decision;
- degree of training – fixed or trainable;
- form of individual component classifiers – common or dissimilar;

- structure – parallel, serial or hierarchical;
- optimisation – separate or simultaneous optimisation of classifiers and combiner.

We assume that there are measurements from C classes and there are L component classifiers.

8.2.1 Feature space

There are several ways of characterising multiple classifier systems according to the feature space, as indeed there are with basic component classifiers. We define three broad categories as follows:

C1. Different feature spaces
This describes the combination of a set of classifiers, each designed on different feature spaces (perhaps using data from different sensors). For example, in a person verification application, several classifier systems may be designed for use with different sensor data (for example, retina scan, facial image, handwritten signature) and we wish to combine the outputs of each system to improve performance.

Figure 8.2 illustrates this situation. A set of L sensors ($S_1, S_2, \ldots, S_L$) provides measurements, $x_1, x_2, \ldots, x_L$, on an object. Associated with each sensor is a classifier ($Cl_1, Cl_2, \ldots, Cl_L$) providing, in this case, estimates of the posterior probabilities of class membership, $p(c|x_i)$ for sensor S_i. The combination rule (denoted Com in the figure), which is itself a classifier defined on a feature space of posterior probabilities, combines these posterior probabilities

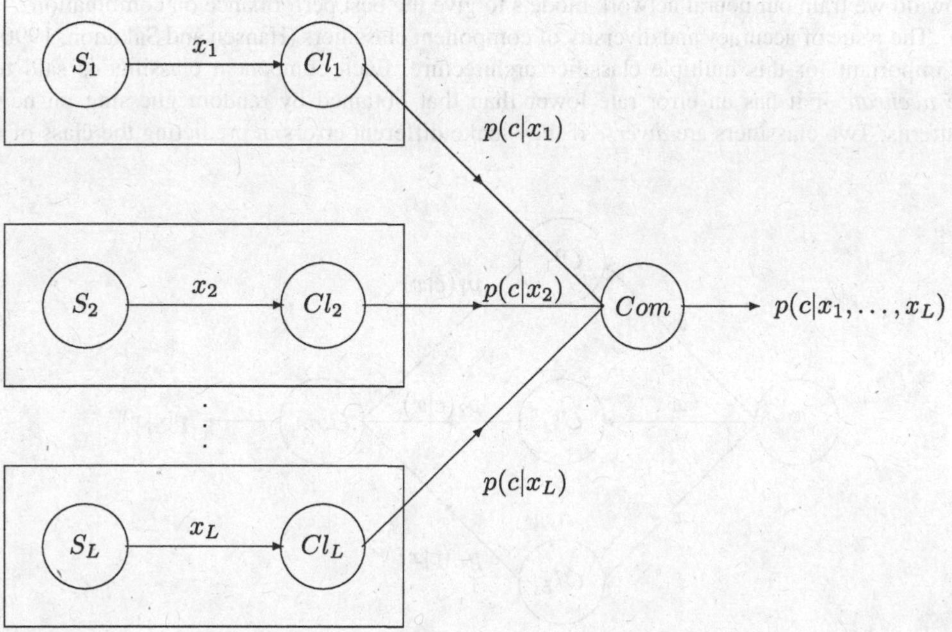

Figure 8.2 Classifier fusion architecture, C1 – component classifiers defined on different feature spaces.

to provide an estimate of $p(c|x_1, \ldots, x_L)$. Thus, the individual classifiers can be thought of as performing a particular form of feature extraction prior to classification by the combiner.

The question usually addressed with this architecture is: Given the component classifiers, what is the best combination rule? This is closely related to dynamic classifier selection and to architectures developed in the data fusion literature (see Section 8.3).

C2. Common feature space

In this case, we have a set of classifiers, $Cl_1, \ldots, Cl_L$, each defined on the same feature space and the combiner attempts to obtain a 'better' classifier through combination (Figure 8.3).

The classifiers can differ from each other in several ways.

1. The classifiers may be of different type, belonging to the set of favourite classifiers of a user: for example, nearest neighbour, neural network, decision tree and linear discriminant.
2. They may be of similar type (for example, all linear discriminant functions or all neural network models) but trained using different training sets (or subsets of a larger training set as in bagging described in Section 8.4.9), perhaps gathered at different times or with different noise realisations added to the input.
3. The classifiers may be of similar type, but with different random initialisation of the classifier parameters (for example, the weights in a neural network of a given architecture) in an optimisation procedure.

In contrast with category C1, a question that we might ask here is: Given the combination rule, what is the best set of component classifiers? Equivalently, in case 2 above for example, how do we train our neural network models to give the best performance on combination?

The issue of accuracy and diversity of component classifiers (Hansen and Salamon, 1990) is important for this multiple classifier architecture. Each component classifier is said to be *accurate* if it has an error rate lower than that obtained by random guessing on new patterns. Two classifiers are *diverse* if they make different errors in predicting the class of a

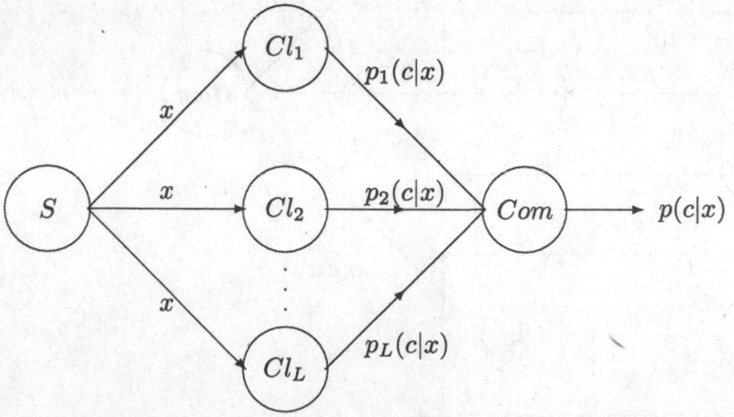

Figure 8.3 Classifier fusion architecture, C2 – component classifiers defined on a common feature space of measurements x made by sensor S.

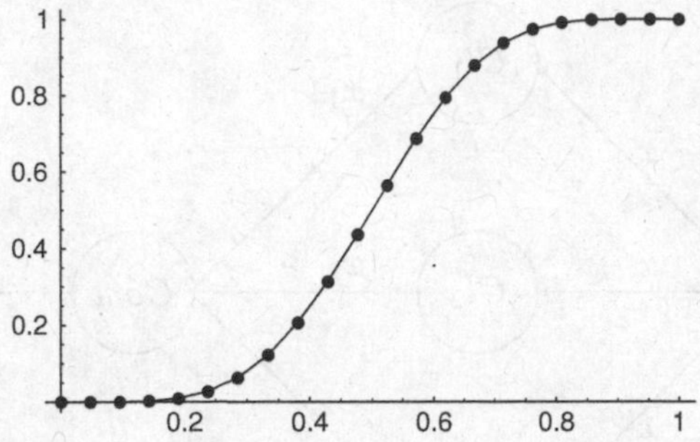

Figure 8.4 Error rate of the ensemble as a function of the error rate of the base classifier, p.

pattern, $\boldsymbol{x}$. To see why both accuracy and diversity are important, consider an ensemble of three classifiers $h_1(\boldsymbol{x})$, $h_2(\boldsymbol{x})$ and $h_3(\boldsymbol{x})$, each predicting a class label as output. If all classifiers produce identical outputs, then there will be no improvement gained by combining the outputs: when h_1 is incorrect, h_2 and h_3 will also be incorrect. However, if the classifier outputs are uncorrelated, then when h_1 is incorrect, h_2 and h_3 may be correct and, if so, a majority vote will give the correct prediction. More specifically, consider L classifiers $h_1, \ldots, h_L$, each with an error rate of $p < \frac{1}{2}$. For the majority vote to be incorrect, we require that $L/2$ or more classifiers be incorrect. The probability that R classifiers are incorrect is

$$\frac{L!}{R!(L-R)!} p^R (1-p)^{L-R}$$

The probability that the majority vote is incorrect is therefore[1]

$$\sum_{R=\lfloor (L+1)/2 \rfloor}^{L} \frac{L!}{R!(L-R)!} p^R (1-p)^{L-R} \tag{8.2}$$

the probability under the binomial distribution that at least $L/2$ are incorrect ($\lfloor . \rfloor$ denotes the integer part). For example, with $L = 11$ classifiers, each with an error rate $p = 0.25$, the probability of six or more classifiers being incorrect is 0.034, which is much less than the individual error rate. Figure 8.4 plots the error rate of the ensemble, given by Equation (8.2), as a function of the error rate of the base classifier p, for $L = 11$.

If the error rates of the individual classifiers exceed 0.5, then the error rate of the majority vote will increase, depending on the number of classes and the degree of correlation between the classifier outputs.

In both cases C1 and C2, each classifier is performing a particular form of feature extraction for inputs to the combiner classifier. If this combiner classifier is not fixed (that is, it is allowed to adapt to the forms of the input), then the general problem is one in which we seek the best forms of feature extractor matched to a combiner (Figure 8.5).

[1] For more than two classes, the majority vote could still produce a correct prediction even if more than $L/2$ classifiers are incorrect.

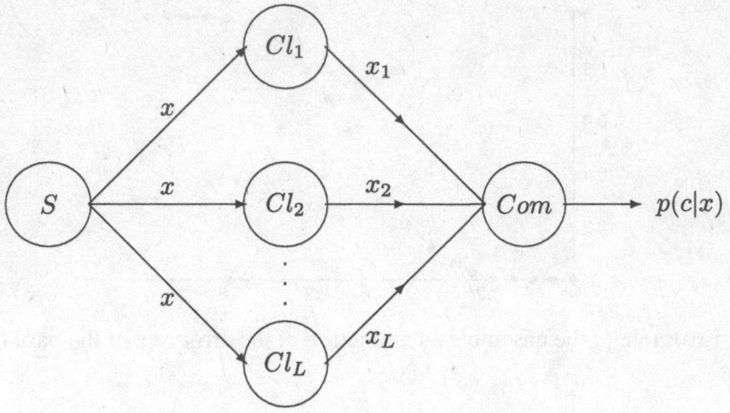

Figure 8.5 Classifier fusion architecture, C2 – component classifiers defined on a common feature space with L component classifiers delivering features x_i.

There is no longer a requirement that the outputs of the classifiers $Cl_1 \ldots Cl_L$ are estimates of posterior probabilities (are positive and sum to unity over classes). Indeed, these may not be the best features. Thus the component classifiers are not classifiers at all and the procedure is essentially the same as many described elsewhere in this book – neural networks, projection pursuit, and so on. In this sense, there is no need for research on classifier combination methods since they impose an unnecessary restriction on the forms of the features input to the combiner.

C3. Repeated measurements

The final category of combination systems arises due to different classification of an object through repeated measurements. This may occur when we have a classifier designed on a feature space giving an estimate of the posterior probabilities of class membership, but in practice several (correlated) measurements may be made on the object (Figure 8.6). An example is that of recognition of aircraft from visual or infrared data.

A probability density function of feature vectors may be constructed using a training dataset. In the practical application, successive measurements are available from the sensor. How can we combine these predictions? This is sometimes described as *multiple observation fusion* or *temporal fusion*.

8.2.2 Level

Combination may occur at different levels of component classifier output.

L1. Data level

Raw sensor measurements are passed to a combiner that produces an estimation of the posterior probabilities of class membership. This simply amounts to defining a classifier on the augmented feature space comprising measurements on all sensor variables. That is, for sensors producing measurements $\boldsymbol{x}, \boldsymbol{y}$ and $\boldsymbol{z}$, a classifier is defined on the feature vector $(\boldsymbol{x}, \boldsymbol{y}, \boldsymbol{z})$. In the *data fusion* literature (see Section 8.3) this is referred to as a *centralised*

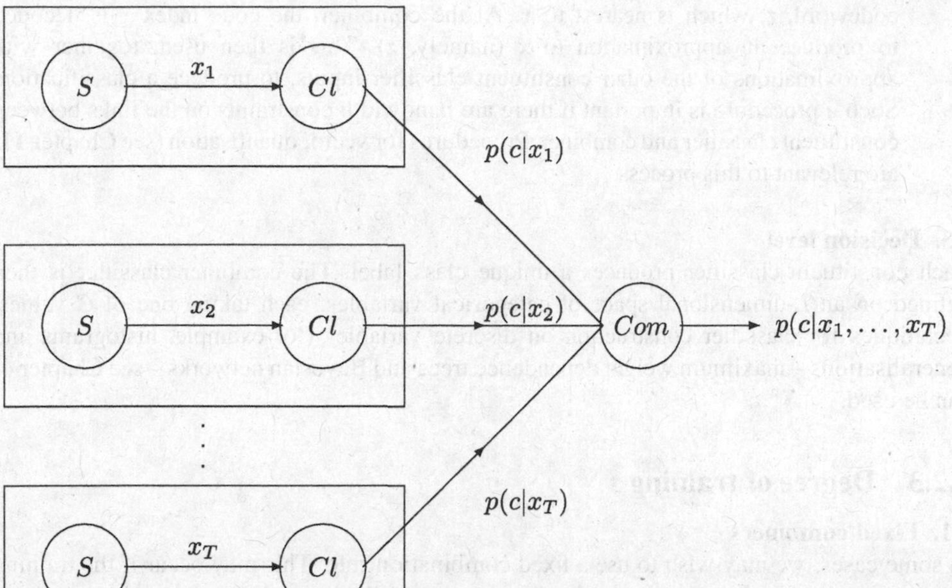

Figure 8.6 Classifier fusion architecture, C3 – repeated measurements on a common feature space. Sensor S produces a sequence of measurements x_i, $i = 1, \ldots, T$, which are input to classifier Cl.

system: all the information collected at distributed sensors is passed to a central processing unit that makes a final decision. A consequence of this procedure is that the combiner must be constructed using data of a high dimensionality. Consequently, it is usual to perform some feature selection or extraction prior to combination.

L2. Feature level

Each constituent classifier (we shall use the term 'classifier' even though the output may not be an estimate of the posterior probabilities of class membership nor a prediction of class) performs some local preprocessing, perhaps to reduce dimensionality. This could be important in some data fusion applications where the communication bandwidth between constituent classifiers and combiner is an important consideration. In data fusion, this is termed a *decentralised* system. The features derived by the constituent classifier for input (transmission) to the combiner can take several forms including the following:

1. A reduced-dimensional representation, perhaps derived using principal components analysis.

2. An estimate of the posterior probabilities of class membership. Thus, each constituent processor is itself a classifier. This is a very specific form of preprocessing that may not be optimum, but it may be imposed by application constraints.

3. A coding of the constituent classifier's input. For an input, $\boldsymbol{x}$, the output of the constituent classifier is the index, y, in a codebook of vectors corresponding to the

codeword, z, which is nearest to x. At the combiner, the code index y is decoded to produce an approximation to x (namely, z). This is then used, together with approximations of the other constituent classifier inputs, to produce a classification. Such a procedure is important if there are bandwidth constraints on the links between constituent classifier and combiner. Procedures for vector quantisation (see Chapter 11) are relevant to this process.

L3. Decision level

Each constituent classifier produces a unique class label. The combiner classifier is then defined on an L-dimensional space of categorical variables, each taking one of C values. Techniques for classifier construction on discrete variables (for example, histograms and generalisations – maximum weight dependence trees and Bayesian networks – see Chapter 4) can be used.

8.2.3 Degree of training

R1. Fixed combiners

In some cases, we may wish to use a fixed combination rule. This may occur if the training data used to design the constituent classifiers are unavailable, or different training sets have been used.

R2. Trainable combiners

Alternatively, the combination rule is adjusted based on knowledge of the training data used to define the constituent classifiers. This knowledge can take different forms:

1. *Probability density functions*. The joint probability density function of constituent classifier outputs is assumed known for each class through knowledge of the distribution of the inputs and the form of the classifiers. For example, in the two-class target detection problem, under the assumption of independent local decisions at each of the constituent classifiers, an optimal detection rule for the combiner can be derived, expressed in terms of the probability of false alarm and the probability of missed detection at each sensor (Chair and Varshney, 1986).
2. *Correlations*. Some knowledge concerning the correlations between constituent classifier outputs is assumed. Again, in the target detection problem under correlated local decisions (correlated outputs of constituent classifiers), Kam *et al*. (1992) expand the probability density function using the Bahadur–Lazarsfeld polynomials to rewrite the optimal combiner rule in terms of conditional correlation coefficients (see Section 8.3).
3. *Training data available*. It is assumed that the outputs of each of the individual constituent classifiers are known for a given input of known class. Thus, we have a set of labelled samples that may be used to train the combiner classifier.

8.2.4 Form of component classifiers

F1. Common form

Classifiers may all be of the same form. For example, they all may be neural networks (multilayer perceptrons) of a given architecture; all linear discriminants or all decision trees.

The particular form may be chosen for several reasons: *interpretability* – it is easy to interpret the classification process in terms of simple rules defined on the input space; *implementability* – the constituent classifiers are easy to implement and do not require excessive computation; *adaptability* – the constituent classifiers are flexible and it is easy to implement diverse classifiers whose combination leads to a lower error rate than any of the individuals.

F2. Dissimilar form
The constituent classifiers may be a collection of neural networks, decision trees, nearest neighbour methods and so on, the set perhaps arising through the analysis of a wide range of classifiers on different training sets. Thus, the classifiers have not necessarily been chosen so that their combination leads to the best improvement.

8.2.5 Structure

The structure of a multiple classifier system is often dictated by a practical application.

T1. Parallel
The results from the constituent classifiers are passed to the combiner together before a decision is made by the combiner.

T2. Serial
Each constituent classifier is invoked sequentially, with the results of one classifier being used by the next one in the sequence, perhaps to set a prior on the classes.

T3. Hierarchical
The classifiers are combined in a hierarchy with the outputs of one constituent classifier feeding as inputs to a parent node, in a similar manner to decision tree classifiers (see Chapter 7). Thus the partition into a single combiner with several constituent classifiers is less apparent, with all classifiers (apart from the leaf and root nodes) taking output of a classifier as input and passing its own outputs as input to another classifier.

8.2.6 Optimisation

Different parts of the combining scheme may be optimised separately or simultaneously, depending on the motivating problem.

O1. Combiner
Optimise the combiner alone. Thus, given a set of constituent classifiers, we determine the combining rule to improve performance the greatest.

O2. Constituent classifiers
Optimise the constituent classifiers. For a fixed combiner rule, and the number and type of constituent classifiers, the parameters of these classifiers are determined to maximise performance of the combined classifier.

O3. Combiner and constituent classifiers
Optimise both the combiner rule and the parameters of the constituent classifiers. In this case, the constituent classifiers may not be classifiers in the strict sense, performing some form of feature extraction. Practical constraints such as limitations on the bandwidth between the constituent classifiers and the combiner may need to be considered.

O4. No optimisation
We are provided with a fixed set of classifiers and use a standard combiner rule that requires no training (see Section 8.4).

8.3 Data fusion

Much of the work on multiple classifier systems reported in the pattern recognition, statistics and machine learning literature has strong parallels, and indeed overlaps substantially, with research on data fusion systems carried out largely within the engineering community (Dasarathy, 1994b; Varshney, 1997; Waltz and Llinas, 1990). In common with the research on classifier combination, different architectures may be considered (for example, serial or parallel) and different assumptions made concerning the joint distribution of classifier outputs, but the final architecture adopted and the constraints under which it is optimised are usually motivated by real problems. One application of special interest is that of distributed detection: detecting the presence of a target using a distributed array of sensors. In this section we review some of the work in this area.

8.3.1 Architectures

Figures 8.7 and 8.8 illustrate the two main architectures for a decentralised distributed detection system – the serial and parallel configurations. We assume that there are L sensors. The observations at each sensor are denoted by y_i and the decision at each sensor by u_i, $i = 1, \ldots, L$, where

$$u_i = \begin{cases} 1 & \text{if 'target present' declared} \\ 0 & \text{if 'target absent' declared} \end{cases}$$

and the final decision is denoted by u_0. Each sensor can be considered as a binary classifier and the problem is termed as one in *decision fusion*. This may be thought a very restrictive model, but one that may arise in some practical situations.

The main drawback with the serial network structure (Figure 8.7) is that it has a serious practical reliability problem.

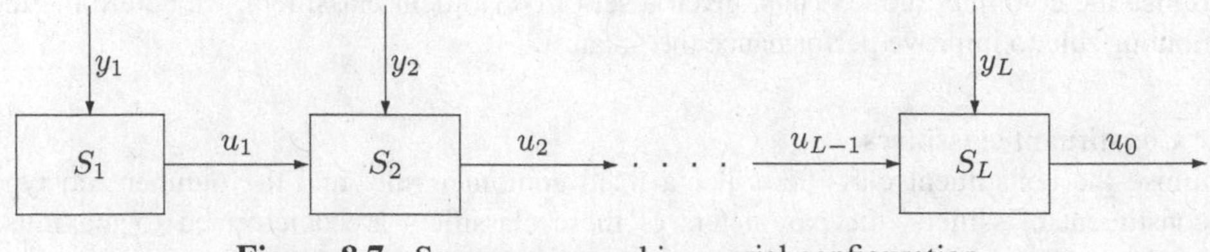

Figure 8.7 Sensors arranged in a serial configuration.

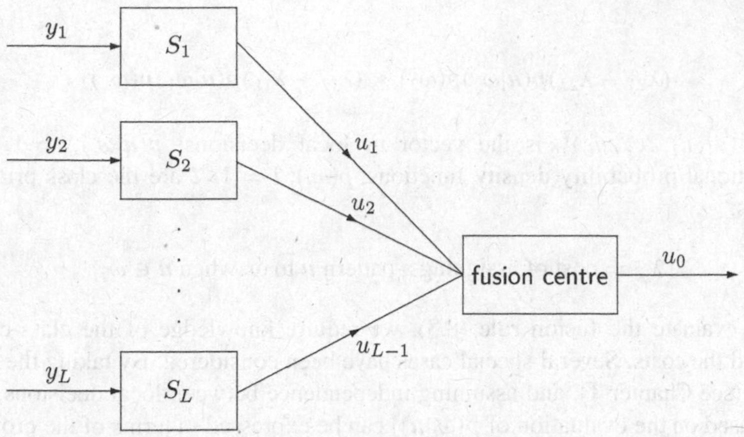

Figure 8.8 Sensors arranged in a parallel configuration.

This problem arises because if there is a link failure between the $(i-1)$th sensor and the ith sensor then all the information used to make the previous decisions would be lost, resulting in the ith sensor becoming effectively the first sensor in the decision process.

The parallel system has been widely considered and is shown in Figure 8.8. Each sensor receives a local observation y_i, $i = 1, \ldots, L$ and produces a local decision u_i which is sent to the fusion centre. At the fusion centre all the local decisions u_i, $i = 1, \ldots, L$ are combined to obtain a global decision u_0. The parallel architecture is far more robust to link failure. A link failure between the ith sensor and the fusion centre does not seriously jeopardize the overall global decision, since it is only the decision of the ith sensor that is lost. The parallel distributed system has been considered under the assumptions of both correlated and independent local decisions, and methods have been proposed for solving the optimal solution in both these cases.

The parallel decision system has also been extended to allow the local decisions to be passed to intermediate fusion centres, each processing all the L local decisions before passing their decisions onto another layer of intermediate fusion centres (Li and Sethi, 1993; Gini, 1997). After K such layers these intermediate fusion centre decisions are passed to the fusion centre and a global decision u_0 is made. This corresponds to the hierarchical model in multiple classifier systems (see Section 8.2).

The parallel architecture has also been used to handle repeated observations. One approach has been to use a memory term that corresponds to the decision made by the fusion centre using the last set of local decisions (Kam *et al.*, 1999). The memory term is used in conjunction with the next set of local decisions to make the next global decision. This memory term therefore allows the fusion centre to take into consideration the decision made on the last set of observations.

8.3.2 Bayesian approaches

We formulate the Bayes rule for minimum risk for the parallel configuration. Let class ω_2 be 'target absent' and class ω_1 be 'target present', then the Bayes' rule for minimum risk is [Equation (1.12)]: declare a target present (class ω_1) if

$$\lambda_{11} p(\omega_1|\boldsymbol{u})p(\boldsymbol{u}) + \lambda_{21} p(\omega_2|\boldsymbol{u})p(\boldsymbol{u}) \leq \lambda_{12} p(\omega_1|\boldsymbol{u})p(\boldsymbol{u}) + \lambda_{22} p(\omega_2|\boldsymbol{u})p(\boldsymbol{u})$$

that is,

$$(\lambda_{21} - \lambda_{22})p(\boldsymbol{u}|\omega_2)p(\omega_2) \leq (\lambda_{12} - \lambda_{11})p(\boldsymbol{u}|\omega_1)p(\omega_1) \tag{8.3}$$

where $\boldsymbol{u} = (u_1, u_2, \ldots, u_L)^T$ is the vector of local decisions; $p(\boldsymbol{u}|\omega_i), i = 1, 2$ are the class–conditional probability density functions; $p(\omega_i)$, $i = 1, 2$ are the class priors and λ_{ji} are the costs:

$$\lambda_{ji} = \text{cost of assigning a pattern } \boldsymbol{u} \text{ to } \omega_i \text{ when } \boldsymbol{u} \in \omega_j$$

In order to evaluate the fusion rule (8.3), we require knowledge of the class-conditional densities and the costs. Several special cases have been considered. By taking the equal cost loss matrix (see Chapter 1), and assuming independence between local decisions, the fused decision [based on the evaluation of $p(\omega_i|\boldsymbol{u})$] can be expressed in terms of the probability of false alarm and the probability of missed detection at each sensor (see the exercises at the end of the chapter).

The problem of how to tackle the likelihood ratios when the local decisions are correlated has been addressed by Kam *et al.* (1992) who showed that by using the Bahadur–Lazarsfeld polynomials to form an expansion of the probability density functions, it is possible to rewrite the optimal data fusion rule in terms of the conditional correlation coefficients.

The Bahadur–Lazarsfeld expansion expresses the density $p(\boldsymbol{x})$ as

$$p(\boldsymbol{x}) = \prod_{j=1}^{L} (p_j^{x_j}(1 - p_j)^{1-x_j}) \times \left[1 + \sum_{i<j} \gamma_{ij} z_i z_j + \sum_{i<j<k} \gamma_{ijk} z_i z_j z_k + \cdots \right]$$

where the γs are the correlation coefficients of the corresponding variables

$$\gamma_{ij} = E[z_i z_j]$$
$$\gamma_{ijk} = E[z_i z_j z_k]$$
$$\gamma_{ij\ldots L} = E[z_i z_j \ldots z_L]$$

and

$$p_i = P(x_i = 1); \quad 1 - p_i = P(x_i = 0)$$

so that

$$E[x_i] = 1 \times p_i + 0 \times (1 - p_i) = p_i$$
$$\text{var}[x_i] = p_i(1 - p_i)$$

and

$$z_i = \frac{x_i - p_i}{\sqrt{\text{var}(x_i)}} = \frac{x_i - p_i}{\sqrt{p_i(1 - p_i)}}$$

Substituting each of the conditional densities in Equation (8.3) by its Bahadur–Lazarsfeld expansion replaces the unknown densities by unknown correlation coefficients (see Exercises).

However, this may simplify considerably under assumptions about the form of the individual detectors (Kam *et al.*, 1992).

8.3.3 Neyman–Pearson formulation

In the Neyman–Pearson formulation, we seek a threshold on the likelihood ratio so that a specified false alarm rate is achieved (see Section 1.5.5). Since the data space is discrete (for an L-dimensional vector, $\boldsymbol{u}$, there are 2^L possible states) the decision rule of Chapter 1 is modified to become:

$$\text{if } \frac{p(\boldsymbol{u}|\omega_1)}{p(\boldsymbol{u}|\omega_2)} \begin{cases} > t & \text{then decide } u_0 = 1 \text{ (target present declared)} \\ = t & \text{then decide } u_0 = 1 \text{ with probability } \epsilon \\ < t & \text{then decide } u_0 = 0 \text{ (target absent declared)} \end{cases} \tag{8.4}$$

where ϵ and t are chosen to achieve the desired false alarm rate.

As an example, consider the case of two sensors, S_1 and S_2, operating with probabilities of false alarm pfa_1 and pfa_2, respectively, and probabilities of detection pd_1 and pd_2. Table 8.1 gives the probability density functions for $p(\boldsymbol{u}|\omega_1)$ and $p(\boldsymbol{u}|\omega_2)$ assuming independence.

There are four values for the likelihood ratio, $p(\boldsymbol{u}|\omega_1)/p(\boldsymbol{u}|\omega_2)$, corresponding to $\boldsymbol{u} = ((0, 0), (0, 1), (1, 0), (1, 1))$. For $pfa_1 = 0.2$; $pfa_2 = 0.4$; $pd_1 = 0.6$ and $pd_2 = 0.7$, these values are 0.25, 0.875, 1.5 and 5.25. Figure 8.9 gives the ROC curve for the combiner, combined using rule (8.4). This is a piecewise linear curve, with four linear segments, each corresponding to one of the values of the likelihood ratio. For example, if we set $t = 0.875$ (one of the values of the likelihood ratio), then $u_0 = 1$ is decided if $\boldsymbol{u} = (1, 0)$ and $\boldsymbol{u} = (1, 1)$ and also for $\boldsymbol{u} = (0, 1)$ with probability ϵ. This gives a probability of detection and a probability of false alarm of (using Table 8.1)

$$pd = pd_1\, pd_2 + (1 - pd_2)\, pd_1 + \epsilon(1 - pd_1)\, pd_2 = 0.6 + 0.28\epsilon$$

$$pfa = pfa_1\, pfa_2 + (1 - pfa_2)\, pfa_1 + \epsilon(1 - pfa_1)\, pfa_2 = 0.2 + 0.32\epsilon$$

a linear variation (as ϵ is varied) of (pfa, pd) values between (0.2, 0.6) and (0.52, 0.88).

Table 8.1 Probability density functions for $p(\boldsymbol{u}|\omega_1)$ (top) and $p(\boldsymbol{u}|\omega_2)$ (bottom) for the case of two sensors S_1 and S_2, operating with probabilities of false alarm pfa_1 and pfa_2, respectively, and probabilities of detection pd_1 and pd_2

		Sensor S_1	
		$u = 0$	$u = 1$
Sensor S_2	$u = 0$	$(1 - pd_1)(1 - pd_2)$	$(1 - pd_2)\, pd_1$
	$u = 1$	$(1 - pd_1)\, pd_2$	$pd_1\, pd_2$

		Sensor S_1	
		$u = 0$	$u = 1$
Sensor S_2	$u = 0$	$(1 - pfa_1)(1 - pfa_2)$	$(1 - pfa_2)\, pfa_1$
	$u = 1$	$(1 - pfa_1)\, pfa_2$	$pfa_1\, pfa_2$

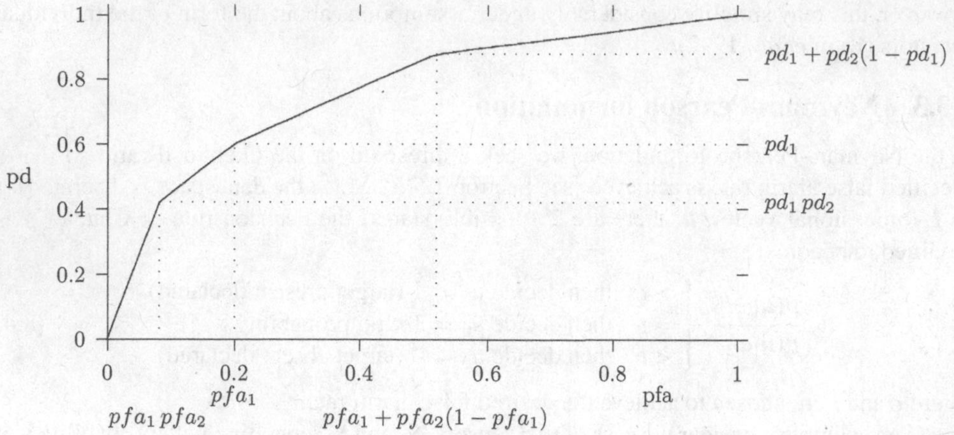

Figure 8.9 ROC curve for two sensors, assuming independence.

It may be possible to achieve a better probability of detection of the combiner, for a given probability of false alarm, by operating the individual sensors at different local thresholds. For L sensors, this is an L-dimensional search problem and requires knowledge of the ROC of each sensor (Viswanathan and Varshney, 1997).

Approaches to the distributed detection problem using the Neyman–Pearson formulation have been proposed for correlated decisions. One approach is to expand the likelihood ratio using the Bahadur–Lazarsfeld polynomials, in a similar manner to the Bayesian formulation of the preceding subsection. The Neyman–Pearson fusion rule can then be expressed as a function of the correlation coefficients.

If the independence assumption is not valid, and it is not possible to estimate the likelihood ratio through other means, then it may still be possible to achieve better performance than individual sensors through a ‘random choice’ fusion system. Consider two sensors S_1 and S_2, with ROC curves shown in Figure 8.10.

For probabilities of false alarm greater than pfa_B, we operate sensor S_2 and for probabilities of false alarm less than pfa_A, we operate sensor S_1. If, for probabilities of false alarm between pfa_A and pfa_B, we operate sensor S_1 with probability of false alarm pfa_A and sensor S_2 at the pfa_B point on its ROC curve, and randomly select sensor S_1 with probability ϵ and sensor S_2 with probability $(1 - \epsilon)$, then the probability of false alarm of the random choice fusion system is $\epsilon pfa_A + (1 - \epsilon)pfa_B$ and the probability of detection is $\epsilon pd_A + (1 - \epsilon)pd_B$. Thus, the best performance is achieved on the convex hull of the two ROC curves.

This differs from the example in Figure 8.9, where the combined output of both sensors is used, rather than basing a decision on a single sensor output.

8.3.4 Trainable rules

One of the difficulties with the Bayesian and the Neyman–Pearson formulations for a set of distributed sensors is that both methods require some knowledge of the probability density of sensor outputs. Often this information is not available and the densities must be estimated using a training set.

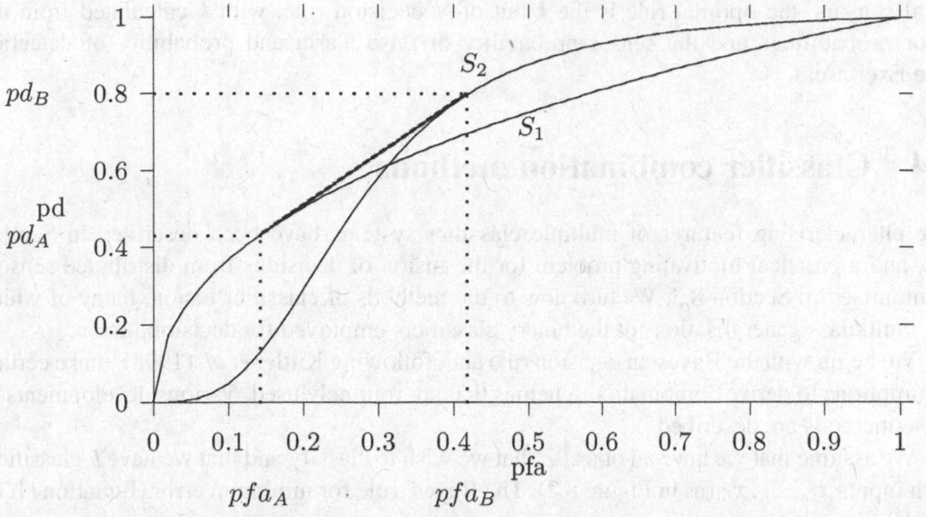

Figure 8.10 ROC curves for sensors S_1 and S_2 and points on the convex hull.

This is simply a problem of classifier design where the classifiers are defined on a feature space comprising the outputs of separate sensors (local decisions). Many of the techniques described elsewhere in this book, suitably adapted for binary variables, may be employed.

8.3.5 Fixed rules

There are a few 'fixed' rules for decision fusion that do not model the joint density of sensor predictions.

AND
Class ω_1 (target present) is declared if all sensors predict class ω_1, otherwise class ω_2 is declared.

OR
Class ω_1 (target present) is declared if at least one of the sensors predicts class ω_1, otherwise class ω_2 is declared.

Majority vote
Class ω_1 (target present) is declared if a majority of the sensors predicts class ω_1, otherwise class ω_2 is declared.

k*-out-of-*N
Class ω_1 (target present) is declared if at least k of the sensors predict class ω_1, otherwise class ω_2 is declared. All of the previous three rules are special cases of this rule.

It is difficult to draw general conclusions about the performance of these rules. For low false alarm rates, there is some evidence to show that the OR rule is inferior to AND and to the majority vote rules in a problem of signal detection in correlated noise. For similar local sensors, the optimal rule is the k-out-of-N decision rule, with k calculated from the prior probabilities and the sensor probability of false alarm and probability of detection (see Exercises).

8.4 Classifier combination methods

The characterising features of multiple classifier systems have been described in Section 8.2, and a practical motivating problem for the fusion of decisions from distributed sensors summarised in Section 8.3. We turn now to the methods of classifier fusion, many of which are multiclass generalisations of the binary classifiers employed for decision fusion.

We begin with the Bayesian decision rule and, following Kittler *et al.* (1998), make certain assumptions to derive combination schemes that are routinely used. Various developments of these methods are described.

We assume that we have an object Z that we wish to classify and that we have L classifiers with inputs $\boldsymbol{x}_1, \ldots, \boldsymbol{x}_L$ (as in Figure 8.2). The Bayes' rule for minimum error [Equation (1.1)] assigns Z to class ω_j if

$$p(\omega_j|\boldsymbol{x}_1, \ldots, \boldsymbol{x}_L) > p(\omega_k|\boldsymbol{x}_1, \ldots, \boldsymbol{x}_L) \quad k = 1, \ldots, C; k \neq j \tag{8.5}$$

or, equivalently [Equation (1.2)], assign Z to class ω_j if

$$p(\boldsymbol{x}_1, \ldots, \boldsymbol{x}_L|\omega_j)p(\omega_j) > p(\boldsymbol{x}_1, \ldots, \boldsymbol{x}_L|\omega_k)p(\omega_k) \quad k = 1, \ldots, C; k \neq j \tag{8.6}$$

This requires knowledge of the class-conditional joint probability densities $p(\boldsymbol{x}_1, \ldots, \boldsymbol{x}_L|\omega_j)$, $j = 1, \ldots, L$, which is assumed to be unavailable.

8.4.1 Product rule

If we assume conditional independence ($\boldsymbol{x}_1, \ldots, \boldsymbol{x}_L$ are conditionally independent given class), then the decision rule (8.6) becomes: assign Z to class ω_j if

$$\prod_{i=1}^{L}(p(\boldsymbol{x}_i|\omega_j))p(\omega_j) > \prod_{i=1}^{L}(p(\boldsymbol{x}_i|\omega_k))p(\omega_k) \quad k = 1, \ldots, C; k \neq j \tag{8.7}$$

or, in terms of the posterior probabilities of the individual classifiers: assign Z to class ω_j if

$$[p(\omega_j)]^{-(L-1)}\prod_{i=1}^{L}p(\omega_j|\boldsymbol{x}_i) > [p(\omega_k)]^{-(L-1)}\prod_{i=1}^{L}p(\omega_k|\boldsymbol{x}_i) \quad k = 1, \ldots, C; k \neq j \tag{8.8}$$

This is the *product rule* and for equal priors simplifies to: assign Z to class ω_j if

$$\prod_{i=1}^{L}p(\omega_j|\boldsymbol{x}_i) > \prod_{i=1}^{L}p(\omega_k|\boldsymbol{x}_i) \quad k = 1, \ldots, C; k \neq j \tag{8.9}$$

Both forms (8.8) and (8.9) have been used in studies. The independence assumption may seem rather severe, but it is one that has been successfully used in many practical problems (Hand and Yu, 2001). The rule requires the individual classifier posterior probabilities, $p(\omega_j|\boldsymbol{x})$, $j = 1, \ldots, C$ to be calculated and these are usually estimated from training data. The main problem with this method is that the product rule is sensitive to errors in the posterior probability estimates, and deteriorates more rapidly than the sum rule (see below) as the estimation errors increase. If one of the classifiers reports that the probability of a sample belonging to a particular class is zero, then the product rule will give a zero probability also, even if the remaining classifiers report that this is the most probable class.

The product rule would tend to be applied where each classifier receives input from different sensors.

8.4.2 Sum rule

The sum rule may be derived from the product rule by making the (rather strong) assumption that

$$p(\omega_k|\boldsymbol{x}_i) = p(\omega_k)(1 + \delta_{ki}) \tag{8.10}$$

where $\delta_{ki} \ll 1$; that is, the posterior probabilities, $p(\omega_k|\boldsymbol{x}_i)$ used in the product rule (8.8) do not deviate substantially from the class priors, $p(\omega_k)$. Then substituting for $p(\omega_k|\boldsymbol{x}_i)$ in the product rule (8.8), neglecting second order and higher terms in δ_{ki}, and using (8.10) leads to the sum rule (see the exercises at the end of the chapter): assign Z to class ω_j if

$$(1 - L)p(\omega_j) + \sum_{i=1}^{L} p(\omega_j|\boldsymbol{x}_i) > (1 - L)p(\omega_k) + \sum_{i=1}^{L} p(\omega_k|\boldsymbol{x}_i) \quad k = 1, \ldots, C; k \neq j \tag{8.11}$$

This is the *sum rule* and for equal priors it simplifies to: assign Z to class ω_j if

$$\sum_{i=1}^{L} p(\omega_j|\boldsymbol{x}_i) > \sum_{i=1}^{L} p(\omega_k|\boldsymbol{x}_i) \quad k = 1, \ldots, C; k \neq j \tag{8.12}$$

The assumption used to derive the sum rule approximation to the product rule, namely that the posterior probabilities are similar to the priors, will be unrealistic in many practical applications. However, it is a rule that is relatively insensitive to errors in the estimation of the posterior probabilities and could be applied to classifiers operating on a common input pattern (Figure 8.3).

In order to implement the above rule, each classifier must produce estimates of the posterior probabilities of class membership. In a comparison of the sum and product rules, Tax *et al.* (2000) concluded that the sum rule is more robust to errors in the estimated posterior probabilities (see also Kittler *et al.*, 1998). The averaging process reduces any effects of overtraining of the individual classifiers and may be thought of as a regularisation process.

It is also possible to applying a weighting to the sum rule to give: assign Z to class ω_j if

$$\sum_{i=1}^{L} w_i p(\omega_j|\boldsymbol{x}_i) > \sum_{i=1}^{L} w_i p(\omega_k|\boldsymbol{x}_i) \quad k = 1, \ldots, C; k \neq j \tag{8.13}$$

where w_i, $i = 1, \ldots, L$ are weights for the classifiers. A key question here is how to choose weights. These may be estimated using a training set to minimise the error rate of the combined classifier. In this case, the same weighting is applied throughout the data space. An alternative is to allow the weights to vary with the location of a given pattern in the data space. An extreme example of this is *dynamic classifier selection* where one weight is assigned the value unity and the remaining weights are zero. For a given pattern, dynamic feature selection attempts to select the best classifier. Thus, the feature space is partitioned into regions with a different classifier for each region.

Dynamic classifier selection has been addressed by Woods *et al.* (1997) who use local regions defined in terms of k-nearest-neighbour regions to select the most accurate classifier (based on the percentage of training samples correctly classified in the region); see also Huang and Suen (1995).

8.4.3 Min, max and median combiners

The max combiner may be derived by approximating the sum of the posterior probabilities in (8.11) by an upper bound, $L \max_i p(\omega_k|\boldsymbol{x}_i)$ to give the decision rule: assign Z to class ω_j if

$$(1 - L)p(\omega_j) + L \max_i p(\omega_j|\boldsymbol{x}_i) > (1 - L)p(\omega_k) + L \max_i p(\omega_k|\boldsymbol{x}_i) \tag{8.14}$$

This is the *max combiner* and for equal priors simplifies to

$$\max_i p(\omega_j|\boldsymbol{x}_i) > \max_i p(\omega_k|\boldsymbol{x}_i) \quad k = 1, \ldots, C; k \neq j \tag{8.15}$$

We can also approximate the product in (8.8) by an upper bound, $\min_i p(\omega_k|\boldsymbol{x}_i)$, to give the decision rule: assign Z to class ω_j if

$$[p(\omega_j)]^{-(L-1)} \min_i p(\omega_j|\boldsymbol{x}_i) > [p(\omega_k)]^{-(L-1)} \min_i p(\omega_k|\boldsymbol{x}_i) \quad k = 1, \ldots, C; k \neq j \tag{8.16}$$

This is the *min combiner* and for equal priors simplifies to: assign Z to class ω_j if

$$\min_i p(\omega_j|\boldsymbol{x}_i) > \min_i p(\omega_k|\boldsymbol{x}_i) \quad k = 1, \ldots, C; k \neq j \tag{8.17}$$

Finally, the *median combiner* is derived by noting that the sum rule calculates the mean of the classifier outputs and that a robust estimate of the mean is the median. Thus, under equal priors, the median combiner is: assign Z to class ω_j if

$$\operatorname*{med}_i p(\omega_j|\boldsymbol{x}_i) > \operatorname*{med}_i p(\omega_k|\boldsymbol{x}_i) \quad k = 1, \ldots, C; k \neq j \tag{8.18}$$

The min, max and median combiners are all easy to implement and require no training.

8.4.4 Majority vote

Among all the classifier combination methods described in this section, the majority vote is one of the easiest to implement. It is applied to classifiers that produce unique class labels as outputs (level L3) and requires no training. It may be considered as an application of the sum rule to classifier outputs where the posterior probabilities, $p(\omega_k|\boldsymbol{x}_i)$, have been 'hardened' (Kittler *et al.*, 1998); that is, $p(\omega_k|\boldsymbol{x}_i)$ is replaced by the binary-valued function, Δ_{ki}, where

$$\Delta_{ki} = \begin{cases} 1 & \text{if } p(\omega_k|\boldsymbol{x}_i) = \max_j p(\omega_j|\boldsymbol{x}_i) \\ 0 & \text{otherwise} \end{cases}$$

which produces decisions at the classifier outputs rather than posterior probabilities. A decision is made to classify a pattern to the class most often predicted by the constituent classifiers. In the event of a tie, a decision can be made according to the largest *prior* class probability (among the tying classes).

An extension to the method is the *weighted majority voting* technique in which classifiers are assigned unequal weights based on their performance. The weights for each classifier may be independent of predicted class, or they may vary across class depending on the performance of the classifier on each class. A key question is how to choose the weights. The weighted majority vote combiner requires the results of the individual classifiers on a training set as training data for the allocation of the weights.

For weights that vary between classifiers but are independent of class, there are $L - 1$ parameters to estimate for L classifiers (we assume that the weights may be normalised to sum to unity). These may be determined by specifying some suitable objective function and an appropriate optimisation procedure. One approach is to define the objective function, F

$$F = R_e - \beta E$$

where R_e is the recognition rate and E is the error rate of the combiner (they do not sum to unity as the individual classifiers may reject patterns – see Chapter 1); β is a user-specified parameter that measures the relative importance of recognition and error rates and is problem dependent (Lam and Suen, 1995). Rejection may be treated as an extra class by the component classifiers and thus the combiner will reject a pattern if the weighted majority of the classifiers also predicts a rejection. In a study of combination schemes applied to a problem in optical character recognition, Lam and Suen (1995) use a genetic optimisation scheme (a scheme that adjusts the weights using a learning method loosely motivated by an analogy to biological evolution) to maximise F and concluded that simple majority voting (all weights equal) gave the easiest and most reliable classification.

8.4.5 Borda count

The Borda count is a quantity defined on the ranked outputs of each classifier. If we define $B_i(j)$ as the number of classes ranked below class ω_j by classifier i, then the Borda count for class ω_j is B_j defined as

$$B_j = \sum_{i=1}^{L} B_i(j),$$

the sum of the number of classes ranked below ω_j by each classifier. A pattern is assigned to the class with the highest Borda count. This combiner requires no training, with the final decision being based on an average ranking of the classes.

8.4.6 Combiners trained on class predictions

The combiners described so far require no training, at least in their basic forms. General conclusions are that the sum rule and median rule can be expected to give better performance than other fixed combiners. We now turn to combiners that require some degree of training and initially we consider combiners acting on discrete variables. Thus, the constituent classifiers deliver class labels and the combiner uses these class predictions to make an improved estimate of class (type L3 combination) – at least that is what we hope.

'Bayesian combiner'
This combiner simply uses the product rule with estimates of the posterior probabilities derived from the classifier predictions of each constituent classifier together with a summary of their performance on a labelled training set.

Specifically, the Bayesian combination rule of Lam and Suen (1995) approximates the posterior probabilities by an estimate based on the results of a training procedure. Let $D^{(i)}$ denote the $C \times C$ confusion matrix (see Chapter 1) for the ith classifier based on the results of a classification of a training set by classifier i. The (j, k) entry, $d_{jk}^{(i)}$, is the number of patterns with true class ω_k that are assigned to ω_j by classifier i. The total number of patterns in class ω_k is

$$n_k = \sum_{l=1}^{C} d_{lk}^{(i)}$$

for any i. The number of patterns assigned to class ω_l is

$$\sum_{k=1}^{C} d_{lk}^{(i)}$$

The conditional probability that a sample $\boldsymbol{x}$ assigned to class ω_l by classifier i actually belongs to ω_k is estimated as

$$p(\omega_k|\text{classifier } i \text{ predicts } \omega_l) = \frac{d_{lk}^{(i)}}{\sum_{k=1}^{C} d_{lk}^{(i)}}$$

Thus, for a given pattern, the posterior probability depends only on the predicted class: two distinct patterns, say $\boldsymbol{x}$ and $\boldsymbol{w}$, having the same predicted class, are estimated as having the same posterior probability. Substituting into the product rule (8.9), equal priors assumed, gives the decision rule: assign the pattern to class ω_j if

$$\prod_{i=1}^{L} \frac{d_{l_i j}^{(i)}}{\sum_{k=1}^{C} d_{l_i k}^{(i)}} > \prod_{i=1}^{L} \frac{d_{l_i m}^{(i)}}{\sum_{k=1}^{C} d_{l_i k}^{(i)}} \qquad m = 1, \ldots, C; m \neq j$$

where ω_{l_i} is the predicted class under classifier i for a given input pattern.

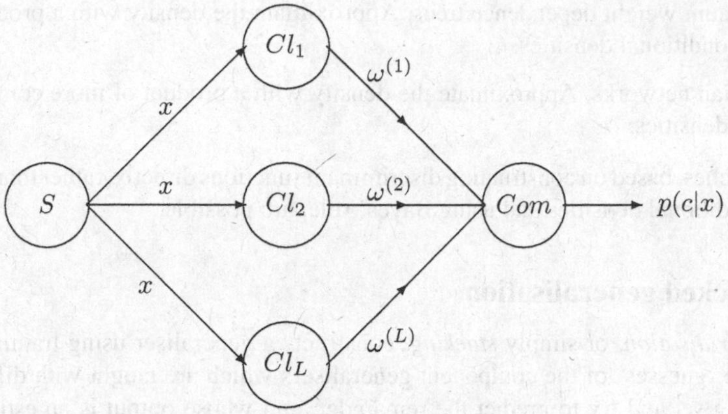

Figure 8.11 Classifier fusion architecture, C2 – component classifiers defined on a common feature space. $\omega^{(i)}$ is the predicted class of classifier Cl_i.

Density estimation in classifier output space

An alternative approach is to regard the L class predictions from the constituent classifiers for an input, $\boldsymbol{x}$, as inputs to a classifier, the combiner, defined on an L-dimensional discrete-valued feature space (Figure 8.11).

Suppose that we have N training patterns ($\boldsymbol{x}_i, i = 1, \ldots, N$) with associated class labels ($y_i, i = 1, \ldots, N$), then the training data for the combiner comprises N L-dimensional vectors ($\boldsymbol{z}_i, i = 1, \ldots, N$) with associated class labels ($y_i, i = 1, \ldots, N$). Each component of $\boldsymbol{z}_i$ is a discrete-valued variable taking 1 of C possible values corresponding to the class label from the component classifier (or 1 of $C + 1$ possible values if a reject option is included in the constituent classifiers).

The combiner is trained using the training set $\{(\boldsymbol{z}_i, y_i), i = 1, \ldots, N\}$ and an unknown pattern $\boldsymbol{x}$ classified by first applying each constituent classifier to obtain a vector of predictions $\boldsymbol{z}$, which is then input to the combiner.

The most obvious approach to constructing the combiner is to estimate class-conditional probabilities, $p(\boldsymbol{z}|\omega_i), i = 1, \ldots, C$, and to classify $\boldsymbol{z}$ to class ω_j if

$$p(\boldsymbol{z}|\omega_j)p(\omega_j) > p(\boldsymbol{z}|\omega_k)p(\omega_k) \quad k = 1, \ldots, C; k \neq j$$

with the priors, $p(\omega_j)$, estimated from the training set (or perhaps using domain knowledge) and the densities, $p(\boldsymbol{z}|\omega_j)$ estimated using a suitable nonparametric density estimation method appropriate for categorical variables.

Perhaps the simplest method is the histogram. This is the approach adopted by Huang and Suen (1995) and termed the *Behaviour-Knowledge Space* method, and also investigated by Mojirsheibani (1999). However, this has the disadvantage of having to estimate and store high order distributions which may be computationally expensive. The multidimensional histogram has C^L cells, which may be large making reliable density estimation difficult in the absence of a large training dataset. In Chapter 4 we described several ways around this difficulty:

1. Independence. Approximate the multivariate density as the product of the univariate estimates.
2. Lancaster models. Approximate the density using marginal distributions.

3. Maximum weight dependence trees. Approximate the density with a product of pairwise conditional densities.

4. Bayesian networks. Approximate the density with a product of more complex conditional densities.

Other approaches, based on constructing discriminant functions directly, rather than estimating the class-conditional densities and using Bayes' rule, are possible.

8.4.7 Stacked generalisation

Stacked generalisation, or simply *stacking*, constructs a generaliser using training data that consist of the 'guesses' of the component generalisers which are taught with different parts of the training set and try to predict the remainder, and whose output is an estimate of the correct class. Thus, in some ways it is similar to the models of the previous section – the combiner is a classifier (generaliser) defined on the outputs of the constituent classifiers – but the training data used to construct the combiner comprise the prediction on held-out samples of the training set.

The basic idea is that the output of the constituent classifiers, termed level 1 data, $\mathcal{L}_1$ (level 0, $\mathcal{L}_0$, is the input level), has information that can be used to construct good combinations of the classifiers. We suppose that we have a set of constituent classifiers, $f_j, j = 1, \ldots, L$, and we seek a procedure for combining them. The level 1 data are constructed as follows:

1. Divide the $\mathcal{L}_0$ data (the training data, $\{(\boldsymbol{x}_i, y_i), i = 1, \ldots, n\}$) into V partitions.

2. For each partition, $v = 1, \ldots, V$,

   (a) Repeat the procedure for constructing the constituent classifiers using a subset of the data: train the constituent classifier j ($j = 1, \ldots, L$) on all the training data apart from partition v to give a classifier denoted, f_j^{-v}.

   (b) Test each classifier, f_j^{-v}, on all patterns in partition v.

This gives a dataset of L predictions on each pattern in the training set. Together with the labels, $\{y_i, i = 1, \ldots, n\}$, these comprise the training data for the combiner.

We must now construct a combiner for the outputs of the constituent classifiers. If the constituent classifiers produce class labels, then the training data for the combiner comprise L-dimensional measurements on categorical variables. Several methods are available to us including those based on histogram estimates of the multivariate density, and variants, mentioned at the end of the previous section; for example, tree-based approaches and neural network methods. Merz (1999) compares an independence model and a multilayer perceptron model for the combiner with an approach based on a multivariate analysis method (*correspondence analysis*). For the multilayer perceptron, when the ith value of the variable occurred, each categorical variable is represented as C binary-valued inputs, with all inputs assigned the value of zero apart from the ith which is assigned a value of one.

8.4.8 Mixture of experts

The *adaptive mixture of local experts* model (Jacobs *et al.*, 1991; Jordan and Jacobs, 1994) is a learning procedure that trains several component classifiers (the 'experts') and a combiner

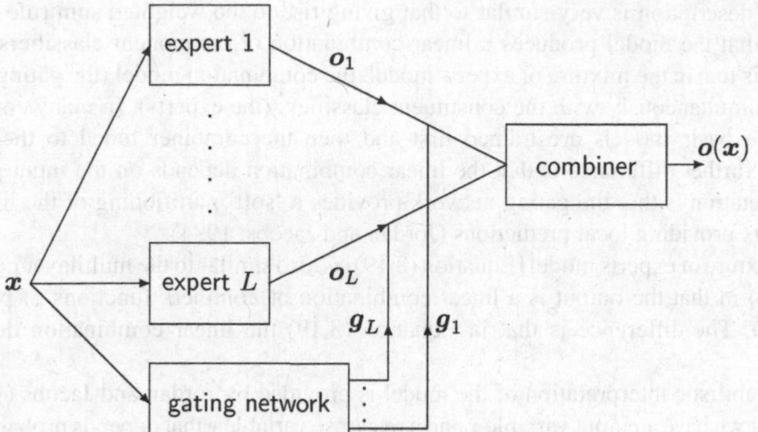

Figure 8.12 Mixture of experts architecture.

(the 'gating function') to achieve improved performance in certain problems. The experts each produce an output vector, $\boldsymbol{o}_i$, $(i = 1, \ldots, L)$, for a given input vector, $\boldsymbol{x}$, and the gating network provides linear combination coefficients for the experts. The gating function may be regarded as assigning a probability to each of the experts, based on the current input (Figure 8.12).

The emphasis of the training procedure is to find the optimal gating function and, for a given gating function, to train each expert to give maximal performance.

In the basic approach, the output of the ith expert, $\boldsymbol{o}_i(\boldsymbol{x})$, is a *generalised linear function* of the input, $\boldsymbol{x}$,

$$\boldsymbol{o}_i(\boldsymbol{x}) = f(\boldsymbol{w}_i \boldsymbol{x})$$

where $\boldsymbol{w}_i$ is a $C \times p$ weight matrix associated with the ith expert and $f(.)$ is a fixed continuous nonlinear function,[2] generally chose to be the logistic function. The gating network is also a generalised linear function, $\boldsymbol{g}$, of its input, with ith component

$$g_i(\boldsymbol{x}) = g(\boldsymbol{x}, \boldsymbol{v}_i) = \frac{\exp(\boldsymbol{v}_i^T \boldsymbol{x})}{\sum_{k=1}^{L} \exp(\boldsymbol{v}_i^T \boldsymbol{x})}$$

for weight vectors $\boldsymbol{v}_i$, $i = 1, \ldots, L$. These outputs of the gating network are used to weight the outputs of the experts to give the overall output, $\boldsymbol{o}(\boldsymbol{x})$, of the mixture of experts architectures as

$$\boldsymbol{o}(\boldsymbol{x}) = \sum_{k=1}^{L} g_k(\boldsymbol{x}) \boldsymbol{o}_k(\boldsymbol{x}) \tag{8.19}$$

[2] We use the shorthand notation that $f(\boldsymbol{u})$, for scalar function f, is the vector $(f(u_1), \ldots, f(u_C))$.

The above description is very similar to that giving rise to the weighted sum rule [Equation (8.13)] in that the model produces a linear combination of component classifiers. The key difference is that in the mixture of experts model, the combination model (the gating network) is trained simultaneously with the constituent classifiers (the experts). In many combination models, the basic models are trained first and then the combiner tuned to these trained models. A further difference is that the linear combination depends on the input pattern, $\boldsymbol{x}$. An interpretation is that the gating network provides a 'soft' partitioning of the input space with experts providing local predictions (Jordan and Jacobs, 1994).

The mixture of experts model [Equation (8.19)] is also similar to the multilayer perceptron[3] (Chapter 6) in that the output is a linear combination of nonlinear functions of projections of the data. The difference is that in Equation (8.19) the linear combination depends on the input.

A probabilistic interpretation of the model is provided by Jordan and Jacobs (1994). We assume that we have an input variable $\boldsymbol{x}$ and a response variable $\boldsymbol{y}$ that depends probabilistically on $\boldsymbol{x}$. The mixing proportions, $g_i(\boldsymbol{x})$, are interpreted as multinomial probabilities associated with the process that maps $\boldsymbol{x}$ to an output $\boldsymbol{y}$. For a given $\boldsymbol{x}$, an output $\boldsymbol{y}$ is generated by selecting an expert according to the values of $g_k(\boldsymbol{x}), k = 1, \ldots, L$, say expert i, and then generating $\boldsymbol{y}$ according to a probability density $p(\boldsymbol{y}|\boldsymbol{x}, \boldsymbol{w}_i)$, where $\boldsymbol{w}_i$ denotes the set of parameters associated with expert i. Therefore, the total probability of generating $\boldsymbol{y}$ from $\boldsymbol{x}$ is the mixture of probabilities of generating $\boldsymbol{y}$ from each component density, where the mixing proportions are the multinomial probabilities $g_k(\boldsymbol{x})$; i.e.

$$p(\boldsymbol{y}|\boldsymbol{x}, \boldsymbol{\Phi}) = \sum_{k=1}^{L} g_k(\boldsymbol{x}) p(\boldsymbol{y}|\boldsymbol{x}, \boldsymbol{w}_k) \tag{8.20}$$

where $\boldsymbol{\Phi}$ is the set of all parameters, including both expert and gating network parameters.

The generating density $p(\boldsymbol{y}|\boldsymbol{x}, \boldsymbol{w}_k)$, can be taken to be one of several popular forms; for a problem in regression, a normal distribution with identical covariance matrices $\sigma^2\boldsymbol{I}$ is often assumed,

$$p(\boldsymbol{y}|\boldsymbol{x}, \boldsymbol{w}_k) \sim \exp\{-\frac{1}{\sigma^2}(\boldsymbol{y} - \boldsymbol{o}_k(\boldsymbol{x}))^T(\boldsymbol{y} - \boldsymbol{o}_k(\boldsymbol{x}))\}$$

For binary classification, the Bernoulli distribution is generally assumed (single output, o_k; univariate binary response variable $y = 0, 1$)

$$p(y|\boldsymbol{x}, \boldsymbol{w}_k) = o_k^y(1 - o_k)^{1-y}$$

and for multiclass problems, the multinomial distribution (C binary variables y_i, $i = 1, \ldots, C$, that sum to unity, where C is the number of classes)

$$p(\boldsymbol{y}|\boldsymbol{x}, \boldsymbol{w}_k) \sim \prod_{i=1}^{C} (o_k^i)^{y_i}$$

[3] The basic multilayer perceptron, with single hidden layer and linear output layer. Further developments of the mixture of experts model to hierarchical models are discussed by Jordan and Jacobs (1994).

(o_k^j is the jth component of $\boldsymbol{o}_k$, the output of the kth expert). Optimisation of the model (8.20) may be achieved via a maximum likelihood approach. Given a training set $\{(\boldsymbol{x}_i, \boldsymbol{y}_i), i = 1, \ldots, n\}$ (in the classification case, $\boldsymbol{y}_i$ would be a C-dimensional vector coding the class – for class ω_j, all entries are zero apart from the jth, which is one), we seek a solution for $\boldsymbol{\Phi}$ for which the log likelihood

$$\sum_t \log \left[\sum_{k=1}^{L} g_k(\boldsymbol{x}_t) p(\boldsymbol{y}_t | \boldsymbol{x}_t, \boldsymbol{w}_k) \right]$$

is a maximum. Jordan and Jacobs (1994) propose an approach based on the EM algorithm (see Chapter 2) for adjusting the parameters $\boldsymbol{w}_k$ and $\boldsymbol{v}_k$. At the stage s of the iteration, the Expectation and Maximisation procedures are as follows:

1. E-step: compute the probabilities

$$h_i^{(t)} = \frac{g(\boldsymbol{x}_t, \boldsymbol{v}_i^{(s)}) p(\boldsymbol{y}_t | \boldsymbol{x}_t, \boldsymbol{w}_i^{(s)})}{\sum_{k=1}^{L} g(\boldsymbol{x}_t, \boldsymbol{v}_k^{(s)}) p(\boldsymbol{y}_t | \boldsymbol{x}_t, \boldsymbol{w}_k^{(s)})}$$

   for $t = 1, \ldots, n; i = 1, \ldots, L$.

2. M-step: solve the following maximisation problem for the parameters of the experts

$$\boldsymbol{w}_i^{(s+1)} = \arg\max_{\boldsymbol{w}_i} \sum_{t=1}^{n} h_i^{(t)} \log[p(\boldsymbol{y}_t | \boldsymbol{x}_t, \boldsymbol{w}_i)]$$

   and for the parameters of the gating network

$$\boldsymbol{V}^{(s+1)} = \arg\max_{\boldsymbol{V}} \sum_{t=1}^{n} \sum_{k=1}^{L} h_k^{(t)} \log[g(\boldsymbol{x}_t, \boldsymbol{v}_k)]$$

where $\boldsymbol{V}$ is the set of all $\boldsymbol{v}_i$.

Procedures for solving these maximisation problems are discussed by Chen *et al*. (1999), who propose a Newton–Raphson method, but other 'quasi-Newton' methods may be used (Press *et al*., 1992).

8.4.9 Bagging

Bagging and boosting (see the following section) are procedures for combining different classifiers generated using the same training set. Bagging or *bootstrap aggregating* (Breiman, 1996) produces replicates of the training set and trains a classifier on each replicate. Each classifier is applied to a test pattern $\boldsymbol{x}$ which is classified on a majority vote basis, ties being resolved arbitrarily. Algorithm 8.1 shows the steps in the bagging algorithm. A *bootstrap sample* is generated by sampling n times from the training set with replacement. This means that some patterns may be used several times, but others may not appear in the sample. This

Algorithm 8.1 The bagging algorithm.

1. Given a training set $(\boldsymbol{x}_i, z_i)$, $i = 1, \ldots, n$, of patterns $\boldsymbol{x}_i$ and labels z_i.
2. For $b = 1, \ldots, B$,
   (a) Generate a bootstrap sample of size n by sampling with replacement from the training set; some patterns will be replicated, others will be omitted.
   (b) Design a classifier, $\eta_b(\boldsymbol{x})$ using the bootstrap sample as training data.
3. Classify a test pattern $\boldsymbol{x}$ by recording the class predicted by $\eta_b(\boldsymbol{x})$, $b = 1, \ldots, B$, and assigning $\boldsymbol{x}$ to the class most represented.

provides a new training set, $\boldsymbol{Y}^b$, of size n. B bootstrap datasets, $\boldsymbol{Y}^b$, $b = 1, \ldots, B$, are generated and a classifier designed for each dataset. The final classifier is that whose output is the class most often predicted by the subclassifiers.

A vital aspect of the bagging technique is that the procedure for producing the classifier is *unstable*. For a given bootstrap sample, a pattern in the training set has a probability of $1 - (1 - 1/n)^n$ of being selected at least once in the n times that patterns are randomly selected from the training set. For large n, this is approximately $1 - 1/e = 0.63$, which means that each bootstrap sample contains only about 63% unique patterns from the training set. This causes different classifiers to be built. If the change in the classifiers is large (that is small changes in a dataset lead to large changes in the predictions), then the procedure is said to be unstable. Bagging of an unstable classifier should result in a better classifier and a lower error rate. However, averaging of a bad classifier can result in a poorer classifier. If the classifier is *stable* – that is, changes in the training dataset lead to little changes in the classifier – then bagging will lead to little improvement.

The bagging procedure reduces the variance of the base classifier and its performance depends on the stability of that base classifier. It is particularly useful in classification problems using neural networks (Chapter 6) and classification trees (Chapter 7) since these are all unstable processes. If the base classifier is stable (robust to minor perturbations in the training set), bagging may not be able to improve the performance. For trees, a negative feature is that there is no longer the simple interpretation as there is with a single tree. Nearest-neighbour classifiers are stable and bagging offers little, if any, improvement.

In studies on *linear classifiers*, Skurichina (2001) reports that bagging may improve the performance on classifiers constructed on critical training sample sizes, but when the classifier is stable, bagging is usually useless. Also, for very large sample sizes, classifiers constructed on bootstrap replicates are similar and combination offers no benefit.

The procedure, as presented in Algorithm 8.1, applies to classifiers whose outputs are class predictions. For classifier methods that produce estimates of the posterior probabilities, $\hat{p}(\omega_j|\boldsymbol{x})$, two approaches are possible. One is to make a decision for the class based on the maximum value of $\hat{p}(\omega_j|\boldsymbol{x})$ and then to use the voting procedure. Alternatively, the posterior probabilities can be averaged over all bootstrap replications, obtaining $\hat{p}_B(\omega_j|\boldsymbol{x})$ and then a decision based on the maximum value of $\hat{p}_B(\omega_j|\boldsymbol{x})$ is made. Breiman (1996) reports a virtually identical misclassification rate for the two approaches in a series of experiments on

Algorithm 8.2 The Adaboost algorithm.

1. Initialise the weights $w_i = 1/n, i = 1, \ldots, n$.
2. For $t = 1, \ldots, T$, (T is the number of boosting rounds)
   (a) Construct a classifier $\eta_t(\boldsymbol{x})$ from the training data with weights $w_i, i = 1, \ldots, n$.
   (b) Calculate e_t as the sum of the weights w_i corresponding to misclassified patterns.
   (c) If $e_t > 0.5$ or $e_t = 0$ then terminate the procedure, otherwise set $w_i = w_i(1 - e_t)/e_t$ for the misclassified patterns and renormalise the weights so that they sum to unity.
3. For a two-class classifier, in which $\eta_t(\boldsymbol{x}) = 1$ implies $\boldsymbol{x} \in \omega_1$ and $\eta_t(\boldsymbol{x}) = -1$ implies $\boldsymbol{x} \in \omega_2$, form a weighted sum of the classifiers, η_t,

$$\hat{\eta} = \sum_{t=1}^{T} \log\left(\frac{1 - e_t}{e_t}\right) \eta_t(\boldsymbol{x})$$

   and assign $\boldsymbol{x}$ to ω_1 if $\hat{\eta} > 0$.

11 datasets. However, bagged estimates of the posterior probabilities are likely to be more accurate than single estimates.

8.4.10 Boosting

Boosting is a procedure for combining or 'boosting' the performance of weak classifiers (classifiers whose parameter estimates are usually inaccurate and give poor performance) in order to achieve a better classifier. It differs from bagging in that it is a *deterministic* procedure and generates training sets and classifiers *sequentially*, based on the results of the previous iteration. In contrast, bagging generates the training sets *randomly* and can generate the classifiers *in parallel*.

Proposed by Freund and Schapire (1996) boosting assigns a weight to each pattern in the training set, reflecting its importance and constructs a classifier using the training set and the set of weights. The weights may be used in several ways.

- The weights may be used by a classifier training procedure to design a classifier that is biased towards higher-weighted patterns. Thus, it requires a classifier that can handle weights on the training samples. The weights are adapted so that the classifier focuses on patterns that are difficult to classify.
- The weights may be used as a sampling distribution to draw samples from the original dataset. Some classifiers may be unable to support weighted patterns. In this case, a subset of the training examples can be sampled according to the distribution of the weights and these examples used to train the classifier in the next stage of the iteration.

The basic boosting procedure is `Adaboost` (Adaptive Boosting; Freund and Schapire, 1996). Algorithm 8.2 presents the basic `Adaboost` algorithm for the binary classification problem.

Initially, all samples are assigned a weight $w_i = 1/n$. At each stage of the algorithm, a classifier $\eta_t(\boldsymbol{x})$ is constructed using the weights w_i (as though they reflect the probability of occurrence of the sample). The weight of misclassified patterns is increased and the weight of correctly classified patterns is decreased. The effect of this is that the higher weight patterns influence the learning classifier more, and thus cause the classifier to focus more on the misclassifications, i.e. those patterns that are nearest the decision boundaries. There is a similarity with support vector machines in this respect (Chapter 6). The error e_t is calculated, corresponding to the sum of the weights of the misclassified samples. These get boosted by a factor $(1 - e_t)/e_t$, increasing the total weight on the misclassified samples (provided that $e_t < 1/2$). This process is repeated and a set of classifiers is generated. The classifiers are combined using a linear weighting whose coefficients are calculated as part of the training procedure.

There are several ways in which the `Adaboost` algorithm has been generalised including how the weights are updated and how the predictions are combined. One generalisation is for the classifiers to deliver a measure of confidence in the prediction. For example, in the two-class case, instead of the output being ± 1 corresponding to one of the two classes, the output is a number in the range $[-1, +1]$. The sign of the output is the predicted class label (-1 or +1) and the magnitude represents the degree of confidence: close to zero is interpreted as low confidence and close to unity as high confidence.

For the multiclass generalisation, Algorithm 8.3 presents the `Adaboost.MH` algorithm (Schapire and Singer, 1999). The basic idea is to expand the training set (of size n) to a training set of size $n \times C$ pairs,

$$((\boldsymbol{x}_i, 1), y_{i1}), ((\boldsymbol{x}_i, 2), y_{i2}), \ldots, ((\boldsymbol{x}_i, C), y_{iC}), \quad i = 1, \ldots, n$$

Thus, each training pattern is replicated C times and augmented with each of the class labels. The new labels for a pattern $(\boldsymbol{x}, l)$ take the values

$$y_{il} = \begin{cases} +1 & \text{if } \boldsymbol{x}_i \in \text{class } \omega_l \\ -1 & \text{if } \boldsymbol{x}_i \notin \text{class } \omega_l \end{cases} \tag{8.21}$$

A classifier, $\eta_t(\boldsymbol{x}, l)$ is trained and the final classifier, $\hat{\eta}(\boldsymbol{x}, l)$, is a weighted sum of the classifiers constructed at each stage of the iteration, with decision: assign $\boldsymbol{x}$ to class j if

$$\hat{\eta}(\boldsymbol{x}, j) \geq \hat{\eta}(\boldsymbol{x}, k) \quad k = 1, \ldots, C; k \neq j$$

Note that the classifier $\eta_t(\boldsymbol{x}, l)$ is defined on a data space of possible mixed variable type: real continuous variables $\boldsymbol{x}$ and categorical variable l. Care will be needed in classifier design (see Chapter 13 for further discussion of designing classifiers for mixed discrete and continuous variables).

In the study of linear classifiers by Skurichina (2001), it is reported that boosting is only useful for large sample sizes, when applied to classifiers that perform poorly. The performance of boosting depends on many factors including training set size, choice of classifier, the way in which the classifier weights are incorporated and the data distribution.

Algorithm 8.3 The `Adaboost.MH` algorithm converts the C class problem into a two-class problem operating on the original training data with an additional 'feature'.

The training data consist of $\{(\boldsymbol{x}_{ij}, y_{ij}), i = 1, \ldots, n;\ j = 1, \ldots, C\}$, where $\boldsymbol{x}_{ij} = (\boldsymbol{x}_i, j)$ and y_{ij} is given by Equation (8.21).

1. Initialise the weights $w_{ij} = 1/(nC), i = 1, \ldots, n; j = 1, \ldots, C$.
2. For $t = 1, \ldots, T$,

   (a) Construct a 'confidence-rated' classifier $\eta_t(\boldsymbol{x}, l)$ from the training data with weights $w_{ij}, i = 1, \ldots, n; j = 1, \ldots, C$.

   (b) Calculate r_t,

   $$r_t = \sum_{i=1}^{n} \sum_{l=1}^{C} w_{il} y_{il} \eta_t(\boldsymbol{x}_i, l)$$

   and α_t,

   $$\alpha_t = \frac{1}{2} \log\left(\frac{1 + r_t}{1 - r_t}\right)$$

   (c) Set $w_{ij} = w_{ij} \exp(-\alpha_t y_{ij} \eta_t(\boldsymbol{x}_i, j))$ and renormalise the weights so that they sum to unity.

3. Set

   $$\hat{\eta}(\boldsymbol{x}, l) = \sum_{t=1}^{T} \alpha_t \eta_t(\boldsymbol{x}, l)$$

   and assign $\boldsymbol{x}$ to ω_j if

   $$\hat{\eta}(\boldsymbol{x}, j) \geq \hat{\eta}(\boldsymbol{x}, k) \quad k = 1, \ldots, C; k \neq j$$

8.4.11 Random forests

Random forests[4] is an ensemble method that combines the predictions made by decision trees using a majority vote decision rule. Each tree is constructed using a bootstrap sample of the data. If there are n patterns in the dataset, n samples are taken, with replacement, to generate a bootstrap set of size n. This results in about two-thirds of the pattern being used to train the classifier and the remaining one-third ($1/e$ as $n \rightarrow \infty$) retained for testing.

[4] Random forests is a trademark of L. Breiman and A. Cutler licensed to Salford Systems.

Algorithm 8.4 The random forests algorithm.

1. Given a training set $(\boldsymbol{x}_i, z_i)$, $i = 1, \ldots, n$, of patterns $\boldsymbol{x}_i$ and labels z_i. Specify the number of trees in the forest, B, and the number of random features to select, m.
2. For $b = 1, \ldots, B$,
   (a) Generate a bootstrap sample of size n by sampling with replacement from the training set; some patterns will be replicated, others will be omitted.
   (b) Design a decision tree classifier, $\eta_b(\boldsymbol{x})$ using the bootstrap sample as training data, randomly selecting at each node in the tree m variables to consider for splitting.
   (c) Classify the nonbootstrap patterns (the 'out-of-bag' data) using the classifier $\eta_b(\boldsymbol{x})$.
3. Assign $\boldsymbol{x}_i$ to the class most represented by the classifiers $\eta_{b'}(\boldsymbol{x})$, where b' refers to the bootstrap samples that do not contain $\boldsymbol{x}_i$.

If the number of features is M, then each tree is constructed using $m \ll M$ features randomly selected at each node of the tree and the best split calculated using these m features. The tree is fully grown without pruning. The procedure is repeated for all trees in the forest using different bootstrap sample of the data. Classification of a pattern in the dataset is achieved by a majority vote using the trees that did not contain the pattern in the bootstrap sample used for their construction (about one-third of the trees). Thus, we see that the approach combines bagging with decision tree classifiers.

The procedure is described in Algorithm 8.4.

Random forests have been shown to give classification accuracies comparable with the best current classifiers on many datasets. They are able to handle data with a very large number of features (such as microarray data). Those features that are important for classification can be determined through the calculation of an importance score for each feature.

8.4.12 Model averaging

8.4.12.1 Introduction

In this section, we address *model* uncertainty (in contrast to *parameter* uncertainty) through a model averaging approach. Suppose that we wish to design a classifier given a set of data and we believe that that data arise from one of a set of models, $M_1, \ldots, M_k$, arising through different assumptions regarding the data probability density functions.

Model selection is the problem of using the data to select one model from the set of candidate models, $M_1, \ldots, M_k$. *Model averaging* (in the context of classification) refers to the process of estimating the class (or the probability of the class given a measurement on an object) for each model M_j and averaging the estimates according to the likelihood of each model.

Many methods of classification ignore model uncertainty, leading to overconfident inferences. In model averaging, in addition to uncertainty in model parameter values (within-model uncertainty), there is also uncertainty in model choice (uncertainty about which model is the best performing) (Gibbons *et al.*, 2008). Bayesian model averaging (BMA) is a particular type of ensemble method that provides a mechanism for accounting for this model uncertainty. It takes model uncertainty into consideration by averaging over posterior distributions of a quantity of interest based on multiple models, weighted by their posterior model probabilities.

A major task in model averaging is in determining the weights for individual models. For forecasting, the predictive abilities of potential models are used to determine the optimal weights [the posterior model probabilities; see Wasserman (2000) who discusses how to derive the weights; also, Gibbons *et al.* (2008)]. Implementing BMA is also discussed in the review by Hoeting *et al.* (1999).

A disadvantage of model averaging methods is that, whereas component models may involve just a few variables, model averages typically involve an order of magnitude more variables (Brown *et al.*, 2002). This may be a drawback if the measurements involve costs or if we are looking for interpretability as an output of the analysis.

A drawback of the Bayesian approach is that it requires assumptions concerning prior information about the distribution of unknown parameters. Other model averaging approaches have been developed (Doppelhofer, 2007).

We seek an implementation of an approach that accounts for model uncertainty which in our case is the choice of variables in variable selection. We work within the linear discriminant analysis framework and combine models (linear discriminants) with different feature subsets.

8.4.12.2 Bayesian model selection and model averaging

Suppose that Y is some quantity of interest. For example, in the classification case, Y is a class prediction at some feature value, x. We assume that we have a set of candidate models, $M_1, \ldots, M_k$, for a training set, D.

The posterior distribution of Y is given by

$$p(Y|D) = \sum_{M_k \in \mathcal{A}} p(Y|D, M_k) p(M_k|D) \tag{8.22}$$

a weighted average of the individual model predictions, with the weights equal to the posterior probability of each model, $p(M_k|D)$. $\mathcal{A}$ is the set of possible models.

The posterior probability of the model M_k may be written

$$p(M_k|D) = \frac{p(D|M_k)p(M_k)}{\sum_{M_l \in \mathcal{A}} p(D|M_l)p(M_l)} \tag{8.23}$$

and if each model M_k has parameters θ_k (in a linear discriminant model it would be the vector of parameters), then we may write, $p(D|M_k)$, the integrated likelihood of model M_k as

$$p(D|M_k) = \int p(D|\theta_k, M_k) p(\theta_k|M_k) d\theta_k \tag{8.24}$$

Evaluation of the integral can be difficult to compute and we seek approximations. We use the *Bayesian Information Criterion* approximation (Raftery, 1995)

$$p(D|M_k) \approx \widehat{m}_k$$

where

$$\log(\hat{m}_k) = \log\left(p(D|M_k, \hat{\theta}_k)\right) - \frac{d_k}{2}\log(n) \tag{8.25}$$

where $\hat{\theta}_k$ is the maximum likelihood estimate of the model parameters for the model M_k, d_k is the number of parameters in the model and n is the number of data samples. Thus, taking all models as equally likely, the posterior probability of the model M_k may be approximated by

$$p(M_k|D) = \frac{\hat{m}_k}{\sum_{M_l \in \mathcal{A}} \hat{m}_l} \tag{8.26}$$

with $\hat{m}_k$ given by Equation (8.25). Substituting Equation (8.26) into Equation (8.22) and using the further approximation,

$$p(Y|D, M_k) \approx p(Y|\hat{\theta}_k)$$

gives the prediction

$$p(Y|D) = \frac{\sum_{M_k \in \mathcal{A}} p(Y|\hat{\theta}_k)\hat{m}_k}{\sum_{M_k \in \mathcal{A}} \hat{m}_k} \tag{8.27}$$

Figure 8.13 lists the stages in the Bayesian model averaging approach. We now need to specify the set of models to include in the summation. Exhaustive summation over all models

1. Compute the maximum likelihood parameters, $\hat{\theta}_k$, for each model, M_k
2. For each model, compute the approximation to the integrated likelihood, $\hat{m}_k$, given by Equation (8.25).
3. For each test sample, evaluate the prediction using the maximum likelihood values, $p(Y|\hat{\theta}_k)$.
4. Calculate the Bayesian model average over all models using Equation (8.27).

Figure 8.13 Stages in Bayesian model averaging.

could render the calculation in Equation (8.27) impractical. For example, within the set of logistic discrimination classifiers, we could consider all possible subsets of variables, but this may be computationally too expensive.

There are several approaches to managing the summation. One approach is to use the 'leaps and bounds' algorithm (Volinsky, 1997) to identify the models to be used in the summation. The strategy described below is due to Madigan and Raftery (1994).

8.4.12.3 Implementation – model selection and search strategy

Determining the set of models

Models that predict the data much less well than the best model in the class are discarded. Thus, models are excluded if they do not belong to the set

$$\mathcal{A}' = \left\{ M_k : \frac{\max_i \{p(M_l|D)\}}{p(M_k|D)} \leq c \right\} \tag{8.28}$$

for a constant c, taken by Madigan and Raftery (1994) to be a value of 20, although other values may be used for some problems.

Also excluded from the summation are models which have a submodel that is better predicted by the data (principle of Occam's razor). Examples of when a model M_0 is considered to be a submodel of M_1 include the following:

- In a linear regression context, the variables that form model M_0 are subsets of the variables that form M_1.
- In a graphical model context, all the links in M_0 are also in M_1.

Thus, models are excluded if they belong to the set

$$\mathcal{B} = \left\{ M_k : \exists M_l \in \mathcal{A}', M_l \subset M_k, \frac{p(M_l|D)}{p(M_k|D)} > 1 \right\} \tag{8.29}$$

Thus, in Equation (8.27), the set $\mathcal{A}$ is taken to be

$$\mathcal{A} = \mathcal{A}' \setminus \mathcal{B}$$

the set of models that belong to $\mathcal{A}'$ but do not belong to $\mathcal{B}$. The next stage is to find the set $\mathcal{A}$ efficiently.

Searching the model space

The approach proceeds by successively comparing the ratio of posterior model probabilities for two models, one of which is a submodel of the other. The basic rule is that if a model is rejected, then so are all of its submodels (since none of the submodels are better predicted by

Algorithm 8.5 BGMS-down algorithm (Madigan and Raftery, 1994).

1. Select a model M from $\mathcal{C}$.
2. Set $\mathcal{C} = \mathcal{C} \setminus \{M\}$; $\mathcal{A} = \mathcal{A} \cup \{M\}$.
3. Select a submodel M_0 of M.
4. Compute the ratio, B, given by

$$B = \log\left(\frac{p(M_0|D)}{p(M|D)}\right)$$

5. If $B > O_R$, then set $A = A \setminus \{M\}$ and if $M_0 \notin \mathcal{C}$, set $\mathcal{C} = \mathcal{C} \cup \{M_0\}$.
6. If $O_L \leq B \leq O_R$, then if $M_0 \notin \mathcal{C}$, set $\mathcal{C} = \mathcal{C} \cup \{M_0\}$.
   (a) If there are more submodels of M, go to step 3.
7. If $\mathcal{C} \neq \phi$, go to step 1.

the data). Two models are compared on the basis of the log of the ratio of the posterior model probabilities.

- If the ratio is greater than some threshold O_R, *i.e.*

$$\log\left(\frac{p(M_0|D)}{p(M_1|D)}\right) > O_R$$

  then this provides evidence for the smaller model, M_0, and M_1 is rejected.

- If the ratio is smaller than some threshold O_L, i.e.

$$\log\left(\frac{p(M_0|D)}{p(M_1|D)}\right) < O_L$$

  [where $O_L = -\log(c)$ with c given by Equation (8.28)] then the model M_0 is rejected.

- For intermediate values of the ratio, both models are considered.

Values of O_L of $-\log(20)$ and O_R of 0 are adopted by Madigan and Raftery (1994), but other values have been proposed for different applications.

We present the BGMS-down and the BGMS-up algorithms of Madigan and Raftery (1994) (see Algorithms 8.5 and 8.6; BMGS, Bayesian Graphical Model Selection). The procedure starts with the down algorithm, and using the models selected as the set of candidate starting models for the up algorithm. Let $\mathcal{C}$ denote a set of starting models and $\mathcal{A}$ denote the set of acceptable models (initially $\mathcal{A} = \phi$).

Algorithm 8.6 BGMS-up algorithm (Madigan and Raftery, 1994).

1. Select a model M from $\mathcal{C}$.
2. Set $\mathcal{C} = \mathcal{C} \setminus \{M\}$; $\mathcal{A} = \mathcal{A} \cup \{M\}$.
3. Select a supermodel M_1 of M; i.e. a model M_1 such that M is a submodel of M_1.
4. Compute the ratio, L, given by

$$L = \log\left(\frac{p(M|D)}{p(M_1|D)}\right)$$

5. If $L < O_L$, then set $A = A \backslash \{M\}$ (if $\{M\} \subset \mathcal{A}$) and if $M_1 \notin \mathcal{C}$, set $\mathcal{C} = \mathcal{C} \cup \{M_1\}$.
6. If $O_L \leq L \leq O_R$, then if $M_1 \notin \mathcal{C}$, set $\mathcal{C} = \mathcal{C} \cup \{M_1\}$.
7. If there are more supermodels of M, go to step 3.
8. If $\mathcal{C} \neq \phi$, go to step 1.

At the conclusion of the algorithm, the set $\mathcal{A}$ contains the potentially acceptable models. We then perform another pass through the set, excluding models that do not belong to the set $\mathcal{A}'$ or belong to the set $\mathcal{B}$ defined by Equations (8.28) and (8.29), respectively.

8.4.12.4 Specification of prior probabilities

The predictions for each individual model are weighted by the posterior probability of the model, given by Equation (8.23). This requires the specification of $p(M_k)$, the prior distribution over competing models. In the analysis of this section, we have assumed as is commonly done that all models are equally likely

$$p(M_k) = \frac{1}{C(\mathcal{A})}$$

where $C(\mathcal{A})$ is the cardinality of set $\mathcal{A}$. Alternative approaches have been considered. For example, in a variable selection context, where each model M_k is defined by a different set of variables, then if we have prior information about the importance of a given variable for classification purposes, the prior probabilities for models incorporating that variable may be greater than those models that do not use that variable. For example, we may take (Hoeting *et al.*, 1999)

$$p(M_k) = \prod_{j=1}^{p} \pi_j^{\delta_{ij}} (1 - \pi_j)^{\delta_{ij}}$$

where $\pi_j \in [0, 1]$ is the prior probability that variable j is important for classification; δ_{ij} is an indicator of whether variable j is included in model M_k. A value of $\pi_j = 0.5$ corresponds to uniform priors; $\pi_j = 1$ implies variable j is included in all the models selected [$p(M_k) = 0$ if variable j is not present].

8.4.12.5 BMA for classification

For the binary classification case, the response variable for a sample in the test set is Y, where $Y = 0$ or $Y = 1$. The posterior probability of $Y = 1$ given the training set is

$$p(Y = 1|D) = \frac{\sum_{M_k \in \mathcal{A}} P(Y = 1|\hat{\theta}_k)\hat{m}_k}{\sum_{M_l \in \mathcal{A}} \hat{m}_k}$$

a weighted sum of component classifiers, with $\hat{m}_k$ given by Equation (8.25).

More generally, for a C-class problem, we sum the posterior class probabilities over all models.

8.4.12.6 BMA for feature selection

For a given class of models (e.g. logistic regression) the family of models is constructed using all possible combinations of the input variables. Thus a supermodel is constructed from a given model by the addition of an extra variable.

When classifying test data, we would like to know which variables are contributing to the classification. Each classifier will use a different combination of variables, but we can calculate the posterior probability that variable is i relevant to the classifier by calculating

$$p(\text{variable } i \text{ is relevant}|D) = \frac{\sum_{M_k \in \mathcal{A}_i} \hat{m}_k}{\sum_{M_l \in \mathcal{A}} \hat{m}_l}$$

where $\mathcal{A}_i$ is the subset of the final set of models that use variable i.

8.4.12.7 Discussion

BMA is not a method of model combination. It is an approach for handling uncertainty in the model under the assumption that the data were generated by one of the models. BMA provides a means of averaging different predictive models, each model based on maximum likelihood estimates of model parameters. The weights applied to each model are the posterior probabilities of each model. Thus, in order to apply the approach we need a predictive model whose parameters are maximum likelihood estimates. Not all classifiers are of this type.

8.4.13 Summary of methods

Table 8.2 lists the combination rules discussed in this section and some of their properties, according to the taxonomy given in Section 8.2. Further discussion is provided by Jain *et al.* (2000) and Sewell (2011).

Table 8.2 Summary of classifier combination methods.

Method	Feature space	Level	Training	Form	Architecture	Optimisation	Comments
Product rule	Different, C1	Feature, L2	Fixed, R1	Common, F1	Parallel, T1	Constituent classifiers, O2	Could be different form; could be common feature selection but different (independent) data realisations; training of constituent classifiers only
Sum rule	Different, C1	Feature, L2	Fixed, R1	Common, F1	Parallel, T1	Constituent classifiers, O2	
Min, max, median	Different, C1	Feature, L2	Fixed, R1	Common, F1	Parallel, T1	Constituent classifiers, O2	
Majority vote	Different, C1	Decision, L3	Fixed, R1	Common, F1	Parallel, T1	Constituent classifiers, O2	
Borda count	Different, C1	Decision, L3	Fixed, R1	Common, F1	Parallel, T1	Constituent classifiers, O2	
Combining class predictors	Common	Decision/ feature	Trainable, R2	Common, F1	Parallel, T1	Combiner + constituent classifiers, O3	
Stacked general	Common	Decision/ feature	Trainable, R2	Common, F1	Parallel, T1	Combiner + constituent classifiers, O3	
Mixture of experts	Common	Feature	Trainable, R2	Dissimilar form	Parallel, T1	Combiner + constituent classifiers, O3	
Bagging	Common	Decision	Fixed	Common, F1	Parallel, T1	Constituent classifiers, O2	Fixed combiner
Boosting	Common	Feature	Trainable	Common, F1	Serial	Constituent classifiers, O2	
Random forests	Common						
Model averaging	Different	Feature	Trainable (weighted sum)	Common, F1 (e.g. logistic)	Parallel, T1	Combiner + constituent classifiers, O3	

8.4.14 Example application study

The problem

This study (Yeung *et al.*, 2005) addresses the problem of selecting a small number of relevant genes for accurate classification of microarray data, essential for the development of diagnostic tests. One of the challenges with using microarray data for classification is the large number of genes (that is, the number of features), which is usually much greater than the number of patterns.

Summary

Yeung *et al.* (2005) develop a BMA approach as a feature selection method for microarray data, using a modification of the traditional BMA implementation of Raftery (1995). Feature selection methods are discussed in detail in Chapter 10. A small set of relevant genes is essential for the development of inexpensive tests. They find that the BMA algorithm developed generally selects fewer features than the existing best feature selection methods. It has the advantages of facilitating biological interpretation.

The data

Three datasets were considered. The breast cancer prognosis dataset (a two-class dataset) consists of primary breast tumour samples hybridised to cDNA (complementary DNA). There are 24481 genes and 97 samples (78 in the training set and 19 in the test set). Thus, the number of features is very much greater than the number of patterns. The samples are divided into two classes: the good prognosis group (those patients who remained disease free for at least 5 years) and the poor prognosis group (those patients who developed distant secondary tumours within 5 years). After preprocessing the data, the training set comprised 95 samples (76 train and 19 test) on 4919 genes.

The leukaemia dataset comprises 72 samples (38 in the training set and 34 in the test set), each consisting (after filtering) of 3051 genes. There are two main classes: patients with acute lymphoblastic leukaemia (ALL) and those with acute myeloid leukaemia (AML). The ALL class could be further subdivided into two ALL subtypes. Therefore, the data can be divided into two or three classes.

The hereditary breast cancer dataset contains 22 samples (from three classes) of 3226 genes.

The model

The BMA algorithm for binary classification was applied to the breast cancer and the two-class leukaemia datasets. The multiclass iterative BMA algorithm was applied to the three-class leukaemia dataset. Yeung *et al.* (2005) use a logistic regression model to predict $P(Y = 1|D, M_k)$ such that

$$\ln\left[P(Y = 1|D, M_k)/P(Y = 0|D, M_k)\right] = \sum_{i=0}^{p} b_i x_i$$

(for multiclass microarray data, the binary logistic regressions were combined).

Training procedure

Yeung *et al.* (2005) use the BMA implementation of Raftery (1995), using Equation (8.23) to compute $p(M_k|D)$ and the Bayesian Information Criterion to approximate $p(D|M_k)$

[Equation (8.25)]. For the hereditary breast cancer dataset, the leave-one-out cross-validation procedure was used in training.

Results

For the breast cancer prognosis dataset, there were 3 classification errors on the test set using 6 selected genes. For the leukaemia dataset, there was one classification error on the test set using 15 selected genes. The hereditary breast cancer dataset produced 6 errors (out of 22) using 13–15 genes (the number varied because of the cross-validation procedure).

8.4.15 Further developments

Many of the methods of classifier combination, even the basic nontrainable methods, are active subjects of research and assessment. For example, further properties of fixed rules (sum, voting, ranking) are presented by Kittler and Alkoot (2001) and Saranli and Demirekler (2001).

Development of stacking to density estimation is reported by Smyth and Wolpert (1999). Further applications of stacking neural network models are given by Sridhar *et al*. (1999).

The basic mixture of experts model has been extended to a tree architecture by Jordan and Jacobs (1994). Termed 'hierarchical mixture of experts', nonoverlapping sets of expert networks are combined using gating networks. Outputs of these gating networks are themselves grouped using a further gating network.

Boosting is classed by Breiman (1998) as an 'adaptive resampling and combining' or *arcing* algorithm. Definitions for the bias and variance of a classifier, C, are introduced and it is shown that

$$e_E = e_B + \text{Bias}(C) + \text{Variance}(C)$$

where e_E and e_B are the expected and Bayes' error rates, respectively (see Chapter 9). Unstable classifiers can have low bias and high variance on a large range of datasets. Combining multiple versions can reduce variance significantly.

Learning a classifier for nominal vector data based on a BMA approach has been described by Sicard *et al*., (2008). Alternative methods of calculating weights in a weighted model averaging approach (for example, frequentist model averaging) are provided by Schomaker *et al*. (2010) and Hjort and Claeskens (2003). EnsembleBMA is an R package for BMA (Fraley *et al*., 2009).

Bayesian model combination (BMC, in contrast to BMA) is an approach proposed by Ghahramani and Kim (2003) for classification. For many classifiers, it is not possible to define the likelihood (for example, rule induction methods) and BMC does not assume that the classifiers are probabilistic. The classifiers may even be human experts.

8.5 Application studies

One of the main motivating applications for research on multiple classifier systems has been the detection and tracking of targets using a large number of different types of sensors. Much of the methodology developed applies to highly idealised scenarios, often failing to take into account practical considerations such as asynchronous measurements, data rates and bandwidth constraints on the communication channels between the sensors and the fusion centre. Nevertheless, methodology developed within the *data fusion* literature is relevant to other practical problems. Example applications include:

- Biomedical data fusion. Various applications include coronary care monitoring and ultrasound image segmentation for the detection of the esophagus (Dawant and Garbay, 1999).
- Airborne target identification (Raju and Sarma, 1991).

Examples of the use of the ensemble methods described in this chapter include the following.

- Biometrics. Chatzis *et al.* (1999) combine the outputs of five methods for person verification, based on image and voice features, in a decision fusion application. Kittler *et al.* (1997) assess a multiple observation fusion (Figure 8.6) approach to person verification. In a writer verification application, Zois and Anastassopoulos (2001) use the Bahadur–Lazarsfeld expansion to model correlated decisions. Prabhakar and Jain (2002) use kernel-based density estimates (Chapter 4) to model the distributions of the component classifier outputs, each assumed to provide a measure of confidence in one of two classes, in a fingerprint verification application.
- Chemical process modelling. Sridhar *et al.* (1996) develop a methodology for stacking neural networks in plant-process modelling applications.
- Remote sensing. In a classification of land cover from remotely sensed data using decision trees, Friedl *et al.* (1999) assess a boosting procedure (see also Chan *et al.*, 2001). In a similar application, Giacinto *et al.* (2000) assess combination methods applied to five neural and statistical classifiers.
- Video indexing. Benmokhtar and Huet (2006) use a classifier fusion (fusion of classifier outputs) approach for automatic semantic-based video content and retrieval (indexing and searching multimedia databases).
- Protein structure prediction. Melvin *et al.* (2008) combine classifiers of different type (nearest neighbour and support vector machine) in a study to predict protein structure from amino acid sequence or structure.

BMA approaches have been exploited in many applications including in the areas of finance (Magnus *et al.*, 2010; Ouysse and Kohn, 2010), weather prediction (Berrocal *et al.*, 2007; Sloughter *et al.*, 2007; Bao *et al.*, 2010; Fraley *et al.*, 2010), political science (Montgomery and Nyhan, 2010), microarray data classification (Yeung *et al.*, 2005; Annest *et al.*, 2009; Abeel *et al.*, 2010), and ecology (Prost *et al.*, 2008).

8.6 Summary and discussion

Combining the results of several classifiers, rather than selecting the best, may offer improved performance. There may be practical motivating problems for this – such as those in distributed data fusion – and many rules and techniques have been proposed and assessed. These procedures differ in several respects: they may be applied at different levels of processing (raw 'sensor' data, feature level, decision level); they may be trainable or fixed; the component classifiers may be similar (for example, all decision trees), or of different forms, developed independently; the structure may be serial or parallel; finally, the combiner alone may be optimised, or jointly with the component classifiers.

In this chapter, we have reviewed the characteristics of combination schemes, a motivating application (distributed sensor detection) and described the properties of some of the more popular methods of classifier combination. Some of the research in this area is motivated by real-world practical applications such as the distributed detection problem and some work on person verification using different identification systems. Often, in applications such as these, the constituent classifier is fixed and an optimal combination is sought. There is no universal best combiner, but simple methods such as the sum, product and median rules can work well.

Of more interest are procedures that simultaneously construct the component classifiers and the combination rule. Unstable classification methods (classification methods for which small perturbations in their training set or construction procedure may result in large changes in the predictor; for example, decision trees) can have their accuracy improved by combining multiple versions of the classifier. Bagging and boosting fall into this category. Bagging perturbs the training set repeatedly and combines by simple voting; boosting reweights misclassified samples and classifiers are combined by weighted voting. Unstable classifiers such as trees can have a high variance that is reduced by bagging and boosting. However, boosting may increase the variance of a stable classifier and be counterproductive. Model averaging methods (such as BMA) offer approaches for handling model uncertainty as well as parameter uncertainty.

8.7 Recommendations

1. If you are combining prescribed classifiers, defined on the same inputs, the sum rule is a good start.
2. For classifiers defined on separate feature spaces, the product rule is a simple one to begin with and one that does not require training.
3. Boosting and bagging are recommended to improve performance of unstable classifiers.

8.8 Notes and references

There is a large amount of literature on combining classifiers. A good starting point is the statistical pattern recognition review by Jain *et al.* (2000); see also Tulyakov *et al.* (2008) and the book by Kuncheva (2004a). Kittler *et al.* (1998) describe a common theoretical framework for some of the fixed combination rules.

Within the defence and aerospace domain, data fusion has received considerable attention, particularly the detection and tracking of targets using multiple distributed sources (Waltz and Llinas, 1990; Dasarathy, 1994b; Varshney, 1997), with benefits in robust operational performance, reduced ambiguity, improved detection and improved system reliability (Harris, 1997).

Stacking originated with Wolpert (1992) and the mixture of experts model with Jacobs *et al.* (1991; see also Jordan and Jacobs, 1994). Combining neural network models is reviewed by Sharkey (1999).

Bagging is presented by Breiman (1996). Comprehensive experiments on bagging and boosting for linear classifiers are described by Skurichina (2001). The first provable polynomial-time boosting algorithm was presented by Schapire. The Adaboost algorithm was introduced by Freund and Schapire (1996, 1999). Improvements to the basic algorithm are given by Schapire and Singer (1999). Empirical comparisons of bagging and boosting are given by Bauer and Kohavi (1999). A statistical view of boosting is provided by Friedman *et al.* (1998).

Raftery (1995) provides a good introduction to Bayesian model averaging; see also the tutorial by Hoeting *et al.* (1999), and comments on that paper, and the review by Wasserman (2000).

Random forests is a trademark of Leo Breiman (Breiman, 2001) and Adele Cutler and is licensed exclusively to Salford Systems. Random forest software is commercially marketed by Salford Systems; open source implementations are available.

Exercises

1. What is the significance of the condition $e_t > 0.5$ in step 2 of the boosting algorithm in Section 8.4.10?
2. Design an experiment to evaluate the boosting procedure. Consider which classifier to use and datasets that may be used for assessment. How would weighted samples be incorporated into the classifier design? How will you estimate generalisation performance? Implement the experiment and describe the results.
3. As above, but assess the bagging procedure as a means of improving classifier performance.
4. Using the expression (8.10) for the posterior probabilities, express the product rule in terms of the priors and δ_{ki}. Assuming $\delta_{ki} \ll 1$, show that the decision rule may be expressed (under certain assumptions) in terms of sums of the δ_{ki}. State your assumptions. Finally, derive (8.11) using (8.10).
5. Given measurements $\boldsymbol{u} = (u_1, \ldots, u_L)$ made by L detectors with probabilities of false alarm pfa_i and probabilities of detection pd_i, $(i = 1, \ldots, L)$, show (assuming independence and equal cost loss matrix)

$$\log\left(\frac{p(\omega_1|\boldsymbol{u})}{p(\omega_2|\boldsymbol{u})}\right) = \log\left(\frac{p(\omega_1)}{p(\omega_2)}\right) + \sum_{S_+} \log\left(\frac{pd_i}{pfa_i}\right) + \sum_{S_-} \log\left(\frac{1 - pd_i}{1 - pfa_i}\right)$$

where S_+ is the set of all detectors such that $u_i = +1$ (target present declared – class ω_1) and S_- is the set of all detectors such that $u_i = 0$ (target absent declared – class ω_2).

Therefore, express the data fusion rule as [see Equation (8.3)]

$$u_0 = \begin{cases} 1 & \text{if } a_0 + \boldsymbol{a}^T \boldsymbol{u} > 0 \\ 0 & \text{otherwise} \end{cases}$$

and determine a_0, $\boldsymbol{a}$.

6. Write a computer program to produce the ROC curve for the L sensor fusion problem [L sensors with probabilities of false alarm pfa_i and probabilities of detection pd_i, ($i = 1, \ldots, L$)] using the decision rule (8.4).

7. Using the Bahadur–Lazarsfeld expansion, derive the Bayesian decision rule in terms of the conditional correlation coefficients,

$$\gamma^i_{ij \ldots L} = \mathrm{E}_i[z_i z_j \ldots z_L] = \int z_i z_j \ldots z_L p(\boldsymbol{u}|\omega_i) d\boldsymbol{u}$$

for $i = 1, 2$.

8. k out of N decision rule. Let $p(\omega_1)$ be the prior probability that a target is present; $p(\omega_2)$ [$=1 - p(\omega_1)$] is the probability that a target is not present. Let p be the probability of detecting a target given that it is present (the probability of detection) and q be the probability of incorrectly detecting a target given that it is not present (the probability of false alarm). For N similar detectors operating independently, write down the probability that a target is present given that at least k detectors fire and show that if all N detectors fire, then the probability that a target is present is

$$\frac{p(\omega_1)}{p(\omega_1) + p(\omega_2)\left(\frac{q}{p}\right)^N}$$

9

Performance assessment

Classifier performance assessment is an important aspect of the pattern recognition cycle. How good is the designed classifier and how well does it compare with competing classification techniques? Error rate is the most widely used performance measure, but it has its shortcomings. Alternatives measures are presented and the use of the receiver operating characteristic for performance assessment described.

9.1 Introduction

The pattern recognition cycle (see Chapter 1) begins with the collection of data and initial data analysis followed, perhaps, by preprocessing of the data and the design of the classification rule. Chapters 2–7 described approaches to classification rule design, beginning with density estimation methods and leading on to techniques that construct a discrimination rule that may be expressed as an interpretable rule. In this chapter we address the issue of measuring classifier performance.

Performance assessment should really be a part of classifier design and not an add-on extra that is considered separately, as it often is. A sophisticated design stage is often followed by a much less sophisticated evaluation stage, perhaps resulting in an inferior rule. The criterion used to design a classifier is often different from that used to assess it, which is different again from a performance measure appropriate for the classifier operating conditions. For example, in constructing a discriminant rule, it is common to choose the parameters of the rule to optimise a square error measure, yet to assess the rule using a different measure of performance, such as error rate.

A related aspect of performance is that of comparing the performance of several classifiers trained on the same dataset. For a practical application, we may implement several classifiers and want to choose the best, measured in terms of error rate or perhaps computational efficiency. In Section 9.3, comparing classifier performance is addressed.

Statistical Pattern Recognition, Third Edition. Andrew R. Webb and Keith D. Copsey.

Table 9.1 2 × 2 confusion matrix.

		True class	
		Positive	Negative
Predicted class	Positive	True positives (*TP*)	False positives (*FP*)
	Negative	False negatives (*FN*)	True negatives (*TN*)

9.2 Performance assessment

Three aspects of performance of a classification rule are addressed. The first is the *discriminability* of a rule (how well it classifies unseen data) and we focus on one particular method, namely the error rate. The second is the *reliability* of a rule. This is a measure of how well it estimates the posterior probabilities of class membership. Finally, the use of the *receiver operating characteristic* (ROC) as an indicator of performance is considered.

9.2.1 Performance measures

Many performance metrics can be calculated from the confusion matrix. Table 9.1 shows a 2 × 2 classifier confusion matrix for two classes (positive and negative). It shows the number of true and false predictions for each of the two classes. The true positives (*TP*) is the number of members of the positive class that are correctly predicted by the classifier to belong to the positive class. The false positives (*FP*) is the number of members of the negative class that are incorrectly predicted to belong to the positive class. Table 9.2 lists some of the common performance metrics derived from the confusion matrix. Many are extendable to multiclass problems (Sing *et al.*, 2007).

Table 9.2 Performance metrics derived from the 2 × 2 confusion matrix; $P = TP + FN$; $N = FP + TN$.

Accuracy (*Acc*)	$\frac{TP+TN}{P+N}$
Error rate (*E*)	$1 - Acc$
False positive rate (*fpr*)	$\frac{FP}{N}$
False alarm rate (identical to *fpr*)	$\frac{FP}{N}$
True positive rate (*tpr*)	$\frac{TP}{P}$
Precision (*Prec*)	$\frac{TP}{TP+FP}$
Recall (*Rec*, identical to *tpr*)	$\frac{TP}{P}$
Sensitivity (*Sens*, identical to *tpr*)	$\frac{TP}{P}$
Specificity (*Spec*)	$\frac{TN}{N}$
F measure	$\frac{2}{\frac{1}{Prec}+\frac{1}{Rec}}$

9.2.2 Discriminability

There are many measures of discriminability (Hand, 1997), the most common being the *misclassification rate* or the *error rate* of a classification rule. Generally, it is very difficult to obtain an analytic expression for the error rate and therefore it must be estimated from the available data. There is a vast amount of literature on error rate estimation, but the error rate suffers from the disadvantage that it is only a single measure of performance, treating all correct classifications equally and all misclassifications with equal weight also (corresponding to a 0/1 loss function – see Chapter 1). In addition to computing the error rate, we may also compute a *confusion* or *misclassification* matrix. The (i, j)th element of this matrix is the number of patterns of class ω_j that are classified as class ω_i by the rule. This is useful in identifying how the error rate is decomposed. A complete review of the literature on error rate estimation deserves a volume in itself, and is certainly beyond the scope of this book. Here, we limit ourselves to a discussion of the more popular types of error rate estimator.

First, let us introduce some notation. Let the training data be denoted by $\boldsymbol{Y} = \{\boldsymbol{y}_i, i = 1, \ldots, n\}$, where the pattern $\boldsymbol{y}_i$ consists of two parts, $\boldsymbol{y}_i^T = (\boldsymbol{x}_i^T, \boldsymbol{z}_i^T)$, where $\{\boldsymbol{x}_i, i = 1, \ldots, n\}$ are the measurements and $\{\boldsymbol{z}_i, i = 1, \ldots, n\}$ are the corresponding class labels, now coded as a vector, $(\boldsymbol{z}_i)_j = 1$ if $\boldsymbol{x}_i \in$ class ω_j and zero otherwise. Let $\omega(\boldsymbol{z}_i)$ be the corresponding categorical class label. Let the decision rule designed using the training data be $\eta(\boldsymbol{x}; \boldsymbol{Y})$ (that is, η is the class to which $\boldsymbol{x}$ is assigned by the classifier designed using $\boldsymbol{Y}$) and let $Q(\omega(\boldsymbol{z}), \eta(\boldsymbol{x}; \boldsymbol{Y}))$ be the loss function

$$Q(\omega(\boldsymbol{z}), \eta(\boldsymbol{x}; \boldsymbol{Y})) = \begin{cases} 0 & \text{if } \omega(\boldsymbol{z}) = \eta(\boldsymbol{x}; \boldsymbol{Y}) \text{ (correct classification)} \\ 1 & \text{otherwise} \end{cases}$$

Apparent error rate

The *apparent error rate*, e_A, or *resubstitution rate* is obtained by using the design set to estimate the error rate,

$$e_A = \frac{1}{n} \sum_{i=1}^{n} Q(\omega(\boldsymbol{z}_i), \eta(\boldsymbol{x}_i; \boldsymbol{Y}))$$

It can be severely optimistically biased, particularly for complex classifiers and a small dataset when there is a danger of overfitting the data – that is, the classifier models the noise on the data rather than its structure. Increasing the number of training samples reduces this bias.

True error rate

The *true error rate* (or *actual error rate* or *conditional error rate*), e_T, of a classifier is the expected probability of misclassifying a randomly selected pattern. It is the error rate on an infinitely large test set drawn from the same distribution as the training data.

Expected error rate

The *expected error rate*, e_E, is the expected value of the true error rate over training sets of a given size, $e_E = \mathrm{E}[e_T]$.

Bayes' error rate

The *Bayes' error rate* or *optimal error rate*, e_B, is the theoretical minimum of the true error rate, the value of the true error rate if the classifier produced the true posterior probabilities of class membership, $p(\omega_i|\boldsymbol{x}), i = 1, \ldots, C$.

9.2.2.1 Holdout estimate

The holdout method splits the data into two mutually exclusive sets, sometimes referred to as the training and test sets. The classifier is designed using the training set and performance evaluated on the independent test set. The method makes inefficient use of the data (using only part of it to train the classifier) and gives a pessimistically biased error estimate (Devijver and Kittler, 1982). However, it is possible to obtain confidence limits on the true error rate given a set of n independent test samples, drawn from the same distribution as the training data. If the true error rate is e_T, and k of the samples are misclassified, then k is binomially distributed

$$p(k|e_T, n) = \text{Bi}(k|e_T, n) \triangleq \binom{n}{k} e_T^k (1 - e_T)^{(n-k)} \tag{9.1}$$

The above expression gives the probability that k samples out of n of an independent test set are misclassified given that the true error rate is e_T. Using Bayes' theorem, we may write the conditional density of the true error rate, given the number of samples misclassified, as

$$p(e_T|k, n) = \frac{p(k|e_T, n)p(e_T, n)}{\int p(k|e_T, n)p(e_T, n)de_T}$$

Assuming $p(e_T, n)$ does not vary with e_T and $p(k|e_T, n)$ is the binomial distribution, we have a beta distribution for e_T,

$$p(e_T|k, n) = \text{Be}(e_T|k + 1, n - k + 1) \triangleq \frac{e_T^k (1 - e_T)^{n-k}}{\int e_T^k (1 - e_T)^{n-k} de_T}$$

where $\text{Be}(x|\alpha, \beta) = [\Gamma(\alpha + \beta)/(\Gamma(\alpha)\Gamma(\beta))]x^{\alpha-1}(1 - x)^{\beta-1}$. The above posterior density provides a complete account of what can be learned given the test error. However, it may be summarised in several ways, one of which is to give an upper and lower bound (a percentage point) on the true error. For a given value of α (for example 0.05), there are many intervals in which e_T lies with probability $1 - \alpha$. These are called $(1 - \alpha)$-credible regions, or Bayesian confidence intervals (O'Hagan, 1994). Among these intervals, the *highest posterior density* (HPD) credible region is the one with the additional property that every point within it has a higher probability than any point outside. It is also the shortest $(1 - \alpha)$-credible region. It is the interval E_α

$$E_\alpha = \{e_T : p(e_T|k, n) \geq c\}$$

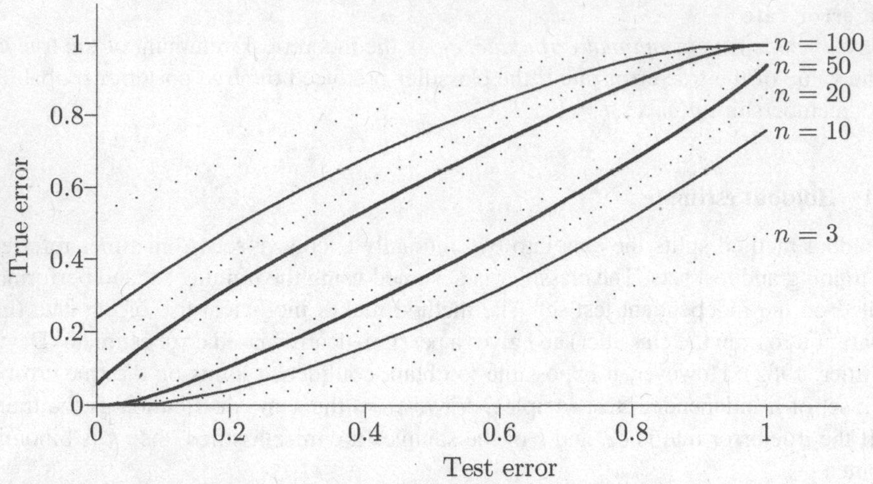

Figure 9.1 HPD credible region limits as a function of test error (number misclassified on test/size of test set) for several values of n, the number of test samples and $\alpha = 0.05$ (i.e. the 95% credible region limits). From top to bottom, the limit lines correspond to $n = 3$, 10, 20, 50, 100, 100, 50, 20, 10, 3.

where c is chosen such that

$$\int_{E_\alpha} p(e_T|k, n)de_T = 1 - \alpha \tag{9.2}$$

For multimodal densities, E_α may be discontinuous. However, for the beta distribution, E_α is a single region with lower and upper bounds $\epsilon_1(\alpha)$ and $\epsilon_2(\alpha)$ (both functions of k and n) satisfying

$$0 \leq \epsilon_1(\alpha) < \epsilon_2(\alpha) \leq 1$$

Figure 9.1 displays the Bayesian confidence intervals as a function of test error for several values of n, the number of samples in the test set, and a value for α of 0.05, i.e. the bounds of the 95% credible region. For example, for 4 out of 20 test samples incorrectly classified, the $(1 - \alpha)$-credible region (for $\alpha = 0.05$) is [0.069, 0.399].

Figure 9.2 plots the maximum length of the 95% HPD credible region (over the test error) as a function of the number of test samples. For example, we can see from the figure that, to be sure of having a HPD interval of less than 0.1, we must have more than 350 test samples.

9.2.2.2 Cross-validation

Cross-validation (also known as the *U-method*, the *leave-one-out estimate* or the *deleted estimate*) calculates the error by using $n - 1$ samples in the design set and tests on the remaining sample. This is repeated for all n subsets of size $n - 1$. For large n, it is computationally

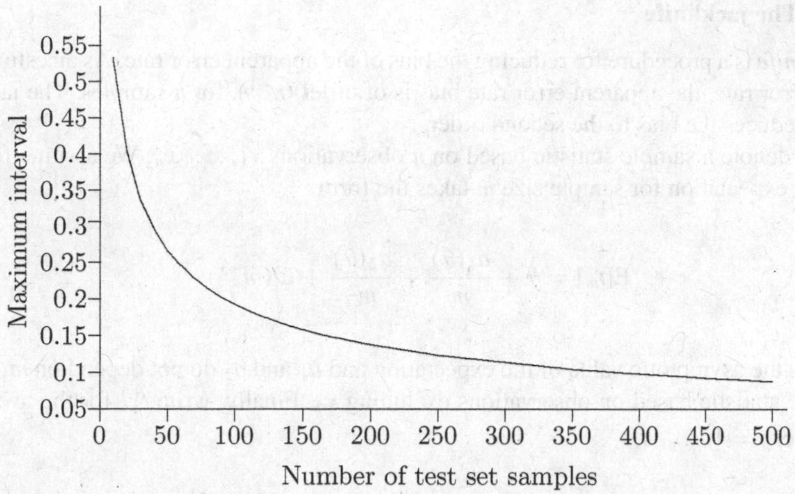

Figure 9.2 Maximum length of the 95% HPD credible region as a function of the number of test samples.

expensive, requiring the design of n classifiers. However, it is approximately unbiased, although at the expense of an increase in the variance of the estimator. Denoting by $\boldsymbol{Y}_j$ the training set with observation $\boldsymbol{x}_j$ deleted, then the cross-validation error is

$$e_{cv} = \frac{1}{n}\sum_{j=1}^{n} Q(\omega(\boldsymbol{z}_j), \eta(\boldsymbol{x}_j, \boldsymbol{Y}_j))$$

One of the disadvantages of the cross-validation approach is that it may involve a considerable amount of computation. However, for discriminant rules based on multivariate normal assumptions, the additional computation can be considerably reduced through the application of the Sherman–Morisson formula (Fukunaga and Kessell, 1971; McLachlan, 1992a):

$$(\boldsymbol{A} + \boldsymbol{u}\boldsymbol{u}^T)^{-1} = \boldsymbol{A}^{-1} - \frac{\boldsymbol{A}^{-1}\boldsymbol{u}\boldsymbol{u}^T\boldsymbol{A}^{-1}}{1 + \boldsymbol{u}^T\boldsymbol{A}^{-1}\boldsymbol{u}} \tag{9.3}$$

for matrix $\boldsymbol{A}$ and vector $\boldsymbol{u}$, which can be used to enable efficient calculation of the covariance matrix.

The *rotation method*, or *ν-fold cross-validation* partitions the training set into ν subsets, training on $\nu - 1$ and testing on the remaining set. This procedure is repeated as each subset is withheld in turn. If $\nu = n$, we have the standard cross-validation, and if $\nu = 2$, we have a variant of the holdout method in which the training set and test set are also interchanged. This method is a compromise between the holdout method and cross-validation, giving reduced bias compared with the holdout procedure, but less computation compared with cross-validation.

9.2.2.3 The jackknife

The *jackknife* is a procedure for reducing the bias of the apparent error rate. As an estimator of the true error rate, the apparent error rate bias is of order (n^{-1}), for n samples. The jackknife estimate reduces the bias to the second order.

Let t_n denote a sample statistic based on n observations $\boldsymbol{x}_1, \ldots, \boldsymbol{x}_n$. We assume for large m that the expectation for sample size m takes the form

$$\mathrm{E}[t_m] = \theta + \frac{a_1(\theta)}{m} + \frac{a_2(\theta)}{m^2} + \mathcal{O}(m^{-3}) \tag{9.4}$$

where θ is the asymptotic value of the expectation and a_1 and a_2 do not depend on m. Let $t_n^{(j)}$ denote the statistic based on observations excluding $\boldsymbol{x}_j$. Finally, write $t_n^{(.)}$ for the average of the $t_n^{(j)}$ over $j = 1, \ldots, n$,

$$t_n^{(.)} = \frac{1}{n}\sum_{j=1}^{n} t_n^{(j)}$$

Then,

$$\begin{aligned} E[t_n^{(.)}] &= \frac{1}{n}\sum_{j=1}^{n}\left(\theta + \frac{a_1(\theta)}{n-1} + \mathcal{O}(n^{-2})\right) \\ &= \theta + \frac{a_1(\theta)}{n-1} + \mathcal{O}(n^{-2}) \end{aligned} \tag{9.5}$$

From (9.4) and (9.5), we may find a linear combination that has bias of order n^{-2},

$$t_J = nt_n - (n-1)t_n^{(.)}$$

t_J is termed the jackknifed estimate corresponding to t_n.

Applying this to error rate estimation, the jackknife version of the apparent error rate, e_J^0, is given by

$$\begin{aligned} e_J^0 &= ne_A - (n-1)e_A^{(.)} \\ &= e_A + (n-1)\left(e_A - e_A^{(.)}\right) \end{aligned}$$

where e_A is the apparent error rate; $e_A^{(.)}$ is given by

$$e_A^{(.)} = \frac{1}{n}\sum_{j=1}^{n} e_A^{(j)}$$

where $e_A^{(j)}$ is the apparent error rate when object j has been removed from the observations,

$$e_A^{(j)} = \frac{1}{n-1} \sum_{k=1,k\neq j}^{n} Q(\omega(z_k), \eta(\boldsymbol{x}_k; \boldsymbol{Y}_j))$$

As an estimator of the expected error rate, the bias of e_J^0 is of order n^{-2}. However, as an estimator of the true error rate, the bias is still of order n^{-1} (McLachlan, 1992a). To reduce the bias of e_J^0 as an estimator of the true error rate to second order, we use

$$e_J = e_A + (n-1)(\tilde{e}_A - e_A^{(\cdot)}) \tag{9.6}$$

where $\tilde{e}_A$ is given by

$$\tilde{e}_A = \frac{1}{n^2} \sum_{j=1}^{n} \sum_{k=1}^{n} Q(\omega(z_k), \eta(\boldsymbol{x}_k; \boldsymbol{Y}_j))$$

The jackknife is closely related to the cross-validation method and both methods delete one observation successively to form bias-corrected estimates of the error rate. A difference is that in cross-validation, the contribution to the estimate is from the deleted sample only, classified using the classifier trained on the remaining set. In the jackknife, the error rate estimate is calculated from all samples, classified using the classifiers trained with each reduced sample set, and from the classifier trained with the full sample set.

9.2.2.4 Bootstrap techniques

The term 'bootstrap' refers to a class of procedures that sample the observed distribution, with replacement, to generate sets of observations that may be used to correct for bias. Introduced by Efron (1979), it has received considerable attention in the literature during the 1980s and 1990s. It provides nonparametric estimates of the bias and variance of an estimator and as a method of error rate estimation, the bootstrap has proved superior to many other techniques. Although computationally intensive, it is a very attractive technique, and there have been many developments of the basic approach, largely by Efron himself [see Efron and Tibshirani (1986) and Hinkley (1988) for a survey of bootstrap methods].

The bootstrap procedure for estimating the bias correction of the apparent error rate is implemented as follows. Let the data be denoted by $\boldsymbol{Y} = \{(\boldsymbol{x}_i^T, z_i^T)^T, i = 1, \ldots, n\}$. Let $\hat{F}$ be the *empirical distribution*. Under joint or mixture sampling it is the distribution with mass $1/n$ at each data point $\boldsymbol{x}_i$, $i = 1, \ldots, n$. Under separate sampling, $\hat{F}_i$ is the distribution with mass $1/n_i$ at point $\boldsymbol{x}_i$ in class ω_i (n_i patterns in class ω_i).

1. Generate a new set of data (the bootstrap sample) $\boldsymbol{Y}^b = \{(\tilde{\boldsymbol{x}}_i^T, \tilde{z}_i^T)^T, i = 1, \ldots, n\}$ according to the empirical distribution.
2. Design the classifier using $\boldsymbol{Y}^b$.
3. Calculate the apparent error rate for this sample and denote it by $\tilde{e}_A$.

4. Calculate the actual error rate for this classifier (regarding the set $\boldsymbol{Y}$ as the entire population) and denote it by $\tilde{e}_c$.

5. Compute $w_b = \tilde{e}_A - \tilde{e}_c$.

6. Repeat steps 1–5 B times.

7. The bootstrap bias of the apparent error rate is

$$W_{\text{boot}} = \text{E}[\tilde{e}_A - \tilde{e}_c]$$

where the expectation is with respect to the sampling mechanism that generates the sets $\boldsymbol{Y}^b$, that is

$$W_{\text{boot}} = \frac{1}{B}\sum_{b=1}^{B} w_b$$

8. The bias-corrected version of the apparent error rate is given by

$$e_A^{(B)} = e_A - W_{\text{boot}}$$

where e_A is the apparent error rate

At step 1, under mixture sampling, n independent samples are generated from the distribution $\hat{F}$, some of these may be repeated in forming the set $\tilde{\boldsymbol{Y}}$ and it may happen that one or more classes are not represented in the bootstrap sample. Under separate sampling, n_i are generated using $\hat{F}_i$, $i = 1, \ldots, C$. Thus, all classes are represented in the bootstrap sample in the same proportions as the original data.

The number of bootstrap samples, B, used to estimate W_{boot} may be limited by computational considerations, but for error rate estimation it can be taken to be of the order of 25–100 (Efron, 1983, 1990; Efron and Tibshirani, 1986).

There are many variants of the basic approach described above (Efron, 1983; McLachlan, 1992a). These include the double bootstrap, the randomised bootstrap and the 0.632 estimator (Efron, 1983).

The 0.632 estimator is a linear combination of the apparent error rate and another bootstrap error estimator, e_0,

$$e_{0.632} = 0.368 e_A + 0.632 e_0$$

where e_0 is an estimator that counts the number of training patterns misclassified that do not appear in the bootstrap sample. The number of misclassified samples is summed over all bootstrap samples and divided by the total number of patterns not in the bootstrap sample. If

A_b is the set of patterns in $\boldsymbol{Y}$, but not in bootstrap sample $\boldsymbol{Y}^b$, then

$$e_0 = \frac{\sum_{b=1}^{B} \sum_{\boldsymbol{x} \in A_b} Q(\omega(\boldsymbol{z}), \eta(\boldsymbol{x}, \boldsymbol{Y}^b))}{\sum_{b=1}^{B} |A_b|}$$

where $|A_b|$ is the cardinality of the set A_b. This estimator gave best performance in Efron's (1983) experiments.

The bootstrap may also be used for parametric distributions in which the samples are generated according to the parametric form adopted, with the parameters replaced by their estimates. The procedure is not limited to estimates of the bias in the apparent error rate and has been applied to other measures of statistical accuracy, though bootstrap calculations for confidence limits require more bootstrap replications, typically 1000–2000 (Efron, 1990). More efficient computational methods aimed at reducing the number of bootstrap replications compared with the straightforward Monte Carlo approach above have been proposed by Davison *et al.* (1986) and Efron (1990). Application to classifiers such as neural networks and classification trees produces difficulties, however, due to multiple local optima of the error surface.

9.2.2.5 Limitations of error rate

There are many other measures of discriminability (see Section 9.2.1) and limiting ourselves to a single measure, such as the error rate, may hide important information as to the behaviour of a rule. The error rate treats all misclassifications equally. For example, in a two-class problem, misclassifying an object from class 0 as class 1 has the same severity as misclassifying an object from class 1 as class 0. Costs of misclassification may be very important in some applications. They may not be known precisely, but rarely can they be considered to be equal.

In addition, in a two-class problem, an object from class 0 that is predicted to be class 1 with a probability of 0.55 contributes equally to error rate as an object from class 0 that is predicted to be class 1 with a probability of 1. In the former case, although the object is misclassified, there is some uncertainty over class membership. Error rate takes no account of this.

Two approaches for overcoming some of these shortcomings are presented in the following two subsections:

1. Performance measures that take account of posterior probabilities of class membership.
2. An ROC methodology. The use of classification error rate (or equivalently, classifier accuracy) as a performance measure is questioned by Provost *et al.* (1997), who argue that a methodology based on ROC analysis is superior.

9.2.3 Reliability

The reliability (termed *imprecision* by Hand, 1997) of a discriminant rule is a measure of how well the posterior probabilities of group membership are estimated by the rule. Thus, we are

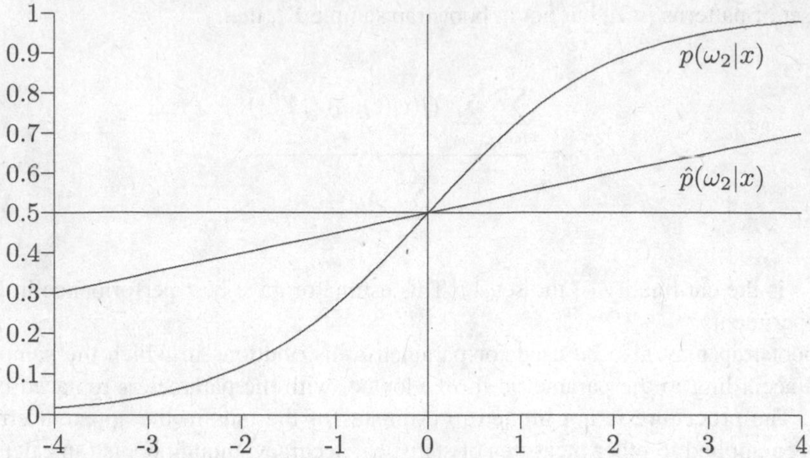

Figure 9.3 Good discriminability, poor reliability (following Hand, 1994).

not simply interested in the class ω_i for which $p(\omega_j|\boldsymbol{x})$ is the greatest, but the value of $p(\omega_j|\boldsymbol{x})$ itself. Of course, we may not easily be able to estimate the reliability for two reasons. The first is that we do not know the true posterior probabilities. Secondly, some discriminant rules do not produce estimates of the posterior probabilities explicitly.

Figure 9.3 illustrates a rule with good discriminability but poor reliability for a two-class problem with equal priors.

An object, $\boldsymbol{x}$, is assigned to class ω_2 if $p(\omega_2|\boldsymbol{x}) > p(\omega_1|\boldsymbol{x})$, or $p(\omega_2|\boldsymbol{x}) > 0.5$. The estimated posterior probabilities $\hat{p}(\omega_i|\boldsymbol{x})$ lead to good discriminability in the sense that the decision boundary is the same as a discriminant rule using the true posterior probabilities [i.e. $\hat{p}(\omega_2|\boldsymbol{x}) = 0.5$ at the same point as $p(\omega_2|\boldsymbol{x}) = 0.5$]. However, the true and estimated posterior probabilities differ.

Why should we want good reliability? Is not good discriminability sufficient? In some cases, a rule with good discriminability may be all that is required. We may be satisfied with a rule that achieves the Bayes' optimal error rate. On the other hand, if we wish to make a decision based on costs, or we are using the results of the classifier in a further stage of analysis, then good reliability is important. Hand (1997) proposes a measure of imprecision obtained by comparing an empirical samples statistic with an estimate of the same statistic computed using the classification function $\hat{p}(\omega_i|\boldsymbol{x})$,

$$R = \sum_{j=1}^{C} \frac{1}{n} \sum_{i=1}^{n} \{\phi_j(\boldsymbol{x}_i)[z_{ji} - \hat{p}(\omega_j|\boldsymbol{x}_i)]\}$$

where $z_{ji} = 1$ if $\boldsymbol{x}_i \in$ class ω_j, zero otherwise and ϕ_j is a function that determines the test statistic [for example, $\phi_j(\boldsymbol{x}_i) = (1 - \hat{p}(\omega_j|\boldsymbol{x}_i))^2$].

Obtaining interval estimates for the posterior probabilities of class membership is another means of assessing the reliability of a rule and is discussed by McLachlan (1992a), both for the multivariate normal class-conditional distribution and in the case of arbitrary class-conditional probability density functions using a bootstrap procedure.

9.2.4 ROC curves for performance assessment

9.2.4.1 Introduction

The ROC curve was introduced in Chapter 1, in the context of the Neyman–Pearson decision rule, as a means of characterising the performance of a two-class discrimination rule and provides a good means of visualising a classifier's performance in order to select a suitable decision threshold. The ROC curve is a plot of the true positive rate on the vertical axis against the false positive rate on the horizontal axis. In the terminology of signal detection theory, it is a plot of the probability of detection against the probability of false alarm, as the detection threshold is varied. Epidemiology has its own terminology: the ROC curve plots the *sensitivity* against $1 - S_e$, where S_e is the specificity.

In practice, the *optimal* ROC curve [the ROC curve obtained from the true class-conditional densities, $p(\boldsymbol{x}|\omega_i)$] is unknown, like error rate. It must be estimated using a trained classifier and an independent test set of patterns with known classes although, in common with error rate estimation, a training set reuse method such as cross-validation or bootstrap methods may be used. Different classifiers will produce different ROC curves characterising performance of the classifiers.

Often however, we may want a single number as a performance indicator of a classifier, rather than a curve, so that we can compare the performance of competing classifier schemes. We begin with an assessment of two-class rules.

In Chapter 1 it was shown that the minimum risk decision rule is defined on the basis of the likelihood ratio [see Equation (1.15)]. Assuming that there is no loss with correct classification, $\boldsymbol{x}$ is assigned to class ω_1 if

$$\frac{p(\boldsymbol{x}|\omega_1)}{p(\boldsymbol{x}|\omega_2)} > \frac{\lambda_{21} p(\omega_2)}{\lambda_{12} p(\omega_1)}, \tag{9.7}$$

where λ_{ji} = cost of assigning a pattern $\boldsymbol{x}$ to ω_i when $\boldsymbol{x} \in \omega_j$, or alternatively

$$p(\omega_1|\boldsymbol{x}) > \frac{\lambda_{21}}{\lambda_{12} + \lambda_{21}} \tag{9.8}$$

and thus corresponds to a single point on the ROC curve determined by the relative costs and prior probabilities. The loss is given by [Equation (1.11)]

$$L = \lambda_{21} p(\omega_2)\epsilon_2 + \lambda_{12} p(\omega_1)\epsilon_1 \tag{9.9}$$

where $p(\omega_i)$ are the class priors and ϵ_i is the probability of misclassifying a class ω_i object. The ROC curve plots $1 - \epsilon_1$ against ϵ_2.

In the ROC curve plane [that is, the $(1 - \epsilon_1, \epsilon_2)$ plane], lines of constant loss [termed *iso-performance lines* by Provost and Fawcett (2001)] are straight lines at gradients of $\lambda_{21} p(\omega_2)/\lambda_{12} p(\omega_1)$ (Figure 9.4), with loss increasing from top left to bottom right in the figure.

Legitimate values for the loss are those for which the loss contours intercept the ROC curve (that is, a possible threshold on the likelihood ratio exists). The solution with minimum loss is that for which the loss contour is tangential with the ROC curve – the point where the

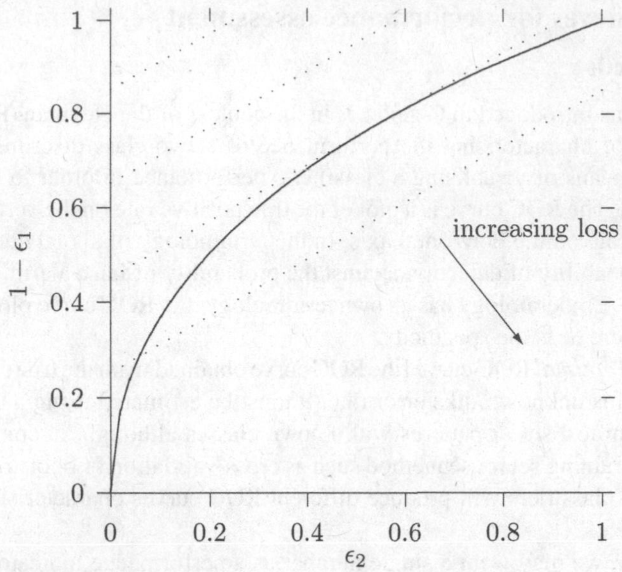

Figure 9.4 ROC curve with selected loss contours (straight lines) superimposed.

ROC curve has gradient $\lambda_{21}p(\omega_2)/\lambda_{12}p(\omega_1)$. There are no other loss contours that intercept the ROC curve with lower loss.

For different values of the relative costs and priors, the loss contours are at different gradients, in general, and the minimum loss occurs at a different point on the ROC curve.

9.2.4.2 Practical considerations

In many cases, even though the misclassification costs (λ_{12} and λ_{21}) are unknown, it is unreasonable to assume equality (leading to the Bayes' rule for minimum error). An alternative strategy is to compare the overall distribution of $\hat{p}(\boldsymbol{x}) \stackrel{\Delta}{=} p(\omega_1|\boldsymbol{x})$ for samples from each of the classes ω_1 and ω_2. We would expect that the values of $p(\omega_1|\boldsymbol{x})$ are greater for samples $\boldsymbol{x}$ from class ω_1 than for samples $\boldsymbol{x}$ from class ω_2. Generally, the larger the difference between the two distributions, the better the classifier. A measure of the separation of these two distributions is the *area* under the ROC curve (AUC). This provides a single numeric value, based on the ROC curve, that ignores the costs, λ_{ij}. Thus, in contrast to the error rate, which assumes equal misclassification costs, it assumes nothing whatever is known about misclassification costs and thus is not influenced by factors that relate to the application of the rule. Both of these assumptions are unrealistic in practice since, usually, something will be known about likely values of the relative cost $\lambda_{12}/\lambda_{21}$. Also, the advantage of the AUC as a measure of performance (namely, that it is independent of the threshold applied to the likelihood ratio) can be a disadvantage when comparing rules. If two ROC curves cross each other, then in general one will be superior for some values of the threshold and the other superior for other values of the threshold. The AUC fails to take this into account. However, an ROC analysis is more than comparing a single number. It can be used to determine regions of the ϵ_2 space where different classifiers dominate. Provost *et al.* (1997) define ranges of the gradient of the tangent line to the ROC convex hull (the convex hull of the ROC curves produced by different classification algorithms) for which a classifier dominates. The slope determines the ranges of costs and class priors for which a particular classifier minimises cost.

9.2.4.3 Interpretation

Let $\hat{p}(\boldsymbol{x}) = p(\omega_1|\boldsymbol{x})$ be the estimated probability that an object $\boldsymbol{x}$ belongs to class ω_1. Let $f(\hat{p}) = f(\hat{p}(\boldsymbol{x})|\omega_1)$ be the probability density function for $\hat{p}$ values for patterns in class ω_1, and $g(\hat{p}) = g(\hat{p}(\boldsymbol{x})|\omega_2)$ be the probability density function for $\hat{p}$ values for patterns in class ω_2. If $F(\hat{p})$ and $G(\hat{p})$ are the cumulative distribution functions, then the ROC curve is a plot of $1 - F(\hat{p})$ against $1 - G(\hat{p})$ (see the exercises at the end of the chapter).

The AUC is given by

$$\int (1 - F(u))dG(u) = 1 - \int F(u)g(u)du \tag{9.10}$$

or alternatively

$$\int G(u)dF(u) = \int G(u)f(u)du \tag{9.11}$$

For an arbitrary point, $\hat{p}(\boldsymbol{x}) = t \in [0, 1]$, the probability that a randomly chosen pattern $\boldsymbol{x}$ from class ω_2 will have a $\hat{p}(\boldsymbol{x})$ value smaller than t is $G(t)$. The density of the $\hat{p}(\boldsymbol{x})$ value from a randomly chosen pattern $\boldsymbol{x}$ from class ω_1 is $f(u)$. Therefore, the probability that a randomly chosen class ω_2 pattern has a smaller $\hat{p}(\boldsymbol{x})$ value than a randomly chosen class ω_1 pattern is $\int G(u)f(u)du$. This is the same as the definition (9.11) for the AUC.

A good classification rule (a rule for which the estimated values of $p(\omega_1|\boldsymbol{x})$ are very different for $\boldsymbol{x}$ from each of the two classes) lies in the upper left triangle. The closer that it gets to the upper corner the better.

A classification rule that is no better than chance produces an ROC curve that follows the diagonal from the bottom left to the top right.

9.2.4.4 Calculating the AUC

The AUC is easily calculated by applying the classification rule to a test set. For a classifier that produces estimates of $p(\omega_1|\boldsymbol{x})$ directly, we can obtain values $\{f_1, \ldots, f_{n_1}; f_i = p(\omega_1|\boldsymbol{x}_i), \boldsymbol{x}_i \in \omega_1\}$ and $\{g_1, \ldots, g_{n_2}; g_i = p(\omega_1|\boldsymbol{x}_i), \boldsymbol{x}_i \in \omega_2\}$ and use these to obtain a measure of how well separated are the distributions of $\hat{p}(\boldsymbol{x})$ for class ω_1 and class ω_2 patterns as shown below (Hand and Till, 2001).

Rank the estimates $\{f_1, \ldots, f_{n_1}, g_1, \ldots, g_{n_2}\}$ in increasing order and let the rank of the ith pattern from class ω_1 within the ranked estimates be r_i. Then there are $r_i - i$ class ω_2 patterns with estimated value of $\hat{p}(\boldsymbol{x})$ less than that of the ith reordered pattern of class ω_1. If we sum over class ω_1 test points, then we see that the number of pairs of points, one from class ω_1 and one from class ω_2, with $\hat{p}(\boldsymbol{x})$ smaller for class ω_2 than the $\hat{p}(\boldsymbol{x})$ value for class ω_1 is

$$\sum_{i=1}^{n_1}(r_i - i) = \sum_{i=1}^{n_1} r_i - \sum_{i=1}^{n_1} i = S_0 - \frac{1}{2}n_1(n_1 + 1)$$

where S_0 is the sum of the ranks of the class ω_1 test patterns. Since there are n_1n_2 pairs, the estimate of the probability that a randomly chosen class ω_2 pattern has a lower estimated probability of belonging to class ω_1 than a randomly chosen class ω_1 pattern is

$$\hat{A} = \frac{1}{n_1 n_2}\left\{S_0 - \frac{1}{2}n_1(n_1+1)\right\}$$

This is equivalent to the AUC and provides an estimate that has been obtained using the rankings alone and has not used threshold values to calculate it.

The standard deviation of the statistic $\hat{A}$ is (Hand and Till, 2001)

$$\sqrt{\frac{\hat{\theta}(1-\hat{\theta}) + (n_1-1)(Q_0-\hat{\theta}^2) + (n_2-1)(Q_1-\hat{\theta}^2)}{n_1 n_2}}$$

where

$$\hat{\theta} = \frac{S_0}{n_1 n_2}$$

$$Q_0 = \frac{1}{6}(2n_1 + 2n_2 + 1)(n_1 + n_2) - Q_1$$

$$Q_1 = \sum_{j=1}^{n_1} (r_j - 1)^2$$

An alternative approach, considered by Bradley (1997), is to construct an estimate of the ROC curve directly for specific classifiers by varying a threshold and then to use an integration rule (for example, the trapezium rule) to obtain an estimate of the area beneath the curve.

9.2.4.5 Cross-validation estimates of the ROC curve

The ROC is a plot of the true positive rate against the false positive rate. Denoting the ratio $\lambda_{21}/(\lambda_{12} + \lambda_{21})$ in Equation (9.8) by t, then the true positive rate is given by

$$tpr(t) = p(p(\omega_1|\boldsymbol{x}) \geq t|\omega_1)$$

the probability that the estimated probability of class ω_1 given the observation $\boldsymbol{x}$ is greater than the threshold t, for x belonging to class ω_1. A sample-based estimate of this, using a test set of n_1 samples of class ω_1 and n_2 samples of class ω_2 is

$$tpr(t) = \frac{1}{n_1} \times \text{number of class } \omega_1 \text{ test samples correctly classified with threshold } t$$

The false positive rate is given by

$$fpr(t) = p(p(\omega_1|\boldsymbol{x}) \geq t|\omega_2)$$

the probability that the estimated probability of class ω_1 given the observation $\boldsymbol{x}$ is greater than the threshold t, for x belonging to class ω_2. A sample-based estimate of this is

$$fpr(t) = \frac{1}{n_2} \times \text{number of class } \omega_2 \text{ test samples incorrectly classified with threshold } t$$

The ROC curve is given by

$$ROC(t) = \{(fpr(t), tpr(t)), t \in [0, 1]\} \tag{9.12}$$

In k-fold cross-validation, the data are divided into k partitions. For each of k runs, the classifier is trained on training data comprising training data in $k-1$ partitions and tested on the remaining partition. The true positive rate is

$$tpr^{cv}(t) = \frac{1}{n_1} \sum_{\boldsymbol{x}_i \in \omega_1} I\left[p_{-k(i)}(\omega_1|\boldsymbol{x}_i) \geq t\right], \quad t \in [0, 1]$$

where $k(i)$ is an indicator function defining the partition to which $\boldsymbol{x}_i$ belongs; $p_{-k(i)}$ is the classifier trained on the set of $k-1$ partitions that excludes the partition $k(i)$; $I[\theta] = 1$ if θ is true and $I[\theta] = 0$ otherwise. The false positive rate is

$$fpr^{cv}(t) = \frac{1}{n_2} \sum_{\boldsymbol{x}_i \in \omega_2} I\left[p_{-k(i)}(\omega_1|\boldsymbol{x}_i) \geq t\right] \quad t \in [0, 1]$$

and the ROC curve given by

$$ROC^{cv}(t) = \{(fpr^{cv}(t), tpr^{cv}(t)), t \in [0, 1]\} \tag{9.13}$$

Adler and Lausen (2009) also provide bootstrap definitions of the ROC curve.

9.2.4.6 Multiclass extensions

There are several extensions of the standard two-class ROC analysis to multiclass problems, including the work of Landgrebe and Duin (2008) who review some of the literature in this area (Ferri *et al.*, 2003; Fieldsend and Everson, 2005; Everson and Fieldsend, 2006; Landgrebe and Duin, 2007).

Further work on the AUC measure includes that of Hand and Till (2001) who develop it to the multiple class classification problem (see also Hajian-Tilaki *et al.*, 1997a,b) and Adams and Hand (1999) who take account of some imprecisely known information on the relative misclassification costs (see also Section 9.3.3).

9.2.5 Population and sensor drift

The basic assumption in classifier design is that the distribution from which the design sample is selected is the same as the distribution from which future objects will arise, i.e. that the training set is representative of the operating conditions. In many applications, this assumption is not valid.

In classifier design, we often have a design or training set that is used to train a classifier; a validation set (used as part of the training process) for model selection or termination of an iterative learning rule; and an independent test set, which is used to measure the generalisation performance of the classifier: the ability of the classifier to generalise to future objects. These datasets are often gathered as part of the same trial and it is common that they are, in fact, different partitions of the same dataset. In many practical situations, the operating conditions may differ from those prevailing at the time of the test data collection, particularly in a sensor data analysis problem. For example, sensor characteristics may drift with time or environmental conditions may change. These effects result in changes to the distributions from which patterns are drawn. This is referred to as *population drift* (Hand, 1997). The circumstances and degree of population drift vary from problem to problem. This presents a difficulty for classifier performance evaluation since the test set may not be representative of the operating conditions and thus the generalisation performance quoted on the test set may be overly optimistic. Designing classifiers to accommodate population drift is problem specific.

9.2.5.1 Population drift

In this section, we review some of the causes of population drift and the approaches that can be taken to mitigate against distributional changes.

Sensor drift
Pattern recognition techniques need to be developed for drifting sensors. An example is an electronic nose, a device that contains a number of different individual sensors whose response characteristics depend on the chemical odour present. These have applications in quality control, bioprocess monitoring and defence. All chemical sensors are affected by drift, stability problems and memory effects. Data processing techniques are required to handle these effects autonomously. This may be simple preprocessing to remove shifts in zero points of responses, and changes in sensitivity handled by a gain control, but there may be more complex effects.

Changes in object characteristics
In medical applications there may be drift in the patient population (changes in patient characteristics) over time. Population drift also occurs in speech recognition when a new speaker is presented. There are various approaches including analysing a standard input from a new speaker and using this to modify stored prototypes. In credit scoring, the behaviour of borrowers is influenced by short-term pressures (for example, Budget announcements by the Chancellor of the Exchequer) and classification rules will need to be changed quite frequently (Kelly *et al.*, 1999). In radar target recognition, classifiers need to be robust to changes in vehicle equipment fit which can give rise to large changes in the radar reflectivity (Copsey and Webb, 2000). In condition monitoring, the healthy state of an engine will change with time. In object recognition in images, it is important that the classifier has some invariance to object pose (translational/rotational invariance). In each of the examples above, it is an advantage if the classification method can be dynamically updated and does not need to be re-computed from scratch (using new sets of training data) as the conditions change.

Environmental changes
The training conditions may only approximate the expected operating conditions and a trained classifier will need some modification (Kraaijveld, 1996). The signal-to-noise ratio of the

operating conditions may be different from the (controlled) training conditions and may possibly be unknown. In order to derive a classifier for noisier operating conditions, several approaches may be adopted including noise injection in the training set and modifying the training procedure to minimise a cost function appropriate for the expected operating conditions (Webb and Garner, 1999). Errors-in-variables models are relevant here. In target recognition, the ambient light conditions may change. The environmental conditions, for example the clutter in which the target is embedded, will differ from the training conditions. Sea clutter is time-varying.

Sensor change

For various reasons, it may not be possible to gather sufficient information with the operating sensor to train a classifier: it might be too expensive or too dangerous. However, measurements can be made in more controlled conditions using a different sensor and a classifier can be designed using this set of measurements. In this type of problem, a classifier needs to be designed using sensor-independent features or a means of translating operating sensor data to the training domain must be developed.

Variable priors and costs

Prior probabilities of class membership are likely to change with time. Thus, although class conditional densities do not change, decision boundaries are altered due to varying priors. This requires high-level modelling, but often there is little data available to model the dependencies and Bayesian networks are constructed using expert opinion (Copsey and Webb, 2002). Costs of misclassification are also variable and unknown. The consequence is that the optimisation criterion used in training (for example, minimum cost) may be inappropriate for the operating conditions.

9.2.5.2 Conclusions

Uncertainties between training and operating conditions mean that there is a limit beyond which it is not worth pushing the development of a classification rule (Hand, 1997). In some cases, population drift is amenable to treatment, but this is problem dependent.

9.2.6 Example application study

The problem

This study (Bradley, 1997) comprises an assessment of AUC as a performance measure on six pattern recognition algorithms applied to datasets characterising medical diagnostic problems.

Summary

The study estimates AUC through an integration of the ROC curve and its standard deviation is calculated using cross-validation.

The data

There are six datasets comprising measurements on two classes:

1. Cervical cancer: 6 features, 117 patterns; classes are normal and abnormal cervical cell nuclei.

2. Post-operative bleeding: 4 features, 113 patterns (after removal of incomplete patterns); classes are normal blood loss and excessive bleeding.
3. Breast cancer: 9 features, 683 patterns; classes are benign and malignant.
4. Diabetes: 8 features, 768 patterns; classes are negative and positive test for diabetes.
5. Heart disease 1: 14 features, 297 patterns; classes are heart disease present and heart disease absent.
6. Heart disease 2: 11 features, 261 patterns; classes are heart disease present and heart disease absent.

Incomplete patterns (patterns for which measurements on some features are missing) were removed from the datasets.

The models
Six classifiers were trained on each dataset:

1. quadratic discriminant function (Chapter 2);
2. *k*-nearest-neighbour (Chapter 4);
3. classification tree (Chapter 7);
4. multiscale classifier method (a development of classification trees);
5. perceptron (Chapter 5);
6. multilayer perceptron (Chapter 6).

The models were trained and classification performance monitored as a threshold was varied in order to estimate the ROC curves. For example, for the *k*-nearest-neighbour classifier, the five nearest neighbours in the training set to a test sample are calculated. If the number of neighbours belonging to class ω_1 is greater than L, where $L = [0, 1, 2, 3, 4, 5]$, then the test sample is assigned to class ω_1, otherwise it is assigned to the second class. This gives six points on the ROC curve.

For the multilayer perceptron, a network with a single output is trained and during testing, it is thresholded at values of $[0, 0.1, 0.2, \ldots, 1.0]$ to simulate different misclassification costs.

Training procedure
A 10-fold cross-validation scheme was used, with 90% of the samples used for training and 10% used in the test set, selected randomly. Thus, for each classifier on each dataset, there are 10 sets of results.

The ROC curve was calculated as a decision threshold was varied for each of the test set partitions and the AUC calculated using trapezoidal integration. The AUC for the rule is taken to be the average of the 10 AUC values obtained from the 10 partitions of the dataset.

9.2.7 Further developments

The aspect of classifier performance that has received most attention in the literature is the subject of error rate estimation. Hand (1997) develops a much broader framework for the assessment of classification rules, defining four concepts.

Inaccuracy
This is a measure of how (in)effective is a classification rule in assigning an object to the correct class. One example is error rate; another is the *Brier* or *quadratic score*, often used as an optimisation criterion for neural networks, defined as

$$\frac{1}{n}\sum_{i=1}^{n}\sum_{j=1}^{C}\left\{z_{ji}-\hat{p}(\omega_j|\boldsymbol{x}_i)\right\}^2$$

where $\hat{p}(\omega_j|\boldsymbol{x}_i)$ is the estimated probability that pattern $\boldsymbol{x}_i$ belongs to class ω_j and $z_{ji} = 1$ if $\boldsymbol{x}_i$ is a member of class ω_j and zero otherwise.

Imprecision
Equivalent to reliability defined in Section 9.2.3, this is a measure of the difference between the estimated probabilities of class membership $\hat{p}(\omega_j|\boldsymbol{x})$ and the (unknown) true probabilities, $p(\omega_j|\boldsymbol{x})$.

Inseparability
This is a measure evaluated using the true probabilities of belonging to a class, and so it does not depend on a classifier. It measures the similarity of the true probabilities of class membership at a point $\boldsymbol{x}$, averaged over $\boldsymbol{x}$. If the probabilities at a point $\boldsymbol{x}$ are similar, then the classes are not separable.

Resemblance
It measures the variation between the true probabilities, conditioned on the estimated ones. Does the predicted classification separate the true classes well? A low value of resemblance is to be hoped for.

In this section we have been unable to do full justice to the elegance of the bootstrap method, and we have simply presented the basic bootstrap approach for the bias correction of the apparent error rate, with some extensions. Further developments may be found in articles by Efron (1983), Efron and Tibshirani (1986), Efron (1990) and McLachlan (1992a).

Provost and Fawcett (2001) propose an approach that uses the convex hull of ROC curves of different classifiers. Classifiers with ROC curves below the convex hull are never optimal (under any conditions on costs or priors) and can be ignored. Classifiers on the convex hull can be combined to produce a better classifier. This idea has been widely used in *data fusion* (see Chapter 8) and is discussed further in Section 9.3.3.

Within the context of classifier ensembles, Barreno *et al.* (2008) discuss an ROC analysis and Kuncheva (2004b) address the issue of changing environments.

9.2.8 Summary

In this section, we have given a rather brief treatment of classification rule performance assessment, covering three measures, discriminability, reliability (or imprecision) and the use of the ROC curve. In particular, we have given emphasis to the error rate of a classifier and schemes for reducing the bias of the apparent error rate, namely cross-validation, the jackknife and the bootstrap methods. These have the advantage over the holdout method in that they do not require a separate test set. Therefore, all the data may be used in classifier design.

The error rate estimators described in this chapter are all nonparametric estimators in that they do not assume a specific form for the class-conditional probability density functions. Parametric forms of error rate estimators, for example based on a normal distribution model for the class-conditional densities, can also be derived. However, although parametric rules may be fairly robust to departures from the true model, parametric estimates of error rates may not be (Konishi and Honda, 1990). Hence our concentration in this chapter on nonparametric forms. For a further discussion of parametric error rate estimators we refer the reader to the book by McLachlan (1992a).

The reliability, or imprecision, of a rule tells us how well we can trust the rule – how close the estimated posterior densities are to the true posterior densities. The area under the ROC curve is a measure that summarises classifier performance over a range of relative costs.

Finally, population drift is a factor that should be considered in all practical classifier design, since it is frequently true that the operating conditions differ from those used to gather data for classifier design.

9.3 Comparing classifier performance

9.3.1 Which technique is best?

Are neural network methods better than 'traditional' techniques? Is the classifier that you develop better than those previously published in the literature? There have been many comparative studies of classifiers and we have referenced these in previous chapters. Perhaps the most comprehensive study is the *Statlog* project (Michie *et al.*, 1994) which provides a study of more than 20 different classification procedures applied to about 20 datasets. Yet comparisons are not easy. Classifier performance varies with the dataset, sample size, dimensionality of the data, and skill of the analyst. There are some important issues to be resolved, as outlined by Duin (1996):

1. An application domain must be defined. Although this is usually achieved by specifying a collection of datasets, these datasets may not be representative of the problem domain that you wish to consider. Although a particular classifier may perform consistently badly on these datasets, it may be particularly suited to the one you have.
2. The skill of the analyst needs to be considered (and removed if possible). Whereas some techniques are fairly well defined (nearest neighbour with a given metric), others require tuning. Can the results of classifications, using different techniques, on a given dataset performed by separate analysts be sensibly compared? If one technique performs better than others, is it due to the superiority of the technique on that dataset or the skill of the implementer in obtaining the best out of a favourite method? In fact, some classifiers are valuable because they have many free parameters and allow a trained analyst to incorporate knowledge into the training procedure. Others are valuable because they are largely automatic and do not require user input. The *Statlog* project was an attempt at developing automatic classification schemes, encouraging minimal tuning.

Related to the second issue above is that often the main contribution to the final performance is the initial problem formulation (abstracting the problem to be solved from the customer, selecting variables and so on), again determined by the skill of the analyst. The classifier may only produce second-order improvements to performance.

In addition, what is the basis on which we make a comparison – error rate, reliability, speed of implementation, speed of testing, etc.?

There is no such thing as a best classifier, but there are several ways in which comparisons may be performed (Duin, 1996):

1. A comparison of experts. A collection of problems is sent to experts who may use whichever technique they feel is appropriate.

2. A comparison of toolsets by nonexperts. Here a collection of toolsets is provided to nonexperts for evaluation on several datasets.

3. A comparison of automatic classifiers (classifiers that require no tuning). This is performed by a single researcher on a benchmark set of problems. Although the results will be largely independent of the expert, they will probably be inferior to those obtained if the expert were allowed to choose the classifier.

9.3.2 Statistical tests

Bounds on the error rate are insufficient when comparing classifiers. Usually, the test sets are not independent – they are common across all classifiers. There are several tests for determining whether one classification rule outperforms another on a particular dataset.

The question of measuring the accuracy of a classification rule using an independent training and test set was discussed in Section 9.2. This can be achieved by constructing a confidence interval or HPD region. Here we address the question: Given two classifiers and sufficient data for a separate test set, which classifier will be more accurate on new test set examples?

Dietterich (1998) assesses five statistical tests, comparing them experimentally to determine the probability of incorrectly detecting a difference between classifier performance when no difference exists (type I error).

Suppose that we have two classifiers, A and B. Let

$$
\begin{aligned}
n_{00} &= \text{number of samples misclassified by both } A \text{ and } B \\
n_{01} &= \text{number of samples misclassified by } A \text{ but not by } B \\
n_{10} &= \text{number of samples misclassified by } B \text{ but not by } A \\
n_{11} &= \text{number of samples misclassified by neither } A \text{ nor } B
\end{aligned}
$$

Compute the z statistic

$$
z = \frac{|n_{01} - n_{10}| - 1}{\sqrt{n_{10} + n_{01}}}
$$

If classifiers A and B have the same performance, The quantity z^2 is χ^2 distributed approximately with one degree of freedom. The null hypothesis (that the classifiers have the same error) can be rejected (with probability of incorrect rejection of 0.05) if $|z| > 1.96$. This is known as McNemar's test or the Gillick test.

9.3.3 Comparing rules when misclassification costs are uncertain

9.3.3.1 Introduction

Error rate, or misclassification rate, discussed in Section 9.2 is often used as a comparison criterion for comparing several classifiers. It requires no choice of costs, making the assumption that misclassification costs are all equal.

An alternative measure of performance is the AUC (also see Section 9.2). This is a measure of the separability of the two distributions $f(\hat{p})$, the probability distribution of $\hat{p} = p(\omega_1|\boldsymbol{x})$ for patterns $\boldsymbol{x}$ in class ω_1 and $g(\hat{p})$, the probability distribution of $\hat{p}$ for patterns $\boldsymbol{x}$ in class ω_2. It has the advantage that it does not depend on the relative costs of misclassification.

There are difficulties with the assumptions behind both of these performance measures. In many, if not most practical applications, the assumptions of equal costs is unrealistic. Also, the minimum loss solution, which requires the specification of costs is not sensible since rarely are costs and priors known precisely. In many real-world environments, misclassification costs and class priors are likely to change over time as the environment may change between design and test. Consequently, the point on the ROC curve corresponding to minimum loss solution [where the threshold on the likelihood ratio is $\lambda_{21}p(\omega_2)/\lambda_{12}p(\omega_1)$ – Equation (9.7)] changes. On the other hand, usually *something* is known about the relative costs and it is therefore inappropriate to summarise over all possible values.

9.3.3.2 ROC curves

Comparing classifiers on the basis of AUC is difficult when the ROC curves cross. Only in the case of one classifier dominating another will the AUC be a valid criterion for comparing different classifiers. If two ROC curves cross, then one curve will be superior for some values of the cost ratio and the other classifier will be superior for different values of the cost ratio. Two approaches for handling this situation are presented here.

LC index

In this approach for comparing two classifiers A and B, the costs of misclassification λ_{12} and λ_{21}, are rescaled so that $\lambda_{12} + \lambda_{21} = 1$ and the loss [Equation (9.9)] calculated as a function of λ_{21} for each classifier. A function, $L(\lambda_{21})$, is defined to take the value $+1$ in regions of the [0, 1] interval for which classifier A is superior (it has a lower value of the loss than classifier B) and -1 in regions for which classifier B is superior. The confidence in any value of λ_{21} is probability density function $D(\lambda_{21})$, defined later, and the LC index is defined as

$$\int_0^1 D(\lambda)L(\lambda)d\lambda$$

which ranges over ± 1, taking positive values when classifier A is more likely to lead to a smaller value of loss than classifier B, and negative values when classifier B is more likely to lead to a smaller value of the loss than classifier A. A value of $+1$ means that A is certain to be a superior classifier since it is superior for all feasible values of λ_{21}.

How do we decide on the set of feasible values of λ_{21}? That is, what form do we choose for the distribution $D(\lambda_{21})$? One proposal is to specify an interval $[a, b]$ for the cost ratio, $\lambda_{12}/\lambda_{21}$, and a most likely value, m and use this to define a unit area triangle with base

$[a, b]$ and apex at m. This is because, it is argued (Adams and Hand, 1999), that experts find it convenient to specify a cost ratio $\lambda_{12}/\lambda_{21}$ and an interval for the ratio.

The ROC convex hull method

In this method a hybrid classification system is constructed from the set of available classifiers. For any value of the cost ratio, the combined classifier will perform at least as well as the best classifier. The combined classifier is constructed to have an ROC curve that is the convex hull of the component classifiers.

Figure 9.5 illustrates the ROC convex hull method. For some values of costs and priors, the slope of the iso-performance lines is such that the optimal classifier (the point on the ROC curve lying to the top left) is classifier B. Line β is the iso-performance line with lowest loss that intercepts the ROC curve of classifier B. For much shallower gradients of the iso-performance lines (corresponding to different values of the priors or costs), the optimal classifier is classifier A. Here, line α is the lowest value iso-performance line (for a given value of priors and costs) that intercepts the ROC curve of classifier A. Classifier C is not optimal for any value of priors or costs. The points on the convex hull of the ROC curves define optimal classifiers for particular values of priors and costs. Provost and Fawcett (2001) present an algorithm for generating the ROC convex hull (see Section 8.3.3).

In practice, we need to store the range of threshold values $[\lambda_{21}p(\omega_2)/\lambda_{12}p(\omega_1)]$ for which a particular classifier is optimal. Thus, the range of the threshold is partitioned into regions, each of which is assigned a classifier, the one that is optimal for that range of thresholds.

Flach and Wu (2005) introduce the term *model repair* to denote approaches that modify given models to obtain better models. They show that it is possible to do better than the

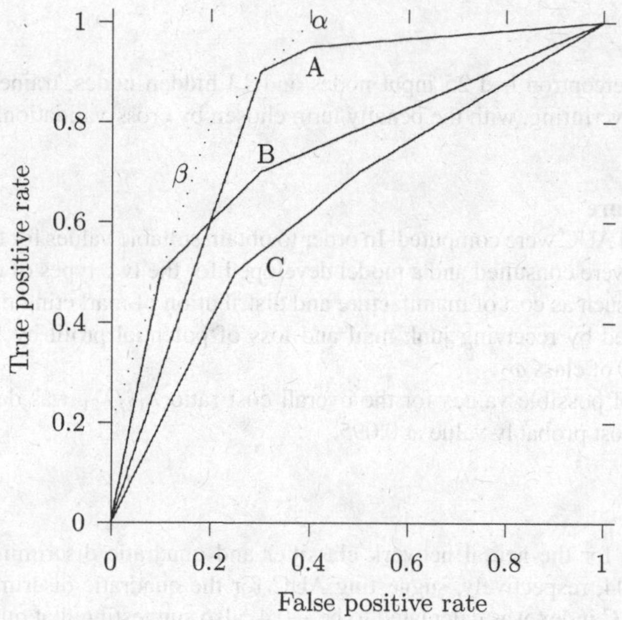

Figure 9.5 ROC convex hull method illustration.

convex hull using a method to repair concavities in an ROC curve using a process of inversion whereby parts of the ROC curve are inverted. Note however, that if we are using ROC curves (either a single curve and repairing concavities or multiple curves and combining them) to derive a better model, then the data used to generate the ROC curves are being used as part of the model training process. There is no guarantee that the ROC curve generated on an independent test set will have the same behaviour.

9.3.4 Example application study

The problem

This study (Adams and Hand, 1999) develops an approach for comparing the performance of two classifiers when misclassification costs are uncertain, but not completely unknown. It is concerned with classifying customers according to their likely response to a promotional scheme.

Summary

The LC index above is evaluated to compare a neural network classifier (Chapter 6) with quadratic discriminant analysis (Chapter 2).

The data

The data comprise 8000 records (patterns) of measurements on 25 variables, mainly describing earlier credit card transaction behaviour. The classes are denoted class ω_1 and class ω_2, with class ω_2 thought likely to return a profit. The priors are set as $p(\omega_1) = 0.87$, $p(\omega_2) = 0.13$.

The model

The multilayer perceptron had 25 input nodes and 13 hidden nodes, trained using 'weight decay' to avoid overfitting, with the penalty term chosen by cross-validation.

Training procedure

The LC index and AUC were computed. In order to obtain suitable values for the ratio of costs, banking experts were consulted and a model developed for the two types of misclassification based on factors such as cost of manufacture and distribution of marketing material, cost due to irritation caused by receiving junk mail and loss of potential profit by failing to mail a potential member of class ω_2.

An interval of possible values for the overall cost ratio $\lambda_{12}/\lambda_{21}$ was derived as [0.065, 0.15], with the most probably value at 0.095.

Results

The AUC values for the neural network classifier and quadratic discriminant analysis are 0.7102 and 0.7244, respectively, suggesting AUC for the quadratic discriminant is slightly preferable. The LC index was calculated to be -0.4, also suggesting that quadratic discriminant analysis is to be preferred.

9.3.5 Further developments

Adams and Hand (2000) present some guidelines for better methodology for comparing classifiers. They identify five common deficiencies in the practice of classifier performance assessment:

1. Assuming equal costs. In many practical applications, the two types of misclassification are not equal.
2. Integrating over costs. The AUC summarises performance over the entire range of costs. It is more likely that something will be known about costs and that a narrower range would be more appropriate.
3. Crossing ROC curves. The AUC measure is only appropriate if one ROC curve dominates over the entire range. If the ROC curves cross, then different classifiers will dominate for different ranges of the misclassification costs.
4. Fixing costs. It is improbable that exact costs can be given in many applications.
5. Variability. Error rate and the AUC measure are sample-based estimates. Standard errors should be given when reporting results.

9.3.6 Summary

There are several ways in which performance may be compared. It is important to use an assessment criterion appropriate to the real problem under investigation. Misclassification costs should be taken into account since they can influence the choice of method. Assuming equal misclassification costs is very rarely appropriate. Usually something can be said about costs, even if they are not known precisely.

9.4 Application studies

That are few application studies that specifically examine different performance metrics. It is more common for application studies to compare different classifiers using a single metric.

Ferri *et al.* (2009) analyse experimentally the behaviour of 18 different performance metrics in several different scenarios. The metrics were classified into three broad families:

- metrics based on a threshold and a qualitative understanding of error (e.g. accuracy);
- metrics based on a probabilistic understanding of error (e.g. Brier score);
- metrics based on ranks (e.g. AUC).

They undertake a comprehensive assessment, identifying important similarities between measures as well as significant differences.

In a case study in breast cancer computer-aided diagnosis, Patel and Markey (2005) compare three performance metrics developed for the extension of the standard two-class ROC analysis to the multiclass situation.

9.5 Summary and discussion

The most common measure of classifier performance assessment is misclassification rate or error rate. We have reviewed the different types of error rate and described procedures for error rate estimation. There are many other performance measures, including reliability – how good is our classifier at estimating the true posterior probabilities – and the area under the receiver operating characteristic (ROC) curve, AUC. Misclassification rate makes the rather strong assumption that misclassification costs are equal. In most practical applications, this is unrealistic. The AUC is a measure averaged over all relative costs and it might be argued that this is equally inappropriate since usually something will be known about relative costs. The LC index was introduced as one attempt to make use of domain knowledge in performance assessment.

One of the greatest sources of uncertainty in classifier performance can be due to the difference between the training conditions and the operating conditions. Some of the factors influencing these differences have been discussed under the concept of population drift. Uncertainty in the operating conditions can lead to greater differences in performance than the difference arising from the choice of classifier. Therefore, spending time understanding population drift may be more beneficial than designing sophisticated classifiers. See Hand (2006) for further discussion.

9.6 Recommendations

1. Use error rate with care. Are the assumptions of equal misclassification costs appropriate for your problem?
2. ROC analysis has the benefit of separating classifier performance from class skew and should be considered as an alternative to error rate.
3. Spend time understanding population drift and how it might apply in your problem, before assessing classifier performance.

9.7 Notes and references

The subject of error rate estimation has received considerable attention. The literature up to 1973 is surveyed in the extensive bibliography of Toussaint (1974) and more recent advances by Hand (1986b) and McLachlan (1987). The holdout method was considered by Highleyman (1962). The leave-one-out method for error estimation is usually attributed to Lachenbruch and Mickey (1968) and cross-validation in a wider context is discussed by Stone (1974).

The number of samples required to achieve good error rate estimates is discussed with application to a character recognition task by Guyon *et al.* (1998).

Quenouille (1949) proposed the method of sample splitting for overcoming bias, later termed the jackknife. The bootstrap procedure as a method of error rate estimation has been widely applied following the pioneering work of Efron (1979, 1982, 1983). Reviews of bootstrap methods are provided by Efron and Tibshirani (1986) and Hinkley (1988). There are several studies comparing the performance of the different bootstrap estimators (Efron, 1983;

Chernick *et al.*, 1985; Fukunaga and Hayes, 1989b; Konishi and Honda, 1990). Davison and Hall (1992) compare the bias and variability of the bootstrap with cross-validation. They find that cross-validation gives estimators with higher variance but lower bias than the bootstrap. The main differences between the estimators are when there is large class overlap, when the bias of the bootstrap is an order of magnitude greater than that of cross-validation.

The 0.632 bootstrap for error rate estimation is investigated by Fitzmaurice *et al.* (1991) and the number of samples required for the double bootstrap by Booth and Hall (1994). The bootstrap has been used to compute other measures of statistical accuracy. The monograph by Hall (1992) provides a theoretical treatment of the bootstrap with some emphasis on curve estimation (including parametric and nonparametric regression and density estimation).

Reliability of posterior probabilities of group membership is discussed in the book by McLachlan (1992a). Hand (1997) also considers other measures of performance assessment.

The use of ROC curves in pattern recognition for performance assessment and comparison is described by Bradley (1997), Hand and Till (2001), Adams and Hand (1999) and Provost and Fawcett (2001). For introduction and reviews of ROC curve analysis see Fawcett (2006), Park *et al.* (2004) and Dubrawski (2004).

Exercises

1. Two hundred labelled samples are used to train two classifiers. In the first classifier, the dataset is divided into training and test sets of 100 samples each and the classifier designed using the training set. The performance on the test set is 80% correct. In the second classifier, the dataset is divided into a training set of 190 samples and a test set of 10 samples. The performance on the test set is 90%.

   Is the second classifier 'better' than the first? Justify your answer.

2. Verify the Sherman–Morisson formula (9.3). Describe how it may be used to estimate the error rate of a Gaussian classifier using cross-validation.

3. Show that Equation (9.8) is an alternative form of Equation (9.7).

4. The ROC curve is a plot of $(1 - \epsilon_1)$, the 'true positive', against ϵ_2, the 'false positive' as the threshold on [see Equation (9.8)]

$$p(\omega_1|\boldsymbol{x})$$

   is varied, where

$$\epsilon_1 = \int_{\Omega_2} p(\boldsymbol{x}|\omega_1)d\boldsymbol{x}$$

$$\epsilon_2 = \int_{\Omega_1} p(\boldsymbol{x}|\omega_2)d\boldsymbol{x}$$

   and Ω_1 is the domain where $p(\omega_1|\boldsymbol{x})$ lies above the threshold.

Show, by conditioning on $p(\omega_1|\boldsymbol{x})$, that the true positive and false positive (for a threshold μ) may be written, respectively, as

$$1 - \epsilon_1 = \int_{\mu}^{1} dc \int_{p(\omega_1|\boldsymbol{x})=c} p(\boldsymbol{x}|\omega_1)d\boldsymbol{x}$$

and

$$\epsilon_2 = \int_{\mu}^{1} dc \int_{p(\omega_1|\boldsymbol{x})=c} p(\boldsymbol{x}|\omega_2)d\boldsymbol{x}$$

The term $\int_{p(\omega_1|\boldsymbol{x})=c} p(\boldsymbol{x}|\omega_1)d\boldsymbol{x}$ is the density of $p(\omega_1|\boldsymbol{x})$ values at c for class ω_1. Hence show that the ROC curve is defined as one minus the cumulative density of $\hat{p} = p(\omega_1|\boldsymbol{x})$ for class ω_1 patterns plotted against one minus the cumulative density for class ω_2 patterns.

5. Generate training data consisting of 25 samples from each of two bivariate normal distributions [means $(-d/2, 0)$ and $(d/2, 0)$ and identity covariance matrix]. Compute the apparent error rate and a bias-corrected version using the bootstrap. Plot both error rates, together with an error rate computed on a test set (of appropriate size) as a function of separation, d. Describe the results.

10

Feature selection and extraction

Reducing the number of useful variables either through feature selection (selecting a subset of the original variables for classifier design) or feature extraction (determining a linear or nonlinear transformation of the original variables to obtain a smaller set) can lead to improved classifier performance and a greater understanding of the data. Approaches to feature selection and feature extraction are introduced. Feature selection methods tied to the classifier type offer improved performance over filter (classifier-independent) methods. The most widely used technique of feature extraction is principal components analysis.

10.1 Introduction

This chapter is concerned with representing data in a reduced number of dimensions. Reasons for doing this may be easier subsequent analysis, improved classification performance through a more stable representation, removal of redundant or irrelevant information or an attempt to discover underlying structure by obtaining a graphical representation of a dataset. Techniques for representing data in a reduced dimension are termed *ordination methods* or *geometrical methods* in the multivariate analysis literature. They include such methods as principal components analysis and multidimensional scaling. In the pattern recognition literature they are termed *feature selection* and *feature extraction* methods and include linear discriminant analysis and methods based on the Karhunen–Loève expansion. Some of the methods are similar, if not identical, in certain circumstances and will be discussed in detail in the appropriate section of this chapter. Here, we approach the topic initially from a pattern recognition perspective and give a brief description of the terms *feature selection* and *feature extraction*.

Given a set of measurements, dimensionality reduction can be achieved in essentially two different ways. The first is to identify those variables that do not contribute to the classification

Statistical Pattern Recognition, Third Edition. Andrew R. Webb and Keith D. Copsey.

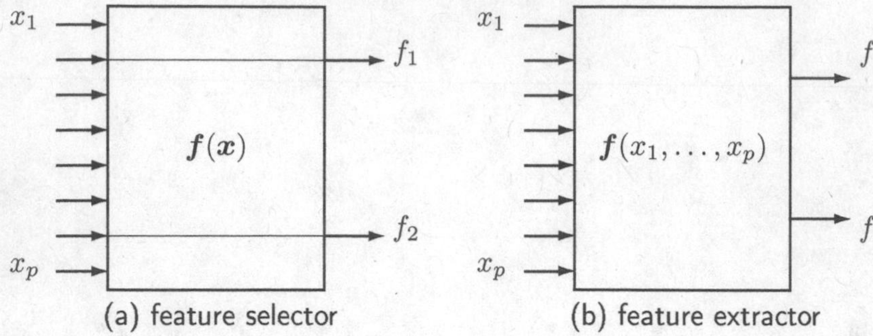

Figure 10.1 Dimensionality reduction by (a) feature selection and (b) feature extraction.

task. In a discrimination problem, we would neglect those variables that do not contribute to class separability. Thus, the task is to seek d features out of the available p measurements (the number of features d must also be determined). This is termed *feature selection in the measurement space* or simply *feature selection* [Figure 10.1(a)]. There are situations other than for discrimination purposes in which it is desirable to select a subset from a larger number of features or variables.[1] Miller (1990) discusses subset selection in the context of regression.

The second approach is to find a *transformation* from the p measurements to a lower-dimensional feature space. The transformation may be performed prior to classifier design – pre-filtering of data – or integrated with a classifier. This is termed *feature selection in the transformed space* or *feature extraction* [Figure 10.1(b)]. This transformation may be a linear or nonlinear combination of the original variables and may be supervised (takes into account class information) or unsupervised. In the supervised case, the task is to find the transformation for which a particular criterion of class separability is maximised.

Feature selection and feature extraction may be regarded as transformation methods where a set of weights is applied to the original variables to obtain transformed variables (Molina *et al.*, 2002). For *feature selection*, the weights are binary valued. For *feature extraction*, the weights are continuous. These weights are learnt from a training dataset. In the binary case, we are interested in retaining a subset of the original features only. In the continuous feature selection problem, all original features are retained.

Both of these approaches require the optimisation of some criterion function, J, which is often based on some measure of distance or dissimilarity between distributions, which in turn may require distances between objects (patterns) to be defined. For feature selection, the optimisation is over the set of all possible subsets of size d, $\mathcal{X}_d$, of the p possible features, $X_1, \ldots, X_p$. Thus we seek the subset $\tilde{X}_d$ for which

$$J(\tilde{X}_d) = \max_{X \in \mathcal{X}_d} J(X)$$

Suboptimal approaches that add features to an existing set are often adopted.

[1] The terms *feature*, *variable* or *attribute* are often used interchangeably.

In feature extraction, the optimisation is performed over all possible transformations of the variables. The class of transformation is usually specified (for example, a linear transformation of the variable set) and we seek the transformation, $\tilde{A}$, for which

$$J(\tilde{A}) = \max_{A \in \mathcal{A}} J(A(X))$$

where $\mathcal{A}$ is the set of allowable transformations. The feature vector is then $\boldsymbol{Y} = \tilde{A}(\boldsymbol{X})$.

10.2 Feature selection

10.2.1 Introduction

There has been a considerable amount of work on feature selection in recent years. In particular, in the bioinformatics area, the analysis of microarray data has been one of the major motivations. However, feature selection is a problem relevant to many applications in pattern recognition and machine learning: text categorisation; remote sensing; drug discovery; marketing; speech processing; and handwritten character recognition. The data are high dimensional and the sample size is often small. A categorisation of feature selection approaches has emerged that was not widely used a decade ago (Molina *et al.*, 2002; Liu and Yu, 2005; Saeys *et al.*, 2007), particularly in the pattern recognition literature.

Reasons for performing feature selection may include:

- increasing predictive accuracy of a classifier;
- removing irrelevant data;
- enhancing learning efficiency – reducing computational and storage requirements;
- reducing the cost of (future) data collection – making measurements on only those variables relevant for discrimination;
- reducing complexity of the resulting classifier description – providing an improved understanding of the data and the model.

Nevertheless, it should be stated that feature selection is not necessarily required as a preprocessing step for classification algorithms to perform well (Guyon, 2008). Many modern pattern recognition algorithms employ regularisation techniques to handle overfitting (e.g. regularised discriminant analysis – Chapter 2; support vector machines – Chapter 6) or average over multiple classifiers (ensemble methods – Chapter 8) and thus preprocessing to remove unnecessary features may not be required.

10.2.1.1 Relevance and redundancy

It is common that a large number of features are not informative – either irrelevant or redundant. Learning can be achieved much more efficiently and effectively with just the relevant and nonredundant features (Yu and Liu, 2004). Irrelevant features are those that do not contribute to a classification rule. Redundant features are those that are strongly correlated. Eliminating redundancy will mean that there is no need to waste effort (time and cost) making measurements on unnecessary variables. Some feature selection approaches will remove irrelevant features but do not handle redundant features very well (since they have similar rankings).

A feature set may be considered to comprise four basic parts (Xie *et al.*, 2006): (1) irrelevant features; (2) redundant features; (3) weakly relevant but nonredundant features; and (4) strongly relevant features. An optimal subset contains features in parts 3 and 4.

Let X be a feature and denote the full set of features by $\mathcal{X}$; denote by S the set of features excluding X ; that is, $S = \mathcal{X} - \{X\}$.

Strong relevance
The feature X is strongly relevant iff

$$p(C|X, S) \neq p(C|S)$$

The distribution of the class predictor depends on the feature, X. Thus, the feature cannot be removed without affecting the distribution of the class predictor.

Weak relevance
The feature X is weakly relevant iff

$$p(C|X, S) = p(C|S)$$

and $\exists\, S' \subset S$ such that $p(C|X, S') \neq p(C|S')$; that is, removal of the feature X from the total set does not affect the class prediction, but X does affect the class predicted based on a subset of the complete set. Thus, the feature is not always necessary, but may be necessary for certain subsets.

Irrelevance
The feature X is irrelevant iff

$$\forall\, S' \subseteq S, \quad p(C|X, S') = p(C|S')$$

The feature X does not affect the class prediction for any subset of the complete set. Thus, the feature is not necessary at all.

Redundancy
A good feature set will include none of the irrelevant features, all of the strongly relevant features and a subset of the weakly relevant features. Feature redundancy is used to determine which of the weakly relevant features to retain. It is defined here using the concept of Markov blanket.

Let $\boldsymbol{M}_X$ be a subset of the features $\boldsymbol{X}$ that does not contain the feature, X. Then, $\boldsymbol{M}_X$ is a **Markov blanket** for X if X is conditionally independent[2] of $\boldsymbol{X} - \boldsymbol{M}_X - \{X\}$ given $\boldsymbol{M}_X$. A Markov blanket means that $\boldsymbol{M}_X$ subsumes the information that X has about all the other features. A graphical interpretation of a Markov blanket can be provided by a Bayesian network.

[2] Consider three variables, A, B and C. We say that A is conditionally independent of B given C (written $A \perp B|C$) to mean $p(A, B|C) = p(A|C)p(B|C)$, or equivalently, $p(A|B, C) = p(A|C)$.

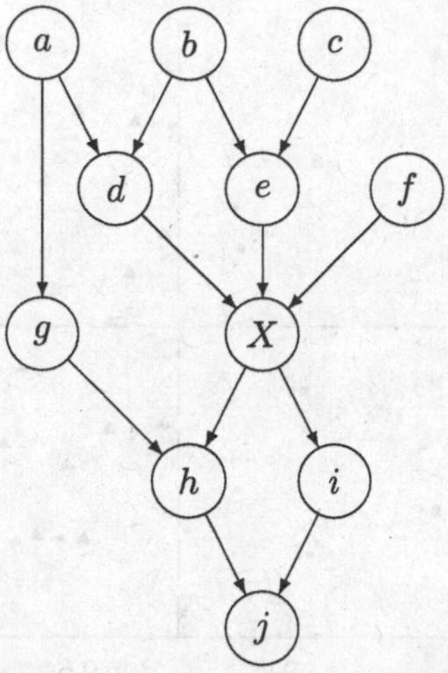

Figure 10.2 A Bayesian network. The Markov blanket, $\boldsymbol{M}_X$, for the node labelled X is the set of nodes $\{d, e, f, g, h, i\}$: the parents, the children and the parents of the children.

A Bayesian network is a statistical model that represents graphically the independencies that hold in a domain (see Chapter 4). For a faithful Bayesian network,[3] the Markov blanket of a feature is unique and is the set of parents, children and spouses (parents of common children) as encoded in the graphical structure of the network (Figure 10.2). If $\boldsymbol{M}_X$ is a Markov blanket of X, then the class variable C is conditionally independent of X given its Markov blanket.

Assume that we have a set of features, $\boldsymbol{G} \subset \boldsymbol{X}$. A feature in $\boldsymbol{G}$ is **redundant** and can be removed from the set *iff* it is weakly relevant and has a Markov blanket in $\boldsymbol{G}$. Thus, the Markov blanket property can be used to remove features that are unnecessary. However, not all features that are retained are necessarily useful: it depends on the objective. The above definitions of relevance and redundancy are cast in terms of the posterior probability of the class variable. If the objective is to minimise classification error, then some of the relevant features may not be necessary. An illustration for a two-class problem (labelled by ▲ and ■) is provided by Figure 10.3 (Guyon, 2008). For the data points above $X_2 = 0.5$, both classes are distributed uniformly over $0 \leq X_1 \leq 1, 0.5 \leq X_2 \leq 1$. Below the line $X_2 = 0.5$, the class ▲ is distributed uniformly over $0.5 \leq X_1 \leq 1, 0 \leq X_2 \leq 0.5$; the class ■ is distributed uniformly over $0 \leq X_1 \leq 0.5, 0 \leq X_2 \leq 0.5$. Clearly, the posterior probability of the class depends on X_2, but the simplest Bayes decision boundary is $X_1 = 0.5$ and does not depend on X_2.

10.2.1.2 Problem statement

Let $\boldsymbol{G}$, be a subset of the variables $\boldsymbol{X}$. Denote by $p(C|\boldsymbol{x})$ the posterior probability of the class variable given a set of measurements, $\boldsymbol{x}$, on the vector of variables, $\boldsymbol{X}$; similarly, denote by $p(C|\boldsymbol{g})$ the posterior probability of the class variable given a set of measurements, $\boldsymbol{g}$, on the vector of variables, $\boldsymbol{G}$. The goal of feature selection is to select the set $\boldsymbol{G}$ so that $p(C|\boldsymbol{x})$ and

[3] A Bayesian network is faithful to a joint distribution, P, *iff* every conditional dependence entailed in the Bayesian network is also present in P.

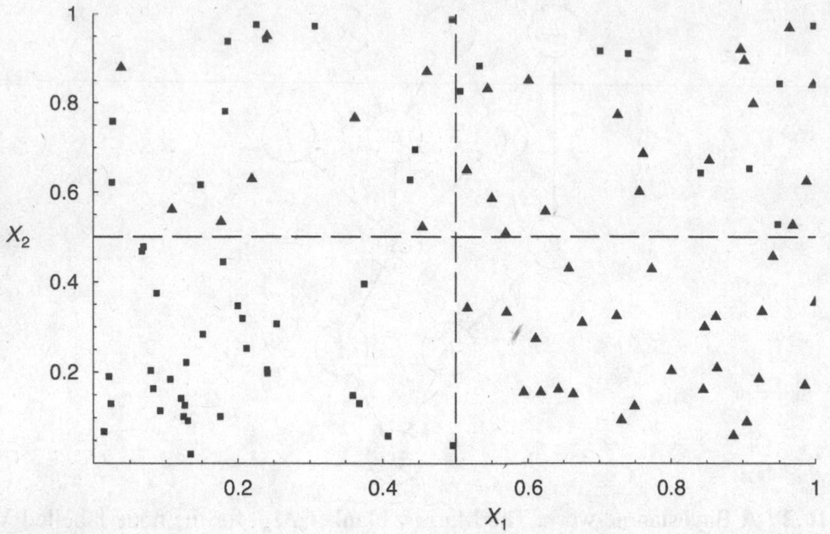

Figure 10.3 Feature X_2 is strongly relevant to the class variable since $p(C|X_1, X_2) \neq p(C|X_1)$. The decision boundary ($X_1 = 0.5$) does not depend on X_2.

$p(C|\boldsymbol{g})$ are as close as possible. For two distributions $\mu(z)$ and $\sigma(z)$, the Kullback–Leibler divergence is defined to be

$$KL(\mu|\sigma) = \sum_z \mu(z) \log\left(\frac{\mu(z)}{\sigma(z)}\right)$$

KL measures the extent of the 'error' in using σ to approximate μ. We use Kullback–Leibler divergence in our feature selection definition and seek the subset of variables such that

$$\begin{aligned}\delta_G(\boldsymbol{x}) &= KL\left(p(C|\boldsymbol{x})|p(C|\boldsymbol{g})\right) \\ &= \sum_C p(C|\boldsymbol{x}) \log\left(\frac{p(C|\boldsymbol{x})}{p(C|\boldsymbol{g})}\right)\end{aligned}$$

is as small as possible. The above is an evaluation for just one measurement vector, therefore we sum over all measurement vectors, or using the unconditional distribution $p(\boldsymbol{x})$, we evaluate

$$\Delta_G = \sum_{\boldsymbol{x}} p(\boldsymbol{x})\delta_G(\boldsymbol{x})$$

where the summation is replaced by an integral for continuous variables.

Unfortunately, it is not practical to implement a feature selection approach directly based on an evaluation of Δ_G since we do not know the distributions involved – we only have a sample of data points from the distributions. Secondly, evaluation of Δ_G for each subset of variables is often infeasible.

For example, suppose that we have a set of measurements on p variables, and we seek the best subset of size d? Ideally, we would evaluate the criterion Δ_G for all possible combinations of d variables selected from p and select that combination for which this criterion is minimised, but the difficulty in obtaining the solution arises because the number of possible subsets is

$$n_d = \frac{p!}{(p-d)!d!}$$

which can be very large even for moderate values of p and d. For example, selecting the best 10 features out of 25 means that 3 268 760 feature sets must be considered, and evaluating the optimality criterion for every feature set in an acceptable time may not be feasible. Therefore, we must consider ways of searching through the space of possible variable sets that reduce the amount of computation, perhaps at the expense of obtaining a suboptimal solution.

Selecting the 'best' subset may be equivalent to selecting the subset with as few variables as possible, although not necessarily if costs are associated with individual variables.

10.2.2 Characterisation of feature selection approaches

There are several characterisations of feature selection methods. A common approach is to organise techniques into three broad categories: filter, wrapper and embedded methods.

Filter methods
These use the statistical properties of the variables to filter out poorly informative variables; this is undertaken prior to a classification stage; thus, feature selection is independent of classifier learning and it relies on various measures of the general characteristics of the training data such as distance and dependency.

Wrapper methods
These are more computationally demanding than filter methods. Subsets are evaluated within a classification algorithm with the predictive accuracy as a measure of the goodness of a feature set, so the approach is classifier dependent. They tend to give better performance than filter methods.

Embedded methods
The search for an optimal set is built into classifier design (rather than being undertaken separately from the classifier design in wrapper methods). The approach is classifier dependent, as in wrapper methods, and can be viewed as a search in the combined space of feature subsets and classifier models. The decision tree classifier (see Chapter 7), which selects features as part of classifier design, is an example of an embedded method.

Filter methods are computationally much more efficient, but usually have poorer performance than wrapper methods. Many heuristic algorithms (e.g. forward and backward selection) have been proposed to reduce the computational complexity of wrapper algorithms. Saeys *et al.* (2007) provide a classification of feature selection techniques in the bioinformatics domain in terms of the filter, wrapper, embedded categories.

There are two key steps in a feature selection process: *evaluation* and *subset generation*:

- **An evaluation measure** is a means of assessing a candidate feature subset. Wrapper and embedded methods tend to use measures based on classifier performance. Filter methods use measures based on the properties of the data.

- **A subset generation method** is a means of generating a subset for evaluation. The procedure may be a simple ranking of individual features or an incremental procedure that adds features to (or removes features from) a current feature subset.

10.2.3 Evaluation measures

In order to choose a good feature set, we require a means of measuring the ability of a feature to contribute to the discriminability of classes, either individually or in the context of other features, already selected. Thus, we require a means of measuring relevance and redundancy. There are essentially two types of measures.

1. **Measures that rely on the general properties of the data**. These include measures that assess the relevance of individual features (for example, simple feature ranking) and measures that are used to eliminate feature redundancy through the estimation of the overlap between the distributions from which the data are drawn and through estimation of Markov blankets. These are optimised to favour those feature sets for which this overlap is minimal (that is, maximise *separability*). All these measures are independent of the final classifier employed (they are used with *filter methods*) and they have the advantage that they are often fairly cheap to implement but they have the disadvantage that the assumptions made in determining the overlap are often crude and may result in a poor estimate of the discriminability.

   Feature ranking Features are ranked by a metric and those that fail to achieve a prescribed score are eliminated. Ranking is performed either individually or in the context of other features.

   Interclass distance A measure of distances between classes is defined based on distances between members of each class.

   Probabilistic distance This is the computation of a probabilistic distance or divergence between class-conditional probability density functions (two classes).

   Probabilistic dependence These measures are multiclass criteria that measure the distance between the class-conditional densities and the mixture probability density function for the data irrespective of the class.

2. **Measures that use a classification rule as part of their evaluation**. In this second approach a classifier is designed using measurements on the reduced feature set and a measure of classifier performance is employed as a separability measure. We then choose the feature sets for which the classifier performs well, according to this measure, on a separate test/validation set. In this approach, the feature set is chosen to match the classifier (these measures are used with a *wrapper* or *embedded method*). A different feature set may result with a different choice of classifier.

Error rate A widely used example is the error rate of the classifier trained with the feature subset, but other metrics that may also be expressed in terms of the number of true positives (*tp*), false positives (*fp*), true negatives (*tn*) and false negatives (*fn*) have been assessed in the context of text classification by Forman (2003).

10.2.3.1 Feature ranking measures

Univariate feature ranking can be used to rank features individually and thus provide a simple filtering method to remove irrelevant or redundant features. The simplest approaches are those measures based on correlation. They are easy to calculate, do not require estimation of probability density functions nor the discretisation of continuous features.

The **Pearson correlation** coefficient measures the degree of linear correlation between two variables. For two variables X and Y with measurements $\{x_i\}$ and $\{y_i\}$ and means $\bar{x}$ and $\bar{y}$ this is given by

$$\rho(X, Y) = \frac{\sum_i (x_i - \bar{x})(y_i - \bar{y})}{\left[\sum_i (x_i - \bar{x})^2 \sum_i (y_i - \bar{y})^2\right]^{\frac{1}{2}}} \tag{10.1}$$

If two variables are completely correlated ($\rho = \pm 1$), then one is redundant and can be removed. However, linear correlation measures may not be able to capture relationships that are nonlinear. Correlation between a feature and a target variable representing class (a categorical variable) requires the coding of the target class as a binary vector. If there are C classes and the class of x_i is ω_k, then y_i is encoded as a C-dimensional binary vector $(0, \ldots, 0, 1, 0, \ldots, 0)$, where the '1' is in the kth position: $y_{ij} = 0, j \neq k, y_{ik} = 1$. We calculate a least-squares regression measure, R, given by

$$R^2 = \frac{\sum_k \left(\sum_i (x_i - \bar{x})(y_{ik} - \bar{y}_k)\right)^2}{\left[\sum_i (x_i - \bar{x})^2 \sum_i \sum_k (y_{ik} - \bar{y}_k)^2\right]} \tag{10.2}$$

where $\bar{y}_k$ is the kth component of the mean target vector. This is the squared magnitude of the vector of correlations between the variable X and the class variables divided by the total sum of squares.

A nonlinear correlation measure is the **mutual information**. The entropy of a (discrete) variable X is given by

$$H(X) = -\sum_x p(x) \log_2(p(x))$$

and the entropy of X after observing Y is defined as

$$H(X|Y) = -\sum_y p(y) \sum_x p(x|y) \log_2(p(x|y))$$

The mutual information is the additional information about X provided by Y and is the decrease in the entropy of X given by

$$\begin{aligned} MI(X|Y) &= H(X) - H(X|Y) \\ &= \sum_{x,y} p(x,y) \log_2 \left[\frac{p(x,y)}{p(x)p(y)} \right] \end{aligned} \tag{10.3}$$

This may be used to rank features. A feature X is more correlated than a feature Z to the class variable Y if

$$MI(X|Y) > MI(Z|Y)$$

A normalised value, termed **symmetrical uncertainty**, is defined as

$$SU(X,Y) = 2\left(\frac{MI(X|Y)}{H(X)+H(Y)}\right)$$

and gives a value in [0, 1]: a value of 1 indicates that knowing the value of one feature will predict the other and a value of 0 indicates independence [as can be seen by Equation (10.3)]. Advantages of mutual information are that the dependency between variables is no longer restricted to be linear and it can handle nominal or discrete features. Disadvantages of the measure are that probability density functions must be estimated and it can be difficult to evaluate for continuous variables, which must be discretised.

Each of the above measures can be used to rank features individually according to their relevance. However, some features may become relevant in the context of others. This is illustrated in Figure 10.4 [after Guyon (2008)]. Therefore, ranking criteria that take into account context are necessary. An example is a family of algorithms called **Relief**. It is based on nearest-neighbour patterns. For each pattern $\boldsymbol{x}_i$, for a prescribed value of K, compute the K closest patterns of the same class $\{\boldsymbol{x}_{H_k(i)}, k = 1, \ldots, K\}$ (the K nearest hits) and the K closest patterns of a different class (the K nearest misses), $\{\boldsymbol{x}_{M_k(i)}, k = 1, \ldots, K\}$. The normalised Relief criterion, evaluated for feature X_j is

$$R_{Relief}(j) = \frac{\displaystyle\sum_{i=1}^{n}\sum_{k=1}^{K} \left| x_{i,j} - x_{M_k(i),j} \right|}{\displaystyle\sum_{i=1}^{n}\sum_{k=1}^{K} \left| x_{i,j} - x_{H_k(i),j} \right|} \tag{10.4}$$

the ratio of the mean distance to the nearest misses to the mean distance to the nearest hits, measured using the feature X_j. Various extensions to the Relief methodology have been proposed (Duch, 2004).

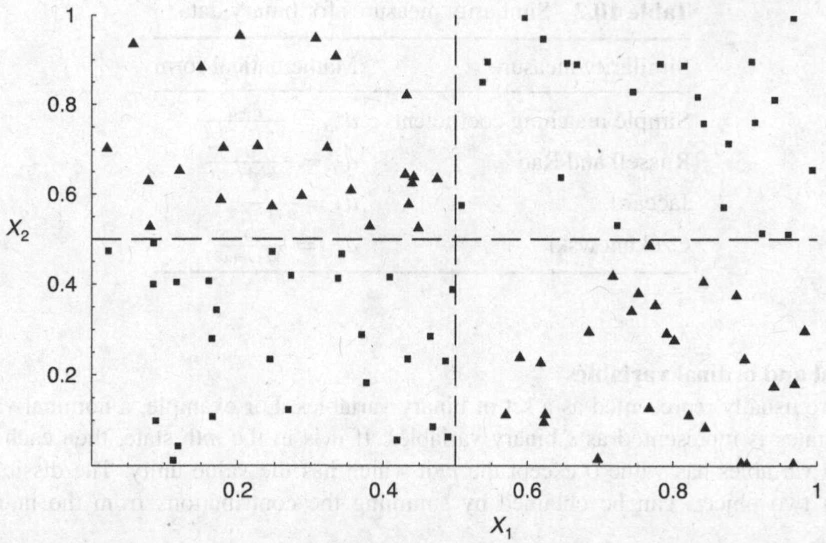

Figure 10.4 Relevance in context. The distribution of X_1 and X_2 are independent of the class variable ($X_1 \perp Y, X_2 \perp Y$), but jointly, the distribution depends on class ($\{X_1, X_2\} \perp\!\!\!\!/\, Y$).

10.2.3.2 Interclass distance

Binary variables

Various dissimilarity measures have been proposed for binary variables. For vectors of binary variables $\boldsymbol{x}$ and $\boldsymbol{y}$ these may be expressed in terms of quantities a, b, c, and d where

a is equal to the number of occurrences of $x_i = 1$ and $y_i = 1$
b is equal to the number of occurrences of $x_i = 0$ and $y_i = 1$
c is equal to the number of occurrences of $x_i = 1$ and $y_i = 0$
d is equal to the number of occurrences of $x_i = 0$ and $y_i = 0$

This is summarised in Table 10.1. Note that $a + b + c + d = p$, the total number of variables (attributes). It is customary to define a similarity measure rather than a dissimilarity measure. Table 10.2 summarises some of the more commonly used similarity measures for binary data.

Table 10.1 Co-occurrence table for binary variables.

		x_i	
		1	0
y_i	1	a	b
	0	c	d

Table 10.2 Similarity measures for binary data.

Similarity measure	Mathematical form
Simple matching coefficient	$d_{sm} = \frac{a+d}{a+b+c+d}$
Russell and Rao	$d_{rr} = \frac{a}{a+b+c+d}$
Jaccard	$d_j = \frac{a}{a+b+c}$
Czekanowski	$d_{Cz} = \frac{2a}{2a+b+c}$

Nominal and ordinal variables

These are usually represented as a set of binary variables. For example, a nominal variable with s states is represented as s binary variables. If it is in the mth state, then each of the s binary variables has value 0 except the mth which has the value unity. The dissimilarity between two objects can be obtained by summing the contributions from the individual variables.

For ordinal variables, the contribution to the dissimilarity between two objects from a single variable does not simply depend on whether or not the values are identical. If the contribution for one variable in state m and one in state l ($m < l$) is δ_{ml}, then we require

$$\begin{aligned} \delta_{ml} &\geq \delta_{ms} \quad \text{for} \quad s < l \\ \delta_{ml} &\geq \delta_{sl} \quad \text{for} \quad s > m \end{aligned}$$

that is, δ_{ml} is monotonic down each row and across each column of the half-matrix of distances between states ($\delta_{14} > \delta_{13} > \delta_{12}$, etc.; $\delta_{14} > \delta_{24} > \delta_{34}$). The values chosen for δ_{ml} depend very much on the problem. For example, we may have a variable describing fruits of a plant that can take the values short, long or very long. We would want the dissimilarity between a plant with very long fruit and one with short fruit to be greater than that between one with long fruit and one with short fruit (all other attributes having equal values). A numeric coding of 1, 2, 3 would achieve this, but so would 1, 10, 100.

Numeric variables

Many dissimilarity measures have been proposed for numeric variables. Table 10.3 gives some of the more common measures. The choice of a particular metric depends on the application. Computational considerations aside, for feature selection and extraction purposes you would choose the metric that gives the best performance (perhaps in terms of classification error on a validation set).

Given a measure of distance, $d(\boldsymbol{x}, \boldsymbol{y})$, between two patterns $\boldsymbol{x}$ and $\boldsymbol{y}$, each from a different class, we define a measure of the separation between two classes, ω_1 and ω_2, as

$$J_{as} = \frac{1}{n_1 n_2} \sum_{i=1}^{n_1} \sum_{j=1}^{n_2} d(\boldsymbol{x}_i, \boldsymbol{y}_j)$$

Table 10.3 Dissimilarity measures for numeric variables (between $\boldsymbol{x}$ and $\boldsymbol{y}$).

	Mathematical form
Euclidean distance	$d_e = \left\{\sum_{i=1}^{p}(x_i - y_i)^2\right\}^{\frac{1}{2}}$
City-block distance	$d_{cb} = \sum_{i=1}^{p} \lvert x_i - y_i \rvert$
Chebyshev distance	$d_{ch} = \max_i \lvert x_i - y_i \rvert$
Minkowski distance of order m	$d_M = \left\{\sum_{i=1}^{p}(x_i - y_i)^m\right\}^{\frac{1}{m}}$
Quadratic distance Q, positive definite	$d_q = \sum_{i=1}^{p}\sum_{j=1}^{p}(x_i - y_i)Q_{ij}(x_j - y_j)$
Canberra distance	$d_{ca} = \sum_{i=1}^{p} \frac{\lvert x_i - y_i \rvert}{x_i + y_i}$
Nonlinear distance	$d_n = \begin{cases} H & d_e > D \\ 0 & d_e \leq D \end{cases}$
Angular separation	$\frac{\sum_{i=1}^{p} x_i y_i}{\left[\sum_{i=1}^{p} x_i^2 \sum_{i=1}^{p} y_i^2\right]^{1/2}}$

for $\boldsymbol{x}_i \in \omega_1$ and $\boldsymbol{y}_j \in \omega_2$. This is the average separation. For $C > 2$ classes, we define the average distance between classes as

$$J = \frac{1}{2}\sum_{i=1}^{C} p(\omega_i)\sum_{j=1}^{C} p(\omega_j)J_{as}(\omega_i, \omega_j)$$

where $p(\omega_i)$ is the prior probability of class ω_i (estimated as $p_i = n_i/n$). Using a Euclidean distance squared for $d(\boldsymbol{x}, \boldsymbol{y})$, then the measure J may be written,

$$J = J_1 = \mathrm{Tr}\{\boldsymbol{S}_W + \boldsymbol{S}_B\} = \mathrm{Tr}\{\hat{\boldsymbol{\Sigma}}\} \tag{10.5}$$

where $\boldsymbol{S}_W$ and $\boldsymbol{S}_B$ are the within and between class scatter matrices (see Notation).

The criterion J_1 is not very satisfactory as a feature selection criterion: it is simply the total variance, which does not depend on class information.

Our aim is to find a set of variables for which the within-class spread is small and the between-class spread is large in some sense. Several criteria have been proposed for achieving this. One popular measure is

$$J_2 = \mathrm{Tr}\left\{\boldsymbol{S}_W^{-1}\boldsymbol{S}_B\right\} \tag{10.6}$$

Another is the ratio of the determinants

$$J_3 = \frac{|\hat{\boldsymbol{\Sigma}}|}{|\boldsymbol{S}_W|} \tag{10.7}$$

the ratio of the total scatter to the within-class scatter.

Finally, a further measure used is

$$J_4 = \frac{\mathrm{Tr}\{S_B\}}{\mathrm{Tr}\{S_W\}} \tag{10.8}$$

As with the probabilistic distance measures below, each of these distance measures may be calculated recursively (see Section 10.2.5.1).

10.2.3.3 Probabilistic distance

Probabilistic distance measures the distance between two distributions, $p(\boldsymbol{x}|\omega_1)$ and $p(\boldsymbol{x}|\omega_2)$, and can be used in feature selection. For example, the divergence is given by

$$J_D(\omega_1, \omega_2) = \int [p(\boldsymbol{x}|\omega_1) - p(\boldsymbol{x}|\omega_2)] \log \left\{ \frac{p(\boldsymbol{x}|\omega_1)}{p(\boldsymbol{x}|\omega_2)} \right\} d\boldsymbol{x}$$

which is the sum of the two Kullback–Leibler divergences $KL(p(\boldsymbol{x}|\omega_1)|p(\boldsymbol{x}|\omega_2))$ and $KL(p(\boldsymbol{x}|\omega_2)|p(\boldsymbol{x}|\omega_1))$. Other measures are given in Table 10.4. A more complete list can be found in the books by Chen (1973) and Devijver and Kittler (1982). All of the measures given in the table have the property that they are maximum when classes are disjoint.

One of the main disadvantages of the probabilistic distance criteria is that they require an estimate of a probability density function and its numerical integration. This restricts their usefulness in many practical situations. However, under certain assumptions regarding the form of the distributions, the expressions can be evaluated analytically. Many of the commonly used distance measures, including those in Table 10.4, simplify for normal distributions. The divergence becomes

$$J_D = \frac{1}{2}(\boldsymbol{\mu}_2 - \boldsymbol{\mu}_1)^T \left(\boldsymbol{\Sigma}_1^{-1} + \boldsymbol{\Sigma}_2^{-1}\right)(\boldsymbol{\mu}_2 - \boldsymbol{\mu}_1) + \mathrm{Tr}\left\{\boldsymbol{\Sigma}_1^{-1}\boldsymbol{\Sigma}_2 + \boldsymbol{\Sigma}_1^{-1}\boldsymbol{\Sigma}_2 - 2\boldsymbol{I}\right\}$$

for normal distributions with means $\boldsymbol{\mu}_1$ and $\boldsymbol{\mu}_2$ and covariance matrices $\boldsymbol{\Sigma}_1$ and $\boldsymbol{\Sigma}_2$.

Table 10.4 Probabilistic distance measures.

Dissimilarity measure	Mathematical form
Chernoff	$J_c = -\log \int p^s(\boldsymbol{x}|\omega_1) p^{1-s}(\boldsymbol{x}|\omega_2)\, d\boldsymbol{x}$
Bhattacharyya	$J_B = -\log \int (p(\boldsymbol{x}|\omega_1) p(\boldsymbol{x}|\omega_2))^{\frac{1}{2}} d\boldsymbol{x}$
Divergence	$J_D = \int [p(\boldsymbol{x}|\omega_1) - p(\boldsymbol{x}|\omega_2)] \log \left(\frac{p(\boldsymbol{x}|\omega_1)}{p(\boldsymbol{x}|\omega_2)} \right) d\boldsymbol{x}$
Patrick–Fischer	$J_P = \left\{ \int [p(\boldsymbol{x}|\omega_1) p(\omega_1) - p(\boldsymbol{x}|\omega_2) p(\omega_2)]^2 d\boldsymbol{x} \right\}^{\frac{1}{2}}$

In a multiclass problem, the pairwise distance measures must be adapted. We may take as our cost function, J, the maximum overlap over all pairwise measures,

$$J = \max_{i,j(i \neq j)} J(\omega_i, \omega_j)$$

or the average of the pairwise measures,

$$J = \sum_{i<j} J(\omega_i, \omega_j) p(\omega_i) p(\omega_j)$$

10.2.3.4 Probabilistic dependence

The probabilistic distance measures are based on discrimination between a pair of classes, using the class-conditional densities to describe each class. Probabilistic dependence measures are multiclass feature selection criteria that measure the distance between the class-conditional density and the mixture probability density function for the data irrespective of class (Figure 10.5).

If $p(\boldsymbol{x}|\omega_i)$ and $p(\boldsymbol{x})$ are identical then we gain no information about class by observing $\boldsymbol{x}$, and the 'distance' between the two distributions is zero. Thus, $\boldsymbol{x}$ and ω_i are independent. If the distance between $p(\boldsymbol{x}|\omega_i)$ and $p(\boldsymbol{x})$ is large, then the observation $\boldsymbol{x}$ is dependent on ω_i. The greater the distance then the greater is the dependence of $\boldsymbol{x}$ on the class ω_i. Table 10.5 gives the probabilistic dependence measures corresponding to the probabilistic distance measures in Table 10.4 (Devijver and Kittler, 1982).

In practice, application of probabilistic dependence measures is limited because, even for normally distributed classes, the expressions given in Table 10.5 cannot be evaluated analytically since the mixture distribution $p(\boldsymbol{x})$ is not normal.

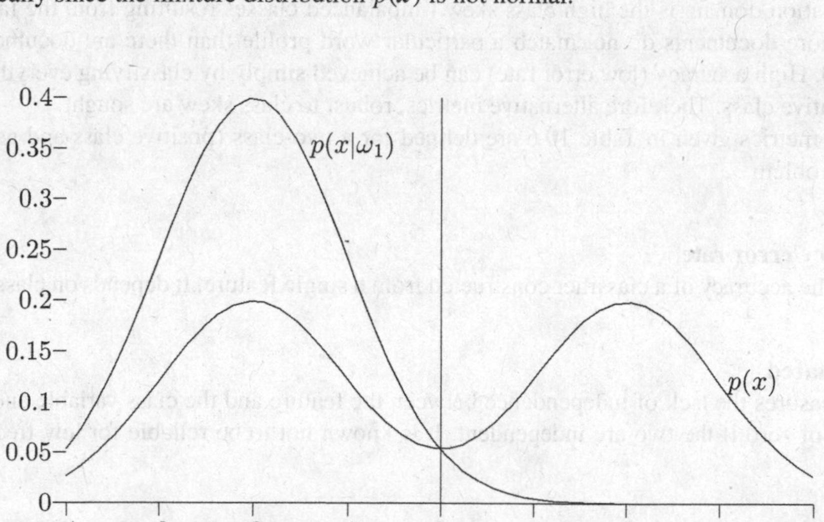

Figure 10.5 Probabilistic dependence.

Table 10.5 Probabilistic dependence measures.

Dissimilarity measure	Mathematical form
Chernoff	$J_c = \sum_{i=1}^{C} p(\omega_i) \left\{ -\log \int p^s(\boldsymbol{x}\|\omega_i) p^{1-s}(\boldsymbol{x})\, d\boldsymbol{x} \right\}$
Bhattacharyya	$J_B = \sum_{i=1}^{C} p(\omega_i) \left\{ -\log \int (p(\boldsymbol{x}\|\omega_i) p(\boldsymbol{x}))^{\frac{1}{2}} d\boldsymbol{x} \right\}$
Joshi	$J_D = \sum_{i=1}^{C} p(\omega_i) \int [p(\boldsymbol{x}\|\omega_i) - p(\boldsymbol{x})] \log \left(\frac{p(\boldsymbol{x}\|\omega_i)}{p(\boldsymbol{x})} \right) d\boldsymbol{x}$
Patrick–Fischer	$J_P = \sum_{i=1}^{C} p(\omega_i) \left\{ \int [p(\boldsymbol{x}\|\omega_i) - p(\boldsymbol{x})]^2 d\boldsymbol{x} \right\}^{\frac{1}{2}}$

10.2.3.5 Error rate

A minimum expected classification error rate is often the main aim in classifier design. Error rate (or misclassification rate) is simply the proportion of samples incorrectly classified. Optimistic estimates of a classifier's performance will result if the data used to design the classifier are naïvely used to estimate the error rate. Such an estimate is termed the apparent error rate. Estimation should be based on a separate test set, but if data are limited in number, we may wish to use all available data in classifier design. Procedures such as the *jackknife* and the *bootstrap* have been developed to reduce the bias of the apparent error rate. An algorithm based on a cross-validation procedure is given in Section 10.2.6. Error rate estimates are discussed in Chapter 9.

Forman (2003) provides a summary of metrics related to error rate that were evaluated as part of a study of machine learning for text classification (see Table 10.6, which lists some of the common metrics for a two-class problem). One of the difficulties with the text classification domain is the high class skew (imbalanced classes resulting from the fact that many more documents do not match a particular word profile than there are documents of interest). High accuracy (low error rate) can be achieved simply by classifying everything as the negative class. Therefore alternative metrics, robust to class skew are sought.

The metrics given in Table 10.6 are defined for a two-class (positive class and negative class) problem.

Accuracy/error rate
This is the accuracy of a classifier constructed from a single feature. It depends on class skew.

Chi-squared
This measures the lack of independence between the feature and the class variable and takes a value of zero if the two are independent. It is known not to be reliable for low frequency classes.

Information gain
This measures the decrease in entropy when the feature is present compared with when it is absent.

Table 10.6 Feature selection metrics.

Name	Definition
Accuracy	$\frac{tp + fp}{tp + tn + fp + fn}$
Error rate	1 - Accuracy
Chi-squared	$\frac{n\,(fp \times fn - tp \times tn)^2}{(tp + fp)\,(tp + fn)\,(fp + tn)\,(tn + fn)}$
Information gain	$e(tp + fn, fp + tn) - \frac{(tp + fp)e(tp, fp) + (tn + fn)e(fn, tn)}{tp + fp + tn + fn}$
	where $e(x, y) = -\frac{x}{x+y}\log_2\frac{x}{x+y} - \frac{y}{x+y}\log_2\frac{y}{x+y}$
Odds ratio	$\frac{tpr}{1 - tpr} \Big/ \frac{fpr}{1 - fpr} = \frac{tp \times tn}{fp \times fn}$
Probability ratio	$\frac{tpr}{fpr}$

tp, true positives; fp, false positives; fn, false negatives; tn, true negatives; $tpr = tp/(tp + fn)$, sample true positive rate; $fpr = fp/(fp + tn)$, sample false positive rate; $precision = tp/(tp + fp)$; $recall = tpr$.

Odds ratio
This has been used to rank documents in information retrieval. It is the odds of a member of the positive class occurring in the positive class divided by the odds of it occurring in the negative class.

Probability ratio
This is the proportion of positive samples predicted correctly divided by the proportion of negative samples predicted as positive.

10.2.4 Search algorithms for feature subset selection

There are three broad categories of search algorithms: complete, sequential and random (Liu and Yu, 2005).

Complete search
These methods guarantee to find the optimal subset of features according to some specified evaluation criterion. Of course, exhaustive search is complete, but the search strategy does not necessarily have to be exhaustive to be complete. For example, a branch and bound method is complete.

Sequential search
Features are added or removed sequentially (sequential forward or backward selection). Such methods are not optimal, but are simple to implement and fast to produce results.

Random search
These include approaches that inject randomness into the above methods and methods that generate the next subset completely at random.

The problem of feature selection is to choose the 'best' possible subset of size d from a set of p features. In this section we consider strategies for doing that – both optimal and suboptimal. The basic approach is to build up a set of d features incrementally, starting with the empty set (a 'bottom-up' method) or to start with the full set of measurements and remove redundant features successively (a 'top-down' approach).

If X_k represents a set of k features or variables then, in a bottom-up approach, the best set at a given iteration, $\tilde{X}_k$, is the set for which the feature (extraction) selection criterion has its maximum value

$$J(\tilde{X}_k) = \max_{X \in \mathcal{X}_k} J(X)$$

The set $\mathcal{X}_k$ of all sets of features at a given step is determined from the set at the previous iteration. This means that the set of measurements at one stage of an iterative procedure is used as a starting point to find the set at the next stage. This does not imply that the sets are necessarily nested ($\tilde{X}_k \subset \tilde{X}_{k+1}$), though they may be.

10.2.5 Complete search – branch and bound

This is an optimal search procedure that does not involve exhaustive search. It is a top-down procedure, beginning with the set of p variables and constructing a tree by deleting variables successively. It relies on one very important property of the feature selection criterion, namely that for two subsets of the variables, X and Y,

$$X \subset Y \Rightarrow J(X) < J(Y) \tag{10.9}$$

That is, evaluating the feature selection criterion on a subset of variables of a given set yields a smaller value of the feature selection criterion. This is termed the *monotonicity property*.

We shall describe the method by way of example (Figure 10.6). Let us assume that we wish to find the best three variables out of a set of five. A tree is constructed whose nodes represent all possible subsets of cardinality 3, 4, and 5 of the total set as follows. Level 0 in the tree contains a single node representing the total set. Level 1 contains subsets of the total set with one variable removed and level 2 contains subsets of the total set with two variables

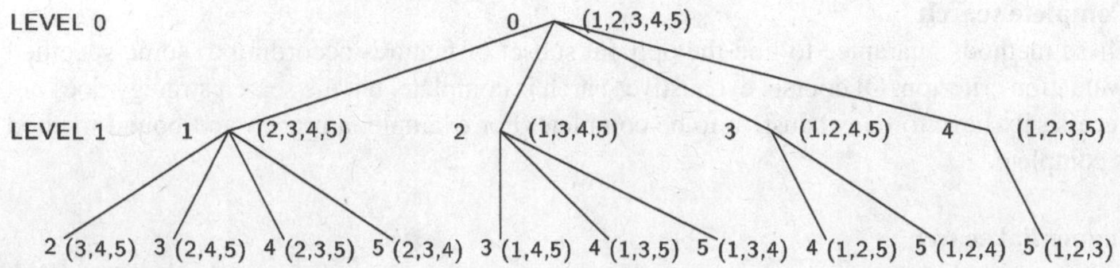

Figure 10.6 Tree figure for branch and bound method.

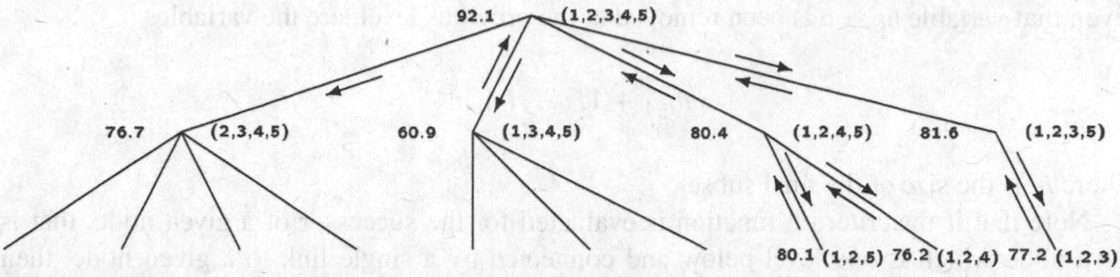

Figure 10.7 Tree figure for branch and bound method with feature selection criterion value at each node.

removed. The numbers to the right of each node in the tree represent a subset of variables. The number to the left represents the variable that has been removed from the subset of the parent node in order to arrive at a subset for the child node. Level 2 contains all possible subsets of five variables of size three. Note that the tree is not symmetrical. This is because removing variables 4 then 5 from the original set (to give the subset {1,2,3}) has the same result as removing variable 5 then variable 4. Therefore, in order for the subsets not to be replicated, we have only allowed variables to be removed in increasing order. This removes unnecessary repetitions in the calculation.

Now we have obtained our tree structure, how are we going to use it? The tree is searched from the least dense part to the part with the most branches (right to left in Figure 10.6).

Figure 10.7 gives a tree structure with values of the criterion J printed at the nodes. Starting at the rightmost set (the set {1,2,3} with a value of $J = 77.2$), the search backtracks to the nearest branching node and proceeds down the rightmost branch evaluating $J(\{1, 2, 4, 5\})$. Since $J(\{1, 2, 4, 5\})$ is larger than $J(\{1, 2, 3\})$, the monotonicity property that all subsets of this set will yield a lower value of the criterion function cannot be utilised to discard the branch and so the search then proceeds down the rightmost branch to $J(\{1, 2, 4\})$, which gives a lower value than the current maximum value of J, J^*, and so is discarded. The set $J(\{1, 2, 5\})$ is next evaluated and retained as the current best value (largest on a subset of three variables), $J^* = 80.1$. $J(\{1, 3, 4, 5\})$ is evaluated next, and since this is less than the current best, the search of the section of the tree originating from this node is not performed. This is because we know from the monotonicity property that all subsets of this set will yield a lower value of the criterion function. The algorithm then backtracks to the nearest branching node and proceeds down the next rightmost branch (in this case, the final branch). $J(\{2, 3, 4, 5\})$ is evaluated, and again since this is lower than the current best value on a subset of three variables, the remaining part of the tree is not evaluated.

Thus, although not all subsets of size 3 are evaluated, the algorithm is optimal since we know by the condition (10.9) that those not evaluated will yield a lower value of J.

From the specific example above, we can see a more general strategy: start at the top level and proceed down the rightmost branch, evaluating the cost function J at each node. If the value of J is less than the current threshold then abandon the search down that particular branch and backtrack to the previous branching node. Continue the search down the next rightmost branch. If, on the search of any branch the bottom level is reached (as it is bound to happen on the initial branch), then if the value of J for this level is larger than the current threshold, the threshold is updated and backtracking begins. Note from Figure 10.6 (and you can verify that this is true in general), that the candidates for removal at level i of the tree, given that variable n_{i-1} has been removed at the previous level, are the variables

$$n_{i-1}+1, \ldots, i+m$$

where m is the size of the final subset.

Note that if the criterion function is evaluated for the successor of a given node, that is for a node which is one level below and connected by a single link to a given node, then during the branch and bound procedure it will be evaluated for all 'brothers and sisters' of that node – that is for all other direct successors of the given node. Now since a node with a low value of J is more likely to be discarded than a node with a high value of J, it is sensible to order these sibling nodes so that those that have lower values have more branches. This is the case in Figure 10.7 where the nodes at level 2 are ordered so that those yielding a smaller value of J have the larger number of branches. Since all the sibling feature sets will be evaluated anyway, this results in no extra computation. This scheme is due to Narendra and Fukunaga (1977).

The quadratic form feature selection criterion used for feature selection both in regression (Miller, 1990) and in classification (Narendra and Fukunaga, 1977) is given by

$$\boldsymbol{x}_k^T \boldsymbol{S}_k^{-1} \boldsymbol{x}_k$$

where $\boldsymbol{x}_k$ is a k-dimensional vector and $\boldsymbol{S}_k$ is a $k \times k$ positive definite matrix when k features are used satisfies (10.9). For example, in a two-class problem, the Mahalanobis distance between two groups with means $\boldsymbol{\mu}_i (i = 1, 2)$ and covariance matrices $\boldsymbol{\Sigma}_i (i = 1, 2)$ is

$$J = (\boldsymbol{\mu}_1 - \boldsymbol{\mu}_2)^T \left(\frac{\boldsymbol{\Sigma}_1 + \boldsymbol{\Sigma}_2}{2} \right)^{-1} (\boldsymbol{\mu}_1 - \boldsymbol{\mu}_2) \tag{10.10}$$

which satisfies the monotonicity criterion. In a multiclass problem, we may take the sum over all pairs of classes as the feature selection criterion. This will also satisfy (10.9) since each component of the sum does. There are many other feature selection criteria satisfying the monotonicity criterion including probabilistic distance measures [for example, Bhattacharyya distance, divergence (Fukunaga, 1990)] and measures based on the scatter matrices (for example, $\text{Tr}\{\boldsymbol{S}_W^{-1}\boldsymbol{S}_B\}$) but not the measures J_3 and J_4, $\text{Tr}\{\boldsymbol{S}_B\}/\text{Tr}\{\boldsymbol{S}_W\}$ or $|\hat{\boldsymbol{\Sigma}}|/|\boldsymbol{S}_W|$.

Many of the search algorithms described in this chapter, both the optimal and the suboptimal algorithms, construct the feature sets at the ith stage of the algorithm from that at the $(i-1)$th stage by the addition or subtraction of a small number of features from the current optimal set. For the normal distribution parametric forms of the probabilistic distance criteria, the value of the criterion function at stage i can be evaluated by updating its value already calculated for stage $i-1$ instead of computing the criterion functions from their definitions. This can result in substantial computational savings.

10.2.5.1 Recursive calculation of separability measures

The normal distribution parametric measures Chernoff, Bhattacharyya, divergence, Patrick–Fischer and Mahalanobis distances, are functions of three basic building blocks of the form

$$x^T S^{-1} x, \quad \text{Tr}\{TS^{-1}\}, \quad |S| \tag{10.11}$$

where S and T are positive definite symmetric matrices and x is a vector of parameters. Thus, to calculate the criteria recursively, we only need to consider each of the building blocks. Let S denote the $k \times k$ positive definite matrix and let $\tilde{S}$ be the matrix with the kth element of the feature vector removed. The matrix S may be written

$$S = \begin{bmatrix} \tilde{S} & y \\ y^T & s_{kk} \end{bmatrix}$$

Assuming that $\tilde{S}^{-1}$ is known, then S^{-1} may be written as

$$S^{-1} = \begin{bmatrix} \tilde{S}^{-1} + \frac{1}{d}\tilde{S}^{-1} yy^T \tilde{S}^{-1} & -\frac{1}{d}\tilde{S}^{-1} y \\ -\frac{1}{d} y^T \tilde{S}^{-1} & \frac{1}{d} \end{bmatrix} \tag{10.12}$$

where $d = s_{kk} - y^T \tilde{S}^{-1} y$. Alternatively, if we know S^{-1}, which may be written as

$$S^{-1} = \begin{bmatrix} A & c \\ c^T & b \end{bmatrix}$$

then we may calculate $\tilde{S}^{-1}$ as

$$\tilde{S}^{-1} = A - \frac{1}{b} cc^T$$

Thus, we can compute the inverse of a matrix if we know the inverse before a feature is added to or deleted from the feature set.

In some cases it may not be necessary to calculate the inverse $\tilde{S}^{-1}$ from S^{-1}. Consider the quadratic form $x^T S^{-1} x$ where x is a k-dimensional vector and $\tilde{x}$ denotes the vector with the kth value removed. This can be expressed in terms of the quadratic form before the kth feature is removed as

$$\tilde{x}^T \tilde{S}^{-1} \tilde{x} = x^T S^{-1} x - \frac{1}{b}[(c^T : b)x]^2 \tag{10.13}$$

where $[c^T : b]$ is the row of S^{-1} corresponding to the feature that is removed. Thus, the calculation of $\tilde{S}^{-1}$ can be deferred until it is confirmed that this feature is to be permanently removed from the candidate feature set.

The second term to consider in (10.11) is $\text{Tr}\{TS^{-1}\}$. We may use the relationship

$$\text{Tr}\{\tilde{T}\tilde{S}^{-1}\} = \text{Tr}\{TS^{-1}\} - \frac{1}{b}(c^T : b)T \begin{pmatrix} c \\ b \end{pmatrix} \tag{10.14}$$

Finally, the determinants satisfy

$$|\boldsymbol{S}| = (s_{kk} - \boldsymbol{y}^T\tilde{\boldsymbol{S}}\boldsymbol{y})|\tilde{\boldsymbol{S}}|$$

10.2.6 Sequential search

There are many problems where suboptimal methods must be used. The branch and bound algorithm may not be computationally feasible (the growth in the number of possibilities that must be examined is still an exponential function of the number of variables) or may not be appropriate if the monotonicity property (10.9) does not hold. Suboptimal algorithms, although not capable of examining every feature combination, will assess a set of potentially useful feature combinations. We consider several techniques, varying in complexity.

10.2.6.1 Best individual *N*

The simplest method for choosing the best *N* features is to assign a score (perhaps a ranking statistic such as mutual information) to each of the features in the original set, $\mathcal{X}$, individually. Thus, the features are ordered so that

$$J(X_1) \geq J(X_2) \geq \cdots \geq J(X_p)$$

and we select as our best set of *N* features the *N* features with the best individual scores:

$$\{X_i | i \leq N\}$$

This requires a specification of the feature subset size, *N*. An alternative approach (Guyon, 2008) is to form nested subsets of features, $S_1 = \{X_1\}, S_2 = \{X_1, X_2\}, \ldots, S_p = \{X_1, \ldots, X_p\}$ and to evaluate the model (a classifier appropriate to the data) performance for each feature subset, choosing the smallest feature subset for which performance is near optimal. This approach selects the feature set size and the features. A cross-validation algorithm (see Chapter 9) is given in Algorithm 10.1.

In some cases this method can produce reasonable feature sets, especially if the features in the original set are uncorrelated since the method ignores multivariate relationships. However, if the features of the original set are highly correlated, the chosen feature set will be suboptimal as some of the features will be adding little discriminatory power (some will be redundant).

10.2.6.2 Sequential forward selection (SFS)

Sequential forward selection (or the method of set addition) is a bottom-up search procedure that adds new features to a feature set one at a time until the final feature set is reached. Suppose we have a set of d_1 features, X_{d_1}. For each of the features ξ_j not yet selected (i.e. in $\mathcal{X} - X_{d_1}$) the criterion function $J_j = J(X_{d_1} + \xi_j)$ is evaluated. The feature that yields the maximum value of J_j is chosen as the one that is added to the set X_{d_1}. Thus, at each stage, the variable is chosen that, when added to the current set, maximises the selection criterion. The feature set is initialised to the null set. When the best improvement makes the feature set worse, or when the maximum allowable number of features is reached, the algorithm terminates. The main disadvantage of the method is that it does not include a mechanism

Algorithm 10.1 Cross-validation algorithm for selection of the best set of individually relevant features.

1. Divide the data into training and test sets.
2. Specify a ranking statistic (e.g. mutual information).
3. Rank the features and form the nested subsets of features, $S_1, S_2, \ldots, S_p$.
4. Cross-validation procedure:
   - Split the training data into (e.g.) 10 equal parts, ensuring that all classes are represented in each part; use nine parts for training and the remaining one part for testing.
   - Train the classifier model for each nested subset of variables, S_h, on each subset, k, of the training data in turn, testing on the remaining part. Obtain the performance (for example, error rate), $CV(h, k), h = 1, \ldots, p; k = 1, \ldots, 10$.
   - Average the result

$$CV(h) = \frac{1}{10} \sum_k CV(h, k)$$

5. Select the smallest feature subset, S_{h^*}, such that $CV(h)$ is optimal or near optimal.
6. Evaluate on test dataset using the feature subset S_{h^*}, training on the entire training set and evaluating performance on the test set.

for deleting features from the feature set once they have been added should further additions render them unnecessary.

Algorithm 10.2 gives a modified version of Algorithm 10.1 for sequential forward selection and the following sequential search algorithms. At each stage of the search, sets of subsets are generated for evaluation using the cross-validation procedure.

10.2.6.3 Generalised sequential forward selection (GSFS)

Instead of adding a single feature at a time to a set of measurements, in the generalised sequential forward selection algorithm r features are added as follows. Suppose we have a set of d_1 measurements, X_{d_1}. All possible sets of size r are generated from the remaining $n - d_1$ features – this gives

$$\binom{n - d_1}{r}$$

sets. For each set of r features, Y_r, the cost function is evaluated for $X_{d_1} + Y_r$ and the set that maximises the cost function is used to increment the feature set. This is more costly than sequential forward selection, but has the advantage that at each stage it is possible to take into account to some degree the statistical relationship between the available measurements.

Algorithm 10.2 Cross-validation algorithm for selection of the best set of features in a sequential search procedure.

1. Divide the data into training and test sets.
2. Specify the search strategy - SFS, GSFS, ...

   At each stage of the algorithm:

   (a) Generate subsets of features for evaluation.

   (b) Cross-validation procedure:

   - Split the training data into (e.g.) 10 equal parts, ensuring that all classes are represented in each part; use nine parts for training and the remaining one part for testing.
   - Train the classifier model for each subset of variables, h, on each subset, k, of the training data in turn, testing on the remaining part. Obtain the performance (for example, error rate), $CV(h, k)$.
   - Average the result

   $$CV(h) = \frac{1}{10} \sum_k CV(h, k)$$

   (c) Select the smallest feature subset, S_{h^*}, such that $CV(h)$ is optimal or near optimal, for the next stage of the search.

3. Evaluate on test dataset using the smallest feature subset, S_{h^*}, from the search procedure giving best performance, training on the entire training set and evaluating performance on the test set.

10.2.6.4 Sequential backward selection (SBS)

Sequential backward selection, or sequential backward elimination, is the top-down analogy to sequential forward selection. Variables are deleted one at a time until d measurements remain. Starting with the complete set, the variable ξ_j is chosen for which $J(\mathcal{X} - \xi_j)$ is the largest (i.e. ξ_j decreases J the least). The new set is $\{\mathcal{X} - \xi_j\}$. This process is repeated until a set of the required cardinality remains. The procedure has the disadvantage over sequential forward selection that it is computationally more demanding since the criterion function J is evaluated over larger sets of variables.

10.2.6.5 Generalised sequential backward selection (GSBS)

If you have read the previous sections, you will not be surprised to learn that generalised sequential backward selection decreases the current set of variables by several variables at a time.

10.2.6.6 Plus l – take away r selection

This is a procedure that allows some backtracking in the feature selection process. If $l > r$, it is a bottom-up procedure. l features are added to the current set using sequential forward selection and then the worst r features are removed using the sequential backward selection procedure. This algorithm removes the problem of nesting since the set of features obtained at a given stage is not necessarily a subset of the features at the next stage of the procedure. If $l < r$ then the procedure is top-down, starting with the complete set of features, removing r, then adding l successively until the required number is achieved.

10.2.6.7 Generalised plus l – take away r selection

The generalised version of the l–r algorithm uses the GSFS and the GSBS algorithms at each stage rather than the SFS and SBS procedures. Kittler (1978) generalises the procedure further by allowing the integers l and r to be composed of several components l_i, $i = 1, \ldots, n_l$, and r_j, $j = 1, \ldots, n_r$ (where n_l and n_r are the number of components), satisfying

$$0 \leq l_i \leq l \qquad 0 \leq r_i \leq r$$

$$\sum_{i=1}^{n_l} l_i = l \qquad \sum_{i=1}^{n_r} r_i = r$$

In this generalisation, instead of applying the generalised sequential forward selection in one step of l variables [denoted GSFS(l)], the feature set is incremented in n_l steps by adding l_i features ($i = 1, \ldots, n_l$) at each increment, i.e. apply GSFS(l_i) successively for $i = 1, \ldots, n_l$. This reduces the computational complexity. Similarly, GSBS(r) is replaced by applying GSBS(r_j), $j = 1, \ldots, n_r$, successively. The algorithm is referred to as the (Z_l, Z_r) algorithm, where Z_l and Z_r denote the sequence of integers l_i and l_j,

$$Z_l = (l_1, l_2, \ldots, l_{n_l})$$
$$Z_r = (r_1, r_2, \ldots, r_{n_r})$$

The suboptimal search algorithms discussed in this subsection and the exhaustive search strategy may be considered as special cases of the (Z_l, Z_r) algorithm (Devijver and Kittler, 1982).

10.2.6.8 Floating search methods

Floating search methods, SFFS (sequential forward floating selection) and SBFS (sequential backward floating selection), may be regarded as a development of the l–r algorithm above in which the values of l and r are allowed to 'float' – that is, they may change at different stages of the selection procedure.

Suppose that at stage k we have a set of subsets $X_1, \ldots, X_k$ of sizes 1 to k, respectively. Let the corresponding values of the feature selection criteria be J_1 to J_k, where $J_i = J(X_i)$, for the feature selection criterion, $J(.)$. Let the total set of features be $\mathcal{X}$. At the kth stage of the SFFS procedure:

1. Select the feature x_j from $Y - X_k$ that increases the value of J the greatest and add it to the current set: $X_{k+1} = X_k + x_j$.
2. Find the feature, x_r, in the current set, X_{k+1}, that reduces the value of J the least; if this feature is the same as x_j then set $J_{k+1} = J(X_{k+1})$; increment k; go to step 1; otherwise remove it from the set to form $X'_k = X_{k+1} - x_r$.
3. Continue removing features from the set X'_k to form reduced sets X'_{k-1} while $J(X'_{k-1}) > J_{k-1}$; $k = k - 1$; until $k = 2$; then continue with step 1.

The algorithm is initialised by setting $k = 0$ and $X_0 = \emptyset$ (the empty set) and using the SFS method until a set of size 2 is obtained.

10.2.7 Random search

Injecting randomness into methods of feature subset selection can be a useful approach in the following cases:

- when deterministic algorithms are susceptible to becoming trapped in local optima of the feature selection criterion;
- when the benefits of obtaining a good solution significantly outweigh the costs of a poor one, i.e. it is worth the additional computational time involved in a random approach;
- when the space of possible feature subsets is large.

There are two sources of randomness in feature selection algorithms

1. Randomise the set of input variables to a classification algorithm. The search procedure may start with a randomly selected subset and either

   (a) follow a sequential procedure, but inject randomness. An example is an approach based on simulated annealing where a new subset is proposed in an interative manner. If the new solution results in a higher value of the feature selection criterion function, J, then it is retained. Otherwise, the new solution is retained with a probability that depends on the difference between the current value of J and the new one, and a parameter called the temperature;

   or

   (b) generate the next subset completely at random, i.e the next subset does not grow or shrink from the current subset.

   An advantage of this approach is that it provides some protection against getting stuck in local minima of the feature selection criterion. A disadvantage is that if only one or a few subsets of variables give good performance, or performance significantly better than other subsets, then the probability of selecting a good solution may be small.
2. Randomise the set of training samples. The dataset used to train the feature subset selection algorithm is a subset of the complete training dataset, generated randomly.

Algorithm 10.3 IAMB algorithm for finding the Markov blanket of a target variable, T.

1. Set the Markov blanket of the target variable to be the empty set, $MB(T) = \phi$.

Growing phase

2. Repeat until there is no change in $MB(T)$.
   (a) Find the variable $v \in \chi - \{T\} - MB(T)$ such that $MI(v, T|MB(T))$ is a maximum.
   (b) If *Not CondInd*$(v, T|MB(T))$ then add v to $MB(T)$.

Shrinking phase

3. For each $v \in MB(T)$
   (a) If *CondInd*$(v, T|MB(T) - \{v\})$ then remove v from $MB(T)$.
4. Return $MB(T)$.

Thus the algorithm selects at random which patterns to use and then performs a traditional deterministic search.

This approach is useful when the number of patterns is very large or the available computation time is short.

Examples of random procedures in feature selection are provided by Stracuzzi (2007). Specific studies include those of Hussein *et al.* (2001) on genetic algorithms (Mitchell, 1997) and Chang (1973) who considers a dynamic programming approach. Monte Carlo methods based on simulated annealing and genetic algorithms are described by Siedlecki and Sklansky (1988), Brill *et al.* (1992) and Chang and Lippmann (1991).

10.2.8 Markov blanket

There has been considerable interest over the past decade in the use of a Markov blanket as a method of feature subset selection as part of a filter method. The Markov blanket of the target variable (the class variable) contains a minimal set of variables. All other variables are conditionally independent of the target variable given the variables in the Markov blanket. Many algorithms for Markov blanket construction have been developed [see Fu and Desmarais (2010) for a review of Markov blanket based feature selection]. Once learnt, the features may be used in a general classifier, or the structure of the Markov blanket exploited in classifier design.

We present the basic IAMB (Incremental Association Markov Blanket) algorithm. It has the advantage that it is simple to implement and predicts a minimal set of features. There have been several developments to address data efficiency issues (e.g. Fast-IAMB: Yaramakala and Margaritis, 2005).

The IAMB algorithm (see Algorithm 10.3) consists of two phases – growing and shrinking. During the growing phase, the features that have a strong dependency with the class variable are identified. The dependency is measured using conditional mutual information. The conditional mutual information is the expected value of the mutual information of two variables given the third and is given by

$$MI(X, Y|Z = z) = \sum_{z} p(z) \sum_{x} \sum_{y} p(x, y|z) \log \left(\frac{p(x, y|z)}{p(x|z)p(y|z)} \right)$$

or

$$MI(X, Y|Z = z) = \sum_{x} \sum_{y} \sum_{z} p(x, y, z) \log \left(\frac{p(x, y, z)}{p(x|z)p(y|z)p(z)} \right)$$

which is the Kullback–Leibler divergence, $KL(p(x, y, z)|p(x|z)p(y|z)p(z))$.

The variable with the strongest dependency is tested to see whether it is conditionally independent of the class variable given the current Markov blanket using the procedure *CondInd*() (there are a number of parametric and nonparametric tests for conditional independence for continuous and categorical variables; Agresti, 1990). If it is not, the Markov blanket is incremented.

The shrinking stage considers each variable, v, in the learnt Markov blanket in turn and removes those that are conditionally independent of the target variable given the variables in the Markov blanket with variable v removed. This shrinking step is necessary to remove false positives (variables that are redundant) from the grown Markov blanket.

The IAMB algorithm has a sound theory, it is simple to implement, but some of the variants proposed achieve better performance on data efficiency (FAST-IAMB: Yaramakala and Margaritis, 2005; parallel-IAMB: Aliferis *et al.*, 2002).

10.2.9 Stability of feature selection

We have reviewed many approaches to feature subset selection. However, the different algorithms proposed do not necessarily produce the same ranking of features for a given dataset. Even for a single method, different feature subsets (each giving similar predictive accuracy) may be identified on different random subsamplings of the data. Such an effect is particularly noticeable for datasets where the number of features greatly exceeds the number of patterns (as in problems involving microarray data). Robustness, or stability, of a feature selection method is the variation in feature subsets, delivered by a feature selection method, due to small changes in the dataset. High reproducibility of features may be as important as high classification accuracy, particularly where interpretation of the output by human experts is required.

10.2.9.1 Measuring stability

A stability measure, s_t, is defined in terms of a similarity between pairs of feature subsets (Saeys *et al.*, 2008)

$$s_t = \frac{2 \sum_{i=1}^{k-1} \sum_{j=i+1}^{k} S(f_i, f_j)}{k(k-1)}$$

where k is the number of feature subsets; f_i is the output of the ith feature selection method (or the output of a feature selection method applied to the ith subset of the data, $1 \leq i \leq k$); and $S(f_i, f_j)$ is a similarity between the outputs f_i and f_j. These outputs may be a feature subset, a set of weights for the features or a ranking of the features.

There is a variety of similarity measures (see the review by He and Yu, 2010) defined for feature selection that produce as outputs sets of features, feature weightings or ranking of the features. We give an example of each type.

Feature subsets

Here f_i is a subset of the original features. The Jaccard coefficient is

$$S(f_i, f_j) = \frac{|f_i \cap f_j|}{|f_i \cup f_j|}$$

the ratio of the number of features common to subsets f_i and f_j to the total number of features in subsets f_i and f_j.

Weights

We now take f_i to be a set of weightings for all of the features resulting from the ith feature selection method. The correlation coefficient is

$$S(f_i, f_j) = \frac{\sum_l \left(f_i^l - \mu_{f_i}\right)\left(f_j^l - \mu_{f_j}\right)}{\sqrt{\sum_l \left(f_i^l - \mu_{f_i}\right)^2 \left(f_j^l - \mu_{f_j}\right)^2}}$$

where f_i^l is the weighting on feature l for feature selection method i and μ_{f_i} is the mean value.

Rank orderings

The Spearman rank correlation coefficient is

$$S(f_i, f_j) = 1 - 6 \sum_l \frac{\left(f_i^l - f_j^l\right)^2}{N\left(N^2 - 1\right)}$$

where f_i^l is the rank of feature l for feature selection method i.

10.2.9.2 Robust feature selection techniques

The stability measures above have been used to assess the stability of individual feature selection methods. Ensemble methods of feature selection, where feature subsets from different feature selection algorithms are combined, have been found to provide a robust approach. Stability of feature selection to variations in the training set has been investigated by Kalousis *et al.* (2005). Bayesian model averaging for gene selection and classification of microarray data (investigating robust approaches to feature selection) has been assessed by Yeung *et al.* (2005) (see also Abeel *et al.*, 2010). See Chapter 8 on ensemble methods.

10.2.10 Example application study

The problem

The application involves intrusion detection in mobile ad hoc networks (Wang *et al.*, 2005).

Summary

The purpose of this study was to investigate techniques for reducing the number of features collected by each node in a network with limited capacity. The aim is to use the features to detect patterns of well-known attacks or normal patterns of behaviour in order to detect potential attacks (as anomalies).

The data

The features comprise a combination of packet types, flow directions and sampling periods. Overall there are 150 features, but 25 were considered in the study. There were 200 patterns.

The model

The data were modelled using a Bayesian network, from which a Markov blanket was extracted. This is not the most efficient approach since the entire Bayesian network is not required – only the Markov blanket of the target node.

Training procedure

The approach taken was to start with a random structure for the Bayesian network and to identify neighbours of the structure by adding or deleting an edge. A *Minimum Description Length* score (see Chapter 13) was evaluated for each neighbour and the current network updated if the score is above that of the current structure. A Markov blanket was extracted from the final structure and the variables used in a classifier algorithm (decision trees and Bayesian network) provided by the WEKA (Witten and Frank, 2005) software. The procedure was run several times and a 10-fold cross-validation used to estimate performance of four classifiers: two classifier types and two feature set sizes (25 and 4, the number of feature selected by the Markov blanket algorithm).

Results

Four features were selected using the Markov blanket approach and the performance of the classifiers using the four features was similar to, but not quite so good as, the performance using all 25 features.

10.2.11 Further developments

Developments of the floating search methods to adaptive floating search algorithms that determine the number of forward (adding) or backward (removing) steps dynamically as the algorithm proceeds are proposed by Somol *et al.* (1999). Other approaches include node pruning in neural networks and methods based on modelling the class densities as finite mixtures of a special type (Pudil *et al.*, 1995).

There is a growing trend towards ensemble methods – combining classifiers based on different feature subsets – as these have demonstrated improved robustness and increased predictive performance. Instead of choosing the one particular feature selection method and accepting its best outcome as the final subset, the results of different methods can be combined.

Alternatively, different solutions produced by a single method can be combined using Bayesian model averaging (see Chapter 8).

10.2.12 Summary

Feature selection is the process of selecting from the original features (or variables) those features that are important for classification, removing irrelevant and redundant variables. A feature selection criterion, J, is defined on subsets of the features and we seek that combination of features for which J is maximised.

In this section we have described some statistical pattern recognition approaches to feature selection, both optimal and suboptimal techniques for maximising J. Some of the techniques (filter approaches) are independent of a specific classifier, but others integrate feature selection with a classification model (wrapper and embedded models). Integrating with a classifier is an additional computational burden if the classifier is complex. Different feature sets may be obtained for different classifiers.

Generally there is a trade-off between algorithms that are computationally feasible, but not optimal, and those that are optimal or close to optimal but are computationally complex even for moderate feature set sizes. Studies of the floating methods suggest that these offer close to optimal performance at an acceptable cost.

Ensemble methods offer improved robustness and are considered in Chapter 8.

10.3 Linear feature extraction

Feature extraction is the transformation of the original data (using all variables) to a dataset with a reduced number of variables.

In the problem of feature selection covered in the previous section, the aim is to select those variables that contain the most discriminatory information. Alternatively, we may wish to limit the number of measurements we make, perhaps on grounds of cost, or we may want to remove redundant or irrelevant information to obtain a reduced complexity classifier.

In feature extraction, all available variables are used and the data are transformed (using a linear or nonlinear transformation) to a reduced dimension space. Thus, the aim is to replace the original variables by a smaller set of underlying variables. There are several reasons for performing feature extraction:

1. to reduce the bandwidth of the input data (with the resulting improvements in speed and reductions in data requirements);
2. to provide a relevant set of features for a classifier, resulting in improved performance, particularly from simple classifiers;
3. to reduce redundancy;
4. to recover new meaningful underlying variables or features that describe the data, leading to greater understanding of the data generation process;
5. to produce a low-dimensional representation (ideally in two dimensions) with minimum loss of information so that the data may easily be viewed and relationships and structure in the data identified.

The techniques covered in this section are data adaptive. The transformations of the original features are dependent on the dataset. Thus we exclude from our treatment preprocessing of the data by fixed linear transformations – the discrete Fourier transform or the discrete cosine transform – and multi-resolution analyses such as the discrete wavelet transform. These are important transformations, particularly for preprocessing signals and images [for a treatment of these transformations, scc Nixon and Aquado (2008) and Gonzalez and Woods (2008)].

The methods described here are to be found in the literature on a diverse range of topics. Many are techniques of *exploratory data analysis* described in the textbooks on multivariate analysis. Sometimes referred to as *geometric methods* or methods of *ordination*, they make no assumption about the existence of groups or clusters in the data. They have found application in a wide range of subjects including ecology, agricultural science, biology and psychology. Geometric methods are sometimes further categorised as being *variable-directed* in that they are primarily concerned with relationships between variables, or *individual-directed* which are primarily concerned with relationships between individuals.

In the pattern recognition literature, the data transformation techniques are termed *feature selection in the transformed space* or *feature extraction* and the techniques may be supervised (make use of class label information) or unsupervised. They may be based on the optimisation of a class separability measure, such as those described in the previous section.

10.3.1 Principal components analysis

10.3.1.1 Introduction

Principal components analysis originated in work by Pearson (1901). It is the purpose of principal components analysis to derive new variables (in decreasing order of importance) that are linear combinations of the original variables and are uncorrelated. Geometrically, principal components analysis can be thought of as a rotation of the axes of the original coordinate system to a new set of orthogonal axes that are ordered in terms of the amount of variation of the original data they account for.

One of the reasons for performing a principal components analysis is to find a smaller group of underlying variables that describe the data. In order to do this, we hope that the first few variables or *components* will account for most of the variation in the original data. However, it does not follow that we will be able to assign an interpretation to these new variables.

Principal components analysis is a variable-directed technique. It makes no assumptions about the existence or otherwise of groupings within the data and so is described as an unsupervised feature extraction technique.

So far, our discussion has been purely descriptive. We have introduced many terms without proper definition. There are several ways in which principal components analysis can be described mathematically, but let us leave aside the mathematics for the time being and continue with a geometrical derivation. We must necessarily confine our illustrations to two dimensions, but nevertheless we shall be able to define most of the attendant terminology and consider some of the problems of a principal components analysis.

In Figure 10.8 are plotted a dozen objects, with the x and y values for each point in the figure representing measurements on each of the two variables. They could represent the height and weight of a group of individuals, for example, in which case one variable would

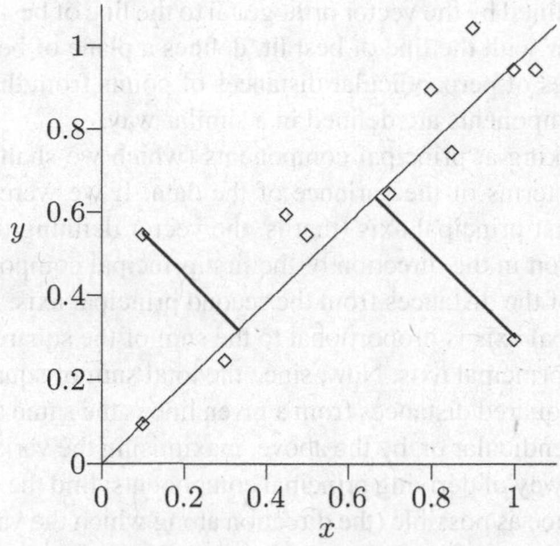

Figure 10.8 Principal components line of best fit.

be measured in metres or centimetres or inches and the other variable in grams or pounds. So the units of measurement may differ.

The problem we want to solve is: What is the best straight line through this set of points? Before we can answer, we must clarify what we mean by best. If we consider the variable x to be an input variable and y a dependent variable so that we wish to calculate the expected value of y given x, $E[y|x]$, then the best (in a least squares sense) regression line of y on x

$$y = mx + c$$

is the line for which the sum of the square distances of points from the line is a minimum, and the distance of a point from the line is the vertical distance.

If y were the *regressor* and x the dependent variable, then the linear regression line is the line for which the sum of squares of horizontal distances of points from the line is a minimum. Of course, this gives a different solution [a good illustration of the two linear regressions on a bivariate distribution is given in Stuart and Ord (1991)].

So we have two lines of best fit and a point to note is that changing the scale of the variables does not alter the predicted values. If the scale of x is compressed or expanded, the slope of the line changes but the predicted value of y does not alter. Principal components analysis produces a single best line and the constraint that it satisfies is that the sum of the squares of the *perpendicular* distances from the sample points to the line is a minimum (Figure 10.8).

A standardisation procedure that is often carried out (and almost certainly if the variables are measured in different units) is to make the variance of each variable unity. Thus the data are transformed to new axes, centred at the centroid of the data sample and in coordinates defined in terms of units of standard deviation. The principal component line of best fit is not invariant to changes of scale.

The variable defined by the line of best fit is the first principal component. The second principal component is the variable defined by the line that is orthogonal with the first and so it is uniquely defined in our two-dimension example. In a problem with higher-dimensional data, it is the variable defined by the vector orthogonal to the line of best fit of the first principal component that, together with the line of best fit, defines a plane of best fit, i.e. the plane for which the sum of squares of perpendicular distances of points from the plane is a minimum. Successive principal components are defined in a similar way.

Another way of looking at principal components (which we shall derive more formally in Section 10.3.1) is in terms of the variance of the data. If we were to project the data in Figure 10.8 on to the first principal axis (that is, the vector defining the first principal component), then the variation in the direction of the first principal component is proportional to the sum of the squares of the distances from the second principal axis. Similarly, the variance along the second principal axis is proportional to the sum of the squares of the perpendicular distances from the first principal axis. Now, since the total sum of squares is a constant, then minimising the sum of squared distances from a given line is the same as maximising the sum of squares from its perpendicular or, by the above, maximising the variance in the direction of the line. This is another way of deriving principal components: find the direction that accounts for as much of the variance as possible (the direction along which the variance is a maximum); the second principal component is defined by the direction orthogonal to the first for which the variance is a maximum, and so on. The variances are the *principal values*.

Principal components analysis produces an orthogonal coordinate system in which the axes are ordered in terms of the amount of variance in the original data for which the corresponding principal components account. If the first few principal components account for most of the variation, then these may be used to describe the data, thus leading to a reduced-dimension representation. It may be possible to interpret the new components as something meaningful in terms of the original variables, but in practice the new components will be difficult to interpret.

10.3.1.2 Derivation of principal components

There are at least three ways in which we can approach the problem of deriving a set of principal components. Let $x_1, \ldots, x_p$ be our set of original variables and let ξ_i, $i = 1, \ldots, p$, be linear combinations of these variables

$$\xi_i = \sum_{j=1}^{p} a_{ij} x_j$$

or

$$\boldsymbol{\xi} = \boldsymbol{A}^T \boldsymbol{x}$$

where $\boldsymbol{\xi}$ and $\boldsymbol{x}$ are vectors of random variables and $\boldsymbol{A}$ is the matrix of coefficients (the columns of $\boldsymbol{A}$ are the vectors $\boldsymbol{a}_i$ with jth component a_{ij}). Then

1. We may seek the orthogonal transformation $\boldsymbol{A}$ yielding new variables ξ_j that have stationary values of their variance. This approach, due to Hotelling (1933), is the one that we choose to present in more detail below.
2. We may seek the orthogonal transformation that gives uncorrelated variables ξ_j.

3. We may consider the problem geometrically and find the line for which the sum of squares of perpendicular distances is a minimum; then the plane of best fit and so on. We used this geometric approach in our two-dimensional illustration above (Pearson, 1901).

Consider the first variable ξ_1:

$$\xi_1 = \sum_{j=1}^{p} a_{1j} x_j$$

We choose $\boldsymbol{a}_1 = (a_{11}, a_{12}, \ldots, a_{1p})^T$ to maximise the variance of ξ_1, subject to the constraint $\boldsymbol{a}_1^T \boldsymbol{a}_1 = |\boldsymbol{a}_1|^2 = 1$. The variance of ξ_1 is

$$\begin{aligned}
\text{var}(\xi_1) &= E\left[\xi_1^2\right] - E\left[\xi_1\right]^2 \\
&= E\left[\boldsymbol{a}_1^T \boldsymbol{x}\boldsymbol{x}^T \boldsymbol{a}_1\right] - E\left[\boldsymbol{a}_1^T \boldsymbol{x}\right] E\left[\boldsymbol{x}^T \boldsymbol{a}_1\right] \\
&= \boldsymbol{a}_1^T \left(E\left[\boldsymbol{x}\boldsymbol{x}^T\right] - E\left[\boldsymbol{x}\right] E\left[\boldsymbol{x}^T\right]\right) \boldsymbol{a}_1 \\
&= \boldsymbol{a}_1^T \boldsymbol{\Sigma} \boldsymbol{a}_1
\end{aligned}$$

where $\boldsymbol{\Sigma}$ is the covariance matrix of $\boldsymbol{x}$ and E[.] denotes expectation. Finding the stationary value of $\boldsymbol{a}_1^T \boldsymbol{\Sigma} \boldsymbol{a}_1$ subject to the constraint $\boldsymbol{a}_1^T \boldsymbol{a}_1 = 1$ is equivalent to finding the unconditional stationary value of

$$f(\boldsymbol{a}_1) = \boldsymbol{a}_1^T \boldsymbol{\Sigma} \boldsymbol{a}_1 - \nu \boldsymbol{a}_1^T \boldsymbol{a}_1$$

where ν is a Lagrange multiplier. (The method of Lagrange multipliers can be found in most textbooks on mathematical methods; Wylie and Barrett, 1995.) Differentiating with respect to each of the components of $\boldsymbol{a}_1$ in turn and equating to zero gives

$$\boldsymbol{\Sigma} \boldsymbol{a}_1 - \nu \boldsymbol{a}_1 = 0$$

For a nontrivial solution for $\boldsymbol{a}_1$ (that is, a solution other than the null vector), $\boldsymbol{a}_1$ must be an eigenvector of $\boldsymbol{\Sigma}$ with ν an eigenvalue. Now $\boldsymbol{\Sigma}$ has p eigenvalues $\lambda_1, \ldots, \lambda_p$, not all necessarily distinct nor all nonzero, but they can be ordered so that $\lambda_1 \geq \lambda_2 \geq \ldots \geq \lambda_p \geq 0$. We must chose one of these for the value of ν. Now, since the variance of $\boldsymbol{\xi}_1$ is

$$\begin{aligned}
\boldsymbol{a}_1^T \boldsymbol{\Sigma} \boldsymbol{a}_1 &= \nu \boldsymbol{a}_1^T \boldsymbol{a}_1 \\
&= \nu
\end{aligned}$$

and we wish to maximise this variance, then we choose ν to be the largest eigenvalue λ_1, and $\boldsymbol{a}_1$ is the corresponding eigenvector. This eigenvector will not be unique if the value of ν is a repeated root of the *characteristic equation*

$$|\boldsymbol{\Sigma} - \nu \boldsymbol{I}| = 0$$

The variable ξ_1 is the first *principal component* and has the largest variance of any linear function of the original variables $x_1, \ldots, x_p$.

The second principal component, $\xi_2 = \boldsymbol{a}_2^T \boldsymbol{x}$, is obtained by choosing the coefficients a_{2i}, $i = 1, \ldots, p$, so that the variance of ξ_2 is maximised subject to the constraint $|\boldsymbol{a}_2| = 1$ *and* that ξ_2 is uncorrelated with the first principal component ξ_1. This second constraint implies

$$E[\xi_2 \xi_1] - E[\xi_2]E[\xi_1] = 0$$

or

$$\boldsymbol{a}_2^T \boldsymbol{\Sigma} \boldsymbol{a}_1 = 0 \tag{10.15}$$

and since $\boldsymbol{a}_1$ is an eigenvector of $\boldsymbol{\Sigma}$, this is equivalent to $\boldsymbol{a}_2^T \boldsymbol{a}_1 = 0$, i.e. $\boldsymbol{a}_2$ is orthogonal to $\boldsymbol{a}_1$.

Using the method of Lagrange's undetermined multipliers again, we seek the unconstrained maximisation of

$$\boldsymbol{a}_2^T \boldsymbol{\Sigma} \boldsymbol{a}_2 - \mu \boldsymbol{a}_2^T \boldsymbol{a}_2 - \eta \boldsymbol{a}_2^T \boldsymbol{a}_1$$

Differentiating with respect to the components of $\boldsymbol{a}_2$ and equating to zero gives

$$2\boldsymbol{\Sigma} \boldsymbol{a}_2 - 2\mu \boldsymbol{a}_2 - \eta \boldsymbol{a}_1 = 0 \tag{10.16}$$

Multiplying by $\boldsymbol{a}_1^T$ gives

$$2\boldsymbol{a}_1^T \boldsymbol{\Sigma} \boldsymbol{a}_2 - \eta = 0$$

since $\boldsymbol{a}_1^T \boldsymbol{a}_2 = 0$. Also, by Equation (10.15), $\boldsymbol{a}_2^T \boldsymbol{\Sigma} \boldsymbol{a}_1 = \boldsymbol{a}_1^T \boldsymbol{\Sigma} \boldsymbol{a}_2 = 0$, therefore $\eta = 0$. Equation (10.16) becomes

$$\boldsymbol{\Sigma} \boldsymbol{a}_2 = \mu \boldsymbol{a}_2$$

Thus, $\boldsymbol{a}_2$ is also an eigenvector of $\boldsymbol{\Sigma}$, orthogonal to $\boldsymbol{a}_1$. Since we are seeking to maximise the variance, it must be the eigenvector corresponding to the largest of the remaining eigenvalues, that is the second largest eigenvalue overall.

We may continue this approach, with the kth principal component $\xi_k = \boldsymbol{a}_k^T \boldsymbol{x}$, where $\boldsymbol{a}_k$ is the eigenvector corresponding to the kth largest eigenvalue of $\boldsymbol{\Sigma}$ and variance equal to the kth largest eigenvalue.

If some eigenvalues are equal, the solution for the eigenvectors is not unique, but it is always possible to find an orthonormal set of eigenvectors for a real symmetric matrix with eigenvalues ≥ 0.

In matrix notation,

$$\boldsymbol{\xi} = \boldsymbol{A}^T \boldsymbol{x} \tag{10.17}$$

where $\boldsymbol{A} = [\boldsymbol{a}_1 | \ldots | \boldsymbol{a}_p]$, the matrix whose columns are the eigenvectors of $\boldsymbol{\Sigma}$.

So now we know how to determine the principal components – by performing an eigenvector decomposition of the symmetric positive definite matrix $\mathbf{\Sigma}$, and using the eigenvectors as coefficients in the linear combination of the original variables. But how do we determine a reduced-dimension representation of some given data? Let us consider the variance.

The sum of the variances of the principal components is given by

$$\sum_{i=1}^{p} \text{var}(\xi_i) = \sum_{i=1}^{p} \lambda_i$$

the sum of the eigenvalues of the covariance matrix $\mathbf{\Sigma}$, equal to the total variance of the original variables. We can then say that the first k principal components account for

$$\sum_{i=1}^{k} \lambda_i \Bigg/ \sum_{i=1}^{p} \lambda_i$$

of the total variance.

We can now consider a mapping to a reduced dimension by specifying that the new components must account for at least a fraction d of the total variance. The value of d would be specified by the user. We then choose k so that

$$\sum_{i=1}^{k} \lambda_i \geq d \sum_{i=1}^{p} \lambda_i \geq \sum_{i=1}^{k-1} \lambda_i$$

and transform the data to

$$\boldsymbol{\xi}_k = \boldsymbol{A}_k^T \boldsymbol{x}$$

where $\boldsymbol{\xi}_k = (\xi_1, \ldots, \xi_k)^T$ and $\boldsymbol{A}_k = [\boldsymbol{a}_1|\ldots|\boldsymbol{a}_k]$ is a $p \times k$ matrix. Choosing a value of d between 70% and 90% preserves most of the information in $\boldsymbol{x}$ (Jolliffe, 1986). Jackson (1991) recommends against the use of this procedure: it is difficult to choose an appropriate value for d – it is very much problem-specific.

An alternative approach is to examine the eigenvalue spectrum and see if there is a point where the values fall sharply before levelling off at small values (the 'scree' test). We retain those principal components corresponding to the eigenvalues before the cut-off point or 'elbow' (Figure 10.9).

However, on occasion the eigenvalues drift downwards with no obvious cutting point and the first few eigenvalues account for only a small proportion of the variance.

It is very difficult to determine the 'right' number of components and most tests are for limited special cases and assume multivariate normality. Jackson (1991) describes a range of procedures and reports the results of several comparative studies.

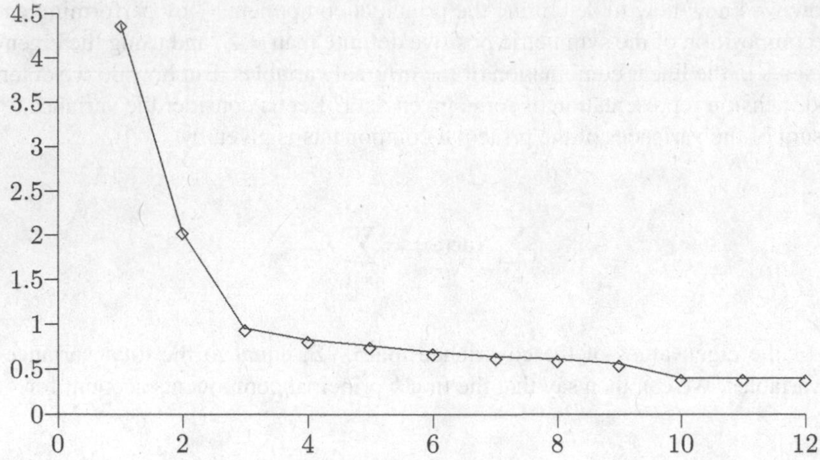

Figure 10.9 Eigenvalue spectrum: a plot of the ordered eigenvalues against eigenvalue number, with an 'elbow' at the third eigenvalue.

10.3.1.3 Remarks

Sampling

The derivation of principal components above has assumed that the covariance matrix $\mathbf{\Sigma}$ is known. In most practical problems, we shall know the sample covariance matrix estimated from a set of sample vectors. We can use the sample covariance matrix to calculate principal components and take these as estimates of the eigenvectors of the covariance matrix $\mathbf{\Sigma}$. Note also, as far as deriving a reduced-dimension representation is concerned, the process is *distribution-free* – a mathematical method with no underlying statistical model. Therefore, unless we are prepared to assume some model for the data, it is difficult to obtain results on how good the estimates of the principal components are.

Standardisation

The principal components are dependent on the scales used to measure the original variables. Even if the units used to measure the variables are the same, if one of the variables has a range of values that greatly exceeds the others, then we expect the first principal component to lie in the direction of this axis. If the units of each variable differ (for example, height, weight), then the principal components will depend on whether the height is measured in feet, inches of centimetres, etc. This does not occur in regression (which is independent of scale) but it does in principal components analysis in which we are minimising a perpendicular distance from a point to a line, plane, etc. and right angles do not change to right angles under changes of scale. The practical solution to this problem is to standardise the data so that the variables have equal range. A common form of standardisation is to transform the data to have zero mean, unit variance so that we find the principal components from the correlation matrix. This gives equal importance to the original variables. We recommend that all data are standardised to zero mean, unit variance. Other forms of standardisation are possible; for example, the data may be transformed logarithmically before a principal components analysis – Baxter

(1995) compares several approaches. The standardisation applied to training data should also be applied to test data.

Mean correction
Equation (10.17) relates the principal components $\boldsymbol{\xi}$ to the observed random vector $\boldsymbol{x}$. In general $\boldsymbol{\xi}$ will not have zero mean. In order for the principal components to have zero mean,[4] they should be defined as

$$\boldsymbol{\xi} = \boldsymbol{A}^T(\boldsymbol{x} - \boldsymbol{\mu}) \tag{10.18}$$

for mean $\boldsymbol{\mu}$. In practice $\boldsymbol{\mu}$ is the sample mean, $\boldsymbol{m}$.

Approximation of data samples
We have seen that in order to represent data in a reduced dimension, we retain only the first few principal components (the number is usually determined by the data). Thus, a data vector $\boldsymbol{x}$ is projected on to the first r (say) eigenvectors of the sample covariance matrix giving

$$\boldsymbol{\xi}_r = \boldsymbol{A}_r^T(\boldsymbol{x} - \boldsymbol{\mu}) \tag{10.19}$$

where $\boldsymbol{A}_r = [\boldsymbol{a}_1|\ldots|\boldsymbol{a}_r]$ is the $p \times r$ matrix whose columns are the first r eigenvectors of the sample covariance matrix and $\boldsymbol{\xi}_r$ is used to denote the measurements on variables $\xi_1, \ldots, \xi_r$ (the first r principal components).

In representing a data point as a point in a reduced dimension, there is usually an error involved and it is of interest to know what the point $\boldsymbol{\xi}_r$ corresponds to in the original space.

The variable $\boldsymbol{\xi}$ is related to $\boldsymbol{x}$ by Equation (10.18) ($\boldsymbol{A}^T = \boldsymbol{A}^{-1}$), giving

$$\boldsymbol{x} = \boldsymbol{A}\boldsymbol{\xi} + \boldsymbol{\mu}$$

If $\boldsymbol{\xi} = (\boldsymbol{\xi}_r, 0)^T$, a vector with its first r components equal to $\boldsymbol{\xi}_r$ and remaining equal to zero, then the point corresponding to $\boldsymbol{\xi}_r$, namely $\boldsymbol{x}_r$, is

$$\begin{aligned}\boldsymbol{x}_r &= \boldsymbol{A}\begin{pmatrix}\boldsymbol{\xi}_r \\ 0\end{pmatrix} + \boldsymbol{\mu} \\ &= \boldsymbol{A}_r\boldsymbol{\xi}_r + \boldsymbol{\mu}\end{aligned}$$

and by Equation (10.19)

$$\boldsymbol{x}_r = \boldsymbol{A}_r\boldsymbol{A}_r^T(\boldsymbol{x} - \boldsymbol{\mu}) + \boldsymbol{\mu}$$

The transformation $\boldsymbol{A}_r\boldsymbol{A}_r^T$ is of rank r and maps the original data distribution to a distribution that lies in an r-dimensional subspace (or on a *manifold* of dimension r) in $\mathbb{R}^p$. The vector $\boldsymbol{x}_r$ is the position the point $\boldsymbol{x}$ maps down to, given in the original coordinate system (the projection

[4] The zero-mean condition on the principal components will be satisfied if the recommended standardisation is performed.

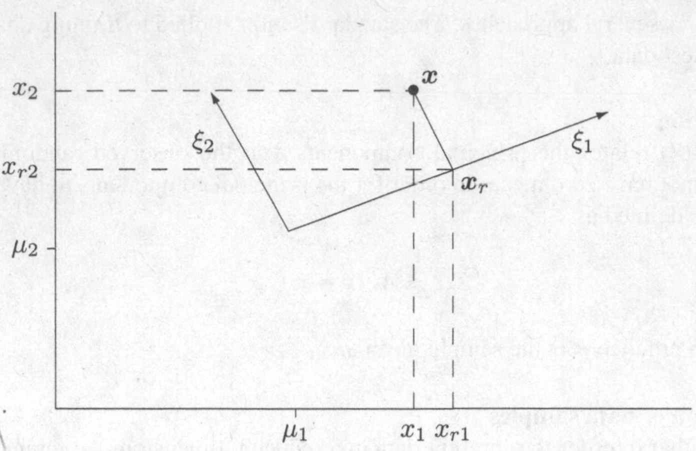

Figure 10.10 Reconstruction from projections: $\boldsymbol{x}$ is approximated by $\boldsymbol{x}_r$ using the first principal component.

of $\boldsymbol{x}$ on to the space defined by the first r principal components). This is illustrated in Figure 10.10 for a projection of a two-dimensional point onto its first principal component.

Singular value decomposition

The right singular vectors of a matrix $\boldsymbol{Z}$ are the eigenvectors of $\boldsymbol{Z}^T\boldsymbol{Z}$ (Figure 5.4). The sample covariance matrix (unbiased estimate) can be written in such a form

$$\frac{1}{n-1}\sum_{i=1}^{n}(\boldsymbol{x}_i - \boldsymbol{m})(\boldsymbol{x}_i - \boldsymbol{m})^T = \frac{1}{n-1}\tilde{\boldsymbol{X}}^T\tilde{\boldsymbol{X}}$$

where $\tilde{\boldsymbol{X}} = \boldsymbol{X} - \boldsymbol{1}\boldsymbol{m}^T$, $\boldsymbol{X}$ is the $n \times p$ data matrix, $\boldsymbol{m}$ is the sample mean and $\boldsymbol{1}$ is an n-dimensional vector of ones. Therefore, if we define

$$\boldsymbol{Z} = \frac{1}{\sqrt{n-1}}\tilde{\boldsymbol{X}} = \frac{1}{\sqrt{n-1}}(\boldsymbol{X} - \boldsymbol{1}\boldsymbol{m}^T)$$

then the right singular vectors of $\boldsymbol{Z}$ are the eigenvectors of the covariance matrix and the singular values are standard deviations of the principal components. Furthermore, setting

$$\boldsymbol{Z} = \frac{1}{\sqrt{n-1}}\tilde{\boldsymbol{X}}\boldsymbol{D}^{-1} = \frac{1}{\sqrt{n-1}}(\boldsymbol{X} - \boldsymbol{1}\boldsymbol{m}^T)\boldsymbol{D}^{-1}$$

where $\boldsymbol{D}$ is a $p \times p$ diagonal matrix with $\boldsymbol{D}_{ii}$ equal to the square root of the variance of the original variables, then the right singular vectors of $\boldsymbol{Z}$ are the eigenvectors of the *correlation* matrix. Thus, given a data matrix, it is not necessary to form the sample covariance matrix in order to determine the principal components.

If the singular value decomposition of $\boldsymbol{X} - \mathbf{1}\boldsymbol{m}^T$ is $\boldsymbol{U}\boldsymbol{S}\boldsymbol{V}^T$, where $\boldsymbol{U} = [\boldsymbol{u}_1| \ldots |\boldsymbol{u}_p]$, $\boldsymbol{S} = \text{Diag}(\sigma_1, \ldots, \sigma_p)$ and $\boldsymbol{V} = [\boldsymbol{v}_1| \ldots |\boldsymbol{v}_p]$, then the $n \times r$ matrix $\boldsymbol{Z}_r$, defined by

$$\boldsymbol{Z}_r = \boldsymbol{U}_r \boldsymbol{\Sigma}_r \boldsymbol{V}_r^T + \mathbf{1}\boldsymbol{m}^T$$

where $\boldsymbol{U}_r = [\boldsymbol{u}_1| \ldots |\boldsymbol{u}_r]$, $\boldsymbol{S}_r = \text{diag}(\sigma_1, \ldots, \sigma_r)$ and $\boldsymbol{V}_r = [\boldsymbol{v}_1| \ldots |\boldsymbol{v}_r]$, is the projection of the original data points on to the hyperplane spanned by the first r principal components and passing through the mean.

Selection of components

There have been many methods proposed for the selection of components in a principal components analysis. There is no single best method as the strategy to adopt will depend on the objectives of the analysis: the set of components that gives a good fit to the data will differ from that which provides good discrimination. Ferré (1995) and Jackson (1993) provide comparative studies. Jackson finds the *broken stick method* one of the most promising, and simple to implement: observed eigenvectors are considered interpretable if their eigenvalues exceed a threshold based on random data; the kth eigenvector is retained if its eigenvalue λ_k exceeds $\sum_{i=k}^{p}(1/i)$. Prakash and Murty (1995) consider a genetic algorithm approach to the selection of components for discrimination. Thus, they calculate principal components as normal, but rather than taking the top k components, a subset from the full set is selected. However, probably the most common approach is the percentage of variance method, retaining eigenvalues that account for approximately 90% of the variance. Another rule of thumb is to retain those eigenvectors with eigenvalues greater than the average (greater than unity for a correlation matrix). For a descriptive analysis, Ferré recommends a rule of thumb method. However, if we do use another more complex approach then we must ask ourselves why are we performing a principal components analysis in the first place? There is no guarantee that a subset of the variables derived will be better for discrimination than a subset of the original variables.

Interpretation

The first few principal components are the most important ones, but it may be very difficult to ascribe meaning to the components. One way this may be done is to consider the eigenvector corresponding to a particular component and select those variables for which the coefficients in the eigenvector are relatively large in magnitude. Then a purely subjective analysis takes place in which the user tries to see what these variables have in common.

An alternative approach is to use an algorithm for orthogonal rotation of axes such as the *varimax* algorithm (Kaiser, 1958, 1959). This rotates the given set of principal axes so that the variation of the squared loadings[5] for a given component is large. This is achieved by making the loadings large on some variables and small on others, though unfortunately it does not necessarily lead to more easily interpretable principal components. Jackson (1991) considers techniques for the interpretation of principal components, including rotation methods.

[5] The loadings are the component values: the weights applied to the original variables.

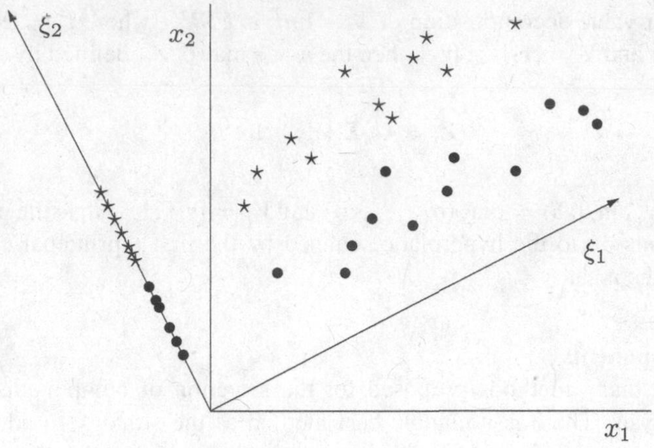

Figure 10.11 Two-group data and the principal axes.

10.3.1.4 Discussion

Principal components analysis is often the first stage in a data analysis and is used to reduce the dimensionality of the data while retaining as much as possible of the variation present in the original dataset.

Principal components analysis takes no account of groups within the data (i.e. it is unsupervised). Although separate groups may emerge as a result of projecting data to a reduced dimension, this is not always the case and dimension reduction may obscure the existence of separate groups.

Figure 10.11 illustrates a dataset in two dimensions with three separate groups and the principal component directions. Projection onto the first eigenvector will remove group separation while projection onto the second retains group separation. Therefore, although dimension reduction may be necessary, the space spanned by the vectors associated with the first few principal components will not necessarily be the best for discrimination.

10.3.1.5 Summary

The stages in performing a principal components analysis are:

1. Form the sample covariance matrix or standardise the data by forming the correlation matrix.
2. Perform an eigendecomposition of the correlation matrix.

OR

1. Standardise the data matrix.
2. Perform a singular value decomposition of the standardised data matrix.

For a reduced-dimension representation of the data, project the data onto the first m eigenvectors, where for example m is chosen using a criterion based on the proportion of variance accounted for.

10.3.2 Karhunen–Loève transformation

The Karhunen–Loève transformation is, in one of its most basic forms, identical to principal components analysis. It is included here because there are variants of the method, occurring in the pattern recognition literature under the general heading of Karhunen–Loève expansion, that incorporate class information in a way in which principal components analysis does not.

The Karhunen–Loève expansion was originally developed for representing a nonperiodic random process as a series of orthogonal functions with uncorrelated coefficients. If $x(t)$ is a random process on $[0, T]$, then $x(t)$ may be expanded as

$$x(t) = \sum_{n=1}^{\infty} x_n \phi_n(t) \tag{10.20}$$

where x_n are random variables and the basis functions ϕ are deterministic functions of time satisfying

$$\int_0^T \phi_n(t)\phi_m^*(t) = \delta_{mn}$$

where ϕ_m^* is the complex conjugate of ϕ_m. Define a correlation function

$$\begin{aligned} R(t, s) &= E\left[x(t)x^*(s)\right] \\ &= E\left[\sum_n \sum_m x_n x_m^* \phi_n(t)\phi_m^*(s)\right] \\ &= \sum_n \sum_m \phi_n(t)\phi_m^*(s) E\left[x_n x_m^*\right] \end{aligned}$$

If the coefficients are uncorrelated ($\mathrm{E}[x_n x_m^*] = \sigma_n^2 \delta_{mn}$) then

$$R(t, s) = \sum_m \sigma_m^2 \phi_m(t)\phi_m^*(s)$$

and multiplying by $\phi_n(s)$ and integrating gives

$$\int R(t, s)\phi_n(s)ds = \sigma_n^2 \phi_n(t)$$

Thus, the functions ϕ_n are the eigenfunctions of the integral equation, with kernel $R(t, s)$ and eigenvalues σ_n^2. We shall not develop the continuous Karhunen–Loève expansion here but proceed straightaway to the discrete case.

If the functions are uniformly sampled, with p samples, then (10.20) becomes

$$\boldsymbol{x} = \sum_{n=1}^{\infty} x_n \boldsymbol{\phi}_n \tag{10.21}$$

and the integral equation becomes

$$\boldsymbol{R}\boldsymbol{\phi}_k = \sigma_k^2 \boldsymbol{\phi}_k$$

where $\boldsymbol{R}$ is now the $p \times p$ matrix with (i, j) element $R_{ij} = \mathrm{E}[x(i)x^*(j)]$. The above equation has only p distinct solutions for $\boldsymbol{\phi}$ and so the summation in (10.21) must be truncated to p. The eigenvectors of $\boldsymbol{R}$ are termed the *Karhunen–Loève coordinate axes* (Devijver and Kittler, 1982).

Apart from the fact that we have assumed zero mean for the random variable $\boldsymbol{x}$, this derivation is identical to that for principal components. Other ways of deriving the Karhunen–Loève coordinate axes are given in the literature, but these correspond to other views of principal components analysis, and so the end result is the same.

In the pattern recognition literature various methods for linearly transforming data to a reduced dimension space defined by eigenvectors of a matrix of second-order statistical moments have been proposed under the umbrella term 'Karhunen–Loève expansion'. These methods could equally be referred to using the term 'generalised principal components' or something similar.

The properties of the Karhunen–Loève expansion are identical to those of the principal components analysis. It produces a set of mutually uncorrelated components, and dimensionality reduction can be achieved by selecting those components with the largest variances. There are many variants of the basic method that incorporate class information or that use different criteria for selecting features.

10.3.2.1 KL1. SELFIC (self-featuring information-compression)

In this procedure, class labels are not used and the Karhunen–Loève feature transformation matrix is $\boldsymbol{A} = [\boldsymbol{a}_1|\ldots|\boldsymbol{a}_p]$, where $\boldsymbol{a}_j$ are the eigenvectors of the sample covariance matrix, $\hat{\boldsymbol{\Sigma}}$, associated with the largest eigenvalues (Watanabe, 1985),

$$\hat{\boldsymbol{\Sigma}}\boldsymbol{a}_i = \lambda_i \boldsymbol{a}_i$$

and $\lambda_1 \geq \ldots \geq \lambda_p$.

This is identical to principal components analysis and is appropriate when class labels are unavailable (unsupervised learning).

10.3.2.2 KL2. Within-class information

If class information is available for the data, then second-order statistical moments can be calculated in a number of different ways. This leads to different Karhunen–Loève coordinate systems. Chien and Fu (1967) propose using the average within-class covariance matrix, $\boldsymbol{S}_W$, as the basis of the transformation. The feature transformation matrix, $\boldsymbol{A}$, is again the matrix of eigenvectors of $\boldsymbol{S}_W$ associated with the largest eigenvalues.

10.3.2.3 KL3. Discrimination information contained in the means

Again, the feature space is constructed from eigenvectors of the averaged within-class covariance matrix, but discriminatory information contained in the class means is used to select the subset of features that will be used in further classification studies (Devijver and Kittler, 1982). For each feature (with eigenvector $\boldsymbol{a}_j$ of $\boldsymbol{S}_W$ and corresponding eigenvalue λ_j) the quantity

$$J_j = \frac{\boldsymbol{a}_j^T \boldsymbol{S}_B \boldsymbol{a}_j}{\lambda_j},$$

where $\boldsymbol{S}_B$ is the between-class scatter matrix, is evaluated and the coordinate axes arranged in descending order of J_j.

10.3.2.4 KL4. Discrimination information contained in the variances

Another means of ordering the feature vectors (eigenvectors $\boldsymbol{a}_j$ of $\boldsymbol{S}_W$) is to use the discriminatory information contained in class variances (Kittler and Young, 1973). There are situations where class mean information is not sufficient to separate the classes and the measure given here uses the dispersion of class-conditional variances. The variance of feature j in the ith class weighted by the prior probability of class ω_i is given by

$$\lambda_{ij} = p(\omega_i)\boldsymbol{a}_j^T \hat{\boldsymbol{\Sigma}}_i \boldsymbol{a}_j$$

where $\hat{\boldsymbol{\Sigma}}_i$ is the sample covariance matrix of class ω_i, and defining $\lambda_j = \sum_{i=1}^{C} \lambda_{ij}$ then a discriminatory measure based on the logarithmic entropy function is

$$H_j = -\sum_{i=1}^{C} \frac{\lambda_{ij}}{\lambda_j} \log\left(\frac{\lambda_{ij}}{\lambda_j}\right)$$

The axes giving low entropy values are selected for discrimination.

A further measure that uses the variances is

$$J_j = \prod_{i=1}^{C} \frac{\lambda_{ij}}{\lambda_j}$$

Both of the above measures reach their maximum when the factors λ_{ij}/λ_j are identical, in which case there is no discriminatory information.

10.3.2.5 KL5. Compression of discriminatory information contained in class means

In the method KL3, the Karhunen–Loève coordinate axes are determined by the eigenvectors of the averaged within-class covariance matrix and the features that are used to represent these data in a reduced-dimension space are determined by ordering the eigenvectors in terms of descending J_j. The quantity J_j is a measure used to represent the discriminatory information contained in the class means. This discriminatory information could be spread over all Karhunen–Loève axes and it is difficult to choose an optimum dimension for the transformed feature space from the values of J_j alone.

The approach considered here (Kittler and Young, 1973) recognises the fact that in a C-class problem, the class means lie in a space of dimension at most $C - 1$, and seeks to find a transformation to a space of at most $C - 1$ giving uncorrelated features.

This is performed in two stages. First of all, a transformation is found that transforms the data to a space in which the averaged within-class covariance matrix is diagonal. This means that the features in the transformed space are uncorrelated, but also any further orthonormal transformation will still produce uncorrelated features for which the *class-centralised vectors* are decorrelated.

If $\boldsymbol{S}_W$ is the average within-class covariance matrix in the original data space, then the transformation that decorrelates the class-centralised vectors is $\boldsymbol{Y} = \boldsymbol{U}^T\boldsymbol{X}$, where $\boldsymbol{U}$ is the matrix of eigenvectors of $\boldsymbol{S}_W$ and the average within-class covariance matrix in the transformed space is

$$\boldsymbol{S}'_W = \boldsymbol{U}^T\boldsymbol{S}_W\boldsymbol{U} = \boldsymbol{\Lambda}$$

where $\boldsymbol{\Lambda} = \text{Diag}(\lambda_1, \ldots, \lambda_n)$ is the diagonal matrix of variances of the transformed features (the eigenvalues of $\boldsymbol{S}_W$). If the rank of $\boldsymbol{S}_W$ is less than p (equal to r, say), then the first stage of dimension reduction is to transform to the r-dimensional space by the transformation $\boldsymbol{U}_r^T\boldsymbol{X}$, $\boldsymbol{U}_r = [\boldsymbol{u}_1|\ldots|\boldsymbol{u}_r]$, so that

$$\boldsymbol{S}'_W = \boldsymbol{U}_r^T\boldsymbol{S}_W\boldsymbol{U}_r = \boldsymbol{\Lambda}_r$$

where $\boldsymbol{\Lambda}_r = \text{Diag}(\lambda_1, \ldots, \lambda_r)$. If we wish the within-class covariance matrix to be invariant to further orthogonal transformations, then it should be the identity matrix. This can be achieved by the transformation $\boldsymbol{Y} = \boldsymbol{\Lambda}_r^{-\frac{1}{2}}\boldsymbol{U}_r^T\boldsymbol{X}$ so that

$$\boldsymbol{S}'_W = \boldsymbol{\Lambda}_r^{-\frac{1}{2}}\boldsymbol{U}_r^T\boldsymbol{S}_W\boldsymbol{U}_r\boldsymbol{\Lambda}_r^{-\frac{1}{2}} = \boldsymbol{I}$$

This is the first stage of dimension reduction, and is illustrated in Figure 10.12.

It transforms the data so that the average within-class covariance matrix in the new space is the identity matrix. In the new space the between-class covariance matrix, $\boldsymbol{S}'_B$, is given by

$$\boldsymbol{S}'_B = \boldsymbol{\Lambda}_r^{-\frac{1}{2}}\boldsymbol{U}_r^T\boldsymbol{S}_B\boldsymbol{U}_r\boldsymbol{\Lambda}_r^{-\frac{1}{2}}$$

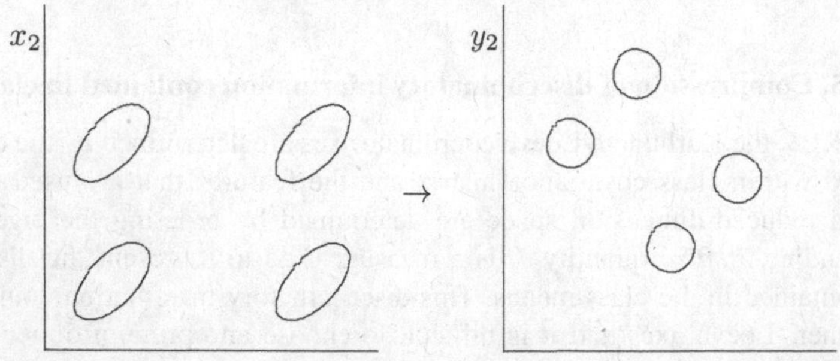

Figure 10.12 Illustration of the first stage of dimension reduction for four groups represented by contours $\boldsymbol{x}^T\hat{\boldsymbol{\Sigma}}_i^{-1}\boldsymbol{x} = \text{constant}$.

where S_B is the between-class covariance matrix in the data space. The second stage of the transformation is to compress the class mean information, i.e. find the orthogonal transformation that transforms the class mean vectors to a reduced dimension. This transformation, V, is determined by the eigenvectors of S'_B

$$S'_B V = V \tilde{\Lambda}$$

where $\tilde{\Lambda} = \text{Diag}(\tilde{\lambda}_1, \dots, \tilde{\lambda}_r)$ is the matrix of eigenvalues of S'_B. There are at most $C - 1$ nonzero eigenvalues and so the final transformation is $Z = V_\nu^T Y$ where $V_\nu = [v_1|\dots|v_\nu]$ and ν is the rank of S_B. The optimal feature extractor is therefore

$$Z = A^T X$$

where the $p \times \nu$ linear transformation A is given by

$$A^T = V_\nu^T \Lambda_r^{-\frac{1}{2}} U_r^T$$

In this transformed space

$$S''_W = V_\nu^T S'_W V_\nu = V_\nu^T V_\nu = I$$
$$S''_B = V_\nu^T S'_B V_\nu = \tilde{\Lambda}_\upsilon$$

where $\tilde{\Lambda}_\nu = \text{diag}(\tilde{\lambda}_1, \dots, \tilde{\lambda}_\nu)$. Thus, the transformation makes the average within-class covariance matrix equal to the identity and the between-class covariance matrix equal to a diagonal matrix (see Figure 10.13).

Usually, all $C - 1$ features are selected, but these can be ordered according to the magnitude of $\tilde{\lambda}_i$ and those with the largest eigenvalues selected. The linear transformation can be found by performing two eigenvector decompositions, first of S_W and then of S'_B, but an alternative approach based on a QR factorisation can be used (Crownover, 1991).

This two-stage process gives a geometric interpretation of linear discriminant analysis described in Chapter 5. The feature vectors (columns of the matrix A) can be shown to be eigenvectors of the generalised symmetric eigenvector equation (Devijver and Kittler, 1982)

$$S_B a = \lambda S_W a$$

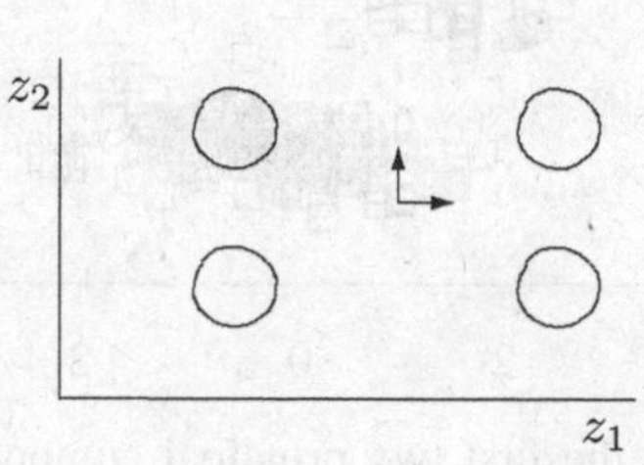

Figure 10.13 Second stage of dimension reduction: orthogonal rotation and projection to diagonalise the between-class covariance matrix.

The first stage of the transformation is simply a rotation of the coordinate axes followed by a scaling (assuming that S_W is full rank). The second stage comprises a projection of the data onto the hyperplane defined by the class means in this transformed space (of dimension at most $C - 1$), followed by a rotation of the axes. If we are to use all $C - 1$ coordinates subsequently in a classifier, then this final rotation is irrelevant since any classifier that we construct should be independent of an orthogonal rotation of the axes. Any set of orthogonal axes within the space spanned by the eigenvectors of S'_B with nonzero eigenvalues could be used. We could simply orthogonalise the vectors $m'_1 - m', \ldots, m'_{C-1} - m'$ where m'_i is the mean of class ω_i in the space defined by the first transformation and m' is the overall mean. However, if we wish to obtain a reduced-dimension display of the data, then it is necessary to perform an eigendecomposition of S'_B to obtain a set of coordinate axes that can be ordered using the values of $\tilde{\lambda}_i$, the eigenvalues of S'_B.

10.3.2.6 Example

Figures 10.14 and 10.15 give two-dimensional projections for simulated oil pipeline data (Bishop, 1998). This synthetic dataset models nonintrusive measurements on a pipe-line transporting a mixture of oil, water and gas. The flow in the pipe takes one out of three possible configurations: horizontally stratified, nested annular or homogeneous mixture flow. The data lie in a 12-dimensional measurement space. Figure 10.15 shows that the Karhunen–Loève transformation, KL5, separates the three classes into (approximately) three spherical clusters. The principal components projection (Figure 10.14) does not separate the classes, but retains some of the structure [for example, class 3 (□) comprises several subgroups].

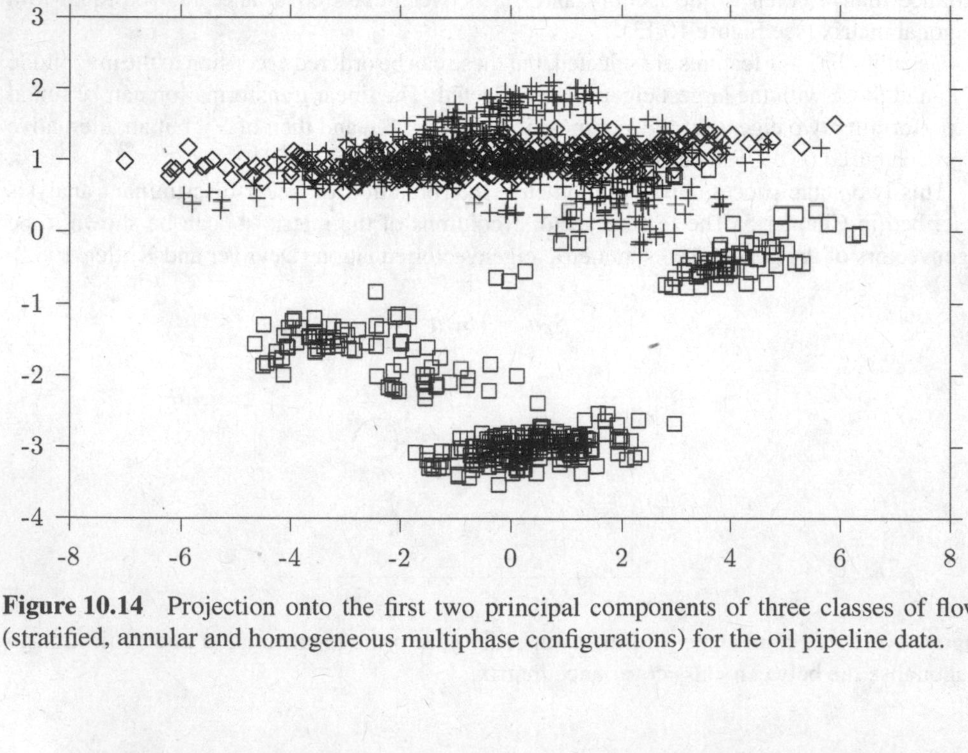

Figure 10.14 Projection onto the first two principal components of three classes of flow (stratified, annular and homogeneous multiphase configurations) for the oil pipeline data.

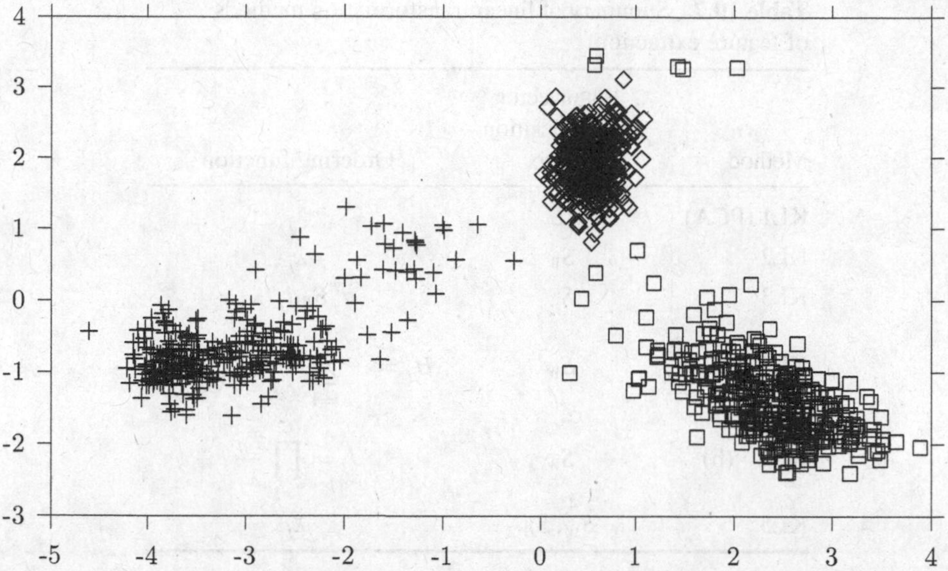

Figure 10.15 Projection (KL5) onto the two linear discriminant directions of three classes of flow (stratified, annular and homogeneous multiphase configurations) for the oil pipeline data.

10.3.2.7 Discussion

All of the above methods have certain features in common. All produce a linear transformation to a reduced-dimension space. The transformations are determined using an eigenvector decomposition of a matrix of second-order statistical moments and produce features or components in the new space that are uncorrelated. The features in the new space can be ordered using a measure of discriminability (in the case of labelled data) or approximation error. These methods are summarised in Table 10.7.

We could add to this list the method of common principal components (Flury, 1988) that also determines a coordinate system using matrices of second-order statistical moments. A reduced-dimension representation of the data could be achieved by ordering the principal components using a measure like H_j or J_j in Table 10.7.

The final method (KL5), derived from a geometric argument, produces a transformation that is identical to the linear discriminant transformation (see Chapter 5), obtained by maximising a discriminability criterion. It makes no distributional assumptions, but a nearest class mean type rule will be optimal if normal distributions are assumed for the classes.

10.3.3 Example application study

The problem

This application involves monitoring changes in land use for planners and government officials (Li and Yeh, 1998).

Table 10.7 Summary of linear transformation methods of feature extraction.

Method	Eigenvector decomposition matrix	Ordering function
KL1 (PCA)	$\hat{\boldsymbol{\Sigma}}$	λ_j
KL2	$\boldsymbol{S}_W$	λ_j
KL3	$\boldsymbol{S}_W$	$\boldsymbol{a}_j^T \boldsymbol{S}_B \boldsymbol{a}_j / \lambda_j$
KL4 – (a)	$\boldsymbol{S}_W$	$H_j = -\sum_{i=1}^{C} \frac{\lambda_{ij}}{\lambda_j} \log\left(\frac{\lambda_{ij}}{\lambda_j}\right)$
KL4 – (b)	$\boldsymbol{S}_W$	$J_j = \prod_{i=1}^{C} \frac{\lambda_{ij}}{\lambda_j}$
KL5	$\boldsymbol{S}_W, \boldsymbol{S}_B$	λ_j

Summary

The application of remote sensing to inventory land resources and to evaluate the impacts of urban developments is addressed. Remote sensing is considered as a fast and efficient means of assessing such developments when detailed 'ground truth' data are unavailable. The aim is to determine the type, amount and location of land use change.

The data

The data consist of satellite images, measured in five wavebands, from two images of the same region measured 5 years apart. A 10-dimensional feature vector is constructed (consisting of pixel measurements in the five wavebands over the two images) and data gathered over the whole region.

The model

A standard principal components analysis is performed.

Training procedure

Each variable is *standardised* to zero mean and unit variance and a principal components analysis performed. The first four principal components account to 97% of the variance. The data are projected on to the subspace characterised by the principal components and subregions (identified from a compressed principal components analysis image) labelled manually according to land use change (16 classes of land use, determined by field data).

10.3.4 Further developments

There are various generalisations of the basic linear approaches to feature extraction described in this section. Common principal components analysis (Flury, 1988), is a generalisation of principal components analysis to the multigroup situation. The common principal components model assumes that the covariance matrices of each group or class, $\mathbf{\Sigma}_i$, can be written

$$\mathbf{\Sigma}_i = \boldsymbol{\beta}\mathbf{\Lambda}_i\boldsymbol{\beta}^T \tag{10.22}$$

where $\mathbf{\Lambda}_i = \text{Diag}(\lambda_{i1}, \ldots, \lambda_{ip})$, the diagonal matrix of eigenvalues of the ith group. Thus, the eigenvectors $\boldsymbol{\beta}$ are common between groups, but the $\mathbf{\Lambda}_i$ are different.

There have been several developments to nonlinear principal components analysis each taking a particular feature of the linear methods and generalising it. The work described by Gifi (1990) applies primarily to categorical data. Other extensions are those of principal curves (Hastie and Stuetzle, 1989; Tibshirani, 1992), and nonlinear principal components based on radial basis functions (Webb, 1996) and kernel functions (Schölkopf *et al.*, 1999). There are more efficient calculations of principal components if the feature dimensionality is much larger than the number of samples (see the work on *eigenfaces*; Turk and Pentland, 1991).

Approaches to principal components analysis have been developed for data that may be considered as curves (Ramsay and Dalzell, 1991; Rice and Silverman, 1991; Silverman, 1995). Principal components for categorical data and functions is discussed by Jackson (1991) who also describes robust procedures.

Independent components analysis (Comon, 1994; Hyvärinen and Oja, 2000) aims to find a linear representation of non-Gaussian data so that the components are statistically independent (or as independent as possible). This is a stronger condition than the uncorrelated variables condition of principal components analysis (they are equivalent for Gaussian random variables only). A linear latent variable model is assumed

$$\boldsymbol{x} = \boldsymbol{A}\boldsymbol{s}$$

where $\boldsymbol{x}$ are the observations, $\boldsymbol{A}$ is a mixing matrix and $\boldsymbol{s}$ is the vector of latent variables. Given T realisations of $\boldsymbol{x}$, the problem is to estimate the mixing matrix $\boldsymbol{A}$ and the corresponding realisations of $\boldsymbol{s}$, under the assumption that the components, s_i, are statistically independent. This technique has found widespread application to problems in signal analysis (medical, financial), data compression, image processing and telecommunications.

10.3.5 Summary

The procedures described in this section construct *linear* transformations based on matrices of first- and second-order statistics:

1. principal components analysis – an unsupervised method based on a correlation or covariance matrix;

2. Karhunen–Loève transformation – an umbrella term to cover transformations based on within and between-class covariance matrices.

Algorithms for their implementation are readily available.

10.4 Multidimensional scaling

Multidimensional scaling (MDS) is a term that is applied to a class of techniques that analyses a matrix of distances or dissimilarities (the proximity matrix) in order to produce a representation of the data points in a reduced-dimension space (termed the *representation space*).

All of the methods of data reduction presented in this chapter so far have analysed the $n \times p$ data matrix $\boldsymbol{X}$ or the sample covariance or correlation matrix. Thus MDS differs in the form of the data matrix on which it operates – it is an *individual-directed* method. Of course, given a data matrix, we could construct a dissimilarity matrix (provided we define a suitable measure of dissimilarity between objects) and then proceed with an analysis using MDS techniques. However, data often arise already in the form of dissimilarities and so there is no recourse to the other techniques. Also, in the methods previously discussed, the data-reducing transformation derived has, in each case, been a linear transformation. We shall see that some forms of multidimensional scaling permit a nonlinear data-reducing transformation (if indeed we do have access to data samples rather than proximities).

There are many types of MDS, but all address the same basic problem: Given an $n \times n$ matrix of dissimilarities and a distance measure (usually Euclidean) find a configuration of n points $\boldsymbol{x}_1, \ldots, \boldsymbol{x}_n$ in the reduced dimension space $\mathbb{R}^e$ $(e < p)$ so that the distance between a pair of points is close in some sense to the dissimilarity between the points. All methods must find the coordinates of the points and the dimension of the space, e. Two basic types of MDS are metric and nonmetric MDS. Metric MDS assumes that the data are quantitative and metric MDS procedures assume a functional relationship between the interpoint distances and the given dissimilarities. Nonmetric MDS assumes that the data are qualitative, having perhaps ordinal significance and nonmetric MDS procedures produce configurations that attempt to maintain the rank order of the dissimilarities.

Metric MDS appears to have been introduced into the pattern recognition literature by Sammon (1969). It has been developed to incorporate class information and has also been used to provide nonlinear transformations for dimension reduction for feature extraction.

We begin our discussion with a description of one form of metric MDS, namely classical scaling.

10.4.1 Classical scaling

Given a set of n points in p-dimensional space, $\boldsymbol{x}_1, \ldots, \boldsymbol{x}_n$, it is straightforward to calculate the Euclidean distance between each pair of points. Classical scaling (or *principal coordinates analysis*) is concerned with the converse problem: Given a matrix of distances, which we assume are Euclidean, how can we determine the coordinates of a set of points in a dimension e (also to be determined from the analysis)? This is achieved via a decomposition of the $n \times n$ matrix $\boldsymbol{T}$, the between-individual sums of squares and products matrix

$$\boldsymbol{T} = \boldsymbol{X}\boldsymbol{X}^T \tag{10.23}$$

where $\boldsymbol{X} = [\boldsymbol{x}_1, \ldots, \boldsymbol{x}_n]^T$ is the $n \times p$ matrix of coordinates. The distance between two individuals i and j is

$$d_{ij}^2 = T_{ii} + T_{jj} - 2T_{ij} \tag{10.24}$$

where

$$T_{ij} = \sum_{k=1}^{p} x_{ik} x_{jk}$$

If we impose the constraint that the centroid of the points $\boldsymbol{x}_i$, $i = 1, \ldots, n$, is at the origin, then Equation (10.24) may be inverted to express the elements of the matrix $\boldsymbol{T}$ in terms of the dissimilarity matrix giving

$$T_{ij} = -\frac{1}{2}\left[d_{ij}^2 - d_{i.}^2 - d_{.j}^2 + d_{..}^2\right] \tag{10.25}$$

where

$$d_{i.}^2 = \frac{1}{n}\sum_{j=1}^{n} d_{ij}^2; \qquad d_{.j}^2 = \frac{1}{n}\sum_{i=1}^{n} d_{ij}^2; \qquad d_{..}^2 = \frac{1}{n^2}\sum_{i=1}^{n}\sum_{j=1}^{n} d_{ij}^2$$

Equation (10.25) allows us to construct $\boldsymbol{T}$ from a given $n \times n$ dissimilarity matrix $\boldsymbol{D}$ (assuming that the dissimilarities are Euclidean distances). All we need to do now is to factorise the matrix $\boldsymbol{T}$ to make it of the form (10.23). Since it is a real symmetric matrix, $\boldsymbol{T}$ can be written in the form

$$\boldsymbol{T} = \boldsymbol{U}\boldsymbol{\Lambda}\boldsymbol{U}^T$$

where the columns of $\boldsymbol{U}$ are the eigenvectors of $\boldsymbol{T}$ and $\boldsymbol{\Lambda}$ is a diagonal matrix of eigenvalues, $\lambda_1, \ldots, \lambda_n$. Therefore we take

$$\boldsymbol{X} = \boldsymbol{U}\boldsymbol{\Lambda}^{\frac{1}{2}}$$

as our matrix of coordinates. If the matrix of dissimilarities is indeed a matrix of Euclidean distances between points in $\mathbb{R}^p$, then the eigenvalues may be ordered

$$\lambda_1 \geq \ldots \geq \lambda_n = 0; \quad \lambda_i = 0, \ i = p+1, \ldots, n$$

If we are seeking a representation in a reduced dimension then we would use only those eigenvectors associated with the largest eigenvalues. Methods for choosing the number of eigenvalues were discussed in relation to principal components analysis. Briefly, we choose the number r so that

$$\sum_{i=1}^{r-1} \lambda_i \; < \; k\sum_{i=1}^{n} \lambda_i \; < \; \sum_{i=1}^{r} \lambda_i$$

for some prespecified threshold, k $(0 < k < 1)$; or use the 'scree test'.

Then we take

$$\boldsymbol{X} = [\boldsymbol{u}_1| \ldots |\boldsymbol{u}_r]\mathrm{Diag}\left(\lambda_1^{\frac{1}{2}} \ldots \lambda_r^{\frac{1}{2}}\right) = \boldsymbol{U}_r\boldsymbol{\Lambda}_r^{\frac{1}{2}}$$

as the $n \times r$ matrix of coordinates, where $\mathbf{\Lambda}_r$ is the $r \times r$ diagonal matrix with diagonal elements λ_i, $i = 1, \ldots, r$.

If the dissimilarities are not Euclidean distances, then $\boldsymbol{T}$ is not necessarily positive semidefinite and there may be negative eigenvalues. Again we may choose the eigenvectors associated with the largest eigenvalues. If the negative eigenvalues are small then this may still lead to a useful representation of the data. In general, the smallest of the set of largest eigenvalues retained should be larger than the magnitude of the most negative eigenvalue. If there is a large number of negative eigenvalues, or some are large in magnitude, then classical scaling may be inappropriate. However, classical scaling appears to be robust to departures from Euclidean distance.

If we were to start with a set of data (rather than a matrix of dissimilarities) and seek a reduced-dimension representation of it using the classical scaling approach (by first forming a dissimilarity matrix and carrying out the procedure above), then the reduced-dimension representation is exactly the same as carrying out a principal components analysis and calculating the component scores, the transformed variable values corresponding to a particular pattern, (provided we have chosen Euclidean distance as our measure of dissimilarity). Thus, there is no point in carrying out classical scaling *and* a principal components analysis on a dataset.

10.4.2 Metric MDS

Classical scaling is one particular form of metric MDS in which an objective function measuring the discrepancy between the given dissimilarities, δ_{ij}, and the derived distances in $\mathbb{R}^e$, d_{ij}, is optimised. The derived distances depend on the coordinates of the samples that we wish to find. There are many forms that the objective function may take. For example, minimisation of the objective function

$$\sum_{1 \leq j < i \leq n} (\delta_{ij}^2 - d_{ij}^2)$$

yields a projection on to the first e principal components if δ_{ij} are exactly Euclidean distances. There are other measures of divergence between the sets $\{\delta_{ij}\}$ and $\{d_{ij}\}$ and the major MDS packages are not consistent in the criterion optimised (Dillon and Goldstein, 1984). One particular measure is

$$S = \sum_{ij} a_{ij}(\delta_{ij} - d_{ij})^2 \tag{10.26}$$

for weighting factors a_{ij}. Taking

$$a_{ij} = \left(\sum_{ij} d_{ij}^2 \right)^{-1}$$

gives $\sqrt{S}$ as similar to Kruskal's stress (Kruskal, 1964a, b), defined in the following section [Equation (10.29)]. There are other forms for the a_{ij} (Sammon, 1969; Koontz and Fukunaga, 1972; de Leeuw and Heiser, 1977; Niemann and Weiss, 1979). The stress is invariant under rigid transformations of the derived configuration (translations, rotations and reflections) and also to uniform stretching and shrinking.

A more general form of (10.26) is

$$S = \sum_{ij} a_{ij}(\phi(\delta_{ij}) - d_{ij})^2 \tag{10.27}$$

where ϕ is a member of a predefined class of functions; for example, the class of all linear functions, giving

$$S = \sum_{ij} a_{ij}(a + b\delta_{ij} - d_{ij})^2 \tag{10.28}$$

for parameters a and b. In general, there is no analytic solution for the coordinates of the points in the representation space. Minimisation of (10.27) can proceed by an alternating least-squares approach [see Gifi (1990)] for further applications of the alternating least squares principle]; that is, by alternating minimisation over ϕ and the coordinates. In the linear regression example (10.28), we would minimise with respect to a and b, for a given initial set of coordinates (and hence the derived distances d_{ij}). Then, keeping a and b fixed, minimise with respect to the coordinates of the data points. This process is repeated until convergence.

The expression (10.27) may be normalised by a function $\tau^2(\phi, \boldsymbol{X})$, that is a function of both the coordinates and the function ϕ. Choices for τ^2 are discussed by de Leeuw and Heiser (1977).

In psychology, in particular, the measures of dissimilarity that arise have ordinal significance at best: their numerical values have little meaning and we are interested only in their order. We can say that one stimulus is larger than another, without being able to attach a numerical value to it. In this case, a choice for the function ϕ above is one that belongs to the class of monotone functions. This is the basis of *nonmetric* multidimensional scaling or *ordinal scaling*.

10.4.3 Ordinal scaling

Ordinal scaling or *nonmetric MDS* is a method of finding a configuration of points for which the rank ordering of the interpoint distance is close to the ranking of the values of the given dissimilarities.

In contrast to classical scaling, there is no analytic solution for the configuration of points in ordinal scaling. Further, the procedure is iterative and requires an initial configuration for the points to be specified. Several initial configurations may be tried before an acceptable final solution is achieved.

The desired requirement that the ordering of the distances in the derived configuration is the same as that of the given dissimilarities is of course equivalent to saying that the distances are a monotonic function of the dissimilarities. Figure 10.16 gives a plot of distances d_{ij} (obtained from a classical scaling analysis) against dissimilarities δ_{ij} for the British town data given in Table 10.8. The numbers in the table are the distances in miles between 10 towns in Britain along recommended routes.

The relationship is clearly not exactly monotonic, though on the whole, the larger the dissimilarity the larger the distance. In ordinal scaling, the coordinates of the points in the

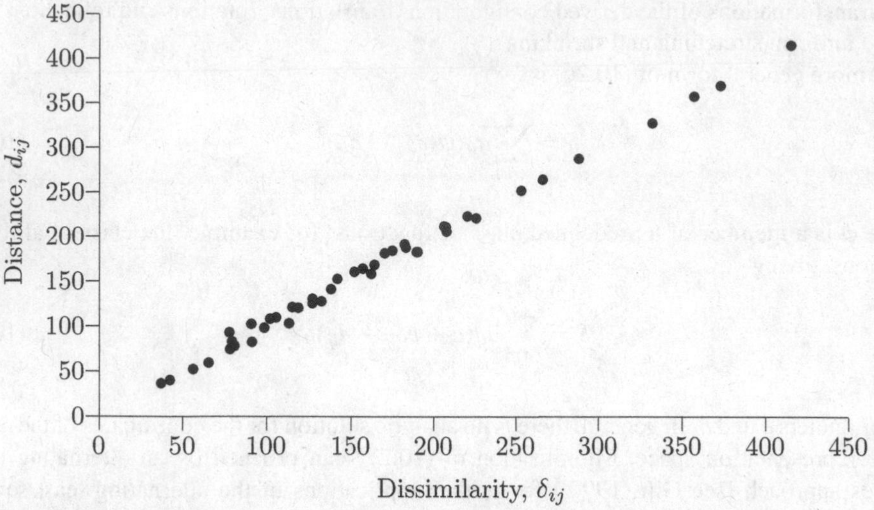

Figure 10.16 Distances versus dissimilarities for the British town data.

representation space are adjusted so as to minimise a cost function that is a measure of the degree of deviation from monotonicity of the relationship between d_{ij} and δ_{ij}. It may not be possible to obtain a final solution that is perfectly monotonic but the final ordering on the d_{ij}s should be 'as close as possible' to that of the δ_{ij}s.

To find a configuration that satisfies the monotonicity requirement, we must first of all specify a definition of monotonicity. Two possible definitions are:

Primary monotone condition

$$\delta_{rs} < \delta_{ij} \Rightarrow \hat{d}_{rs} \leq \hat{d}_{ij}$$

Secondary monotone condition

$$\delta_{rs} \leq \delta_{ij} \Rightarrow \hat{d}_{rs} \leq \hat{d}_{ij}$$

Table 10.8 Dissimilarities between 10 towns in the British Isles (measured as distances in miles along recommended routes).

London (LON)	0.0									
Birmingham (B)	111	0.0								
Cambridge (C)	55	101	0.0							
Edinburgh (E)	372	290	330	0.0						
Hull (H)	171	123	124	225	0.0					
Lincoln (LIN)	133	85	86	254	39	0.0				
Manchester (M)	184	81	155	213	96	84	0.0			
Norwich (N)	112	161	62	360	144	106	185	0.0		
Scarborough (SCA)	214	163	167	194	43	81	105	187	0.0	
Southampton (SOU)	77	128	130	418	223	185	208	190	266	0.0
	LON	B	C	E	H	LIN	M	N	SCA	SOU

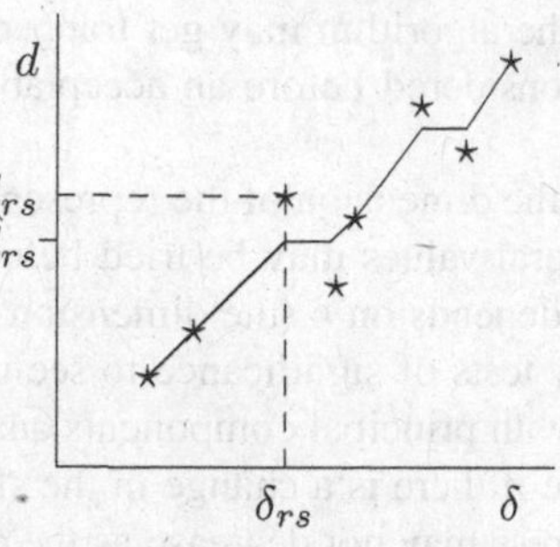

Figure 10.17 Least-squares monotone regression line.

where $\hat{d}_{rs}$ is the point on the fitting line (see Figure 10.17) corresponding to δ_{rs}. The $\hat{d}_{rs}$ are termed the *disparities* or the *pseudo-distances*. The difference between these two conditions is the way in which ties between the δs are treated. In the secondary monotone condition, if $\delta_{rs} = \delta_{ij}$ then $\hat{d}_{rs} = \hat{d}_{ij}$, whereas in the primary condition there is no constraint on $\hat{d}_{rs}$ and $\hat{d}_{ij}$ if $\delta_{rs} = \delta_{ij}$: $\hat{d}_{rs}$ and $\hat{d}_{ij}$ are allowed to differ (which would give rise to vertical lines in Figure 10.17). The secondary condition is usually regarded as too restrictive, often leading to convergence problems.

Given the above definition, we can define a goodness of fit as

$$S_q = \sum_{i<j} (d_{ij} - \hat{d}_{ij})^2$$

and minimising gives the primary (or secondary) least-squares monotone regression line. (In fact, it is not a line, only being defined at points δ_{ij}.)

An example of a least-squares monotone regression is given in Figure 10.17. The least-squares condition ensures that the sum of squares of vertical displacements from the line is a minimum. Practically, this means that for sections of the data where d is actually a monotonic function of δ then the line passes through the points. If there is a decrease, the value taken is the mean of a set of samples.

The quantity S_q is a measure of the deviation from monotonicity, but it is not invariant to uniform dilation of the geometric configuration. This can be removed by normalisation and the normalised *stress*, given by

$$S = \sqrt{\frac{\sum_{i<j} (d_{ij} - \hat{d}_{ij})^2}{\sum_{i<j} d_{ij}^2}} \tag{10.29}$$

used as the measure of fit. (In some texts the square root factor is omitted in the definition.)

Since S is a differentiable function of the desired coordinates, we use a nonlinear optimisation scheme (Press *et al.*, 1992) which requires an initial configuration of data points to be specified. In practice, it has been found that some of the more sophisticated methods do not work so well as steepest descents (Chatfield and Collins, 1980).

The initial configuration of data points could be chosen randomly, or as the result of a principal coordinates analysis (classical scaling). Of course, there is no guarantee of finding the global minimum of S and the algorithm may get trapped in poor local minima. Several initial configurations may be considered before an acceptable value of the minimum stress is achieved.

In the algorithm, a value of the dimension of the representation space, e, is required. This is unknown in general and several values may be tried before a 'low' value of the stress is obtained. The minimum stress depends on n (the dimension of the dissimilarity matrix) and e and it is not possible to apply tests of significance to see if the 'true' dimension has been found (this may not exist). As with principal components analysis, we may plot the stress as a function of dimension and see if there is a change in the slope (elbow in the graph). If we do this, we may find that the stress may not decrease as the dimension increases, because of the problem of finding poor local minima. However, it can be made to do so by initialising the solution for dimension e by that obtained in dimension $e - 1$ (extra coordinates of zero are added).

The summations in the expression for the stress are over all pairwise distances. If the dissimilarity matrix is asymmetric we may include both the pairs (i, j) and (j, i) in the summation. Alternatively, we may carry out ordinal scaling on the symmetric matrix of dissimilarities with

$$\delta^*_{rs} = \delta^*_{sr} = \frac{1}{2}(\delta_{rs} + \delta_{sr})$$

Missing values in δ_{ij} can be accommodated by removing the corresponding indices from the summation in estimating the stress (10.29).

10.4.4 Algorithms

Most implementations of MDS algorithms use standard gradient methods. There is some evidence that sophisticated nonlinear optimisation schemes do not work so well (Chatfield and Collins, 1980). Siedlecki *et al.* (1988) report that steepest descents out-performed conjugate gradients on a metric MDS optimisation problem, but better performance was given by the coordinate descent procedure of Niemann and Weiss (1979).

One approach to minimising the objective function, S, is to use the principle of majorisation (de Leeuw and Heiser, 1977; Heiser, 1991, 1994) as part of the alternating least-squares process. Given the current values of the coordinates, say θ_t, an upper bound, $W(\theta_t, \theta)$, for the criterion function is defined. It is usually a quadratic form with a single minimum as a function of the coordinates θ. It has the property that $W(\theta_t, \theta)$ is equal to the value of the objective function at θ_t and greater than it everywhere else. Minimising $W(\theta_t, \theta)$ with respect to θ yields a value θ_{t+1} at which the objective function is lower. A new majorising function $W(\theta_{t+1}, \theta)$ is defined and the process repeated (Figure 10.18).

This generates a sequence of estimates $\{\theta_t\}$ for which the objective function decreases and converges to a local minimum.

All algorithms start with an initial configuration and converge to a local minimum. It has been reported that the secondary definition of monotonicity is more likely to get trapped in poor local minima than the primary definition (Gordon, 1999). We recommend that you repeat your experiments for several starting configurations.

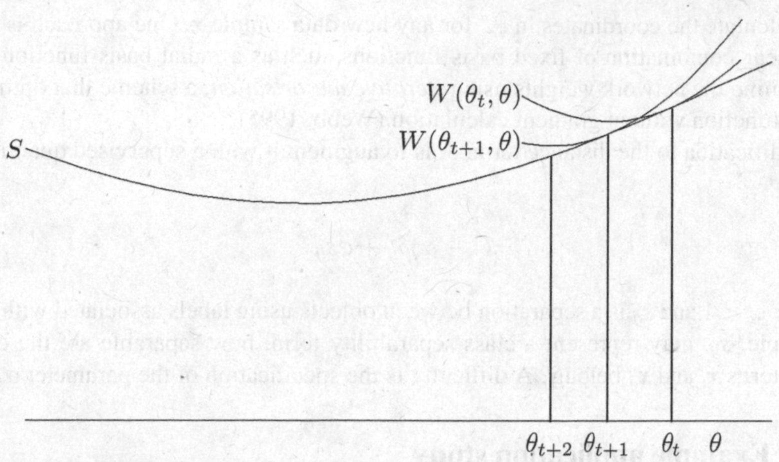

Figure 10.18 Illustration of iterative majorisation principle: minimisation of $S(\theta)$ is achieved through successive minimisations of the majorisation functions, W.

10.4.5 MDS for feature extraction

There are several obstacles in applying MDS to the pattern recognition problem of feature extraction that we are addressing in this chapter. The first is that usually we are not presented with an $n \times n$ matrix of dissimilarities, but with an $n \times p$ matrix of observations. Although in itself this is not a problem since we can form a dissimilarity matrix using some suitable measure (e.g. Euclidean distance), the number of patterns n can be very large (in some cases thousands). The storage of an $n \times n$ matrix may present a problem. Further, the number of adjustable parameters is $n' = e \times n$ where e is the dimension of the derived coordinates. This may prohibit the use of some nonlinear optimisation methods, particularly those of the quasi-Newton type which either calculate, or iteratively build up, an approximation to the inverse Hessian matrix of size $n' \times n' = e^2 n^2$.

Even if these problems can be overcome, MDS does not readily define a transformation that, given a new data sample $\boldsymbol{x} \in \mathbb{R}^p$, produces a result $\boldsymbol{y} \in \mathbb{R}^e$. Further calculation is required.

One approach to this problem is to regard the transformed coordinates $\boldsymbol{y}$ as a nonlinear parametrised function of the data variables

$$\boldsymbol{y} = \boldsymbol{f}(\boldsymbol{x}; \theta)$$

for parameters, θ. In this case,

$$d_{ij} = \left|\boldsymbol{f}(\boldsymbol{x}_i; \theta) - \boldsymbol{f}(\boldsymbol{x}_j; \theta)\right|$$
$$\delta_{ij} = \left|\boldsymbol{x}_i - \boldsymbol{x}_j\right|$$

and we may minimise the criterion function, for example (10.26), *with respect to the parameters*, θ, of $\boldsymbol{f}$ rather than with respect to the coordinates of the data points in the transformed space. This is termed *multidimensional scaling by transformation*. Thus the number of parameters can be substantially reduced. Once the iteration has converged, the function $\boldsymbol{f}$ can be used to calculate the coordinates in $\mathbb{R}^e$ for any new data sample $\boldsymbol{x}$. One approach is to model f as a linear combination of fixed basis functions such as a radial basis function network and determine the network weights using *iterative majorisation*, a scheme that optimises the objective function without gradient calculation (Webb, 1995).

A modification to the distance term, δ, is to augment it with a supervised quantity giving a distance

$$(1-\alpha)\delta_{ij} + \alpha s_{ij} \tag{10.30}$$

where $0 < \alpha < 1$ and s_{ij} is a separation between objects using labels associated with the data. For example, s_{ij}, may represent a class separability term: how separable are the classes to which patterns $\boldsymbol{x}_i$ and $\boldsymbol{x}_j$ belong? A difficulty is the specification of the parameter α.

10.4.6 Example application study

The problem

The application involves an exploratory data analysis to investigate relationships between research assessment ratings of UK higher education institutions (Lowe and Tipping, 1996).

Summary

Several methods of feature extraction, both linear and nonlinear, were applied to high-dimensional data records from the 1992 UK Research Assessment Exercise and projections to two dimensions produced.

The data

Institutions supply information on research activities within different subjects in the form of quantitative indicators of their research activity, such as the number of active researchers, postgraduate students, values of grants and numbers of publications. Together with some qualitative data (example publications), this forms part of the data input to committees which provide a research rating, on a scale of 1–5. There are over 4000 records for all subjects, but the analysis concentrated on three subjects: physics, chemistry and biological sciences.

Preprocessing included the removal of redundant and repeated variables, accumulating indicators that were given for a number of years and standardisation of variables. The training set consisted of 217 patterns each with 80 variables.

The model

Several models were assessed. These included a principal components analysis, an MDS (metric) [termed a Sammon mapping by Lowe and Tipping (1996)] and an MDS by transformation modelled as a radial basis function network.

Training procedure

For the MDS by transformation, the dissimilarity was augmented with a subjective quantity as in (10.30) where s_{ij} is a separation between objects based on the subjective research rating.

Since the objective function is no longer quadratic (as commonly used in the radial basis function classification model; Chapter 6), an analytic matrix inversion routine cannot be used for the weights of the radial basis function. A conjugate gradients nonlinear function minimisation routine was used to minimise the stress criterion.

10.4.7 Further developments

Within the pattern recognition literature, there have been several attempts to use MDS techniques for feature extraction both for exploratory data analysis and classification purposes (Sammon, 1969; also see comments by Kruskal, 1971; Koontz and Fukunaga, 1972; Cox and Ferry, 1993).

Approaches that model the nonlinear dimension-reducing transformation as a radial basis function network are described by Webb (1995) and Lowe and Tipping (1996). Mao and Jain (1995) model the transformation as a multilayer perceptron. A comparative study of neural network feature extraction methods was carried out by Lerner *et al.* (1999).

10.4.8 Summary

Multidimensional scaling is a name given to a range of techniques that analyse dissimilarity matrices and produce coordinates of points in a 'low' dimension. Three approaches to MDS have been presented:

Classical scaling
This assumes that the dissimilarity matrix is Euclidean though it has been shown to be robust if there are small departures from this condition. An eigenvector decomposition of the dissimilarity matrix is performed and the resulting set of coordinates is identical to the principal components analysis scores (to within an orthogonal transformation) if indeed the dissimilarity matrix is a matrix of Euclidean distances. Therefore, in this case, there would be nothing to be gained over a principal components analysis in using this technique as a method of feature extraction, given an $n \times p$ data matrix $\boldsymbol{X}$.

Metric scaling
This method regards the coordinates of the points in the derived space as parameters of a stress function that is minimised. This method allows nonlinear reductions in dimensionality. The procedure assumes a functional relationship between the interpoint distances and the given dissimilarities.

Nonmetric scaling
As with metric scaling, a criterion function (stress) is minimised but the procedure assumes that the data are qualitative, having perhaps ordinal significance at best.

10.5 Application studies

There are very many applications of feature selection methods (and indeed, feature selection methods will have been employed in many of the other application studies considered in this book). Recent examples include the following are:

- Bioinformatics: for example, sequence analysis, microarray analysis and mass spectra analysis. Microarray data are characterised by thousands of genes and small sample sizes (Ahmad *et al.*, 2008; see review by Saeys *et al.*, 2007).
- Text classification (Forman, 2003; Liu and Yu, 2005).

- Image retrieval, customer relationship management, intrusion detection and genomic analysis (Liu and Yu, 2005).
- 'Upcoming domains' (Saeys *et al.*, 2007) include single nucleotide polymorphism analysis, text and literature analysis.

Feature extraction application studies include:

- EEG. Jobert *et al.* (1994) use principal components analysis to produce a two-dimensional representation of spectral data (sleep electroencephalogram – EEG) to view time-dependent variation.
- Positron emission tomography. Pedersen *et al.* (1994) use principal components analysis for data visualisation purposes on dynamic PET images to enhance clinically interesting information.
- Remote sensing. Eklundh and Singh (1993) compare principal components analysis using correlation and covariance matrices in the analysis of satellite remote sensing data. The correlation matrix gives improvement to the signal-to-noise ratio.
- Calibration of near-infrared spectra (Oman *et al.*, 1993).
- Structure–activity relationships. Darwish *et al.* (1994) apply principal components analysis (14 variables, nine compounds) in a study to investigate the inhibitory effect of benzine derivatives.
- Target classification. Liu *et al.* (1994) use principal components analysis for feature extraction in a classification study of materials design.
- Face recognition. Principal components analysis has been used in several studies on face recognition to produce 'eigenfaces'. The weights that characterise the expansion of a given facial image in terms of these eigenfaces are features used for face recognition and classification (see the review by Chellappa *et al.*, 1995).
- Speech. Pinkowski (1997) uses principal components analysis for feature extraction on a speaker-dependent dataset consisting of spectrograms of 80 sounds representing 20 speaker-dependent words containing English semivowels.

Applications of MDS and Sammon mappings include:

- Medical. Ratcliffe *et al.* (1995) use MDS to recover three-dimensional localisation of sonomicrometry transducer elements glued to excised and living ovine hearts. The inter-element distances were measured by the sonomicrometry elements by sequentially activating a single array element followed by eight receiver elements (thus giving inter-transducer distances).
- Bacterial classification. Bonde (1976) uses nonmetric MDS to produce two- and three-dimensional plots of groups of organisms (using a steepest descent optimisation scheme).
- Chemical vapour analysis. For a potential application of an 'artificial nose' (an array of 14 chemical sensors) to atmosphere pollution monitoring, cosmetics, food and defence applications, Lowe (1993) considers an MDS approach to feature extraction, where the dissimilarity matrix is determined by class (concentration of the substance).

10.6 Summary and discussion

In this chapter we have considered a variety of techniques for mapping data to a reduced dimension. There are many more that we have not covered, and we have tried to point to the literature where some of these may be found. A comprehensive account of data transformation techniques requires a volume in itself and in this chapter we have only been able to consider some of the more popular multivariate methods. In common with the following chapter on clustering, many of the techniques are used as part of data preprocessing in an exploratory data analysis.

The techniques vary in their complexity – both from mathematical ease of understanding and numerical ease of implementation points of view. Most methods of feature extraction produce linear transformations, but nonmetric MDS is nonlinear. Some use class information, others are unsupervised, although there are both variants of the Karhunen–Loève transformation. Some techniques, although producing a linear transformation, require the use of a nonlinear optimisation procedure to find the parameters. Others are based on eigenvector decomposition routines, perhaps performed iteratively.

To some extent, the separation of the classifier design into two processes of feature extraction and classification is artificial, but there are many reasons, some of which were enumerated in Section 10.3, why dimension reduction may be advisable.

Some methods of feature selection (wrapper/embedded methods) are tied to classifier type whilst others (filter approaches) are independent of a specific classifies and the main conclusions that can be drawn from the feature selection review are as follows:

- Filter methods are the most computationally efficient.
- Feature selection methods tied to classifier type (wrapper/embedded methods) offer improved performance over filter methods.
- Embedded methods, where feature selection is integral to model selection, are likely to provide optimal performance.
- It is important to handle irrelevance and redundancy. Therefore, a multivariate approach is required.
- Cross-validation should be performed as part of the feature selection process.
- Model averaging provides a robust solution.

10.7 Recommendations

If explanation is required of the variables that are used in a classifier, then a feature selection process, as opposed to a feature extraction process, is recommended for dimension reduction. For feature selection, the probabilistic criteria for estimating class separability are complicated, involving estimation of probability density functions and their numerical integration. Even the simple error rate measure is not easy to evaluate for nonparametric density functions. Therefore, we recommend using the following:

- The parametric form of the probabilistic distance measures assumes normal distributions. These have the advantage for feature selection that the value of the criteria for a given set of features may be used in the evaluation of the criteria when an additional feature is included. This reduces the amount of computation in some of the feature set search algorithms.

- The interclass distance measures, J_1 to J_4 [Equations (10.5), (10.6), (10.7) and (10.8)].
- Use an embedded or wrapper method (with error rate estimation). Wrapper and embedded methods of feature selection are optimised for the classifier type and offer improved performance.
- Implement feature selection within the cross-validation loop, where a robust method of classifier design, incorporating cross-validation, is being implemented.
- Combine feature selection methods. Typical feature selection methods ignore model uncertainty and select a single best feature set for use in classifier design. Model uncertainty can be handled by averaging over multiple models (that use potentially overlapping subsets of features).

Which algorithms should you employ for feature extraction? Whatever your problem, always start with the simplest approach which for feature extraction is, in our view, a principal components analysis of your data. This will tell you whether your data lie on a linear subspace in the space spanned by the variables and a projection on to the first two principal components, and displaying your data may reveal some interesting and unexpected structure.

It is recommended to apply principal components analysis to standardised data for feature extraction, and to consider it particularly when dimensionality is high. Use a simple heuristic to determine the number of principal components to use, in the first instance. For class-labelled data, use linear discriminant analysis (KL5) for a reduced-dimension representation.

If you believe there to be nonlinear structure in the data, then techniques based on MDS (for example, MDS by transformation) are straightforward to implement. Try several starting conditions for the parameters.

10.8 Notes and references

A comprehensive review of developments in feature selection is covered in the books by Liu and Motoda (2007) and Guyon *et al.* (2006) [see also Guyon (2008) and Saeys *et al.* (2007)]. For a review of filter methods see Duch (2004). For feature selection for classification, see Dash and Liu (1997).

The branch and bound method (referred to as complete search in this chapter) has been used in many areas of statistics (Hand, 1981b). It was originally proposed for feature subset selection by Narendra and Fukunaga (1977) and receives a comprehensive treatment by Devijver and Kittler (1982) and Fukunaga (1990). Hamamoto *et al.* (1990) evaluate the branch and bound algorithm using a recognition rate measure that does not satisfy the monotonicity condition. Krusińska (1988) describes a semioptimal branch and bound algorithm for feature selection in mixed variable discrimination.

Stepwise procedures have been considered by many authors: Whitney (1971) for the sequential forward selection algorithm; Michael and Lin (1973) for the basis of the plus l – take away r algorithm (the l–r algorithm); and Stearns (1976) for the l–r algorithm.

Floating search methods were introduced by Pudil *et al.* [1994b; see also Pudil *et al.* (1994a) for their assessment with nonmonotonic criterion functions and Kudo and Sklansky (2000) for a comparative study]. Error-rate-based procedures are described by McLachlan (1992a). Ganeshanandam and Krzanowski (1989) also use error rate as the selection criterion. Within the context of regression, the book by Miller (1990) gives very good accounts of variable selection.

Many of the standard feature extraction techniques may be found in most textbooks on multivariate analysis. Descriptions (with minimal mathematics) may be found in Reyment *et al.* (1984, chapter 3) and Clifford and Stephenson (1975, chapter 13).

Thorough treatments of principal components analysis are given in the books by Jolliffe (1986) and Jackson (1991), the latter providing a practical approach and giving many worked examples and illustrations. Common principal components and related methods are described in the book by Flury (1988).

Multidimensional scaling is described in textbooks on multivariate analysis (Chatfield and Collins, 1980; Dillon and Goldstein, 1984) and more detailed treatments are given in the books by Schiffman *et al.* (1981) and Jackson (1991). Cox and Cox (1994) provide an advanced treatment, with details of some of the specialised procedures. An extensive treatment of nonmetric MDS can be found in the collection of articles edited by Lingoes *et al.* (1979). There are many computer packages available for performing scaling. The features of some of these are given by Dillon and Goldstein (1984) and Jackson (1991).

Many of the techniques described in this chapter are available in standard statistical software packages.

Exercises

Numerical routines for matrix operations, including eigendecomposition, can be found in many numerical packages. Press *et al.* (1992) give descriptions of algorithms.

1. Consider the divergence,

$$J_D = \int (p(\boldsymbol{x}|\omega_1) - p(\boldsymbol{x}|\omega_2))\log\left(\frac{p(\boldsymbol{x}|\omega_1)}{p(\boldsymbol{x}|\omega_2)}\right) d\boldsymbol{x}$$

   where $\boldsymbol{x} = (x_1, \ldots, x_p)^T$. Show that under conditions of independence, J_D may be expressed as

$$J_D = \sum_{j=1}^{p} J_j(x_j)$$

2. Suppose the p-variate random variable $\boldsymbol{x}$ has covariance matrix $\boldsymbol{\Sigma}$ with eigenvalues $\{\lambda_i\}$ and orthonormal eigenvectors $\{\boldsymbol{a}_i\}$. Show that the identity matrix is given by

$$\boldsymbol{I} = \boldsymbol{a}_1\boldsymbol{a}_1^T + \cdots + \boldsymbol{a}_p\boldsymbol{a}_p^T$$

   and that

$$\boldsymbol{\Sigma} = \lambda_1\boldsymbol{a}_1\boldsymbol{a}_1^T + \cdots + \lambda_p\boldsymbol{a}_p\boldsymbol{a}_p^T$$

The latter result is called the *spectral decomposition* of $\mathbf{\Sigma}$ (Chatfield and Collins, 1980).

3. Given a set of n measurements on p variables, describe the stages in performing a principal components analysis for dimension reduction.
4. Let X_1 and X_2 be two random variables with covariance matrix

$$\mathbf{\Sigma} = \begin{bmatrix} 9 & \sqrt{6} \\ \sqrt{6} & 4 \end{bmatrix}$$

Obtain the principal components. What is the percentage of total variance explained by each component?

5. Athletic records for 55 countries comprise measurements made on eight running events. These are the country's record times for: (1) 100 m (s); (2) 200 m (s); (3) 400 m (s); 800 m (min); (5) 1500 m (min); (6) 5000 m (min); (7) 10 000 m (min); (8) marathon (min).

Describe how a principal components analysis may be used to obtain a two-dimensional representation of the data.

The results of a principal components analysis are shown below (Everitt and Dunn, 1991). Interpret the first two principal components.

	$PC1 \times \lambda_1$	$PC2 \times \lambda_2$
100 m	0.82	0.50
200 m	0.86	0.41
400 m	0.92	0.21
800 m	0.87	0.15
1500 m	0.94	−0.16
5000 m	0.93	−0.30
10 000 m	0.94	−0.31
Marathon	0.87	−0.42
Eigenvalue	6.41	0.89

What is the percentage of the total variance explained by the first principal component? State any assumptions you make.

6. Four measurements are made on each of a random sample of 500 animals. The first three variables were different linear dimensions, measured in centimetres, while the fourth variable was the weight of the animal measured in grammes. The sample covariance matrix was calculated and its four eigenvalues were found to be 14.1, 4.3, 1.2 and 0.4. The eigenvectors corresponding to the first and second eigenvalues were:

$$\boldsymbol{u}_1^T = [0.39, 0.42, 0.44, 0.69]$$
$$\boldsymbol{u}_2^T = [0.40, 0.39, 0.42, -0.72]$$

Comment on the use of the sample covariance matrix for the principal components analysis for these data. What is the percentage of variance in the original data accounted for by the first two principal components? Describe the results.

Suppose the data were stored by recording the eigenvalues and eigenvectors together with the 500 values of the first and second principal components and the mean values for the original variables. Show how to reconstruct the original covariance matrix and an approximation to the original data (Chatfield and Collins, 1980).

7. Given that

$$\boldsymbol{S} = \begin{bmatrix} \tilde{\boldsymbol{S}} & \boldsymbol{y} \\ \boldsymbol{y}^T & s_{kk} \end{bmatrix}$$

and assuming that $\tilde{\boldsymbol{S}}^{-1}$ is known, verify that $\boldsymbol{S}^{-1}$ is given by (10.12). Conversely, with $\boldsymbol{S}$ given by the above and assuming that $\boldsymbol{S}^{-1}$ is known,

$$\boldsymbol{S}^{-1} = \begin{bmatrix} \boldsymbol{A} & \boldsymbol{c} \\ \boldsymbol{c}^T & b \end{bmatrix}$$

show that the inverse of $\tilde{\boldsymbol{S}}$ (the inverse of $\boldsymbol{S}$ after the removal of a feature) can be written as

$$\tilde{\boldsymbol{S}}^{-1} = \boldsymbol{A} - \frac{1}{b}\boldsymbol{c}\boldsymbol{c}^T$$

8. For a given symmetric matrix $\boldsymbol{S}$ of known inverse (of the above form) and symmetric matrix $\boldsymbol{T}$, verify (10.14), where $\tilde{\boldsymbol{T}}$ is the submatrix of $\boldsymbol{T}$ after the removal of a feature. Hence, show that the feature extraction criterion $\text{Tr}(\boldsymbol{S}_W^{-1}\boldsymbol{S}_B)$, where $\boldsymbol{S}_W$ and $\boldsymbol{S}_B$ are the within and between-class covariance matrices, satisfies the monotonicity property.

9. How could you use floating search methods for radial basis function centre selection? What are the possible advantages and disadvantages of such methods compared with random selection or k-means, for example?

10. Suppose we take classification rate using a nearest class mean classifier as our feature selection criterion. Show by considering the two distributions,

$$p(\boldsymbol{x}|\omega_1) = \begin{cases} 1 & 0 \leq x_1 \leq 1,\ 0 \leq x_2 \leq 1 \\ 0 & \text{otherwise} \end{cases}$$

$$p(\boldsymbol{x}|\omega_2) = \begin{cases} 1 & 1 \leq x_1 \leq 2,\ -0.5 \leq x_2 \leq 0.5 \\ 0 & \text{otherwise} \end{cases}$$

where $\boldsymbol{x} = (x_1, x_2)^T$, that classification rate does not satisfy the monotonicity property.

11. Derive the relationship (10.25) expressing the elements of the sum of squares and products matrix in terms of the elements of the dissimilarity matrix from (10.24) and the definition of T_{ij} and zero-mean data.

12. Given n p-dimensional measurements (in the $n \times p$ data matrix $\boldsymbol{X}$, zero mean, $p < n$) show that a low-dimensional representation in $r < p$ dimensions obtained by constructing the sums of squares and products matrix, $\boldsymbol{T} = \boldsymbol{X}\boldsymbol{X}^T$, and performing a principal coordinates analysis, results in the same projection as principal components (to within an orthogonal transformation).

13. For two classes normally distributed, $N(\boldsymbol{\mu}_1, \boldsymbol{\Sigma})$ and $N(\boldsymbol{\mu}_2, \boldsymbol{\Sigma})$ with common covariance matrix, $\boldsymbol{\Sigma}$, show that the divergence

$$J_D(\omega_1, \omega_2) = \int [p(\boldsymbol{x}|\omega_1) - p(\boldsymbol{x}|\omega_2)] \log \left\{ \frac{p(\boldsymbol{x}|\omega_1)}{p(\boldsymbol{x}|\omega_2)} \right\} d\boldsymbol{x}$$

is given by

$$(\boldsymbol{\mu}_2 - \boldsymbol{\mu}_1)^T \boldsymbol{\Sigma}^{-1} (\boldsymbol{\mu}_2 - \boldsymbol{\mu}_1)$$

the *Mahalanobis distance*.

14. Describe how multidimensional scaling solutions that optimise stress can always be constructed that result in a decrease of stress with increasing dimension of the representation space.

11

Clustering

Clustering methods are used for data exploration and to provide prototypes for use in supervised classifiers. Methods that operate both on dissimilarity matrices and feature vector measurements on individuals are described, each implicitly imposing its own structure on the data. Mixtures explicitly model the data structure. Spectral clustering methods exploit the eigenstructure of a similarity matrix to perform clustering.

11.1 Introduction

Cluster analysis is the grouping of individuals in a population in order to discover structure in the data. In some sense, we would like the individuals within a group to be close or similar to one another, but dissimilar from individuals in other groups.

Clustering is fundamentally a collection of methods of data exploration. One often uses a method to see if natural groupings are present in the data. If groupings do emerge, these may be named and their properties summarised. For example, if the clusters are compact, then it may be sufficient for some purposes to reduce the information on the original dataset to information about a small number of groups, in some cases representing a group of individuals by a single pattern. The results of a cluster analysis may produce identifiable structure that can be used to generate hypotheses (to be tested on a separate dataset) to account for the observed data.

It is difficult to give a universal definition of the term 'cluster'. All of the methods described in this chapter can produce a *partition* of the dataset – a division of the dataset into mutually nonoverlapping groups. However, different methods will often yield different groupings since each implicitly imposes a structure on the data. Also, the techniques will produce groupings even when there is no 'natural' grouping in the data. The term 'dissection' is used when the data consist of a single homogeneous population that one wishes to partition. Clustering techniques may be used to obtain dissections, but the user must be aware that a structure is being imposed on the data that may not be present. This does not matter in some applications.

Statistical Pattern Recognition, Third Edition. Andrew R. Webb and Keith D. Copsey.

Before attempting a classification, it is important to understand the problem you are wishing to address. Different classifications, with consequently different interpretations, can be imposed on a sample and the choice of variables is very important. For example, there are different ways in which books may be grouped on your bookshelf – by subject matter or by size – and different classifications will result from the use of different variables. Each classification may be important in different circumstances, depending on the problem under consideration. Once you understand your problem and data, you must choose your method carefully. An inappropriate match of method to data can give results that are misleading.

There is a vast literature on clustering. Some of the more useful texts are given in Section 11.11. There is a wide range of application areas, sometimes with conflicting terminology. This has led to methods being rediscovered in different fields of study. Much of the early work was in the fields of biology and zoology, but clustering methods have also been applied in the fields of psychology, archaeology, linguistics and signal processing.

The topics discussed in this chapter are:

1. *Hierarchical methods*: methods that derive a clustering from a given dissimilarity matrix.

2. *Quick partitions*: methods for obtaining a partition as an initialisation to more elaborate approaches.

3. *Mixture models*: models that express the probability density function as a sum of component densities.

4. *Sum-of-squares methods*: methods that minimise a sum-of-squares error criterion, including k-means, fuzzy k-means, vector quantisation and stochastic vector quantisation.

5. *Spectral clustering*: methods that use the eigenvectors of a graph Laplacian to embed the data into a space that captures the underlying structure.

6. *Cluster validity*: addressing the problem of model selection.

11.2 Hierarchical methods

Hierarchical clustering procedures are the most commonly used means for summarising data structure. A hierarchical tree is a nested set of partitions represented by a tree diagram or *dendrogram* (Figure 11.1). Sectioning a tree at a particular level produces a partition into g disjoint groups. If two groups are chosen from different partitions (the results of partitioning at different levels) then either the groups are disjoint or one group wholly contains the other.

An example of a hierarchical classification is the classification of the animal kingdom. Each species belongs to a series of nested clusters of increasing size with a decreasing number of common characteristics. In producing a tree diagram like that in Figure 11.1, it is necessary to order the points so that branches do not cross. This ordering is somewhat arbitrary, but does not alter the structure of the tree, only its appearance. There is a numerical value associated with each position up the tree where branches join. This is a measure of the distance or dissimilarity between two merged clusters. There are many different measures of distances between clusters and these give rise to different hierarchical structures, as we shall see in

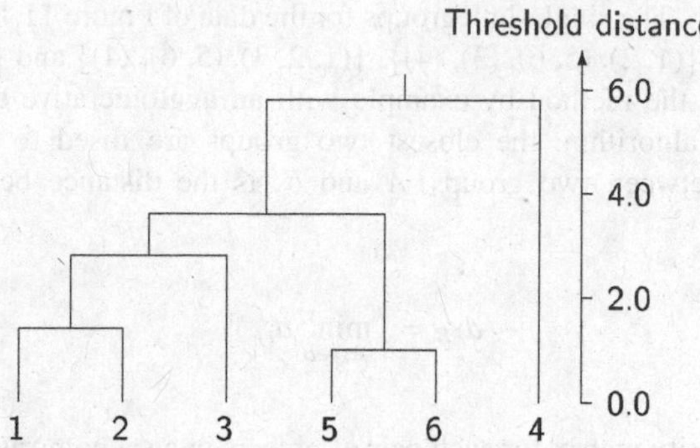

Figure 11.1 Dendrogram.

later sections of this chapter. Sectioning a tree partitions the data into a number of clusters of comparable homogeneity (as measured by the clustering criterion).

There are several different algorithms for finding a hierarchical tree. An *agglomerative algorithm* begins with n subclusters, each containing a single data point and at each stage merges the two most similar groups to form a new cluster, thus reducing the number of clusters by one. The algorithm proceeds until all the data fall within a single cluster. A *divisive algorithm* operates by successively splitting groups beginning with a single group and continuing until there are n groups, each of a single individual. Generally, divisive algorithms are computationally inefficient (except where most of the variables are binary attribute variables).

From the tree diagram, a new set of distances between individuals may be defined with the distance between individual i and individual j being the distance between the two groups that contain them, when these two groups are amalgamated (i.e. the distance level of the lowest link joining them). Thus, the procedure for finding a tree diagram may be viewed as a transformation of the original set of dissimilarities d_{ij} to a new set $\hat{d}_{ij}$ where $\hat{d}_{ij}$ satisfy the *ultrametric inequality*

$$\hat{d}_{ij} \leq \max(\hat{d}_{ik}, \hat{d}_{jk}) \qquad \text{for all objects } i, j, k$$

This means that the distances between three groups can be used to define a triangle that is either equilateral or isosceles (either the three distances are the same or two are equal and the third smaller – see Figure 11.1, for example). A transformation $D : d \rightarrow \hat{d}$ is termed an *ultrametric transformation*. All of the methods in this section produce a clustering from a given dissimilarity matrix.

11.2.1 Single-link method

The single-link method is one of the oldest methods of cluster analysis. It is defined as follows. Two objects a and b belong to the same single-link cluster at level d if there exists a chain of intermediate objects $i_1, \ldots, i_{m-1}$ linking them such that all the distances

$$d_{i_k, i_{k+1}} \leq d \qquad \text{for } k = 0, \ldots, m-1$$

where $i_0 = \text{a}$ and $i_m = \text{b}$. The single-link groups for the data of Figure 11.1 for a threshold of $d = 2.0$, 3.0 and 5.0 are {(1, 2), (5, 6), (3), (4)}, {(1, 2, 3), (5, 6), (4)} and {(1, 2, 3, 5, 6), (4)}.

We shall illustrate the method by example with an agglomerative algorithm in which, at each stage of the algorithm, the closest two groups are fused to form a new group where the distance between two groups, A and B, is the distance between their closest members, i.e.

$$d_{AB} = \min_{i \in A, j \in B} d_{ij} \tag{11.1}$$

Consider the dissimilarity matrix for each pair of objects in a set comprising six individuals:

	1	2	3	4	5	6
1	0	4	13	24	12	8
2		0	10	22	11	10
3			0	7	3	9
4				0	6	18
5					0	8.5
6						0

The closest two groups (which contain a single object each at this stage) are those containing the individuals 3 and 5. These are fused to form a new group {3, 5} and the distances between this new group and the remaining groups calculated according to Equation (11.1) so that $d_{1,(3,5)} = \min\{d_{13}, d_{15}\} = 12$; $d_{2,(3,5)} = \min\{d_{23}, d_{25}\} = 10$, $d_{4,(3,5)} = 6$, $d_{6,(3,5)} = 8.5$ giving the new dissimilarity matrix:

	1	2	(3, 5)	4	6
1	0	4	12	24	8
2		0	10	22	10
(3, 5)			0	6	8.5
4				0	18
6					0

The closest two groups now are those containing objects 1 and 2; therefore these are fused to form a new group (1, 2). We now have four clusters (1, 2), (3, 5), 4 and 6. The distance between the new group and the other three clusters is calculated: $d_{(1,2)(3,5)} = \min\{d_{13}, d_{23}, d_{15}, d_{25}\} = 10$; $d_{(1,2)4} = \min\{d_{14}, d_{24}\} = 22$; $d_{(1,2)6} = \min\{d_{16}, d_{26}\} = 8$. The new dissimilarity matrix is:

	(1, 2)	(3, 5)	4	6
(1, 2)	0	10	22	8
(3, 5)		0	6	8.5
4			0	18
6				0

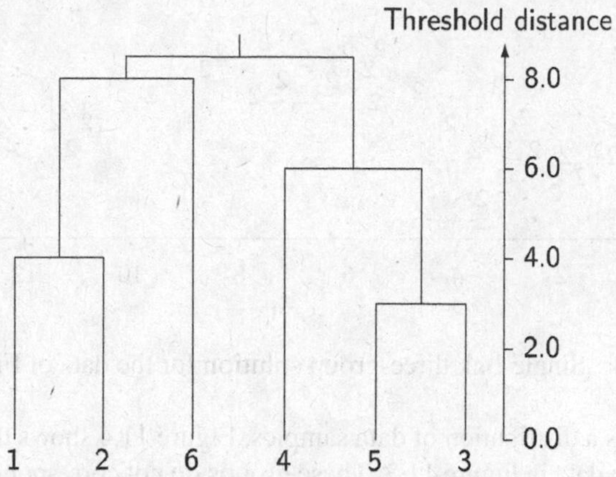

Figure 11.2 Single-link dendrogram.

The closest two groups are now those containing 4 and (3, 5). These are fused to form (3, 4, 5) and a new dissimilarity matrix calculated. This is shown with the result of fusing the next two groups:

	(1, 2)	(3, 4, 5)	6
(1, 2)	0	10	8
(3, 4, 5)		0	8.5
6			0

	(1, 2, 6)	(3, 4, 5)
(1, 2, 6)	0	8.5
(3, 4, 5)		0

The single-link dendrogram is given in Figure 11.2.

The above agglomerative algorithm for a single-link method illustrates the fact that it takes only a single link to join two distinct groups and that the distance between two groups is the distance of their closest neighbours. Hence the alternative name of *nearest-neighbour method*. A consequence of this joining together by a single link is that some groups can become elongated with some distant points, having little in common, being grouped together because there is a chain of intermediate objects. This drawback of *chaining* is illustrated in Figures 11.3 and 11.4.

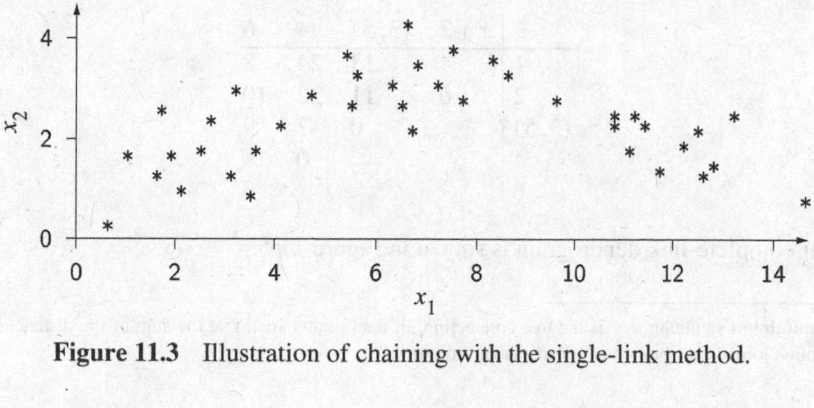

Figure 11.3 Illustration of chaining with the single-link method.

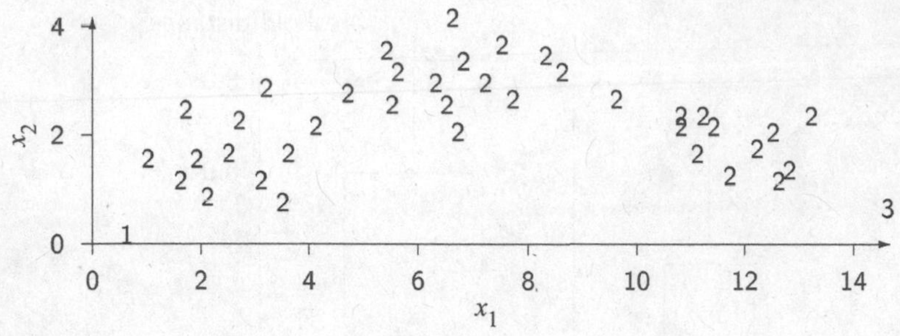

Figure 11.4 Single-link three-group solution for the data of Figure 11.3.

Figure 11.3 shows a distribution of data samples. Figure 11.4 shows the single-link three-group solution for the data in Figure 11.3. These groups do not correspond to those suggested by the data in Figure 11.3.

There are many algorithms for finding a single-link tree. Some are agglomerative, like the procedure described above, some are divisive; some are based on an ultrametric transformation and others generate the single-link tree via the minimum spanning tree[1] [see Rohlf (1982) for a review of algorithms]. The algorithms vary in their computational efficiency, storage requirements and ease of implementation. Sibson's (1973) algorithm uses the property that only local changes in the reduced dissimilarity result when two clusters are merged and it has been extended to the complete-link method discussed in the following section. It has computational requirements $\mathcal{O}(n^2)$, for n objects. More time-efficient algorithms are possible if knowledge of the metric properties of the space in which the data lie is taken into account. In such circumstances, it is not necessary to compute all dissimilarity coefficients. Also, preprocessing the data to facilitate searches for nearest neighbours can reduce computational complexity.

11.2.2 Complete-link method

In the *complete-link* or *furthest-neighbour method* the distance between two groups A and B is the distance between the two furthest points, one taken from each group

$$d_{AB} = \max_{i \in A, j \in B} d_{ij}$$

In the example used to illustrate the single-link method, the second stage dissimilarity matrix (after merging the closest groups 3 and 5 using the complete-link rule above) becomes:

	1	2	(3, 5)	4	6
1	0	4	13	24	8
2		0	11	22	10
(3, 5)			0	7	9
4				0	18
6					0

The final complete-link dendrogram is shown in Figure 11.5.

[1] The minimum spanning tree is the tree connecting all data points such that the sum of the distances between pairs of points joined by a branch of the tree is a minimum.

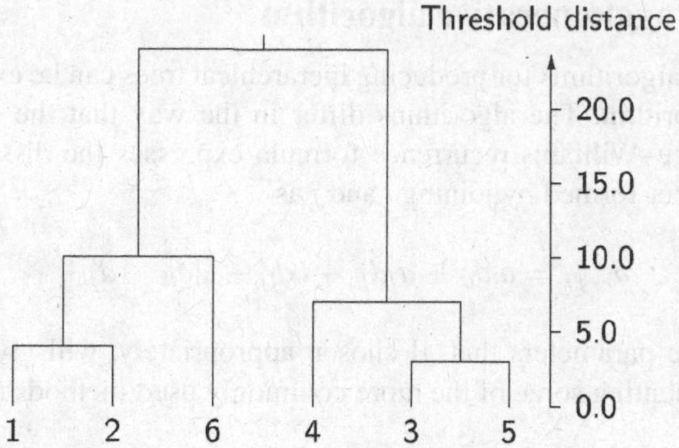

Figure 11.5 Complete-link dendrogram.

At each stage, the closest groups are merged of course. The difference between this method and the single-link method is the measure of distance between groups. The groups found by sectioning the complete-link dendrogram at level h have the property that $d_{ij} < h$ for all members in the group. The method concentrates on the internal *cohesion* of groups in contrast to the single-link method, which seeks isolated groups. Sectioning a single-link dendrogram at a level h gives groups with the property that they are separated from each other by at least a 'distance' h.

Defays (1977) provides an algorithm for the complete-link method using the same representation as Sibson. It should be noted that the algorithm is sensitive to the ordering of the data, and consequently has several solutions. Thus it provides only an approximate complete link clustering.

11.2.3 Sum-of-squares method

The sum-of-squares method is appropriate for the clustering of points in Euclidean space. The aim is to minimise the total within-group sum of squares. *Ward's hierarchical clustering method* (Ward, 1963) uses an agglomerative algorithm to produce a set of hierarchically nested partitions that can be represented by a dendrogram. However, the *optimal* sum-of-squares partitions for different numbers of groups are not necessarily hierarchically nested. Thus Ward's algorithm is suboptimal.

At each stage of the algorithm, the two groups that produce the smallest increase in the total within-group sum of squares are amalgamated. The dissimilarity between two groups is defined to be the increase in the total sum of squares that would result if they were amalgamated. The updating formula for the dissimilarity matrix is

$$d_{i+j,k} = \frac{n_k + n_i}{n_k + n_i + n_j} d_{ik} + \frac{n_k + n_j}{n_k + n_i + n_j} d_{jk} - \frac{n_k}{n_k + n_i + n_j} d_{ij}$$

where $d_{i+j,k}$ is the distance between the amalgamated groups $i + j$ and the group k and n_i is the number of objects in group i. Initially, each group contains a single object and the element of the dissimilarity matrix, d_{ij}, is the squared Euclidean distance between the ith and the jth object (Everitt *et al.*, 2011).

11.2.4 General agglomerative algorithm

Many agglomerative algorithms for producing hierarchical trees can be expressed as a special case of a single algorithm. The algorithms differ in the way that the dissimilarity matrix is updated. The Lance–Williams recurrence formula expresses the dissimilarity between a cluster *k* and the cluster formed by joining *i* and *j* as

$$d_{i+j,k} = a_i d_{ik} + a_j d_{jk} + b d_{ij} + c|d_{ik} - d_{jk}|$$

where a_i, b and c are parameters that, if chosen appropriately, will give an agglomerative algorithm for implementing some of the more commonly used methods (Table 11.1).

Centroid distance
This defines the distance between two clusters to be the distance between the cluster means or (centroids).

Median distance
When a small cluster is joined to a larger one, the centroid of the result will be close to the centroid of the larger cluster. For some problems this may be a disadvantage. This measure attempts to overcome this by defining the distance between two clusters to be the distance between the medians of the clusters.

Group average link
In the group average method, the distance between two clusters is defined to be the average of the dissimilarities between all pairs of individuals, one from each group

$$d_{AB} = \frac{1}{n_i n_j} \sum_{i \in A, j \in B} d_{ij}$$

11.2.5 Properties of a hierarchical classification

What desirable properties should a hierarchical clustering method possess? It is difficult to write down a set of properties on which everyone will agree. What might be a set of

Table 11.1 Special cases of the general agglomerative algorithm.

	a_i	b	c
Single link	$\frac{1}{2}$	0	$-\frac{1}{2}$
Complete link	$\frac{1}{2}$	0	$\frac{1}{2}$
Centroid	$\frac{n_i}{n_i+n_j}$	$-\frac{n_i n_j}{(n_i+n_j)^2}$	0
Median	$\frac{1}{2}$	$-\frac{1}{4}$	0
Group average link	$\frac{n_i}{n_i+n_j}$	0	0
Ward's method	$\frac{n_i+n_k}{n_i+n_j+n_k}$	$-\frac{n_k}{n_i+n_j+n_k}$	0

commonsense properties to one person may be the extreme position of another. Jardine and Sibson (1971) suggest a set of six mathematical conditions that an ultrametric transformation should satisfy; for example, that the results of the method should not depend on the labelling of the data. They show that the single-link method is the only one to satisfy all their conditions and recommend this method of clustering. However, this method has its drawbacks (as do all methods) which has led people to question the plausibility of the set of conditions proposed by Jardine and Sibson. We shall not list the conditions here but refer to Jardine and Sibson (1971) and Williams *et al.* (1971) for further discussion.

11.2.6 Example application study

The problem

This application involves the analysis of microarray gene expression data of *Toxoplasma gondii* (Gautam *et al.*, 2010). *Toxoplasma gondii* is a parasite that can infect a wide range of warm-blooded animals, including humans, and develop into a chronic infection that cannot be eliminated by currently used drugs. The aim of the study is to find out new possible drug targets (key molecules) through microarray data analysis.

The data

Microarray data were provided by the Stanford Microarray Database. The data were filtered to 327 genes for analysis.

The model

The assumption is that if the expression of one gene is similar to another, then it is likely that they are related in function. Thus, a method for grouping expression data is implemented. A hierarchical clustering method was implemented (the complete-link method) and results compared with *k*-means clustering (Section 11.5).

Results

For the *Toxoplasma gondii* dataset, five main clusters were identified using the complete-link method. One particular cluster was identified as important for further investigation since it included surface antigens (SAGs), known to be important. Most of the SAG gene family members present in this cluster were also present in a single cluster produced by the *k*-means clustering algorithm. The study identified some possible drug targets for the treatment of toxoplasmosis, since the genes have an important role in the immune system.

11.2.7 Summary

The concept of having a hierarchy of nested clusters was developed primarily in the biological field and may be inappropriate to model the structure in some data. Each hierarchical method imposes its own structure on the data. The single-link method seeks isolated clusters, but is generally not favoured, even though it is the only one satisfying the conditions proposed by Jardine and Sibson. It is subject to the chaining effect, which can result in long straggly groups. This may be useful in some circumstances if the clusters sought are not homogeneous, but it can mean that distinct groups are not resolved because of intermediate points present between the groups. The group average, complete-link and Ward's method tend to concentrate on internal cohesion, producing homogeneous, compact (often spherical) groups.

The centroid and median methods may lead to *inversions* (a reduction in the dissimilarity between an object and a cluster when the cluster increases in size) which make the dendrogram difficult to interpret. Also, ties in the dissimilarities may lead to multiple solutions (*nonuniqueness*) of which the user should be aware (Morgan and Ray, 1995).

Hierarchical agglomerative methods are one of the most common clustering techniques employed. Divisive algorithms are less popular, but efficient algorithms have been proposed, based on recursive partitioning of the cluster with largest diameter (Guénoche *et al.*, 1991).

11.3 Quick partitions

Many of the techniques described subsequently in this chapter are implemented by algorithms that require an initial partition of the data. The normal mixture methods require initial estimates of means and covariance matrices. These could be sample-based estimates derived from an initial partition. The k-means algorithm also requires an initial set of means. Hierarchical vector quantisation (see Section 11.5.3) requires initial estimates of the code vectors. In the context of discrimination, radial basis functions, introduced in Chapter 4, require initial estimates for 'centres'. These could be derived using the quick partition methods of this section or be the result of a more principled clustering approach (which in turn may need to be initialised).

Let us suppose that we have a set of n data samples and we wish to find an initial partition into k groups, or to find k seed vectors. We can always find a seed vector, given a group of objects, by taking the group mean. Also, we can partition a set, given k vectors, using a nearest-neighbour assignment rule. There are many heuristic partition methods. We shall consider a few of them.

Random k selection
We wish to have k different vectors, so we select one randomly from the whole dataset, the another from the remaining $n - 1$ samples in the dataset, and so on. In a supervised classification problem, these vectors should be spread across all classes ideally.

Variable division
Choose a single variable. This may be selected from one of the measured variables or be a linear combination of variables; for example, the first principal component. Divide it into k equal intervals that span the range of the variable. The data are partitioned according to which bin they fall in and k seed vectors are found from the means of each group.

Leader algorithm
The leader cluster algorithm (Hartigan, 1975; Späth, 1980) partitions a dataset such that for each group there is a leader object and all other objects within the group are within a distance T of the leading example.

Figure 11.6 illustrates a partition in two dimensions. The first data point, A, is taken as the centre of the first group. Successive data points are examined. If they fall inside the circle centred at A of radius T then they are assigned to group 1. The first data sample examined to fall outside the circle, say at B, is taken as the leader of the second group. Further data points are examined to see if they fall within the first two clusters. The first one to fall outside, say at C, is taken as the centre of the third cluster and so on.

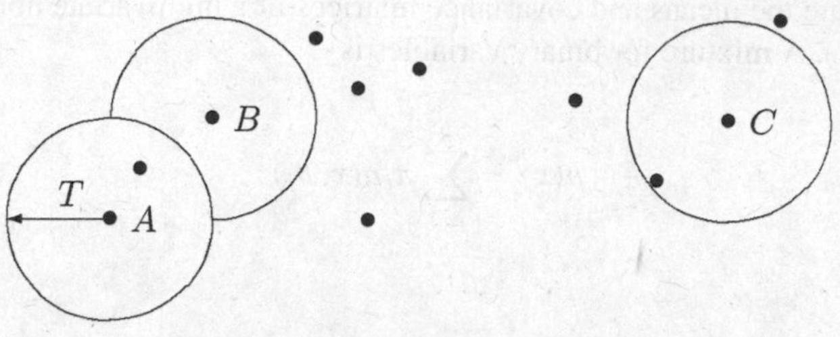

Figure 11.6 Leader clustering.

Points to note about the algorithm:

1. All cluster centres are at least a distance T from each other.
2. It is fast, requiring only one pass through the dataset.
3. It can be applied to a given dissimilarity matrix.
4. It is dependent on the ordering of the dataset. The first point is always a cluster leader. Also, initial clusters tend to be larger than later ones.
5. The distance T is specified, not the number of clusters.

11.4 Mixture models

11.4.1 Model description

In the mixture method of clustering, each different group in the population is assumed to be described by a different probability distribution. These different probability distributions may belong to the same family but differ in the values they take for the parameters of the distribution. Alternatively, the population may comprise sums of different component densities (for modelling different effects such as a signal and noise). The population is said to be described by a *finite mixture distribution* of the form

$$p(\boldsymbol{x}) = \sum_{i=1}^{g} \pi_i p(\boldsymbol{x}; \boldsymbol{\theta}_i)$$

where π_i are the mixing proportions ($\sum_{i=1}^{g} \pi_i = 1$) and $p(\boldsymbol{x}; \boldsymbol{\theta}_i)$ is a p-dimensional probability function depending on a parameter vector $\boldsymbol{\theta}_i$. There are three sets of parameters to estimate: the values of π_i, the components of the vectors $\boldsymbol{\theta}_i$ and the value of g, the number of groups in the population.

Many forms of mixture distributions have been considered and there are many methods for estimating their parameters. An example of a mixture distribution for continuous variables is the mixture of normal distributions

$$p(\boldsymbol{x}) = \sum_{i=1}^{g} \pi_i p(\boldsymbol{x}; \boldsymbol{\Sigma}_i, \boldsymbol{\mu}_i)$$

where $\boldsymbol{\mu}_i$ and $\boldsymbol{\Sigma}_i$ are the means and covariance matrices of a multivariate normal distribution (see Section 2.3.1). A mixture for binary variables is

$$p(\boldsymbol{x}) = \sum_{i=1}^{g} \pi_i p(\boldsymbol{x}; \boldsymbol{\theta}_i)$$

where

$$p(\boldsymbol{x}; \boldsymbol{\theta}_j) = \prod_{l=1}^{p} \theta_{jl}^{x_l} (1 - \theta_{jl})^{1-x_l}$$

is the multivariate Bernoulli density. The value of θ_{jl} is the probability that variable l in the jth group is unity.

Maximum likelihood procedures for estimating the parameters of normal mixture distributions were given in Chapter 2. Other examples of continuous and discrete mixture distributions, and methods of parameter estimation, can be found in Everitt and Hand (1981) and Titterington *et al.* 1985). Also, in some applications, variables are often of a mixed type – both continuous and discrete.

The usual approach to clustering using finite mixture distributions is first of all to specify the form of the component distributions, $p(\boldsymbol{x}, \boldsymbol{\theta}_i)$. Then the number of clusters, g, is prescribed. The parameters of the model are now estimated and the objects are grouped on the basis of their estimated posterior probabilities of group membership; that is the object $\boldsymbol{x}$ is assigned to group i if

$$\pi_i p(\boldsymbol{x}; \boldsymbol{\theta}_i) \geq \pi_j p(\boldsymbol{x}; \boldsymbol{\theta}_j) \qquad \text{for all } j \neq i,\ j = 1, \ldots, g$$

Clustering using a normal mixture model may be achieved by using the EM algorithm described in Chapter 2, to which we refer for further details.

The main difficulty with the method of mixtures concerns the number of components, g (see Chapter 2). This is the question of model selection we return to many times in this book. Many algorithms require g to be specified before the remaining parameters can be estimated. Several test statistics have been put forward. Many apply to special cases such as assessing the question as to whether or not the data come from a single component distribution or a two-component mixture. However, others have been proposed based on likelihood ratio tests (Everitt and Hand, 1981, chapter 5; Titterington *et al.*, 1985, chapter 5).

Another problem with a mixture model approach is that there may be many local minima of the likelihood function and several initial configurations may have to be tried before a satisfactory clustering is produced. In any case, it is worthwhile trying several initialisations, since agreement between the resulting classifications lends more weight to the chosen solution.

11.4.2 Example application study

The problem

This application considers clustering of gene expression data for gene function discovery (Dai *et al.*, 2009). The assumption is that genes having similar expression patterns should have similar cellular functions. Therefore, a clustering approach is developed to cluster gene profiles.

Summary
The study develops a model for data arising from multiple sources. It is applied to real data providing biologically plausible results.

The data
Real and simulated datasets were used to assess the approach. The real data comprised mouse protein DNA binding data and gene expression data. There were 1775 genes in both datasets, which were filtered to provide 673 genes for analysis.

The model
An approach based on mixture models is developed. The mixture components are modelled as a product of two independent terms. That is, if $\boldsymbol{x} = [\boldsymbol{y}^T, \boldsymbol{z}^T]^T$, then

$$p(\boldsymbol{x}) = p(\boldsymbol{y})p(\boldsymbol{z})$$

and $p(\boldsymbol{y})$ is a Gaussian distribution and $p(\boldsymbol{z})$ is a beta distribution.

The motivation for this is that the observation $\boldsymbol{x}$ arises from two data sources, each generated from a different probability distribution.

Training procedure
The model parameters were determined by maximising the likelihood using the EM algorithm (see Chapter 2). Three forms of EM algorithm are considered: the standard EM algorithm, an approximated EM and a hybrid EM. The latter two approaches were developed to reduce the computational requirement.

Four model selection criteria (to determine the number of mixture components) were assessed: Akaike Information Criterion (AIC); a modified AIC (AIC3); the Bayesian Information Criterion (BIC); and the integrated classification likelihood-BIC (ICL-BIC).

Results
The approach yielded more biologically plausible results, independent of the model selection criteria, than using simpler mixture models. Three important groups of genes were identified.

11.5 Sum-of-squares methods

Sum-of-squares methods find a partition of the data that maximises a predefined clustering criterion based on the within-class and between-class scatter matrices. The methods differ in the choice of clustering criterion optimised and the optimisation procedure adopted. However, the problem all methods seek to solve is: *Given a set of n data samples, partition the data into g clusters so that the clustering criterion is optimised.*

Most methods are suboptimal. Computational requirements prohibit optimal schemes, even for moderate values of n. Therefore we require methods that, although producing a suboptimal partition, give a value of the clustering criterion that is not much worse than the optimal one. First of all, let us consider the various criteria that have been proposed.

11.5.1 Clustering criteria

Let the n data samples be $\boldsymbol{x}_1, \ldots, \boldsymbol{x}_n$. The sample covariance matrix, $\hat{\boldsymbol{\Sigma}}$, is given by

$$\hat{\boldsymbol{\Sigma}} = \frac{1}{n} \sum_{i=1}^{n} (\boldsymbol{x}_i - \boldsymbol{m})(\boldsymbol{x}_i - \boldsymbol{m})^T$$

where $\boldsymbol{m} = \frac{1}{n}\sum_{i=1}^{n} \boldsymbol{x}_i$, the sample mean. Let there be g clusters. The *within-class scatter matrix* or *pooled within-group scatter matrix* is

$$\boldsymbol{S}_W = \frac{1}{n} \sum_{j=1}^{g} \sum_{i=1}^{n} z_{ji} (\boldsymbol{x}_i - \boldsymbol{m}_j)(\boldsymbol{x}_i - \boldsymbol{m}_j)^T,$$

the sum of the sums of squares and cross-products (scatter) matrices over the g groups, where $z_{ji} = 1$ if $\boldsymbol{x}_i \in$ group j, 0 otherwise, $\boldsymbol{m}_j = \frac{1}{n_j}\sum_{i=1}^{n} z_{ji}\boldsymbol{x}_i$ is the mean of cluster j and $n_j = \sum_{i=1}^{n} z_{ji}$, the number in cluster j. The between-class scatter matrix is

$$\boldsymbol{S}_B = \hat{\boldsymbol{\Sigma}} - \boldsymbol{S}_W = \sum_{j=1}^{g} \frac{n_j}{n} (\boldsymbol{m}_j - \boldsymbol{m})(\boldsymbol{m}_j - \boldsymbol{m})^T$$

and describes the scatter of the cluster means about the total mean.

The most popular optimisation criteria are based on univariate functions of the above matrices and are similar to the criteria given in Chapter 10 on feature selection and extraction. The two areas of clustering and feature selection are very much related. In clustering we are seeking clusters that are internally cohesive but isolated from other clusters. We do not know the number of clusters. In feature selection or extraction, we have labelled data from a known number of groups or classes and we seek a transformation that makes the classes distinct. Therefore one that transforms the data into isolated clusters will achieve this.

1. Minimisation of $\mathrm{Tr}(\boldsymbol{S}_W)$

The trace of $\boldsymbol{S}_W$ is the sum of the diagonal elements

$$\begin{aligned}\mathrm{Tr}(\boldsymbol{S}_W) &= \frac{1}{n} \sum_{j=1}^{g} \sum_{i=1}^{n} z_{ij} |\boldsymbol{x}_i - \boldsymbol{m}_j|^2 \\ &= \frac{1}{n} \sum_{j=1}^{g} S_j\end{aligned}$$

where $S_j = \sum_{i=1}^{n} z_{ji} |\boldsymbol{x}_i - \boldsymbol{m}_j|^2$, the within-group sum of squares for group j. Thus, the minimisation of $\mathrm{Tr}(\boldsymbol{S}_W)$ is equivalent to minimising the total within-group sum of squares about the g centroids. Clustering methods that minimise this quantity are sometimes referred to as sum-of-squares or *minimum-variance* methods. They tend to produce clusters that are hyperspherical in shape. The criterion is not invariant to the scale of the axes and usually some form of standardisation of the data must be performed prior to application of the method. Alternatively, criteria that are not invariant to linear transformations of the data may be employed.

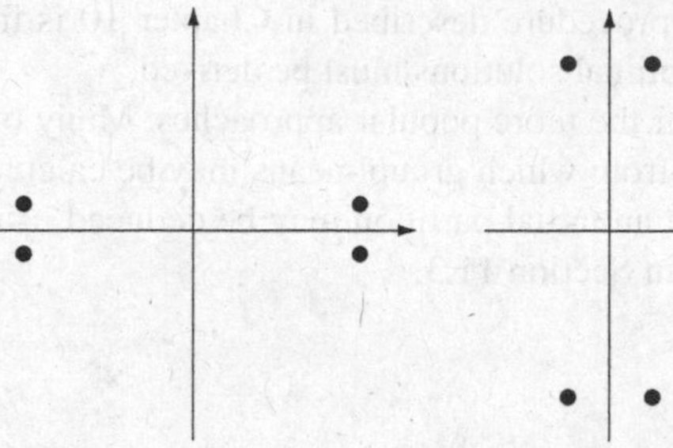

Figure 11.7 Example of effects of scaling in clustering.

2. Minimisation of $|S_W|/|\hat{\Sigma}|$
This criterion is invariant to nonsingular linear transformations of the data. For a given dataset, it is equivalent to finding the partition of the data that minimises $|S_W|$ (the matrix $\hat{\Sigma}$ is independent of the partition).

3. Maximisation of $\mathrm{Tr}(S_W^{-1} S_B)$
This is a generalisation of the sum-of-squares method in that the clusters are no longer hyperspherical, but hyperellipsoidal. It is equivalent to minimising the sum of squares under the Mahalanobis metric. It is also invariant to nonsingular transformations of the data.

4. Minimisation of $\mathrm{Tr}(\hat{\Sigma}^{-1} S_W)$
This is identical to minimising the sum of squares for data that have been normalised to make the total scatter matrix equal to the identity.

Note that the two examples in Figure 11.7 would be clustered differently by the sum-of-squares method (criterion 1 above). However, since they only differ from each other by a linear transformation, they must both be local optima of a criterion invariant to linear transformations. Thus, it is not necessarily an advantage to use a method that is invariant to linear transformations of the data since structure may be lost. The final solution will depend very much on the initial assignment of points to clusters.

11.5.2 Clustering algorithms

The problem we are addressing is one in combinatorial optimisation. We seek a nontrivial partition of n objects into g groups for which the chosen criterion is optimised. However, to find the optimum partition requires the examination of every possible partition. The number of nontrivial partitions of n objects into g groups is

$$\frac{1}{g!}\sum_{i=1}^{g}(-1)^{g-i}\binom{g}{i}i^n$$

with the final term in the summation being most significant if $n \gg g$. This increases rapidly with the number of objects. For example, there are $2^{59} - 1 \approx 6 \times 10^{17}$ partitions of 60 objects into two groups. This makes exhaustive enumeration of all possible subsets infeasible. In fact, even the branch and bound procedure described in Chapter 10 is impractical for moderate values of n. Therefore, suboptimal solutions must be derived.

We now describe some of the more popular approaches. Many of the procedures require initial partitions of the data, from which group means may be calculated, or initial estimates of group means (from which an initial partition may be deduced using a nearest-class-mean rule). These were discussed in Section 11.3.

11.5.2.1 *k*-means

The aim of the *k-means* (which also goes by the names of the *c-means* or *iterative relocation* or *basic ISODATA*) algorithm is to partition the data into k clusters so that the within-group sum of squares (criterion 1) is minimised. The simplest form of the k-means algorithm is based on alternating two procedures. The first is one of assignment of objects to groups. An object is usually assigned to the group to whose mean it is closest in the Euclidean sense. The second procedure is the calculation of new group means based on the assignments. The process terminates when no movement of an object to another group will reduce the within-group sum of squares. Let us illustrate with a very simple example.

Consider the two-dimensional data shown in Figure 11.8. Let us set $k = 2$ and choose two vectors from the dataset as initial cluster mean vectors. Those selected are points 5 and 6. We now cycle through the dataset and allocate individuals to groups A and B represented by the initial vectors 5 and 6 respectively. Individuals 1, 2, 3, 4 and 5 are allocated to A and individual 6 to B. New means are calculated and the within-group sum of squares is evaluated, giving 6.4. The results of this iteration are summarised in Table 11.2.

The process is now repeated, using the new mean vectors as the reference vectors. This time, individuals 1, 2, 3, and 4 are allocated to group A and 5 and 6 to group B. The within-group sum-of-squares has now decreased to 4.0. A third iteration produces no change in the within-group sum-of-squares.

The iterative procedure of allocating objects to groups on a nearest-group-mean basis, followed by recalculation of group means, gives the version of the k-means called HMEANS by Späth (1980). It is also termed Forgy's method or the basic ISODATA method.

There are two main problems with HMEANS. It may lead to empty groups and it may lead to a partition for which the sum-square error could be reduced by moving an individual from one group to another. Thus the partition of the data by HMEANS is not necessarily one

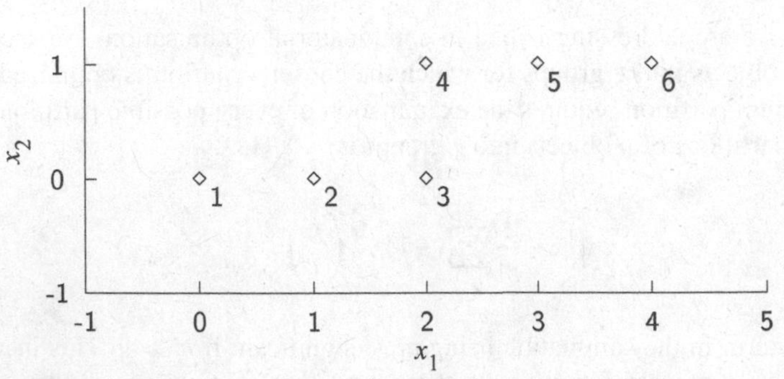

Figure 11.8 Data to illustrate the k-means procedure.

Table 11.2 Summary of k-means iterations.

	Group A		Group B		
Step	Membership	Means	Membership	Means	Tr($\boldsymbol{W}$)
1	1, 2, 3, 4, 5	(1.6, 0.4)	6	(4.0, 1.0)	6.4
2	1, 2, 3, 4	(1.25, 0.25)	5, 6	(3.5, 1.0)	4.0
3	1, 2, 3, 4	(1.25, 0.25)	5, 6	(3.5, 1.0)	4.0

for which the within-group sum of squares is a minimum [see Selim and Ismail (1984a) for a treatment of the convergence of this algorithm].

For example, in Figure 11.9 four data points and two groups are illustrated. The means are at positions (1.0, 0.0) and (3.0, 1.0), with a sum-square error of 4.0. Repeated iterations of the algorithm HMEANS will not alter that allocation. However, if we allocate object 2 to the group containing objects 3 and 4, the means are now at (0.0, 0.0) and (8/3, 2/3), and the sum-square error is reduced to 10/3. This suggests an iterative procedure that cycles through the data points and allocates each to a group for which the within-group sum of squares is reduced the most. Allocation takes place on a sample-by-sample basis, rather than after a pass through the entire dataset. An individual $\boldsymbol{x}_i$ (in group l) is assigned to group r if

$$\frac{n_l}{n_l - 1} d_{il}^2 > \frac{n_r}{n_r + 1} d_{ir}^2$$

where d_{il} is the distance to the lth centroid and n_l is the number in group l. The greatest decrease in the sum-square error is achieved by choosing the group for which $n_r d_{ir}^2/(n_r + 1)$ is a minimum. This is the basis of the k-means algorithm.

There are many variants of the k-means algorithm to improve efficiency of the algorithm in terms of computing time and of achieving smaller error. Some algorithms allow new clusters to be created and existing ones deleted during the iterations. Others may move an object to another cluster on the basis of the best improvement in the objective function. Alternatively, the first encountered improvement during the pass through the dataset could be used.

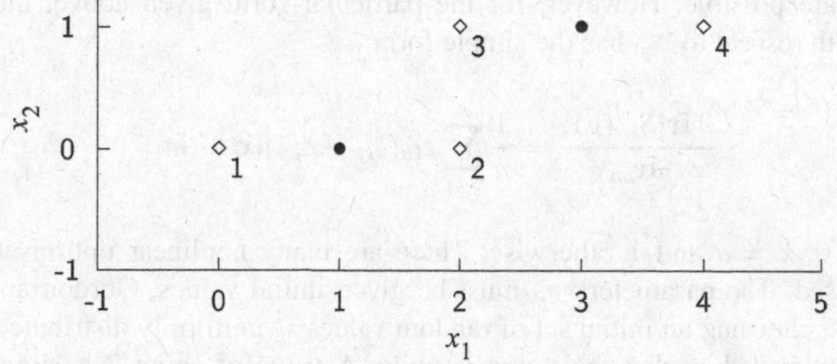

Figure 11.9 Data to illustrate the k-means local optimum.

11.5.2.2 Nonlinear optimisation

The within-groups sum-of-squares criterion may be written in the form

$$\mathrm{Tr}(S_W) = \frac{1}{n}\sum_{i=1}^{n}\sum_{k=1}^{g} z_{ki}\sum_{j=1}^{p}(x_{ij} - m_{kj})^2 \tag{11.2}$$

where x_{ij} is the jth coordinate of the ith point ($i = 1, \ldots, n$; $j = 1, \ldots, p$); m_{kj} is the jth coordinate of the mean of the kth group and $z_{ki} = 1$ if the ith point belongs to the kth group and 0 otherwise. The mean quantities m_{kj} may be written as

$$m_{kj} = \frac{\sum_{i=1}^{n} z_{ki}x_{ij}}{\sum_{i=1}^{n} z_{ki}} \tag{11.3}$$

To obtain an optimal partition, we must find the values of z_{ki} (either 0 or 1) for which the expression (11.2) is a minimum.

The approach of Gordon and Henderson (1977) is to regard the $g \times n$ matrix $\mathbf{Z}$ with (ij) element z_{ij} as consisting of real-valued quantities (as opposed to binary quantities) with the property

$$\sum_{k=1}^{g} z_{ki} = 1 \quad \text{and} \quad z_{ki} \geq 0 \quad (i = 1, \ldots, n; k = 1, \ldots, g) \tag{11.4}$$

Minimisation of the expression (11.2) with respect to z_{ki}, ($i = 1, \ldots, n; k = 1, \ldots, g$), subject to the constraints above, yields a final solution for $\mathbf{Z}$ with elements that are all 0 or 1 (even though they are not constrained to be binary-valued). Therefore we can obtain a partition by minimising (11.2) subject to the constraints (11.4) and assigning objects to groups on the basis of the values z_{ik}. Thus, m_{kj} is not equal to a group mean until the iteration has converged.

The problem can be transformed to one of unconstrained optimisation by writing z_{ji} as

$$z_{ji} = \frac{\exp(\nu_{ji})}{\sum_{k=1}^{g}\exp(\nu_{ki})} \quad (j = 1, \ldots, g; i = 1, \ldots, n)$$

that is, we regard $\mathrm{Tr}\{\boldsymbol{S}_W\}$ as a nonlinear function of parameters ν_{ki}, $i = 1, \ldots, n$; $k = 1, \ldots, g$, and seek a minimum of $\mathrm{Tr}\{\boldsymbol{S}_W(\nu)\}$. Other forms of transformation to unconstrained optimisation are possible. However, for the particular form given above, the gradient of $\mathrm{Tr}\{\boldsymbol{S}_W(\nu)\}$ with respect to ν_{ab} has the simple form

$$\frac{\partial\,\mathrm{Tr}\{\boldsymbol{S}_W(\nu)\}}{\partial \nu_{ab}} = \frac{1}{n}\sum_{k=1}^{g} z_{kb}(\delta_{ka} - z_{ab})|\boldsymbol{x}_b - \boldsymbol{m}_k|^2 \tag{11.5}$$

where $\delta_{ka} = 0$, $k \neq a$ and 1 otherwise. There are many nonlinear optimisation schemes that can be used. The parameters ν_{ij} must be given initial values. Gordon and Henderson (1977) suggest choosing an initial set of random values z_{ji} uniformly distributed in the range $[1, 1 + a]$ and scaled so that their sum is unity. A value of about 2 is suggested for the parameter a. Then ν_{ji} is initialised to $\nu_{ji} = \log(z_{ji})$.

11.5.2.3 Fuzzy *k*-means

The partitioning methods described so far in this chapter have the property that each object belongs to one group only, though the mixture model can be regarded as providing degrees of cluster membership. Indeed, the early work on fuzzy clustering was closely related to multivariate mixture models. The basic idea of the fuzzy clustering method is that patterns are allowed to belong to all clusters with different degrees of membership. The first generalisation of the *k*-means algorithm was presented by Dunn (1974). The *fuzzy k-means* (or fuzzy *c*-means) algorithm attempts to find a solution for parameters y_{ji} ($i = 1, \ldots, n; j = 1, \ldots, g$) for which

$$J_r = \sum_{i=1}^{n} \sum_{j=1}^{g} y_{ji}^r |\boldsymbol{x}_i - \boldsymbol{m}_j|^2 \tag{11.6}$$

is minimised subject to the constraints

$$\sum_{j=1}^{g} y_{ij} = 1 \quad 1 \leq i \leq n$$

$$y_{ji} \geq 0 \quad i = 1, \ldots, n; \;\; j = 1, \ldots, g$$

The parameter y_{ji} represents the *degree of association* or *membership function* of the ith pattern or object with the jth group. In the above expression (11.6), r is a scalar termed the *weighting exponent* which controls the 'fuzziness' of the resulting clusters ($r \geq 1$) and $\boldsymbol{m}_j$ is the 'centroid' of the jth group

$$\boldsymbol{m}_j = \frac{\sum_{i=1}^{n} y_{ji}^r \boldsymbol{x}_i}{\sum_{i=1}^{n} y_{ji}^r} \tag{11.7}$$

A value of $r = 1$ gives the same problem as the nonlinear optimisation scheme presented earlier. In that case, we know that a minimum of (11.6) gives values for the y_{ji} that are either 0 or 1.

The basic algorithm is iterative and can be stated as follows (Bezdek, 1981):

1. Select r ($1 < r < \infty$); initialise the membership function values y_{ji}, $i = 1, \ldots, n; j = 1, \ldots, g$.
2. Compute the cluster centres $\boldsymbol{m}_j$, $j = 1, \ldots, g$, according to (11.7).
3. Compute the distances d_{ij}, $i = 1, \ldots, n; j = 1, \ldots, g$, where $d_{ij} = |\boldsymbol{x}_i - \boldsymbol{m}_j|$.
4. Compute the membership function according to

   If $d_{il} = 0$ for some l, $y_{li} = 1$, and $y_{ji} = 0$, for all $j \neq l$

   $$\text{otherwise } y_{ji} = \frac{1}{\sum_{k=1}^{g} \left(\frac{d_{ij}}{d_{ik}}\right)^{\frac{2}{r-1}}}$$

5. If not converged, go to step 2.

As $r \to 1$, this algorithm tends to the basic *k*-means algorithm. Improvements to this basic algorithm, resulting in faster convergence, are described by Kamel and Selim (1994).

Several stopping rules have been proposed (Ismail, 1988). One is to terminate the algorithm when the relative change in the centroid values becomes small; that is, terminate when

$$D_z \triangleq \left\{ \sum_{j=1}^{g} |\boldsymbol{m}_j(k) - \boldsymbol{m}_j(k-1)|^2 \right\}^{\frac{1}{2}} < \epsilon$$

where $\boldsymbol{m}_j(k)$ is the value of the jth centroid on the kth iteration and ϵ is a user-specified threshold. Alternative stopping rules are based on changes in the membership function values, y_{ji}, or the cost function, J_r. Another condition based on the local optimality of the cost function is given by Selim and Ismail (1986). It is proposed to stop when

$$\max_{1 \leq i \leq n} \alpha_i < \epsilon$$

where

$$\alpha_i = \max_{1 \leq j \leq g} y_{ji}^{r-1} |\boldsymbol{x}_i - \boldsymbol{m}_j|^2 - \min_{1 \leq j \leq g} y_{ji}^{r-1} |\boldsymbol{x}_i - \boldsymbol{m}_j|^2$$

since at a local minimum, $\alpha_i = 0$, $i = 1, \ldots, n$.

11.5.2.4 Complete search

Complete search of the space of partitions of n objects into g groups is impractical for all but very small datasets. The branch and bound method (described in a feature subset selection context in Chapter 10) is one approach for finding the partition that results in the minimum value of the clustering criterion, without exhaustive enumeration. Nevertheless, it may still be impractical. Koontz *et al.* (1975) have developed an approach that extends the range of problems to which branch and bound can be applied. The criterion they seek to minimise is $\mathrm{Tr}\{\boldsymbol{S}_W\}$. Their approach is to divide the dataset into 2^m independent sets. The branch and bound method is applied to each set separately and then sets are combined in pairs (to give 2^{m-1} sets) and the branch and bound method applied to each of these combined sets, using the results obtained from the branch and bound application to the constituent parts. This is continued until the branch and bound procedure is applied to the entire set. This hierarchical approach results in a considerable saving in computer time.

Other approaches based on global optimisation algorithms such as simulated annealing have also been proposed. Simulated annealing is a stochastic relaxation technique in which a randomly selected perturbation to the current configuration is accepted or rejected probabilistically. Selim and Al-Sultan (1991) apply the method to the minimisation of $\mathrm{Tr}\{\boldsymbol{S}_W\}$. Generally, the method is slow, but it can lead to effective solutions.

11.5.3 Vector quantisation

Vector quantisation is not a *method* of producing clusters or partitions of a dataset but rather an *application* of many of the clustering algorithms already presented. Indeed, many clustering techniques have been rediscovered in the vector quantisation literature. On the other hand, there are some important algorithms in the vector quantisation literature that are not found in the standard texts on clustering. This section is included in the section on optimisation methods

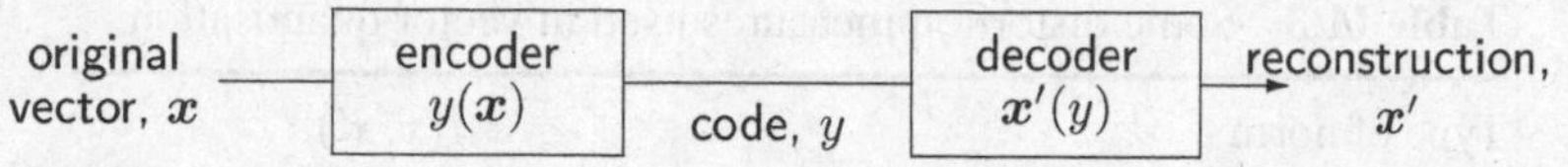

Figure 11.10 The encoding–decoding operation in vector quantisation.

since in vector quantisation a distortion measure (often, but by no means exclusively, based on the Euclidean distance) is optimised during training. A comprehensive and very readable account of the fundamentals of vector quantisation is given by Gersho and Gray (1992).

Vector quantisation is the encoding of a p-dimensional vector $\boldsymbol{x}$ as one from a *codebook* of g vectors, $z_1, \ldots, z_g$, termed the *code vectors* or the *codewords*. The purpose of vector quantisation is primarily to perform data compression. A vector quantiser consists of two components: an *encoder* and a *decoder* (Figure 11.10).

The encoder maps an input vector $\boldsymbol{x}$ to a scalar variable, y, taking discrete values $1, \ldots, g$. After transmission of the *index*, y, the inverse operation of reproducing an approximation $\boldsymbol{x}'$ to the original vector takes place. This is termed decoding and is a mapping from the index set $\mathcal{I} = \{1, \ldots, g\}$ to the codebook $\mathcal{C} = \{z_1, \ldots, z_g\}$. Codebook design is the problem of determining the codebook entries given a set of training samples. From a clustering point of view, we may regard the problem of codebook design as one of clustering the data and then choosing a representative vector for each cluster. These vectors could be cluster means for example and they form the entries in the codebook. They are indexed by integer values. Then the code vector for a given input vector $\boldsymbol{x}$ is the representative vector, say z of the cluster to which $\boldsymbol{x}$ belongs. Membership of a cluster may be determined on a nearest-to-cluster-mean basis. The distortion or error in approximation is then $d(\boldsymbol{x}, z)$, the distance between $\boldsymbol{x}$ and z (Figure 11.11).

The problem in vector quantisation is to find a set of codebook vectors that characterise a dataset. This is achieved by choosing the set of vectors for which a distortion measure between an input vector, $\boldsymbol{x}$, and its quantised vector, $\boldsymbol{x}'$, is minimised. Many distortion measures have been proposed, the most common being based on the square error measure giving average distortion, D_2,

$$\begin{aligned} D_2 &= \int p(\boldsymbol{x}) d(\boldsymbol{x}, \boldsymbol{x}') d\boldsymbol{x} \\ &= \int p(\boldsymbol{x}) \left\| \boldsymbol{x}'(y(\boldsymbol{x})) - \boldsymbol{x} \right\|^2 d\boldsymbol{x} \end{aligned} \tag{11.8}$$

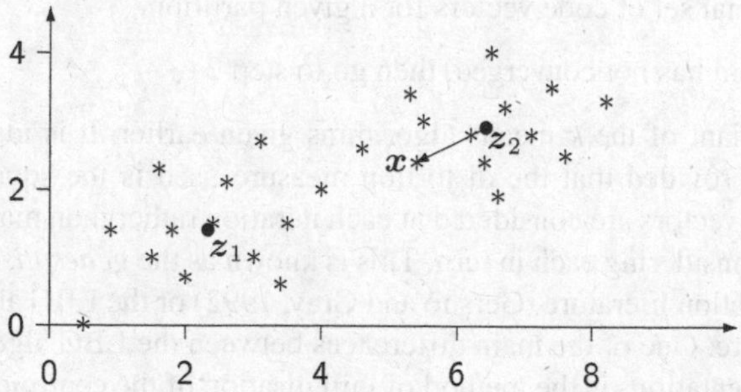

Figure 11.11 Vector quantisation distortion for two code vectors. The reconstruction of $\boldsymbol{x}$ after encoding and decoding is z_2, the nearest code vector. The distortion is $d(\boldsymbol{x}, z_2)$.

Table 11.3 Some distortion measures used in vector quantisation.

Type of norm	$d(\boldsymbol{x}, \boldsymbol{x}')$
L_2, Euclidean	$\lvert\boldsymbol{x}' - \boldsymbol{x}\rvert$
L_ν	$\{\sum_{i=1}^{p} \lvert\boldsymbol{x}' - \boldsymbol{x}\rvert^\nu\}^{\frac{1}{\nu}}$
Minkowski	$\max_{1\leq i\leq p} \lvert x_i' - x_i\rvert$
Quadratic (for positive definite symmetric $\boldsymbol{B}$)	$(\boldsymbol{x}' - \boldsymbol{x})^T \boldsymbol{B} (\boldsymbol{x}' - \boldsymbol{x})$

where $p(\boldsymbol{x})$ is the probability density function over samples $\boldsymbol{x}$ used to train the vector quantiser and $\|.\|$ denotes the norm of a vector. Other distortion measures are given in Table 11.3.

For a finite number of training samples, $\boldsymbol{x}_1, \ldots, \boldsymbol{x}_n$, we may write the distortion as

$$D = \sum_{j=1}^{g} \sum_{\boldsymbol{x}_i \in S_j} d(\boldsymbol{x}_i, \boldsymbol{z}_j)$$

where S_j is the set of training vectors for which $y(\boldsymbol{x}) = j$, i.e. those that map onto the jth code vector, $\boldsymbol{z}_j$. For a given set of code vectors $\boldsymbol{z}_j$, the partition that minimises the average distortion is constructed by mapping each $\boldsymbol{x}_i$ to the $\boldsymbol{z}_j$ for which $d(\boldsymbol{x}_i, \boldsymbol{z}_j)$ is a minimum over all $\boldsymbol{z}_j$ – i.e. choosing the minimum distortion or nearest-neighbour code vector. Alternatively, for a given partition the code vector of a set S_j, $\boldsymbol{z}_j$, is defined to be the vector for which

$$\sum_{\boldsymbol{x}_i \in S_j} d(\boldsymbol{u}, \boldsymbol{x}_i)$$

is a minimum with respect to $\boldsymbol{u}$. This vector is called the centroid (for the square error measure it is the mean of the vectors $\boldsymbol{x}_i$).

This suggests an iterative algorithm for a vector quantiser:

1. Initialise the code vectors.
2. Given a set of code vectors, determine the minimum distortion partition.
3. Find the optimal set of code vectors for a given partition.
4. If the algorithm has not converged, then go to step 2.

This is clearly a variant of the k-means algorithms given earlier. It is identical to the basic k-means algorithm provided that the distortion measure used is the square error distortion since all the training vectors are considered at each iteration rather than making an adjustment of code vectors by considering each in turn. This is known as the *generalised Lloyd algorithm* in the vector quantisation literature (Gersho and Gray, 1992) or the LBG algorithm in the data compression literature. One of the main differences between the LBG algorithm and some of the k-means implementations is the method of initialisation of the centroid vectors. The LBG algorithm (Linde *et al.*, 1980) given below starts with a one-level quantiser (a single cluster) and after obtaining a solution for the code vector $\boldsymbol{z}$, 'splits' the vector $\boldsymbol{z}$ into two close vectors

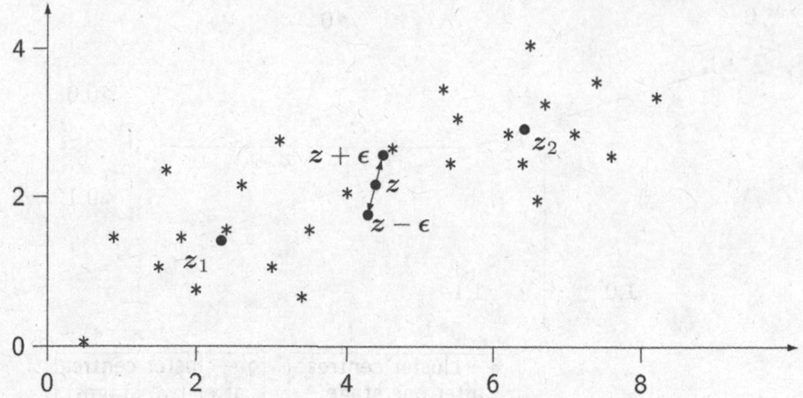

Figure 11.12 LBG algorithm illustration. z denotes the group centroid; z_1 and z_2 denote the code vectors for a two-level quantiser.

that are used as seed vectors for a two-level quantiser. This is run until convergence and a solution is obtained for the two-level quantiser. Then these two codewords are split to give four seed vectors for a four-level quantiser. The process is repeated so that finally quantisers of $1, 2, 4, \ldots, N$ levels are obtained (Figure 11.12). The steps in the algorithm are:

1. Initialise a code vector z_1 to be the group mean; initialise ϵ.
2. Given a set of m code vectors, 'split' each vector z_i to form $2m$ vectors, $z_i + \epsilon$ and $z_i - \epsilon$. Set $m = 2m$; relabel the code vectors as x'_i, $i = 1, \ldots, m$.
3. Given the set of code vectors, determine the minimum distortion partition.
4. Find the optimal set of code vectors for a given partition.
5. Repeat steps 3 and 4 until convergence.
6. If $m \neq N$, the desired number of levels, go to step 2.

Although it appears that all we have achieved with the introduction of vector quantisation in this chapter is yet another version of the k-means algorithm, the vector quantisation framework allows us to introduce two important concepts: that of tree-structured codebook search that reduces the search complexity in vector quantisation and that of topographic mappings in which a topology is imposed on the code vectors.

11.5.3.1 Tree-structured vector quantisation

Tree-structured vector quantisation is a way of structuring the codebook in order to reduce the amount of computation required in the encoding operation. It is a special case of the classification trees or decision trees discussed in a discrimination context in Chapter 7. Here we shall consider *fixed-rate* coding, in which there are the same number of bits used to represent each code vector. *Variable-rate coding*, which allows pruning of the tree, will not be addressed. Pruning methods for classification trees are described in Chapter 7, and in the vector quantisation context by Gersho and Gray (1992).

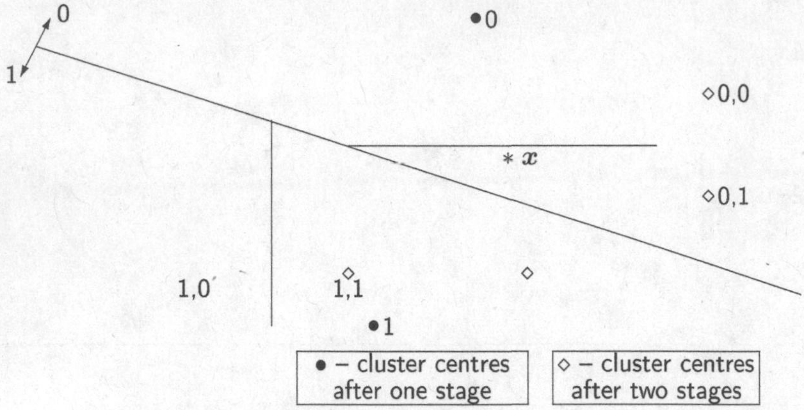

Figure 11.13 Tree-structured vector quantisation.

We shall begin our description of tree-structured vector quantisation with a simple *binary tree* example. The first stage in the design procedure is to run the k-means algorithm on the entire dataset to partition the set into two parts. This leads to two code vectors (the means of each cluster) (Figure 11.13).

Each group is considered in turn and the k-means algorithm applied to each group, partitioning each group into two parts again. This second stage then produces four code vectors and four associated clusters. The mth stage produces 2^m code vectors. The *total* number of code vectors produced in an m-stage design algorithm is $\sum_{i=1}^{m} 2^i = 2^{m+1} - 2$. This process produces a hierarchical clustering in which two clusters are disjoint or one wholly contains the other.

Encoding of a given vector, $\boldsymbol{x}$, proceeds by starting at the root of the tree (labelled A_0 in Figure 11.14) and comparing $\boldsymbol{x}$ with each of the two level 1 code vectors, identifying the nearest. We then proceed along the branch to A_1 and compare the vector $\boldsymbol{x}$ with the two code vectors at this level which were generated from members of the training set in this group. Thus there are m comparisons in an m-stage encoder. This compares with 2^m code vectors at the final level. Tree-structured vector quantisation may not be optimal in the sense that the

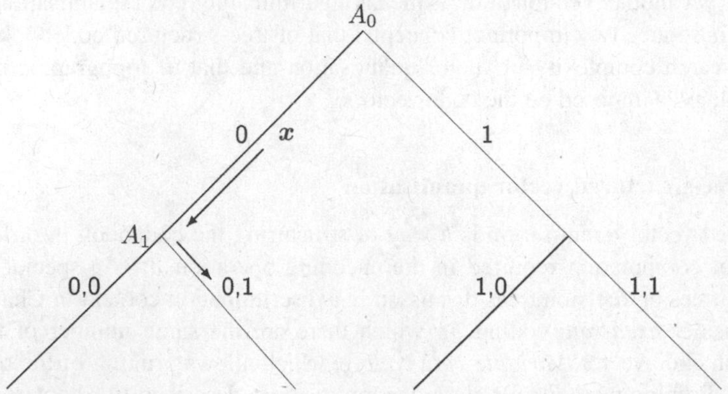

Figure 11.14 Tree-structured vector quantisation tree.

nearest neighbour of the final level code vectors is not necessarily found (the final partition in Figure 11.13 does not consist of nearest-neighbour regions). However, the code has the property that it is a progressively closer approximation as it is generated and the method can lead to a considerable saving in encoding time.

11.5.3.2 Self-organising feature maps

Self-organising feature maps are a special kind of vector quantisation in which there is an ordering or topology imposed on the code vectors. The aim of self-organisation is to represent high-dimensional data as a low-dimensional array of numbers (usually a one- or two-dimensional array) that captures the structure in the original data. Distinct clusters of data points in the data space will map to distinct clusters of code vectors in the array, although the converse is not necessarily true: separated clusters in the array do not necessarily imply separated clusters of data points in the original data. In some ways, self-organising feature maps may be regarded as a method of exploratory data analysis in keeping with those described in Chapter 10. The basic algorithm has the k-means algorithm as a special case.

Figures 11.15 and 11.16 illustrate the results of the self-organising feature map algorithm applied to data in two dimensions.

- In Figure 11.15, 50 data samples are distributed in three groups in two dimensions and we have used a self-organisation process to obtain a set of nine ordered cluster centres in one dimension. By a set of ordered cluster centres we mean that centre z_i is close in some sense to z_{i-1} and z_{i+1}. In the k-means algorithm, the order that the centres are stored in the computer is quite arbitrary and depends on the initialisation of the procedure.
- In Figure 11.16, the data (not shown) comprise 500 samples drawn from a uniform distribution over a square ($[-1 \leq x_1, x_2 \leq 1]$) and do not lie on (or close to) a one-dimensional manifold in the two-dimensional space. Again, we have imposed a one-dimensional topology on the cluster centres, which are joined by straight lines. In this case, we have obtained a space-filling curve.

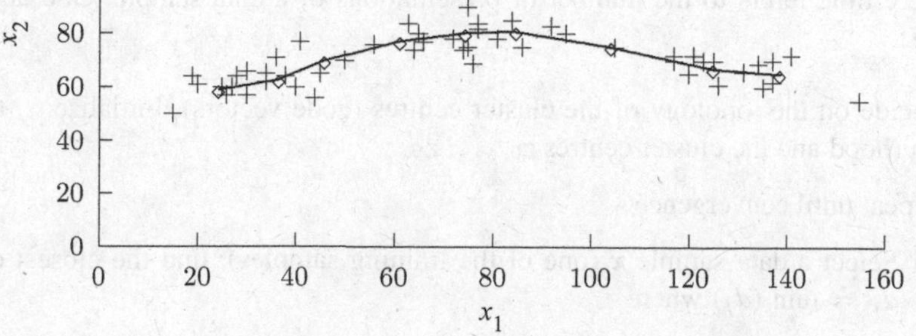

Figure 11.15 Topographic mapping. Adjacent cluster centres (◇) in the stored array of code vectors are joined.

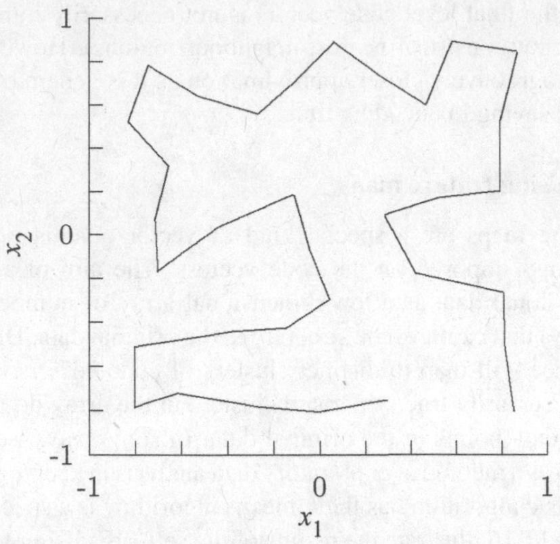

Figure 11.16 Topographic mapping for data uniformly distributed over a square. Thirty-three centres are determined and adjacent cluster centres in the stored array are joined.

In Figures 11.15 and 11.16, a transformation to a reduced dimension is achieved using a topographic mapping in which there is an ordering on the cluster centres. Each point in the data space is mapped to the ordered index of its nearest cluster centre. The mapping is nonlinear, and for purposes of illustration we considered mappings to one dimension only. If the data do lie on a reduced-dimension manifold within a high-dimensional space, then it is possible for topographic mappings to capture the structure in the data and present it in a form that may aid interpretation. In a supervised classification problem, it is possible to label each cluster centre with a class label according to the majority of the objects for which that cluster centre is the nearest. Of course we can do this even if there were no ordering on the cluster centres, but the ordering does allow the relationships between classes (according to decision boundaries) to be viewed easily.

The algorithm for determining the cluster centres may take many forms. The basic approach is to cycle through the dataset and adjust the cluster centres in the neighbourhood (suitably defined) of each data point. The algorithm is often presented as a function of time, where time refers to the number of presentations of a data sample. One algorithm is as follows:

1. Decide on the topology of the cluster centres (code vectors). Initialise α, the neighbourhood and the cluster centres $z_1, \ldots, z_N$.

2. Repeat until convergence:

   (a) Select a data sample $\boldsymbol{x}$ (one of the training samples), find the closest centre: let $d_{j^*} = \min_j (d_j)$ where

$$d_j = |\boldsymbol{x} - \boldsymbol{z}_j| \quad j = 1, \ldots, N$$

(b) Update the code vectors in the neighbourhood, $\mathcal{N}_{j^*}$ of code vector z_{j^*}

$$z(t+1) = z(t) + \alpha(t)(x(t) - z(t)) \quad \text{for all centres } z \in \mathcal{N}_{j^*}$$

where α is a learning rate that decreases with iteration number, t ($0 \leq \alpha \leq 1$).

(c) Decrease the neighbourhood and the learning parameter, α.

In order to apply the algorithm, an initial set of code vectors, the learning rate $\alpha(t)$ and the change with t of the neighbourhoods must be chosen.

Definition of topology

The choice of topology of the cluster centres requires some prior knowledge of the data structure. For example, if you suspect circular topology in your data, then the topology of your cluster centres should reflect this. Alternatively, if you wish to map your data onto a two-dimensional surface, then a regular lattice structure for the code vectors may be sufficient.

Learning rate

The learning rate, α, is a slowly decreasing function of t. It is suggested by Kohonen (1989) that it could be a linear function of t, stopping when α reaches 0, but there are no hard and fast rules for choosing $\alpha(t)$. It could be linear, inversely proportional to t or exponential. Haykin (1994) describes two phases: the *ordering* phase, of about 1000 iterations, when α starts close to unity and decreases, but remains above 0.1; and the *convergence* phase when α decreases further and is maintained at a small value – 0.01 or less – for typically thousands of iterations.

Initialisation of code vectors

Code vectors z_i are initialised to $\boldsymbol{m} + \boldsymbol{\epsilon}_i$ where $\boldsymbol{m}$ is the sample mean and $\boldsymbol{\epsilon}_i$ is a vector of small random values.

Decreasing the neighbourhood

The topological neighbourhood $\mathcal{N}_j$ of a code vector z_j is itself a function of t and decreases as the number of iterations proceeds. Initially, the neighbourhood may cover most of the code vectors ($z_{j-r}, \ldots, z_{j-1}, z_{j+1}, \ldots, z_{j+r}$ for large r), but towards the end of the iterations, it covers the nearest (topological) neighbours, z_{j-1} and z_{j+1} only. Finally it shrinks to zero. The problem is how to initialise the neighbourhood and how to reduce it as a function of t. During the *ordering* phase, the neighbourhood is decreased to cover only a few neighbours.

An alternative approach proposed by Luttrell (1989) is to fix the neighbourhood size and to start off with a few code vectors. The algorithm is run until convergence, and then the number of vectors is increased by adding vectors intermediate to those already calculated. The process is repeated and continued until a mapping of the desired size has been grown. Although the neighbourhood size is fixed, it starts off by covering a large area (since there are few centres) and the physical extent is reduced as the mapping grows. Specifically, given a data sample $\boldsymbol{x}$, if the nearest neighbour is z^*, then all code vectors z in the neighbourhood of z^* are updated according to

$$z \rightarrow z + \pi(z, z^*)(\boldsymbol{x} - z) \tag{11.9}$$

where π (> 0) is a function that depends on the position of z in the neighbourhood of z^*. For example, with a one-dimensional topology, we may take

$$\pi(z, z^*) = \begin{cases} 0.1 & \text{for } z = z^* \\ 0.01 & \text{for } z \text{ a topographic neighbour of } z^* \end{cases}$$

The Luttrell algorithm for a one-dimensional topographic mapping is as follows:

1. Initialise two code vectors, z_1 and z_2; set $m = 2$. Define the neighbourhood function π; set the number of updates per code vector, u.
2. Repeat until the distortion is small enough or the maximum number of code vectors is reached:
   (a) For $j = 1$ to $m \times u$ do
      - Sample from the dataset $\boldsymbol{x}_1, \ldots, \boldsymbol{x}_n$, say $\boldsymbol{x}$.
      - Determine the nearest-neighbour code vector, say z^*.
      - Update the code vectors according to: $z \rightarrow z + \pi(z, z^*)(\boldsymbol{x} - z)$.

   (b) Define $2m - 2$ new code vectors (z_1 remains unaltered): for $j = m - 1$ down to 1 do
      - $z_{2j+1} = z_{j+1}$
      - $z_{2j} = \dfrac{z_j + z_{j+1}}{2}$

   (c) Set $m = 2m - 1$.

Topographic mappings have received widespread use as a means of exploratory data analysis (Kraaijveld *et al.*, 1992; Roberts and Tarassenko, 1992). They have also been misused and applied when the ordering of the resulting cluster centres is irrelevant in any subsequent data analysis and a simple k-means approach could have been adopted. An assessment of the method and its relationship to other methods of multivariate analysis is provided by Murtagh and Hernández-Pajares (1995). Luttrell (1989) has derived the basic learning algorithm from a vector quantisation approach assuming a minimum distortion (Euclidean) and a robustness to noise on the codes. This puts the approach on a firmer mathematical footing. Also, the requirement for *ordered* cluster centres is demonstrated for a hierarchical vector quantiser.

11.5.3.3 Learning vector quantisation

Vector quantisation or clustering (in the sense of partitioning a dataset, not seeking meaningful groupings of objects) is often performed as a preprocessor to supervised classification. There are several ways in which vector quantisers or self-organising maps have been used with labelled training data. In the radar target classification example of Luttrell (1995), each class is modelled separately using a self-organising map. Test data are classified by comparing each pattern with the prototype patterns in each of the self-organising maps (codebook entries) and classifying on a nearest-neighbour rule basis.

An alternative approach that uses vector quantisers in a supervised way is to model the whole of the training data with a single vector quantiser (rather than each class separately).

Each training pattern is assigned to the nearest code vector, which is then labelled with the class of the majority of the patterns assigned to it. A test pattern is then classified using a nearest-neighbour rule using the labelled codebook entries.

Learning vector quantisation is a supervised generalisation of vector quantisation that takes account of class labels *in the training process*. The basic algorithm is as follows:

1. Initialise cluster centres (or code vectors), $z_1, \ldots, z_N$, and labels of cluster centres, $\omega_1, \ldots, \omega_N$.
2. Select a sample $\boldsymbol{x}$ from the training dataset with associated class $\omega_{\boldsymbol{x}}$ and find the closest centre: let $d_{j^*} = \min_j (d_j)$ where

$$d_j = |\boldsymbol{x} - \boldsymbol{z}_j| \quad j = 1, \ldots, N$$

with corresponding centre z_{j^*} and class ω_{j^*}.
3. If $\omega_{\boldsymbol{x}} = \omega_{j^*}$ then update the nearest vector, z_{j^*}, according to

$$z_{j^*}(t+1) = z_{j^*}(t) + \alpha(t)(\boldsymbol{x}(t) - z_{j^*}(t))$$

where $0 < \alpha_t < 1$ and decreases with t, starting at ≈ 0.1.
4. If $\omega_{\boldsymbol{x}} \neq \omega_{j^*}$ then update the nearest vector, z_{j^*}, according to

$$z_{j^*}(t+1) = z_{j^*}(t) - \alpha(t)(\boldsymbol{x}(t) - z_{j^*}(t))$$

5. Go to step 1 and repeat until several passes have been made through the dataset.

Correct classification of a pattern in the dataset leads to a refinement of the codeword in the direction of the pattern. Incorrect classification leads to a movement of the codeword away from the training pattern.

11.5.3.4 Stochastic vector quantisation

Stochastic vector quantisation (Luttrell, 1999) is a generalisation of the standard approach in which an input vector, $\boldsymbol{x}$, is encoded as a *vector* of code indices, $\boldsymbol{y}$, (rather than as a *single* code index) that are stochastically sampled from a probability distribution $p(\boldsymbol{y}|\boldsymbol{x})$ that depends on the input vector, $\boldsymbol{x}$. The decoding operation that produces a reconstruction, $\boldsymbol{x}'$, is also probabilistic, with $\boldsymbol{x}'$ being a sample drawn from $p(\boldsymbol{x}|\boldsymbol{y})$ given by

$$p(\boldsymbol{x}|\boldsymbol{y}) = \frac{p(\boldsymbol{y}|\boldsymbol{x})p(\boldsymbol{x})}{\int p(\boldsymbol{y}|\boldsymbol{z})p(\boldsymbol{z})d\boldsymbol{z}} \tag{11.10}$$

One of the key factors motivating the development of the stochastic vector quantisation approach is that of scalability to high dimensions. A problem with standard vector quantisation is that the codebook grows exponentially in size as the dimensionality of the input vector is increased, assuming that the contribution to the reconstruction error from each dimension is held constant. This means that such vector quantisers are not appropriate for encoding extremely high-dimensional input vectors, such as images. An advantage of using the stochastic approach is that it automates the process of splitting high-dimensional input vectors into low-dimensional blocks before encoding them, because minimising the mean Euclidean reconstruction error can encourage different stochastically sampled code indices to become associated with different input subspaces.

11.5.4 Example application study

The problem

This application involves a study into the detection of breast lesions in magnetic resonance imaging (MRI) imagery and the discrimination of malignant breast lesions from benign lesions (Lee *et al.*, 2007).

Summary

A k-means algorithm (Section 11.5.2) was applied to measured data obtained from a set of MRI images. The approach appeared to be useful, in detecting heterogeneous patterns within a malignant tumour, but failed when malignant lesions presented patterns similar to benign tumours. It was concluded that morphological analysis is required.

The data

The data comprised a set of five MRI breast images from each of 13 patients. For each patient, signal measurements were made on a voxel-by-voxel basis from the malignant tumour area. The total number of patterns was 1735; each pattern was five-dimensional.

The model

A k-means clustering approach was adopted.

Training procedure

Classification was achieved by labelling cluster centroids and a pattern from a random set of images assigned to the nearest reference centroid.

Results

The results showed that it is not possible to discriminate between benign and malignant lesions using the presented approach. There are two possible reasons:

- Further features, for example generated through a morphological analysis, are required.
- A supervised approach is likely to offer improved performance rather than using an unsupervised approach with a postprocessing stage that labels the centroid vectors with class labels.

11.5.5 Further developments

Procedures for reducing the computational load of the k-means algorithm are discussed by Venkateswarlu and Raju (1992). Further developments of k-means procedures to other metric spaces (with l_1 and l_∞ norms) are described by Bobrowski and Bezdek (1991). Juan and Vidal (1994) propose a fast k-means algorithm (based on the approximating and eliminating search algorithm, AESA) for the case when data arise in the form of a dissimilarity. That is, the data cannot be represented in a suitable vector space (without performing a multidimensional scaling procedure), though the dissimilarity between points is available. Termed the *k-centroids* procedure, it determines the 'most centred sample' as a centroid of a cluster. Popescu-Borodin (2008) develops a fast k-means algorithm for the quantisation of still images (see also Chen *et al.*, 2008).

There have been many developments of the basic fuzzy clustering approach and many algorithms have been proposed. Sequential approaches are described by de Màntaras and Aguilar-Martín (1985). In 'semi fuzzy' or 'soft' clustering (Selim and Ismail, 1984b; Ismail, 1988) patterns are considered to belong to some, though not necessarily all, clusters. In thresholded fuzzy clustering (Kamel and Selim, 1991) membership values below a threshold are set to zero, with the remaining being normalised and an approach that performs a fuzzy classification with prior assumptions on the number of clusters is reported by Gath and Geva (1989). Developments of the fuzzy clustering approach to data arising in the form of dissimilarity matrices are described by Hathaway and Bezdek (1994).

11.5.6 Summary

The techniques described in this section minimise a square error objective function. The k-means procedure is a special case of all of the techniques. It is widely used in pattern recognition, forming the basis for many supervised classification techniques. It produces a 'crisp' coding of the data in that a pattern belongs to one cluster only.

Fuzzy k-means is a development that allows a pattern to belong to more than one cluster. This is controlled by a membership function. The clusters resulting from a fuzzy k-means approach are softly overlapping, in general, with the degree of overlap controlled by a user-specified parameter. The learning procedure determines the partition.

The vector quantisation approaches are application driven. The aim is to produce a crisp coding and algorithms such as tree-structured vector quantisation are motivated by the need for fast coding. Self-organising feature maps produce an ordered coding.

11.6 Spectral clustering

In Section 11.2 we examined approaches to cluster analysis that exploited the structure of a (dis-)similarity matrix to produce a clustering. In Section 11.5, the k-means algorithm and variants for clustering patterns in a high-dimensional space was presented. *Spectral clustering* refers to a class of techniques that brings together these two concepts by exploiting the eigenstructure of a similarity matrix, or other matrices derived from it, to partition a set of patterns into disjoint clusters such that patterns in the same cluster have high similarity and patterns in different clusters have low similarity. It is a heuristic that often works well in practice.

11.6.1 Elementary graph theory

Many complex datasets arise as a result of interactions between entities of various kinds. For example, social networks of interactions between people; networks of computers; online purchasing between people and vendors. The use of pattern recognition techniques to analyse such transactional data is considered further in Chapter 12. Such complex data can be

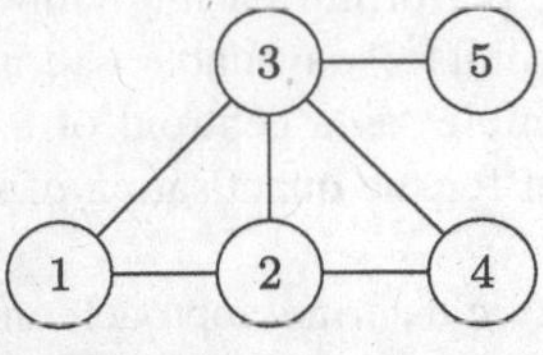

	1	2	3	4	5
1	0	1	1	0	0
2	1	0	1	1	0
3	1	1	0	1	1
4	0	1	1	0	0
5	0	0	1	0	0

Figure 11.17 Undirected graph illustration with its adjacency matrix.

represented using the mathematical concept of a graph with nodes representing entities of interest (people, computers, shops) and the edges a transaction (email message, purchase, telephone call) between those nodes. Spectral clustering also exploits some results from graph theory and therefore here we introduce some basic graph theory notation.

A graph, G, comprises a set of vertices (or nodes) and edges, $G = \{V, E\}$. The structure of the graph is described by the *adjacency matrix*, A, where (also, see Section 12.2)

$$A_{ij} = 1 \quad \text{if node } i \text{ and node } j \text{ are connected}$$

$$A_{ij} = 0 \quad \text{if node } i \text{ and node } j \text{ are not connected}$$

Figure 11.17 gives an illustration of a graph and its adjacency matrix. The graph may be directed (the edges are directed from one node to an other) or undirected (there is no direction associated with the edge between two nodes). For undirected graphs, the adjacency matrix is symmetrical. In addition, there may be weights associated with the edges in a graph. These weights may represent the degree of interaction between nodes (for example, amount of email traffic between two computers, or level of purchasing from an online shop) and may change with time. In this case, the binary adjacency matrix is replaced by a (time-varying) weighted adjacency matrix.

The graph Laplacian is defined in terms of the adjacency matrix as

$$L = D - A$$

where D is a diagonal matrix, $D = \text{Diag}(d_1, \ldots, d_n)$, where

$$d_i = \sum_{j=1}^{n} A_{ij}$$

which, for a binary adjacency matrix, is the degree of vertex i. The (unnormalised graph) Laplacian has the following properties:

1. L is symmetric and positive semi-definite.
2. All eigenvalues are real and non-negative. The smallest eigenvalue is zero. Thus the eigenvalues can be ordered:

$$\lambda_n \geq \lambda_{n-1} \geq \cdots \geq \lambda_1 = 0$$

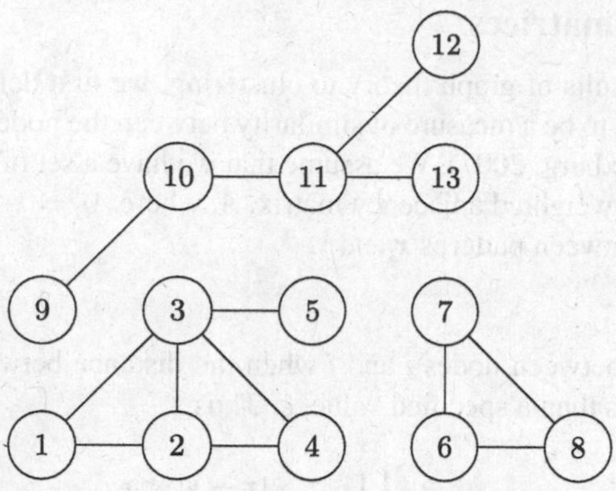

Figure 11.18 Graph with disjointed components.

3. The multiplicity of the eigenvalue with value 0 is equal to the number of connected components of the graph. The corresponding eigenvectors provide indicator vectors to show which component a given node belongs to. For example, for the graph shown in Figure 11.18, the eigenvalue $\lambda = 0$ of the Laplacian has multiplicity 3, indicating three separate components. The corresponding eigenvectors are shown in Table 11.4. The table shows that nodes 1, 2, 3, 4 and 5 form one component; nodes 6, 7 and 8 form a second; and nodes 9, 10, 11, 12 and 13 form a third. Thus, the rows of the eigenvector matrix provide an indication of which group a node belongs to.

This definition extends to a weighted adjacency matrix, for which a normalised graph Laplacian is defined (see Section 11.6.5). The rows of the eigenvectors corresponding to zero eigenvalues can be used to identify disjoint components.

Table 11.4 Eigenvectors v_1, v_2, v_3 corresponding to the eigenvalue $\lambda = 0$ for the graph of Figure 11.18

	v_1	v_2	v_3
1	0	0	1
2	0	0	1
3	0	0	1
4	0	0	1
5	0	0	1
6	0	1	0
7	0	1	0
8	0	1	0
9	1	0	0
10	1	0	0
11	1	0	0
12	1	0	0
13	1	0	0

11.6.2 Similarity matrices

In order to apply the results of graph theory to clustering, we first define the weights on the edge between two nodes to be a measure of similarity between the nodes. This may be defined in several ways (von Luxburg, 2007). We assume that we have a set of patterns, $\{\boldsymbol{x}_1, \ldots, \boldsymbol{x}_n\}$, and we wish to define a weighted adjacency matrix, A, where $A_{ij} = s(\boldsymbol{x}_i, \boldsymbol{x}_j)$, and $s(\boldsymbol{x}, \boldsymbol{y})$ is a measure of similarity between patterns $\boldsymbol{x}$ and $\boldsymbol{y}$.

ϵ-neighbourhood
Here, edges are drawn between nodes i and j when the distance between the corresponding patterns, $\boldsymbol{x}_i$ and $\boldsymbol{x}_j$ is less than a specified value, ϵ. Thus

$$s(\boldsymbol{x}, \boldsymbol{y}) = \begin{cases} 1 & |\boldsymbol{x} - \boldsymbol{y}| < \epsilon \\ 0 & |\boldsymbol{x} - \boldsymbol{y}| \geq \epsilon \end{cases}$$

Fully connected graph
Each node is connected to every other node with an edge weight that is the similarity of the patterns associated with each of the nodes. There are different definitions of a similarity function, $s(\boldsymbol{x}, \boldsymbol{y})$. For example, the Gaussian similarity function for modelling local neighbourhood relationships is given by

$$s(\boldsymbol{x}, \boldsymbol{y}) = \exp\left(-\frac{|\boldsymbol{x} - \boldsymbol{y}|^2}{2\sigma^2}\right) \tag{11.11}$$

for parameter σ controlling the width of the neighbourhood. The weight on an edge connecting node i to node j is $A_{ij} = s(\boldsymbol{x}_i, \boldsymbol{x}_j)$.

***k*-nearest neighbour**
There are two basic models. The first is the *k-nearest-neighbour graph*:

$$A_{ij} = \begin{cases} s(\boldsymbol{x}_i, \boldsymbol{x}_j) & \boldsymbol{x}_i \text{ is one of the } k \text{ nearest neighbours of } \boldsymbol{x}_j \\ & \text{OR } \boldsymbol{x}_j \text{ is one of the } k \text{ nearest neighbours of } \boldsymbol{x}_I \\ 0 & \text{otherwise} \end{cases}$$

In this model, node i is connected to node j if node j is one of the k-nearest neighbours of node i or node i is one of the k-nearest neighbours of node j. The weight on the edge is the similarity between the patterns associated with the nodes.

The second model, the *mutual k-nearest-neighbour graph* is described by the weighted adjacency matrix:

$$A_{ij} = \begin{cases} s(\boldsymbol{x}_i, \boldsymbol{x}_j) & \boldsymbol{x}_i \text{ is one of the } k \text{ nearest neighbours of } \boldsymbol{x}_j \\ & \text{AND } \boldsymbol{x}_j \text{ is one of the } k \text{ nearest neighbours of } \boldsymbol{x}_I \\ 0 & \text{otherwise} \end{cases}$$

11.6.3 Application to clustering

We have seen that the graph Laplacian can be used to group nodes in a graph through the analysis of the eigenvectors corresponding to the $\lambda = 0$ eigenvalues: nodes in the same graph component are grouped together; nodes in disjoint components are not. This is also the case for a Laplacian constructed from a weighted adjacency matrix.

This suggests an approach to clustering patterns:

1. Construct a matrix of similarities between patterns;
2. Construct the graph Laplacian based on such a matrix of similarities;
3. Analyse the eigenvectors corresponding to the smallest eigenvalues. In general, there will not be completely disjoint components, but components that are 'almost disjoint'. In this case, the smallest eigenvalues will not be zero.

11.6.4 Spectral clustering algorithm

The basic spectral clustering algorithm is given in Algorithm 11.1. Step one constructs a similarity matrix. There are several ways of constructing this given patterns. It may be a simple transformation of a distance between patterns, or one of the measures given in Section 11.6.2. Given this similarity matrix, the graph Laplacian is calculated. Step 3 calculates the eigenvectors and eigenvalues of the Laplacian. Only the eigenvectors corresponding to the k lowest eigenvalues are required. These are then used to define an $n \times k$ data matrix, V. The final step is to use a k-means algorithm (Section 11.5) to apply to the n k-dimensional patterns defined by the rows.

Figure 11.19 shows the results of a spectral clustering applied to some points in two dimensions. A Gaussian similarity function was used [Equation (11.11)]. Points in different clusters are labelled with different symbols.

11.6.5 Forms of graph Laplacian

For a weighted adjacency matrix, a normalised graph Laplacian is used (see Section 12.2). A symmetric form (Ng *et al.*, 2002) is given by

$$L_{sym} = D^{-\frac{1}{2}} L D^{-\frac{1}{2}} = I - D^{-\frac{1}{2}} A D^{-\frac{1}{2}}$$

Algorithm 11.1 Spectral clustering.

---

1. Form a matrix, S, of similarities, using a definition of similarity between patterns, such as those given in Section 11.6.2.
2. Compute the graph Laplacian from the similarity matrix.
3. Perform an eigendecomposition of the Laplacian, selecting the $n \times k$ matrix of eigenvectors, $V = [v_1, \ldots, v_k]$, corresponding to the k smallest eigenvalues.
4. Perform the k-means algorithm on the n k-dimensional patterns corresponding to the rows of the eigenvector matrix to cluster the patterns into k clusters.

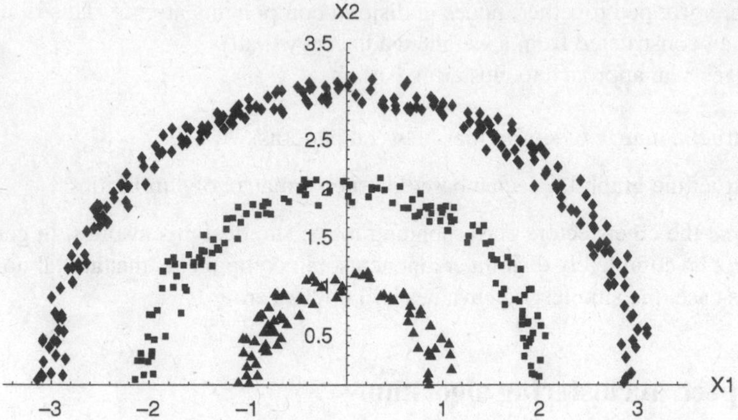

Figure 11.19 Clustering of points in two dimensions: a value of $k = 3$ was taken. Points in different clusters are highlighted by different symbols.

A nonsymmetric form (proposed by Shi and Malik, 2000) is

$$L_{nonsym} = D^{-1}L = I - D^{-1}A$$

These normalised forms have the following properties:

1. The normalised Laplacians are positive semidefinite.
2. All eigenvalues are real and non-negative. The smallest eigenvalue is zero. Thus the eigenvalues can be ordered:

$$\lambda_n \geq \lambda_{n-1} \geq \cdots \geq \lambda_1 = 0$$

3. The multiplicity of the eigenvalue with value 0 is equal to the number of connected components of the graph. For L_{nonsym}, the eigenspace is spanned by eigenvectors of the form $\mathbf{1}_i$, where the vector has a '1' in the jth position if node j is in component i and zero otherwise. The eigenvectors of L_{sym} are $D^{-1}\mathbf{1}_i$.
4. The eigenvectors, v, and eigenvalues, λ, of L_{nonsym} satisfy the generalised symmetric eigenvector equation $Lv = \lambda Dv$.

A clustering algorithm for a normalised graph Laplacian, L_{nonsym}, is given in Algorithm 11.2.

11.6.6 Example application study

The problem

This application involves recovering signals from a linear mixture using partial knowledge of the mixing process and the signals (blind signal separation) using a spectral clustering approach (Bach and Jordan, 2006). This is applied to the problem of one-microphone blind source separation of speech.

Algorithm 11.2 Spectral clustering for normalised Laplacian, L_{nonsym}.

1. Form a matrix, S, of similarities, using a definition of similarity between patterns, such as those given in Section 11.6.2.
2. Compute the graph Laplacian from the similarity matrix.
3. Solve the generalised symmetric eigenvector equation

$$Lv = \lambda Dv$$

   and select the $n \times k$ matrix of eigenvectors, $V = [v_1, \ldots, v_k]$, corresponding to the k smallest eigenvalues.
4. Perform the k-means algorithm on the n k-dimensional patterns corresponding to the rows of the eigenvector matrix to cluster the patterns into k clusters.

Summary

Blind signal separation is a problem that occurs in many fields, such as radar signal processing, speech processing and imaging. An extreme case is when there is only one sensor and two or more signals to estimate. The aim of this study was to separate speech mixtures from a single microphone without specific speaker models. A cost function is derived with the feature that minimisation with respect to the partition of the data leads to a spectral clustering algorithm. Alternatively, for labelled training data (i.e. the partition is known), minimising with respect to the (parameters of a) similarity matrix produces an algorithm for learning the similarity matrix.

The data

The training data comprise speech from four male and female speakers of duration 3 s. For testing, mixes of speakers different from those in the training data were used.

The model

The speech separation problem is formulated as a segmentation problem in the time-frequency plane (speech is represented as a spectrogram). There are many methods for segmenting image-like data, but speech segments have their own characteristics and therefore speech-specific features are derived. These features are used to define parametrised similarity matrices.

Training procedure

There are two stages to deriving a partition of a spectrogram. The first is to use labelled training data from datasets of known partitions to learn the parameters of the similarity matrix. The next stage uses these parameters in an algorithm to cluster previously unseen datasets.

Results

The study showed that a mixture of two English speakers could be successfully demixed, with the segmentation algorithm producing audible signals of reasonable quality.

11.6.7 Further developments

Developments include approaches to address the application of spectral clustering for large datasets. Spectral clustering requires the computation of the eigenvectors of an $n \times n$ matrix, which is an operation of the order of n^3. Therefore, it is infeasible for problems where n is very large. A fast approximate spectral clustering approach is described by Yan *et al.* (2009). An approach for online applications, in which the spectral clustering is updated for small changes in the dataset, is presented by Ning *et al.* (2007). Zare *et al.* (2010) describe a data reduction method, whereby the original data are sampled to produce a smaller set, which is then clustered.

11.6.8 Summary

Spectral clustering uses the eigenvectors of a graph Laplacian for embedding the data into a space that captures the underlying structure. It requires the calculation of the smallest k eigenvectors of an $n \times n$ matrix. The clusters that the approach produces are not always convex sets (as in k-means), but there are several parameters to be set. These include the parameters of the similarity function (ϵ for ϵ-neighbourhood, σ for fully connected graph, k for k nearest neighbour) as well as the form of the graph Laplacian. The normalised form, L_{nonsym} is recommended by von Luxburg (2007).

11.7 Cluster validity

11.7.1 Introduction

Cluster validity is the procedure by which the results of a clustering are evaluated. A clustering algorithm will partition a dataset of objects even if there are no natural clusters within the data. Different clustering methods (or the same method with different parameter settings) may produce very different classifications. How do we know whether the structure is a property of the dataset and not imposed by the particular method that we have chosen? This is the problem of *cluster validity*, which is one that is full of difficulties and rarely straightforward. In some applications of clustering techniques we may not be concerned with groupings in the dataset. For example, in vector quantisation we may be concerned with the average distortion in reconstructing the original data or in the performance of any subsequent analysis technique. This may be measured by the error rate in a discrimination problem or diagnostic performance in image reconstruction (see the examples in the following section). In these situations, clustering is simply a means of obtaining a partition, not to discover structure in the data.

Yet, if we are concerned with discovering groupings within a dataset, how do we validate the clustering? For two-dimensional data, the user can easily verify the result visually. For higher-dimensional data, a simple approach is to view the clustering in a low-dimensional representation of the data. Linear and nonlinear data projection methods have been discussed in Chapter 10. Alternatively, we may perform several analyses using different clustering methods and compare the resulting classifications to see whether the derived structure is an artifact of a particular method. More formal procedures may also be applied and we discuss some approaches in this section.

There are three approaches to cluster validity based on three types of clustering criteria:

Internal criteria
The clustering results are evaluated in terms of quantities that involve the dataset; for example, the dissimilarity matrix.

External criteria
The clustering is evaluated using information that was not used in performing the clustering; for example, a user-specified structure.

Relative criteria
These are criteria to evaluate a clustering by comparing it with that produced by different algorithms (or the same algorithm with different parameters).

The first two cluster validity approaches are based on statistical tests using models for different types of *a priori* structure. The approach based on relative criteria aims to find the best clustering scheme according to a defined metric.

11.7.2 Statistical tests

11.7.2.1 Hypothesis testing

Internal and external criteria approaches are based on statistical hypothesis testing in which the null hypothesis, H_0, concerns the randomness of the data and the test statistic, q, is calculated using the given dataset. In many cases, the probability density function of the test statistic under the given hypothesis is unknown. A sample-based approach is adopted whereby the probability density function is calculated from samples obtained via a Monte Carlo approach.

11.7.2.2 Monte Carlo

For a one-sided test (in which a high value of the test statistic means evidence against the null hypothesis – a right-sided test), we seek the probability that extreme values of the statistic occur by chance under the null hypothesis,

$$Pr(q \geq Q|H_0)$$

where Q is the observed value of the test statistic on the dataset.

In a Monte Carlo approach, sample datasets, X_i, $i = 1, \ldots, r$, are generated and the value of q calculated for each dataset, to produce a set of sample values, $\{q_i, i = 1, \ldots, r\}$. If the significance level is ρ, then the decision is

Reject the null hypothesis if q is greater than $(1 - \rho)r$ of the values, q_i.

Similarly, for a left-sided test. For a two-sided test, for which the significance level is

$$Pr(|q| \geq Q)$$

then the decision is

Reject the null hypothesis if q is greater than $(1 - \rho/2)r$ of the values, q_i or if q is less than $(\rho/2)r$ of the values, q_i.

11.7.2.3 Null models

Gordon (1999) identifies five types of null model associated with pattern or dissimilarity matrices for the complete absence of structure in a dataset (see also Gordon 1994b, 1996a).

- *Poisson model* (Bock, 1985) or *random position hypothesis* (Jain and Dubes, 1988). Objects are represented as points uniformly distributed in some region A of the p-dimensional data space. The main problem with this model is the specification of A. Standard definitions include the unit hypercube and the hypersphere.
- *Unimodal model* (Bock, 1985). The variables have a unimodal distribution. The difficulty here is the specification of this density function. Standard definitions include a spherical multivariate normal.
- *Random permutation model*. In this model, the entries in each column of an $n \times p$ data matrix, X, are independently permuted giving $(n!)^{p-1}$ essentially different matrices, each regarded as equally likely.
- *Random dissimilarity matrix* (Ling, 1973) or *random graph hypothesis* (Jain and Dubes, 1988). This is based on data arising in the form of dissimilarities. The elements of the (lower triangle of the) dissimilarity matrix are ranked in random order, all orderings being regarded as equally likely. Viewed as a graph, with the nodes representing patterns and the values on the edges representing dissimilarities between patterns, the edges below a specified threshold are inserted into the graph in random order. One of the problems with this model is that if objects i and j are close (d_{ij} is small), you would expect d_{ik} and d_{jk} to have similar ranks for each object k, but such correlation is not accounted for.
- *Random labels model* (Jain and Dubes, 1988). This assumes that all possible permutations of the labels associated with each pattern (obtained as a result of a clustering) are equally likely. The observed value of the test statistic is compared with the distribution obtained by randomly permuting the labels.

11.7.3 Absence of class structure

Tests for the absence of class structure address the question:

Are the data $\boldsymbol{x}_1, \ldots, \boldsymbol{x}_n$ sampled from a homogeneous population?

Thus, the null hypothesis is a statement about the randomness of the data. There have been many test statistics proposed (for a review see Gordon, 1998).

For the Poisson model, example test statistics are:

- the number of interpoint distances less than a specified threshold;
- the largest nearest-neighbour distance within a set of objects.

For the random dissimilarity matrix model, example test statistics include:

- the minimum number of edges required to connect a random graph (the number of edges before the graph comprises a single component);
- the number of components in the graph when it contains a specified number of edges.

11.7.4 Validity of individual clusters

Determining whether individual clusters are valid requires the specification of properties of a cluster that an ideal cluster is expected to possess. Such properties of a cluster will be dataset dependent. An alternative approach is to define a statistic and examine its distribution under some null model that the data does not have class structure. This is the approach we present here. The question we address is:

Is a cluster $\mathcal{C}$ (of size c) defined by

$$\mathcal{C} = \{i : d_{ij} < d_{ik} \text{ for all } j \in \mathcal{C}, k \notin \mathcal{C}\}$$

a valid cluster?

Gordon (1999) describes a Monte Carlo approach to individual cluster validation based on a U statistic:

$$U_{ijkl} = \begin{cases} 0 & \text{if } d_{ij} < d_{kl} \\ \frac{1}{2} & \text{if } d_{ij} = d_{kl} \\ 1 & \text{if } d_{ij} > d_{kl} \end{cases}$$

and

$$U = \sum_{(i,j)\in W} \sum_{(k,l)\in B} U_{ijkl}$$

for subsets W and B of ordered pairs $(i, j) \in W$ and $(k, l) \in B$. W is taken to be those pairs where objects i and j both belong to the cluster $\mathcal{C}$ and B comprises pairs where one element belongs to $\mathcal{C}$ and the other does not.

The basic algorithm is defined as follows:

1. Evaluate U for the cluster $\mathcal{C}$; denote it by U^*.
2. Generate a random $n \times p$ pattern matrix and cluster it using the same algorithm used to produce $\mathcal{C}$.
3. Calculate $U(k)$ for each cluster of size k ($k = 2, \ldots, n-1$) (arising through the partitioning of a dendrogram, for example). If there is more than one of a given size, select one of that size randomly.
4. Repeat steps 2 and 3 until there are $(m-1)$ values of $U(k)$ for each value of k.
5. If U^* is less than the jth smallest value of $U(c)$, the null hypothesis of randomness is rejected at the $100(j/m)\%$ level of significance.

Note that the values of $U(k)$ are independent of the dataset.

Gordon takes $m = 100$ and evaluates the above approach under both the Poisson and unimodal (spherical multivariate normal) models. The clusterings using Ward's method are assessed on four datasets. Results for the approach are encouraging. Further refinements could include the use of other test statistics and developments of the null models.

11.7.5 Hierarchical clustering

Here we address the question of the internal validation of a hierarchical clustering:

> Is the hierarchical clustering an accurate summary of the data?

The *cophenetic correlation* coefficient is a measure of the agreement between a set of dissimilarities and a hierarchical classification derived from the data. It is the correlation between the dissimilarities, d_{ij}, and the ultrametric distances, $\hat{d}_{ij}$

$$\frac{\sum_{i<j}\left(d_{ij}-\bar{d}\right)\left(\hat{d}_{ij}-\bar{\hat{d}}\right)}{\left[\sum_{i<j}\left(d_{ij}-\bar{d}\right)^2\sum_{i<j}\left(\hat{d}_{ij}-\bar{\hat{d}}\right)^2\right]^{\frac{1}{2}}}$$

It is a measure of how well the hierarchical classification represented by the dendrogram preserves inter-point distances.

The distribution of this statistic is difficult to calculate and a Monte Carlo approach under a Poisson null model, for example, may be adopted.

11.7.6 Validation of individual clusterings

Here we address the validation of an individual clustering, not a hierarchical clustering, produced by a clustering algorithm. The question we ask is:

> Is the clustering, $\mathcal{C}$, of data into m clusters consistent with information in the dataset X?

Define an $n \times n$ matrix Y:

$$Y_{ij} = \begin{cases} 1 & \boldsymbol{x}_i \text{ and } \boldsymbol{x}_j \text{ belong to different clusters} \\ 0 & \text{otherwise} \end{cases}$$

Then, the Γ statistic,

$$\Gamma(D, Y) = \frac{1}{M}\sum_{i=1}^{N-1}\sum_{j=i+1}^{N} d_{ij}Y_{ij} \tag{11.12}$$

or the normalised Γ statistic, $\hat{\Gamma}$,

$$\hat{\Gamma}(D, Y) = \frac{1}{M}\frac{\sum_{i=1}^{N-1}\sum_{j=i+1}^{N}\left(d_{ij}-\mu_D\right)\left(Y_{ij}-\mu_Y\right)}{\sigma_D\sigma_Y} \tag{11.13}$$

is used to measure the agreement between the dissimilarity matrix, D, and the matrix Y, where μ_D and μ_Y are the means and σ_D and σ_Y are the standard deviations of the D and Y matrix elements and $M = N(N-1)/2$.

A Monte Carlo approach under the Poisson null model may be taken. For each dataset generated, the clustering algorithm that was used to produce the clustering $\mathcal{C}$ is applied to the dataset, X, and the statistic calculated. The decision to reject, or not, the null hypothesis at a given significant level, is based on the distribution of the statistics.

11.7.7 Partitions

The questions we seek to address here are those requiring external validation tests and are of the form:

> Does the clustering produced by an algorithm agree with an externally prescribed partition?
>
> Does the proximity matrix of X agree with a predetermined partition?

The approach adopted is the standard one: define statistics that measure the degree to which a partition agrees with either a clustering or a proximity matrix and use a Monte Carlo approach for the estimation of the probability density function of the statistic. Denoting the clustering by $\mathcal{C} = \{C_1, \ldots, C_m\}$ and the independent partition by $\mathcal{P} = \{P_1, \ldots, P_s\}$ (Theodoridis and Koutroumbas, 2009), and introducing the quantities a, b, c and d as

> a is the number of pairs of pattern vectors that belong to the same cluster in $\mathcal{C}$ and to the same group in $\mathcal{P}$
>
> b is the number of pairs of pattern vectors that belong to the same cluster in $\mathcal{C}$ and to different groups in $\mathcal{P}$
>
> c is the number of pairs of pattern vectors that belong to different clusters in $\mathcal{C}$ and to the same group in $\mathcal{P}$
>
> d is the number of pairs of pattern vectors that belong to different clusters in $\mathcal{C}$ and to different groups in $\mathcal{P}$

then the total number of distinct pairs is $M = a + b + c + d$ and we define the following statistics:

Rand

$$R = (a + d)/M$$

Jaccard

$$J = a/\,(a + b + c)$$

Both statistics have the property that they lie between 0 and 1 and the larger their value then the closer the agreement between $\mathcal{C}$ and $\mathcal{P}$.

The Γ and $\hat{\Gamma}$ statistics may also be used. Defining $X_{ij} = 1$ if the patterns $\boldsymbol{x}_i$ and $\boldsymbol{x}_j$ are in the same cluster in $\mathcal{C}$ and zero otherwise; and $Y_{ij} = 1$ if the patterns $\boldsymbol{x}_i$ and $\boldsymbol{x}_j$ are in the same group in $\mathcal{P}$ and zero otherwise, then $\Gamma(X, Y)$ and $\hat{\Gamma}(X, Y)$ have the property that the larger the magnitude the closer the agreement between $\mathcal{C}$ and $\mathcal{P}$.

11.7.8 Relative criteria

Relative criteria are used to compare the results of different clustering algorithms and seek to address the question:

> Which clustering, out of a set produced by applying a clustering algorithm with different parameter values to a dataset X, best fits the data, X?

The basis of the approach is the definition of a validity index.

11.7.8.1 Validity indices

Dunn and Dunn-like indices

The Dunn index is defined as

$$D = \min_{i=1,\ldots,g} \left\{ \min_{j=i+1,\ldots,g} \left(\frac{d\left(C_i, C_j\right)}{\max_{k=1,\ldots,g} \left(\text{diam}\left(C_k\right)\right)} \right) \right\}$$

where $d\left(C_i, C_j\right) = \min_{x \in C_i, y \in C_j} \{d(x, y)\}$ is the distance between two clusters defined as the minimum interpoint distance, with the members of the pair coming from different clusters; and $\text{diam}\left(C_i\right) = \max_{x,y \in C_i} \{d(x, y)\}$ is the cluster diameter. It is large for compact and well-separated clusters.

The Dunn index can be sensitive to noise (the cluster diameter can be large in a noisy environment). Variations have been proposed using different definitions of cluster distance and cluster diameter that are more robust to noise.

Davies–Bouldin index

The Davies–Bouldin index for a clustering of g clusters is defined as

$$DB = \frac{1}{g} \sum_{i=1}^{g} R_i$$

where

$$R_i = \max_{j=1,\ldots,g, j \neq i} \left(R_{ij}\right)$$

and R_{ij} is a measure of within-cluster spread to between-cluster distance given by

$$R_{ij} = \frac{s_i + s_j}{d_{ij}}$$

where d_{ij} is the distance between two clusters, defined as the distance between the cluster centroids, v_i, and s_i is a measure of the spread of cluster C_i

$$d_{ij} = d(v_i, v_j), \quad s_i = \frac{1}{\|C_i\|} \sum_{x \in C_i} d(x, v_i)$$

and $\|C_i\|$ is the number of patterns in cluster C_i. The Davies–Bouldin index measures average of the similarity between a cluster and its most similar one. Small values of the index are indicative of compact and separated clusters.

Γ statistic

The Γ statistic and its normalised form, $\hat{\Gamma}$, given by Equations (11.12) and (11.13), can be used to measure the relationship between the dissimilarity matrix, D, and the matrix Q, whose (i, j) element is the distance between the cluster centres to which patterns $\boldsymbol{x}_i$ and $\boldsymbol{x}_j$ belong.

$$Q_{ij} = d\left(v_{c_i}, v_{c_j}\right)$$

where c_i is the index of the cluster that contains the pattern $\boldsymbol{x}_i$. Large values are an indication of compact clusters.

Hierarchical clustering indices

In assessing particular hierarchical schemes we must consider how well the structure in the original data can be described by a dendrogram. However, since the structure in the data is not known (this is precisely what we are trying to determine) and since each clustering is simply a method of exploratory data analysis that imposes its own structure on the data, this is a difficult question to address. One approach is to examine various measures of distortion. Many measures have been proposed [see, for example, Cormack (1971) for a summary]. They are based on differences between the dissimilarity matrix d and d^*, the ultrametric dissimilarity coefficient, where d^*_{ij} is the distance between the groups containing i and j when the groups are amalgamated. Jardine and Sibson (1971) propose several goodness of fit criteria. One scale-free measure of classifiability is defined by

$$\Delta_1 = \frac{\sum_{i<j} |d_{ij} - d^*_{ij}|}{\sum_{i<j} d_{ij}}$$

Small values of Δ_1 are indicative that the data are amenable to the classification method that produced d^*.

There are many other measures of distortion, both for hierarchical and nonhierarchical schemes. Milligan (1981) performed an extensive Monte Carlo study of thirty internal criterion measures applied to the results of hierarchical clusterings, although the results may also apply to nonhierarchical methods.

11.7.9 Choosing the number of clusters

The problem of deciding how many clusters are present in the data is one common to all clustering methods. If we regard the number of clusters, g, as a model parameter then the cluster validity indices may be used to compare the value of the index function for different values of g. Plotting the validity index against g, there may be a value for g for which a significant local change in the index occurs. When applied to hierarchical schemes, these procedures for determining the number of clusters g are sometimes referred to as *stopping rules*. Many intuitive schemes have been proposed for hierarchical methods; for example, we may look at the plot of fusion level against the number of groups, g (Figure 11.20), and look for a flattening of the curve showing that little improvement in the description of the data structure is to be gained above a particular value of g. Defining $\alpha_j, j = 0, \ldots, n-1$, to be the fusion level corresponding to the stage with $n-j$ clusters, Mojena (1977) proposes a stopping rule that selects the number of groups as g where α_{n-g} is the lowest value of α for which

$$\alpha_{n-g} > \overline{\alpha} + k s_\alpha$$

where $\overline{\alpha}$ and s_α are the mean and the unbiased standard deviation of the fusion levels α; k is a constant, suggested by Mojena to be in the range 2.75–3.5.

Milligan and Cooper (1985) examine thirty procedures applied to classifications of datasets containing 2, 3, 4, or 5 distinct nonoverlapping clusters by four hierarchical schemes. They

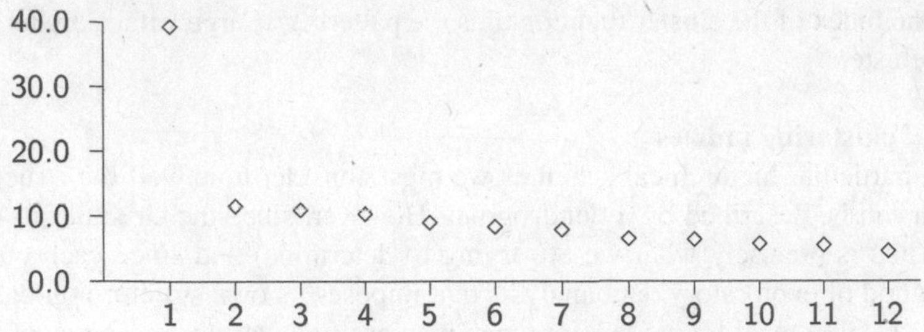

Figure 11.20 Fusion level against the number of clusters.

find that Mojena's rule performs poorly with only two groups present in the data and the best performance is for 3, 4 or 5 groups with a value of k of 1.25. One of the better criteria that Milligan and Cooper (1985) assess is that of Calinski and Harabasz (1974). The number of groups is taken to be the value of g that corresponds to the maximum of C, given by

$$C = \frac{\text{Tr}(S_B)}{\text{Tr}(S_W)}\left(\frac{n-g}{g-1}\right)$$

This is evaluated further by Atlas and Overall (1994) who compare it with a split-sample replication rule of Overall and Magee (1992). This gave improved performance over the Calinski and Harabasz criterion.

Dubes (1987) also reports the results of a Monte Carlo study on the effectiveness of two internal criterion measures in determining the number of clusters. Jain and Moreau (1987) propose a method to estimate the number of clusters in a dataset using the bootstrap technique. A clustering criterion based on the within and between-group scatter matrices (developing a criterion of Davies and Bouldin, 1979) is proposed and the k-means and three hierarchical algorithms are assessed. The basic method of determining the number of clusters using a bootstrap approach can be used with any cluster method.

Several authors have considered the problem of testing for the number of components of a normal mixture (see Chapter 2). Ismail (1988) reports the results of cluster validity studies within the context of soft clustering and lists nine validity functionals that provide useful tools in determining cluster structure (see also Pal and Bezdek, 1995).

It is not reasonable to expect a single statistic to be suitable for all problems in cluster validity. Many different factors are involved and since clustering is essentially a method of exploratory data analysis, we should not put too much emphasis on the results of a single classification, but perform several clusterings using different algorithms and measures of fit.

11.8 Application studies

Applications of hierarchical methods of cluster analysis include the following:

- Flight monitoring. Eddy *et al.* (1996) consider single-link clustering of large datasets (more than 40 000 observations) of high-dimensional data relating to aircraft flights over the USA.

- Clinical data. D'Andrea *et al.* (1994) apply the nearest centroid method to data relating to adult children of alcoholics.

- In a comparative study of seven methods of hierarchical cluster analysis on 20 datasets, Morgan and Ray (1995) examine the extent of inversions in dendrograms and nonuniqueness. They conclude that inversions are expected to be encountered and recommend against the use of the median and centroid methods. Also, nonuniqueness is a real possibility for many datasets.

The k-means clustering approach is widely used as a preprocessor to supervised classification to reduce the number of prototypes:

- Coal petrography. In a study to classify the different constituents (macerals) of coal (Mukherjee *et al.*, 1994), the k-means algorithm was applied to training images (training vectors consist of RGB level values) to determine four clusters (identified as one of known types – vitrinite, inertinite, exinite and background). These clusters are labelled and test images are classified using the labelled training vectors.

- Crop classification. Conway *et al.* (1991) use a k-means algorithm to segment synthetic aperture radar images as part of a study into crop classification. k-means is used to identify sets of image regions that share similar attributes prior to labelling. Data were gathered from a field of five known crop types and could be clearly separated into two clusters – one containing the broad-leaved crops and the other the narrow-leaved crops.

k means is also used for image and speech coding applications (see below).

Examples of fuzzy c-means applications are:

- Medical diagnosis. Li *et al.* (1993) use a fuzzy c-means algorithm for image segmentation in a study on automatic classification and tissue labelling of two-dimensional magnetic resonance human brain images.

- Acoustic quality control. Meier *et al.* (1994) describe the application of fuzzy c-means to cluster six-dimensional feature vectors as part of a quality control system of ceramic tiles. The signals are derived by hitting the tiles and digitising and filtering the recorded signal. The resulting classes are interpreted as good or bad tiles.

- Water quality. Mukherjee *et al.* (1995) compared fuzzy c-means with two alternative approaches to image segmentation in a study to identify and count bacterial colonies from images.

See also the survey on fuzzy clustering by Yang (1993) for further references to applications of fuzzy c-means.

Bayesian approaches to mixture modelling:

- Some applications of mixture models are given in Chapter 2. Dellaportas (1998) considers the application of mixture modelling to the classification of neolithic ground stone tools. A Bayesian methodology is adopted and developed in three main ways to apply to data (147 measurements on four variables) consisting of variables of mixed type – continuous and categorical; to handle missing values and measurement errors (errors in variables) in the continuous variables. Gibbs sampling is used to generate samples from the posterior densities of interest and classification to one of two classes is based on the mean of the indicator variable, z_i. After a 'burn-in' of 4000 iterations, 4000 further samples were used as the basis for posterior inference.

Examples of self-organising feature map applications include:

- Engineering applications. Kohonen *et al.* (1996) review the self-organising map algorithm and describe several engineering applications including fault detection, process analysis and monitoring, computer vision, speech recognition, robotic control and in the area of telecommunications.
- Human protein analysis. Ferrán *et al.* (1994) use a self-organising map to cluster protein sequences into families. Using 1758 human protein sequences, they cluster using two-dimensional maps of various sizes and label the nodes in the grid using proteins belonging to known sequences.
- Radar target classification. Stewart *et al.* (1994) develop a self-organising map and a learning vector quantisation approach to radar target classification using turntable data of four target types. The data consist of 33-dimensional feature vectors (33 range gates) and 36 000 patterns per target were used. Performance as a function of the number of cluster centres is reported with the learning vector quantisation performance better than a simplistic nearest-neighbour algorithm.
- Fingerprint classification. Halici and Ongun (1996), in a study on automatic fingerprint classification, use a self-organising map, and one modified by preprocessing the feature vectors by combining them with 'certainty' vectors that encode uncertainties in the fingerprint images. Results show an improvement on previous studies using a multilayer perceptron on a database of size 2000.

Sum-of-squares method application:

- Language disorder. Powell *et al.* (1979) use a normal mixture approach and a sum-of-squares method in a study of 86 aphasic cases referred to a speech therapy unit. Four groups are found, which are labelled as severe, high–moderate, low–moderate and mild aphasia.

Vector quantisation has been widely applied as a preprocessor in many studies:

- Speech recognition. Zhang *et al.* (1994) assess three different vector quantisers (including the LBG algorithm and an algorithm based on normal mixture modelling) as preprocessors for a hidden Markov model based recogniser in a small speech recognition problem. They found that the normal mixture model gave the best performance of the subsequent classifier. See also Bergh *et al.* (1985).
- Medical diagnosis. Cosman *et al.* (1993) assess the quality of tree-structured vector quantisation images by the diagnostic performance of radiologists in a study on lung tumour and lymphadenopathy identification. Initial results suggest that a 12 bits per pixel (bpp) computerised tomography chest scan image can be compressed to between 1 bpp and 2 bpp with no significant change in diagnostic accuracy: subjective quality seems to degrade sooner than diagnostic accuracy falls off.

- Speaker recognition. Speaker recognition approaches are reviewed by Furui (1997). Vector quantisation methods are used to compress the training data and produce codebooks of representative feature vectors characterising speaker-specific features. A codebook is generated for each speaker by clustering training feature vectors. At the speaker recognition stage, an input utterance is quantised using the codebook of each speaker and recognition performed by assigning the utterance to the speaker whose codebook produces minimum distortion. A tutorial on vector quantisation for speech coding is given by Makhoul *et al.* (1985).

Spectral clustering applications include the following:

- Analysis of flow cytometry (measurements of microscopic particles in a fluid) data. Zare *et al.* (2010) develop a modification of the spectral clustering approach to large datasets and apply it to flow cytometry data as an example of data containing potentially hundreds of thousands of data points.
- Monitoring blog communities. Ning *et al.* (2007) develop an incremental algorithm that extends the existing offline spectral clustering algorithms to evolving data such as that occurring in the real-time monitoring of the evolving communities of blogs and their links.
- Tissue classification (Crum, 2009).
- Language analysis. Brew and Schulte im Walde (2002) apply spectral methods to the clustering of German verbs. The distance between two 'verb frames' is calculated using different measures and transformed to a similarity matrix using a Gaussian similarity function [Equation (11.11)]. The clustering solution gave as good alignment with the distance measure as a 'gold standard' human solution.

11.9 Summary and discussion

In this chapter we have covered a wide range of techniques for partitioning a dataset. This has included approaches based on cluster analysis methods and vector quantisation methods. Although both approaches have much in common – they both produce a dissection of a given dataset – there are differences. In cluster analysis, we tend to look for 'natural' groupings in the data that may be labelled in terms of the subject matter of the data. In contrast, the vector quantisation methods are developed to optimise some appropriate criterion from communication theory. One area of common ground we have discussed in this chapter is that of optimisation methods with specific implementations in terms of the *k*-means algorithm in cluster analysis and the LBG algorithm in vector quantisation.

As far as cluster analysis or classification is concerned, there is no single best technique. Different clustering methods can yield different results and some methods will fail to detect obvious clusters. The reason for this is that each method implicitly forces a structure on the given data. For example, the sum-of-squares methods will tend to produce hyperspherical clusters. Also, the fact that there is a wide range of available methods partly stems from the lack of a single definition of the word cluster. There is no universal agreement as to what constitutes a cluster and so a single definition is insufficient.

A further difficulty with cluster analysis is in deciding the number of clusters present. This is a tradeoff between parsimony and some measure of increase in within-cluster homogeneity. This problem is partly due to the difficulty in deciding what a cluster actually is and partly because clustering algorithms tend to produce clusters even when applied to random data.

Both of the above difficulties may be overcome to some degree by considering several possible classifications or comparing classifications on each half of a dataset (McIntyre and Blashfield, 1980; Breckenridge, 1989). The interpretation of these is more important than a rigid inference of the number of groups. But which methods should we employ? There are advantages and disadvantages of all the approaches we have described. The optimisation methods tend to require a large amount of computer time (and consequently may be infeasible for large datasets, though this is becoming less critical these days). Of the hierarchical methods, the single link is preferred by many users. It is the only one to satisfy the Jardine–Sibson conditions, yet with noisy data it can join separate clusters (chaining effect). It is also invariant under monotone transformations of the dissimilarity measure. Ward's method is also popular. The centroid and median methods should be avoided since inversions may make the resulting classification difficult to interpret.

There are several aspects of cluster analysis that we have mentioned only briefly in this chapter and we must refer the reader to the literature on cluster analysis for further details. An important problem is the choice of technique for mixed mode data (i.e. containing both numerical and categorical measurements). Everitt and Merette (1990) (see also Everitt, 1988) propose a finite mixture model approach for clustering mixed mode data, but computational considerations may mean that it is not practically viable when the datasets contain a large number of categorical variables.

The techniques described in this chapter all apply to the clustering of objects. However, there may be some situations where clustering of variables, or simultaneous clustering of objects and variables, is required. In clustering of variables, we seek subsets of variables that are so highly correlated that each can be replaced by any one of the subset, or perhaps a (linear or nonlinear) combination of the members. Many of the techniques described in this chapter can be applied to the clustering of variables and therefore we require a measure of similarity or dissimilarity between variables. Of course, techniques for feature extraction (for example, principal components analysis) perform this process.

Another point to reiterate about cluster analysis is that it is essentially an exploratory method of multivariate data analysis providing a description of the measurements. Once a solution or interpretation has been obtained then the investigator must re-examine and assess the dataset. This may allow further hypotheses (perhaps concerning the variables used in the study, the measures of dissimilarity and the choice of technique) to be generated. These may be tested on a new sample of individuals.

Both cluster analysis and vector quantisation are means of reducing a large amount of data to a form in which it is either easier to describe or represent in a machine. Clustering of data may be performed prior to supervised classification. For example, the number of stored prototypes in a k-nearest-neighbour classifier may be reduced by a clustering procedure. The new prototypes are the cluster means, and the class of the new prototype is decided on a majority basis of the members of the cluster. A development of this approach that adjusts the decision surface by modifying the prototypes is *learning vector quantisation* (Kohonen, 1989).

Self-organising maps may be viewed as a form of *constrained classification*: clustering in which there is some form of constraint on the solution. In this particular case, the constraint is an ordering on the cluster centres. Other forms of constraint may be that objects within a cluster are required to comprise a spatially contiguous set of objects (for example, in some texture segmentation applications). This is an example of *contiguity constrained clustering* [Murtagh, 1992; Gordon, 1999; see also Murtagh (1995), for an application to the outputs of the self-organising map]. Other forms of constraint may be on the topology of the dendrogram or the size or composition of the classes. We refer to Gordon (1996b) for a review.

Spectral clustering has become a popular method in recent years. It is simple to implement and methods to overcome its computational burden when applied to large datasets have been developed (Ning *et al.*, 2007; Yan *et al.*, 2009).

Gordon (1994, 1996a) provides reviews of approaches to cluster validation; see also Bock (1989), Jain and Dubes (1988), Theodoridis and Koutroumbas (2009) and the papers by Kovács *et al.* (2006) and Halkidi *et al.* (2001).

11.10 Recommendations

There is a large number of techniques to choose from and the availability of computer packages means that analyses can be readily performed. Nevertheless, there are some general guidelines to follow when carrying out a classification.

1. Detect and remove outliers. Many clustering techniques are sensitive to the presence of outliers. Therefore, some of the techniques discussed in Chapter 13 should be used to detect and possibly remove these outliers.
2. Plot the data in two dimensions if possible in order to understand structure in the data. It might be helpful to use the first two principal components (see Chapter 10).
3. Carry out any preprocessing of the data. This may include a reduction in the number of variables or standardisation of the variables to zero mean and unit variance.
4. If the data are not in the form of a dissimilarity matrix then a dissimilarity measure must be chosen (for some techniques) and a dissimilarity matrix formed.
5. Choose an appropriate technique. Of the hierarchical methods, some studies favour the use of the group average link method, but the single-link method gives solutions that are invariant to a monotone transformation measure. The single-link method is the only one to satisfy all the conditions laid down by Jardine and Sibson (1971) and is their preferred method. We recommend against the use of the centroid and median methods since inversions are likely to arise.
6. Evaluate the method. Assess the results of the clustering method you have employed. How do the clusters differ? We recommend that you split the dataset into two parts and compare the results of clustering on each subset. Similar results would suggest that useful structure has been found. Also, use several methods and compare results. With many of the methods some parameters must be specified and it is worthwhile carrying out clustering over a range of parameter values to assess stability.
7. In using vector quantisation as a preprocessor for a supervised classification problem, model each class separately rather than model the whole dataset and label the resulting codewords.

8. Spectral clustering should be considered when clusters may not be compact/concave.

9. If you require some representative prototypes, we recommend using the k-means algorithm.

Finally, we reiterate that cluster analysis is usually the first stage in an analysis and unquestioning acceptance of the results of clustering is extremely unwise.

11.11 Notes and references

There is a vast literature on cluster analysis. A very good starting point is the book by Everitt *et al.* (2011). Now in its fifth edition, this book is a mainly nonmathematical account of the most common clustering techniques, together with their advantages and disadvantages. The practical advice is supported by several empirical investigations. Another good introduction to methods and assessments of classification is the book by Gordon (1999). McLachlan and Basford (1988) discuss the mixture model approach to clustering in some detail.

Of the review papers, that by Cormack (1971) is worth reading and provides a good summary of the methods and problems of cluster analysis. The article by Diday and Simon (1976) gives a more mathematical treatment of the methods, together with descriptive algorithms for their implementation.

Several books have an orientation towards biological and ecological matters. Jardine and Sibson (1971) give a mathematical treatment. The book by Sneath and Sokal (1973) is a comprehensive account of cluster analysis and the biological problems to which it can be applied. The book by Clifford and Stephenson (1975) is a nonmathematical general introduction to the ideas and principles of numerical classification and data analysis, though it does not cover many of the approaches described in this chapter, concentrating on hierarchical classificatory procedures. The book by Jain and Dubes (1988) has a pattern recognition emphasis. McLachlan (1992b) reviews cluster analysis in medical research.

A more specialist book is that of Zupan (1982). This monograph is concerned with the problem of implementing hierarchical techniques on large datasets.

The literature on fuzzy techniques in cluster analysis is reviewed by Bezdek and Pal (1992). This book contains a collection of some of the important papers on fuzzy models for pattern recognition, including cluster analysis and supervised classifier design, together with fairly extensive bibliographies. A survey of fuzzy clustering and its applications is provided by Yang (1993). An interesting probabilistic perspective of fuzzy methods is provided by Laviolette *et al.* (1995).

Tutorials and surveys of self-organising maps are given by Kohonen (1990, 1997), Kohonen *et al.* (1996) and Ritter *et al.* (1992).

There are various books on techniques for implementing methods, which give algorithms in the form of FORTRAN code, or pseudo-code. The books by Anderberg (1973), Hartigan (1975), Späth (1980) and Jambu and Lebeaux (1983) all provide a description of a clustering algorithm, FORTRAN source code and a supporting mathematical treatment, sometimes with case studies. The book by Murtagh (1985) covers more recent developments and is also concerned with implementation on parallel machines.

An introduction to spectral clustering is provided by von Luxburg (2007), Spielman and Teng (1996) and Chung (1997).

Exercises

Dataset 1: Generate $n = 500$ samples (x_i, y_i), $i = 1, \ldots, n$, according to

$$x_i = \frac{i}{n}\pi + n_x$$

$$y_i = \sin\left(\frac{i}{n}\pi\right) + n_y$$

where n_x and n_y are normally distributed with mean 0.0 and variance 0.01.

Dataset 2: Generate n samples from a multivariate normal (p variables) of diagonal covariance matrix $\sigma^2\boldsymbol{I}$, $\sigma^2 = 1$ and zero mean. Take $n = 40$, $p = 2$.

Dataset 3: This consists of data comprising two classes: class ω_1 is distributed as $0.5N((0, 0), \boldsymbol{I}) + 0.5N((2, 2), \boldsymbol{I})$ and class $\omega_2 \sim N((2, 0), \boldsymbol{I})$ (generate 500 samples for training and test sets, $p(\omega_1) = p(\omega_2) = 0.5$).

1. Is the square of the Euclidean distance a metric? Does it matter for any clustering algorithm?

2. Observations on six variables are made for seven groups of canines and given in Table 11.5 (Manly, 1986; Krzanowski and Marriott, 1994). Construct a dissimilarity matrix using Euclidean distance after standardising each variable to unit variance. Carry out a single-link cluster analysis.

3. Compare the single-link method of clustering with k-means, discussing computational requirements, storage, and applicability of the methods.

4. A mixture of two normals divided by a normal density having the same mean and variance as the mixed density is always bimodal. Prove this for the univariate case.

5. Implement a k-means algorithm and test it on two-dimensional normally distributed data (dataset 2 with $n = 500$). Also, use the algorithm within a tree-structured vector quantiser and compare the two methods.

6. Using dataset 1, code the data using the Luttrell self-organising feature map algorithm (Section 11.5.3). Plot the positions of the centres for various numbers of code vectors.

Table 11.5 Data on mean mandible measurements (from Manly, 1986).

Group	x_1	x_2	x_3	x_4	x_5	x_6
Modern Thai dog	9.7	21.0	19.4	7.7	32.0	36.5
Golden jackal	8.1	16.7	18.3	7.0	30.3	32.9
Chinese wolf	13.5	27.3	26.8	10.6	41.9	48.1
Indian wolf	11.5	24.3	24.5	9.3	40.0	44.6
Cuon	10.7	23.5	21.4	8.5	28.8	37.6
Dingo	9.6	22.6	21.1	8.3	34.4	43.1
Prehistoric dog	10.3	22.1	19.1	8.1	32.3	35.0

Compute the average distortion as a function of the number of code vectors. How would you modify the algorithm for data having circular topology?

7. Using dataset 1, construct a tree-structured vector quantiser, partitioning the clusters with the largest sum-square error at each stage. Compute the average distortion.

8. Using dataset 2, cluster the data using Ward's method and Euclidean distance. Using Gordon's approach for identifying genuine clusters (unimodal null model), how many clusters are valid at the 5% level of significance?

9. Implement a learning vector quantisation algorithm on dataset 3. Plot performance as a function of the number of cluster centres. What would be the advantages and disadvantages of using the resulting cluster centres as centres in a radial basis function network?

10. Show that the single-link dendrogram is invariant to a nonlinear monotone transformation of the dissimilarities.

11. For a distance between two clusters A and B of objects given by $d_{AB} = |\boldsymbol{m}_A - \boldsymbol{m}_b|^2$, where $\boldsymbol{m}_A$ is the mean of the objects in cluster A, show that the formula expressing the distance between a cluster k and a cluster formed by joining i and j is

$$d_{i+j,k} = \frac{n_i}{n_i + n_j} d_{ik} + \frac{n_j}{n_i + n_j} d_{jk} - \frac{n_i n_j}{(n_i + n_j)^2} d_{ij}$$

where there are n_i objects in group i. This is the update rule for the centroid method.

12. Show that, if the number of clusters in a clustering is m and the number of groups in a partition is s and $s \neq m$, the Rand and Jaccard coefficients (Section 11.7.7) are less than 1.

12

Complex networks

Complex networks can be used to describe the behaviour of a wide range of systems of importance in nature and society. These include social networks describing the pattern of contact between groups of people, biological networks (for example, the food web and the network of protein–protein interactions characterising a cell) and computer networks such as the Internet. Datasets from these domains differ from those considered elsewhere in this book. They are not 'flat' (measurements on a fixed set of variables for a set of individuals), but comprise measurements that describe the *relationship* between individuals (often pairs of individuals) and can be represented using the mathematical concept of a graph. Analysis of the data is undertaken to understand network structure, including determining the presence of groups or communities, finding patterns in the network structure and the identification of anomalous behaviour. This chapter presents a brief introduction to complex networks and introduces the topics of community detection and link prediction.

12.1 Introduction

Complex networks is the name given to a multidisciplinary area, bringing together concepts from physics, mathematics and statistics, computer science and the social sciences, that relates to interactions between individual entities or groups of individuals. It is an area of research that has been of increasing importance over the past decade or so, as data acquisition systems have enabled large datasets of real networks to be gathered. These datasets differ from those we have used in this book to develop supervised or unsupervised classifiers: they are not simply measurements on a fixed set of features for a given number of individuals where the aim has been to determine a rule to classify unknown patterns using the dataset to train the rule (supervised classification); or to discover groups in the data (unsupervised classification or clustering). Rather, these datasets comprise measurements on the interactions between (often,

Statistical Pattern Recognition, Third Edition. Andrew R. Webb and Keith D. Copsey.

but not exclusively) pairs of individuals. Examples of networks that arise through interactions or transactions between individuals are (Albert and Barabási, 2002; Newman, 2003):

- the Internet – a network of routers and computers connected by physical wires;
- the World Wide Web – a network of pages, accessible over the Internet;
- social networks – a network of people, or groups of people, linked by friendships, business relationships and family ties;
- online transactions network – a network of shoppers and the Internet sites that they purchase from;
- telephone call network – a network of telephone company subscribers linked by the calls they make to each other;
- the cell – a network of protein–protein interactions characterising a cell;
- citation network – a network of articles, with a link from article A to article B if article A cites article B.

Investigations to understand such networks – their structure, growth and dynamics – has led to their description using the mathematical concept of a graph (introduced in Chapter 11), where the nodes in the graph represent individual entities (for example, people, computers, web pages, articles), which are joined if there is an interaction between them. An example is shown in Figure 12.1, which displays a social network of frequent associations between dolphins in a community living off Doubtful Sound, New Zealand (Lusseau and Newman, 2004).

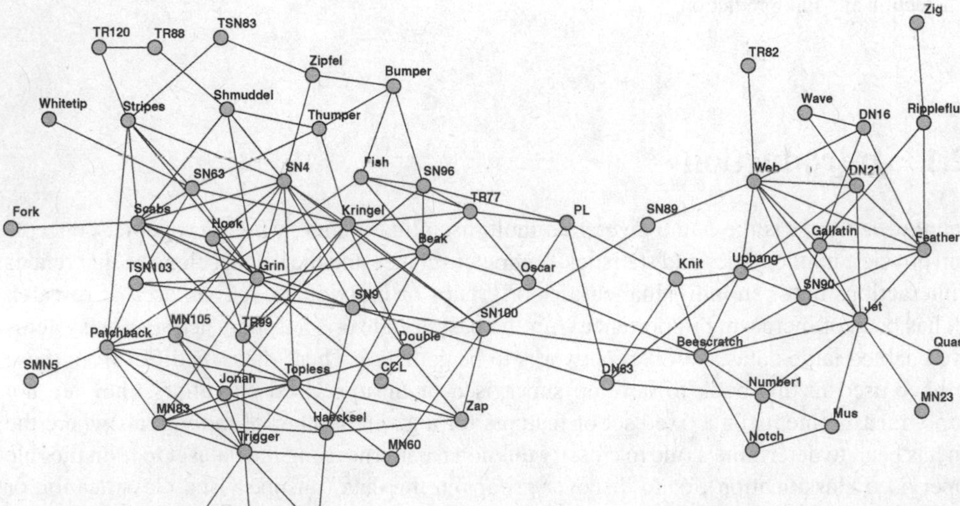

Figure 12.1 An undirected social network of frequent associations between 62 dolphins in a community living off Doubtful Sound, New Zealand (Lusseau and Newman, 2004).

12.1.1 Characteristics

The datasets describing the networks above typically have the following characteristics:

- Large size. Networks can have millions, if not billions, of nodes, particularly some social networks and the World Wide Web. Graphical displays can be difficult, if not impossible, to interpret. Automatic procedures are required to discover structure in the data.
- Sparseness. The networks are sparse, most nodes having few connections to other nodes.
- Weighted edges. The connections between nodes are often weighted. These weights describe the degree of interaction between the nodes. For example, these weights could describe the amount of email traffic between people, the cost of an Internet purchase or the length of a telephone call.
- Temporal dependence of weights. The strength of a connection between nodes is time varying: weights change with time as friendships form, purchases are made or people work together on a project.
- Growing/shrinking network. Nodes join and leave a network at different times, as new Internet sites are created and old ones deleted, or as new social groups develop. Thus, most networks are dynamic.
- Attributed graphs. Associated with each node may be a set of attributes or features (for example, the age of a person, gender, location, qualifications). The nodes may belong to different classes (purchaser and vendor) and edges may exist only between nodes of different classes.[1] Similarly, there may be features associated with the links between nodes (for example, the type of purchase from an online store).

More generally, the datasets may be described by a hypergraph, a generalisation of a graph where the edges connect any number of vertices. A hypergraph has many of the properties above, but we restrict our treatment to edges connecting two nodes.

12.1.2 Properties

There are some properties of real-world networks that are common to networks of different types. Here we summarise three common properties.

12.1.2.1 The small-world effect

The *small-world effect* refers to the property that in most, often large, networks, pairs of vertices are often connected by a short pathlength, where the pathlength between two vertices is the number of edges along the shortest path. The most popular example of this is the study by Milgram (1967), who investigated the distribution of pathlengths in an acquaintance network, where two individuals are connected if they were on first name terms with each other. The average separation between any two individuals was six, leading to the concept of 'six degrees of separation'.

---

[1] A *bipartite* graph is a graph where the nodes belong to one of two classes and there are no edges joining nodes of the same class.

Defining d_{ij} to be the shortest distance between node i and node j in an undirected network (i.e. the *geodesic*[2] *distance*), then the mean geodesic distance is given by

$$l = \frac{1}{\frac{1}{2}n(n+1)} \sum_{i \geq j} d_{ij} \tag{12.1}$$

l is often quite small – much smaller than the number of nodes in the network. If there are separate components in the network – parts of the network that are disconnected from other parts – then d_{ij} will not be defined for some pairs of nodes. Therefore, the mean geodesic is modified to be the mean distance between nodes for which a connecting path exists.

An alternative measure is the harmonic mean, defined by

$$l^{-1} = \frac{1}{\frac{1}{2}n(n+1)} \sum_{i \geq j} d_{ij}^{-1} \tag{12.2}$$

The implications of the small-world effect are that the spread of information can be very rapid: for a rumour or a virus hoax, or for a disease to spread.

12.1.2.2 Clustering

In social networks, groups or *communities* form according to factors such as the interests, age or occupation of the members. This tendency to form communities (groups of nodes with a high degree of within group connectivity and a low between-group connectivity) is not only a property of social networks and the identification of communities in networks is an important research topic.

The degree to which nodes in a network cluster together to form a community can be quantified by the clustering coefficient. For a node i with k_i neighbours, then the ratio of the number of edges between the neighbours, E_i, to the maximum number of edges,

$$C_i = \frac{E_i}{\frac{1}{2}k_i(k_i - 1)}$$

is a measure of the clustering of node i. The clustering of the whole network is the average

$$C = \frac{1}{n} \sum_{i=1}^{n} \frac{E_i}{\frac{1}{2}k_i(k_i - 1)}$$

Albert and Barabási (2002) provide several alternative definitions.

12.1.2.3 Degree distributions

Nodes in a network have different numbers of connections to other nodes – the *node degree*. The distribution of node degree, is denoted by $P(k)$, the probability that a randomly selected

[2] The *geodesic* is the shortest path between two nodes in a network.

node has degree k. For a random graph,[3] the distribution of node degree is binomial, or Poisson in the limit of a large graph. However, for most large networks, the degree distribution deviates significantly from Poisson, particularly in the tail of the distribution. There is a long tail – there are nodes whose values of the degree are far greater than the mean. For many real-world networks, the distributions are found to follow a power law in their tails

$$P(k) \propto k^{-\alpha}$$

for some exponent α. Networks with power law degree distributions are referred to as *scale-free* networks.

12.1.2.4 Other properties

There are other properties of real-world networks that are explored in many studies. These include:

- Network resilience – the robustness of the network to the removal of nodes from the network (or the removal of the means by which two adjacent nodes communicate, leading to the removal of an edge).
- Mixing patterns – the pattern of links between nodes of different types.
- Degree correlations – the correlation between the degree of neighbouring nodes. (Are high degree nodes preferentially attached to other high degree nodes, or low degree nodes?)

Newman (2003) discusses these and others in detail.

12.1.3 Questions to address

There are many questions that have been posed of datasets that may be represented graphically as a network. Some of these are of a more exploratory nature, relating to the structure of the network, particularly when the network is difficult to visualise graphically due to its large size. Others are more application specific. Listed below are some of the properties of a network that we may wish to discover.

- Anomalous nodes. Do any of the nodes have an unusual pattern of behaviour? They may have a particularly high degree or have a pattern of connectivity to other nodes that is unusual. Alternatively, the temporal behaviour of the node may be unusual. The behaviour of a node may change with time (for example, the rate of interaction with neighbouring nodes), resulting in behaviour that is different from past behaviour.
- Significant edges. These are connections between two nodes whose presence is vital to the network's function.
- Anomalous edges. These are edges, or communication channels between nodes, that are striking by their presence (or, alternatively, an edge may be absent where one is expected).

[3] This is a graph in which edges are placed randomly – an edge between nodes occurs with a specified probability, p.

- Influential nodes. These are nodes whose removal results in a significant change in network behaviour. For example, an influential node may be the only common node to two communities. Its removal causes the communities to become disjointed and hence lack a communication route between them.
- Central nodes. Which nodes are the most central? These are nodes that communicate with many others. Which nodes are outliers, with few connections, or connections to others for a short period of time only?
- Community structure. This is the clustering of nodes into groups or communities for which there is high inter-community connectivity and low between-community connectivity.
- Community evolution. How the community structure changes with time – the addition or removal of nodes or edges, the change in the strength of connection between nodes – may be of interest for some applications.

12.1.4 Descriptive features

What features should we use to describe the complex networks described by datasets of transactions between individuals? This depends on the questions that we are seeking to address, just as in the supervised classification of objects the features employed are (ideally) those important for discrimination. For very large networks, for which the structure is difficult to ascertain from a graphical display, the questions may be more exploratory in nature about the structure of the network. Examples of features that we may measure on the complete network include the degree distribution, network clustering coefficient and the network diameter (the largest geodesic distance between two nodes).

In other cases, the question may be more specifically about particular nodes or nodes of a characteristic type: we may seek nodes with certain characteristics. This is a supervised problem, where the characteristics are specified by the user or learnt from a design set. Features characterising individual nodes or groups of nodes include centrality measures,[4] node clustering coefficient and community structure measures (size, diameter, community clustering coefficient – see Section 12.3.1).

12.1.5 Outline

The topics of this chapter lie at the intersection of the research areas of theory and algorithms for the analysis of graphs and pattern recognition. There are broadly two main themes within this area:

1. the use of graph-based methods in pattern recognition applications;
2. the application of data analysis methods of the type described in this book to data comprising measurements of interactions between entities.

---

[4] A centrality measure describes the importance of nodes in a network [see Section 12.2; for their application in social networks, see Wasserman and Faust (1994)].

The focus of this chapter is on the latter theme: what patterns (structure) are present in the network? How can we find it? The former theme is addressed in the book by Marchette (2004). We exclude from our treatment important topics such as methods for the analysis of time evolving networks (dynamic graphs) and the development of network models (random graphs and developments). Section 12.8 provides references to more extensive treatments of complex networks.

12.2 Mathematics of networks

A network may be represented mathematically as a graph, $G = \{V, E\}$, comprising a set of *nodes* or *vertices*, $V = \{v_1, \ldots, v_n\}$, and a set of edges, $E = \{e_1, \ldots, e_m\}$, where each edge, e_j, is a pair of vertices

$$e_j = \{v_{j1}, v_{j2}\}$$

where $v_{j1} \neq v_{j2}$ and $v_{j1}, v_{j2} \in V$.

For a graph $G = \{V, E\}$, the *neighbours* of a node $v \in V$ are those nodes connected to v by an edge $e \in E$.

12.2.1 Graph matrices

The structure of the graph is described by the symmetric $n \times n$ adjacency matrix, $\mathbf{A}$, where

$$A_{ij} = \begin{cases} 1 & \text{if } v_i \text{ and } v_j \text{ are connected by an edge} \\ 0 & \text{otherwise} \end{cases}$$

A *directed* graph (Figure 12.2) is one in which the direction of the edges is important (for example, a graph of email traffic between employees of a company where the direction is

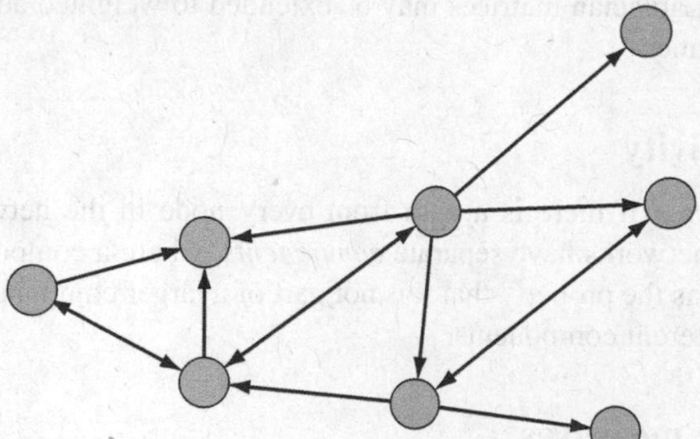

Figure 12.2 A directed graph in which some of the edges are unidirectional and some bidirectional.

from sender to recipient), the adjacency matrix is not symmetric and

$$A_{ij} = \begin{cases} 1 & \text{if there is an edge from } v_j \text{ to } v_i \\ 0 & \text{otherwise} \end{cases}$$

The *graph Laplacian* is the difference between the adjacency matrix, $\boldsymbol{A}$, and the diagonal degree matrix, $\boldsymbol{D}$

$$\boldsymbol{L} = \boldsymbol{D} - \boldsymbol{A} \tag{12.3}$$

i.e.

$$L_{ij} = \begin{cases} k_i & i = j \\ -A_{ij} & i \neq j \end{cases}$$

where k_i is the number of edges connected to node i. There are two common forms of normalised graph Laplacian (see Chapter 11); the symmetric form,

$$\boldsymbol{L}_{sym} = \boldsymbol{D}^{-\frac{1}{2}}\boldsymbol{L}\boldsymbol{D}^{-\frac{1}{2}} = \boldsymbol{I} - \boldsymbol{D}^{-\frac{1}{2}}\boldsymbol{A}\boldsymbol{D}^{-\frac{1}{2}} \tag{12.4}$$

and the nonsymmetric form,

$$\boldsymbol{L}_{nonsym} = \boldsymbol{D}^{-1}\boldsymbol{L} = \boldsymbol{I} - \boldsymbol{D}^{-1}\boldsymbol{A} \tag{12.5}$$

These normalised forms have the properties:

1. The normalised Laplacians are positive semidefinite.
2. All eigenvalues are real and non-negative. The smallest eigenvalue is zero.
3. The multiplicity of the eigenvalue with value 0 is equal to the number of connected components of the graph.
4. The eigenvectors, $\boldsymbol{v}$, and eigenvalues, λ, of $\boldsymbol{L}_{nonsym}$ satisfy the generalised symmetric eigenvector equation $\boldsymbol{L}\boldsymbol{v} = \lambda \boldsymbol{D}\boldsymbol{v}$.

The definition of all Laplacian matrices may be extended to weighted adjacency matrices in a straightforward manner.

12.2.2 Connectivity

A network is *connected* if there is a path from every node in the network to every other node. Disconnected networks have separate *components*, where a component is a connected group of nodes that has the property that it is not part of a larger component. There is no path between nodes in different components.

12.2.3 Distance measures

The geodesic distance between two nodes is the length of the shortest path between them, where the pathlength between two nodes is the number of edges traversed in going from one node to the other. For disconnected graphs, the geodesic distance between nodes in separate components cannot be calculated and is conventionally taken to be infinity.

The maximum geodesic distance between any pair of nodes in a connected graph is the *graph diameter*.

12.2.4 Weighted networks

Many networks are *weighted* – they have strengths associated with the edges connecting pairs of nodes. For example, the rate of email traffic between two friends; the cost of a purchase by a buyer from an online store. Weighted networks may be represented by a weighted adjacency matrix,

$$A_{ij} = \begin{cases} \text{weight of connection between } v_j \text{ and } v_i \\ 0 \text{ if no connection} \end{cases}$$

Weights are usually non-negative; the larger the value of a weight on an edge then the closer are the two nodes.

12.2.5 Centrality measures

Degree centrality
The *degree* of a node is the number of neighbours of that node. The degree of node v_i, k_i, is expressed in terms of the adjacency matrix as

$$k_i = \sum_{j=1}^{n} A_{ij}$$

The *degree matrix*, $\boldsymbol{D}$, is the diagonal matrix of degrees, $\boldsymbol{D} = \text{Diag}[k_1, \ldots, k_n]$. For a directed network, each node has an *in-degree* and an *out-degree*, corresponding to the number of incoming or outgoing edges.

Betweenness centrality
For a given node, v, the betweenness centrality is a measure of how many geodesic paths from different pairs of nodes in the network pass through v. Defining σ_{st} as the number of shortest paths between node s and node t, and $\sigma_{st}(v)$ as the number of those shortest paths that pass through node v [for many pairs of nodes in real-world networks, $\sigma_{st} = 1$ and $\sigma_{st}(v) = 0$ or 1, i.e. the single shortest path either excludes v or includes 1], then the normalised betweenness centrality for node v (for an undirected graph) is given by

$$C_B(v) = \frac{2}{(n-1)(n-2)} \sum_{\text{all pairs } s,t} \frac{\sigma_{st}(v)}{\sigma_{st}} \tag{12.6}$$

where $(n-1)(n-2)/2$ is the number of pairs of vertices not including v.

Eigenvector centrality
The components of the eigenvector of the adjacency matrix corresponding to the largest eigenvalue are used as centrality indices. The centrality of a node is proportional to the sum of the centralities of the nodes to which it is connected. Thus, a node is more 'important' if its neighbours are important. Let $\boldsymbol{x}$ denote the vector of centralities, then we may express this statement as

$$x_i = \lambda^{-1} \sum_{j=1}^{n} A_{ij} x_j \quad i = 1, \ldots, n$$

for a constant of proportionality λ^{-1}, or

$$\boldsymbol{A}\boldsymbol{x} = \lambda \boldsymbol{x}$$

The eigenvector corresponding to the largest eigenvalue has the property that its components are all positive (by the Perron–Frobenius theorem which states that a real matrix with positive entries has the largest real eigenvalue whose corresponding eigenvector has positive components).

This definition of eigenvector centrality extends to weighted adjacency matrices.

Closeness centrality
The closeness centrality is a measure of the distance of a node to the other nodes in the network. Denoting the geodesic distance between a node i and a node j by $d_g(i, j)$, then the closeness of the node i to other nodes in the network is defined as the reciprocal of the sum of the geodesic distances

$$C_l(i) = \frac{1}{\sum_j d_g(i, j)} \tag{12.7}$$

12.2.6 Random graphs

A random graph is one in which the number of vertices, the number of edges and the placement of those edges are determined in some random manner. The simplest model and one widely studied is due to Erdös and Rényi (1959, 1960) and denoted $G_{n,p}$. Each edge occurs independently with probability p. The probability p_k that a vertex has degree exactly k is given by the Binomial distribution

$$p_k = \binom{n-1}{k} p^k (1-p)^{n-1-k}$$

For $n \gg kz$, where z is the average degree of a vertex, $(n-1)p$, then

$$p_k = \frac{z^k e^{-z}}{k!}$$

which is the Poisson distribution.

12.3 Community detection

Detecting the presence of communities – clusters of nodes with a high intracluster connectivity and a low intercluster connectivity – is important in many application areas such as protein discovery, marketing, sociology, security and others where a graphical representation of the data is a natural one. The identification of high order structures in the data provides insights into the network's organisation leading, for example, to the discovery of functionally related proteins, groups of people sharing common interests or potential markets for new products. The term *graph clustering* is also used to refer to the task of grouping vertices in a graph, taking account of graph structure.

In this section, we review some of the approaches for identifying communities in a network, both global approaches that use the entire vertex set and local approaches that use a particular node as a seed to identify a local community.

12.3.1 Clustering methods

Many of the techniques of cluster analysis, described in Chapter 11, may be applied to discover communities in networks.

12.3.1.1 Hierarchical methods

Hierarchical clustering algorithms are often based on the analysis of a matrix of similarities or dissimilarities between pairs of objects. Therefore, since we seek to group vertices with high similarity, we require a measure of inter-node similarity or dissimilarity between each pair of vertices. A similarity measure, s_{ij}, can often be converted to a dissimilarity, d_{ij}, by $d_{ij} = k - s_{ij}$, for some constant k.

Similarity measures may be structural (they take account of the patterns of connections of the nodes) or attribute-based (they use the properties of the nodes). Structural similarity indices may be classified into three broad categories (Lü and Zhou, 2010):

- Local indices. These use the properties of the network in the vicinity of the two nodes whose similarity is required.
- Global indices. These use the complete adjacency matrix or weighted adjacency matrix.
- Quasi-local indices. These use the network in the neighbourhood of the two nodes, but with a 'broader horizon' than local indices.

Tables 12.1 and 12.2 list some of the local and global indices. Further measures and an evaluation in the context of link prediction (see Section 12.4) are given by Lü and Zhou (2010).

Hierarchical methods require some measure of the dissimilarity between groups or clusters of nodes: for example, the single link method takes the distance between a node and a cluster as the smallest distance between the node and any member of the cluster; the complete link method uses the greatest distance between a node and any member of the cluster as the distance between a node and a cluster (see Chapter 11). Different algorithms, and different measures of (dis-)similarity will lead to the identification of different community structure. Some networks do exhibit hierarchical structure, but a hierarchical algorithm may impose

Table 12.1 Similarity measures for graph vertices.

Local measures	
$\lvert\Gamma(i) \cap \Gamma(j)\rvert$	**Common neighbours**. $\Gamma(i)$ is the set of neighbours of vertex i. This measures the size of the overlap between the neighbourhood of vertex i and the neighbourhood of j. Many local measures are variants of this
$\frac{\lvert\Gamma(i) \cap \Gamma(j)\rvert}{\lvert\Gamma(i) \cup \Gamma(j)\rvert}$	**Jaccard index**
$\frac{\lvert\Gamma(i) \cap \Gamma(j)\rvert}{\sqrt{k_i k_j}}$	**Salton index**. k_i is the degree of node i
$\frac{\sum_k (A_{ik} - \mu_i)(A_{jk} - \mu_j)}{n\sigma_i\sigma_j}$	Correlation between rows of the adjacency matrix; $\mu_i = \frac{1}{n}\sum_j A_{ij}$; $\sigma_i = \sqrt{\sum_j (A_{ij} - \mu_i)^2/n}$
Global measures	
$(\boldsymbol{I} - \beta \boldsymbol{A})^{-1} - \boldsymbol{I}$	**Katz index**. $\boldsymbol{I}$ is the identity matrix, $\boldsymbol{A}$ the adjacency matrix and β a decay factor. This measures a weighted sum of all paths connecting pairs of vertices
$\frac{1}{L_{ii}^\dagger + L_{jj}^\dagger - 2L_{ij}^\dagger}$	**Average commute time**. $L_{ij}^\dagger$ are the components of the pseudo-inverse of the Laplacian. This is proportional to the reciprocal of the average number of steps by a random walker in moving between the two nodes
Proximity, measured using random walk with restart $\boldsymbol{r}_i = p\left(1 - (1-p)\boldsymbol{P}^T\right)^{-1}\boldsymbol{e}_i$	**Random walk with restart**. A particle is considered to start at node i and move to its neighbours with probability proportional to the edge weights from node i. At each step, the particle may return to node i with some restart probability p. The *proximity* of node j from node i is defined as the steady-state probability, r_{ij}, that the particle will be on node j. $\boldsymbol{e}_i$ is the unit vector with 1 in the ith position and 0 elsewhere; $P_{ij} = 1/k_i$ if nodes i and j are connected

Table 12.2 Dissimilarity measures for graph vertices.

Geodesic distance	The length of the shortest path between two nodes
$d_{ij} = \sqrt{\sum_{k \neq i,j} (A_{ik} - A_{jk})^2}$	Two vertices are structurally equivalent if they have the same neighbours

artificial structure in some cases. Single nodes connected to a community can often remain isolated, rather than part of the community to which they are joined.

12.3.1.2 *k*-medoids method

The *graph k-medoids* algorithm is the community detection algorithm corresponding to the *k*-means clustering algorithm (Chapter 11). It has the disadvantage compared with the hierarchical approach in that it requires the specification of the number of communities, *k*. The algorithm proceeds by initialising the communities by randomly selecting *k* seed nodes. All nodes are then assigned to the cluster of the nearest seed node (measured in geodesic distance). Since the geodesic distance is an integer (there is an integer number of steps between nodes), there may be several seed nodes at the same distance from a given node, v. In this case, v is assigned randomly to one of the nearest seed nodes. The next step is to calculate the *closeness centrality* for each node in each cluster. The closeness centrality of a cluster is a measure of the distance of a node to the other nodes in the cluster. If node i is in cluster k and we denote the geodesic distance between a node i and a node j by $d_g(i, j)$, then the closeness of the node i to other nodes in its cluster is defined as the reciprocal of the sum of the geodesic distances

$$C_l(i) = \frac{1}{\sum\limits_{j \in \text{ cluster } k} d_g(i, j)} \tag{12.8}$$

This is a generalisation of the network closeness centrality given by Equation (12.7).

The new community centre nodes are those with the smallest closeness centrality in each cluster. The procedure iterates until a stable solution is obtained. This may not be convergence, since the random nature of the assignment of nodes to clusters in the case of equal geodesic distances can result in some instability. Therefore, the algorithm may be considered stable when there is less than x% change in the number of cluster medoids. Rattigan *et al.* (2007) suggest a value of x of between 1 and 3.

The algorithm is summarised in Algorithm 12.1.

Algorithm 12.1 *k*-medoids algorithm.

---

1. Initialise set of vertices as seeds for cluster centres.
2. Assign nodes to the cluster of the nearest cluster centre (using geodesic distance – ties are resolved randomly).
3. For each cluster, calculate a new set of cluster centres. These are the nodes in each cluster with the lowest value of closeness [Equation (12.8)].
4. If the algorithm has stabilised, exit, otherwise go to step 2.

12.3.1.3 Spectral methods

The algorithm for a spectral clustering approach to community detection in networks is based on an eigendecomposition of the graph Laplacian and is given by Algorithms 11.1 and 11.2. There are three forms for the Laplacian that may be employed: the unnormalised form, $\boldsymbol{L} = \boldsymbol{D} - \boldsymbol{A}$ [Equation (12.3)] and its normalised variants, $\boldsymbol{L}_{sym}$ [Equation (12.4)] and $\boldsymbol{L}_{nonsym}$ [Equation (12.5)]. If the nodes degrees are similar, then the results of spectral clustering are similar for the unnormalised and normalised forms of the Laplacian. However, if there is a large variation in node degree, the normalised forms are more reliable with the nonsymmetric form, $\boldsymbol{L}_{nonsym}$, preferred (Fortunato, 2010).

12.3.2 Girvan–Newman algorithm

The betweenness centrality of a node is a measure of the influence of the node in a network and is defined as the number of shortest paths that run through the node [Equation (12.6) gives a normalised version and allows for multiple shortest paths]. Girvan and Newman (2002; Newman and Girvan, 2004) generalise this concept and define *edge betweenness* as the number of shortest paths between vertices that run along a particular edge.

The basic idea is that for a network consisting of several communities, loosely connected by a few edges, then the shortest paths between nodes in different communities must run along these edges. Hence, they will have a high edge betweenness and can be identified (Figure 12.3) and removed, revealing the underlying community structure of the network.

The algorithm is an example of a divisive approach: beginning with a complete network and removing edges one at a time (Algorithm 12.2). It can be computationally expensive for large networks.

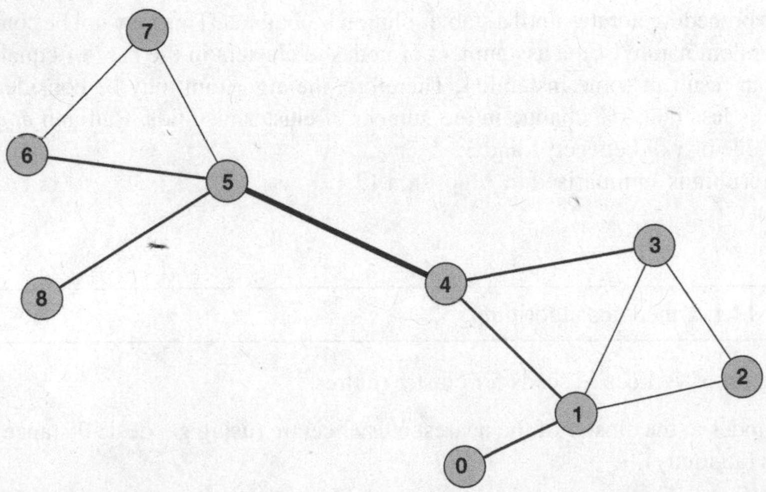

Figure 12.3 A network of two communities (nodes 0–4 and nodes 5–8), where the thickness of the line joining nodes is proportional to the edge betweenness. The line between nodes 4 and 5 is on all paths between nodes in the two different communities and so it has a high edge betweenness.

Algorithm 12.2 Girvan–Newman divisive algorithm for partitioning networks into communities.

---

1. For all edges in the network, calculate the betweenness.
2. Remove the edge with the largest value of betweenness.
3. Recalculate the edge betweenness for the edges affected by the removal.
4. Repeat from step 2 until no edges remain.

Newman and Girvan (2004) develop the approach further by considering alternative definitions of edge betweenness.

1. Geodesic edge betweenness – based on the number of shortest paths that run along an edge (as in Figure 12.3).
2. Random walk betweenness – based on the number of times that a random walk between a pair of vertices passes along a particular edge.
3. Current flow betweenness – based on the current flow in a circuit with a resistor placed on each edge.

Separate communities may not emerge until several edges have been removed, but the algorithms always produce a partition of the network into a nested set of communities, which may be represented in the form of a dendrogram. However, the choice of the best set of communities to describe the network structure requires a measure of the quality of community structure. Girvan and Newman (2002) define a quantity termed *modularity*, and select the structure with the largest value of modularity.

Consider the division of a network with vertex set V into k communities, $V_1, \ldots, V_k$ ($\bigcup_{i=1}^{k} V_k = V$). Define $A(S, T)$ as

$$A(S,T) = \sum_{i \in S,\, j \in T} A_{ij}$$

the number of edges between the two communities S and T. Then the modularity, Q, is defined as

$$Q = \sum_{i=1}^{k} \left[\frac{A(V_i, V_i)}{A(V, V)} - \left(\frac{A(V_i, V)}{A(V, V)} \right)^2 \right] \qquad (12.9)$$

where the first term in brackets in the summation, $\frac{A(V_i,V_i)}{A(V,V)}$, is the fraction of edges that connect vertices within community V_i; the term $\frac{A(V_i,V)}{A(V,V)}$ is the the fraction of edges that connect to vertices in community V_i.

Equation (12.9) may be simplified to

$$Q = \frac{1}{2m} \sum_{i=1}^{n} \sum_{j=1}^{n} \left(A_{ij} - \frac{k_i k_j}{2m} \right) \delta(C_i, C_j) \tag{12.10}$$

where k_i is the degree of node i, m is the number of edges in the network, C_i is the community to which node i belongs and $\delta(C_i, C_j) = 1$ if nodes i and j are in the same community.

Q has the property that it is bounded below by 0, when the number of within-community edges is no better than random. The maximum is attained when $Q = 1$, indicating strong community structure.

12.3.3 Modularity approaches

Girvan and Newman (2002) introduced modularity as a community quality measure to be used as a stopping criterion for their divisive algorithm (Algorithm 12.2), with high values of modularity indicating good partitions of the network into communities. More recently, there has been a class of community detection algorithms proposed that attempt to maximise modularity directly. Even for moderately sized graphs, there is a large number of ways of partitioning the graph and so exhaustive optimisation of modularity is not feasible. Therefore, approximate methods have been proposed [see Fortunato (2010) for a review].

Here, we present an approach due to Newman (2006) based on the eigendecomposition of a modularity matrix, $\boldsymbol{B}$, defined by

$$B_{ij} = A_{ij} - \frac{k_i k_j}{2m}$$

where A_{ij} are the components of the adjacency matrix and k_i is the degree of the ith node; m is the number of edges. For the partition of the network into two components, we define an index vector, $\boldsymbol{s}$, with components

$$s_i = \begin{cases} +1 & \text{if vertex } i \text{ belongs to group 1} \\ -1 & \text{if vertex } i \text{ belongs to group 2} \end{cases}$$

Then the modularity [Equation (12.10)] may be written as

$$Q = \frac{1}{4m} \boldsymbol{s}^T \boldsymbol{B} \boldsymbol{s} \tag{12.11}$$

Maximising Q subject to a normalisation constraint on the vector $\boldsymbol{s}$ ($\boldsymbol{s}^T \boldsymbol{s} = n$) gives $\boldsymbol{s}$ proportional to $\boldsymbol{u}_1$, the eigenvector of $\boldsymbol{B}$ corresponding to the largest eigenvalue.[5] However, there is a further restriction on the vector $\boldsymbol{s}$: its components are constrained to be ± 1, which means that $\boldsymbol{s}$

[5] The eigenvectors, $\boldsymbol{u}_i$, of the $n \times n$ real symmetric matrix $\boldsymbol{B}$ are orthogonal; the eigenvalues, λ_i, are real and can be positive or negative. They can be ordered so that $\lambda_1 \geq \lambda_2 \geq \cdots \geq \lambda_n$.

cannot usually be taken to be $\boldsymbol{u}_1$. Newman and Girvan (2004) report that a good approximation that works well is achieved by setting

$$s_i = \begin{cases} +1 & \text{if } u_{1i} \geq 0 \\ -1 & \text{if } u_{1i} < 0 \end{cases}$$

The simplest approach for generalising this method to more than two communities is through repeated division of the communities discovered into two, whereby each of the two communities is split separately and the modularity (computed from the full adjacency matrix of the original graph) calculated. If a partition of a 'parent' community into two 'child' communities fails to increase the modularity, then the original parent community is retained. Repeated subdivision is not an optimal approach for finding a set of communities, but works well in practice.

A modularity maximisation algorithm will partition the network into communities, but a high value of modularity is not necessarily a good indicator of community structure. The modularity maximum is an indicator of structure only if it is appreciably larger than the modularity maximum of random graphs of the same size and the same expected degree sequence [6] (Fortunato, 2010).

12.3.4 Local modularity

The approaches to community detection presented above are global in that they require knowledge of the entire structure of the network. For large graphs, such approaches become computationally demanding, if not infeasible. However, we may still wish to discover local structure in a particular region of a network, of the communities surrounding a given vertex; for example, the social network of a particular individual. This requirement has led to the development of methods for identifying local community structure in the vicinity of a seed node.

Clauset (2005) introduces a definition of *local modularity* as a measure of local community structure. Suppose that we have a subset, $\mathcal{C}$, of a graph and we have complete knowledge of the connections of all the nodes in $\mathcal{C}$. Some of the nodes in $\mathcal{C}$ will be connected only to nodes within $\mathcal{C}$; others will be connected to nodes outside $\mathcal{C}$ about which we have imperfect knowledge. Denote this set of vertices external to $\mathcal{C}$ but connected to $\mathcal{C}$ by $\mathcal{U}$ (Figure 12.4). Within the set $\mathcal{C}$, there is a set of boundary nodes, $\mathcal{B}$, which have at least one connection to nodes in $\mathcal{U}$. The local modularity, R, is defined as

$$R = \frac{I}{T}$$

where T is the number of edges with one or more endpoints in $\mathcal{B}$ and I is the number of those edges with neither endpoint in $\mathcal{U}$ ($T = 12$ and $I = 7$ in Figure 12.4). Thus, R is the fraction of boundary edges internal to the set $\mathcal{C}$ and is a measure of the sharpness of the boundary.

The steps of the Clauset algorithm are given in Algorithm 12.3. Initially, the seed node v_0 (a node about which we wish to discover further information regarding community membership) is placed in $\mathcal{C}$ and its neighbours in $\mathcal{U}$. At each step, the members of $\mathcal{U}$ are examined and the

[6] The degree sequence of an undirected graph is the monotonic, nonincreasing sequence of its vertex degrees.

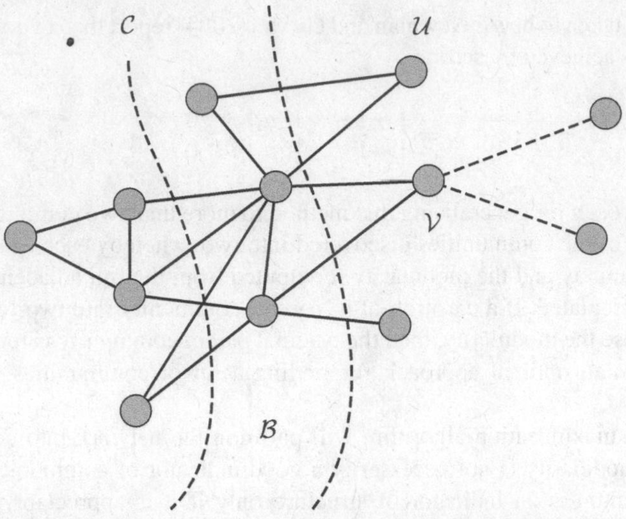

Figure 12.4 Regions of a network for the definition of local modularity. The two nodes on the right-hand side do not belong to $\mathcal{U}$ until the node v is added to $\mathcal{C}$.

node, v, that produces the largest increase in R (or the smallest decrease, since R may decrease with the addition of a node to $\mathcal{C}$) is added to the set $\mathcal{C}$ (and the set $\mathcal{B}$ if v is a boundary node of the new set $\mathcal{C}$). Newly discovered nodes (the neighbours of v not already in $\mathcal{C}$) are added to $\mathcal{U}$ (Figure 12.4). The algorithm produces a function, $R(t)$, the modularity of the community centred on v_0 at time step t and a corresponding set of nested communities. Local peaks in $R(t)$ are examined for potential community boundaries.

Algorithm 12.3 Clauset local modularity algorithm for finding nested community structure around node v_0.

Specify a value for k, the maximum community size. Specify v_0.

1. Add v_0 to $\mathcal{C}$.
2. Add the neighbours of v_0 to the set $\mathcal{U}$.
3. Set $\mathcal{B} = v_0$.
4. **While** $|\mathcal{C}| < k$ **do**
   (a) For each $v_j \in \mathcal{U}$ compute the change in R, denoted ΔR_j, if v_j were to be added to $\mathcal{C}$.
   (b) Find the v_j for which ΔR_j is a maximum.
   (c) Add that v_j to $\mathcal{C}$ and its new neighbours to $\mathcal{U}$.
   (d) Update R and the set $\mathcal{B}$.

12.3.5 Clique percolation

The algorithms presented in previous sections have sought to partition a network into disjoint communities, or to find isolated communities within a network: each vertex is assigned to a single community. The communities discovered do not overlap (apart for the case where a community is a complete subset of another in a hierarchical approach). In many practical situations, this assumption is unrealistic. Real communities do overlap: a person may belong to a community of colleagues at work, a member of a local sports group and be one of the parents of children at a local school. These communities may share several members. In this section, we consider an approach for detecting overlapping communities.

The clique percolation method (CPM) was proposed by Palla *et al.* (2005a). We begin with the following definitions:

***k*-clique:** a complete (i.e. fully connected) subgraph with k vertices (Figure 12.5).

***k*-clique adjacency:** two k-cliques are adjacent if they share $k - 1$ vertices (i.e. they differ by a singe node). Adjacent 4-cliques are shown in Figure 12.6.

***k*-clique community:** the largest connected subgraph obtained by the union of adjacent k-cliques. There are two overlapping 4-clique communities in the network of Figure 12.6, one comprising 6 vertices and the second 7 vertices.

The stages in finding the k-clique community are first to find the *maximal cliques* (the cliques that are not subgraphs of larger cliques) and to construct the $n_c \times n_c$ clique-clique overlap matrix whose (i, j) component is the number of vertices shared by clique i and clique j. For k-clique community detection, the elements of the matrix less than $k - 1$ are set to zero, and the remaining entries analysed to find connected components [Palla *et al.* (2005b) for further details].

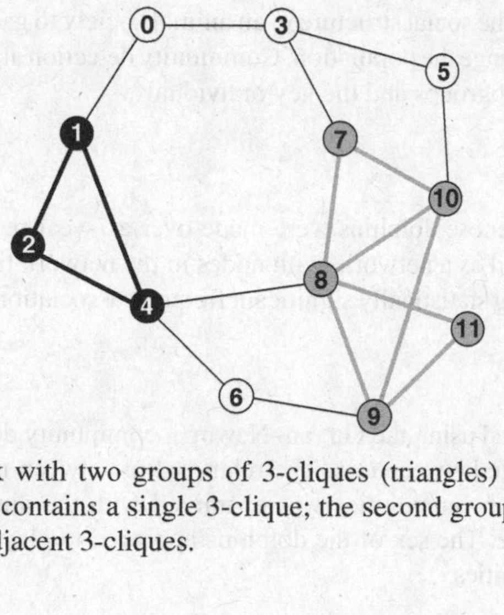

Figure 12.5 A network with two groups of 3-cliques (triangles) highlighted. One group (black nodes 1, 2 and 4) contains a single 3-clique; the second group (grey nodes 7, 8, 9, 10 and 11) contains three adjacent 3-cliques.

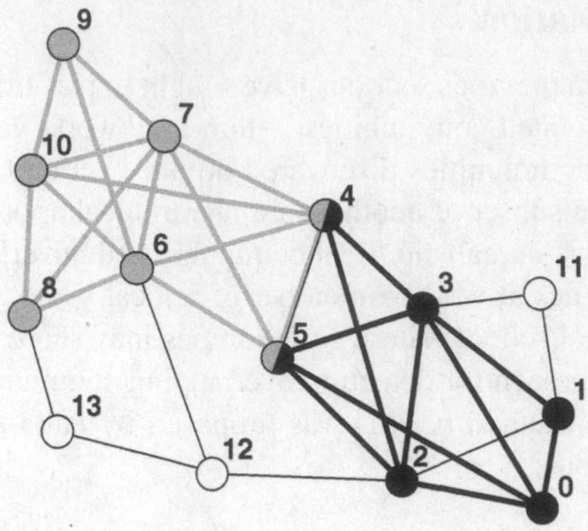

Figure 12.6 Two overlapping groups of adjacent 4-cliques. One group contains the 4-cliques (0, 1, 2, 3), (0, 2, 3, 5) and (2, 3, 4, 5); the second contains the 4-cliques (4, 5, 6, 7), (4, 6, 7, 10), (6, 7, 8, 10) and (6, 7, 9, 10).

Development of CPM to directed and weighted graphs is described by Farkas *et al.* (2007) and Palla *et al.* (2007). Downloadable software is provided by Adamcsek *et al.* (2006).

12.3.6 Example application study

The problem
The application involves the identification of communities and subcommunities within a dolphin population living off the coast of New Zealand (Lusseau and Newman, 2004).

Summary
The aim is to investigate the social structure in an animal society to gain a better understanding in order to help better manage the population. Community detection algorithms were employed to identify groups and subgroups and the key individuals.

The data
Observations of 62 bottlenose dolphins were made over a 7-year period from 1994 to 2001. The data were represented as a network, with nodes in the network being individual dolphins and the edges representing statistically significant frequent associations between dolphin pairs.

The approach
The network was dissected using the Girvan–Newman community detection algorithm based on the calculation of edge betweenness. The relationship between members of the subcommunities and between subcommunities was examined with the aim of understanding how the subcommunities arise. The sex of the dolphins appeared to play an important role in the definition of subcommunities.

Results
Two communities and four subcommunities were identified. Betweenness centrality was used to identify 'boundary individuals' who act as links between subcommunities and appear important to the social cohesion of the group.

12.3.7 Further developments

The approaches presented have largely been described in the context of undirected and unweighted networks. Development to directed, weighted (including networks with both positive and negative weights) and bipartite networks have been addressed by a number of authors including Yook *et al.* (2001) and Leicht and Newman (2008).

Alternative approaches, based on probabilistic models have been considered by Newman and Leicht (2007) (mixture models) and Lusseau *et al.* (2008) (incorporating uncertainty).

The computational issues, particularly for divisive algorithms, have been addressed by Rattigan *et al.* (2007) and Tyler *et al.* (2003).

The significance of community structure (that is, could the structure discovered have arisen by chance?) is reviewed by Fortunato (2010). Related to this is the model selection – how many communities are there in a network? A Bayesian approach is presented by Hofman and Wiggins (2008).

12.3.8 Summary

The techniques described in this section aim to find community structure within complex networks using structural information. With appropriate definitions of similarity or dissimilarity, many of the techniques of cluster analysis can be exploited, including hierarchical and spectral methods. A measure of the quality of community structure is *modularity*, and techniques that maximise modularity directly have been developed.

The application of such approaches to very large graphs raises computational issues, and a *local modularity* approach that attempts to discover the community surrounding a node of interest has been presented. Finally, the clique percolation method was introduced as an example of an approach for detecting overlapping clusters. Fortunato (2010) and Schaeffer (2007) provide thorough reviews of approaches to community detection.

12.4 Link prediction

Link prediction is the problem of predicting a link between two nodes in a network based on the existing observed links in the network and the properties of the nodes. Reasons for link prediction include (Lü and Zhou, 2010):

- to extract missing information. This may occur in cases when not all links are observed due to incomplete data collection mechanisms.
- to identify anomalous interactions. There may be links between individuals in a social network, for example, that may be unusual or unexpected.

- to evaluate models of network evolution. Here, link prediction is used to predict the future occurrence of links in an evolving network and the metrics on link prediction accuracy used to assess the performance of different models.

Link prediction may be used in many diverse application areas. Examples include:

- Social networks – predicting friendships.
- Marketing – identifying potential markets for products.
- Security – predicting anomalous links.

12.4.1 Approaches to link prediction

One of the approaches to link prediction is to view the problem as one in binary classification. The two classes comprise pairs of nodes that are linked and pairs of nodes that are not linked, taken from a network at time t, or sampled from part of a larger network. Attributes of these pairs of nodes are used as inputs to a classifier which is then tested on the network at a later time or on an unseen part of a large network. One of the difficulties with a classification approach is that there is a very large class skew, with many more examples of nonlinked pairs, and this imbalance grows as a network evolves and becomes larger. This can lead to poor model performance with the prior probability of a link usually being very small.

Another approach is to calculate the area under the receiver operating characteristic (ROC) curve (AUC, see Chapter 9) based on thresholds applied to node similarity or proximity indices. This measure is independent of the degree of class skew.

12.4.1.1 Similarity-based approaches

This approach requires a measure of similarity between pairs of nodes. It could be a structural index, defined globally or locally (Table 12.1), or one that takes account of node attributes. Let E be the set of edges in the network and U be the set of all $n(n-1)/2$ possible edges, so that the missing links are in the set $U - E$.

Independent train and test set

A training set, E^T, is randomly selected from E. This defines a graph which is used to calculate the similarity between pairs of nodes in $E - E^T$ (edges) and $U - E$ (the missing links). These similarities are ranked (low value to high) to produce a list of edges, together with their similarity and set label,

$$\{(i, j), s_{ij}, \omega_{ij}\}$$

where s_{ij} is the similarity between node i and node j and ω_{ij} is the class label for the pair (i, j) (either edge or no edge).

Cross-validation

The set of edges, E, is randomly partitioned into k (for example, a value of 10) subsets. $k - 1$ subsets are used as training data and the similarity calculated on the pairs of edges in the remaining set and on a subset of $U - E$ (also partitioned into k subsets). This is repeated

Algorithm 12.4 Link prediction using an independent train and test set.

1. Select a subset of edges, E^T, from the set E.
2. Calculate the similarity between all pairs of nodes defined by the edges in the sets $E - E^T$ and $U - E$ using the graph defined by the set of edges in E^T.
3. Rank all edges based on the similarity.
4. For a given threshold, s, calculate the proportion of edges in $E - E^T$ greater than the threshold [and denoted by $E(s)$] and the proportion in $U - E$ greater than the threshold [denoted by $U(s)$].
5. Plot $E(s)$ against $U(s)$ as s varies and calculate the area under the curve.

Algorithm 12.5 Link prediction with cross-validation assessment.

1. Partition the set of edges, E, into k subsets, $E_l, l = 1, \ldots, k$. Partition the set $U - E$ into k subsets, $U_l, l = 1, \ldots, k$.
2. For $l = 1, \ldots, k$
   - Calculate the similarity between all pairs or nodes in the subsets E_l and U_l using the graph defined by the set of edges in remaining subsets $E_i, i \neq l$.
3. Rank all edges based on the similarity.
4. For a given threshold, s, calculate the proportion of edges in $E - E^T$ greater than the threshold [and denoted by $E(s)$] and the proportion in $U - E$ greater than the threshold [denoted by $U(s)$].
5. Plot $E(s)$ against $U(s)$ as s varies and calculate the area under the curve.

for each set of $k - 1$ subsets in turn. This again produces an ordered list of pairs of nodes, with similarity and label.

Area under the ROC curve (AUC)

For a given threshold value for the similarity, s, the proportion of members in the test set or the cross-validation set with similarity greater than the threshold is plotted against the proportion in the set $U - E$ that is greater than the threshold, as the threshold, s, is varied. This produces an ROC curve (see Algorithms 12.4 and 12.5). The AUC is a measure calculated from it that is independent of class skew (see Chapter 9). If the similarity is a monotonic function of the probability that an edge is present[7] given the pair of nodes, then this approach produces the ROC of the Bayesian classifier without explicitly calculating probability density functions.

[7] That is, the greater the similarity then the greater the probability that an edge is present (or is likely to be present as the network evolves).

12.4.2 Example application study

The problem

This application involves the detection of anomalous email as an example of a security informatics application, where it is important to detect unusual communication patterns in various channels of communication (Huang and Zeng, 2006).

Summary

An email communication anomaly detection framework is evaluated that is based on the development of probabilistic models of email traffic in directed and weighted networks. These models are then used to identify anomalous email traffic.

The data

The data comprise the Enron email corpus, a large-scale email collection from a real organisation over a period of more than three and a half years (Shetty and Adibi, 2005). The dataset contained 252 759 emails from 151 Enron employees. Only emails between these 151 employees were considered (emails to/from contacts outside this set were excluded).

The approach

Preprocessing reduced the dataset to 40 489 emails. Data in the time periods $t-1, t-2, \ldots, t-g$ were used to make a prediction on possible email communications (i.e. possible links) in the time period t. The time periods were taken to be 1 week duration. For the data, a directed and weighted graph describing the email traffic between individuals was generated. Three link prediction approaches, adapted to handle weighted and directed networks, were implemented:

- Preferential attachment – based on sender degree and recipient degree.
- Spreading activation – this uses the idea that if there is a link $A \rightarrow B$ and $B \rightarrow C$, then $A \rightarrow C$ might be likely.
- Generative model – a probabilistic process is assumed to generate the data and its parameters are estimated using the EM algorithm.

Emails of the previous 8 weeks (i.e. $g = 8$) were used to predict the links of the current week.

Results

The generative model gave a better fit to that data than the other two models: individuals sending emails to a large number of senders do not necessarily send to people who themselves either send or receive a large number of emails (preferential attachment assumption); the transitivity property – if A sends to B and B sends to C then A sends to C (spreading activation) – is not prominent in emails. For the generative model, there were weeks when there was a significant departure of data from the prediction. Further investigation revealed unusual email patterns in those weeks.

12.4.3 Further developments

Link prediction is an active area of research within a broader area of link data mining (Getoor and Diehl, 2005) and only a simple approach based on a measure of similarity in unweighted and undirected networks has been presented in this section. Further approaches include the extension of similarity measures to directed and weighted networks, the development of methods to predict both the presence of an edge and the weight on the edge and the development of probabilistic models.

12.5 Application studies

Examples of application studies of community detection algorithms include:

- a study of a network of mobile phone users (Palla *et al.*, 2007) using the clique percolation method;
- identification of protein complexes and functional modules in molecular networks (Spirin and Mirny, 2003);
- hierarchical decomposition of metabolic networks (Holme *et al.*, 2003), using a modification of the Girvan–Newman algorithm for bipartite networks;
- understanding scientific communication in the citation patterns of journals (Rosvall and Bergstrom, 2008);
- analysis of the bidding behaviour of users on eBay using a community detection algorithm that maximises modularity (Jin *et al.*, 2007);
- identification of community structure in the networks of committee and subcommittee assignments in the United States House of Representatives using methods of hierarchical clustering and quantifying using modularity (Porter *et al.*, 2007).

12.6 Summary and discussion

There are many datasets that arise through interactions between pairs or groups of individuals. Data of this type can be represented using the mathematical concept of a graph where the nodes represent individuals (people, computers, shops, purchasers, telephone subscribers, web sites, and so on) and the edges in the graph relate to some 'transaction' between them (purchase, telephone call, email, meeting, for example). The networks from which the graphs are abstracted are dynamic – the number of nodes and the degree of interaction between them change with time. This chapter has reviewed the properties of such networks and focused on the application of data analysis methods to the data arising from the networks. In particular, two main areas of interest have been presented: community detection, which largely employs unsupervised pattern recognition techniques to discover communities in networks; and link prediction, which is supervised. Modelling the dynamics of evolving networks is outside the scope of this chapter.

Complex networks is an area where there have been many developments over the past decade and is currently receiving considerable research attention. There is great scope to apply the tools of statistical pattern recognition to this domain.

12.7 Recommendations

For community detection, many of the methods of cluster analysis may be employed given suitable definitions of the similarity between individuals.

- Hierarchical methods are a good starting point. They are straightforward to implement using the similarity measures of Table 12.1. Spectral methods have achieved success and fast implementations developed for large networks.
- If the community in the vicinity of a specific node is required, and it is not possible to analyse the complete graph due to its large size, than a local method should be considered (Clauset, 2005).
- For overlapping clusters, clique percolation is a popular method.

12.8 Notes and references

Good introductions to complex networks are provided in the books by Newman (2010) and Easley and Kleinberg (2010); see also the reviews by Albert and Barabási (2002) and the graph mining review by Chakrabarti and Faloutsos (2006). Within the context of community detection, a recent extensive review is that of Fortunato (2010); see also Schaeffer (2007). A survey of link prediction in complex networks is provided by Lü and Zhou (2010).

Exercises

1. Show that the modularity defined by Equation (12.9) can be written in the form given by Equation (12.10).
2. Derive the solution for $\boldsymbol{s}$ that maximises the modularity Q [Equation (12.11)] as proportional to the largest eigenvector of $\boldsymbol{B}$.
3. Show that the vector $(1, 1, \ldots, 1)$ is an eigenvector of the graph Laplacian. What is the corresponding eigenvalue?
4. Show that the normalised graph Laplacians [Equations (12.4) and (12.5)] are positive semidefinite.

13

Additional topics

Issues in classifier design are addressed. The first concerns model selection – choosing the appropriate type and complexity of classifier. The second concerns problems with data – mixed variables, outliers and missing values. Finally, the Vapnik-Chervonenkis dimension, a measure of the capacity of a learning algorithm, is introduced. This can provide guidance on classifier design.

13.1 Model selection

In many areas of pattern recognition we are faced with the problem of model selection; that is, how complex should we allow our model to be, measured perhaps in terms of the number of free parameters to estimate? The optimum complexity of the model depends on the quantity and the quality of the training data. If we choose a model that is too complex, then we may be able to model the training data very well (and also any noise on the training data), but it is likely to have poor generalisation performance on unseen data, drawn from the same distribution as the training set was drawn from (thus the model *over-fits* the data). If the model is not complex enough, then it may fail to model structure in the data adequately. Model selection is inherently a part of the process of determining optimum model parameters. In this case, the complexity of the model is a parameter to determine. As a consequence, many model selection procedures are based on optimising a criterion that penalises a goodness of fit measure by a model complexity measure.

The problem of model selection arises with many of the techniques described in this book. Some examples are:

1. How many components in a mixture model should be chosen to model the data adequately? (Chapter 2)
2. How is the optimum tree structure found in a decision-tree approach to discrimination? This is an example in determining an appropriate number of basis functions in an

Statistical Pattern Recognition, Third Edition. Andrew R. Webb and Keith D. Copsey.

expansion, where the basis functions in this case are hyperrectangles, with sides parallel to the coordinate axes. (Chapter 7)

3. How many hidden units do we take in a multilayer perceptron, or centres in a radial basis function network? (Chapter 6)
4. How many features do we select on our model? (Chapter 10)
5. How many clusters describe the dataset? (Chapter 11)

In this section, we give some general procedures that have been widely used for model selection within classification problems. Anders and Korn (1999) examine model selection procedures in the context of neural networks, comparing five strategies in a simulation study.

13.1.1 Separate training and test sets

In the separate training and test set approach, both training and test sets are used for model selection. A test set used in this way is often termed the *validation set*. The training set is used to optimise a goodness of fit criterion and the performance recorded on the validation set. As the complexity of the model is increased, it is expected that performance on the training set will improve (as measured in terms of the goodness of fit criterion), while the performance on the validation set will begin to deteriorate beyond a certain model complexity.

This is onc approach used for the classification and regression tree training models (Breiman *et al.*, 1984). However, the preferred approach is to grow a large tree to over-fit the data, and prune it back until the performance on a separate validation set fails to improve. A similar approach may be taken for neural network models (Reed, 1993).

The separate training and validation set procedure may also be used as part of the optimisation process for a model of a given complexity, particularly when the optimisation process is carried out iteratively, as in a nonlinear optimisation scheme. The values of the parameters of the model are chosen, not as the ones that minimise the given criterion on the training set, but those for which the validation set performance is best. Thus, as training proceeds, the performance on the validation set is monitored (by evaluating the goodness of fit criterion using the validation data) and training is terminated when the validation set performance begins to deteriorate.

Note that in this case the validation set is not an independent test set that may be used for error rate estimation. It is part of the training data. A third dataset is required for an independent estimate of the error rate.

13.1.2 Cross-validation

Cross-validation as a method of error rate estimation was described in Chapter 9. It is a simple idea. The dataset of size n samples is partitioned into two parts. The model parameters are estimated using one set (by minimising some optimisation criterion) and the goodness-of-fit criterion evaluated on the second set. The usual version of cross-validation is the simple leave-one-out method in which the second set consists of a single sample. The cross-validation estimate of the goodness-of-fit criterion, CV, is then the average over all possible training sets of size $n - 1$.

As a means of determining an appropriate model, the cross-validation error is calculated for each member of the family of candidate models, $\{M_k, k = 1, \ldots, K\}$, and the model $M_{\hat{k}}$ chosen where

$$\mathrm{CV}(\hat{k}) \leq CV(k) \text{ for all } k$$

Cross-validation tends to over-fit when selecting a correct model; that is, it chooses an over-complex model for the dataset. There is some evidence that multifold cross-validation, when $d > 1$ samples are deleted from the training set, does better than simple leave-one-out cross-validation for model selection purposes (Zhang, 1993). Using cross-validation to select a classification method is discussed further by Schaffer (1993).

13.1.3 The Bayesian viewpoint

In the Bayesian approach, prior knowledge about models, M_k, and parameters, $\boldsymbol{\theta}_k$, is incorporated into the model-selection process. Given a dataset $\boldsymbol{X}$, the distribution of the models may be written using Bayes' theorem as

$$\begin{aligned} p(M_k|\boldsymbol{X}) &\propto p(\boldsymbol{X}|M_k)\, p(M_k) \\ &= p(M_k) \int p(\boldsymbol{X}|M_k, \boldsymbol{\theta}_k)\, p(\boldsymbol{\theta}_k|M_k)\, d\boldsymbol{\theta}_k \end{aligned}$$

and we are therefore required to specify a prior distribution $p(M_k, \boldsymbol{\theta}_k)$. If a single model is required, we may choose $M_{\hat{k}}$ where

$$p(M_{\hat{k}}|\boldsymbol{X}) \geq p(M_k|\boldsymbol{X}) \text{ for all } k$$

However, in a Bayesian approach, all models are considered. Over-complex models are penalised since their ability to model a wide range of datasets (through their complexity) results in the assignment of low likelihood to any single dataset. This effect is referred to as Occam's razor. Additionally, the prior model probabilities $p(M)$ can provide a regularisation effect, penalising over-complex models. Model averaging as an approach to handling model uncertainty was discussed in Chapter 8 as an example of ensemble methods.

13.1.4 Akaike's Information Criterion

Akaike (1973, 1974, 1977, 1981, 1985) used ideas from information theory to suggest a model selection criterion. A good introduction to the general principles is given by Bozdogan (1987; see also Sclove, 1987).

Suppose that we have a family of candidate models $\{M_k, k = 1, \ldots, K\}$, with the kth model depending on a parameter vector $\boldsymbol{\theta}_k = (\theta_{k,1}, \ldots, \theta_{k,\epsilon(k)})^T$, where $\epsilon(k)$ is the number of free parameters of model k. Then the information criterion proposed by Akaike (AIC) is given by

$$\mathrm{AIC}(k) = -2\log[L(\hat{\boldsymbol{\theta}}_k)] + 2\epsilon(k) \tag{13.1}$$

where $\hat{\boldsymbol{\theta}}_k$ is the maximum likelihood estimate of $\boldsymbol{\theta}_k$, $L[.]$ is the likelihood function, and the model M_k is chosen as that model, $M_{\hat{k}}$, where

$$\text{AIC}(\hat{k}) \leq AIC(k) \text{ for all } k$$

Equation (13.1) represents the unbiased estimate of minus twice the expected log likelihood,

$$-2\text{E}[\log(p(\boldsymbol{X}_n|\boldsymbol{\theta}_k))]$$

where $\boldsymbol{X}_n$ is the set of observations $\{\boldsymbol{x}_1, \ldots, \boldsymbol{x}_n\}$, characterised by $p(\boldsymbol{x}|\boldsymbol{\theta})$.

There are a number of difficulties in applying (13.1) in practice. The main problem is that the correction for the bias of the log-likelihood, $\epsilon(k)$, is only valid asymptotically. Various other corrections have been proposed that have the same asymptotic performance, but different finite sample performance.

13.1.5 Minimum description length

The *minimum description length* (MDL) is another way to incorporate model complexity in an information criterion. This is illustrated in Figure 13.1. Person A and Person B are given a dataset X, a set of measurements $\boldsymbol{x}_i$ on n individuals. Person A is also given the class labels z_i of each pattern and develops a classification model or *Theory*, T, to summarise the relationship between $\boldsymbol{x}$ and z.

Person A now wants to transmit to Person B the class labels of the patterns, $\boldsymbol{x}_i$. Person A could transmit the set of labels of all the patterns, or transmit the model in a compact form. However, the model may not be perfect, so if he/she transmits the model, he/she must also transmit information concerning those data samples $\boldsymbol{x}_i$ incorrectly labelled by the model.

If the cost of encoding the model is $L(M)$ and the cost of encoding the training set given the model is $L(X|M)$ (*i.e.* the cost of encoding the mislabelled patterns), then the total *description length* is

$$L(M) + L(X|M)$$

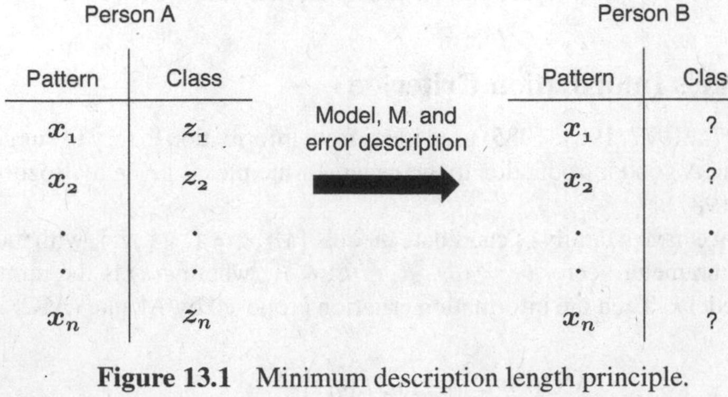

Figure 13.1 Minimum description length principle.

The MDL principle states that we seek the model that minimises the total description length.

A difficulty is in encoding the model in the most efficient way. It depends on the shared knowledge of the people A and B. Encoding the errors is usually more straightforward, but there are sophisticated ways of encoding errors (which reduces the need for a model). For further details, see Grünwald (2007).

13.2 Missing data

Many classification techniques assume that we have a set of observations with measurements made on each of p variables. However, missing values are common. For example, questionnaires may be returned incomplete; in an archaeological study, it may not be possible to make a complete set of measurements on an artifact because of missing parts; in a medical problem, a complete set of measurements may not be made on a patient, perhaps owing to forgetfulness by the physician or being prevented by the medical condition of the patient.

How missing data are handled depends on a number of factors: how much is missing; why data are missing; whether the missing values can be recovered; are values missing in both the design and test set? There are several approaches to this problem.

1. We may omit all incomplete vectors from our analysis. This may be acceptable in some circumstances, but not if there are many observations with missing values. For example, in the head injury study of Titterington *et al.* (1981) referred to in Chapter 2, 206 out of the 500 training patterns and 199 out of the 500 test patterns have at least one observation missing. Neglecting an observation because perhaps one out of 100 variables has not been measured means that we are throwing away potentially useful information for classifier design. Also, in an incomplete observation, it may be that the variables that have been measured are the important ones for classification anyway.

2. We may use all the available information. The way that we would do this depends on the analysis that we are performing. In estimating means and covariances, we would use only those observations for which measurements have been made on the relevant variables. Thus, the estimates would be made on different numbers of samples. This can give poor results and may lead to covariance matrices that are not positive definite. Other approaches must be used to estimate principal components when data are missing (Jackson, 1991). In clustering, we would use a similarity measure that takes missing values into account. In density estimation using an independence assumption, the marginal density estimates will be based on different numbers of samples.

3. We may substitute for the missing values and proceed with our analysis as if we had a complete dataset.

There are many approaches to missing value estimation, varying in sophistication and computational complexity.

The simplest and perhaps the crudest method is to substitute mean values of the corresponding components. This has been used in many studies. In a supervised classification problem, class means may be substituted for the missing values in the training set and the sample mean for the missing values in the test set since in this case we do not know the class.

A thorough treatment of missing data in statistical analysis is given by Little and Rubin (1987). A review in the context of regression is given by Little (1992) and a discussion in the classification context by Liu *et al.* (1997).

13.3 Outlier detection and robust procedures

We now consider the problem of detecting outliers in multivariate data. This is one of the aims of *robust statistics*. Outliers are observations that are not consistent with the rest of the data. They may be genuine observations [termed *discordant observations* by Beckman and Cook (1983)] that are surprising to the investigator. Perhaps they may be the most valuable, indicating certain structure in the data that shows deviations from normality. Alternatively, outliers may be *contaminants*, errors caused by copying and transferring the data. In this situation, it may be possible to examine the original data source and correct for any transcription errors.

In both of the above cases, it is important to detect the outliers and to treat them appropriately. Many of the techniques we have discussed in this book are sensitive to outlying values. If the observations are atypical on a single variable, it may be possible to apply univariate outlier detection methods to the variable. Outliers in multivariate observations can be difficult to detect, particularly when there are several outliers present. A classical procedure is to compute the Mahalanobis distance for each sample $\boldsymbol{x}_i$ $(i = 1, \ldots, n)$

$$D_i = \left\{ (\boldsymbol{x}_i - \boldsymbol{m})^T \hat{\boldsymbol{\Sigma}}^{-1} (\boldsymbol{x}_i - \boldsymbol{m}) \right\}^{\frac{1}{2}}$$

where $\boldsymbol{m}$ is the sample mean and $\hat{\boldsymbol{\Sigma}}$ the sample covariance matrix. Outliers may be identified as those samples yielding large values of the Mahalanobis distance. This approach suffers from two problems in practice:

1. *Masking*. Multiple outliers in a cluster will distort $\boldsymbol{m}$ and $\hat{\boldsymbol{\Sigma}}$, attracting $\boldsymbol{m}$ and inflating $\hat{\boldsymbol{\Sigma}}$ in their direction, thus giving lower values for the Mahalanobis distance.
2. *Swamping*. This refers to the effect that a cluster of outliers may have on some observations that are consistent with the majority. The cluster could cause the covariance matrix to be distorted so that high values of D are found for observations that are not outliers.

One way of overcoming these problems is to use robust estimates for the mean and covariance matrices. Different estimators have different *breakdown points*, the fraction of outliers that they can tolerate. Rousseeuw (1985) has proposed a *minimum volume ellipsoid* (MVE) estimator that has a high breakdown point of approximately 50%. It can be computationally expensive, but approximate algorithms have been proposed.

Outlier detection and robust procedures have represented an important area of research, investigated extensively in the statistical literature. Chapter 1 of Hampel *et al.* (1986) gives a good introduction and background of robust procedures. Robust estimates of mean and covariance matrices are reviewed by Rocke and Woodruff (1997); further procedures for the detection of outliers in the presence of appreciable masking are given by Atkinson and Mulira (1993). Krusińska (1988) reviews robust methods within the context of discrimination.

13.4 Mixed continuous and discrete variables

In many areas of pattern recognition involving multivariate data, the variables may be of mixed type, comprising perhaps continuous, ordered categorical, unordered categorical and binary variables. If the discrete variable is ordered, and it is important to retain this information, then the simplest approach is to treat the variable as continuous and to use a technique developed for multivariate continuous data. Alternatively, a categorical variable with k states can be coded as $k-1$ dummy binary variables. All take the value zero except the jth if the observed categorical variable is in the jth state, $j = 1, \ldots, k-1$. All are zero if the variable is in the kth state. This allows some of the techniques developed for mixtures of binary and continuous variables to be used (with some modifications, since not all combinations of binary variables are observable).

The above approaches attempt to distort the data to fit the model. Alternatively, we may apply existing methods to mixed variable data with little or no modification. These include

1. nearest-neighbour methods (Chapter 4) with a suitable choice of metric;
2. independence models where each univariate density estimate is chosen to be appropriate for the particular variable (Chapter 4);
3. kernel methods using product kernels, where the choice of kernel depends on the variable type (Chapter 4);
4. dependence tree models and Bayesian networks where the conditional densities are modelled appropriately (Chapter 4);
5. recursive partitioning methods such as CART and MARS (Chapter 7);
6. rule induction approaches (Chapter 7).

The *location model*, introduced by Olkin and Tate (1961), was developed specifically with mixed variables in mind and applied to a two-class discriminant analysis problem by Chang and Afifi (1974). Consider the problem of classifying a vector $\boldsymbol{v}$ which may be partitioned into two parts, $\boldsymbol{v} = (\boldsymbol{z}^T, \boldsymbol{y}^T)^T$, where $\boldsymbol{z}$ is a vector of r binary variables and $\boldsymbol{y}$ is a vector of p continuous variables. The random vector $\boldsymbol{z}$ gives rise to 2^r different cells. We may order the cells such that, given a measurement $\boldsymbol{z}$, the cell number of $\boldsymbol{z}$, cell$(\boldsymbol{z})$, is given by

$$\text{cell}(\boldsymbol{z}) = 1 + \sum_{i=1}^{r} z_i 2^{i-1}$$

The location model assumes a multivariate normal distribution for $\boldsymbol{y}$, whose mean depends on the state of $\boldsymbol{z}$ and the class from which $\boldsymbol{v}$ is drawn. It also assumes that the covariance matrix is the same for both classes and for all states. Thus, given a measurement $(\boldsymbol{z}, \boldsymbol{y})$ such that $m = \text{cell}(\boldsymbol{z})$, the probability density for class ω_j $(j = 1, 2)$ is

$$p(\boldsymbol{y}|\boldsymbol{z}) = \frac{1}{(2\pi)^{p/2}|\boldsymbol{\Sigma}|}\exp\left\{-\frac{1}{2}\left(\boldsymbol{y}-\boldsymbol{m}_j^m\right)^T \boldsymbol{\Sigma}^{-1}\left(\boldsymbol{y}-\boldsymbol{m}_j^m\right)\right\}$$

Thus, the means of the distribution depend on z and the class. Then, if the probability of observing z in cell m for class ω_j is p_{jm}, then $\boldsymbol{v}$ may be assigned to ω_1 if

$$\left(\boldsymbol{m}_1^m - \boldsymbol{m}_2^m\right)^T \boldsymbol{\Sigma}^{-1}\left(\boldsymbol{y} - \frac{1}{2}\left(\boldsymbol{m}_1^m + \boldsymbol{m}_2^m\right)\right) \geq \log(p_{2m}/p_{1m})$$

The maximum likelihood estimates for the parameters p_{jm}, $\boldsymbol{m}_j^m$ and $\boldsymbol{\Sigma}$ are

$$\hat{p}_{jm} = \frac{n_{jm}}{n}$$

$$\hat{\boldsymbol{m}}_j^m = \sum_{i=1}^{n} v_{imj} \frac{1}{n_{jm}} \boldsymbol{y}_i$$

$$\boldsymbol{\Sigma} = \frac{1}{n} \sum_{j=1}^{2} \sum_{m=1}^{k} \sum_{i=1}^{n} v_{imj} \left(\boldsymbol{y}_i - \hat{\boldsymbol{m}}_j^m\right)\left(\boldsymbol{y}_i - \hat{\boldsymbol{m}}_j^m\right)^T$$

where $v_{imj} = 1$ if $\boldsymbol{y}_i$ is in cell m of class ω_j, 0 otherwise; and n_{jm} is the number of observations in cell m of class ω_j equal to $\sum_i v_{imj}$.

If the sample size n is very large relative to the number of cells, then these naïve estimates may be sufficient. However, in practice there will be too many parameters to estimate. Some of the cells may not be populated giving poor estimates for $\hat{p}_{jm}$ and no estimates for $\hat{\boldsymbol{m}}_j^m$. There have been several developments of the basic approach. For a review see Krzanowski (1993).

13.5 Structural risk minimisation and the Vapnik-Chervonenkis dimension

13.5.1 Bounds on the expected risk

There are general bounds in statistical learning theory (Vapnik, 1998) that govern the relationship between the capacity of a learning system and its performance and thus can provide some guidance in the design of such systems.

We assume that we have a training set of independently and identically distributed samples, $\{(\boldsymbol{x}_i, y_i), i = 1, \ldots, n\}$ drawn from a distribution $p(\boldsymbol{x}, y)$. We wish to learn a mapping $\boldsymbol{x} \rightarrow y$ and we have a set of possible functions (classifiers) indexed by parameters $\boldsymbol{\alpha}$, namely $f(\boldsymbol{x}; \boldsymbol{\alpha})$. A particular choice for $\boldsymbol{\alpha}$ results in a particular classifier or *trained machine*. We would like to choose $\boldsymbol{\alpha}$ to minimise the classification error. We assume a two-class problem. If y_i takes the value $+1$ for patterns in class ω_1 and -1 for patterns in class ω_2, then the expected value of the test error (the true error – see Chapter 9) is

$$\mathcal{R}(\boldsymbol{\alpha}) = \frac{1}{2} \int |y - f(\boldsymbol{x}; \boldsymbol{\alpha})| \, p(\boldsymbol{x}, y) dx dy$$

This is sometimes termed the *expected risk* (note that differs from the definition of risk in Chapter 1). This is not known in general, but we may estimate it based on a training set, to give the *empirical risk*,

$$\mathcal{R}_K(\boldsymbol{\alpha}) = \frac{1}{2n}\sum_{i=1}^{n}|y_i - f(\boldsymbol{x}_i; \boldsymbol{\alpha})|$$

For any value of η, $0 \leq \eta \leq 1$, then the following bound of statistical learning theory holds (Vapnik, 1998)

$$\mathcal{R}(\boldsymbol{\alpha}) \leq \mathcal{R}_K + \sqrt{\frac{h(\log(2n/h)+1) - \log(\eta/4)}{n}} \tag{13.2}$$

with probability $1 - \eta$, where h is a non-negative integer called the *Vapnik–Chervonenkis (VC) dimension*. The first term on the right-hand side of the inequality above depends on the particular function f chosen by the training procedure. The second term, the *VC confidence*, depends on the class of functions.

13.5.2 The VC dimension

The VC dimension is a property of the set of functions $f(\boldsymbol{x}; \boldsymbol{\alpha})$. If a given set of m points can be labelled in all possible 2^m ways using the functions $f(\boldsymbol{x}; \boldsymbol{\alpha})$, then the set of points is said to be *shattered* by the set of functions. That is, for any labelling of the set of points, there is a function $f(\boldsymbol{x}; \boldsymbol{\alpha})$ that correctly classifies the patterns.

The VC dimension of a set of functions is defined as the maximum number of training points that can be shattered. Note that if the VC dimension is m, then there is at least one set of m points that can be shattered, but not necessarily every set of m points can be shattered. For example, if $f(\boldsymbol{x}; \boldsymbol{\alpha})$ is the set of all lines in the plane, then every set of two points can be shattered, and most sets of three (Figure 13.2), but no sets of four points can be shattered by a linear model. Thus the VC dimension is 3. More generally, the VC dimension of a set of hyperplanes in r-dimensional Euclidean space is $r + 1$.

The inequality (13.2) shows that the risk may be controlled through a balance of optimising a fit to the data and the capacity of functions used in learning. In practice, we would consider sets of models, f, with each set of a fixed VC dimension. For each set, minimise the empirical

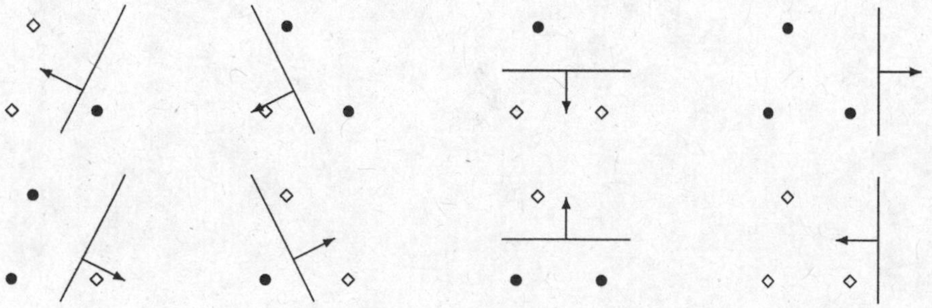

Figure 13.2 Shattering three points in two dimensions. There are $2^3 = 8$ possible labellings.

risk and choose the model over all sets for which the sum of the empirical risk and VC confidence is a minimum. However, the inequality (13.2) is only a guide. There may be models with equal empirical risk but with different VC dimensions. The one with higher VC dimension does not necessarily have poorer performance. A nearest-neighbour classifier has zero empirical risk (any labelling of a set of points will be correctly classified) and infinite VC dimension.

References

S. Abe. *Support Vector Machines for Pattern Classification*. Springer, second edition, 2010.

T. Abeel, T. Helleputte, Y. Van de Peer, P. Dupont, and Y. Saeys. Robust biomarker identification for cancer diagnosis with ensemble feature selection methods. *Bioinformatics*, 26(3):392–398, 2010.

I.S. Abramson. On bandwidth variation in kernel estimates – a square root law. *The Annals of Statistics*, 10:1217–1223, 1982.

Y.S. Abu-Mostafa, A.F. Atiya, M. Magdon-Ismail, and H. White, editors. *IEEE Transactions on Neural Networks*, 12(4): 2001. Special issue on 'Neural Networks in Financial Engineering'.

B. Adamcsek, G. Palla, I.J. Farkas, I. Derényi, and T. Vicsek. CFinder: locating cliques and overlapping modules in biological networks. *Bioinformatics*, 22(8):1021–1023, 2006.

N.M. Adams and D.J. Hand. Comparing classifiers when the misallocation costs are uncertain. *Pattern Recognition*, 32:1139–1147, 1999.

N.M. Adams and D.J. Hand. Improving the practice of classifier performance assessment. *Neural Computation*, 12:305–311, 2000.

W. Adler and B. Lausen. Bootstrap estimated true and false positive rates and ROC curves. *Computational Statistics and Data Analysis*, 53:718–729, 2009.

S. Aeberhard, D. Coomans, and O. de Vel. Improvements to the classification performance of RDA. *Chemometrics*, 7:99–115, 1993.

S. Aeberhard, D. Coomans, and O. de Vel. Comparative analysis of statistical pattern recognition methods in high dimensional settings. *Pattern Recognition*, 27(8):1065–1077, 1994.

A. Agresti. *Categorical Data Analysis*. Wiley Series in Probability and Mathematical Statistics. John Wiley & Sons, Ltd, 1990.

F.K. Ahmad, N. Md. Norwawi, S. Denis and N.H. Othman. A review of feature selection techniques via gene expression profiles – in *International Symposium on Information Technology*, pages 1–7, 2008.

J. Aitchison, J.D.F. Habbema, and J.W. Kay. A critical comparison of two methods of statistical discrimination. *Applied Statistics*, 26:15–25, 1977.

H. Akaike. Information theory and an extension of the maximum likelihood principle. In B.N. Petrov and B.F. Csaki, editors, *Second International Symposium on Information Theory*, pages 267–281. Academiai Kiado, 1973.

H. Akaike. A new look at the statistical model identification. *IEEE Transactions on Automatic Control*, 19:716–723, 1974.

H. Akaike. On entropy maximisation principle. In P.R. Krishnaiah, editor, *Proceedings of the Symposium on Applications of Statistics*, pages 27–47. North Holland, 1977.

H. Akaike. Likelihood of a model and information criteria. *Journal of Econometrics*, 16:3–14, 1981.

H. Akaike. Prediction and entropy. In A.C. Atkinson and S.E. Fienberg, editors, *A Celebration of Statistics*, pages 1–24. Springer-Verlag, 1985.

M. Al-Alaoui. A new weighted generalized inverse algorithm for pattern recognition. *IEEE Transactions on Computers*, 26(10):1009–1017, 1977.

M. Aladjem. Parametric and nonparametric linear mappings of multidimensional data. *Pattern Recognition*, 24(6):543–553, 1991.

M. Aladjem and I. Dinstein. Linear mappings of local data structures. *Pattern Recognition Letters*, 13:153–159, 1992.

A. Albert and E. Lesaffre. Multiple group logistic discrimination. *Computers and Mathematics with Applications*, 12A(2):209–224, 1986.

J. Albert. *Bayesian Computation with R*. Springer, 2009.

R. Albert and A.-L. Barabási. Statistical mechanics of complex networks. *Review of Modern Physics*, 74(1):47–97, 2002.

C. Aliferis, I. Tsamardinos, and A. Statnikov. Large-scale feature selection using Markov blanket induction for the prediction of protein-drug binding. Technical Report DSL TR-02-06, Department of Biomedical Informatics, Vanderbilt University, Nashville, 2002.

M.K.S. Alsmadi, K.B. Omar, and S.A. Noah. Back propagation algorithm: the best algorithm among the multi-layer perceptron algorithm. *IJCSNS International Journal of Computer Science and Network Security*, 9(4):378–383, 2009.

M.R. Anderberg. *Cluster Analysis for Applications*. Academic Press, 1973.

U. Anders and O. Korn. Model selection in neural networks. *Neural Networks*, 12:309–323, 1999.

J.A. Anderson. Diagnosis by logistic discriminant function: further practical problems and results. *Applied Statistics*, 23:397–404, 1974.

J.A. Anderson. Logistic discrimination. In P.R. Krishnaiah and L.N. Kanal, editors, *Handbook of Statistics*, volume 2, pages 169–191. North Holland, 1982.

J.J. Anderson. Normal mixtures and the number of clusters problem. *Computational Statistics Quarterly*, 2:3–14, 1985.

H.C. Andrews. *Introduction to Mathematical Techniques in Pattern Recognition*. Wiley Interscience, 1972.

C. Andrieu, N. De Freitas, and Doucet A. Robust full Bayesian learning for radial basis networks. *Neural Computation*, 13(10):2359–2407, 2001.

C. Andrieu and A. Doucet. Joint Bayesian model selection and estimation of noisy sinusoids via reversible jump MCMC. *IEEE Transactions on Signal Processing*, 47(10):2667–2676, 1999.

C. Andrieu, A. Doucet, and R. Holenstein. Particle Markov chain Monte Carlo methods. *Journal of the Royal Statistical Society Series B*, 72(3):269–342, 2010.

C. Andrieu and J. Thoms. A tutorial on adaptive MCMC. *Statistics and Computing*, 18(4):343–373, 2008.

S.P. Andrieu, N. de Freitas, A. Doucet, and M.I. Jordan. An introduction to MCMC for machine learning. *Machine Learning*, 50:5–43, 2003.

A. Annest, R.E. Bumgarner, A.E. Raftery, and K.Y. Yeung. Iterative Bayesian model averaging: a method for the application of survival analysis to high-dimensional microarray data. *BMC Bioinformatics*, 10(1):72, 2009.

C. Apté, R. Sasisekharan, S. Seshadri, and S.M. Weiss. Case studies of high-dimensional classification. *Journal of Applied Intelligence*, 4:269–281, 1994.

M.S. Arulampalam, S. Maskell, Gordon N., and T. Clapp. A tutorial on particle filters for online nonlinear/non-Gaussian Bayesian tracking. *IEEE Transactions on Signal Processing*, 50(2):174–188, 2002.

T. Ashikaga and P.C. Chang. Robustness of Fisher's linear discriminant function under two-component mixed normal models. *Journal of the American Statistical Association*, 76:676–680, 1981.

A.C. Atkinson and H.-M. Mulira. The stalactite plot for the detection of multivariate outliers. *Statistics and Computing*, 3:27–35, 1993.

L. Atlas, J. Connor, D. Park, M. El-Sharkawi, R. Marks, A. Lippman, R. Cole, and Y. Muthusamy. A performance comparison of trained multilayer perceptrons and trained classification trees. In *Proceedings of the 1989 IEEE Conference on Systems, Man and Cybernetics*, pages 915–920. IEEE, 1989.

R.S. Atlas and J.E. Overall. Comparative evaluation of two superior stopping rules for hierarchical cluster analysis. *Psychometrika*, 59(4):581–591, 1994.

H. Attias. A variational Bayesian framework for graphical models. *Advances in Neural Information Processing Systems*, 12:209–215, 2000.

G.A. Babich and O.I. Camps. Weighted Parzen windows for pattern classification. *IEEE Transactions on Pattern Analysis and Machine Intelligence*, 18(5):567–570, 1996.

F.R. Bach and M.I. Jordan. Learning spectral clustering, with application to speech segmentation. *Journal of Machine Learning Research*, 7:1963–2001, 2006.

C. Bahlmann, B. Haasdonk, and H. Burkhardt. On-line handwriting recognition with support vector machines – a kernel approach. In *Proceedings of the 8th International Workshop on Frontiers in Handwriting Recognition (IWFHR'02)*, pages 49–54. IEEE, 2002.

M. Bahrololum and M. Khaleghi. Anomaly intrusion detection system using Gaussian mixture model. In *Third International Conference on Convergence and Hybrid Information Technology*, pages 1162–1167. IEEE, 2008.

F. Bajramovic, F. Mattern, N. Butko, and J. Denzler. A comparison of nearest neighbor search algorithms for generic object recognition. In *Proceedings of ACIVS 2006*, pages 1186–1197. Springer–Verlag, 2006.

N. Balakrishnan and K. Subrahmaniam. Robustness to nonnormality of the linear discriminant function: mixtures of normal distributions. *Communications in Statistics – Theory and Methods*, 14(2):465–478, 1985.

L. Bao, T. Gneiting, E.P. Grimit, P. Guttorp, and A.E. Raftery. Bias correction and Bayesian model averaging for ensemble forecasts of surface wind direction. *Monthly Weather Review*, 138(5):1811–1821, 2010.

M. Barreno, A.A. Cárdenas, and J.D. Tygar. Optimal ROC curve for a combination of classifiers. In *Advances in Neural Information Processing Systems (NIPS)*, pages 57–64. The MIT Press, 2008.

A.R. Barron and R.L. Barron. Statistical learning networks: a unifying view. In E.J. Wegman, D.T. Gantz, and J.J. Miller, editors, *Symposium on the Interface: Statistics and Computing Science*, pages 192–203. American Statistical Association, 1988.

E. Bauer and R. Kohavi. An empirical comparison of voting classification algorithms: bagging, boosting and variants. *Machine Learning*, 36:105–139, 1999.

C. Bavoux, G. Burneleau, and V. Bretagnolle. Gender determination in the Western marsh harrier (*Circus aeruginosus*) using morphometrics and discriminant analysis. *The Journal of Raptor Research*, 40(1):57–64, 2006.

M.J. Baxter. Standardisation and transformation in principal component analysis with applications to archaeometry. *Applied Statistics*, 4(4):513–527, 1995.

M.J. Beal. *Variational Algorithms for Approximate Bayesian Inference*. PhD thesis, The Gatsby Computational Neuroscience Unit, University College London, 2003.

M.A. Beaumont, J.-M. Cornuet, J.M. Marin, and C.P. Robert. Adaptivity for ABC algorithms: the ABC-PMC scheme. *Biometrika*, 96(4):983–990, 2009.

M.A. Beaumont, W. Zhang, and D.J. Balding. Approximate Bayesian computation in population genetics. *Genetics*, 162:2025–2035, 2002.

R.J. Beckman and R.D. Cook. Outlier..........s. *Technometrics*, 25(2):119–163, 1983.

M.D. Bedworth, L. Bottou, J.S. Bridle, F. Fallside, L. Flynn, F. Fogelman, K.M. Ponting, and R.W. Prager. Comparison of neural and conventional classifiers on a speech recognition problem. In *IEE International Conference on Artificial Neural Networks*, pages 86–89. IEEE, 1989.

Y. Bengio, J.M. Buhmann, M. Embrechts, and J.M. Zurada, editors. *IEEE Transactions on Neural Networks*, 11(3): 2000. Special issue on 'Neural Networks for Data Mining and Knowledge Discovery'.

R. Benmokhtar and B. Huet. Classifier fusion: combination methods for semantic indexing in video content. In *Artificial Neural Networks – ICANN 2006*, volume 4132 of *Lecture Notes in Computer Science*, pages 65–74. Springer, 2006.

H. Bensmail and G. Celeux. Regularized Gaussian discriminant analysis through eigenvalue decomposition. *Journal of the American Statistical Association*, 91:1743–1748, 1996.

S.S. Bentow. *A Markov Chain Monte Carlo Method for Approximating 2-Way Contingency Tables with Applications in the Stability Analysis of Ecological Ordination*. PhD thesis, University of California, Los Angeles, 1999.

A.F. Bergh, F.K. Soong, and L.R. Rabiner. Incorporation of temporal structure into a vector-quantization-based preprocessor for speaker-independent, isolated-word recognition. *AT&T Technical Journal*, 64(5):1047–1063, 1985.

J.M. Bernardo and A.F.M. Smith. *Bayesian Theory*. John Wiley & Sons, Ltd, 1994.

V.J. Berrocal, A.E. Raftery, and T. Gneiting. Combining spatial statistical and ensemble information in probabilistic weather forecasts. *Monthly Weather Review*, 135(4):1386–1402, 2007.

J. Besag. Markov chain Monte Carlo for statistical inference. Working Paper, Centre for Statistics and the Social Sciences, University of Washington, USA, 2000.

J.C. Bezdek. *Pattern Recognition with Fuzzy Objective Function Algorithms*. Plenum Press, 1981.

J.C. Bezdek and S.K. Pal, editors. *Fuzzy Models for Pattern Recognition. Methods that Search for Structure in Data*. IEEE Press, 1992.

N. Bhatia and Vandana. Survey of nearest neighbor techniques. *International Journal of Computer Science and Information Security*, 8(2):302–305, 2010.

C.M. Bishop. Curvature-driven smoothing: a learning algorithm for feedforward networks. *IEEE Transactions on Neural Networks*, 4(5):882–884, 1993.

C.M. Bishop. *Neural Networks for Pattern Recognition*. Oxford University Press, 1995.

C.M. Bishop. Variational principal components. In *Proceedings of the Ninth International Conference on Artificial Neural Networks, ICANN'99*, pages 505–514. IEEE, 1999.

C.M. Bishop. *Pattern Recognition and Machine Learning*. Springer, 2007.

C.M. Bishop, M. Svensén, and C.K.I. Williams. GTM: the generative topographic mapping. *Neural Computation*, 10:215–234, 1998.

P. Bladon, P.S. Day, T. Hughes, and P. Stanley. High-level fusion using Bayesian networks: applications in command and control. *Information Fusion for Command Support, Meeting Proceedings RTO-MP-IST-055*, 2006.

J.L. Blue, G.T. Candela, P.J. Grother, R. Chellappa, and C.L. Wilson. Evaluation of pattern classifiers for fingerprint and OCR applications. *Pattern Recognition*, 27(4):485–501, 1994.

L. Bobrowski and J.C. Bezdek. *c*-means clustering with the l_1 and l_∞ norms. *IEEE Transactions on Systems, Man, and Cybernetics*, 21(3):545–554, 1991.

H.H. Bock. On some significance tests in cluster analysis. *Journal of Classification*, 2:77–108, 1985.

H.H. Bock. Probabilistic aspects in cluster analysis. In O. Opitz, editor, *Conceptual and Numerical Analysis of Data*, pages 12–44. Springer-Verlag, 1989.

G.J. Bonde. Kruskal's non-metric multidimensional scaling – applied in the classification of bacteria. In J. Gordesch and P. Naeve, editors, *Proceedings in Computational Statistics*, pages 443–449. Physica-Verlag, 1976.

J.G. Booth and P. Hall. Monte Carlo approximation and the iterated bootstrap. *Biometrika*, 81(2):331–340, 1994.

P. Borini and R.C. Guimarães. Noninvasive classification of liver disease in asymptomatic and oligosymptomatic male alcoholics. *Brazilian Journal of Medical and Biological Research*, 36:1367–1373, 2003.

Z.I. Botev, J.F. Grotowski, and D.P. Kroese. Kernel density estimation via diffusion. *The Annals of Statistics*, 38(5):2916–2957, 2010.

C. Bouveyron, C. Brunet, and V. Vigneron. Classification of high-dimensional data for cervical cancer detection. In M. Verleysen, editor, *Proceedings of ESANN 2009, European Symposium on Artificial Neural Networks – Advances in Computational Intelligence and Learning*, pages 361–366, 2009.

A.W. Bowman. A comparative study of some kernel-based nonparametric density estimators. *Journal of Statistical Computation and Simulation*, 21:313–327, 1985.

H. Bozdogan. Model selection and Akaike's information criterion (AIC): the general theory and its analytical extensions. *Psychometrika*, 52(3):345–370, 1987.

H. Bozdogan. Choosing the number of component clusters in the mixture-model using a new informational complexity criterion of the inverse-Fisher information matrix. In O. Optiz, B. Lausen, and R. kia, editors, *Information and Classification*, pages 40–54. Springer-Verlag, 1992, 1993.

A.P. Bradley. The use of the area under the ROC curve in the evaluation of machine learning algorithms. *Pattern Recognition*, 30(7):1145–1159, 1997.

J.N. Breckenridge. Replicating cluster analysis: method, consistency, and validity. *Multivariate Behavioral Research*, 24(32):147–161, 1989.

L. Breiman. Bagging predictors. *Machine Learning*, 26(2):123–140, 1996.

L. Breiman. Arcing classifiers. *The Annals of Statistics*, 26(3):801–849, 1998.

L. Breiman. Random forests. *Machine Learning*, 45(1):5–32, 2001.

L. Breiman and J.H. Friedman. Discussion on article by Loh and Vanichsetakul: 'Tree-structured classification via generalized discriminant analysis'. *Journal of the American Statistical Association*, 83:715–727, 1988.

L. Breiman, J.H. Friedman, R.A. Olshen, and C.J. Stone. *Classification and Regression Trees*. Wadsworth International Group, 1984.

L. Breiman, W. Meisel, and E. Purcell. Variable kernel estimates of multivariate densities. *Technometrics*, 19(2):135–144, 1977.

R.P. Brent. Fast training algorithms for multilayer neural nets. *IEEE Transactions on Neural Networks*, 2(3):346–354, 1991.

C. Brew and S. Schulte im Walde. Spectral clustering for German verbs. In *Proceedings of the Conference on Empirical Methods in Natural Language Processing (EMNLP)*, pages 117–124. Association for Computational Linguistics, 2002.

F.Z. Brill, D.E. Brown, and W.N. Martin. Fast genetic selection of features for neural network classifiers. *IEEE Transactions on Neural Networks*, 3(2):324–328, 1992.

S.P. Brooks. Markov chain Monte Carlo method and its application. *The Statistician*, 7:69–100, 1998.

D.S. Broomhead and D. Lowe. Multi-variable functional interpolation and adaptive networks. *Complex Systems*, 2(3):269–303, 1988.

D.E. Brown, V. Corruble, and C.L. Pittard. A comparison of decision tree classifiers with backpropagation neural networks for multimodal classification problems. *Pattern Recognition*, 26(6):953–961, 1993.

M. Brown, H.G. Lewis, and S.R. Gunn. Linear spectral mixture models and support vector machines for remote sensing. *IEEE Transactions on Geoscience and Remote Sensing*, 38(5):2346–2360, 2000.

P.J. Brown. *Measurement, Regression, and Calibration*. Clarendon Press, 1993.

P.J. Brown, M. Vannucci, and T. Fearn. Bayes model averaging with selection of regressors. *Journal of the Royal Statistical Society Series B*, 64(3): 519–536, 2002.

S.B. Bull and A. Donner. The efficiency of multinomial logistic regression compared with multiple group discriminant analysis. *Journal of the American Statistical Association*, 82:1118–1121, 1987.

W.L. Buntine. Learning classification trees. *Statistics and Computing*, 2:63–73, 1992.

W.L. Buntine. A guide to the literature on learning probabilistic networks from data. *IEEE Transactions on Knowledge and Data Engineering*, 8(2):195–210, 1996.

W.L. Buntine and A.S. Weigend. Bayesian back-propagation. *Complex Systems*, 5:603–643, 1991.

R. Burbidge, M. Trotter, B. Buxton, and S. Holden. Drug design by machine learning: support vector machines for pharmaceutical data analysis. *Computers and Chemistry*, 26:5–14, 2001.

C.J.C. Burges. A tutorial on support vector machines for pattern recognition. *Data Mining and Knowledge Discovery*, 2:121–167, 1998.

P.R. Burrell and B.O. Folarin. The impact of neural networks in finance. *Neural Computing and Applications*, 6:193–200, 1997.

L.J. Buturović. Improving *k*-nearest neighbor density and error estimates. *Pattern Recognition*, 26(4):611–616, 1993.

R.B. Calinski and J. Harabasz. A dendrite method for cluster analysis. *Communications in Statistics*, 3:1–27, 1974.

B. Cao, D. Zhan, and X. Wu. Application of SVM in financial research. In *Proceedings of 2009 International Joint Conference on Computational Sciences and Optimization*, pages 507–511. IEEE, 2009.

L. Cao and F.E.H. Tay. Financial forecasting using support vector machines. *Neural Computing and Applications*, 10:184–192, 2001.

R. Cao, A. Cuevas, and W.G. Manteiga. A comparative study of several smoothing methods in density estimation. *Computational Statistics and Data Analysis*, 17:153–176, 1994.

G. Casella and E.I. George. Explaining the Gibbs sampler. *The American Statistician*, 46(3):167–174, 1992.

G. Casella and C.P. Robert. Rao-Blackwellisation of sampling schemes. *Biometrika*, 83(1):81–94, 1996.

G.C. Cawley and N.L.C Talbot. Sparse Bayesian learning and the relevance multi-layer perceptron network. In *International Joint Conference on Neural Networks (IJCNN-2005)*, pages 1320–1324. IEEE, 2005.

G. Celeux and G. Govaert. A classification EM algorithm for clustering and two stochastic versions. *Computational Statistics and Data Analysis*, 14:315–332, 1992.

G. Celeux and G. Govaert. Gaussian parsimonious clustering models. *Pattern Recognition*, 28(5):781–793, 1995.

G. Celeux and A. Mkhadri. Discrete regularized discriminant analysis. *Statistics and Computing*, 2(3):143–151, 1992.

G. Celeux and G. Soromenho. An entropy criterion for assessing the number of clusters in a mixture model. *Journal of Classification*, 13:195–212, 1996.

N.N. Čencov. Evaluation of an unknown distribution density from observations. *Soviet Math.*, 3:1559–1562, 1962.

S. Chainey, L. Tompson, and S. Uhlig. The utility of hotspot mapping for predicting spatial patterns of crime. *Security Journal*, 21:4–28, 2008.

Z. Chair and P.R. Varshney. Optimal data fusion in multiple sensor detection systems. *IEEE Transactions on Aerospace and Electronic Systems*, 22:98–101, 1986.

D. Chakrabarti and C. Faloutsos. Graph mining: laws, generators, and algorithms. *ACM Computing Surveys*, 38(1): Article no. 2, 2006.

C.-W.J. Chan, C. Huang, and R. DeFries. Enhanced algorithm performance for land cover classification from remotely sensed data using bagging and boosting. *IEEE Transactions on Geoscience and Remote Sensing*, 39(3):693–695, 2001.

C.-Y. Chang. Dynamic programming as applied to feature subset selection in a pattern recognition system. *IEEE Transactions on Systems, Man and Cybernetics*, 3(2):166–171, 1973.

C.C. Chang and C.J. Lin. LIBSVM: a library for support vector machines. *ACM Transactions on Intelligent Systems and Technology*, 2(3):27:1–27:27, 2011.

E.I. Chang and R.P. Lippmann. Using genetic algorithms to improve pattern classification performance. In R.P. Lippmann, J.E. Moody, and D.S. Touretzky, editors, *Advances in Neural Information Processing Systems*, volume 3, pages 797–803. Morgan Kaufmann, 1991.

E.I. Chang and R.P. Lippmann. A boundary hunting radial basis function classifier which allocates centers constructively. In S.J. Hanson, J.D. Cowan, and C.L. Giles, editors, *Advances in Neural Information Processing Systems*, volume 5, pages 139–146. Morgan Kaufmann, 1993.

P.C. Chang and A.A. Afifi. Classification based in dichotomous and continuous variables. *Journal of the American Statistical Association*, 69:336–339, 1974.

C. Chatfield and A.J. Collins. *Introduction to Multivariate Analysis*. Chapman and Hall, 1980.

V. Chatzis, A.G. Borş, and I. Pitas. Multimodal decision-level fusion for person authentication. *IEEE Transactions on Systems, Man, and Cybernetics – Part A: Systems and Humans*, 29(6):674–680, 1999.

R. Chellappa, K. Fukushima, A.K. Katsaggelos, S.-Y. Kung, Y. LeCun, N.M. Nasrabadi, and T. Poggio, editors. *IEEE Transactions on Image Processing*, 7(8): 1998. Special issue on 'Applications of Artificial Neural Networks to Image Processing'.

R. Chellappa, C.L. Wilson, and S. Sirohey. Human and machine recognition of faces: a survey. *Proceedings of the IEEE*, 83(5):705–740, 1995.

C.H. Chen. *Statistical Pattern Recognition*. Hayden Book Co., 1973.

J. Chen, H. Huang, S. Tian, and Y. Qu. Feature selection for text classification with naïve Bayes. *Expert Systems with Applications*, 36:5432–5435, 2009.

J.S. Chen and E.K. Walton. Comparison of two target classification schemes. *IEEE Transactions on Aerospace and Electronic Systems*, 22(1):15–22, 1986.

K. Chen, L. Xu, and H. Chi. Improved learning algorithms for mixture of experts in multiclass classification. *Neural Networks*, 12:1229–1252, 1999.

L.-F. Chen, H.-Y.M. Liao, M.-T Ko, J.-C. Lin, and G.-J. Yu. A new LDA-based face recognition system which can solve the small sample size problem. *Pattern Recognition*, 33:1713–1726, 2000.

M.-H. Chen, Q.-M. Shao, and J.G. Ibrahim. *Monte Carlo Methods in Bayesian Computation*. Springer Series in Statistics. Springer, 2000.

S. Chen, E.S. Chng, and K. Alkadhimi. Regularised orthogonal least squares algorithm for constructing radial basis function networks. *International Journal of Control*, 64(5):829–837, 1996.

S. Chen, C.F.N. Cowan, and P.M. Grant. Orthogonal least squares learning algorithm for radial basis function networks. *IEEE Transactions on Neural Networks*, 2(2):302–309, 1991.

S. Chen, P.M. Grant, and C.F.N. Cowan. Orthogonal least-squares algorithm for training multioutput radial basis function networks. *IEE Proceedings, Part F*, 139(6):378–384, 1992.

S. Chen, S. Gunn, and C.J. Harris. Decision feedback equaliser design using support vector machines. *IEE Proceedings on Vision, Image and Signal Processing*, 147(3):213–219, 2000.

T. Chen and H. Chen. Approximation capability to functions of several variables, nonlinear functionals, and operators by radial basis function neural networks. *IEEE Transactions on Neural Networks*, 6(4):904–910, 1995.

T.-W. Chen, Y.-L. Chen, and S.Y. Chen. Fast image segmentation based on k-means clustering with histograms in HSV color space. In *IEEE 10th Workshop on Multimedia Signal Processing*, pages 322–325. IEEE, 2008.

X. Chen, X. Liu, and Y. Jia. Discriminative structure selection method of Gaussian mixture models with its application to handwritten digit recognition. *Neurocomputing*, 74:954–961, 2011.

B. Cheng and D.M. Titterington. Neural networks: a review from a statistical perspective (with discussion). *Statistical Science*, 9(1):2–54, 1994.

Y.-Q. Cheng, Y.-M. Zhuang, and J.-Y. Yang. Optimal Fisher discriminant analysis using rank decomposition. *Pattern Recognition*, 25(1):101–111, 1992.

M.R. Chernick, V.K. Murthy, and C.D. Nealy. Application of bootstrap and other resampling techniques: evaluation of classifier performance. *Pattern Recognition Letters*, 3:167–178, 1985.

S. Chib and E. Greenberg. Understanding the Metropolis–Hastings algorithm. *The American Statistician*, 49(4):327–335, 1995.

Y.T. Chien and K.S. Fu. On the generalized Karhunen–Loève expansion. *IEEE Transactions on Information Theory*, 13:518–520, 1967.

E.F. Chinganda and K. Subrahmaniam. Robustness of the linear discriminant function to nonnormality: Johnson's system. *Journal of Statistical Planning and Inference*, 3:69–77, 1979.

S.C. Chiou and R.S. Tsay. A copula-based approach to option pricing and risk assessment. *Journal of Data Science*, 6:273–301, 2008.

P.A. Chou. Optimal partitioning for classification and regression trees. *IEEE Transactions on Pattern Analysis and Machine Intelligence*, 13(4):340–354, 1991.

C.K. Chow. On optimum recognition error and reject tradeoff. *IEEE Transactions on Information Theory*, 16(1):41–46, 1970.

C.K. Chow and C.N. Liu. Approximating discrete probability distributions with dependence trees. *IEEE Transactions on Information Theory*, 14(3):462–467, 1968.

M.-Y. Chow, editor. *IEEE Transactions on Industrial Electronics*, 40(2): 1993. Special issue on 'Applications of Intelligent Systems to Industrial Electronics'.

F.R.K. Chung. *Spectral Graph Theory*, volume 92 of *Regional Conference Series in Mathematics*. Conference Board of the Mathematical Sciences, 1997.

P. Clark and T. Niblett. The CN2 induction algorithm. *Machine Learning*, 3:261–283, 1989.

A. Clauset. Finding local community structure in networks. *Physical Review E*, 72(2):026132, 2005.

R.T. Clemen and T. Reilly. Correlations and copulas for decision and risk analysis. *Management Science*, 45:208–224, 1999.

H.T. Clifford and W. Stephenson. *An Introduction to Numerical Classification*. Academic Press, 1975.

W.W. Cohen. Fast effective rule induction. In *Proceedings of the Twelth International Conference on Machine Learning*, pages 115–123. Morgan Kaufmann, 1995.

P. Comon. Independent component analysis, a new concept? *Signal Processing*, 36:287–314, 1994.

A.G. Constantinides, S. Haykin, Y.H. Hu, J.-N. Hwang, S. Katagiri, S.-Y. Kung, and T.A. Poggio, editors. *IEEE Transactions on Signal Processing*, 45(11): 1997. Special issue an 'Neural Networks for Signal Processing'.

J.A. Conway, L.M.J. Brown, N.J. Veck, and R.A. Cordey. A model-based system for crop classification from radar imagery. *GEC Journal of Research*, 9(1):46–54, 1991.

G.F. Cooper and E. Herskovits. A Bayesian method for the induction of probabilistic networks from data. *Machine Learning*, 9:309–347, 1992.

K.D. Copsey and A.R. Webb. Bayesian approach to mixture models for discrimination. In F.J. Ferri, J.M. Iñesta, A. Amin, and P. Pudil, editors, *Advances in Pattern Recognition*, volume 1876 of *Lecture Notes in Computer Science*, pages 491–500. Springer, 2000.

K.D. Copsey and A.R. Webb. Bayesian networks for incorporation of contextual information in target recognition systems. In T. Caelli, A. Amin, R.P.W. Duin, D. de Ridder, and M. Kamel, editors, *Structural, Syntactic, and Statistical Pattern Recognition, Lecture Notes in Computer Science*, pages 709–717. Springer, 2002.

A. Corduneanu and C. M. Bishop. Variational Bayesian model selection for mixture distributions. In *Proceedings of the 8th International Conference on Artificial Intelligence and Statistics*, pages 27–34. Morgan Kaufmann, 2001.

R.M. Cormack. A review of classification (with discussion). *Journal of the Royal Statistical Society Series A*, 134:321–367, 1971.

C. Cortes and V. Vapnik. Support-vector networks. *Machine Learning*, 20:273–297, 1995.

P.C. Cosman, C. Tseng, R.M. Gray, R.A. Olshen, L.E. Moses, H.C. Davidson, C.J. Bergin, and E.A. Riskin. Tree-structured vector quantization of CT chest scans: image quality and diagnostic accuracy. *IEEE Transactions on Medical Imaging*, 12(4):727–739, 1993.

R. Courant and D. Hilbert. *Methods of Mathematical Physics*. John Wiley & Sons, Ltd, 1959.

T.M. Cover and P.E. Hart. Nearest neighbour pattern classification. *IEEE Transactions on Information Theory*, 13:21–27, 1967.

R.G. Cowell, A.P. Dawid, T. Hutchinson, and D.J. Spiegelhalter. A Bayesian expert system for the analysis of an adverse drug reaction. *Artificial Intelligence in Medicine*, 3:257–270, 1991.

M.K. Cowles and B.P. Carlin. Markov chain Monte Carlo convergence diagnostics: a comparative review. *Journal of the American Statistical Association*, 91(434):883–904, 1996.

T.F. Cox and M.A.A. Cox. *Multidimensional Scaling*. Chapman and Hall, 1994.

T.F. Cox and G. Ferry. Discriminant analysis using non-metric multidimensional scaling. *Pattern Recognition*, 26(1):145–153, 1993.

T.F. Cox and K.F. Pearce. A robust logistic discrimination model. *Statistics and Computing*, 7:155–161, 1997.

P. Craven and G. Wahba. Smoothing noisy data with spline functions. *Numerische Mathematik*, 31:317–403, 1979.

S.L. Crawford. Extensions to the CART algorithm. *International Journal of Man Machine Studies*, 31:197–217, 1989.

N. Cristianini and J. Shawe-Taylor. *An Introduction to Support Vector Machines*. Cambridge University Press, 2000.

R.M. Crownover. A least squares approach to linear discriminant analysis. *SIAM Journal on Scientific and Statistical Computing*, 12(3):595–606, 1991.

W.R. Crum. Spectral clustering and label fusion for 3D tissue classification: Sensitivity and consistency analysis. *Annals of the British Machine Vision Association*, 6:1–12, 2009.

S.P. Curram and J. Mingers. Neural networks, decision tree induction and discriminant analysis: an empirical comparison. *Journal of the Operational Research Society*, 45(4):440–450, 1994.

X. Dai, T. Erkkilä, O. Yli-Harja, and H. Lähdesmäki. A joint finite mixture model for clustering genes from independent Gaussian and beta distributed data. *BMC Bioinformatics*, 10(1):165, 2009.

L.M. D'Andrea, G.L. Fisher, and T.C. Harrison. Cluster analysis of adult children of alcoholics. *The International Journal of Addictions*, 29(5):565–582, 1994.

A. Darwiche. *Modeling and Reasoning with Bayesian Networks*. Cambridge University Press, 2009.

Y. Darwish, T. Cserháti, and E. Forgács. Use of principal component analysis and cluster analysis in quantitative structure-activity relationships: a comparative study. *Chemometrics and Intelligent Laboratory Systems*, 24:169–176, 1994.

B.V. Dasarathy. NN concepts and techniques. an introductory survey. In B.V. Dasarathy, editor, *Nearest Neighbour Norms: NN Pattern Classification Techniques*, pages 1–30. IEEE Computer Society Press, 1991.

B.V. Dasarathy. Minimal consistent set (MCS) identification for nearest neighbour decision systems design. *IEEE Transactions on Systems, Man, and Cybernetics*, 24(3):511–517, 1994a.

B.V. Dasarathy. *Decision Fusion*. IEEE Computer Society Press, 1994b.

M. Dash and H. Liu. Feature selection for classification. *Intelligent Data Analysis*, 1:131–156, 1997.

D.L. Davies and D. Bouldin. A cluster separation measure. *IEEE Transactions on Pattern Analysis and Machine Intelligence*, 1:224–227, 1979.

A.C. Davison and P. Hall. On the bias and variability of bootstrap and cross-validation estimates of error rate in discrimination problems. *Biometrika*, 79:279–284, 1992.

A.C. Davison, D.V. Hinkley, and E. Schechtman. Efficient bootstrap simulation. *Biometrika*, 73(3):555–556, 1986.

M. Davy, C. Doncarli, and J.-Y. Tourneret. Classification of chirp signals using hierarchical Bayesian learning and MCMC methods. *IEEE Transactions on Signal Processing*, 50(2): 377–388, 2002.

B.M. Dawant and C. Garbay, editors. *IEEE Transactions on Biomedical Engineering*, 46(10): 1999. Special topic section on 'Biomedical Data Fusion'.

N.E. Day and D.F. Kerridge. A general maximum likelihood discriminant. *Biometrics*, 23:313–323, 1967.

D. Defays. An efficient algorithm for a complete link method. *The Computer Journal*, 20(4):364–366, 1977.

J.G. De Gooijer, B.K. Ray, and K. Horst. Forecasting exchange rates using TSMARS. *Journal of International Money and Finance*, 17(3):513–534, 1998.

J. de Leeuw and W.J. Heiser. Convergence of correction matrix algorithms for multidimensional scaling. In J.C. Lingoes, editor, *Geometric Representations of Relational Data*, pages 735–752. Mathesis Press, 1977.

P. Dellaportas. Bayesian classification of neolithic tools. *Applied Statistics*, 47(2):279–297, 1998.

P. Del Moral, A. Doucet, and A. Jasra. Sequential Monte Carlo samplers. *Journal of the Royal Statistical Society Series B*, 68(3):411–436, 2006.

R.L. de Màntaras and J. Aguilar-Martín. Self-learning pattern classification using a sequential clustering technique. *Pattern Recognition*, 18(3/4):271–277, 1985.

S. Demarta and A.J. McNeil. The t copula and related copulas. *International Statistical Review*, 73:111–129, 2005.

A.P. Dempster, N.M. Laird, and D.B. Rubin. Maximum likelihood from incomplete data via the EM algorithm. *Journal of the Royal Statistical Society Series B*, 39:1–38, 1977.

D.G.T. Denison, B.K. Mallick, and A.F.M. Smith. A Bayesian CART algorithm. *Biometrika*, 85(2):363–377, 1998a.

D.G.T. Denison, B.K. Mallick, and A.F.M. Smith. Bayesian MARS. *Statistics and Computing*, 8:337–346, 1998b.

L. De Raedt, P. Frasconi, K. Kersting, and S.H. Muggleton, editors. *Probabilistic Inductive Logic Programming*. Springer, 2008.

T. Deselaers, G. Heigold, and H. Ney. Speech recognition with state-based nearest neighbour classifiers. *Interspeech 2007 Conference, Antwerp, Belgium*, 2007.

De Vel, A. Anderson, M. Corney, and G. Mohay. Mining e-mail content for author identification forensics. *SIGMOD Record*, 30(4):55–64, 2001.

P.A. Devijver. Relationships between statistical risks and the least-mean-square error criterion in pattern recognition. In *Proceedings of the First International Joint Conference on Pattern Recognition*, pages 139–148. 1973.

P.A. Devijver and J. Kittler. *Pattern Recognition, A Statistical Approach*. Prentice-Hall, Inc., 1982.

L. Devroye. *Non-uniform Random Variate Generation*. Springer-Verlag, 1986.

L. Devroye and L. Györfi. *Nonparametric Density Estimation. The L_1 View*. John Wiley & Sons, Ltd, 1985.

E. Diday and J.C. Simon. Clustering analysis. In K.S. Fu, editor, *Digital Pattern Recognition*, pages 47–94. Springer-Verlag, 1976.

X. Didelot, R.G. Everitt, A.M. Johansen, and D.J. Lawson. Likelihood-free estimation of model evidence. *Bayesian Analysis*, 6:49–76, 2011.

J. Diederich. Authorship attribution with support vector machines. *Applied Intelligence, Special Issue: Neural Networks and Machine Learning for Natural Language Processing*, 19(1-2):109–123, 2006.

T.G. Dietterich. Approximate statistical tests for comparing supervised classification learning algorithms. *Neural Computation*, 10:1895–1923, 1998.

P.J. Diggle and P. Hall. The selection of terms in an orthogonal series density estimator. *Journal of the American Statistical Association*, 81:230–233, 1986.

T. Dillon, P. Arabshahi, and R.J. Marks, editors. *IEEE Transactions on Neural Networks*, volume 8(4). 1997. Special issue on 'Everyday Applications of Neural Networks'.

W.R. Dillon and M. Goldstein. *Multivariate Analysis Methods and Applications*. John Wiley & Sons, Ltd, 1984.

A. Djouadi and E. Bouktache. A fast algorithm for the nearest-neighbor classifier. *IEEE Transactions on Pattern Analysis and Machine Intelligence*, 19(3):277–282, 1997.

P. Domingos and M. Pazzani. On the optimality of the simple Bayesian classifier under zero-one loss. *Machine Learning*, 29:103–130, 1997.

R.D. Dony and S. Haykin. Neural network approaches to image compression. *Proceedings of the IEEE*, 83(2):288–303, 1995.

G. Doppelhofer. Model averaging. In L. Blune and S. Durkauf, editors, *The New Palgrave Dictionary of Economics*. Palgrave Macmillan, second edition, 2007.

M. Dorey and P. Joubert. Modelling copulas: an overview. The Staple Inn Actuarial Society, 2007.

A. Doucet, N. De Freitas, and N. Gordon, editors. *Sequential Monte Carlo Methods in Practice*. Springer-Verlag, 2001.

D.C. Dracopoulos and P.L. Rosin, editors. *Neural Computing and Applications*, 7(3): 1998. Special issue on 'Machine Vision Using Neural Networks'.

K.C. Drake, Y. Kim, T.Y. Kim, and O.D. Johnson. Comparison of polynomial network and model-based target recognition. In *Proceedings of SPIE*, volume 2333, pages 2–11. SPIE, 1994.

H. Drucker and Y. Le Cun. Improving generalization performance using double backpropagation. *IEEE Transactions on Neural Networks*, 3(6):991–997, 1992.

R.C. Dubes. How many clusters are best? – an experiment. *Pattern Recognition*, 20(6):645–663, 1987.

A. Dubrawski. A framework for evaluating predictive capability of classifiers using receiver operating characteristic (ROC) approach: a brief introduction. Technical Report, Auton Laboratory, Carnegie Mellon University, 2004.

W. Duch. Filter methods. In I. Guyon, S. Gunn, M. Nikravesh, and L. Zadeh, editors, *Feature Extraction: Foundations and Applications*, pages 89–118. Springer, 2004.

J. Duchene and S. Leclercq. An optimal transformation for discriminant analysis and principal component analysis. *IEEE Transactions on Pattern Analysis and Machine Intelligence*, 10(6):978–983, 1988.

J. Duchon. Interpolation des fonctions de deux variables suivant le principe de la flexion des plaques minces. *RAIRO Analyse Numérique*, 10(12):5–12, 1976.

R.O. Duda, P.E. Hart, and D.G. Stork. *Pattern Classification*. John Wiley & Sons, Ltd, second edition, 2001.

S.A. Dudani. The distance-weighted *k*-nearest-neighbour rule. *IEEE Transactions on Systems, Man and Cybernetics*, 6(4):325–327, 1976.

D. Duffy, B. Yuhas, A. Jain, and A. Buja. Empirical comparisons of neural networks and statistical methods for classification and regression. In B. Yuhas and N. Ansari, editors, *Neural Networks in Telecommunications*, pages 325–349. Kluwer Academic, 1994.

R.P.W. Duin. On the choice of smoothing parameters for Parzen estimators of probability density functions. *IEEE Transactions on Computers*, 25:1175–1179, 1976.

R.P.W. Duin. A note on comparing classifiers. *Pattern Recognition Letters*, 17:529–536, 1996.

J.C. Dunn. A fuzzy relative of the ISODATA process and its use in detecting compact well-separated clusters. *Journal of Cybernetics*, 3(3):32–57, 1974.

D. Easley and J. Kleinberg. *Networks, Crowds, and Markets: Reasoning About a Highly Connected World*. Cambridge University Press, 2010.

J.E. Eck, S. Chainey, J.G. Cameron, M. Leitner, and R.E. Wilson. Mapping crime: understanding hot spots. US Department of Justice, National Institute of Justice Special Report, August 2005.

W.F. Eddy, A. Mockus, and S. Oue. Approximate single linkage cluster analysis of large data sets in high-dimensional spaces. *Computational Statistics and Data Analysis*, 23:29–43, 1996.

B. Efron. Bootstrap methods: another look at the jackknife. *Annals of Statistics*, 7:1–26, 1979.

B. Efron. *The Jackknife, the Bootstrap, and Other Resampling Plans*. Society for Industrial and Applied Mathematics, 1982.

B. Efron. Estimating the error rate of a prediction rule: Improvement on cross-validation. *Journal of the American Statistical Association*, 78:316–331, 1983.

B. Efron. More efficient bootstrap computations. *Journal of the American Statistical Association*, 85:79–89, 1990.

B. Efron and R.J. Tibshirani. Bootstrap methods for standard errors, confidence intervals, and other measures of statistical accuracy (with discussion). *Statistical Science*, 1:54–77, 1986.

L. Eklundh and A. Singh. A comparative analysis of standardised and unstandardised principal components analysis in remote sensing. *International Journal of Remote Sensing*, 14(7):1359–1370, 1993.

A. Elgammal, R. Duraiswami, D. Harwood, and L.S. Davis. Background and foreground modeling using nonparametric kernel density estimation for visual surveillance. *Proceedings of the IEEE*, 90(7):1151–1163, 2002.

M. El-Telbany, M. Warda, and M. El-Borahy. Mining the classification rules for Egyptian rice diseases. *The International Arab Journal of Information Technology*, 3(4):303–307, 2006.

J. Elzinga and D.W. Hearn. The minimum covering sphere problem. *Management Science*, 19(1):96–104, 1972.

P. Embrechts, F. Lindskog, and A. McNeil. Modelling dependence with copulas and applications to risk management. Technical Report, Department of Mathematics ETHZ, Zurich, September 2001.

G.G. Enas and S.C. Choi. Choice of the smoothing parameter and efficiency of *k*-nearest neighbor classification. *Computers and Mathematics with Applications*, 12A(2):235–244, 1986.

P. Erdös and A. Rényi. On random graphs. I. *Publicationes Mathematicae Debrecen*, 6:290–297, 1959.

P. Erdös and A. Rényi. On the evolution of random graphs. *Publications of the Mathematical Institute of the Hungarian Academy of Sciences*, 5:17–61, 1960.

F. Esposito, D. Malerba, and G. Semeraro. A comparative analysis of methods for pruning decision trees. *IEEE Transactions on Pattern Analysis and Machine Intelligence*, 19(5):476–491, 1997.

B.S. Everitt. A Monte Carlo investigation of the likelihood ratio test for the number of components in a mixture of normal distributions. *Multivariate Behavioral Research*, 16:171–180, 1981.

B.S. Everitt. A finite mixture model for the clustering of mixed-mode data. *Statistics and Probability Letters*, 6:305–309, 1988.

B.S. Everitt and G. Dunn. *Applied Multivariate Data Analysis*. Edward Arnold, 1991.

B.S. Everitt and D.J. Hand. *Finite Mixture Distributions*. Chapman and Hall, 1981.

B.S. Everitt, S. Landau, and M. Leese. *Cluster Analysis*. Edward Arnold, fourth edition, 2001.

B.S. Everitt, S. Landau, M. Leese, and D. Stahl. *Cluster Analysis*. Wiley-Blackwell, fiifth edition, 2011.

B.S. Everitt and C. Merette. The clustering of mixed-mode data: a comparison of possible approaches. *Journal of Applied Statistics*, 17:283–297, 1990.

R.G. Everitt and R.H. Glendinning. A statistical approach to the problem of restoring damaged and contaminated images. *Pattern Recognition*, 42(1):115–125, 2009.

R.M. Everson and J.E. Fieldsend. Multi-class ROC analysis from a multi-objective optimisation perspective. *Pattern Recognition Letters*, 27(8):918–927, 2006.

J.A. Falconer, B.J. Naughton, D.D. Dunlop, E.J. Roth, D.C. Strasser, and J.M. Sinacore. Predicting stroke in patient rehabilitation outcome using a classification tree approach. *Archives of Physical Medicine and Rehabilitation*, 75:619–625, 1994.

T.H. Falk, H. Shatkay, and W.-Y. Chan. Breast cancer prognosis via Gaussian mixture regression. In *Proceedings of the Canadian Conference on Electrical and Computer Engineering (CCECE 2006)*, pages 987–990. IEEE, 2006.

R.-E. Fan, P.-H. Chen, and Lin C.-J. Working set selection using second order information for training support vector machines. *Journal of Machine Learning Research*, 6:1889–1918, 2005.

A. Faragó and G. Lugosi. Strong universal consistency of neural network classifiers. *IEEE Transactions on Information Theory*, 39(4):1146–1151, 1993.

I. Farkas, D. Ábel, G. Palla, and T. Vicsek. Weighted network modules. *New Journal of Physics*, 9(6):180, 2007.

T. Fawcett. An introduction to ROC analysis. *Pattern Recognition Letters*, 27(8):861–874, 2006.

C. Feng and D. Michie. Machine learning of rules and trees. In D. Michie, D.J. Spiegelhalter, and C.C. Taylor, editors, *Machine Learning, Neural and Statistical Classification*, pages 50–83. Ellis Horwood, 1994.

J. Fernández de Cañete and A.B. Bulsari, editors. *Neural Computing and Applications*, 9(3): 2000. Special issue on 'Neural Networks in Process Engineering'.

E.A. Ferrán, B. Pflugfelder, and P. Ferrara. Self-organized neural maps of human protein sequences. *Protein Science*, 3:507–521, 1994.

L. Ferré. Selection of components in principal components analysis: a comparison of methods. *Computational Statistics and Data Analysis*, 19:669–682, 1995.

C. Ferri, J. Hernández-Orallo, and R. Modroiu. An experimental comparison of performance measures for classification. *Pattern Recognition Letters*, 30:27–38, 2009.

C. Ferri, J. Hernández-Orallo, and M.A. Salido. Volume under the ROC surface for multi-class problems. In *Machine Learning: ECML 2003*, volume 2837 of *Lecture Notes in Computer Science*, pages 108–120. Springer, 2003.

F.J. Ferri, J.V. Albert, and E. Vidal. Considerations about sample-size sensitivity of a family of nearest-neighbor rules. *IEEE Transactions on Systems, Man and Cybernetics*, 29(4):667–672, 1999.

F.J. Ferri and E. Vidal. Small sample size effects in the use of editing techniques. In *Proceedings of the 11th IAPR International Conference on Pattern Recognition*, pages 607–610. IEEE, 1992a.

F.J. Ferri and E. Vidal. Colour image segmentation and labeling through multiedit-condensing. *Pattern Recognition Letters*, 13:561–568, 1992b.

J.E. Fieldsend and R.M. Everson. Formulation and comparison of multi-class ROC surfaces. In *Proceedings of the ICML 2005 Workshop on ROC Analysis in Machine Learning*, pages 41–48. 2005.

G.M. Fitzmaurice, W.J. Krzanowski, and D.J. Hand. A Monte Carlo study of the 632 bootstrap estimator of error rate. *Journal of Classification*, 8:239–250, 1991.

P.A. Flach and S. Wu. Repairing concavities in ROC curves. In *Proceedings of the 19th International Joint Conference on Artificial intelligence*, pages 702–707. Morgan Kaufmann, 2005.

R. Fletcher. *Practical Methods of Optimization*. John Wiley & Sons, Ltd, 1988.

B. Flury. A hierarchy of relationships between covariance matrices. In A.K. Gupta, editor, *Advances in Multivariate Analysis*. D. Reidel Publishing Co., 1987.

B. Flury. *Common Principal Components and Related Multivariate Models*. John Wiley & Sons, Ltd, 1988.

D.H. Foley and J.W. Sammon. An optimal set of discriminant vectors. *IEEE Transactions on Computers*, 24(3):281–289, 1975.

G. Forman. An extensive empirical study of feature selection metrics for text classification. *Journal of Machine Learning Research*, 3:1289–1305, 2003.

S. Fortunato. Community detection in graphs. *Physics Reports*, 486:75–174, 2010.

C. Fraley, A.E. Raftery, and T. Gneiting. Calibrating multimodal forecast ensembles with exchangeable and missing members using Bayesian model averaging. *Monthly Weather Review*, 138(1):190–202, 2010.

C. Fraley, A.E. Raftery, T. Gneiting, and J.M. Sloughter. ensembleBMA: an R package for probabilistic forecasting using ensembles and Bayesian model averaging. Technical Report 516R, Department of Statistics, University of Washington, 2009.

E.W. Frees and E.A. Valdez. Understanding relationships using copulas. *North American Actuarial Journal*, 2(1):1–25, 1998.

S. French and J.Q. Smith. *The Practice of Bayesian Analysis*. Arnold, 1997.

Y. Freund and R. Schapire. Experiments with a new boosting algorithm. In *Proceedings of the 13th International Conference on Machine Learning*, pages 256–285. Morgan Kaufman, 1996.

Y. Freund and R. Schapire. A short introduction to boosting. *Journal of the Japanese Society for Artificial Intelligence*, 14(5):771–780, 1999.

M.A. Friedl, C.E. Brodley, and A.H. Strahler. Maximising land cover classification accuracies produced by decision trees at continental to global scales. *IEEE Transactions on Geoscience and Remote Sensing*, 37(2):969–977, 1999.

J.H. Friedman. Exploratory projection pursuit. *Journal of the American Statistical Association*, 82:249–266, 1987.

J.H. Friedman. Regularized discriminant analysis. *Journal of the American Statistical Association*, 84:165–175, 1989.

J.H. Friedman. Multivariate adaptive regression splines. *Annals of Statistics*, 19(1):1–141, 1991.

J.H. Friedman. Estimating functions of mixed ordinal and categorical variables using adaptive splines. In S. Morgenthaler, E.M.D. Ronchetti, and W.A. Stahel, editors, *New Directions in Statistical Data Analysis and Robustness*, pages 73–113. Birkhäuser-Verlag, 1993.

J.H. Friedman. Flexible metric nearest neighbor classification. Report, Department of Statistics, Stanford University, CA, 1994.

J.H. Friedman, J.L. Bentley, and R.A. Finkel. An algorithm for finding best matches in logarithmic expected time. *ACM Transactions on Mathematical Software*, 3(3):209–226, 1977.

J.H. Friedman, T.J. Hastie, and R.J. Tibshirani. Additive logistic regression: a statistical view of boosting. *Annals of Statistics*, 28(2): 337–407, 1998.

J.H. Friedman and W. Stuetzle. Projection pursuit regression. *Journal of the American Statistical Association*, 76:817–823, 1981.

J.H. Friedman, W. Stuetzle, and A. Schroeder. Projection pursuit density estimation. *Journal of the American Statistical Association*, 79:599–608, 1984.

J.H. Friedman and J.W. Tukey. A projection pursuit algorithm for exploratory data analysis. *IEEE Transactions on Computers*, 23(9):881–889, 1974.

N. Friedman, D. Geiger, and M. Goldszmidt. Bayesian network classifiers. *Machine Learning*, 29:131–163, 1997.

N. Friedman and D. Koller. Being Bayesian about Bayesian network structure: a Bayesian approach to structure discovery in Bayesian networks. *Machine Learning*, 50(1–2):95–125, 2003.

H. Fröhlich and A. Zell. Efficient parameter selection for support vector machines in classification and regression via model-based global optimization. In *Proceedings of the International Joint Conference on Neural Networks (IJCNN)*, pages 1431–1438. IEEE, 2005.

K.S. Fu. *Sequential Methods in Pattern Recognition and Machine Learning*. Academic Press, 1968.

S. Fu and M.C. Desmarais. Markov blanket based feature selection: a review of past decade. In S.I. Ao, L. Gelman, D.W.L. Hukins, A. Hunter, and A.M. Korsunsky, editors, In *Proceedings of the World Congress on Engineering*, volume 1, pages 302–308. Newswood Ltd, 2010.

K. Fukunaga. *Introduction to Statistical Pattern Recognition*. Academic Press, Inc., second edition, 1990.

K. Fukunaga and T.E. Flick. An optimal global nearest neighbour metric. *IEEE Transactions on Pattern Analysis and Machine Intelligence*, 6:314–318, 1984.

K. Fukunaga and R.R. Hayes. The reduced Parzen classifier. *IEEE Transactions on Pattern Analysis and Machine Intelligence*, 11(4):423–425, 1989a.

K. Fukunaga and R.R. Hayes. Estimation of classifier performance. *IEEE Transactions on Pattern Analysis and Machine Intelligence*, 11(10):1087–1101, 1989b.

K. Fukunaga and D.M. Hummels. Bias of nearest neighbor error estimates. *IEEE Transactions on Pattern Analysis and Machine Intelligence*, 9(1):103–112, 1987a.

K. Fukunaga and D.M. Hummels. Bayes error estimation using Parzen and *k*-nn procedures. *IEEE Transactions on Pattern Analysis and Machine Intelligence*, 9(5):634–643, 1987b.

K. Fukunaga and D.M. Hummels. Leave-one-out procedures for nonparametric error estimates. *IEEE Transactions on Pattern Analysis and Machine Intelligence*, 11(4):421–423, 1989.

K. Fukunaga and D.L. Kessell. Estimation of classification error. *IEEE Transactions on Computers*, 20:1521–1527, 1971.

K. Fukunaga and P.M. Narendra. A branch and bound algorithm for computing *k*-nearest neighbors. *IEEE Transactions on Computers*, 24(7):750–753, 1975.

T.S. Furey, N. Vristianini, N. Duffy, D.W. Bednarski, M. Schummer, and D. Haussler. Support vector machine classification and validation of cancer tissue samples using microarray expression data. *Bioinformatics*, 16(10):906–914, 2000.

S. Furui. Recent advances in speaker recognition. *Pattern Recognition Letters*, 18:859–872, 1997.

D. Gamerman and H.F. Lopes. *Markov Chain Monte Carlo: Stochastic Simulation for Bayesian Inference*. Texts in Statistical Science. Chapman and Hall/CRC, second edition, 2006.

S. Ganeshanandam and W.J. Krzanowski. On selecting variables and assessing their performance in linear discriminant analysis. *Australian Journal of Statistics*, 31(3):433–447, 1989.

I. Gath and A.B. Geva. Unsupervised optimal fuzzy clustering. *IEEE Transactions on Pattern Analysis and Machine Intelligence*, 11(7):773–781, 1989.

B. Gautam, P. Katara, S. Singh, and R. Farmer. Drug target identification using gene expression microarray data of *Toxoplasma gondii*. *International Journal of Biometrics and Bioinformatics*, 4(3):113–124, 2010.

S. Geisser. Posterior odds for multivariate normal classifications. *Journal of the Royal Statistical Society Series B*, 26:69–76, 1964.

A.E. Gelfand. Gibbs sampling. *Journal of the American Statistical Association*, 95:1300–1304, 2000.

A.E. Gelfand and A.F.M. Smith. Sampling-based approaches to calculating marginal densities. *Journal of the American Statistical Association*, 85:398–409, 1990.

S.B. Gelfand and E.J. Delp. On tree structured classifiers. In I.K. Sethi and A.K. Jain, editors, *Artificial Neural Networks and Statistical Pattern Recognition*, pages 51–70. North Holland Publishing Company, 1991.

A. Gelman. Implementing and monitoring convergence. In W.R. Gilks, S. Richardson, and D.J. Spiegelhalter, editors, *Markov Chain Monte Carlo in Practice*, pages 131–143. Chapman and Hall, 1996.

A. Gelman, J.B. Carlin, H.S. Stern, and D.B. Rubin. *Bayesian Data Analysis*. Chapman and Hall/CRC, second edition, 2004.

A. Gersho and R.M. Gray. *Vector Quantization and Signal Compression*. Kluwer Academic, 1992.

L. Getoor and C.P. Diehl. Link mining: a survey. *SIGKDD Explorations Newsletter*, 7(2):3–12, 2005.

L. Getoor and B. Taskar, editors. *Introduction to Statistical Relational Learning*. The MIT Press, 2007.

J. Geweke. Bayesian inference in econometric models using Monte Carlo integration. *Econometrica*, 57(6):1317–1339, 1989.

Z. Ghahramani and M.J. Beal. Variational inference for Bayesian mixtures of factors analysers. *Advances in Neural Information Processing Systems*, 12:449–455, 1999.

Z. Ghahramani and H.-C. Kim. Bayesian model combination. Technical Report, Gatsby Computational Neuroscience Unit, University College London, 2003.

F. Giacinto, G. Roli and L. Bruzzone. Combination of neural and statistical algorithms for supervised classification of remote-sensing images. *Pattern Recognition Letters*, 21:385–397, 2000.

J.M. Gibbons, G.M. Cox, A.T.A. Wood, J. Craigon, S.J. Ramider, D. Tarsikono, and N.M.T. Crout. Applying (Baysian) model averaging to mechanistic models: an example and comparison of methods. *Environmental Modelling and Software*, 23(8):973–985, 2008.

A. Gifi. *Nonlinear Multivariate Analysis*. John Wiley & Sons, Ltd, 1990.

W.R. Gilks, S. Richardson, and D.J. Spiegelhalter, editors. *Markov Chain Monte Carlo in Practice*. Chapman and Hall, 1996.

F. Gini. Optimal multiple level decision fusion with distributed sensors. *IEEE Transaction on Aerospace and Electronic Systems*, 33(3):1037–1041, 1997.

M. Girvan and M.E.J. Newman. Community structure in social and biological networks. *Proceedings of the National Academy of Sciences of the United States of America*, 99(12):7821–7826, 2002.

G. Glonek, T. Staniford, M. Rumsewicz, O. Mazonka, J. McMahon, D. Fletcher, and M. Jokic. Range safety application of kernel density estimation. Technical Report Defence Science and Technology Organisation, Australia, DSTO-TR-2292, 2010.

F. Glover. Tabu search – Part I. *ORSA Journal on Computing*, 1(3):190–206, 1989.

F. Glover. Tabu search – Part II. *ORSA Journal on Computing*, 2(1):4–32, 1990.

D.E. Goldberg. *Generic Algorithms in Search, Optimization and Machine Learning*. Addison Wesley, 1989.

G.H. Golub, M. Heath, and G. Wahba. Generalised cross-validation as a method of choosing a good ridge parameter. *Technometrics*, 21:215–223, 1979.

R.C. Gonzalez and R.E. Woods. *Digital Image Processing*. Pearson Education, third edition, 2008.

R.M. Goodman and P. Smyth. Decision tree design using information theory. *Knowledge Acquisition*, 2:1–19, 1990.

A.D. Gordon. Clustering algorithms and cluster validity. In P. Dirschedl and R. Ostermann, editors, *Computational Statistics*, pages 497–512. Physica-Verlag, 1994.

A.D. Gordon. Null models in cluster validation. In W. Gaul and D. Pfeifer, editors, *From Data to Knowledge: Theoretical and Practical Aspects of Classification, Data Analysis and Knowledge Organization*, pages 32–44. Springer-Verlag, 1996a.

A.D. Gordon. A survey of constrained classification. *Computational Statistics and Data Analysis*, 21:17–29, 1996b.

A.D. Gordon. Cluster validation. In C. Hayashi, N. Ohsumi, K. Yajima, Y. Tanaka, H.-H. Bock, and Y. Baba, editors, *Data Science, Classification, and Related Methods*, pages 22–39. Springer-Verlag, 1998.

A.D. Gordon. *Classification*. Chapman and Hall, second edition, 1999.

A.D. Gordon and J.T. Henderson. An algorithm for Euclidean sum of squares classification. *Biometrics*, 33:355–362, 1977.

N.J. Gordon, D.J. Salmond, and A.F.M. Smith. Novel approach to nonlinear/non-Gaussian Bayesian state estimation. *IEE Proceedings-F*, 140(2):107–113, 1993.

P.J. Green. Reversible jump MCMC computation and Bayesian model determination. *Biometrika*, 82:711–732, 1995.

P.J. Green and B.W. Silverman. *Nonlinear Regression and Generalized Linear Models. A Roughness Penalty Approach*. Chapman and Hall, 1994.

T. Greene and W. Rayens. Partially pooled covariance estimation in discriminant analysis. *Communications in Statistics*, 18(10):3679 3702, 1989.

A. Grelaud, C.P. Robert, J.-M. Marin, F. Rodolphe, and J.-F. Taly. ABC likelihood-free methods for model choice in Gibbs random fields. *Bayesian Analysis*, 4(2):317–336, 2009.

P.D. Grünwald, editor. *The Minimum Description Length Principle*. The MIT Press, 2007.

A. Guénoche, P. Hansen, and B. Jaumard. Efficient algorithms for divisive hierarchical clustering with the diameter criterion. *Journal of Classification*, 8:5–30, 1991.

Y. Guo, T. Hastie, and R. Tibshirani. Regularized discriminant analysis and its application in microarrays. *Biostatistics*, 8(1):86–100, 2007.

I. Guyon. Practical feature selection: from correlation to causality. In F. Fogelman-Soulié, D. Perrotta, J. Pislorski, and R. Steinberger, editors, *Mining Massive Data Sets for Security*, pages 27–43. IOS Press, 2008.

I. Guyon, S. Gunn, M. Nikravesh, and L. Zadeh, editors. *Feature Extraction. Foundations and Applications*. Springer, 2006.

I. Guyon, J. Makhoul, R. Schwartz, and V. Vapnik. What size test set gives good error rate estimates? *IEEE Transactions on Pattern Analysis and Machine Intelligence*, 20(1):52–64, 1998.

I. Guyon and D.G. Stork. Linear discriminant and support vector classifiers. In A. Smola, P. Bartlett, B. Schölkopf, and C. Schuurmans, editors, *Large Margin Classifiers*, pages 147–169. The MIT Press, 1999.

I. Guyon, J. Weston, S. Barnhill, and V. Vapnik. Gene selection for cancer classification using support vector machines. *Machine Learning*, 46:389–422, 2002.

Z. Haikun, W. Liguang, and Z. Weican. Kernel density estimation applied to tropical cyclones genesis in Northwestern Pacific. *2009 International Conference on Environmental Science and Information Application Technology*, 2009.

K.O. Hajian-Tilaki, J.A. Hanley, L. Joseph, and J.-P. Collett. A comparison of parametric and non-parametric approaches to ROC analysis of quantitative diagnostic tests. *Medical Decision Making*, 17(1):94–102, 1997a.

K.O. Hajian-Tilaki, J.A. Hanley, L. Joseph, and J.-P. Collett. Extension of receiver operating characteristic analysis to data concerning multiple signal detection. *Academic Radioloy*, 4(3):222–229, 1997b.

U. Halici and G. Ongun. Fingerprint classification through self-organising feature maps modified to treat uncertainties. *Proceedings of the IEEE*, 84(10):1497–5112, 1996.

M. Halkidi, Y. Batistakis, and M. Vazirgiannis. On clustering validation techniques. *Journal of Intelligent Information Systems*, 17(2/3):107–145, 2001.

P. Hall. *The Bootstrap and Edgeworth Expansion*. Springer-Verlag, 1992.

P. Hall, T.-C. Hu, and J.S. Marron. Improved variable window kernel estimates of probability densities. *Annals of Statistics*, 23(1):1–10, 1995.

Y. Hamamoto, Y. Fujimoto, and S. Tomita. On the estimation of a covariance matrix in designing Parzen classifiers. *Pattern Recognition*, 29(10):1751–1759, 1996.

Y. Hamamoto, T. Kanaoka, and S. Tomita. On a theoretical comparison between the orthonormal discriminant vector method and discriminant analysis. *Pattern Recognition*, 26(12):1863–1867, 1993.

Y. Hamamoto, Y. Matsuura, T. Kanaoka, and S. Tomita. A note on the orthonormal discriminant vector method for feature extraction. *Pattern Recognition*, 24(7):681–684, 1991.

Y. Hamamoto, S. Uchimura, Y. Matsuura, T. Kanaoka, and S. Tomita. Evaluation of the branch and bound algorithm for feature selection. *Pattern Recognition Letters*, 11:453–456, 1990.

Y. Hamamoto, S. Uchimura, and S. Tomita. A bootstrap technique for nearest neighbor classifier design. *IEEE Transactions on Pattern Analysis and Machine Intelligence*, 19(1):73–79, 1997.

F.R. Hampel, E.M. Ronchetti, P.J. Rousseuw, and W.A. Stahel. *Robust Statistics. The Approach Based on Influence Functions*. John Wiley & Sons, Ltd, 1986.

E.-H. Han, G. Karypis, and V. Kumar. Text categorisation using weight adjusted k-nearest neighbor classification. In *Advances in Knowledge Discovery and Data Mining, volume 2035 of Lecture Notes in Computer Science*, pages 53–65. Springer, 2001.

J. Han and M. Kamber. *Data Mining. Concepts and Techniques*. Morgan Kaufmann, second edition, 2006.

L. Han, Y. Wang, and S.H. Bryant. Developing and validating predictive decision tree models from mining chemical structure fingerprints and high-throughput screening data in pubchem. *BMC Bioinformatics*, 9(1):401, 2008.

D.J. Hand. *Discrimination and Classification*. John Wiley & Sons, Ltd, 1981a.

D.J. Hand. Branch and bound in statistical data analysis. *The Statistician*, 30:1–13, 1981b.

D.J. Hand. *Kernel Discriminant Analysis*. Research Studies Press, Herts, UK, 1982.

D.J. Hand. Recent advances in error rate estimation. *Pattern Recognition Letters*, 4:335–346, 1986b.

D.J. Hand. Statistical methods in diagnosis. *Statistical Methods in Medical Research*, 1(1):49–67, 1992.

D.J. Hard. Assessing classification rules. *Journal of Applied Statistics*, 21(3): 3–16, 1994.

D.J. Hand. *Construction and Assessment of Classification Rules*. John Wiley & Sons, Ltd, 1997.

D.J. Hand. Classifier technology and the illusion of progress (with discussion). *Statistical Science*, 21(1):1–29, 2006.

D.J. Hand, N.M. Adams, and M.G. Kelly. Multiple classifier systems based on interpretable linear classifiers. In J. Kittler and F. Roli, editors, *Proceedings of the 2001 Workshop on Multiple Classifiers Systems*, pages 136–147. Springer-Verlag, 2001.

D.J. Hand and B.G. Batchelor. Experiments on the edited condensed nearest neighbour rule. *Information Sciences*, 14:171–180, 1978.

D.J. Hand and R.J. Till. A simple generalisation of the area under the ROC curve for multiple class classification problems. *Machine Learning*, 45:171–186, 2001.

D.J. Hand and K. Yu. Idiot's Bayes – not so stupid after all? *International Statistical Review*, 69(3):385–398, 2001.

P.L. Hansen and P. Salamon. Neural network ensembles. *IEEE Transactions on Pattern Analysis and Machine Intelligence*, 12:993–1001, 1990.

L.S. Harkins, J.M. Sirel, P.J. McKay, R.C. Wylie, D.M. Titterington, and R.M. Rowan. Discriminant analysis of macrocytic red cells. *Clinical and Laboratory Haematology*, 16:225–234, 1994.

C.J. Harris, A. Bailey, and T.J. Dodd. Multi-sensor data fusion in defence and aerospace. *The Aeronautical Journal*, 102:229–244, 1997.

J.D. Hart. On the choice of a truncation point in Fourier series density estimation. *Journal of Statistical Computation and Simulation*, 21:95–116, 1985.

P.E. Hart. The condensed nearest neighbor rule. *IEEE Transactions on Information Theory*, 14:515–516, 1968.

J.A. Hartigan. *Clustering Algorithms*. John Wiley & Sons, Ltd, 1975.

V. Hasselblad. Estimation of parameters for a mixture of normal distributions. *Technometrics*, 8:431–444, 1966.

T. Hastie, S. Rosset, R. Tibshirani, and J. Zhu. The entire regularisation path for the support vector machine. *Journal of Machine Learning Research*, 5:1391–1415, 2004.

T. Hastie and R. Tibshirani. Discriminant adaptive nearest neighbour classification. *IEEE Transactions on Pattern Analysis and Machine Intelligence*, 18(6):607–616, 1996.

T.J. Hastie, A. Buja, and R.J. Tibshirani. Penalized discriminant analysis. *Annals of Statistics*, 23(1):73–102, 1995.

T.J. Hastie and W. Stuetzle. Principal curves. *Journal of the American Statistical Association*, 84:502–516, 1989.

T.J. Hastie and R.J. Tibshirani. Discriminant analysis by Gaussian mixtures. *Journal of the Royal Statistical Society Series B*, 58(1):155–176, 1996.

T.J. Hastie, R.J. Tibshirani, and A. Buja. Flexible discriminant analysis by optimal scoring. *Journal of the American Statistical Association*, 89:1255–1270, 1994.

T.J. Hastie, R.J. Tibshirani, and J.H. Friedman. *The Elements of Statistical Learning: Data Mining, Inference, and Prediction*. Springer, 2001.

R.J. Hathaway and J.C. Bezdek. Nerf *c*-means: non-Euclidean relational fuzzy clustering. *Pattern Recognition*, 27(3):429–437, 1994.

S. Haykin. *Neural Networks. A Comprehensive Foundation*. Macmillan College Publishing Inc., 1994.

S. Haykin, W. Stehwien, C. Deng, P. Weber, and R. Mann. Classification of radar clutter in an air traffic control environment. *Proceedings of the IEEE*, 79(6):742–772, 1991.

Z. He and W. Yu. Stable feature selection for biomarker discovery. *Computational Biology and Chemistry*, 34(4):215–225, 2010.

D. Heath, S. Kasif, and S. Salzberg. Induction of oblique decision trees. *Journal of Artificial Intelligence Research*, 2(2):1–32, 1993.

D. Heckerman, D. Geiger, and D.M. Chickering. Learning Bayesian networks: the combination of knowledge and statistical data. *Machine Learning*, 20:197–243, 1995.

W.J. Heiser. A generalized majorization method for least squares multidimensional scaling of pseudodistances that may be negative. *Psychometrika*, 56(1):7–27, 1991.

W.J. Heiser. Convergent computation by iterative majorization: theory and applications in multidimensional data analysis. In W.J. Krzanowski, editor, *Recent Advances in Descriptive Multivariate Analysis*, pages 157–189. Clarendon Press, 1994.

W.E. Henley and D.J. Hand. A *k*-nearest-neighbour classifier for assessing consumer credit risk. *The Statistician*, 45(1):77–95, 1996.

W.H. Highleyman. The design and analysis of pattern recognition experiments. *Bell System Technical Journal*, 41:723–744, 1962.

D.V. Hinkley. Bootstrap methods. *Journal of the Royal Statistical Society Series B*, 50(3):321–337, 1988.

N.L. Hjort and G. Claeskens. Frequentist model average estimators. *Annals of the American Statistical Society*, 98(464):879–899, 2003.

N.L. Hjort and I.K. Glad. Nonparametric density estimation with a parametric start. *Annals of Statistics*, 23(3):882–904, 1995.

N.L. Hjort and M.C. Jones. Locally parametric nonparametric density estimation. *Annals of Statistics*, 24(4):1619–1647, 1996.

Y.-C. Ho and A.K. Agrawala. On pattern classification algorithms. Introduction and survey. *Proceedings of the IEEE*, 56(12):2101–2114, 1968.

Y.-C. Ho and R.L. Kashyap. An algorithm for linear inequalities and its applications. *IEEE Transactions on Electronic Computers*, 14(5):683–688, 1965.

J.A. Hoeting, D. Madigan, A.E. Raffery, and C.T. Volinsky. Bayesian model averaging: a tutorial. *Statistical Science*, 14(4): 382–417, 1999.

J.M. Hofman and C.H. Wiggins. A Bayesian approach to network modularity. *Physical Review Letters*, 100(25):258701, 2008.

P. Holme, M. Huss, and H. Jeong. Subnetwork hierarchies of biochemical pathways. *Bioinformatics*, 19(4):532–538, 2003.

C.C. Holmes and D.G.T. Denison. Classification with Bayesian MARS. *Machine Learning*, 50:159–173, 2003.

C.C. Holmes and B.K. Mallick. Bayesian radial basis functions of variable dimension. *Neural Computation*, 10:1217–1233, 1998.

L. Holmström and P. Koistinen. Using additive noise in back-propagation training. *IEEE Transactions on Neural Networks*, 3(1):24–38, 1992.

L. Holmström, P. Koistinen, J. Laaksonen, and E. Oja. Neural and statistical classifiers – taxonomy and two case studies. *IEEE Transactions on Neural Networks*, 8(1):5–17, 1997.

L. Holmström and S.R. Sain. Multivariate discrimination methods for top quark analysis. *Technometrics*, 39(1):91–99, 1997.

Z.-Q. Hong and J.-Y. Yang. Optimal discriminant plane for a small number of samples and design method of classifier on the plane. *Pattern Recognition*, 24(4):317–324, 1991.

K. Hornik. Some new results on neural network approximation. *Neural Networks*, 6:1069–1072, 1993.

D. Hosseinzadeh and S. Krishnan. Gaussian mixture modeling of keystroke patterns for biometric applications. *IEEE Transactions on Systems, Man, and Cybernetics - Part C: Applications and Reviews*, 38(6):816–826, 2008.

H. Hotelling. Analysis of a complex of statistical variables into principal components. *Journal of Educational Psychology*, 24:417–444, 1933.

C.W. Hsu, C.C. Chang, and C.J. Lin. A practical guide to support vector classification. Technical Report, Department of Computer Science, National Taiwan University, Taipei, 2003.

C.W. Hsu and C.J. Lin. A comparison on methods for multi-class support vector machines. *IEEE Transactions on Neural Networks*, 13(2):415–425, 2002.

S. Hua and Z. Sun. A novel method of protein secondary structure prediction with high segment overlap measure: support vector machine approach. *Journal of Molecular Biology*, 308:397–407, 2001.

C.-L. Huang and C.-J. Wang. A GA-based feature selection and parameters optimization for support vector machines. *Expert Systems with Applications*, 31(2):231–240, 2006.

Y.S. Huang and C.Y. Suen. A method of combining multiple experts for the recognition of unconstrained handwritten numerals. *IEEE Transactions on Pattern Analysis and Machine Intelligence*, 17(1):90–94, 1995.

Z. Huang and D. Zeng. A link prediction approach to anomalous email detection. In *IEEE International Conference on Systems, Man and Cybernetics*, pages 1131–1136. IEEE, 2006.

P.J. Huber. Projection pursuit (with discussion). *Annals of Statistics*, 13(2):435–452, 1985.

D.R. Hush, W. Horne, and J.M. Salas. Error surfaces for multilayer perceptrons. *IEEE Transactions on Systems, Man, and Cybernetics*, 22(5):1151–1161, 1992.

F. Hussein, N. Khanna, and R. Ward. Genetic algorithms for feature selection and weighting, a review and study. In *Proceedings of the Sixth International Conference on Document Analysis and Recognition*, pages 1240–1244. IEEE, 2001.

J.-N. Hwang, S.-R. Lay, and A. Lippman. Nonparametric multivariate density estimation: a comparative study. *IEEE Transactions on Signal Processing*, 42(10):2795–2810, 1994.

A. Hyvärinen and E. Oja. Independent component analysis: algorithms and applications. *Neural Networks*, 13:411–430, 2000.

S. Ingrassia. A comparison between the simulated annealing and the EM algorithms in normal mixture decompositions. *Statistics and Computing*, 2:203–211, 1992.

M.A. Ismail. Soft clustering: algorithms and validity of solutions. In M.M. Gupta and T. Yamakawa, editors, *Fuzzy Computing*, pages 445–472. Elsevier, 1988.

S.G. Iyengar, P.K. Varshney, and T. Damarla. A parametric copula based framework for multimodal signal processing. In *Proceedings of the IEEE International Conference on Acoustics, Speech, and Signal Processing (ICASSP'09)*, pages 1893–1896. IEEE, 2009.

A.J. Izenman. Recent developments in nonparametric density estimation. *Journal of the American Statistical Association*, 86:205–223, 1991.

T.S. Jaakkola. Tutorial on variational approximation methods. *Advanced Mean Field Methods: Theory and Practice,* The MIT Press, 2000.

D.A. Jackson. Stopping rules in principal components analysis: a comparison of heuristical and statistical approaches. *Ecology*, 74(8):2204–2214, 1993.

J.E. Jackson. *A User's Guide to Principal Components*. John Wiley & Sons, Ltd, 1991.

R.A. Jacobs, M.I. Jordan, S.J. Nowlan, and G.E. Hinton. Adaptive mixtures of local experts. *Neural Computation*, 3:79–87, 1991.

A.K. Jain and R.C. Dubes. *Algorithms for Clustering Data*. Prentice-Hall, 1988.

A.K. Jain, R.P.W. Duin, and J. Mao. Statistical pattern recognition: a review. *IEEE Transactions on Pattern Analysis and Machine Intelligence*, 22(1):4–37, 2000.

A.K. Jain and J.V. Moreau. Bootstrap technique in cluster analysis. *Pattern Recognition*, 20(5):547–568, 1987.

S. Jain and R.K. Jain. Discriminant analysis and its application to medical research. *Biomedical Journal*, 36(2):147–151, 1994.

M. Jambu and M.-O. Lebeaux. *Cluster Analysis and Data Analysis*. North Holland, 1983.

G.M. James and T.J. Hastie. Functional linear discriminant analysis for irregularly samples curves. *Journal of the Royal Statistical Society Series B*, 63(3):533–550, 2001.

M. Jamshidian and R.I. Jennrich. Conjugate gradient acceleration of the EM algorithm. *Journal of the American Statistical Association*, 88:221–228, 1993.

M. Jamshidian and R.I. Jennrich. Acceleration of the EM algorithm by using quasi-Newton methods. *Journal of the Royal Statistical Society Series B*, 59(3):569–587, 1997.

N. Jardine and R. Sibson. *Mathematical Taxonomy*. John Wiley & Sons, Ltd, 1971.

A. Jasra and P. Del Moral. Sequential Monte Carlo methods for option pricing. *Stochastic Analysis and Applications*, 29(2):292–316, 2011.

A. Jasra, D.A. Stephens, A. Doucet, and T. Tsagaris. Inference for Lévy driven stochastic volatility models via adaptive sequential Monte Carlo. *Scandinavian Journal of Statistics*, 38(1):1–22, 2011.

F.V. Jensen. *Introduction to Bayesian Networks*. Springer, 1997.

F.V. Jensen. *Bayesian Networks and Decision Graphs*. Statistics for Engineering and Information Science. Springer, 2002.

B. Jeon and D.A. Landgrebe. Fast Parzen density estimation using clustering-based branch and bound. *IEEE Transactions on Pattern Analysis and Machine Intelligence*, 16(9):950–954, 1994.

J. Jiang. Image compression with neural networks – a survey. *Signal Processing: Image Communication*, pages 737–760, 1999.

Q. Jiang and W. Zhang. An improved method for finding nearest neighbours. *Pattern Recognition Letters*, 14:531–535, 1993.

R.-K. Jin, D.C. Parkes, and P.J. Wolfe. Analysis of bidding networks in eBay: aggregate preference identification through community detection. In *Proceedings of the AAAI Workshop on Plan, Activity and Intent Recognition (PAIR)*, pages 66–73. 2007.

T. Joachims. Text categorization with support vector machines: Learning with many relevant features. In *Proceedings of ECML-98, 10th European Conference on Machine Learning*, pages 137–142. 1998.

T. Joachims. Structured output prediction with support vector machines. In *Structural, Syntactic, and Statistical Pattern Recognition, volume 4109 of Lecture Notes in Computer Science*, pages 1–7. Springer, 2006.

M. Jobert, H. Escola, E. Poiseau, and P. Gaillard. Automatic analysis of sleep using two parameters based on principal component analysis of electroencephalography spectral data. *Biological Cybernetics*, 71:197–207, 1994.

H. Joe. Asymptotic efficiency of the two-stage estimation method for copula-based models. *Journal of Multivariate Analysis*, 94:401–419, 2005.

N. Johnson and D. Hogg. Representation and synthesis of behaviour using Gaussian mixtures. *Image and Vision Computing*, 20:889–894, 2002.

I.T. Jolliffe. *Principal Components Analysis*. Springer-Verlag, 1986.

M.C. Jones and H.W. Lotwick. A remark on algorithm AS176. Kernel density estimation using the fast Fourier transform. *Applied Statistics*, 33:120–122, 1984.

M.C. Jones, J.S. Marron, and S.J. Sheather. A brief survey of bandwidth selection for density estimation. *Journal of the American Statistical Association*, 91:401–407, 1996.

M.C. Jones, I.J. McKay, and T.-C. Hu. Variable location and scale kernel density estimation. *Annals of the Institute of Statistical Mathematics*, 46(3):521–535, 1994.

M.C. Jones and S.J. Sheather. Using non-stochastic terms to advantage in kernel-based estimation of integrated squared density derivatives. *Statistics and Probability Letters*, 11:511–514, 1991.

M.C. Jones and R. Sibson. What is projection pursuit? (with discussion). *Journal of the Royal Statistical Society Series A*, 150:1–36, 1987.

M.C. Jones and D.F. Signorini. A comparison of higher order bias kernel density estimators. *Journal of the American Statistical Association*, 92:1063–1073, 1997.

M. Jordan. *Learning in Graphical Models*. The MIT Press, 1998.

M.I. Jordan, Z. Ghahramani, T.S. Jaakola, and L.K. Saul. An introduction to variational methods for graphical models. *Machine Learning*, 37:183–233, 1999.

M.I. Jordan and R.A. Jacobs. Hierarchical mixtures of experts and the EM algorithm. *Neural Computation*, 6:181–214, 1994.

A. Juan and E. Vidal. Fast *k*-means-like clustering in metric spaces. *Pattern Recognition Letters*, 15:19–25, 1994.

B.-H. Juang and L.R. Rabiner. Mixture autoregressive hidden Markov models for speech signals. *IEEE Transactions on Acoustics, Speech and Signal Processing*, 33(6):1404–1413, 1985.

H.F. Kaiser. The varimax criterion for analytic rotation in factor analysis. *Psychometrika*, 23:187–200, 1958.

H.F. Kaiser. Computer program for varimax rotation in factor analysis. *Educational and Psychological Measurement*, 19:413–420, 1959.

D. Kalles and T. Morris. Efficient incremental induction of decision trees. *Machine Learning*, 24:231–242, 1996.

J. Kalousis, A.and Prados and M. Hilario. Stability of feature selection algorithms. In *Proceedings of the Fifth IEEE International Conference on Data Mining*, ICDM '05, pages 218–225. IEEE, 2005.

M. Kam, C. Rorres, W. Chang, and X. Zhu. Performance and geometric interpretation for decision fusion with memory. *IEEE Transactions on Systems, Man and Cybernetics*, 29(1):52–62, 1999.

M. Kam, Q. Zau, and W.S. Gray. Optimal data fusion of correlated local decisions in multiple sensor detection systems. *IEEE Transactions on Aerospace and Electronic Systems*, 28(3):916–920, 1992.

M.S. Kamel and S.Z. Selim. A thresholded fuzzy *c*-means algorithm for semi-fuzzy clustering. *Pattern Recognition*, 24(9):825–833, 1991.

M.S. Kamel and S.Z. Selim. New algorithms for solving the fuzzy clustering problem. *Pattern Recognition*, 27(3):421–428, 1994.

B. Kamgar-Parsi and L. Kanal. An improved branch and bound algorithm for computing *k*-nearest neighbors. *Pattern Recognition Letters*, 3(1):7–12, 1985.

N.B. Karayiannis and G.W. Mi. Growing radial basis neural networks: merging supervised and unsupervised learning with network growth techniques. *IEEE Transactions on Neural Networks*, 8(6):1492–1506, 1997.

N.B. Karayiannis and A.N. Venetsanopolous. Efficient learning algorithms for neural networks (ELEANNE). *IEEE Transactions on Systems, Man and Cybernetics*, 23(5):1372–1383, 1993.

R.J. Karunamuni and T. Alberts. On boundary correction in kernel density estimation. *Statistical Methodology*, 2(3):191–212, 2005.

R.L. Kashyap. Algorithms for pattern classification. In J.M. Mendel and K.S. Fu, editors, *Adaptive, Learning and Pattern Recognition Systems. Theory and Applications*, pages 81–113. Academic Press, 1970.

R.E. Kass, B.P. Carlin, A. Gelman, and R.M. Neal. Markov chain Monte Carlo in practice: a roundtable discussion. *The American Statistician*, 52(2):93–100, 1998.

J.W. Kay. Comments on paper by Esposito *et al. IEEE Transactions on Pattern Analysis and Machine Intelligence*, 19(5):492–493, 1997.

S.S. Keerthi, S.K. Shevade, C. Bhattacharyya, and K.R.K. Murthy. Improvements to Platt's SMO algorithm for SVM classifier design. *Neural Computation*, 13:637–649, 2001.

M.G. Kelly, D.J. Hand, and N.M. Adams. The impact of changing populations on classifier performance. In *Proceedings of the 5th ACM SIGKDD Conference*, pages 367–371. ACM, 1999.

J. Kennedy and R.C. Eberhart. Particle swarm optimization. In *Proceedings of IEEE Conference on Neural Networks*, pages 1942–1948. IEEE, 1995.

R.D. Keppel, K.M. Brown, and K. Welch. *Forensic Pattern Recognition: from Fingerprints to Toolmarks*. Prentice-Hall, 2006.

G. Kim, M.J. Silvapulle, and P. Silvapulle. Comparison of semiparametric and parametric methods for estimating copulas. *Computational Statistics and Data Analysis*, 51(6):2836–2850, 2007.

S.-B. Kim, K.-S. Han, H.-C. Rim, and S.H. Myaeng. Some effective techniques for naïve Bayes text classification. *IEEE Transactions on Knowledge and Data Engineering*, 18(11):1457–1466, 2006.

S.C. Kim and T.J. Kang. Texture classification and segmentation using wavelet packet frame and Gaussian mixture model. *Pattern Recognition*, 40:1207–1221, 2007.

A.J. Kinderman and J.F. Monahan. Computer generation of random variables using the ratio of uniform deviates. *ACM Transactions on Mathematical Software (TOMS)*, 3(3):257–260, 1977.

C.A. Kirkwood, B.J. Andrews, and P. Mowforth. Automatic detection of gait events: a case study using inductive learning techniques. *Journal of Biomedical Engineering*, 11:511–516, 1989.

J. Kittler. Une généralisation de quelques algorithmes sous-optimaux de recherche d'ensembles d'attributs. In *Proceedings of Congrès AFCET/IRIA Reconnaissance des Formes et Traitement des Images*, pages 678–686, 1978.

J. Kittler and F.M. Alkoot. Relationship of sum and vote fusion strategies. In J. Kittler and F. Roli, editors, *Proceedings of the 2001 Workshop on Multiple Classifiers Systems*, pages 339–348. Springer-Verlag, 2001.

J. Kittler, M. Hatef, R.P.W. Duin, and J. Matas. On combining classifiers. *IEEE Transactions on Pattern Analysis and Machine Intelligence*, 20(3):226–239, 1998.

J. Kittler and P.C. Young. A new approach to feature selection based on the Karhunen–Loève expansion. *Pattern Recognition*, 5:335–352, 1973.

J. Kittler, J. Matas, K. Jonsson, and Ramos Sánchez, M.U. Combining evidence in personal identity verification systems. *Pattern Recognition Letters*, 18:845–852, 1997.

U.B. Kjaerulff and A.L. Madsen. *Bayesian Networks and Influence Diagrams: A Guide to Construction and Analysis*. Information Science and Statistics. Springer, 2010.

J. Klemelä. *Smoothing of Multivariate Data: Density Estimation and Visualization*. John Wiley & Sons, Ltd, 2009.

T. Kohonen. *Self-organization and Associative Memory*. Springer-Verlag, third edition, 1989.

T. Kohonen. The self-organizing map. *Proceedings of the IEEE*, 78:1464–1480, 1990.

T. Kohonen. *Self-organizing Maps*. Springer-Verlag, second edition, 1997.

T. Kohonen, E. Oja, O. Simula, A. Visa, and J. Kangas. Engineering applications of the self-organising map. *Proceedings of the IEEE*, 84(10):1358–1384, 1996.

D. Koller and N. Friedman. *Probabilistic Graphical Models: Principals and Techniques*. The MIT Press, 2009.

T. Komviriyavut, P. Sangkatsanee, N. Wattanapongsakorn, and C. Charnsripinyo. Network intrusion detection and classification with decision tree and rule based approaches. In *Proceedings of the 9th International Conference on Communications and Information Technologies*, pages 1046–1050. IEEE, 2009.

S. Konishi and M. Honda. Comparison of procedures for estimation of error rates in discriminant analysis under nonnormal populations. *Journal of Statistical Computation and Simulation*, 36:105–115, 1990.

W.L.G. Koontz and K. Fukunaga. A nonlinear feature extraction algorithm using distance transformation. *IEEE Transactions on Computers*, 21(1):56–63, 1972.

W.L.G. Koontz, P.M. Narendra, and K. Fukunaga. A branch and bound clustering algorithm. *IEEE Transactions on Computers*, 24(9):908–915, 1975.

T. Koski and J.M. Noble. *Bayesian Networks: an Introduction*. Series in Probability and Statistics. John Wiley & Sons, Ltd, 2009.

F. Kovács, C. Legány, and A. Babos. Cluster validity measurement techniques. In *6th International Symposium of Hungarian Researchers on Computational Intelligence*, pages 388–393. 2006.

M.A. Kraaijveld. A Parzen classifier with an improved robustness against deviations between training and test data. *Pattern Recognition Letters*, 17:679–689, 1996.

M.A. Kraaijveld, J. Mao, and A.K. Jain. A non-linear projection method based on Kohonen's topology preserving maps. In *11th International Conference on Pattern Recognition*. 1992.

D.E. Kreithen, S.D. Halversen, and G.J. Owirka. Discriminating targets from clutter. *The Lincoln Laboratory Journal*, 6(1):25–51, 1993.

M. Kristan, A. Leonardis, and Skočaj. Multivariate online kernel density estimation with Gaussian kernels. *Pattern Recognition*, 44(10–11):2630–2642, 2011.

R.A. Kronmal and M. Tarter. The estimation of probability densities and cumulatives by Fourier series methods. *Journal of the American Statistical Association*, 63:925–952, 1962.

E. Krusińska. Robust methods in discriminant analysis. *Rivista di Statistica Applicada*, 21(3):239–253, 1988.

J.B. Kruskal. Multidimensional scaling by optimizing goodness-of-fit to a nonmetric hypothesis. *Psychometrika*, 29:1–28, 1964a.

J.B. Kruskal. Nonmetric multidimensional scaling: a numerical method. *Psychometrika*, 29(2):115–129, 1964b.

J.B. Kruskal. Comments on 'A nonlinear mapping for data structure analysis'. *IEEE Transactions on Computers*, 20:1614, 1971.

W.J. Krzanowski. The location model for mixtures of categorical and continuous variables. *Journal of Classification*, 10(1):25–49, 1993.

W.J. Krzanowski, P. Jonathan, W.V. McCarthy, and M.R. Thomas. Discriminant analysis with singular covariance matrices: methods and applications to spectroscopic data. *Applied Statistics*, 44(1):101–115, 1995.

W.J. Krzanowski and F.H.C. Marriott. *Multivariate Analysis. Part 1: Distributions, Ordination and Inference*. Edward Arnold, London, 1994.

W.J. Krzanowski and F.H.C. Marriott. *Multivariate Analysis. Part 2: Classification, Covariance Structures and Repeated Measurements*. Edward Arnold, London, 1996.

A. Krzyzak. Classification procedures using multivariate variable kernel density estimate. *Pattern Recognition Letters*, 1:293–298, 1983.

M. Kudo and J. Sklansky. Comparison of algorithms that select features for pattern classifiers. *Pattern Recognition*, 33:25–41, 2000.

L. Kuncheva. *Combining Pattern Classifiers: Methods and Algorithms*. Wiley-Blackwell, 2004a.

L.I. Kuncheva. Classifier ensembles for changing environments. In F. Roli, J. Kittler, and T. Windeatt, editors, *Multiple Classifier Systems*, volume 3077 of *Lecture Notes in Computer Science*, pages 1–15. Springer, 2004b.

C.-K. Kwoh and D.F. Gillies. Using hidden nodes in Bayesian networks. *Artificial Intelligence*, 88:1–38, 1996.

T.-Y. Kwok and D.-Y. Yeung. Use of bias term in projection pursuit learning improves approximation and convergence properties. *IEEE Transactions on Neural Networks*, 7(5):1168–1183, 1996.

P.A. Lachenbruch and M.R. Mickey. Estimation of error rates in discriminant analysis. *Technometrics*, 10:1–11, 1968.

P.A. Lachenbruch, C. Sneeringer, and L.T. Revo. Robustness of the linear and quadratic discriminant function to certain types of non-normality. *Communications in Statistics*, 1(1):39–56, 1973.

L. Lam and C.Y. Suen. Optimal combinations of pattern classifiers. *Pattern Recognition Letters*, 16:945–954, 1995.

C.G. Lambert, S.E. Harrington, C.R. Harvey, and A. Glodjo. Efficient on-line nonparametric kernel density estimation. *Algorithmica*, 25(1):37–57, 1999.

J. Lampinen and A. Vehtari. Bayesian approach for neural networks – review and case studies. *Neural Networks*, 14:257–274, 2001.

G. Landeweerd, T. Timmers, E. Gersema, M. Bins, and M. Halic. Binary tree versus single level classification of white blood cells. *Pattern Recognition*, 16:571–577, 1983.

T.C.W. Landgrebe and R.P.W. Duin. Approximating the multiclass ROC by pairwise analysis. *Pattern Recognition Letters*, 28(13):1747–1758, 2007.

T.C.W. Landgrebe and R.P.W. Duin. Efficient multiclass ROC approximation by decomposition via confusion matrix perturbation analysis. *IEEE Transactions on Pattern Analysis and Machine Intelligence*, 30:810–822, 2008.

R.O. Lane. Non-parametric Bayesian super-resolution. *IET Radar, Sonar and Navigation*, 4(4):639–648, 2010.

R.O. Lane, M. Briers, and K. Copsey. Approximate Bayesian computation for source term estimation. In *Mathematics in Defence 2009*. 2009.

K. Lange. A gradient algorithm locally equivalent to the EM algorithm. *Journal of the Royal Statistical Society Series B*, 57(2):425–437, 1995.

P. Langley and H.A. Simon. Applications of machine learning and rule induction. *Communications of the ACM*, 38:55–64, 1995.

F. Lauer and G. Bloch. Incorporating prior knowledge in support vector machines for classification: a review. *Neurocomputing*, 71(7–9):1578–1594, 2008.

S.L. Lauritzen and D.J. Spiegelhalter. Local computations with probabilities on graphical structures and their application to expert systems (with discussion). *Journal of the Royal Statistical Society Series B*, 50:157–224, 1988.

M. Lavine and M. West. A Bayesian method for classification and discrimination. *The Canadian Journal of Statistics*, 20(4):451–461, 1992.

M. Laviolette, J.W. Seaman, J.D. Barrett, and W.H. Woodall. A probabilistic and statistical view of fuzzy methods (with discussion). *Technometrics*, 37(3):249–292, 1995.

E.L. Lawler and D.E. Wood. Branch-and-bound methods: a survey. *Operations Research*, 14(4):699–719, 1966.

M. Lázaro, I. Santamaría, and C. Pantaleón. A new EM-based training algorithm for RBF networks. *Neural Networks*, 16:69–77, 2003.

J.R. Leathwick, D. Rowe, J. Richardson, J. Elith, and T. Hastie. Using multivariate adaptive regression splines to predict the distributions of New Zealand's freshwater diadromous fish. *Freshwater Biology*, 50:2034–2052, 2005.

Y. Le Cun, B. Boser, J. Denker, D. Henderson, R. Howard, W. Hubbard, and L. Jackel. Backpropagation applied to hand-written zip code recognition. *Neural Computation*, 1(4):541–551, 1989.

C.-C. Lee, S.-S. Huang, and Shih C.-Y. Facial affect recognition using regularized discriminant analysis-based algorithms. *EURASIP Journal on Advances in Signal Processing*, 2010:1:1–1:10, 2010.

J.S. Lee, M.R. Grunes, and R. Kwok. Classification of multi-look polarimetric SAR imagery based on complex Wishart distribution. *International Journal of Remote Sensing*, 15(11):2299–2311, 1994.

P.M. Lee. *Bayesian Statistics: an Introduction*. Arnold, third edition, 2004.

S.H. Lee, J.H. Kim, K.G. Kim, S.J. Park, and W.K. Moon. K-means clustering and classification of kinetic curves on malignancy in dynamic breast MRI. *IFMBE Proceedings*, 14(15):2536–2539, 2007.

E.A. Leicht and M.E.J. Newman. Community structure in directed networks. *Physical Review Letters*, 100(11):118703, 2008.

B. Lerner, H. Guterman, M. Aladjem, and I. Dinstein. A comparative study of neural network based feature extraction paradigms. *Pattern Recognition Letters*, 20:7–14, 1999.

M. Leshno, V.Y. Lin, A. Pinkus, and S. Schocken. Multilayer feedforward networks with a non-polynomial activation function can approximate any function. *Neural Networks*, 6:861–867, 1993.

D.D. Lewis. Naïve Bayes at forty: the independence assumption in information retrieval. In *Proceedings of ECML-98, 10th European Conference on Machine Learning*, pages 4–15. 1998.

C. Li, D.B. Goldgof, and L.O. Hall. Knowledge-based classification and tissue labeling of MR images of human brain. *IEEE Transactions on Medical Imaging*, 12(4):740–750, 1993.

D.X. Li. On default correlation: a copula function approach. *The Journal of Fixed Income*, 9(4):43–54, 2000.

T. Li and I.K. Sethi. Optimal multiple level decision fusion with distributed sensors. *IEEE Transaction on Aerospace and Electronic Systems*, 29(4):1252–1259, 1993.

X. Li and A.G.O. Yeh. Principal component analysis of stacked multi-temporal images for the monitoring of rapid urban expansion in the Pearl River Delta. *International Journal of Remote Sensing*, 19(8):1501–1518, 1998.

H.-T. Lin, C.-J. Lin, and Weng R.C. A note on Platt's probabilistic outputs for support vector machines. *Machine Learning*, 68:267–276, 2007.

S.-W. Lin, K.-C. Ying, S.-C. Chen, and Z.-J. Lee. Particle swarm optimization for parameter determination and feature selection of support vector machines. *Expert Systems with Applications*, 35:1817–1824, 2008.

Y. Lin, Y. Lee, and G. Wahba. Support vector machines for classification in nonstandard situations. *Machine Learning*, 46(1–3):191–202, 2002.

Y. Linde, A. Buzo, and R.M. Gray. An algorithm for vector quantizer design. *IEEE Transactions on Communications*, 28(1):84–95, 1980.

B.G. Lindsay and P. Basak. Multivariate normal mixtures: a fast consistent method of moments. *Journal of the American Statistical Association*, 88:468–476, 1993.

R.F. Ling. A probability theory for cluster analysis. *Journal of the American Statistical Association*, 68:159–164, 1973.

J.C. Lingoes, E.E. Roskam, and I. Borg, editors. *Geometric Representations of Relational Data. Readings in Multidimensional Scaling*, volume 20(2). Mathesis Press, 1979.

R.J.A. Little. Regression with missing X's: a review. *Journal of the American Statistical Association*, 87:1227–1237, 1992.

R.J.A. Little and D.B. Rubin. *Statistical Analysis with Missing Data*. John Wiley & Sons, Ltd, 1987.

B. Liu. *Web Data Mining: Exploring Hyperlinks, Contents, and Usage Data*. Springer, 2006.

C. Liu and D.B. Rubin. The ECME algorithm: a simple extension of EM and ECM with faster monotone convergence. *Biometrika*, 81(4):633–648, 1994.

H. Liu and H. Motoda, editors. *Computational Methods of Feature Selection*. Chapman and Hall/CRC, 2007.

H. Liu and L. Yu. Toward integrating feature selection algorithms for classification. *IEEE Transactions on Knowledge and Data Engineering*, 17(4):491–502, 2005.

H.-L. Liu, N.-Y. Chen, W.-C. Lu, and X.-W. Zhu. Multi-target classification pattern recognition applied to computer-aided materials design. *Analytical Letters*, 27(11):2195–2203, 1994.

J.S. Liu, W.H. Wong, and A. Kong. Covariance structure of the Gibbs sampler with applications to the comparisons of estimators and augmentation schemes. *Biometrika*, 81(1):27–40, 1994.

K. Liu, Y.Q. Cheng, and J.-Y. Yang. A generalized optimal set of discriminant vectors. *Pattern Recognition*, 25(7):731–739, 1992.

K. Liu, Y.Q. Cheng, and J.-Y. Yang. Algebraic feature extraction for image recognition based on an optimal discriminant criterion. *Pattern Recognition*, 26(6):903–911, 1993.

T. Liu, A.W. Moore, and Gray A. New algorithms for efficient high-dimensional nonparametric classification. *Journal of Machine Learning Research*, 7:1135–1158, 2006.

W. Liu, J. Tian, and X. Chen. RDA for automatic airport recognition on FLIR image. In *Proceedings of the 7th World Congress on Intelligent Control and Automation*, pages 5966–5969. IEEE, 2008.

W.Z. Liu and A.P. White. A comparison of nearest neighbour and tree-based methods of non-parametric discriminant analysis. *Journal of Statistical Computation and Simulation*, 53:41–50, 1995.

W.Z. Liu, A.P. White, S.G. Thompson, and M.A. Bramer. Techniques for dealing with missing values in classification. In X. Liu, P. Cohen, and M. Berthold, editors, *Advances in Intelligent Data Analysis*. Springer-Verlag, 1997.

K. Lock and Gelman A. Bayesian combination of state polls and election forecasts. *Political Analysis*, 18:337–348, 2010.

H. Lodhi, C. Saunders, J. Shawe-Taylor, N. Cristianini, and C. Watkins. Text classification using string kernels. *Journal of Machine Learning Research*, 2:419–444, 2002.

A.M. Logar, E.M. Corwin, and W.J.B. Oldham. Performance comparisons of classification techniques for multi-font character recognition. *International Journal of Human-Computer Studies*, 40:403–423, 1994.

W.-L. Loh. On linear discriminant analysis with adaptive ridge classification rules. *Journal of Multivariate Analysis*, 53:264–278, 1995.

W.-Y. Loh and N. Vanichsetakul. Tree-structured classification via generalized discriminant analysis (with discussion). *Journal of the American Statistical Association*, 83:715–727, 1988.

M. Loog and R.P.W. Duin. Linear dimensionality reduction via a heteroscedastic extension of LDA: the Chernoff criterion. *IEEE Transactions on Pattern Analysis and Machine Intelligence*, 26(6):732–739, 2004.

D. Lowe. Novel 'topographic' nonlinear feature extraction using radial basis functions for concentration coding on the 'artificial nose'. In *3rd IEE International Conference on Artificial Neural Networks*, pages 95–99. IEE, 1993.

D. Lowe, editor. *IEE Proceedings on Vision, Image and Signal Processing*, 141(4): 1994. Special issue on 'Applications of Artificial Neural Networks'.

D. Lowe. Radial basis function networks. In M.A. Arbib, editor, *The Handbook of Brain Theory and Neural Networks*, pages 779–782. The MIT Press, 1995a.

D. Lowe. On the use of nonlocal and non-positive definite basis functions in radial basis function networks. In *IEE International Conference on Artificial Neural Networks*, pages 206–211. IEE, 1995b.

D. Lowe and M. Tipping. Feed-forward neural networks and topographic mappings for exploratory data analysis. *Neural Computing and Applications*, 4:83–95, 1996.

D. Lowe and A.R. Webb. Exploiting prior knowledge in network optimization: an illustration from medical prognosis. *Network*, 1:299–323, 1990.

D. Lowe and A.R. Webb. Optimized feature extraction and the Bayes decision in feed-forward classifier networks. *IEEE Transactions on Pattern Analysis and Machine Intelligence*, 13(4):355–364, 1991.

J. Lu, K.N. Plataniotis, and A.N. Venetsanopoulos. Regularized discriminant analysis for the small sample size problem in face recognition. *Pattern Recognition Letters*, 24(16):3079–3087, 2003.

J. Lu, K.N. Plataniotis, and A.N. Venetsanopoulos. Regularization studies of linear discriminant analysis in small sample size scenarios with application to face recognition. *Pattern Recognition Letters*, 26:181–191, 2005.

L. Lü and T. Zhou. Link prediction in complex networks: a survey. *Physica A: Statistical Mechanics and its Applications*, 390(6):1150–1170, 2010.

A.J. Lunn, A. Thomas, N. Best, and D. Spiegelhalter. WinBUGS – a Bayesian modelling framework: contents, structure and extensibility. *Statistics and Computing*, 10:325–337, 2000.

D. Lusseau and M.E.J. Newman. Identifying the role that animals play in their social networks. *Proceedings of the Royal Society of London B*, 271(S6):S477–S481, 2004.

D. Lusseau, H. Whitehead, and S. Gero. Applying network methods to the study of animal social structures. *Animal Behaviour*, 75:1809–1815, 2008.

S.P. Luttrell. Hierarchical vector quantisation. *IEE Proceedings Part I*, 136(6):405–413, 1989.

S.P. Luttrell. Using self-organising maps to classify radar range profiles. In *Proceedings of the 4th International Conference on Artificial Neural Networks*, pages 335–340. IEE, 1995.

S.P. Luttrell. Self-organised modular neural networks for encoding data. In A.J.C. Sharkey, editor, *Combining Artificial Neural Nets: Ensemble and Modular Multi-Net Systems*, pages 235–263. Springer-Verlag, 1999.

M.A. Lyons, R.S.H. Yang, A.N. Mayeno, and B. Reisfeld. Computational toxicology of chloroform: Reverse dosimetry using Bayesian inference, Markov chain Monte Carlo simulation, and human biomonitoring data. *Environmental Health Perspectives*, 116(8):1040–1043, 2008.

D.J.C. MacKay. Probable networks and plausible predictions – a review of practical Bayesian methods for supervised neural networks. *Network: Computation in Neural Systems*, 6:469–505, 1995.

D. Madigan and A.E. Raftery. Model selection and accounting for model uncertainty in graphical models using Occam's window. *Journal of the American Statistical Association*, 89:1535–1546, 1994.

M. Magee, R. Weniger, and D. Wenzel. Multidimensional pattern classification of bottles using diffuse and specular illumination. *Pattern Recognition*, 26(11):1639–1654, 1993.

J.R. Magnus, O. Powell, and P. Prüfer. A comparison of two model averaging techniques with an application to growth empirics. *Journal of Econometrics*, 154:139–153, 2010.

J. Makhoul, S. Roucos, and H. Gish. Vector quantization in speech coding. *Proceedings of the IEEE*, 73(11):1511–1588, 1985.

B.F.J. Manly. *Multivariate Statistical Methods, a Primer*. Chapman and Hall, 1986.

J. Mao and A.K. Jain. Artificial neural networks for feature extraction and multivariate data projection. *IEEE Transactions on Neural Networks*, 6(2):296–317, 1995.

D.J. Marchette. *Random Graphs for Statistical Pattern Recognition*. John Wiley & Sons, Ltd, 2004.

M. Marinaro and S. Scarpetta. On-line learning in RBF neural networks: a stochastic approach. *Neural Networks*, 13:719–729, 2000.

P. Marjoram, J. Molitor, V. Plagnol, and S. Tavaré. Markov chain Monte Carlo without likelihoods. *Proceedings of the National Academy of Sciences of the United States of America*, 100(26):15324–15328, 2003.

J.S. Marron. Automatic smoothing parameter selection: a survey. *Empirical Economics*, 13:187–208, 1988.

A.D. Marrs. An application of reversible-jump MCMC to multivariate spherical Gaussian mixtures. *Advances in Neural Information Processing Systems*, 10:577–583, 1998.

G. Martinelli, J. Eidsvik, R. Hauge, and M.D. Forland. Bayesian networks for prospect analysis in the North Sea. Technical Report of Department of Mathematical Sciences, NTNU, Trondheim, 2010.

S.R. Maskell. A Bayesian approach to fusing uncertain, imprecise and conflicting information. *Information Fusion Journal*, 9(2):259–277, April 2008.

K. Matsuoka. Noise injection into inputs in back-propagation learning. *IEEE Transactions on Systems, Man, and Cybernetics*, 22(3):436–440, 1992.

A. McCallum and K. Nigam. A comparison of event models for naïve Bayes text classification. In *Proceedings of AAAI/ICML-98 Workshop on Learning for Text Categorization*, pages 41–48. AAAI Press, 1998.

C.A. McGrory and D.M. Titterington. Variational Bayesian analysis for hidden Markov models. *Australian and New Zealand Journal of Statistics*, 51(2):227–244, 2009.

R.M. McIntyre and R.K. Blashfield. A nearest-centroid technique for evaluating the minimum-variance clustering procedure. *Multivariate Behavioral Research*, 2:225–238, 1980.

S.J. McKenna, S. Gong, and Y. Raja. Modelling facial colour and identity with Gaussian mixtures. *Pattern Recognition*, 31(12):1883–1892, 1998.

G.J. McLachlan. Error rate estimation in discriminant analysis: recent advances. In A.K. Gupta, editor, *Advances in Multivariate Analysis*. D. Reidel Publishing Company, 1987.

G.J. McLachlan. *Discriminant Analysis and Statistical Pattern Recognition*. John Wiley & Sons, Ltd, 1992a.

G.J. McLachlan. Cluster analysis and related techniques in medical research. *Statistical Methods in Medical Research*, 1(1):27–48, 1992b.

G.J. McLachlan and K.E. Basford. *Mixture Models: Inference and Applications to Clustering*. Marcel Dekker, 1988.

G.J. McLachlan and T. Krishnan. *The EM Algorithm and Extensions*. John Wiley & Sons, Ltd, 1996.

G.J. McLachlan and D. Peel. *Finite Mixture Models*. John Wiley & Sons, Ltd, 2000.

W. Meier, R. Weber, and H.-J. Zimmermann. Fuzzy data analysis – methods and industrial applications. *Fuzzy Sets and Systems*, 61:19–28, 1994.

J. Meinguet. Multivariate interpolation at arbitrary points made simple. *Journal of Applied Mathematics and Physics (ZAMP)*, 30:292–304, 1979.

M.R. Melchiori. Which Archimedean copula is the right one? *YieldCurve.com*, 2003.

F. Melgani and L. Bruzzone. Classification of hyperspectral remote sensing images with support vector machines. *IEEE Transactions on Geoscience and Remote Sensing*, 42(8):1778–1790, 2004.

I. Melvin, J. Weston, C.S. Leslie, and W.S. Noble. Combining classifiers for improved classification of proteins from sequence or structure. *BMC Bioinformatics*, 9:389, 2008.

X.-L. Meng and D.B. Rubin. Recent extensions to the EM algorithm. In J.M. Bernado, J.O. Berger, A.P. Dawid, and A.F.M. Smith, editors, *Bayesian Statistics*, volume 4, pages 307–320. Oxford University Press, 1992.

X.-L. Meng and D.B. Rubin. Maximum likelihood estimation via the ECM algorithm: a general framework. *Biometrika*, 80(2):267–27, 1993.

X.-L. Meng and D. van Dyk. The EM algorithm – an old folk-song sung to a fast new tune (with discussion). *Journal of the Royal Statistical Society Series B*, 59(3):511–567, 1997.

K.L. Mengersen, C.P. Robert, and C. Guihenneuc-Jouyaux. MCMC convergence diagnostics: a review. In J.M. Bernardo, J.O. Berger, A.P. Dawid, and A.F.M. Smith, editors, *Bayesian Statistics*, pages 399–432. Oxford University Press, 1998.

C.L. Merz. Using correspondence analysis to combine classifiers. *Machine Learning*, 36:33–58, 1999.

A. Meyer-Baese. *Pattern Recognition in Medical Imaging*. Academic Press, 2003.

C.A. Micchelli. Interpolation of scattered data: distance matrices and conditionally positive definite matrices. *Constructive Approximation*, 2:11–22, 1986.

M. Michael and W.-C. Lin. Experimental study of information measure and inter-intra class distance ratios on feature selection and orderings. *IEEE Transactions on Systems, Man and Cybernetics*, 3(2):172–181, 1973.

D. Michie, D.J. Spiegelhalter, and C.C. Taylor. *Machine Learning, Neural and Statistical Classification*. Ellis Horwood Limited, 1994.

M.L. Micó, J. Oncina, and E. Vidal. A new version of the nearest-neighbour approximating and eliminating search algorithm (AESA) with linear preprocessing time and memory requirements. *Pattern Recognition*, 15:9–17, 1994.

T. Mikosch. Copulas: tales and facts. *Extremes*, 9:3–20, 2006.

J. Milgram, M. Cheriet, and R. Sabourin. 'One against one' or 'one against all': which one is better for handwriting recognition with SVMs? In G. Lorette, editor, *Proceedings of 10th International Workshop on Frontiers in Handwriting Recognition*. Suvisoft, 2006.

S. Milgram. The small world problem. *Psychology Today*, 2:60–67, 1967.

A.J. Miller. *Subset Selection in Regression*. Chapman and Hall, 1990.

G.W. Milligan. A Monte Carlo study of thirty internal measures for cluster analysis. *Psychometrika*, 46(2):187–199, 1981.

G.W. Milligan and M.C. Cooper. An examination of procedures for determining the number of clusters in a data set. *Psychometrika*, 50(2):159–179, 1985.

J. Mingers. An empirical comparison of pruning methods for decision tree inductions. *Machine Learning*, 4:227–243, 1989.

M.L. Minsky and S.A. Papert. *Perceptrons. An Introduction to Computational Geometry*. The MIT Press, 1988.

T.M. Mitchell. *Machine Learning*. McGraw-Hill, 1997.

C. Miyajima, Y. Nishiwaki, K. Ozawa, T. Wakita, K. Itou, K. Takeda, and F. Itakura. Driver modeling based on driving behavior and its evaluation in driver identification. *Proceedings of the IEEE*, 95(2):427–437, 2007.

A. Mkhadri. Shrinkage parameter for the modified linear discriminant analysis. *Pattern Recognition Letters*, 16:267–275, 1995.

A. Mkhadri, G. Celeux, and A. Nasroallah. Regularization in discriminant analysis: an overview. *Computational Statistics and Data Analysis*, 23:403–423, 1997.

R. Mojena. Hierarchical grouping methods and stopping rules: an evaluation. *Computer Journal*, 20:359–363, 1977.

M. Mojirsheibani. Combining classifiers via discretization. *Journal of the American Statistical Association*, 94(446):600–609, 1999.

L.C. Molina, L. Belanche, and A. Nebot. Feature selection algorithms: a survey and experimental evaluation. In *Proceedings of the IEEE International Conference on Data Mining*, pages 306–313. IEEE, 2002.

M. Montemerlo, S. Thrun, D. Koller, and B. Wegbreit. FastSLAM 2.0: an improved particle filtering algorithm for simultaneous localization and mapping that provably converges. In *Proceedings of the Sixteenth International Joint Conference on Artificial Intelligence (IJCAI)*. 2003.

J. Montgomery and B. Nyhan. Bayesian model averaging: theoretical developments and practical applications. *Political Science*, 18(2):245–270, 2010.

A.W. Moore. *Efficient Memory-based Learning for Robot Control: An introductory tutorial on kd-trees*. PhD thesis, Computer Laboratory, University of Cambridge, 1991.

M.A. Moran and B.J. Murphy. A closer look at two alternative methods of statistical discrimination. *Applied Statistics*, 28(3):223–232, 1979.

F. Moreno-Seco, L. Micó, and J. Oncina. Extending LAESA fast nearest neighbour algorithm to find the k nearest neighbours. *Springer Lecture Notes in Computer Science*, 2396:718–724, 2002.

B.J.T. Morgan and A.P.G. Ray. Non-uniqueness and inversions in cluster analysis. *Applied Statistics*, 44(1):117–134, 1995.

N. Morgan and H.A. Bourlard. Neural networks for the statistical recognition of continuous speech. *Proceedings of the IEEE*, 83(5):742–772, 1995.

D.P. Mukherjee, D.K. Banerjee, B. Uma Shankar, and D.D. Majumder. Coal petrography: a pattern recognition approach. *International Journal of Coal Geology*, 25:155–169, 1994.

D.P. Mukherjee, A. Pal, S.E. Sarma, and D.D. Majumder. Water quality analysis: a pattern recognition approach. *Pattern Recognition*, 28(2):269–281, 1995.

S. Mukkamala, A.H. Sung, A. Abraham, and V. Ramos. Intrusion detection using adaptive regression splines. In *6th International Conference on Enterprise Information Systems, EIS'04*, pages 26–33. Kluwer Academic Press, 2004.

D.J. Munro, O.K. Ersoy, M.R. Bell, and J.S. Sadowsky. Neural network learning of low-probability events. *IEEE Transactions on Aerospace and Electronic Systems*, 32(3):898–910, 1996.

P.M. Murphy and D.W. Aha. UCI repository of machine learning databases. Technical Report, http://www.ics.uci.edu/ mlearn/MLRepository.html, UCI, 1995.

F. Murtagh. *Multidimensional Clustering Algorithms*. Physica-Verlag, 1985.

F. Murtagh. Contiguity-constrained clustering for image analysis. *Pattern Recognition Letters*, 13:677–683, 1992.

F. Murtagh. Interpreting the Kohonen self-organizing feature map using contiguity-constrained clustering. *Pattern Recognition Letters*, 16:399–408, 1995.

F. Murtagh and M. Hernández-Pajares. The Kohonen self-organizing map method: an assessment. *Journal of Classification*, 12(2):165–190, 1995.

S.K. Murthy. Automatic construction of decision trees from data: a multidisciplinary survey. *Data Mining and Knowledge Discovery*, 2:345–389, 1998.

S.K. Murthy, S. Kasif, and S. Salzberg. A system for induction of oblique decision trees. *Journal of Artificial Intelligence Research*, 2:1–32, 1994.

M.T. Musavi, W. Ahmed, K.H. Chan, K.B. Faris, and D.M. Hummels. On the training of radial basis function classifiers. *Neural Networks*, 5:595–603, 1992.

J.P. Myles and D.J. Hand. The multi-class metric problem in nearest neighbour discrimination rules. *Pattern Recognition*, 23(11):1291–1297, 1990.

I.T. Nabney. *NETLAB: Algorithms for Pattern Recognition*. Springer, 2001.

I.T. Nabney, editor. *NETLAB Algorithms for Pattern Recognition*. Springer, 2002.

E.A. Nadaraya. *Nonparametric Estimation of Probability Densities and Regression Curves*. Kluwer Academic, 1989.

P.M. Narendra and K. Fukunaga. A branch and bound algorithm for feature subset selection. *IEEE Transactions on Computers*, 26:917–922, 1977.

R. Neal. Annealed importance sampling. *Statistics and Computing*, 11(2):125–139, April 2001.

R.M. Neal. Slice sampling. *The Annals of Statistics*, 31(3):705–767, 2003.

R.E Neapolitan, editor. *Learning Bayesian Networks*. Series in Artificial Intelligence. Prentice Hall, 2003.

M. Neil, D. Marquez, and N. Fenton. Using Bayesian networks to model the operational risk to information technology infrastructure in financial institutions. *Journal of Financial Transformation*, 22:131–138, 2008.

M. Neil, M. Tailor, and D. Marquez. Inference in hybrid Bayesian networks using dynamic discretisation. *Statistics and Computing*, 17(3):219–233, 2007.

M.E.J. Newman. The structure and function of complex networks. *SIAM Review*, 45(2):167–256, 2003.

M.E.J. Newman. Finding community structure in networks using the eigenvectors of matrices. *Physical Review E*, 74:036184, 2006.

M.E.J. Newman. *Networks. An Introduction*. Oxford University Press, 2010.

M.E.J. Newman and M. Girvan. Finding and evaluating community structure in networks. *Physical Review E*, 69(5):026113, 2004.

M.E.J. Newman and E.A. Leicht. Mixture models and exploratory analysis in networks. *Proceedings of the National Academy of Sciences of the United States of America*, 104(23):9564–9569, 2007.

A.Y. Ng, M.I. Jordan, and Y. Weiss. On spectral clustering: analysis and an algorithm. In T. Dieterrich, S. Becker, and Z. Ghahramani, editors, *Advances in Neural Information Processing Systems*, pages 849–856. The MIT Press, 2002.

H. Niemann and R. Goppert. An efficient branch-and-bound algorithm nearest neighbour classifier. *Pattern Recognition Letters*, 7(2):67–72, 1988.

H. Niemann and J. Weiss. A fast-converging algorithm for nonlinear mapping of high dimensional data to a plane. *IEEE Transactions on Computers*, 28(2):142–147, 1979.

N.J. Nilsson. *Learning Machines: Foundations of Trainable Pattern-Classifying Systems*. McGraw-Hill, 1965.

H. Ning, W. Xu, Y. Chi, Y. Gong, and T. Huang. Incremental spectral clustering with application to monitoring of evolving blog communities. In *SIAM International Conference on Data Mining*, pages 261–272. SIAM, 2007.

M. Nixon and A.S. Aquado. *Feature Extraction and Image Processing*. Academic Press, second edition, 2008.

I. Ntzoufras. *Bayesian Modeling Using WinBUGS*. Wiley Series in Computational Statistics. John Wiley & Sons, Ltd, 2009.

A. O'Hagan. *Bayesian Inference*. Edward Arnold, 1994.

T. Okada and S. Tomita. An optimal orthonormal system for discriminant analysis. *Pattern Recognition*, 18(2):139–144, 1985.

J.J. Oliver and D.J. Hand. Averaging over decision trees. *Journal of Classification*, 13(2):281–297, 1996.

I. Olkin and R.F. Tate. Multivariate correlation models with mixed discrete and continuous variables. *Annals of Mathematical Statistics*, 22:92–96, 1961.

S.D. Oman, T. Naes, and A. Zube. Detecting and adjusting for non-linearities in calibration of near-infrared data using principal components. *Journal of Chemometrics*, 7:195–212, 1993.

S.M. Omohundro. Five balltree construction algorithms. Technical Report, International Computer Science Institute, 1989.

T.J. O'Neill. Error rates of non-Bayes classification rules and the robustness of Fisher's linear discriminant function. *Biometrika*, 79(1):177–184, 1992.

M.J.L. Orr. Regularisation in the selection of radial basis function centers. *Neural Computation*, 7:606–623, 1995.

E. Osuna, R. Freund, and F. Girosi. Training support vector machines: an application to face detection. In *Proceedings of 1997 IEEE Computer Society Conference on Computer Vision and Pattern Recognition*, pages 130–136. Computer Society Press, 1997.

R. Ouysse and R. Kohn. Bayesian variable selection and model averaging in the arbitrage pricing theory model. *Computational Statistics and Data Analysis*, 54(12):3249–3268, 2010.

J.E. Overall and K.N. Magee. Replication as a rule for determining the number of clusters in a hierarchical cluster analysis. *Applied Psychological Measurement*, 16:119–128, 1992.

N.R. Pal and J.C. Bezdek. On cluster validity for the fuzzy *c*-means model. *IEEE Transactions on Fuzzy Systems*, 3(3):370–379, 1995.

G. Palla, A.-L. Barabási, and T. Vicsek. Quantifying social group evolution. *Nature*, 446:664–667, 2007.

G. Palla, I. Derényi, I. Farkas, and T. Vicsek. Uncovering the overlapping community structure of complex networks in nature and society. *Nature*, 435:814–818, 2005a.

G. Palla, I. Derényi, I. Farkas, and T. Vicsek. Uncovering the overlapping community structure of complex networks in nature and society. Supplementary information. *Nature*, 435:814–818, 2005b.

Y.-H. Pao. *Adaptive Pattern Recognition and Neural Networks*. Addison-Wesley, 1989.

J. Park and I.W. Sandberg. Approximation and radial-basis-function networks. *Neural Computation*, 5:305–316, 1993.

S.H. Park, J.M. Goo, and C.-H. Jo. Receiver operating characteristic (ROC) curve: practical review for radiologists. *Korean Journal of Radiology*, 5:11–18, 2004.

E. Parzen. On estimation of a probability density function and mode. *Annals of Mathematical Statistics*, 33:1065–1076, 1962.

A.C. Patel and M.K. Markey. Comparison of three-class classification performance metrics: a case study in breast cancer CAD. In M.P. Eckstein and Y. Jiang, editors, *Proceedings SPIE Medical Imaging 2005: Image Perception, Observer Performance and Technology Assessment*, volume 5749, pages 581–589. SPIE, 2005.

M. Pawlak. Kernel classification rules from missing data. *IEEE Transactions on Information Theory*, 39(3):979–988, 1993.

J. Pearl. *Probabilistic Reasoning in Intelligent Systems: Networks of Plausible Inference*. Morgan Kaufmann, 1988.

K. Pearson. On lines and planes of closest fit to systems of points in space. *Philosophical Magazine*, 2:559–572, 1901.

F. Pedersen, M. Bergström, E. Bengtsson, and B. Langström. Principal component analysis of dynamic positron emission tomography images. *European Journal of Nuclear Medicine*, 21(12):1285–1292, 1994.

D. Peel and G.J. McLachlan. Robust mixture modelling using the t distribution. *Statistics and Computing*, 10(4):339–348, 2000.

G. Peters. *Topics in Sequential Monte Carlo Samplers*. MSc Dissertation, Cambridge University, 2005.

P.J. Phillips, H. Moon, S.A. Rizvi, and P.J. Rauss. The FERET evaluation methodology for face-recognition algorithms. *IEEE Transactions on Pattern Analysis and Machine Intelligence*, 22(10):1090–1104, 2000.

B. Pinkowski. Principal component analysis of speech spectrogram images. *Pattern Recognition*, 30(5):777–787, 1997.

J. Platt. Fast training of support vector machines using sequential minimal optimisation. In B. Schölkopf, C.J.C. Burges, and A.J. Smola, editors, *Advances in Kernel Methods: Support Vector Learning*, pages 185–208. The MIT Press, 1998.

J.C. Platt. Probabilities for SV machines. In A.J. Smola, P. Bartlett, B. Schölkopf, and D. Schuurmans, editors, *Advances in Large Margin Classifiers*, pages 61–74. The MIT Press, 2000.

N. Popescu-Borodin. Fast k-means image quantization algorithm and its application to iris segmentation. *Buletin Stiintific - Universitatea de Pitesti. Seria Matematica si Informatica*, 14:1–18, 2008.

M.A. Porter, P.J. Mucha, M.E.J. Newman, and A.J. Friend. Community structure in the United States House of Representatives. *Physica A*, 386:414–438, 2007.

O. Pourret, P. Naim, and B. Marcot, editors. *Bayesian Networks: A Practical Guide to Applications*. Statistics in Practice. John Wiley & Sons, Ltd, 2008.

G.E. Powell, E. Clark, and S. Bailey. Categories of aphasia: a cluster-analysis of Schuell test profiles. *British Journal of Disorders of Communication*, 14(2):111–122, 1979.

M.J.D. Powell. Radial basis functions for multivariable interpolation: a review. In J.C. Mason and M.G. Cox, editors, *Algorithms for Approximation*, pages 143–167. Clarendon Press, 1987.

S. Prabhakar and A.K. Jain. Decision-level fusion in fingerprint verification. *Pattern Recognition*, 35:861–874, 2002.

M. Prakash and M.N. Murty. A genetic approach for selection of (near-) optimal subsets of principal components for discrimination. *Pattern Recognition Letters*, 16:781–787, 1995.

S.J. Press and S. Wilson. Choosing between logistic regression and discriminant analysis. *Journal of the American Statistical Association*, 73:699–705, 1978.

W.H. Press, B.P. Flannery, S.A. Teukolsky, and W.T. Vetterling. *Numerical Recipes. The Art of Scientific Computing*. Cambridge University Press, second edition, 1992.

J.K. Pritchard, M.T. Seielstad, A. Perex-Lezaun, and M.W. Feldman. Population growth of human Y chromosomes: a study of Y chromosome microsatellites. *Molecular Biology and Evolution*, 16(12):1791–1798, 1999.

L. Prost, D. Makowski, and M.-H. Jeuffroy. Comparison of stepwise selection and Bayesian model averaging for yield gap analysis. *Ecological Modelling*, 219:66–76, 2008.

F. Provost and T. Fawcett. Robust classification for imprecise environments. *Machine Learning*, 42:203–231, 2001.

F. Provost, T. Fawcett, and R. Kohavi. The case against accuracy estimation for comparing induction algorithms. In *Proceedings of the Fifteenth International Conference on Machine Learning*, pages 445–453. Morgan Kaufmann, 1997.

D. Psaltis, R.R. Snapp, and S.S. Venkatesh. On the finite sample performance of the nearest neighbor classifier. *IEEE Transactions on Information Theory*, 40(3):820–837, 1994.

P. Pudil, F.J. Ferri, J. Novovičová, and J. Kittler. Floating search methods for feature selection with nonmonotonic criterion functions. In *Proceedings of the International Conference on Pattern Recognition*, volume 2, pages 279–283. IEEE, 1994a.

P. Pudil, J. Novovičová, N Choakjarernwanit, and J. Kittler. Feature selection based on the approximation of class densities by finite mixtures of special type. *Pattern Recognition*, 28(9):1389–1398, 1995.

P. Pudil, J. Novovičová, and J. Kittler. Floating search methods in feature selection. *Pattern Recognition Letters*, 15:1119–1125, 1994b.

M. H. Quenouille. Approximate tests of correlation in time series. *Journal of the Royal Statistical Society Series B*, 11:68–84, 1949.

J.R. Quinlan. Simplifying decision trees. *International Journal of Man Machine Studies*, 27:221–234, 1987.

J.R. Quinlan. Learning logical definitions from relations. *Machine Learning*, 5(3):239–266, 1990.

J.R. Quinlan. *C4.5: Programs for Machine Learning*. Morgan Kaufmnann, 1993.

J.R. Quinlan and R.L. Rivest. Inferring decision trees using the minimum description length principle. *Information and Computation*, 80:227–248, 1989.

L.R. Rabiner, B.-H. Juang, S.E. Levinson, and M.M. Sondhi. Recognition of isolated digits using hidden Markov models with continuous mixture densities. *AT&T Technical Journal*, 64(4):1211–1234, 1985.

A.E. Raftery. Bayesian model selection in social research (with discussion). In P.V. Marsden, editor, *Sociological Methodology 1995*, pages 111–196. Blackwell, 1995.

A.E. Raftery and S.M. Lewis. Implementing MCMC. In W.R. Gilks, S. Richardson, and D.J. Spiegelhalter, editors, *Markov Chain Monte Carlo in Practice*, pages 115–130. Chapman and Hall, 1996.

S. Raju and V.V.S. Sarma. Multisensor data fusion and decision support for airborne target identification. *IEEE Transactions on Systems, Man, and Cybernetics*, 21(5):1224–1230, 1991.

A. Ramalingam and S. Krishnan. Gaussian mixture modeling of short-time Fourier transform features for audio fingerprinting. *IEEE Transactions on Information Forensics and Security*, 1(4):457–463, 2006.

V. Ramasubramanian and K.K. Paliwal. Fast nearest-neighbor search algorithms based on approximation-elimination search. *Pattern Recognition*, 33:1497–1510, 2000.

S. Ramaswamy, P. Tamayo, R. Rifkin, S. Mukherjee, C.-H. Yeang, M. Angelo, C. Ladd, M. Reich, E. Latulippe, J.P. Mesirov, T. Poggio, W. Gerald, M. Loda, E.S. Lander, and T.R. Golub. Multiclass cancer diagnosis using tumor gene expression signatures. *Proceedings of the National Academy of Sciences of the United States of America*, 98(26):15149–15154, 2001.

J.O. Ramsay and C.J. Dalzell. Some tools for functional data analysis (with discussion). *Journal of the Royal Statistical Society Series B*, 53:539–572, 1991.

M.B. Ratcliffe, K.B. Gupta, J.T. Streicher, E.B. Savage, D.K. Bogen, and L.H. Edmunds. Use of sonomicrometry and multidimensional scaling to determine the three-dimensional coordinates of multiple cardiac locations: feasibility and initial implementation. *IEEE Transactions on Biomedical Engineering*, 42(6):587–598, 1995.

S. Rattanasiri, D. Böhning, P. Rojanavipart, and S. Athipanyakom. A mixture model application in disease mapping of malaria. *Southeast Asian Journal of Tropical Medicine and Public Health*, 35(1):38–47, 2004.

M.J. Rattigan, M. Maier, and D. Jensen. Graph clustering with network structure indices. In *ICML '07: Proceedings of the 24th International Conference on Machine Learning*, pages 783–790. ACM, 2007.

S.J. Raudys. Scaled rotation regularisation. *Pattern Recognition*, 33:1989–1998, 2000.

W. Rayens and T. Greene. Covariance pooling and stabilization for classification. *Computational Statistics and Data Analysis*, 11:17–42, 1991.

R.A. Redner and H.F. Walker. Mixture densities, maximum likelihood and the EM algorithm. *SIAM Review*, 26(2):195–239, 1984.

R. Reed. Pruning algorithms – a survey. *IEEE Transactions on Neural Networks*, 4(5):740–747, 1993.

A.-P.N. Refenes, A.N. Burgess, and Y. Bentz. Neural networks in financial engineering: a study in methodology. *IEEE Transactions on Neural Networks*, 8(6):1222–1267, 1997.

J. Remme, J.D.F. Habbema, and J. Hermans. A simulative comparison of linear, quadratic and kernel discrimination. *Journal of Statistical Computation and Simulation*, 11:87–106, 1980.

S. Renals. Nearest neighbours and the kD-tree, Course notes for informatics 2B, algorithms data structures and learning. Technical Report, The University of Edinburgh School of Informatics, 2007.

J.D.M. Rennie, L. Shih, J. Teevan, and D.R. Karger. Tackling the poor assumptions of naïve Bayes text classifiers. In *Proceedings of the Twentieth International Conference on Machine Learning (ICML-2003)*, Pages 616–623. IEEE, 2003.

M. Revow, C.K.I. Williams, and G.E. Hinton. Using generative models for handwritten digit recognition. *IEEE Transactions on Pattern Analysis and Machine Intelligence*, 18(6):592–606, 1996.

R.A. Reyment, R.E. Blackith, and N.A. Campbell. *Multivariate Morphometrics*. Academic Press, second edition, 1984.

D.A. Reynolds, T.F. Quatieri, and R.B. Dunn. Speaker verification using adapted Gaussian mixture models. *Digital Signal Processing*, 10:19–41, 2000.

J.A. Rice and B.W. Silverman. Estimating the mean and covariance structure nonparametrically when the data are curves. *Journal of the Royal Statistical Society Series B*, 53:233–243, 1991.

S. Richardson and P.J. Green. On Bayesian analysis of mixtures with an unknown number of components (with discussion). *Journal of the Royal Statistical Society Series B*, 59(4):731–792, 1997.

S. Richardson and P.J. Green. Corrigendum: On Bayesian analysis of mixtures with an unknown number of components. *Journal of the Royal Statistical Society Series B*, 60(3):661, 1998.

B.D. Ripley. *Stochastic Simulation*. John Wiley & Sons, Ltd, 1987.

B.D. Ripley. Neural and related methods of classification. *Journal of the Royal Statistical Society Series B*, 56(3), 1994.

B.D. Ripley. *Pattern Recognition and Neural Networks*. Cambridge University Press, 1996.

E.A. Riskin and R.M. Gray. A greedy tree growing algorithm for the design of variable rate vector quantizers. *IEEE Transactions on Signal Processing*, 39(11):2500–2507, 1991.

B. Ristic, B. La Scala, M. Morelande, and N. Gordon. Statistical analysis of motion patterns in AIS data: anomaly detection and motion prediction. In *Proceedings of 11th International Conference on Information Fusion*, pages 1–7. 2008.

H. Ritter, T. Martinetz, and K. Schulten. *Neural Computation and Self-Organizing Maps: an Introduction*. Addison-Wesley, 1992.

C.P. Robert. *The Bayesian Choice: From Decision-Theoretic Foundations to Computational Implementation*. Springer Texts in Statistics. Springer, second edition, 2001.

C.P. Robert and G. Casella. *Monte Carlo Statistical Methods*. Springer Texts in Statistics. Springer, 2004.

C.P. Robert and G. Casella. *Introducing Monte Carlo Methods with R*. Springer, 2009.

C.P. Robert, J.-M. Cornuet, J.-M. Marin, and N.S. Pillai. Lack of confidence in ABC model choice. *Proceedings of the National Academy of Sciences of the United States of America*, submitted.

G.O. Roberts. Markov chain concepts related to sampling algorithms. In W.R. Gilks, S. Richardson, and D.J. Spiegelhalter, editors, *Markov Chain Monte Carlo in Practice*, pages 45–57. Chapman and Hall, 1996.

S. Roberts and L. Tarassenko. Analysis of the sleep EEG using a multilayer network with spatial organisation. *IEE Proceedings Part F*, 139(6):420–425, 1992.

D.M. Rocke and D.L. Woodruff. Robust estimation of multivariate location and shape. *Journal of Statistical Planning and Inference*, 57:245–255, 1997.

K. Roeder and L. Wasserman. Practical Bayesian density estimation using mixtures of normals. *Journal of the American Statistical Association*, 92(439):894–902, 1997.

S.K. Rogers, J.M. Colombi, C.E. Martin, J.C. Gainey, K.H. Fielding, T.J. Burns, D.W. Ruck, M. Kabrisky, and M. Oxley. Neural networks for automatic target recognition. *Neural Networks*, 8(7/8):1153–1184, 1995.

F.J. Rohlf. Single-link clustering algorithms. In P.R. Krishnaiah and L.N. Kanal, editors, *Handbook of Statistics*, volume 2, pages 267–284. North Holland, 1982.

M. Rosenblatt. Remarks on some nonparametric estimates of a density function. *The Annals of Mathematical Statistics*, 27:832–835, 1956.

M. Rosvall and C.T. Bergstrom. Maps of random walks on complex networks reveal community structure. *Proceedings of the National Academy of Sciences of the United States of America*, 105(4):1118–1123, 2008.

M.W. Roth. Survey of neural network technology for automatic target recognition. *IEEE Transactions on Neural Networks*, 1(1):28–43, 1990.

P.J. Rousseeuw. Multivariate estimation with high breakdown point. In W. Grossmann, G. Pflug, I. Vincze, and Wertz W, editors, *Mathematical Statistics and Applications*, pages 283–297. Reidel, 1985.

D.E. Rumelhart, G.E. Hinton, and R.J. Williams. Learning internal representation by error propagation. In D.E. Rumelhart, J.L. McClelland, and the PDP Research Group, editors, *Parallel Distributed Processing: Explorations in the Microstructure of Cognition*, volume 1, pages 318–362. The MIT Press, 1986.

Y. Saeys, T. Abeel, and Y. Van de Peer. Robust feature selection using ensemble feature selection techniques. In *ECML PKDD*, volume 5212 of *Lecture Notes in Artificial Intelligence*, pages 313–325. Springer, 2008.

Y. Saeys, I. Inza, and P. Larranaga. A review of feature selection techniques in bioinformatics. *Bioinformatics*, 23(19):2507–2517, 2007.

S.R. Safavian and D.A. Landgrebe. A survey of decision tree classifier methodology. *IEEE Transactions on Systems, Man, and Cybernetics*, 21(3):660–674, 1991.

A. Samal and P.A. Iyengar. Automatic recognition and analysis of human faces and facial expressions: a survey. *Pattern Recognition*, 25:65–77, 1992.

J.W. Sammon. A nonlinear mapping for data structure analysis. *IEEE Transactions on Computers*, 18(5):401–409, 1969.

A. Sankar and R.J. Mammone. Combining neural networks and decision trees. In *Applications of Neural Networks II*, volume 1469, pages 374–383. SPIE, 1991.

A. Saranli and M. Demirekler. A statistical framework for rank-based multiple classifier decision fusion. *Pattern Recognition*, 34:865–884, 2001.

K. Saravanan. An efficient detection mechanism for intrusion detection systems using rule learning method. *International Journal of Computer and Electrical Engineering*, 1(4):503–506, 2009.

S. Schaack, A. Mauthofer, and U. Brunsmann. Stationary video-based pedestrian recognition for driver assistance systems. In *Proceedings of 21st International Technical Conference on the Enhanced Safety of Vehicles (ESV)*, Paper no. 09-0276. 2009.

S.E. Schaeffer. Graph clustering. *Computer Science Review*, 1(1):27–64, 2007.

C. Schaffer. Selecting a classification method by cross-validation. *Machine Learning*, 13:135–143, 1993.

R. Schalkoff. *Pattern Recognition. Statistical Structural and Neural*. John Wiley & Sons, Ltd, 1992.

R.E. Schapire. The strength of weak learnability. *Machine Learning*, 5(2):197–227, 1990.

R.E. Schapire and Y. Singer. Improved boosting algorithms using confidence-rated predictions. *Machine Learning*, 37:297–336, 1999.

S.S. Schiffman, M.L. Reynolds, and F.W. Young. *An Introduction to Multidimensional Scaling*. Academic Press, 1981.

B. Schölkopf and A.J. Smola. *Learning with Kernels. Support Vector Machines, Regularization, Optimization and Beyond*. The MIT Press, 2001.

B. Schölkopf, A.J. Smola, and K. Müller. Kernel principal component analysis. In B. Schölkopf, C.J.C. Burges, and A.J. Smola, editors, *Advances in Kernel Methods – Support Vector Learning*, pages 327–352. The MIT Press, 1999.

B. Schölkopf, A.J. Smola, R.C. Williamson, and P.L. Bartlett. New support vector algorithms. *Neural Computation*, 12:1207–1245, 2000.

B. Schölkopf, K.-K. Sung, C.J.C. Burges, F. Girosi, P. Niyogi, T. Poggio, and V. Vapnik. Comparing support vector machines with Gaussian kernels to radial basis function classifiers. *IEEE Transactions on Signal Processing*, 45(11):2758–2765, 1997.

C. Schölzel and P. Friederichs. Multivariate non-normally distributed random variables in climate research – introduction to the copula approach. *Nonlinear Processes in Geophysics*, 15:761–772, 2008.

M. Schomaker, A.T.K. Wan, and C. Heumann. Frequentist model averaging with missing observations. *Computational Statistics and Data Analysis*, 54(12):3336–3347, 2010.

J.R. Schott. Dimensionality reduction in quadratic discriminant analysis. *Computational Statistics and Data Analysis*, 16:161–174, 1993.

G. Schwarz. Estimating the dimension of a model. *The Annals of Statistics*, 6(2):461–464, 1978.

F. Schwenker, H.A. Kestler, and G. Palm. Three learning phases for radial-basis-function networks.. *Neural Networks*, 14:439–458, 2001.

S.L. Sclove. Application of model selection criteria to some problems in multivariate analysis. *Psychometrika*, 52(3):333–343, 1987.

D.W. Scott. *Multivariate Density Estimation. Theory, Practice and Visualization*. John Wiley & Sons, Ltd, 1992.

D.W. Scott, A.M. Gotto, J.S. Cole, and G.A. Gorry. Plasma lipids as collateral risk factors in coronary artery disease – a study of 371 males with chest pains. *Journal of Chronic Diseases*, 31:337–345, 1978.

G. Sebestyen and J. Edie. An algorithm for non-parametric pattern recognition. *IEEE Transactions on Electronic Computers*, 15(6):908–915, 1966.

S.Z. Selim and K.S. Al-Sultan. A simulated annealing algorithm for the clustering problem. *Pattern Recognition*, 24(10):1003–1008, 1991.

S.Z. Selim and M.A. Ismail. K-means-type algorithms: a generalized convergence theorem and characterization of local optimality. *IEEE Transactions on Pattern Analysis and Machine Intelligence*, 6(1):81–87, 1984a.

S.Z. Selim and M.A. Ismail. Soft clustering of multidimensional data: a semi-fuzzy approach. *Pattern Recognition*, 17(5):559–568, 1984b.

S.Z. Selim and M.A. Ismail. On the local optimality of the fuzzy isodata clustering algorithm. *IEEE Transactions on Pattern Analysis and Machine Intelligence*, 8(2):284–288, 1986.

P.S. Sephton. Cointegration tests on MARS. *Computational Economics*, 7:23–35, 1994.

S.B. Serpico, L. Bruzzone, and F. Roli. An experimental comparison of neural and statistical non-parametric algorithms for supervised classification of remote-sensing images. *Pattern Recognition Letters*, 17:1331–1341, 1996.

I.K. Sethi and J.H. Yoo. Design of multicategory multifeature split decision trees using perceptron learning. *Pattern Recognition*, 27(7):939–947, 1994.

M. Sewell. Ensemble learning. RN/11/02. Technical Report, University College London, 2011.

S. Shah and P.S. Sastry. New algorithms for learning and pruning oblique decision trees. *IEEE Transactions on Systems, Man, and Cybernetics Part C*, 29(4):494–505, 1999.

A.J.C. Sharkey. Multi-net systems. In A.J.C. Sharkey, editor, *Combining Artificial Neural Nets. Ensemble and Modular Multi-net Systems*, pages 1–30. Springer-Verlag, 1999.

J.W. Shavlik, R.J. Mooney, and G.G. Towell. Symbolic and neural learning algorithms: an experimental comparison. *Machine Learning*, 6:111–143, 1991.

S.J. Sheather and M.C. Jones. A reliable data-based bandwidth selection method for kernel density estimation. *Journal of the Royal Statistical Society Series B*, 53:683–690, 1991.

J. Shetty and J. Adibi. The Enron email dataset database schema and brief statistical report. Information Sciences Institute Technical Report, University of Southern California, 2005.

J. Shi and J. Malik. Normalized cuts and image segmentation. *IEEE Transactions on Pattern Analysis and Machine Intelligence*, 22(8):888–905, 2000.

L. Shuhui, D. C. Wunsh, E.A. O'Hair, and M.G. Giesselmann. Using neural networks to estimate wind turbine turbine power generation. *IEEE Transactions on Energy Conversion*, 16(3):276–282, September 2001.

R. Sibson. SLINK: an optimally efficient algorithm for the single-link cluster method. *The Computer Journal*, 16(1):30–34, 1973.

R. Sicard, T. Artières, and E. Petit. Learning iteratively a classifier with the Bayesian model averaging principle. *Pattern Recognition*, 41:930–938, 2008.

W. Siedlecki, K. Siedlecka, and J. Sklansky. An overview of mapping techniques for exploratory pattern analysis. *Pattern Recognition*, 21(5):411–429, 1988.

W. Siedlecki and J. Sklansky. On automatic feature selection. *International Journal of Pattern Recognition and Artificial Intelligence*, 2(2):197–220, 1988.

B.W. Silverman. Kernel density estimation using the fast Fourier transform. *Applied Statistics*, 31:93–99, 1982.

B.W. Silverman. *Density Estimation for Statistics and Data Analysis*. Chapman and Hall, 1986.

B.W. Silverman. Incorporating parametric effects into functional principal components analysis. *Journal of the Royal Statistical Society Series B*, 57(4):673–689, 1995.

P.K. Simpson, editor. *IEEE Journal of Oceanic Engineering*, 17: 1992. Special issue on 'Neural Networks for Oceanic Engineering'.

T. Sing, O. Sander, N. Beerenwinkel, and T. Lengauer. The ROCR package. Technical Report, http://rocr.bioinf.mpi-sb.mpg.de, 2007.

A. Skabar. Application of Bayesian MLP techniques to predicting mineralization potential from geoscientific data. In *Artificial Neural Networks: Formal Models and their Applications - ICANN, volume 3697 of Springer Lecture Notes in Computer Science*, pages 963–968. Springer, 2005.

A. Sklar. Fonctions de répartition à *n* dimensions et leurs marges. *Publications of the Institute of Statistics of the University of Paris*, 8:229–231, 1959.

A. Sklar. Random variables, joint distribution functions and copulas. *Kybernetika*, 9(6):449–460, 1973.

M. Skurichina. *Stabilizing Weak Classifiers*. Technical University of Delft, 2001.

J.M. Sloughter, A.E. Raftery, T. Gneiting, and C. Fraley. Probabilistic quantitative precipitation forecasting using Bayesian model averaging. *Monthly Weather Review*, 135(9):3209–3220, 2007.

A.F.M. Smith and A.E. Gelfand. Bayesian statistics without tears: a sampling-resampling perspective. *The American Statistician*, 46(2):84–88, 1992.

S.J. Smith, M.O. Bourgoin, K. Sims, and H.L. Voorhees. Handwritten character classification using nearest neighbour in large databases. *IEEE Transactions on Pattern Analysis and Machine Intelligence*, 16(9):915–919, 1994.

P. Smyth and Wolpert. Linearly combining density estimators via stacking. *Machine Learning*, 36:59–83, 1999.

P.H.A. Sneath and R.R. Sokal. *Numerical Taxonomy*. Freeman, 1973.

J.V.B. Soares, J.J.G. Leandro, R.M. Cesar Jr., and H.F. Jelinek. Retinal vessel segmentation using the 2-D Gabor wavelet and supervised classification. *IEEE Transactions on Medical Imaging*, 25(9):1214–1222, 2006.

A.H.S. Solberg, G. Storvik, R. Solberg, and E. Volden. Automatic detection of oil spills in ERS SAR images. *IEEE Transactions on Geoscience and Remote Sensing*, 37(4):1916–1924, 1999.

P. Sollich. Bayesian methods for support vector machines: evidence and predictive class probabilities. *Machine Learning*, 46(1-3):21–52, 2002.

P. Somol, P. Pudil, J. Novovičová, and P. Paclík. Adaptive floating search methods in feature selection. *Pattern Recognition Letters*, 20:1157–1163, 1999.

T. Sorsa, H.N. Koivo, and H. Koivisto. Neural networks in process fault diagnosis. *IEEE Transactions on Systems, Man and Cybernetics*, 21(4):815–825, 1991.

H. Späth. *Cluster Analysis Algorithms for Data Reduction and Classification of Objects*. Ellis Horwood Limited, 1980.

D.J. Spiegelhalter, A.P. Dawid, T.A. Hutchinson, and R.G. Cowell. Probabilistic expert systems and graphical modelling: a case study in drug safety. *Philosophical Transactions of the Royal Society of London*, 337:387–405, 1991.

D.A. Spielman and S.-H. Teng. Spectral partitioning works: planar graphs and finite element meshes. In *Proceedings of the 37th Annual Symposium on Foundations of Computer Science*, pages 96–105. IEEE Computer Society Press, 1996.

V. Spirin and L.A. Mirny. Protein complexes and functional modules in molecular networks. *Proceedings of the National Academy of Sciences of the United States of America*, 100(21):12123–12128, 2003.

R.F. Sproull. Refinements to nearest-neighbour searching in k-dimensional trees. *Algorithmica*, 6:579–589, 1991.

D.V. Sridhar, E.B. Bartlett, and R.C. Seagrave. An information theoretic approach for combining neural network process models. *Neural Networks*, 12:915–926, 1999.

D.V. Sridhar, R.C. Seagrave, and E.B. Bartlett. Process modeling using stacked neural networks. *Process Systems Engineering*, 42(9):387–405, 1996.

C. Staelin. Parameter selection for support vector machines. Technical Report, HP Laboratories Israel, 2002.

F. Stäger and M. Agarwal. Three methods to speed up the training of feedforward and feedback perceptrons. *Neural Networks*, 10(8):1435–1443, 1997.

A. Stassopoulou, M. Petrou, and J. Kittler. Bayesian and neural networks for geographic information processing. *Pattern Recognition Letters*, 17:1325–1330, 1996.

S.D. Stearns. On selecting features for pattern classifiers. In *Proceedings of the 3rd International Conference on Pattern Recognition*, pages 71–75. 1976.

M. Stephens. *Bayesian Methods for Mixtures of Normal Distributions*. PhD thesis, Magdalen College, University of Oxford, 1997.

M. Stephens. Bayesian analysis of mixture models with an unknown number of components – an alternative to reversible jump methods. *Annals of Statistics*, 28(1):40–74, 2000.

J. Stevenson. Multivariate statistics VI. The place of discriminant function analysis in psychiatric research. *Nordic Journal of Psychiatry*, 47(2):109–122, 1993.

C. Stewart, Y.-C. Lu, and V. Larson. A neural clustering approach for high resolution radar target classification. *Pattern Recognition*, 27(4):503–513, 1994.

G.W. Stewart. *Introduction to Matrix Computation*. Academic Press, Inc., 1973.

C. Stone, M. Hansen, C. Kooperberg, and Y. Truong. Polynomial splines and their tensor products (with discussion). *Annals of Statistics*, 25(4):1371–1470, 1997.

M. Stone. Cross-validatory choice and assessment of statistical predictions. *Journal of the Royal Statistical Society Series B*, 36:111–147, 1974.

D.J. Stracuzzi. Randomized feature selection. In H. Liu and H. Motoda, editors, *Computational Methyods of Feature Selection*. Chapman and Hall/CRC, 2007.

A. Stuart and J.K. Ord. *Kendall's Advanced Theory of Statistic*, volume 2. Edward Arnold, fifth edition, 1991.

R.G. Sumpter, C. Getino, and D.W. Noid. Theory and applications of neural computing in chemical science. *Annual Reviews of Physical Chemistry*, 45:439–481, 1994.

B.D. Sutton and G.J. Steck. Discrimination of Caribbean and Mediterranean fruit fly larvae (Diptera: Tephritidae) by cuticular hydrocarbon analysis. *Florida Entomologist*, 77(2):231–237, 1994.

A. Swarnkar and K.R. Niazi. CART for online security evaluation and preventive control of power systems. In *Proceedings of the 5th WSEAS/IASME International Conference on Electric Power Systems, High Voltages, Electric Machines*, pages 378–383. 2005.

K.S. Swarup, R. Mastakar, and K.V. Prasad Reddy. Decision tree for steady state security assessment and evaluation of power systems. In *Proceedings of the 2005 International Conference on Intelligent Sensing and Information Processing*, pages 211–216. 2005.

M.A. Tahir, A. Bouridane, and F. Kurugollu. Simultaneous feature selection and feature weighting using hybrid tabu search/k-nearest neighbor classifier. *Pattern Recognition Letters*, 28:438–446, 2007.

P.-N. Tan, M. Steinbach, and V. Kumar. *Introduction to Data Mining*. Pearson Education, 2005.

L. Tarassenko. *A Guide to Neural Computing Applications*. Arnold, 1998.

S. Tavaré, D.J. Balding, R.C. Griffiths, and P. Donnelly. Inferring coalescence times from DNA sequence data. *Genetics*, 145:505–518. 1997.

D.M.J. Tax, M. van Breukelen, R.P.W. Duin, and J. Kittler. Combining multiple classifiers by averaging or multiplying? *Pattern Recognition*, 33:1475–1485, 2000.

G.R. Terrell and D.W. Scott. Variable kernel density estimation. *Annals of Statistics*, 20(3):1236–1265, 1992.

S. Theodoridis and K. Koutroumbas. *Pattern Recognition*. Academic Press, fourth edition, 2009.

S. Theodoridis, A. Pikrakis, K. Koutroumbas, and D. Cavouras. *Introduction to Pattern Recognition: A Matlab Approach*. Academic Press, 2010.

C.W. Therrien. *Decision, Estimation and Classification. An Introduction to Pattern Recognition and Related Topics*. John Wiley & Sons, Ltd, 1989.

H.H. Thodberg. A review of Bayesian neural networks with application to near infrared spectroscopy. *IEEE Transactions on Neural Networks*, 7(1):56–72, 1996.

C.E. Thomaz, D.F. Gillies, and R.Q. Feitosa. A new covariance estimate for Bayesian classifiers in biometric recognition. *IEEE Transactions on Circuits and Systems for Video Technology*, 14(2):214–223, 2004.

Q. Tian, Y. Fainman, and S.H. Lee. Comparison of statistical pattern-recognition algorithms for hybrid processing. II. Eigenvector-based algorithm. *Journal of the Optical Society of America A*, 5(10):1670–1682, 1988.

R.J. Tibshirani. Principal curves revisited. *Statistics and Computing*, 2(4):183–190, 1992.

L. Tierney. Markov chains for exploring posterior distributions. *Annals of Statistics*, 22(4):1701–1762, 1994.

D.M. Titterington. A comparative study of kernel-based density estimates for categorical data. *Technometrics*, 22(2):259–268, 1980.

D.M. Titterington and G.M. Mill. Kernel-based density estimates from incomplete data. *Journal of the Royal Statistical Society Series B*, 45(2):258–266, 1983.

D.M. Titterington, G.D. Murray, L.S. Murray, D.J. Spiegelhalter, A.M. Skene, J.D.F. Habbema, and G.J. Gelpke. Comparison of discrimination techniques applied to a complex data set of head injured patients (with discussion). *Journal of the Royal Statistical Society Series A*, 144(2):145–175, 1981.

D.M. Titterington, A.F.M. Smith, and U.E. Makov. *Statistical Analysis of Finite Mixture Distributions*. John Wiley & Sons, Ltd, 1985.

R. Todeschini. *k*-nearest neighbour method: the influence of data transformations and metrics. *Chemometrics and Intelligent Laboratory Systems*, 6:213–220, 1989.

T. Toni and M.P.H. Stumpf. Simulation-based model selection for dynamical systems in systems and population biology. *Bioinformatics*, 26(1):104–110, 2010.

T. Toni, D. Welch, N. Strelkowa, A. Ipsen, and M.P.H. Stumpf. Approximate Bayesian computation scheme for parameter inference and model selection in dynamical systems. *Journal of the Royal Society Interface*, 6(31):187–2002, 2009.

J.T. Tou and R.C. Gonzales. *Pattern Recognition Principles*. Addison-Wesley, 1974.

G.T. Toussaint. Bibliography on estimation of misclassification. *IEEE Transactions on Information Theory*, 20(4):472–479, 1974.

P.K. Trivedi and D.M. Zimmer. Copula modeling: an introduction for practitioners. *Foundations and Trends® in Econometrics*, 1(1):1–111, 2005.

S. Tulyakov, S. Jaeger, V. Govindaraju, and D. Doermann. Review of classifier combination methods. In S. Marinai and H. Fujisawa, editors, *Studies in Computational Intelligence: Machine Learning in Document Analysis and Recognition*, pages 361–386. Springer, 2008.

M. Turk and A. Pentland. Eigenfaces for recognition. *Journal of Cognitive Neuroscience*, 3(1):71–86, 1991.

J.R. Tyler, D.M. Wilkinson, and B.A. Huberman. *Email as spectroscopy: automated discovery of community structure within organizations*, pages 81–96. Kluwer, 2003.

D.G. Tzikas, C.A. Likas, and N.P. Galatsanos. The variational approximation for Bayesian inference. *IEEE Signal Processing Magazine*, 25(6):131–146, 2008.

J.K. Uhlmann. Satisfying general proximity/similarity queries with metric trees. *Information Processing Letters*, 40:175–179, 1991.

R. Unbehauen and F.-L. Luo, editors. *Signal Processing*, 64:1998. Special issue on 'Neural Networks'.

D. Valentin, H. Abdi, A.J. O'Toole, and G.W. Cottrell. Connectionist models of face processing: a survey. *Pattern Recognition*, 27(9):1209–1230, 1994.

R.S. Valiveti and B.J. Oommen. On using the chi-squared metric for determining stochastic dependence. *Pattern Recognition*, 25(11):1389–1400, 1992.

R.S. Valiveti and B.J. Oommen. Determining stochastic dependence for normally distributed vectors using the chi-squared metric. *Pattern Recognition*, 26(6):975–987, 1993.

F. van der Heiden, R.P.W. Duin, D. de Ridder, and D.M.J. Tax. *Classification, Parameter Estimation and State Estimation: an Engineering Approach Using MATLAB*. Wiley-Blackwell, 2004.

R. van der Heiden and F.C.A. Groen. The Box-Cox metric for nearest neighbour classification improvement. *Pattern Recognition*, 30(2):273–279, 1997.

P.P. van der Smagt. Minimisation methods for training feedforward networks. *Neural Networks*, 7(1):1–11, 1994.

T. Van Gestel, J.A.K. Suykens, D.-E. Baestaens, A. Lambrechts, G. Lanckriet, B.V. Vandaele, B. De Moor, and J. Vandewalle. Financial time series prediction using least squares support vector machines within the evidence framework. *IEEE Transactions on Neural Networks*, 12(4):809–821, 2001.

T. Van Gestel, J.A.K. Suykens, G. Lanckriet, A. Lambrechts, B. De Moor, and J. Vandewalle. A Bayesian framework for least squares support vector machine classifiers, Gaussian processes and kernel Fisher discriminant analysis. *Neural Computation*, 14(5):1115–1147, 2002.

V.N. Vapnik. *Statistical Learning Theory*. John Wiley & Sons, Ltd, 1998.

P.K. Varshney. *Distributed Detection and Data Fusion*. Springer-Verlag, 1997.

N.B. Venkateswarlu and P.S.V.S.K. Raju. Fast ISODATA clustering algorithms. *Pattern Recognition*, 25(3):335–345, 1992.

G.G. Venter. Tails of copulas. *2001 ASTIN Colloquium*. 2001.

E. Vidal. An algorithm for finding nearest neighbours in (approximately) constant average time. *Pattern Recognition Letters*, 4(3):145–157, 1986.

E. Vidal. New formulation and improvements of the nearest-neighbour approximating and eliminating search algorithm (AESA). *Pattern Recognition Letters*, 15:1–7, 1994.

R. Viswanathan and P.K. Varshney. Distributed detection with multiple sensors: part 1 – fundamentals. *Proceedings of the IEEE*, 85(1):54–63, 1997.

F. Vivarelli and C.K.I. Williams. Comparing Bayesian neural network algorithms for classifying segmented outdoor images. *Neural Networks*, 14:427–437, 2001.

C.T. Volinsky. *Bayesian Model Averaging for Censored Survival Models*. PhD thesis, University of Washington, Seattle, 1997.

U. von Luxburg. A tutorial on spectral clustering. *Statistics and Computing*, 17(4):395–416, 2007.

P.W. Wahl and R.A. Kronmal. Discriminant functions when covariances are unequal and sample sizes are moderate. *Biometrics*, 33:479–484, 1977.

E. Waltz and J. Llinas. *Multisensor Data Fusion*. Artech House, 1990.

M.P. Wand and M.C. Jones. Multivariate plug-in bandwidth selection. *Computational Statistics*, 9:97–116, 1994.

M.P. Wand and M.C. Jones. *Kernel Smoothing*. Chapman and Hall, 1995.

X. Wang, T.L. Lin, and J. Wong. Feature selection in intrusion detection system over mobile ad-hoc network. Technical Report, Computer Science, Iowa State University, 2005.

J.H. Ward. Hierarchical grouping to optimise an objective function. *Journal of the American Statistical Association*, 58:236–244, 1963.

L. Wasserman. Bayesian model selection and model averaging. *Journal of Mathematical Psychology*, 44(1):92–107, 2000.

S. Wasserman and K. Faust. *Social Network Analysis*. Cambridge University Press, 1994.

S. Watanabe. *Pattern Recognition: Human and Mechanical*. John Wiley & Sons, Ltd, 1985.

A.R. Webb. Functional approximation in feed-forward networks: a least-squares approach to generalisation. *IEEE Transactions on Neural Networks*, 5(3):363–371, 1994.

A.R. Webb. Multidimensional scaling by iterative majorisation using radial basis functions. *Pattern Recognition*, 28(5):753–759, 1995.

A.R. Webb. An approach to nonlinear principal components analysis using radially-symmetric kernel functions. *Statistics and Computing*, 6:159–168, 1996.

A.R. Webb. Gamma mixture models for target recognition. *Pattern Recognition*, 33:2045–2054, 2000.

A.R. Webb and P.N. Garner. A basis function approach to position estimation using microwave arrays. *Applied Statistics*, 48(2):197–209, 1999.

A.R. Webb and D. Lowe. A hybrid optimisation strategy for feed-forward adaptive layered networks. DRA Memo 4193, DERA, 1988.

A.R. Webb, D. Lowe, and M.D. Bedworth. A comparison of nonlinear optimisation strategies for feed-forward adaptive layered networks. DRA Memo 4157, DERA, 1988.

W.G. Wee. Generalized inverse approach to adaptive multiclass pattern recognition. *IEEE Transactions on Computers*, 17(12):1157–1164, 1968.

L. Wehenkel and M. Pavella. Decision tree approach to power systems security assessment. *International Journal of Electrical Power and Energy Systems*, 15(1):13–36, 1993.

M. West. Modelling with mixtures. In J.M. Bernardo, J.O. Berger, A.P. Dawid, and A.F.M. Smith, editors, *Bayesian Statistics*, pages 503–524. Oxford University Press, 1992.

N. Weymaere and J.-P. Martens. On the initialization and optimization of multilayer perceptrons. *IEEE Transactions on Neural Networks*, 5(5):738–751, 1994.

A.W. Whitney. A direct method of nonparametric measurement selection. *IEEE Transactions on Computers*, 20:1100–1103, 1971.

C.K.I. Williams and X. Feng. Combining neural networks and belief networks for image segmentation. In *Proceedings of the 1998 IEEE Signal Processing Society Workshop on Neural Networks for Signal Processing*. IEEE, 1998.

W.T. Williams, G.N. Lance, M.B. Dale, and H.T. Clifford. Controversy concerning the criteria for taxonomic strategies. *Computer Journal*, 14:162–165, 1971.

D. Wilson. Asymptotic properties of NN rules using edited data. *IEEE Transactions on Systems, Man and Cybernetics*, 2(3):408–421, 1972.

J. Winn and C.M. Bishop. Variational message passing. *Journal of Machine Learning Research*, 6:661–694, 2005.

J.M. Winn. *Variational Message Passing and its Applications*. PhD thesis, Inference Group, Cavendish Laboratory, University of Cambridge, 2004.

I.H. Witten and E. Frank. *Data Mining: Practical Machine Learning Tools and Techniques*. Morgan Kaufmann, second edition, 2005.

J.H. Wolfe. A Monte Carlo study of the sampling distribution of the likelihood ratio for mixtures of multinormal distributions. Technical Bulletin STB 72-2, Naval Personnel and Training Research Laboratory, San Diego, 1971.

D.H. Wolpert. Stacked generalization. *Neural Networks*, 5(2):241–260, 1992.

S.K.M. Wong and F.C.S. Poon. Comments on 'Approximating discrete probability distributions with dependence trees'. *IEEE Transactions on Pattern Analysis and Machine Intelligence*, 11(3):333–335, 1989.

K. Woods, W.P. Kegelmeyer, and K. Bowyer. Combination of multiple classifiers using local accuracy estimates. *IEEE Transactions on Pattern Analysis and Machine Intelligence*, 19(4):405–410, 1997.

J. Wray and G.G.R. Green. Neural networks, approximation theory, and finite precision computation. *Neural Networks*, 8(1):31–37, 1995.

T.-F. Wu, C.-J. Lin, and R. C. Weng. Probability estimates for multi-class classification by pairwise coupling. *Journal of Machine Learning Research*, 5:975–1005, 2004.

X. Wu and K. Zhang. A better tree-structured vector quantizer. In J.A. Storer and J.H. Reif, editors, *Proceedings Data Compression Conference*, pages 392–401. IEEE Computer Society Press, 1991.

C.R. Wylie and L.C. Barrett. *Advanced Engineering Mathematics*. McGraw-Hill, sixth edition, 1995.

Z.-X. Xie, Q.H. Hu, and D.-R. Yu. Improved feature selection algorithm based on SVM and correlation. In *Advances in Neural Networks - ISNN 2006*, pages 1373–1380. Springer, 2006.

D. Yan, L. Huang, and M.I. Jordan. Fast approximate spectral clustering. In *Proceedings of the 15th ACM SIGKDD International Conference on Knowledge Discovery and Data Mining, KDD '09*, pages 907–916. ACM, 2009.

H. Yan. Handwritten digit recognition using an optimised nearest neighbor classifier. *Pattern Recognition Letters*, 15:207–211, 1994.

M.-S. Yang. A survey of fuzzy clustering. *Mathematical and Computer Modelling*, 18(11):1–16, 1993.

Y. Yang and X. Liu. A re-examination of text categorization methods. In *Proceedings of SIGIR'99, 22nd Annual International ACM SIGIR Conference on Research and Development in Information Retrieval*, pages 42–49. ACM, 1999.

D. Yaramakala, S. Margaritis. Speculative Markov blanket discovery for optimal feature selection. In *Fifth IEEE International Conference on Data Mining* (ICDM'05), pages 809–812. IEEE, 2005.

R. Yasdi, editor. *Neural Computing and Applications*, 9(4):2000. Special issue on 'Neural Computing in Human-Computer Interaction'.

K.Y. Yeung, R.E. Bumgarner, and A.E. Raftery. Bayesian model averaging: development of an improved multi-class, gene selection and classification tool for microarray data. *Bioinformatics*, 21:2394–2402, 2005.

S.H. Yook, H. Jeong, and A.-L. Barabási. Weighted evolving networks. *Physical Review Letters.*, 86(25):5835–5838, 2001.

T.Y. Young and T.W. Calvert. *Classification, Estimation and Pattern Recognition*. Elselvier, 1974.

L. Yu and H. Liu. Efficient feature selection via analysis of relevance and redundancy. *Journal of Machine Learning Research*, 5:1205–1224, 2004.

H. Zare, P. Shooshtari, A. Gupta, and R.R. Brinkman. Data reduction for spectral clustering to analyze high throughput flow cytometry. *BMC Bioinformatics*, 11(403), 2010.

R. Zentgraf. A note on Lancaster's definition of higher-order interactions. *Biometrika*, 62(2):375–378, 1975.

I. Zezula. On multivariate Gaussian copulas. *Journal of Statistical Planning and Inference*, 139:3942–3946, 2009.

G.P. Zhang. Neural networks for classification: a survey. *IEEE Transactions on Systems, Man, and Cybernetics – Part C: Applications and Reviews*, 30(4):451–462, 2000.

P. Zhang. Model selection via multifold cross validation. *Annals of Statistics*, 21(1):299–313, 1993.

T. Zhang and F. J. Oles. Text categorization based on regularized linear classification methods. *Information Retrieval*, 4(1):5–31, 2001.

X. Zhang, M.L. King, and R.J. Hyndman. A Bayesian approach to bandwidth selection for multivariate kernel density estimation. *Compuational Statistics and Data Analysis*, 50(11):3009–3031, 2006.

Y. Zhang, C.J.S. de Silva, R. Togneri, M. Alder, and Y. Attikiouzel. Speaker-independent isolated word recognition using multiple hidden Markov models. *IEEE Proceedings on Vision, Image and Signal Processing*, 141(3):197–202, 1994.

Q. Zhao, J.C. Principe, V.L. Brennan, D. Xu, and Z. Wang. Synthetic aperture radar automatic target recognition with three strategies of learning and representation. *Optical Engineering*, 39(5):1230–1244, 2000.

Y. Zhao and C.G. Atkeson. Implementing projection pursuit learning. *IEEE Transactions on Neural Networks*, 7(2):362–373, 1996.

E.N. Zois and V. Anastassopoulos. Fusion of correlated decisions for writer verification. *Pattern Recognition*, 34:47–61, 2001.

J. Zupan. *Clustering of Large Data Sets*. Research Studies Press, 1982.

Index

activation function, 277
adjacency matrix, 532, 561
Akaike information criterion, 53, 583
application studies
  classifier combination, 399
  clustering, 546
  data fusion, 399
  decision trees, 340, 350
  feature selection and extraction, 493
  MARS, 357
  mixture models, 64, 547
  neural networks, 548
  nonparametric methods of density estimation, 213
  normal-based linear and quadratic discriminant rule, 63
  rule induction, 357
  support vector machines, 315
approximate Bayesian computation, 137–144

back-propagation, 302–305
bagging, 385, 402
ball trees, 170
Bayes
  decision rule, *see* decision rule
  error, *see* error, Bayes
Bayesian information criterion, 53
Bayesian learning methods, 70–87
  analytic, 73–87
  conjugacy, 73
  recursive, 72
Bayesian multinet, 189
Bayesian networks, 186–190
Bayesian sampling schemes, 87–126
  importance sampling, 92–95
  Markov chain Monte Carlo, 95–116
  ratio of uniforms, 90
  rejection sampling, 89
  sequential Monte Carlo samplers, 119–126
boosting, 387
bootstrap
  in cluster validity, 546
  in error rate estimation, *see* error-rate estimation, bootstrap
Borda count, 379
branch and bound
  in clustering, 516, 520
  in density estimation, 203
  in feature selection, 450–454, 496
  in kd-trees, 163
  in nearest-neighbour classification, 163–174

CART, *see* decision trees, CART
chain rule, 187
change of variables, 91

class-conditional probability density
function, 7
classification trees, *see* decision trees
classifier combination methods, *see*
ensemble methods
classifier fusion, 362
classifier generalisation, 6
classifier performance assessment,
404–432
clique percolation method, 573
clustering
agglomerative methods, 203
application studies, 546
cluster validity, 538–546
hierarchical methods, 502–510
agglomerative algorithm, 503
complete-link method, 506–507
divisive algorithm, 503, 510
general agglomerative algorithm,
508
inversions, 510, 547
nonuniqueness, 510, 547
single-link method, 503–506
sum-of-squares method, 507
mixture models, 511–513
quick partitions, 510–511
spectral clustering, 531–538
sum-of-squares methods, 513–531
clustering criteria, 514
complete search, 520
fuzzy *k*-means, 519–520
k-means, 49, 280, 516–517
nonlinear optimisation, 518
stochastic vector quantisation, 529–531
vector quantisation, 520–531
community detection, *see* complex
networks, community detection
comparative studies, 30
approximation elimination search
algorithms, 178
comparing performance, 424
decision trees, 357, 358
feature selection, 497
fuzzy *c*-means, 547
hierarchical clustering methods, 547
kernel bandwidth estimators, 200
kernel methods, 201, 203, 216
linear discriminant analysis – for small
sample size, 248
MARS, 358
maximum weight dependence trees, 186
naïve Bayes, 216
neural networks, 315
neural networks model selection, 582
nonlinear feature extraction, 493
nonlinear optimisation algorithms, 313
normal-based models, 64
number of mixture components, 53
principal components analysis, 469, 473
RBF learning, 290
regularised discriminant analysis, 45, 46
tree-pruning methods, 339
complete link, *see* clustering, hierarchical
methods, complete-link method
complex networks, 555–580
community detection, 565–575
link prediction, 575–579
properties, 557
condensing of nearest-neighbour training
set, 176
conditional risk, 14
confusion matrix, 5, 40, 406
conjugate gradients, 305
conjugate priors, 73
copulas, *see* nonparametric discrimination,
copulas
covariance matrix
maximum likelihood estimate, 37
structures
common principal components, 45
proportional, 45
unbiased estimate, 80
cross-validation, 327, 582
error rate, *see* error-rate estimation,
cross-validation
curse of dimensionality, 181

data
dimension, 4
head injury patient, 39
training set, 3
data fusion, 361, 366, 367, 370–376
data mining, 2, 315
data visualisation, 433, 463

decision rule, 7
  Bayes
    for minimum error, 8–15, 18
    for minimum risk, 13–15
  minimax, 18
  Neyman–Pearson, 15, 373
decision surfaces, 7
decision theory, 7–20
decision trees, 323–342
  CART, 330, 332, 334
  construction, 326, 337–339
  definition, 333
  pruning algorithm, 330–332
  splitting rules, 327–330
deduction, 3
dendrogram, 502
density estimation
  nonparametric, 150–218
    expansion by basis functions, 204
    properties, 150
  parametric
    estimative, 31
    predictive, 35, 70
  semiparametric, 203
design set, *see* data, training set
detailed balance, 97, 105
discriminability, *see* performance assessment, discriminability
discriminant functions, 20–26
discrimination
  normal-based models, 35–46
    linear discriminant function, 41
    quadratic discriminant function, 36
  regularised discriminant analysis, *see* regularised discriminant analysis
distance measures
  binary variables, 443
  distance between distributions
    Bhattacharyya, 452
    Chernoff, 452
    divergence, 446, 452
    Kullback–Leibler, 126, 184
    Mahalanobis, 271, 452, 500
    multiclass measures, 447
    Patrick–Fischer, 452
  nominal and ordinal variables, 443
divergence, 446, 452

editing of nearest-neighbour training set, 174
EM algorithm, *see* estimation, maximum likelihood, EM algorithm
ensemble methods, 361–402
error rate, 406
  apparent, 406, 448
  Bayes, 10, 407
  estimation, *see* error-rate estimation
  expected, 406
  for feature selection, 448
  true, 406
error-rate estimation
  bootstrap, 411–413, 448
  cross-validation, 408
  holdout, 407
  jackknife, 410, 448
errors-in-variables models, 289
estimation
  maximum likelihood
    definition, 34
    EM algorithm, 55
Expectation Maximisation algorithm, 49

false negative rate, 16
false positive rate, 16
feature extraction, 433, 434, 463–495
  Karhunen–Loève transformation, 475–481
    Kittler–Young, 477
    SELFIC, 476
  multidimensional scaling, 484–493
    by transformation, 491
    classical scaling, 484
    metric multidimensional scaling, 486
    ordinal scaling, 487
  principal components analysis, 98, 464–474
feature selection, 433, 435–463
  algorithms, 449–463
    branch and bound, 450
    Markov blanket, 459
    suboptimal methods, 454–463
  criteria, 440
    error rate, 448
    probabilistic distance, 446
    scatter matrices, 443
  embedded methods, 439

feature selection (*Continued*)
filter methods, 439
floating search methods, 457
redundancy, 436
relevance, 435
stability, 460
wrapper methods, 439
features, 2
forward propagation, 303
fuzzy *k*-means clustering, *see* clustering, sum-of-squares methods, fuzzy *k*-means

Gaussian classifier, *see* discrimination, normal-based models, 35–40
geometric methods, 433
Gibbs sampling, *see* Markov chain Monte Carlo algorithms, Gibbs sampling
Gini criterion, 328
graph, 188, 532–533, 561
graph centrality measures, 563–564
graph diameter, 563

Hermite polynomial, 206

imprecision, *see* performance assessment, imprecision
independent components analysis, 483
induction, 3
information gain, 329
intrinsic dimensionality, 4, 285

k-means clustering, *see* clustering, sum-of-squares methods, *k*-means
k-nearest-neighbour methods, *see* nonparametric discrimination, nearest-neighbour methods
Karhunen–Loéve transformation, *see* feature extraction, Karhunen–Loéve transformation
Karush–Kuhn–Tucker conditions, 252
kernel methods, 88, 194
Kullback–Leibler divergence, 126, 128, 129, 184, 438, 446, 460

Laplacian, 532, 535–536, 562
latent variables, 483
LBG algorithm, 522
learning vector quantisation, 528
likelihood ratio, 9
linear discriminant analysis, 221–249
error correction procedure
multiclass, 237
two-class, 223
Fisher's criterion
multiclass, 238
two-class, 227
for feature extraction, *see* feature extraction, linear discriminant anlysis
least mean-squared-error procedures, 228, 241–246
multiclass algorithms, 236–249
perceptron criterion, 223–227
support vector machines, *see* support vector machines
two-class algorithms, 222–236
linear discriminant function, *see* discrimination, normal models, normal-based linear discriminant function
generalised, 24
piecewise, 23
link prediction, *see* complex networks, link prediction
logistic discrimination, 263–268
loss matrix, 14
equal cost, 14
Luttrell algorithm, 528

machine learning, 322
Mahalanobis distance, *see* distance measures, distance between distributions, Mahalanobis
Markov blanket, 436, 459
Markov chain Monte Carlo methods, 95–116
Gibbs sampling, 95–103
Metropolis–Hastings, 103–107, 114
practical example, 111–114
reversible jump Markov chain Monte Carlo, 108
slice sampling, 109–111
MARS, *see* multivariate adaptive regression splines

MCMC algorithms, *see* Markov chain Monte Carlo algorithms
Mercer's condition, 293
Metropolis–Hasting, *see* Markov chain Monte Carlo algorithms, Metropolis–Hasting
minimum description length, 339, 345, 350, 462, 584
minimum-distance classifier, 22, 241
misclassification matrix, *see* confusion matrix
missing data, 585–586
mixture models, 46–63
  discriminant analysis, 48
  EM algorithm, 55
  in cluster analysis, *see* clustering, mixture models
  in discriminant analysis, *see* normal mixture models
  number of components, 52
  sampling from, 48
mixture of experts model, 382–385
mixture sampling, 264
model averaging, 390–396
model selection, 6, 53, 135, 286, 293, 581–585
modularity, 569, 570–572
monotonicity property, 450
multidimensional scaling, *see* feature extraction, multidimensional scaling
multilayer perceptron, 26, 275, 298–314
multivariate adaptive regression splines, 351–356
mutual information, 441

naïve Bayes, 40, 181
nearest class mean classifier, 23, 42
neural networks, 274–321
  model selection, 582
nonparametric discrimination
  copulas, 207–213, 216
  expansion by basis functions, 204
  histogram approximations, 181
    Bayesian networks, 186–190
    independence, 181
    Lancaster models, 182
    maximum weight dependence trees, 183–190
  histogram method, 180, 216
    variable cell, 181
  kernel methods, 194–204, 216
    choice of kernel, 201
    choice of smoothing parameter, 199
    multivariate kernel, 198
    product kernels, 198
    variable kernel, 200
  nearest-neighbour algorithms, 159–174
    ball-trees, 170–174
    kd-trees, 163–170
    LAESA, 159–163
  nearest-neighbour methods, 152–180, 216
    choice of *k*, 180
    choice of metric, 157
    condensing, 176
    discriminant adaptive nearest neighbour classification, 177
    editing techniques, 174
    for RBF initialisation, 281
    k-nearest-neighbour decision rule, 156
normal (Gaussian) distribution, 36
normal mixture models, 47, 67
  cluster analysis, 511
  discriminant analysis, 48
  EM acceleration, 62
  EM algorithm, 49, 55, 67
  for RBF initialisation, 280
  number of components, 52
normal-based quadratic discriminant function, 36

oblique decision trees, 341
ordination methods, 433
outlier detection, 586
overfitting, 6

parametric density estimation, *see* density estimation, parametric
particle filters, 93
pattern
  definition, 2
perceptron, *see* linear discriminant analysis, perceptron criterion

performance assessment
  discriminability, 405, 406
  imprecision, 413
  reliability, 405, 413
population and sensor drift, 7, 202, 245, 289, 317, 419–421
primary monotone condition, 488
principal components analysis, *see* feature extraction, principal components analysis
principal coordinates analysis, 484
probabilistic expert systems, 189
probability
  a posteriori, 8
  a priori, 8
probability density function
  Dirichlet, 86
  gamma, 74
  normal, 36
  Poisson, 74
  student, 81
  Wishart, 80
projection pursuit, 26, 313
pseudo-distances, 489

quadratic discriminant function, *see* discrimination, normal models, quadratic discriminant function

radial basis function network, 26, 269, 275–290
random forests, 389
random graph, 564
Rao-Blackwellisation, 88
receiver operating characteristic, 17, 415–422
  area under the curve, 416–419
recursive partitioning, 322
regression, 21, 26–28
regularisation, 246, 284
regularised discriminant analysis, 42–67
reject option, 7, 12, 15
reliability, *see* performance assessment, reliability
representation space, 484
robust procedures, 586
ROC, *see* receiver operating characteristic
rule induction, 342–351

scale-free networks, 559
scree test, 469, 485
secondary monotone condition, 488
self-organising feature maps, 525–531
Sherman–Morisson formula, 409
simulated annealing
  in clustering, 520
single-link, *see* clustering, hierarchical methods, single-link method
small-world effect, 557
softmax, 284
spectral clustering, 531–538
stacked generalisation, 382
stochastic vector quantisation, *see* clustering, sum-of-squares methods, stochastic vector quantisation
stress, 489
support vector machines, 249–263, 291–298, 388
  application studies, 268, 315
  canonical hyperplanes, 250
  linear
    multiclass algorithms, 247, 256
    two-class algorithms, 249–256, 262
  nonlinear, 291–298
surrogate splits, 339
Sylvester's determinant theorem, 84

text classification, 190
training set, *see* data, training set, 7
tree-structured vector quantisation, 523

ultrametric dissimilarity coefficient, 545
ultrametric inequality, 503

validation set, 582
variables of mixed type, 587–588
variational Bayes, 126–137
varimax rotation, 473
VC dimension, 589, 590
vector quantisation, *see* clustering, sum-of-squares methods, vector quantisation